TERRESTRIAL PLANT ECOLOGY

THIRD EDITION

Michael G. Barbour
University of California, Davis

Jack H. Burk
California State University, Fullerton

Wanna D. Pitts
Emeritus, San Jose State University

Frank S. Gilliam
Marshall University

Mark W. Schwartz
University of California, Davis

An imprint of Addison Wesley Longman, Inc.

Menlo Park, California • Reading, Massachusetts • New York • Harlow, England
Don Mills, Ontario • Sydney • Mexico City • Madrid • Amsterdam

This third edition is dedicated to the career and memory of William Dwight Billings, 1910–1997: President of the Ecological Society of America, Fellow of the American Academy of Arts and Sciences, father of the discipline of plant physiological ecology, author of more than 125 publications, and mentor to more than 50 Ph.D. students. Directly or indirectly, he has influenced every one of the coauthors of this book.

Publisher: Jim Green
Sponsoring Editor: Elizabeth Fogarty
Managing Editor: Laura Kenney
Production Editor: Janet Vail
Copyeditor: Anita Wagner
Composition/Art: G&S Typesetters
Index: Nancy Ball, Nota Bene Indexing

Cover and part opener illustrations are from Alan E. Bessett and William K. Chapman, eds., 1992. *Plants and Flowers: 1,761 Illustrations for Artists and Designers.* New York: Dover Publications, Inc.

Library of Congress Cataloging-in-Publication Data

Terrestrial plant ecology / Michael G. Barbour . . . [et al.]. — 3rd ed.
    p. cm.
  Includes bibliographical references (p.) and index.
  ISBN 0-8053-0004-x
  1. Plant ecology. 2. Plant ecology—North America.
  I. Barbour, Michael G.
  QK901.T49   1998
  581.7—dc21   98-30070
                                      CIP

ISBN 0-8053-0004-x
 2 3 4 5 6 7 8 9 10—MA—02 01 00 99

Benjamin/Cummings, an imprint of Addison Wesley Longman, Inc.
2725 Sand Hill Road
Menlo Park, CA 94025

TERRESTRIAL PLANT ECOLOGY

THIRD EDITION

Of Related Interest
from the Benjamin/Cummings Series in the Life Sciences

General Biology

N. A. Campbell, J. B. Reece, and L. G. Mitchell *Biology*, Fifth Edition (1999)

N. A. Campbell, L. G. Mitchell, and J. B. Reece *Biology: Concepts and Connections*, Second Edition (1997)

J. Dickey *Laboratory Investigations for Biology* (1995)

R. J. Ferl and R. A. Wallace *Biology: The Realm of Life*, Third Edition (1996)

J. Hagen, D. Allchin, and F. Singer *Doing Biology* (1996)

A. Jones, R. Reed, and J. Weyers *Practical Skills in Biology*, Second Edition (1998)

R. J. Kosinski *Fish Farm: A Simulation of Commercial Aquaculture* (1993)

A. Lawson and B. D. Smith *Studying for Biology* (1995)

J. G. Morgan and M. E. B. Carter *Investigating Biology: A Laboratory Manual for Biology*, Third Edition (1999)

J. Pechenik *A Short Guide to Writing Biology*, Third Edition (1997)

G. Sackheim *Introduction to Chemistry for Biology Students*, Sixth Edition (1999)

R. M. Thornton *The Chemistry of Life* CD ROM (1998)

R. A. Wallace *Biology: The World of Life*, Seventh Edition (1997)

R. A. Wallace, G. P. Sanders, and R. J. Ferl *Biology: The Science of Life*, Fourth Edition (1996)

Ecology

C. J. Krebs *Ecological Methodology*, Second Edition (1999)

C. J. Krebs *Ecology*, Fourth Edition (1994)

C. J. Krebs *The Message of Ecology* (1987)

E. R. Pianka *Evolutionary Ecology*, Fifth Edition (1994)

R. L. Smith *Ecology and Field Biology*, Fifth Edition (1996)

R. L. Smith and T. M. Smith *Elements of Ecology*, Fourth Edition (1998)

Evolution

E. C. Minkoff *Evolutionary Biology* (1983)

P. Skelton *Evolution: A Biological and Paleontological Approach* (1998)

Marine Biology and Oceanography

M. Lerman *Marine Biology: Environment, Diversity, and Ecology* (1986)

J. W. Nybakken *Marine Biology: An Ecological Approach*, Fourth Edition (1997)

D. A. Ross *Introduction to Oceanography* (1995)

H. V. Thurman and H. H. Webber *Marine Biology*, Second Edition (1991)

Environmental Science (Supplements)

J. Bowers *Sustainability and Environmental Economics* (1998)

J. Graves and D. Reavey *Global Environmental Change: Plants, Animals and Communities* (1995)

A. R. W. Jackson and J. M. Jackson *Environmental Science: The Natural Environment and Human Impact* (1996)

W. Levy and C. Hallowell *Green Perspectives: Thinking and Writing about Nature and the Environment* (1994)

W. J. Makofske and E. F. Karlin *Technology and the Global Environment Issues* (1995)

I. F. Spellerberg *Conservation Biology* (1996)

Plant Physiology

D. Dennis, D. H. Turpin, D. D. Lefebvre, and D. B. Layzell *Plant Metabolism*, Second Edition (1997)

Zoology

C. L. Harris *Concepts in Zoology*, Second Edition (1996)

PREFACE

A decade of research separates our second and third editions. Important advances have been made in the way ecologists think about photosynthesis, plant populations, the plant community, landscape dynamics, vegetation classification, and conservation. Such fields of study and tools as landscape ecology, conservation biology, restoration ecology, ecohistory, and GIS were relatively unexplored at the time of our last edition. The pages of ecological journals today are not only more numerous than they were 10 years ago but are also devoted to a new array of subjects.

It is no longer possible for the three original coauthors to cover the range of topics that characterizes modern plant ecology, and for that reason two more coauthors have been added. Dr. Mark Schwartz has added depth to text sections on mechanisms that allow plants to coexist in communities, the conservation of endangered species, and the principles of restoration ecology. Dr. Frank Gilliam has added depth to text sections on vegetation sampling and quantitative analysis and the use of these methods to deduce patterns of vegetation change over time. In addition, both have considerable research and teaching experience on the forests of eastern North America, and their contributions have broadened the geographic focus on this edition.

Our writing philosophy has remained unchanged. We want the book to be inclusive, covering the entire breadth of modern plant ecology; to blend classic topics with the results of new research; to summarize concisely and simply yet present conflicting evidence and opinions when consensus viewpoints do not exist; to be clear, using as little jargon as possible; and to be a reasonably short textbook instead of an encyclopedic reference work.

Significant revisions and updating have been incorporated in every chapter, but major revisions have been made to chapters on population biology, species interactions, succession, nutrient cycling, soil, fire, photosynthesis, and the vegetation of North America. Although this edition is longer than the last, we successfully kept the increase in length to a modest level. Some sections, tables, figures, and references from the last edition have been shortened or eliminated, either because the new material was better or the older material now requires less elaboration.

We thank those readers, students, and colleagues who encouraged us to prepare a third edition. It is an honor and privilege to have been a source of accurate and stimulating information for students for nearly 20 years. We hope the third edition will be a valuable resource for the next decade, well into the 21st century, in this way serving an entire generation of teachers and students.

We are grateful to the many reviewers who criticized drafts of individual chapters, clusters of chapters, or the entire book. Their thoughtful comments have improved the breadth, accuracy, and logic of the book immeasurably. Any unfortunate errors of fact or interpretation that remain in the book are due to our own carelessness and are not the fault of reviewers.

We deeply appreciate the sacrifices made by family members who granted us the release time we required to complete a project of this scale. We especially mention our spouses, Valerie Whitworth, Patricia Burk, Jerry Pitts, Laura Gilliam, and Sharon Strauss.

Thanks are extended also to Leonard Deutsch, Graduate School Dean, Marshall University, for providing logistical support to Frank Gilliam for work on the book.

We also thank Lisa Woo-Bloxberg and the editorial staff of Addison Wesley Longman for their patience with us, their consistent encouragement, their thorough review procedures, and their professional dedication to producing the best book we were jointly capable of creating.

MGB, JHB, WDP, FSG, and MWS

Reviewers for the Third Edition

James K. Agee, University of Washington

Jay E. Anderson, Idaho State University

Robert S. Boyd, Auburn University

C. John Burk, Smith College

Gary K. Clambey, North Dakota State University

Ross C. Clark, Eastern Kentucky University

Robert Curry, California State University, Monterey

Randy Dahlgren, University of California, Davis

Thelma H. Dalmas, Longwood College

Roger del Moral, University of Washington

Diane DeSteven, University of Wisconsin, Milwaukee

Peggy L. Fiedler, San Francisco State University

Richard W. Fonda, Western Washington University

Irwin Forseth, University of Maryland

Norma Fowler, University of Texas, Austin

David Gill, California State University, Fullerton

David C. Hartnett, Kansas State University

Karl E. Holte, Idaho State University

Jon E. Keeley, Occidental College, NSF

Alan K. Knapp, Kansas State University

James O. Luken, Northern Kentucky University

Brian C. McCarthy, Ohio University

A. David McGuire, University of Alaska, Fairbanks

Robert S. Nowak, University of Nevada, Reno

William J. Platt, Louisiana State University

Daniel D. Richter, Duke University

Kristina A. Schierenbeck, California State University, Fresno

Randal J. Southard, University of California, Davis

Dirk R. Walters, California Polytechnic State University, San Luis Obispo

Thomas R. Wentworth, North Carolina State University

Truman P. Young, University of California, Davis

CONTENTS

PART   IV

ENVIRONMENTAL FACTORS   375

CHAPTER   14

LIGHT AND TEMPERATURE   377

CHAPTER   15

PHOTOSYNTHESIS   411

CHAPTER   16

FIRE   441

CHAPTER   17

SOIL   473

TERRESTRIAL
PLANT
ECOLOGY

THIRD EDITION

PART I

BACKGROUND AND BASIC CONCEPTS

The origins of informal plant ecology go back hundreds of years to plant geographers, taxonomists, and naturalists who were people of energy and insight, such as Humboldt, De Candolle, and Darwin. In contrast, the science of ecology has had a very brief, intensive history, beginning less than a century ago. In that brief period, certain energetic men and women have had such a profound effect on the developing science that many current research projects and viewpoints are simply extensions of their work.

In this part, we will examine some of that history and some of those personalities, and we will present a common language to use in our study of plant ecology. There are many approaches to plant ecology, but each one attempts to answer the same basic question: How do plants cope with their environment? Some of the approaches we will discuss are paleoecology, phytosociology, community dynamics, systems ecology, and autecology.

CHAPTER 1

INTRODUCTION

Plant ecologists try to discover an underlying order to vegetation. They do this at progressively finer scales for the same reasons that biologists, chemists, and physicists pursue their worlds to the level of DNA, hydrogen bonds, and subatomic particles: There seems to be a human need to know the complete story, to explain the past, and to predict the future. What threads link plants to each other and to their environment? How flexible are these threads, and how inter-meshed? How do plants "solve" the problems of reproduction, dispersal, germi-nation in a suitable site, competition, and the acquisition of energy and nutrients? How can they withstand unfavorable periods of fire, flood, drought, or storm?

What can plants tell us by their presence, vigor, or abundance about the past, present, and future course of their habitat? Can plants be used as a scientific tool to analyze the intricacies of the environment or to test hypotheses about evolution?

What can plants tell us about our management hopes for the land? Once a for-est is cut, what will replace it, how long will the process take, and how can we most efficiently manipulate this process? Once domestic livestock graze at a given den-sity for a given time, what will the vegetation look like and how many animals will it then be able to support? Once topsoil is removed by strip mining, what plants should be introduced to stabilize the remodeled landscape? Once brushfields have been sprayed, burned, and replanted as grasslands, what will happen to watershed quality, soil nutrient levels, and the rate of siltation behind nearby dams? What is the residence time of herbicides in soils and are there side effects on nontargeted organisms? How many backpackers can use an alpine trail without altering adja-cent vegetation? If fire or flood is a natural catastrophe that must recur with a cer-tain frequency to maintain particular vegetation types, how can we incorporate such regular disasters into state, private, and federal park management plans? How can we know that we are managing a particular landscape or a rare species within it sustainably? (See Fiedler and Jain 1992.)

All of these questions, and more, are being investigated by plant ecologists. Some researchers are more interested in generating basic information that has to do with the description of vegetation or the biology of component species. Others are more interested in applying that basic information to management problems.

Applied plant ecologists may be called range managers, foresters, restoration specialists, conservation biologists, weed biologists, or agronomists, but they are all plant ecologists and share pleasure in discovering the subtle ways in which plants arc adapted to their environment. Their objective is very close to a formal definition of **ecology:** the study of organisms in relation to their natural environment.

Ecology, Environment, and Vegetation

The word *ecology* was coined more than 100 years ago by the German zoologist Ernst Haeckel. He spelled the word *oekologie,* but ecologists soon dropped the first *o,* to the annoyance and confusion of some purists. The word comes from the Greek roots *oikos,* "home," and *logos,* "the study of." Thus, oekologie translates loosely as the study of organisms in their home, their environment.

Environment is the summation of all **biotic** (living) and **abiotic** (nonliving) factors that surround and potentially influence an organism; it is the organism's habitat. Examples of biotic factors include competition, mutualism, allelopathy, herbivory, and other interactions between organisms, which are described in Chapters 6 and 7. Abiotic factors include all chemical and physical aspects of the environment that influence a plant's growth and distribution.

The environment extends along a continuum between two extremes: the macroenvironment and the microenvironment. The **macroenvironment** is the prevailing regional environment and the **microenvironment** is the environment close enough to an object to be influenced by it (Figure 1-1a). The microenvironment may be quite different from the macroenvironment. For example, the microenvironment beneath a forest canopy is different from the macroenvironment above it in such traits as humidity, wind speed, and light intensity; the microenvironment beneath a rock in desert soil may be cooler and moister than other parts of the macroenvironment; the microenvironment just 1 mm above a leaf surface may differ in wind speed, humidity, and temperature from the macroenvironment 10 mm away. Each organ or part of a plant is exposed to a different microenvironment, as shown in Figure 1-1b. Obviously, the microenvironment is what a plant responds to, and so the microenvironment will be emphasized in this book. At this microenvironmental scale, ecologists are interested in knowing how physiological and metabolic processes allow plants to succeed within their immediate environment.

Plant ecology is concerned not only with individual plants and plant species, but also with vegetation. **Vegetation** consists of all the plant species in a region (the flora) and the ways those species are spatially or temporally distributed. If the region is large, its vegetation will consist of several prominent plant communities. Each vegetation type can be characterized by the **life form** or **growth form** of its dominant plants (the largest, most abundant, characteristic plants). Examples of growth forms include annual herbs, broadleaf evergreen trees, drought-deciduous

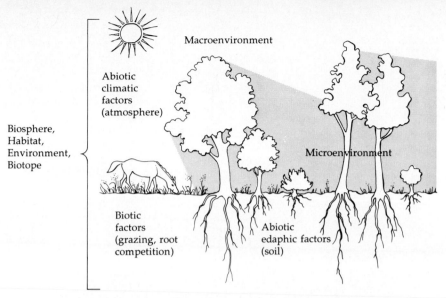

Macroenvironment

Abiotic climatic factors (atmosphere)

Biosphere, Habitat, Environment, Biotope

Microenvironment

Biotic factors (grazing, root competition)

Abiotic edaphic factors (soil)

Figure 1-1a   Components of the environment.

Figure 1-1b   Temperature of the microenvironment (°C) near different parts of an alpine plant in sun and shade on a July day when the temperature of the macroenvironment was 11.7°C. Redrawn from Tikhomirov (1963); also cited in Larcher (1995).

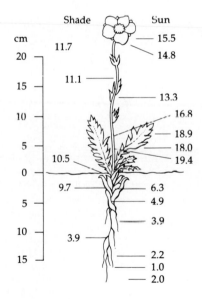

Shade   Sun

cm
20   11.7   15.5
                        14.8
15   11.1
                        13.3
10   16.8
5   18.9
                        18.0
      10.5   19.4
0
      9.7   6.3
5   4.9
                        3.9
10   3.9
15   2.2
                        1.0
                        2.0

shrubs, plants with bulbs or rhizomes, needleleaf evergreen trees, perennial bunch-grass, and dwarf shrubs (Figure 1-2). Growth forms may include any or all of the following, depending on the context: (a) the life span, woodiness, and size of a taxon, for example, annual, perennial, herb, herbaceous perennial, woody perennial, shrub, tree, or vine; (b) the degree of independence of a taxon, for example, green and rooted to the ground, parasitic, saprophytic, or epiphytic; (c) the

(a)

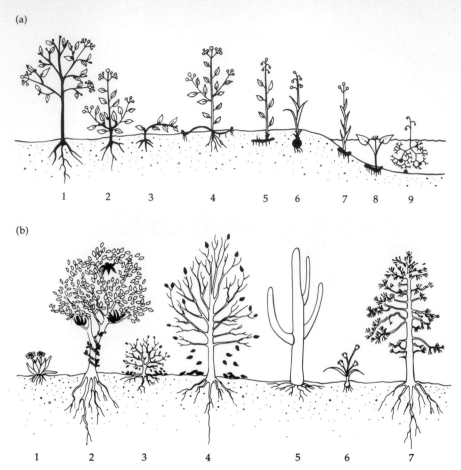

(b)

Figure 1-2 Examples of life forms (growth forms). (a) A scheme of classification by Raunkiaer (1934) based on the location of overwintering perennating parts such as buds, bulbs, or seeds, shown in solid dark form. Number 1 represents the phanerophyte category, 2–3 are chamaephytes, 4 is a hemicryptophyte, and 5–9 are cryptophytes (also called geophytes or hydrophytes). Redrawn from C. Raunkiaer, 1934, *The Life Forms of Plants and Statistical Plant Geography,* courtesy of Clarendon Press, Oxford. (b) Life forms based on other criteria such as length of life, succulence, and leaf traits. Number 1 is an annual herb, 2 is a broadleaf evergreen tree with a liana (woody vine), 3 is a drought-deciduous shrub, 4 is a winter-deciduous broadleaf tree, 5 is a stem succulent, 6 is a bulbous herbaceous perennial, and 7 is a needleleaf evergreen.

morphology of a taxon, for example, stem succulent, leaf succulent, rosette form, spinescent, or pubescent; (d) the leaf traits of a taxon, for example, large, small, sclerophyllous, evergreen, winter-deciduous, drought-deciduous, needleleaf, or broadleaf; (e) the location of perennating buds, as defined by Raunkiaer (1934); and (f) **phenology,** the timing of life-cycle events in relation to environmental cues.

Vegetation is also characterized by the architecture of its canopy layers. Dif-

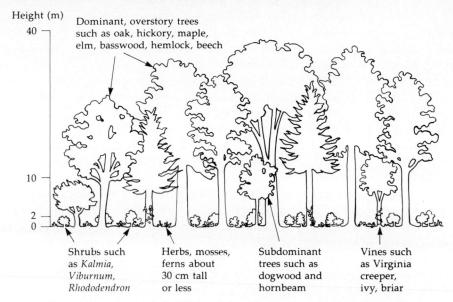

Height (m)

40 — Dominant, overstory trees such as oak, hickory, maple, elm, basswood, hemlock, beech

10 —

2 —
0 —

Shrubs such as *Kalmia, Viburnum, Rhododendron*

Herbs, mosses, ferns about 30 cm tall or less

Subdominant trees such as dogwood and hornbeam

Vines such as Virginia creeper, ivy, briar

Figure 1-3 Physiognomy of a typical eastern deciduous forest. The dominant overstory trees may reach 60 m or more but typically are 30–40 m tall.

ferent forest types have one to four tree canopy layers. Architecture and life form both contribute to the **physiognomy** (outer appearance) of vegetation, and each vegetation type has its own characteristic physiognomy. A **vegetation type** extends over a large region, and its name usually describes its location and then its physiognomy. For example, "eastern deciduous forest" (Figure 1-3) is a vegetation type that extends throughout the southeastern portion of North America, roughly from the Atlantic coast west to 95°W longitude, and from 45 to 30°N latitude. "Eastern" refers to its relative location within North America and "deciduous forest" refers to dominance by deciduous trees dense enough to form a closed forest canopy.

The physiognomy of eastern deciduous forest is typically four-layered. The overstory may reach up to 65 m but is more typically 35 m tall. It is a closed, interlocking canopy of broad-leaved winter-deciduous tree species, typically oaks (*Quercus*), maples (*Acer*), and hickory (*Carya*). Beneath the overstory is a second tree canopy that is much more open and consists of smaller trees about 5–10 m tall, such as dogwood (*Cornus*). A third layer is made up of scattered shrubs, often members of the Ericaceae, about 1–2 m tall. A fourth layer consists of seasonally active patches of perennial grasses, broad-leaved herbs, mosses, and ferns. Vines such as the Virginia creeper (*Parthenocissus quinque-folia*) pass through several layers.

Other extensive vegetation types in North America include boreal forest, desert scrub, montane mixed conifer forest, intermountain steppe, arctic tundra, and northwestern temperate rain forest. Each possesses a unique physiognomy

and inhabits a unique regional environment. All of these, including eastern deciduous forest, are described in Chapter 20. The plant biogeographer A. V. Küchler
(1964) mapped and described more than 100 vegetation types in the United States,
but more regional, modern, and intensive studies would probably identify several
hundred. Indeed, Küchler himself identified about 50 types in California alone
in a later study (1997). A fascinating ecological phenomenon is the similarity of
vegetation types in similar macroenvironments scattered around the world. It is
as though a particular physiognomy has been selected for in similar but isolated
habitats, suggesting convergent evolution among vegetation types and their component species. We will return to this subject later in the book.

Within each vegetation type are local variants that differ only in the identity of the dominant species. For example, only portions of the eastern deciduous
forest are dominated by oak and hickory; other portions are dominated by beech,
maple, basswood, magnolia, or hemlock. Some moist, protected valleys support
such a rich mix of trees that no single species can be identified as a dominant. Each
variant exhibits the physiognomy of the vegetation type but differs in the identity
of the overstory dominant species. These variants are usually called **series.** Each
series is named after its dominant species. There is a sugar maple series, a beech-
maple series, a southern magnolia series, an oak-hickory series, a hemlock series,
and a mixed tree series. Common names or scientific names can be used, thus a
sugar maple series is equivalent to an *Acer saccharum* series. Obviously, there will
be more series than vegetation types in a given region. In California there are approximately 250 series (Sawyer and Keeler-Wolf 1995).

The ultimate classification level is an **association.** An association is defined in
reference to many of its co-occurring species, rather than just to the dominant one
or two species. Within a given series, very local conditions will permit different
understory species to coexist with the same overstory dominants. Thus, several
hectares of a sugar maple forest may have an understory particularly rich in certain species of shrubs and herbs, whereas several adjacent acres—perhaps with
shallower soil or on a different slope face—will exhibit different species of shrubs
and herbs. Each local variant is an association. As formally defined in an international botanical congress early this century, an association has the following attributes: (a) it has a relatively fixed floristic composition, (b) it exhibits a relatively
uniform physiognomy, and (c) it occurs in a relatively consistent type of localized
habitat. The same species tend to recur together wherever a particular habitat
repeats itself. Associations are usually named after their more characteristic species—a species relatively restricted to that association and no other. These characteristic species seldom are dominants but rather are understory species of only
moderate abundance. Many countries of the world have made progress this century in documenting, mapping, and classifying all of their associations. North
America, for a variety of reasons, has lagged in this effort. We can only guess at the
number of associations that might be present in any region (perhaps 2000 in California alone). Our ignorance is unfortunate because the distribution of most of our
rare and endangered plants is best described at the association level rather than at
the series or vegetation type levels.

Community is a general term that can be used as a synonym for any unit, from
a regional vegetation type to a narrower series or to a very local association. It has

the same sort of use as the term **taxon** does for taxonomy: Taxon can refer to a family, genus, species, or variety.

Specializations Within Plant Ecology

Community Ecology (Synecology)

One large segment of plant ecology has followed directly from plant geography (Figure 1-4). This is **synecology,** and some of its many synonyms include community ecology, phytosociology, geobotany, vegetation science, and vegetation ecology.

One phase of synecology is **plant sociology,** the description and mapping of vegetation types and communities. In the past 50 years there has been a proliferation of standard methods for sampling vegetation and treating and analyzing sampling data. With these standard methods, valid conclusions can be drawn and vegetation from all over the world can be compared on an equal basis. (The description of *past* vegetation types and associations, as they have existed through geologic time, is part of the field called **paleoecology.**)

Another phase of synecology is the examination of **community dynamics,** which includes processes such as the transfer of nutrients and energy between members, the antagonistic or symbiotic relationships between members, and the process and causes of **succession** (community change over time). If one studies a series of communities across a landscape that are related by being phases of the same succession—that is, they have been disturbed at various times in the past and represent everything from the earliest phases of recovery to complete recovery—then this level of investigation is part of a relatively new science called **landscape ecology.** Landscape ecologists often use remotely sensed data because the physical area included in their studies is relatively large. The data are usually obtained from orbiting satellites that are equipped to photograph the vegetated surface of the landscape below. The photographic mosaic of community types can be laid over maps of soil, topography, and disturbance history to tease out relationships. The process of overlaying mapped information creates a geographic information system (GIS), and it is becoming a powerful ecological tool.

The most encompassing scale of study combines plants, animals, their environment, and their exchange of energy and nutrients with that environment. Such an integrated entity is called an **ecosystem.** Often, the study of ecosystems involves a degree of mathematical abstraction. Complex formulae and computer programs summarize, simulate, or model the ecosystem in an approach called **systems ecology** (Golley 1993).

Population Ecology (Autecology)

Another large segment of plant ecology deals with the adaptations and behavior of individual species or populations in relation to their environment. Subdivisions of this segment (population ecology) include population dynamics and

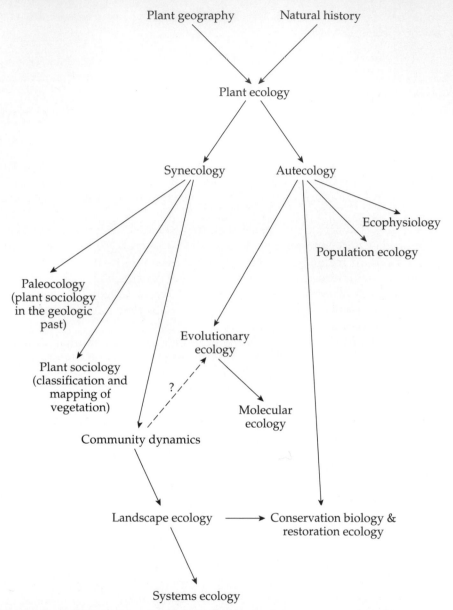

Figure 1-4   The relationship among specializations within the field of plant ecology.

demography (the regulation of population size), genecology (population genetics and speciation), and ecophysiology (linkage of metabolism with microenvironment). Autecologists try to explain the *why* of a given species' distribution: What phenological, physiological, morphological, behavioral, or genetic traits seem most important to the species' continued success in a given habitat? They try to illustrate the pervasive influence of the environment at the population, organismic, and sub-organismic levels. There is some overlap with evolutionary ecology when autecologists attempt to summarize all this as a species' adaptive pattern for survival. Autecology flows easily into other specialties outside the field of ecology proper, such as physiology, genetics, evolution, biophysics, and systematics (taxonomy).

Ecological evolutionists, for example, try to deduce how plants and animals have coevolved such complicated interactions as pollination, seed dispersal, and herbivory/resistance. They try to measure the ecological value ("fitness") of different patterns of carbon allocation to reproduction, leaves, stems, and roots. They try to measure the impact of species richness on community stability and ask whether species numbers per community might be limited by some sort of natural laws or rules. Are species packed together (assembled) in communities in some predictable way? The evidence so far is inconclusive (Wilson and Whittaker 1995). Evolutionists and geneticists presume that evolution occurs only within populations. Some ecological evolutionists, however, believe that communities may also evolve, becoming more efficient at nutrient or energy cycling over evolutionary time or more resistant to environmental stresses. Devising tests of such hypotheses has so far been a difficult and unrewarding process.

Population and landscape ecologists have recently collaborated in creating the new fields of conservation biology and restoration ecology (Fiedler and Jain 1992, Simberloff 1988). **Conservation biology** is the study of sustainable wildland management and the maintenance of biological diversity. How should preserves be situated to maximize diversity and minimize the probability of species extinction? How can we tell, by short-term monitoring, if a given endangered population is stable, declining, or increasing? **Restoration ecology** develops the technology of re-creating or restoring degraded habitats, communities, and ecosystems. It also prescribes management activities and monitoring necessary to maintain the ecological health of restored vegetation.

A Final Word on Specialization

Plant ecology itself may be thought of as a specialization within ecology. Some scientists and educators criticize the division of ecology into plant and animal ecology, arguing that the division is artificial and damaging to an understanding of the interdependence that permeates ecosystems. Plant ecology as a discrete science is partly an artificial creation, but no more so than any other science. There are many ways in which chemistry is linked to physiology, or physiology to behavior, or soils to geology, or mathematics to economics, yet these fields are accepted as reasonable specializations. We all specialize, and in this manner progress is made. One person cannot master all of plant ecology, let alone plant and animal ecology both.

In addition, we think a case could be made that the inherent differences in structure, behavior, and function between plants and animals are so profound that many principles of plant ecology cannot be translated into principles of animal ecology, and vice versa. For example, most plants are intimately associated with soil and geological substrates, whereas most animals are not. Plants have **modular growth,** producing repeating units of leaves, stems, roots, and reproductive organs, each somewhat independent of the others; animal bodies are more integrated. Plants exhibit greater degrees of plasticity in size, phenology, and life span than do most animal species. Plants occupy fewer trophic niches than animals, meaning that they compete more intensely for only a handful of environmental resources. Plants have several unique biochemical pathways relating to food acquisition (photosynthesis), food storage (starch), bodybuilding (cellulose and lignin), and chemical defense from herbivores (a wide variety of complex molecules). Plants also have **indeterminate growth,** retaining meristematic tissue throughout their life span.

Even plant ecology and animal ecology have become fragmented. Entire books, courses, agency programs, and research careers have been built on such narrower topics as insect ecology, microbial ecology, pollination ecology, systems ecology, and agroecology. The answer to the charge that we artificially fragment ecology is not that we should specialize less, but that we should communicate more. The excellent reviews of the history of ecology in North America by Egerton (1976), McIntosh (1976, 1980, 1985), and Golley (1993) are a good beginning in this direction.

CHAPTER 2

A BRIEF HISTORY
OF PLANT ECOLOGY

The route a science takes in its development is determined by the personal traits, interests, culture, and social surroundings of certain people along the way perhaps as much as it is determined by the facts, hypotheses, or approaches that each person has contributed. Good science may be impersonal, dispassionate, and unbiased, but most scientists are not. Consequently, this brief history of plant ecology is flavored not only by people's ideas but also by the people themselves.

Foundations in Plant Geography

Prior to the Nineteenth Century

Plant ecology has been studied informally and pragmatically since the beginning of the human race. Gatherers and hunters mastered a knowledge of the distribution of the wild food and forage plants that sustained both them and the prey they hunted. Tribal healers and shamans learned the narrow habitat requirements of rare species that had healing, narcotic, or hallucinogenic qualities; no doubt being able to find such plants was as important as knowing how to use them. Aristotle and Theophrastus, ca. 300 B.C., may have been the first to write about plant geography and plant ecology, but preliterate peoples throughout the world must certainly have had an understanding of these topics well before that time.

Some of the earliest formal ecology papers, dating back to the seventeenth century, were concerned with the succession of communities surrounding gradually filling lakes and bogs, and the term *succession* was used in its modern context by the beginning of the nineteenth century (Clements 1916). However, the real development of plant ecology came through books on plant geography written by people trained as taxonomists or general botanists. World exploration, especially in the nineteenth century, molded an increasingly ecological viewpoint. Carl Ludwig Willdenow (1765–1812) was the pioneer of this line of thought. He was an early plant geographer who noted that similar climates produce similar vegetation types, even in regions thousands of kilometers apart, such as southern Africa and Australia.

The Nineteenth Century

Willdenow's teaching at the University of Göttingen, in Germany, greatly influenced a wealthy young Prussian student, Friedrich Heinrich Alexander von Humboldt (1769–1859) (Figure 2-1), who studied botany to round out his education in higher mathematics, natural sciences, and chemistry. Not long after graduating, Humboldt met Johann Forster, who had accompanied James Cook on his world voyage of discovery. Forster's stories made Humboldt determined to visit the new world tropics. A decade later Humboldt met a young French botanist named Aimé Bonpland, who had a similar desire. Plans crystallized, and they received the permission and protection of King Carlos IV of Spain to travel in what is now Latin America. They took with them the best equipment of the day for measuring latitude, elevation, temperature, humidity, and other physical factors. For five years they traveled from steamy lowland rain forest to cold alpine paramo and from arid desert to thorn scrub. They explored Cuba, Venezuela, Ecuador, Peru, Mexico, and the Orinoco and Amazon Rivers; they climbed nearly to the top of Mt. Chimborazo (to 5900 m); they collected 60,000 plant specimens. On his way

Figure 2-1   Friedrich Heinrich Alexander von Humboldt, 1769–1859. Courtesy of the Hunt Institute for Botanical Documentation, Carnegie-Mellon University, Pittsburgh, PA.

back to Europe in 1804 Humboldt was a house guest of President Jefferson in Washington. Jefferson himself was keenly interested in plant responses to climate and studied the phenology of garden plants along latitudinal gradients. Jefferson had only recently sent Lewis and Clark west, and he questioned Humboldt on the nature of his discoveries (Billings 1985). Summaries of Humboldt's travels can be found in Botting (1973) and McIntyre (1985).

Humboldt returned to France and began to write his monumental 30-volume work, *Voyage aux régions equinoxiales*. The first 14 volumes were devoted to botany, and in those he coined the term *association*, described vegetation in terms of physiognomy, correlated the distribution of vegetation types with environmental factors, and described the synergistic effects of some physical factors (for example, elevation, latitude, and temperature). His statement, "In the great chain of causes and effects no thing and no activity should be regarded in isolation," is a striking preview of our modern view of interdependence within communities and ecosystems. Near the end of his life he wrote a five-volume encyclopedia, *Kosmos*, which attempted to describe and explain the entire universe. Humboldt was one of the last Renaissance people, attempting to master all the knowledge of his time.

Humboldt's study of plant geography was furthered by Schouw, De Candolle, Kerner, and Grisebach, among others. J. E. Schouw (1789–1852), a professor at the University of Copenhagen, methodically described the effects of major environmental factors on plant distribution in an 1822 book, emphasizing the role of temperature. This search for single most important factors is still with us today, but more and more we understand the interdependence of all factors. Schouw popularized the procedure of naming associations by combining the dominant genus with the suffix *-etum.* Thus, *Quercetum* is an oak woodland association (*Quercus + etum*) and *Pinetum* is a pine forest association (*Pinus + etum*). Some modern schemes of association nomenclature still use this concept.

Alphonse Luis Pierre Pyramus De Candolle (1806–1893) was an herbarium taxonomist, an armchair plant geographer, but his access to vast plant collections led him to try to "discern the laws of plant distribution." Like Schouw, he chose to study temperature. He summed temperatures according to a formula so useful that his 1874 data later became the basis for Köppen's famous classification of climate, published half a century later (see Gates 1972 and Ackerman 1941).

August Grisebach (1814–1879) traveled widely and described more than 50 major vegetation types in very modern physiognomic terms, relating their distribution to various climatic factors. He succeeded Willdenow as professor of botany at Göttingen.

Anton Kerner von Marilaun (1831–1898) studied medicine at the University of Vienna but gave up practice after he experienced a cholera epidemic as an intern. He turned to botany as a less traumatic career, became professor of botany at the University of Innsbruck, and then was commissioned by the Hungarian government to describe the vegetation in parts of eastern Hungary and Transylvania. The book that resulted, *Plant Life of the Danube Basin* (1863), has fortunately been translated into English, so a larger audience can now appreciate the beauty of his vegetation descriptions and his clear understanding of succession. Later, Kerner became one of the first experimental ecologists. He established several transplant gardens at various elevations in the Tirolean Alps, in Vienna at 180 m, at his villa at 1200 m, and at 2200 m. In each garden he grew alpine and lowland forms of more than 300 species together. He found that some variations within a species were fixed and heritable, but other variations were nonheritable modifications induced by the environment (Kerner 1895).

Other biologists who contributed to the development of plant ecology in the mid- to late-nineteenth century include Oscar Drude (1890 and 1896), Adolf Engler (1903), George Marsh (1864), Asa Gray (1889), and Charles Darwin. The theory of evolution and the evidence that Darwin marshaled to support it in his book *Origin of Species* (1859) are inherently ecological (see Harper 1967, Hagen 1992). Darwin (1809–1882) revolutionized all of biology with his theory of evolution by natural selection. One of the greatest geneticists of the twentieth century, Theodosius Dobzhansky, once wrote that "nothing makes sense in biology except in the light of evolution." This is particularly true for ecology. Darwin was not only the father of evolutionary biology, he could also be called the first modern ecologist. His interest in ecology included experimental and descriptive research in plant ecology that stands up today as a model of insight and thoroughness. His voluminous work was published in book-length monographs on plant movement, insectivorous plants, and pollination ecology. Because he advanced the field of ecology, it is fitting that his own explorations were inspired by reading the works of Humboldt, an earlier pioneer in ecology.

The Establishment of Plant Ecology
Apart from Plant Geography

The plant geography period culminated with the publication of several extremely important, innovative books between 1895 and 1916. At the end of those 20 years, plant ecology was firmly established as a science in its own right, and most of our modern research directions had been initiated. The major contributors were Warming, Schimper, Ramensky, and Paczoski in Europe, and Merriam, Cowles, and Clements in America.

Johannes Eugenius Bülow Warming (1841–1924)

As a young man in Holland, Johannes Warming (Figure 2-2) was offered the chance to help a paleontologist conduct a field study near Lagoa Santa, Brazil. It was an area of tropical woodland-savanna, 42 days' travel northwest of Rio de Janeiro. He spent 3 years there, and 30 years later, when the last of his 2600 plant specimens had finally been identified or described, he wrote a book (1892) on the vegetation. Its classical ecological organization could serve as a model for any vegetation study done today. Introductory chapters on geology, soils, and climate are followed by sections on each of the major vegetation types and communities. He discussed dominants and subdominants, the adaptive value of various life

Figure 2-2 Johannes Eugenius Bülow Warming, 1841–1924. Courtesy of the Hunt Institute for Botanical Documentation, Carnegie-Mellon University, Pittsburgh, PA.

forms, the effect of fire on community composition and succession, and the phenology of communities and taxa.

In the meantime, he had returned to teach at the University of Copenhagen as a successor to Schouw. He organized what may have been the world's first ecology course and was recognized as an outstanding teacher. He published his lecture notes in 1895, in Danish, as the world's first plant ecology text. The book had an immediate impact on many botanists throughout the Western world. It was later translated into German, Polish, Russian, and finally English (1909). Warming synthesized plant morphology, physiology, taxonomy, and biogeography into a coherent science for the first time. He concluded that soil affects vegetation more than climate does, and he emphasized both moisture and temperature as prime climatic factors. He coined such useful, and still commonly used, terms as **halo-, hydro-, meso-,** and **xerophyte,** meaning plants of saline, wet, moist, and dry habitats, respectively.

Andreas Franz Wilhelm Schimper (1856–1901)

Andreas Schimper was born into a celebrated family of German botanists. He studied geology and botany at the University of Strassburg, and later taught plant histology, physiology, ecology, and geography at the University of Bonn. He traveled extensively in the tropics and concluded that the basic work of plant geography and taxonomy would soon be completed; he was not quite correct. Based on his conclusion, however, the goal he set for himself was to explain the *causes* of regional differences in floras and vegetation. Near the end of his short life, he published *Plant Geography upon a Physiological Basis* (1898 and later in English in 1903).

The book's title is somewhat misleading; he stressed morphological features of presumed adaptive value and presented little of plant physiology in the modern sense. However, many of his semiphysiological conclusions are still accepted today, and Billings (1985) calls his work the real beginning of plant physiological ecology.

Like Warming, Schimper gave weight to both climatic and edaphic (soil) factors, and among climatic factors he emphasized temperature and moisture. He appeared to borrow heavily from Warming's text and figures (Warming's book had been translated into German two years before), but nowhere did he give Warming so much as a footnote of credit. It is possible that Schimper came to Warming's conclusions independently, but some of Warming's supporters suspect Schimper of plagiarism (Goodland 1975).

Jozef Paczoski (1864–1941) and Leonid Ramensky (1884–1953)

The ideas of Jozef Paczoski (pronounced pach · ós · ky) spread slowly because they were published in a Slavic rather than a western European language and they were not quickly translated (see Maycock 1967). He has only belatedly been credited in Russia and North America as the father of **phytosociology** (which he defined in his 1891 and 1896 papers as all the sociological relationships of plants). He showed how plants modify the habitat, creating their own microenvironment. He discussed the role of competition, the causes of succession, the role of fire, the interdependence of species in a community, the continuum nature of community boundaries, and such physiological adaptations as shade tolerance. He later published a text on phytosociology (1921) and founded the world's first phytosociology department at the University of Poznan, Poland. Many current phytosociology practices, terms, and concepts in Europe were conceived by him long before their current popularity.

Ramensky (also written Ramenskii) has also only recently been appreciated by Western plant ecologists (Rabotnov 1953 and 1978, Sobolev and Utekhin 1978, McIntosh 1983a). His concepts of the individuality of species and the continuum of vegetation predate those of Gleason and Whittaker by half a century. He developed methods of gradient analysis and, like Paczoski, showed how communities grade into one another (he coined the term **phytocoenosis**). He also expressed community composition in a table form much like that later popularized by Braun-Blanquet. Ramensky also prefigured the autecological *C-S-R* and *r-K* categories of Grime, MacArthur, and Pianka by dividing plants into three groups: "violent" (competitors, *K*-strategists), "patient" (stress-tolerators), and "exploring" (ruderals, *r*-strategists). See Chapter 5 for more details.

Clinton Hart Merriam (1855–1942)

Clinton Merriam received an M.D. degree from Columbia University, but later devoted himself to biology (especially zoology) and served as chief of the U.S. Biological Survey from 1885 to 1910. During that time he visited many parts of the western United States, writing with a naturalist's eye about new species of mammals and the distribution of vegetation types, and with an anthropologist's eye about Indian groups in California. He was a founder of the National Geographic Society.

As his large expeditions with long supply trains moved slowly through the West, he was impressed with the similarities of elevational zones of vegetation from one mountain range to another. He believed that temperature, especially warmth in the growing season, was the determining environmental factor, and he developed formulae to sum degrees of warmth much as Grisebach had done. He then correlated each vegetation type to values of warmth, enlarged the vegetation types into life zones (which he named Boreal, Transitional, Canadian, Hudsonian, Tropical, etc.) and extrapolated the distribution of his life zones across the entire North American continent (1890, 1894, and 1898; Figure 2-3 is a small example).

Merriam had a strong measure of self-confidence, for he later wrote, "in its broader aspects the study of the geographic distribution of life in North America is completed. The primary regions and their subdivisions have been defined and mapped, the problems involved in the control of distribution have been solved, and the laws themselves have been formulated." Although some naturalists and biology books still use this life-zone terminology and concept, many of his assumptions, calculations, and conclusions have been rightfully challenged (see, for example, Daubenmire 1938), and his latitudinal extrapolations are not widely accepted. Nevertheless, Merriam had a strong impact on American plant ecology.

Henry Chandler Cowles (1869–1939)

Henry Cowles was a geologist-turned-botanist who investigated plant succession on sand dunes around Lake Michigan from the perspective of a geologist. A leading botanist of that time, John Coulter, was his major professor. Cowles had a gentle, amiable personality that attracted many students to him. Gleason wrote in 1940, "he was a man of infectious gaiety and high spirits, of infinite humanity and humor." The University of Chicago, where he taught, became a center of plant ecology during the first two decades of the twentieth century. He was an organizer of the Ecological Society of America in 1915 and served as its president three years later. Most professional ecologists in the United States today belong to this society. It publishes research journals and sponsors scientific meetings at which ecologists can exchange information and ideas.

His Ph.D. dissertation and later papers on dune succession, from 1898 to 1911, emphasized the dynamic nature of vegetation, as Warming and Schimper somehow had not. This dynamic aspect of ecology attracted botanists who until then had thought of plant ecology as merely the mapping and description of static pieces of vegetation. Cowles had an impact on American plant ecology mainly through his students, who adopted his emphasis on succession and applied it throughout the midwestern and eastern United States. The pioneer animal ecologists Charles C. Adams and Victor Shelford were colleagues of Cowles at Chicago and they were strongly influenced by his work. However, the University of Chicago "school" of plant ecology was soon overshadowed by the Nebraska "school" of Clements.

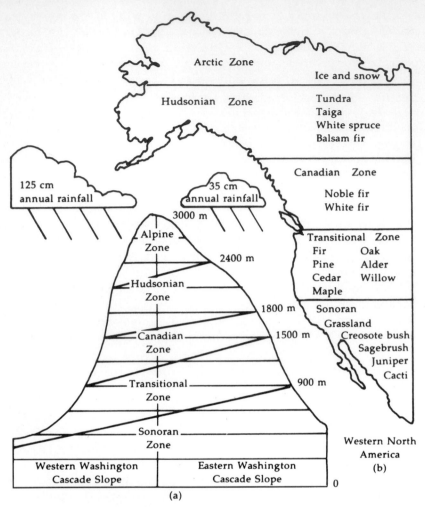

Figure 2-3   Example of Merriam's life zones (a) in the Cascade Mountains of Washington and (b) diagrammatically extended across the western states. Adapted from *Living Systems* by J. Ford and J. Monroe. Copyright 1971 by Ford and Monroe. Reprinted by permission of Harper and Row, Publishers, Inc.

Frederick Edward Clements (1874–1945)

Frederick Clements (Figure 2-4) provided a geographic balance to Cowles; he was born, raised, and educated in Nebraska and traveled widely in the western United States. Clements's life and contributions have been well described by Tobey (1981). During his student days at the University of Nebraska, he was prodded by Professor of Botany C. E. Bessey to expand his classical undergraduate education by adding a large dose of field botany. Together with some other students, Clements attempted to collect, identify, and describe the distribution of every

Figure 2-4   Frederick Edward Clements, 1874–1945. Courtesy of the Hunt Institute for Botanical Documentation, Carnegie-Mellon University, Pittsburgh, PA.

plant species in the state, from algae to oaks. He collected hundreds of plants and identified many new species of fungi, ultimately writing a book in 1898, *The Phytogeography of Nebraska*, which received wide recognition.

Clements described much of the vegetation of North America, naming regional formations and associations, local variants, and seral stages with great authority, if not always with great precision. He wrote about the causes of succession, the use of certain species as environmental indicators, and the methods for documenting succession and identifying associations. His classical background meshed with his philosophical, precise, rigid personality to produce large, comprehensive books (e.g., 1916 and 1920) filled with many new terms and with conclusions that have a dogmatic tinge. He defined a plant association in terms of an organism in order to illustrate the interdependent nature of an association's component species (see Chapters 8 and 11). The wealth of information, the new terms, the philosophical sweep of completeness, and the assertive conclusions confused some readers and offended others, and thus many of his ideas have been criticized unjustly. Much of his plant community work retains validity today.

In 1917 he was hired by the Carnegie Institution of Washington, D.C., to direct research at a coastal laboratory at Santa Barbara, California, during the winter and an alpine laboratory at Pikes Peak, Colorado, during the summer. He devoted the rest of his life to full-time research on the causes of plant distribution, experimenting with transplant gardens much as Kerner had done before him, yet arriving at completely different conclusions, which are not considered valid today (see Clements and Hall 1921). The Carnegie Institution was an important source of support for the young science of plant ecology (McIntosh 1983b, Billings 1985). Frederick Coville, who did pioneering botanical work in Death Valley, convinced the institution to establish a desert laboratory for the study of southwestern desert vegetation. In 1903 the laboratory was erected just west of Tucson, Arizona. Very productive researchers, such as William Cannon, Daniel MacDougal, Forrest Shreve, and Heinrich Walter, worked out of that laboratory during the same years that Clements worked at the other two Carnegie laboratories.

Clements performed valuable public service work with the Soil Conservation Service during the Dust Bowl of the mid-1930s, and he wrote one of the first American plant ecology textbooks with John E. Weaver in 1929. He was actively engaged in research until a few weeks before his death at the age of 70.

His personality was quite different from that of Cowles. It was "powerful" according to Tansley (1947a), "decidedly puritan, even ascetic . . . and his manner was apt to be tinged with a certain arrogance. . . . [Nevertheless] he had that best of all senses of humour which enables a man to laugh at himself. . . . [He was] by far the greatest individual creator of the modern science of vegetation."

Clements's ideas about vegetation have been called both holistic and organismic. Where did his theory of vegetation come from? Organicism, according to the *Random House Dictionary of the English Language,* is "The view that some systems resemble organisms in having parts that function in relation to the whole to which they belong . . . [also a] view of society as an autonomous entity analogous to and following the same developmental pattern as a biological organism." Clements applied this philosophy to plant communities. He believed that the component species of a plant community were tightly interdependent and that they functioned as a unit. Plant communities, then, were not simply convenient human abstractions; they were as natural as species.

Communities, Clements also wrote, are more than the sum of their parts. These are the same words that the *Random House Dictionary* employs to define holism: a theory that wholes are more than the mere sum of their parts. The word *holism* was invented by the South African philosopher and politician Jan Christiaan Smuts early in this century (Smuts 1926), but the concept of holism has long been at the base of human thought about the surrounding world. Holistic theories explain behavior by adding an element beyond the mere atoms, organs, or parts of which an individual—or an ecosystem—is composed. Each layer of increasing complexity has emergent properties that make it unique beyond the sum of the parts of which it is composed. In contrast, reductionists argue that a sufficient explanation for the behavior of any object or any ecosystem comes from a dissection and study of the object's component parts. Behavior can be understood and predicted once we know all the component details. Scientists are more often reductionists than holists.

Several historians (Tobey 1981, Hagen 1992, Worster 1985), as well as the South African ecologist John Phillips (1936), credit the writings of nineteenth-century sociologist Lester Ward (1883) with leading Clements to his organismic view of nature. Ward equated human social systems with organisms: "Society is simply a compound organism whose acts exhibit the resultant of all the individual forces which its members exert. The acts obey fixed laws. Objectively viewed, society is a natural object." Ward's second edition of *Dynamic Sociology* was published only a few years before Clements first wrote of his own organismic hypothesis, and the book's title was remarkably parallel with Clements's description of his own research as "dynamic ecology." Although Clements never admitted any links between Ward and himself, the parallelism with Ward is striking. Clements did, however, publicly acknowledge support for his biological ideas from Smuts.

Clements's organicism and holism were simply extensions of commonly accepted nineteenth-century ideas. Ecologists who preceded Clements, such as Humboldt, Kerner, Haeckel, Warming, McMillan, and Cowles, all to some extent employed organic metaphors for the plant and animal communities they studied (Barbour 1995).

The Second Generation of Plant Ecologists

By 1925 there were many plant ecologists in Europe and America, contemporaries of the above pioneers, each asking his or her own particular questions and expanding the field of plant ecology in a particular direction.

William Cooper, Edgar Transeau, and Emma Lucy Braun, all students of Cowles, described forests and bogs and paths of succession in the Midwest. Cooper also traveled long distances, for those days, to examine succession behind retreating glaciers in Alaska, on sand dunes along the Pacific coast, and in California chaparral. Transeau introduced ecology into general botany texts and foreshadowed our current interest in ecosystem productivity and energy transfer by working out an energy budget of a cornfield in 1926. E. Lucy Braun became famous for her descriptions of the virgin deciduous forests of eastern North America. She and her sister traveled extensively through the Appalachian Mountains during Prohibition and the Depression, gaining acceptance by the local people when male strangers might have been rebuffed.

A new theory of vegetation, completely antithetical to Clements's view, grew in acceptance during this time. It was originally formulated in 1917 by Henry Allen Gleason (Figure 2-5), born and raised in a rural Illinois setting. Like Clements, he was a schoolboy collector of wild plants. While an undergraduate at the University of Illinois, he took a biology course from two pioneering ecologists—Charles Adams and Stephen Forbes—and this experience stimulated him toward a career in ecology and plant systematics. Mathematics also attracted him, and he made some of the earliest attempts to quantify biotic diversity. His hypothesis was called the "individualistic" or "continuum" concept (Micolson 1990, McIntosh 1975). According to this hypothesis, plant communities are not real, natural units; they are merely human abstractions. Collections of co-occurring species are not obligately interdependent; rather, each species spreads out into a variety of habitats as an independent entity. Where Clements saw predictability, uniformity, cooperation, stability, simplicity, and certainty, Gleason saw only chaos, individualism, competition, a blur of continuous change, complexity, and fuzzy zones defined by probability (see Chapter 8 for more details).

At first, few ecologists agreed with Gleason. Many, in fact, argued vehemently

Figure 2-5   Henry Allen Gleason, 1882–1975. Courtesy of the Hunt Institute for Botanical Documentation, Carnegie-Mellon University, Pittsburgh, PA.

with him. He republished versions of his theory in 1926 and 1939. Not until 30 years after his original paper did a prominent ecologist cite his theory in print favorably (Cain 1947). New field research in Wisconsin forests (as later summarized by Fralish et al. 1993) and the Great Smoky Mountains (Whittaker 1956) tended to support Gleason's idea of the continuum. Incredibly, the majority of ecologists adopted his theory as part of a scientific minirevolution within a dozen years. In 1959, the Ecological Society of America awarded him the title "eminent ecologist" specifically because of his individualistic hypothesis. Gleason (1953) recalled his unpopular years as lonely. "To ecologists I was anathema. Not one believed my ideas; not one would even argue the matter. . . . For ten years, or thereabout, I was an ecological outlaw, sometimes referred to as 'a good man gone wrong.'" Gleason's ideas have since become standard parts of current ecology books, now often taken with as much faith as Clements's ideas had been taken before.

Forrest Shreve and William Cannon labored in the deserts, Cannon relating morphology and anatomy to habitat in the tradition of Schimper and Warming, and Shreve describing the major communities of the warm deserts of North America as a research associate of the Carnegie Institution from 1908 to 1945 (Bowers 1988). Shreve drew one of the first published vegetation maps of North America in 1917, and he was a member of the small group Cowles brought together to found the Ecological Society of America. His wife, Edith Shreve, wrote some excellent articles on plant-water relations (e.g., see her 1923 article in the *Botanical Gazette*).

Figure 2-6 Robert H. Whittaker, 1920–1980. From W. E. Westman and R. K. Peet. Reproduced by permission from Dr. W. Junk, Publisher, *Vegetatio* 48: 97–122, 1982.

Major synecological contributions were made by Robert H. Whittaker from the 1940s through the 1970s (see Figure 2-6). Raised during the Depression, Whittaker developed a strong work ethic. He turned his keen intellect and energy to a wide diversity of topics: the classification of communities; development of techniques such as ordination and gradient analysis, which permitted complex vegetation patterns to be related to equally complex environmental factors; the measurement of species diversity and an assessment of its significance; studies of the process and driving forces of succession; comparative studies of biomass and plant productivity; and an analysis of the roles of inhibitory metabolic compounds (allelochemics) in the ecosystem. He also proposed a five-kingdom classification for organisms, an approach since widely adopted by biology texts and instructors.

Whittaker concentrated on zoology and entomology in his undergraduate work but combined botany and zoology as a graduate student at the University of Illinois, working under both Vestal (a botanist) and Kendeigh (a zoologist). Interestingly, the botany department would not admit him as a botany graduate student because it was felt he lacked appropriate background courses. (Similar mistakes can be made anywhere.) He completed his Ph.D. in two and one-half years, far below today's normal period of five to six years. For a first-person account of his difficult times as a graduate student, read his essay in Jensen and Salisbury's 1972 textbook. Whittaker was aggressive in his challenge of many ecological ideas accepted as dogma at the time (especially Clementsian views on vegetation). Perhaps because of the unease this raised in colleagues who felt challenged, Whittaker was released from his first academic position without tenure. He went on, however, to serve at several universities, completing his career at Cornell, where he died prematurely of cancer in 1980. In view of his enormous importance to plant ecology on this continent, many view him as a second Clements. Unlike Clements, however, his ideas were accepted worldwide, and Whittaker did much to bring together plant ecologists of Europe and North America in a common approach to the study of vegetation (Westman and Peet 1982).

In Europe, Christen Raunkiaer succeeded Warming as professor of botany at Copenhagen. He developed a life form classification that is still widely used (see Figure 1-2a) and a quantitative method of sampling vegetation whereby data could be treated statistically without bias. In Sweden, Göte Turesson experimentally brought plants together in a common garden to examine the genetic basis of variation. His method was not much different from those of Bonnier (1895) and Kerner (1895) before him, but his conclusions were much more general, powerful, and well documented. He described the ecological variation within species as ecotypes (see Chapter 3), and this new concept of the "ecological species" triggered a long series of experiments by others on the subtle ways in which genotypes are selected by local environments. In the United States, one of Clements's associates, Hall, began to establish a transect through California for the purpose of studying plant variation. After his death, Clausen, Keck, and Hiesey, from the Carnegie Institution, collaborated to extend that work into a classic ecotype study that further generalized Turesson's ideas.

Sir Arthur Tansley investigated the vegetation of Britain, founded the British Ecological Society, coined the term *ecosystem*, called for more physiological investigations in field studies, and later led a conservation movement in England decades before a movement equally strong developed in the United States. In his first presidential address to the British Ecological Society (1914), he was not hesitant to put the importance of the new science of ecology above the older, traditional fields of study. Tansley contributed a great deal to the philosophy and process of ecological research, and he called for a more experimental approach even though he himself was not adept at such an approach. He encouraged an international exchange of views and fostered that with field trips on three continents between 1911 and 1930 (Cooper 1957).

In Switzerland and later in France, Josias Braun-Blanquet (Figure 2-7), following the path laid down by Kerner, developed his methods of community sampling, data reduction, and association nomenclature that dominate much of plant community ecology today. Eduard Rübel made many other early contributions in Switzerland, and cooperation between Rübel and Braun-Blanquet led to the development of an approach to plant synecology called the Zurich-Montpellier School of Phytosociology.

Braun-Blanquet began his career as a bank clerk, but his strong interest in alpine plants drove him to become a self-made plant ecologist. Winning the acceptance and esteem of more traditional botanists by the quality of his work (and to some extent by the pleasant force of his personality), he eventually received a doctorate and in 1930 founded a research station at Montpellier, France, called Station Internationale de Géobotanique Méditerrènne et Alpine (SIGMA). SIGMA remains a center for vegetation research to this day. Although Braun-Blanquet's methods of vegetation analysis and classification are widely accepted throughout the world, other schools or approaches to synecology have developed in the Czech Republic, Germany, Scandinavia, and Russia (see Whittaker 1962).

Figure 2-7 Josias Braun-Blanquet, 1884–1980. Courtesy of the Hunt Institute for Botanical Documentation, Carnegie-Mellon University, Pittsburgh, PA.

Ecophysiological research expanded in the 1940s and 1950s (Billings 1985). Frits Went supervised the construction at the California Institute of Technology of an elaborate growth chamber called a phytotron, which permitted research on the roles of temperature and photoperiod on plant behavior. Gas exchange chambers, capable of measuring whole plant or leaf photosynthesis, were developed in the 1950s, and the first portable one was taken into the mountains of Wyoming in the summer of 1958 by Harold Mooney, Ed Clebsch, and Dwight Billings (Billings et al. 1966). Within a few years, several groups in the United States and Europe had mobile laboratories for the measurement of transpiration, plant-water stress, and photosynthesis.

In Wales, beginning in the 1950s, John Harper developed the area of plant demography into an exciting science that attracted many autecologists throughout the world. His examination of the dynamics of plant populations focused on weedy species, but the concepts he formulated in his 1977 book have been applied by others to wildland plants in natural habitats. As a result of his work, weed science has turned in a more biological, ecological direction. Integrated pest management, rather than reliance on herbicides or mechanical methods, has gained acceptance (Radosevich and Holt 1984). Autecological research on weedy species has contributed a great deal of basic understanding in the field of population ecology, in particular the mechanisms of competition and other forms of interference such as allelopathy (see Chapter 6). This research has also been used by community ecologists trying to understand why some communities are more resistant to invasions than others (Rejmanek 1989, 1996).

Internationalism characterized two decades of novel research alliances promoted by the International Biological Program (IBP), created in 1960. Hundreds of scientists—animal ecologists, plant ecologists, mathematicians, biochemists, climatologists, systematists—studied different aspects of a single ecosystem in the hope that the pieces could be put together and large questions answered. IBP projects were located in arctic tundra, boreal forests, deserts, grasslands, and marine seabeds, with study sites on several continents. The IBP formally ended in 1974, but publications resulting from its work continued to appear into the 1980s.

Frank Golley (1993) pointed out that the IBP was a new chapter for ecology in the sense that large, multidisciplinanry teams of ecologists were assembled for the first time. More than 1800 U.S. scientists participated in the IBP. The models for this team approach were the engineering teams of NASA, which accomplished enormous technological advances that took humans to the moon in the 1960s and 1970s. Ecology at that time became "big science" with large nationally supported budgets, just like physics, medicine, and engineering. Was the new mode of research fruitful? Golley reached a mixed conclusion. General ecosystem theory did not advance, he concluded, but basic ecological knowledge and applied knowledge about environmental problems grew significantly, setting the stage for the modern environmental age. Today, relatively large budgets from the National Science Foundation (NSF) and other federal agencies support ecosystem research at a network of sites that represent major vegetation types. The NSF has made a commitment to fund monitoring and research at these sites for decades into the future. The establishment of this network was surely a consequence of the IBP.

The Environmental Age

In 1969, passage of the National Environmental Policy Act (NEPA) by the U.S. Congress required that future federal activities that would have a significant effect on the environment be postponed until an adequate environmental impact statement could be written and commented on by the public and other agencies. Since then, many states have adopted similar legislation for nonfederal projects. Some impact statements are written by federal, state, or local agencies, but many others are written by private consulting companies who hire ecologists and other specialists. New magazines, technical journals, and many books have been published for this audience of applied ecology professionals.

Ecologists are more and more being seen by the public as problem solvers. As a consequence, the research areas of landscape ecology, conservation biology, restoration ecology, and environmental science have become well established in a remarkably short time. As already defined in Chapter 1, landscape ecology is the study of vegetation processes at a landscape scale (on the order of thousands of hectares), with special attention paid to the role of natural disturbance in creating different patches of vegetation within that landscape. Landscape ecologists investigate spatial patterns at a scale between that of the local association and that of the regional ecosystem, where interactions between contiguous associations, series, and ecosystems become evident (Forman and Godron 1981). The nature of human-related problems in the environment is often revealed at this scale, as is the nature of possible solutions. **Environmental science** is the study of human-mediated disturbance to the environment, such as the stresses and changes caused by pollutants, urbanization, and agricultural activities. Environmental scientists propose policy-level actions necessary to correct the disturbances. Conservation biologists, restoration ecologists, environmental toxicologists, atmospheric scientists, soil scientists, and hydrologists are all involved in this field, as are sociologists, economists, and planners. Environmental science is the place where politics and science mingle. Currently, the mix is rich and provocative, spinning off such movements as environmental justice, deep ecology, green parties, ecofeminism, bioregionalism, ecohistory, and environmental ethics (see, for example, Zimmerman et al. 1993).

Doing ecological research in the environmental age must be interdisciplinary because it requires approaches beyond the scope of individual scientists or even single institutions. Doing this research well will require public education about environmental issues because only an educated citizenry will have the collective political will to change wasteful habits, outmoded political units, and underfunded research programs. To this end, the Ecological Society of America (Lubchenko 1991, Christensen et al. 1996, Levin 1996) has published conceptual papers called "The sustainable biosphere initiative," "The scientific basis for ecosystem management," and "Economic growth and environmental quality," which are intended to educate planners, decision makers, and scientists. To ecologists, **sustainability** means management of human and natural resources in such a way that biological diversity and ecosystem health remain relatively steady into the future, despite the demands of a growing human population and increases in standards of living. Defined this way, sustainability is quite a challenge, one the human species has never met even with the ecosystems we know the best (fisheries, farms, forests). One ecologist (Ludwig 1993) has gone so far as to label our belief in achieving sustainability magic, not science, because "magic is based on the notion that hope cannot fail," whereas science is based on results and experience.

Major forms of pollutants that will surely receive continued attention include: increasing carbon dioxide concentration in the atmosphere, acid deposition from oxides of sulfur and nitrogen (SO_x, NO_x), ozone and hydrocarbons from internal combustion engines, and toxic waste. In addition, the continued losses of tropical forests and other vegetation types have a major ecological impact that will require assessment. The biological consequences include the loss of gene sets (species) as valuable genetic resources. Apparently, we still need to generate and adopt a life ethic—a credo that all living organisms are important because they share the unique anomaly we call life, quite apart from any homocentric uses we may attribute to them.

Basic ecological research will also continue and should be encouraged. Applied ecology depends on continuing work in basic ecology, and basic research does often lead to practical applications (McIntosh 1974). Also, collaboration among different kinds of ecologists is adding new foci. In Chapter 1, we mentioned that evolutionary ecology has been enriched by both population and community ecologists; conservation biology and restoration ecology have been products of both population and landscape ecologists; ecophysiologists, demographers, and genecologists are jointly creating a new understanding of population biology. An example of a modern synthetic approach is the work of David Tilman (1982, 1986), who attempts to explain such community traits as species richness, patterns of dominance, and pathways of succession on the basis of simply knowing how individual species compete for resources. Another example is the work of Marcel Rejmanek (1996), which represents a new field we could call **molecular ecology** (see Figure 1-4). He has concluded that the invasiveness (aggressive weediness) of a species can be predicted by knowing its DNA content (genome size); the lower the genome size, the more potentially invasive the species.

It is unfair to cite only two individuals to represent the exceptionally rich research period we are in at present, because the richness of knowledge is being created by hundreds of bright, curious, and energetic plant ecologists. Our apologies to those who cannot be mentioned because of a lack of space here. All aspects of basic research should be encouraged to continue. None is unimportant; none is completely understood. As Pielou (1981) has cautioned us, conceptual models of the ecological world are still of limited use; they can be improved only by incorporating additional data based on observations of, and experiments with, nature.

PART II

THE SPECIES AS AN ECOLOGICAL UNIT

An autecological question is asked in this part: How do the members of just one species in a community cope with their environment? An incidental, secondary question is also asked: What is an ecological species? A taxonomic species is made up of individuals and populations that may be genetically heterogeneous. An ecological species, however, is a genetically more homogeneous collection of plants adapted to one particular set of microenvironmental conditions.

Part of an organism's environment is composed of adjacent plants and animals, which may or may not be members of the same species. Interactions between pairs of organisms lie along every possible part of a continuum between obligate and incidental, and between mutually beneficial and mutually detrimental. These interactions, whether mediated by chemical or physical factors, can affect the spatial distribution of individuals; in fact, a striking distribution pattern may be an ecologist's first clue to the existence of an interaction.

CHAPTER 3

THE SPECIES IN THE ENVIRONMENTAL COMPLEX

The surface of the earth is a network of environmental factors that vary in both space and time. These environmental factors determine the direction of evolution of plant species and they are correlated with the patterns of plant life on the planet. We will consider the ways that environmental extremes and the gradients between them are related to the physiological tolerances of species, as well as how plant evolution reflects variations in and predictability of the environmental complex. Every plant species is a dynamic set of individuals able to respond to an ever-changing environment. We will consider examples of how these sedentary organisms are able to adjust to variations in the environment, not only as individuals but also on the level of organs, such as leaves, which react physiologically and developmentally.

Environmental Factors and Plant Distribution

Later chapters will review the importance of certain environmental factors one by one, but there are some general principles that we will discuss in this chapter: the law of the minimum, the theory of tolerance, and the holocoenotic concept of the environment.

The "Law" of the Minimum

In 1840, the agriculturist and physiologist Justus von Liebig wrote that the yield of any crop depended on the soil nutrient most limited in amount. This conclusion has come to be called the **law of the minimum.** The factors covered have been expanded, so that a loose definition of the law would now be: The growth and/or distribution of a species is dependent on the one environmental factor most critically in demand.

The validity of the law has been shown in many parts of the world. The poor growth of some clover pastures in parts of Australia, for example, was shown to be a result of deficiencies in the micronutrients copper, zinc, or molybdenum. Addition of only 6–8 kg ha^{-1} of copper or zinc sulfate every 4 to 10 years increased plant growth 300% and increased the wool harvest from sheep grazing the vegetation even more. As little as 140 g ha^{-1} of sodium molybdate, applied every 5 to 10 years, increased pasture yield six- to sevenfold (Moore 1970). In England, the range of certain calcicoles (plants on calcium-rich soil with basic pH) abruptly ends when soil pH drops below pH 5 (Grubb et al. 1969, Rorison 1969), and in California the abundance of typical chaparral shrubs declines (sometimes to zero) when the substrate changes to serpentinite, which has an exceptionally low level of calcium (Walker 1954).

Two limitations to the law of the minimum have become evident, however. First, organisms have an upper tolerance limit to every factor as well as a lower limit. Second, most factors act in concert rather than in isolation; a low level of one factor can sometimes be partially compensated for by appropriate levels of other factors, or the influence of one factor may be magnified as other factors reach their maximum or minimum limits.

The Theory of Tolerance

The pioneer American animal ecologist Victor Shelford (1913) noted weaknesses in Liebig's concept and proposed a modification, which has come to be called the **theory of tolerance.** Ronald Good, a plant geographer, later elaborated on it (1931, 1953). Each and every plant species is able to exist and reproduce successfully only within a definite range of environmental conditions (Figure 3-1). In general importance, Good rated climatic factors above edaphic (soil) factors, and both of these above biotic factors (competition, etc.). Tolerance ranges may be broad for some factors and narrow for others, and they may vary in their relative width according to the phenological stage of the species. Tolerance ranges cannot be determined from morphology; they are related to physiological features that must be measured experimentally. Tolerance ranges may change in the course of evolution, but this process is so slow that environmental change is typically accompanied by plant migration rather than a change in tolerance.

Some ecologists disagree strongly with Good's second conclusion, believing that edaphic or biotic factors are more important than climatic ones, depending on the species under discussion. There have been a few experiments that dramatically show how the tolerance range or optimum for a physical factor is modified by competition (Harper 1964, Ellenberg 1958). For example, when the common weedy annuals wild radish (*Raphanus raphanistrum*) and spurrey (*Spergula arvensis*) are grown in separate pots in controlled conditions, their growth curves exhibit similar pH tolerance ranges and optima. Both optima are at pH 5–6 (Figure 3-2). When grown together, however, the optimum of spurrey shifts to pH 4 and its

Tolerance ranges for:

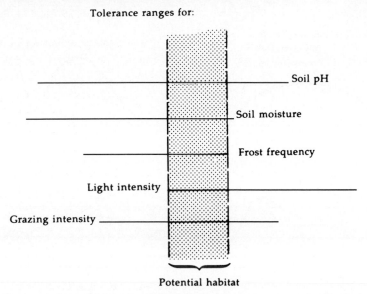

Figure 3-1 According to the theory of tolerance, the range of habitats for a given species is the sum of tolerance limits for each environmental factor. The extent of each horizontal line represents the tolerance range for that factor. The stippled area is the region of overlap for all tolerance ranges, which represents the potential habitat.

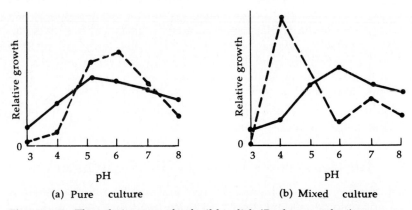

Figure 3-2 The relative growth of wild radish (*Raphanus raphanistrum*, solid line) and spurrey (*Spergula arvensis*, dashed line) as a function of substrate pH (a) when grown alone and (b) when grown in competition with each other. From H. Ellenberg, "Bodenreaktion (einschlieblich Kaltfrage)." In W. Rhuland (editor), *Handbuch der Pflanzenphysiologie*, vol. 4, 1958. By permission of Springer-Verlag.

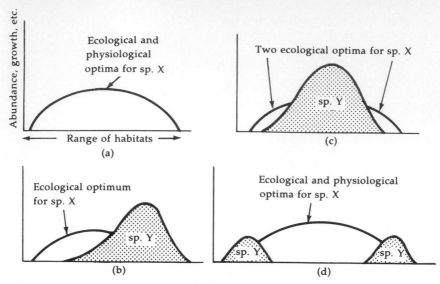

Figure 3-3   Examples of how the ecological optimum may be different from the physiological optimum due to competition. In (a), species X grows alone and laboratory experiments determine that its physiological optimum lies in the center of the curve. In nature, it faces competition from other species that displace it from habitats it could grow in alone; consequently its ecological optimum is shifted or its ecological range is truncated (b, c, d). From H. Walter. 1979. *Vegetation of the Earth and Ecological Systems of the Geo-biosphere,* 2nd ed. By permission of Springer-Verlag.

range for good growth becomes narrow, while the optimum for radish shifts slightly towards pH 6 and its tolerance range remains much as when grown alone.

The conditions under which a species can exist and grow best in isolation are its **potential (physiological) range** and **potential (physiological) optimum,** respectively. These may differ from its observed **ecological range** and observed **ecological optimum** in nature, where the species grows in competition with other species (Figure 3-3). The role of competition in plant distribution is very important. Undoubtedly, other biotic interactions, such as herbivory and pollination, also affect the distribution of species.

The Holocoenotic Concept of the Environment

Roughly 100 years after Humboldt wrote that everything was somehow interconnected and interdependent, Karl Friederich gave that ecosystem attribute a name: **holocoenotic** (Billings 1952) (holocoen is a synonym for ecosystem). The holocoenotic concept is a natural climax to the modifications others have applied to Liebig's law. The holocoenotic concept states that it is impossible to isolate the importance of single environmental factors to the distribution or abundance of a species, because the factors are interdependent and synergistic. Therefore, single-factor ecology (the search for one all-important environmental factor that best determines plant distribution) is short-sighted and naive, according to this concept.

This concept does not mean that all factors are necessarily equal—only that they are interactive. Certain factors in any ecosystem are of overriding importance, such as moisture in desert scrub, fire in prairies, or moving sand in coastal dune scrub. Billings (1970) calls these important factors **trigger factors.**

The Taxonomic Species

Although the species is at the heart of our classification scheme, there is no unanimity on a working definition for species. We will use the following definition, synthesized from several sources. A **species** consists of groups of genetically, morphologically, and ecologically similar natural populations that may or may not be interbreeding but that are reproductively isolated from other such groups.

Three aspects of classification are combined in this definition: (1) external appearance (morphology), (2) breeding behavior, and (3) habitat distinctiveness. The last aspect is clearly third in importance to most taxonomists, and morphology or reproductive isolation receives the most weight in deciding what a species is. Traditional taxonomists weight morphology heavily, but biosystematists give more weight to reproductive isolation.

Most biologists believe that living organisms do not vary continuously over the whole range but fall into more or less well-defined groups, which are commonly called species. Traditional taxonomists adhere to this philosophy of discrete species. A traditional taxonomist examines primarily plant morphology, searching for a few conservative, genetically controlled traits that consistently allow the separation of plants into well-defined groups. Traditional taxonomists do not limit themselves to the examination of only a few traits at the start of a study. They examine many morphological, anatomical, and chemical characters from many specimens but eventually select only a few morphological characters to serve in defining the various species. The characters selected are those that show discontinuities and thus are most helpful in separating species. However, the selection process is somewhat subjective.

Biosystematists—in contrast to traditional taxonomists—are interested in determining natural biotic units: populations of plants that maintain their distinctiveness because of biological barriers that genetically isolate them from other populations. These isolating barriers may be due to breeding behavior (time of flowering, type of pollinator), habitat or geographic isolation, or inability to form fertile hybrids with closely related groups (perhaps because of chromosomal differences or pollen incompatibility with the stigma and style). This particular approach to defining species has been called the biological species concept, or BSC. The BSC has been replacing the older, more traditional morphological species concept since the early 1900s and appears to be the more widely accepted of the two at this time (Mayer 1992).

Conflicts arise between biosystematists and traditional taxonomists because the natural biotic units do not always correspond to well-defined groups. Two populations of the same traditional species may prove, upon crossing in the greenhouse, to yield no offspring or infertile offspring; thus they belong to two different biosystematic species. The traditionalist argues that such nonvisible traits as crossability are theoretically important, but they cannot be interpreted in the field with a hand lens and thus are of no practical importance. Also, greenhouse crosses may not be a valid imitation of crossing frequency in nature. One further difficulty with the biosystematic approach is that crossability is seldom all or nothing, so subjective decisions still have to be made. For example, if populations A and B are 78% interfertile, are A and B in the same species?

Phylogenetic systematists attempt to deduce evolutionary (phylogenetic) relationships by noting the presence of certain traits each taxonomic group exhibits, and then linking pairs and clusters of taxonomic groups in a diagram based on the number of traits shared. The traits selected are only those that systematists agree reflect evolutionary time. A given trait is either in a primitive, hence ancestral, state (mathematically equivalent to 0) or in an advanced, hence derived, state (rated 1). The resulting diagram (called a cladogram) illustrates two things at once: (1) the evolutionary relatedness of the groups (the closer together they are on the cladogram, the greater the number of traits in common); and (2) the direction of evolution through time (the most recently evolved groups are at the tips of the cladogram).

As an abstract example, consider Table 3-1, which lists six taxonomic groups of plants, A–F, and six traits, 1–6. Notice that all six groups possess the derived version of trait 1. We will assume in this example that no other plants in the world

Table 3-1   Cladistic analysis of six taxonomic groups (A–F) and six traits (1–6). Each taxon either shows an advanced (1) or primitive (0) version of the trait. At this stage of classification, the taxa might represent provisional populations, species, or higher categories. All six groups belong to a single clade, meaning they share a common ancestor, and that ancestor had an advanced version of trait 1.

| Taxonomic group | Trait | | | | | |
	1	2	3	4	5	6
A	1	0	0	0	0	0
B	1	1	1	1	0	0
C	1	1	1	0	1	1
D	1	0	1	0	0	0
E	1	1	1	0	0	0
F	1	1	1	0	1	1

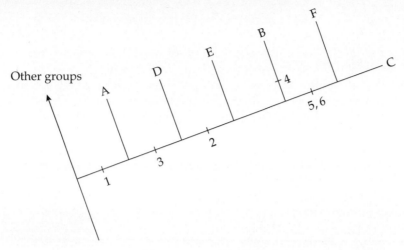

Figure 3-4   Cladogram of data presented in Table 3-1. One clade, with taxonomic groups A–F, is represented as a branch from the main line of all other plants (arrow pointing up and to the left). The most highly evolved taxa appear at the distal tips of the clade (C and F); they show the advanced form of traits 1–3 and 5–6. Taxon B is most closely related to them, and taxon A is the least related to them.

have it; therefore we are dealing with a clade—a group of organisms descended from a common ancestor. This clade is represented in our cladogram as though it were a branch coming off from the rest of the plant kingdom (Figure 3-4). Our six groups will be arranged on this branch in a way that signifies evolutionary relationship. Groups C and F have the most derived traits and they will be placed at the outermost tips of the cladogram (Figure 3-4). They represent the most recently derived groups in the clade, the two that have evolved farthest from the ancestral type. Group B has the next highest number of derived traits; it is placed further down the branch, closer to the ancestral type. Group E is next, followed by D and then by A. Hash marks on the cladogram indicate that every group above the mark possesses the advanced form of that trait; those below the mark lack it. For the purposes of taxonomy, the cladogram in Figure 3-4 could be interpreted in several ways. Each of the six groups might be classified as six separate species; groups C and F could be retained as subspecies within the same species, giving only five species; or A, D, E, and B could similarly be lumped into one species, giving only two species in the clade.

   All of these approaches to taxonomy are partly arbitrary, leading to classifications that are partly natural and partly human artifacts. Furthermore, the spatial scale at which taxonomic lines are drawn are generally very broad and regional. Plant ecologists often work at finer scales. Can the average large, somewhat arbitrary taxonomic species be redefined or subdivided to make it a better, more natural ecological tool?

Variation Within Species

Linnaeus and taxonomists after him recognized that some taxonomic species were not homogeneous: their member plants varied in height, leaf size, flowering time, or other attributes, with change in light intensity, latitude, elevation, or other site characteristics. It was thought that such differences within a species were plastic, not heritable responses. The early transplant gardens of Kerner (1895) in the Tirol supported this view.

This conclusion was challenged by the experiments of the botanist Göte Turesson in the early twentieth century. He hypothesized that many variations within a species were heritable and were of adaptive value to particular habitats within the species' range. During the 1920s he attempted to arrive at, as he put it, "an ecological understanding of the Linnaean species." At first he examined plants from Sweden only, but later he studied species that ranged all over Europe. For each species (usually a perennial), he brought back vegetative material or seeds from different habitats or regions and grew the plants in a test garden near his home at Åkarp, Sweden. He reasoned that if the morphological or phenological differences noted in the field were retained in the garden, then the traits were heritable and genetically based.

Table 3-2 summarizes a typical result, using the herbaceous perennial hawkweed (*Hieraceum umbellatum*), which in southern Sweden grows on coastal sand dunes, on rocky headlands, and in inland fields and woodlands. The field differences were retained in the test garden. Were these genetically distinct types of plants different species? Turesson made all the crosses and found all the types to be interfertile. Therefore, the types were technically part of only a single species.

Turesson called these entities ecotypes. An **ecotype** is the product of the genetic response of a population to a habitat. It is a population or group of populations distinguished by morphological and/or physiological characters, interfertile with other ecotypes of the same species, but usually prevented from natural in-

Table 3-2   Some morphological and phenological traits of hawkweed (*Hieraceum umbellatum*) ecotypes, as revealed in Turesson's uniform garden. From Briggs and Walters. 1969. *Plant Variation and Evolution.* McGraw-Hill, New York.

	Ecotypes		
Traits	Woodland	Field	Dune
Habit	Erect	Prostrate	Intermediate
Leaves	Broad	Intermediate	Narrow
Pubescence	Absent	Present	Absent
Autumn dormancy	Present	Present	Absent

terbreeding by ecological barriers. (Although Turesson wrote in English, it is a very dense form of English and is best read in the "translation" written by Turrill in 1946.)

It is important to outline all the elements that were part of ecotypes as Turesson saw them: (a) they were genetically based, (b) their distinctiveness could be morphological, physiological, phenological, or all three, (c) they occurred in distinctive habitat types, (d) the genetic differences were adaptations to the different habitats, (e) they were potentially interfertile with other ecotypes of the same species, and (f) they were discrete entities, with clear differences separating one ecotype from another.

At the same time that Turesson was describing ecotypes in over 50 common European species, three biologists were reporting similar results with western North American perennial plants. In 1922, Jens Clausen, a geneticist and cytologist, David Keck, a taxonomist, and William Hiesey, a physiological ecologist, established a 323-km-long study transect in California, supported by the Carnegie Institution. The transect extended from near sea level at Stanford University (just south of San Francisco), across the Coast Ranges, through California's Central Valley, up the gradual west face of the Sierra Nevada to timberline at 3000 m elevation, and partly down the steep, relatively arid east face of the range (Figure 3-5). Despite great environmental diversity along this transect, Clausen, Keck, and Hiesey were able to find about 180 species whose ranges extended over much or all of the transect.

Each species was collected at a variety of locations along the transect, brought back to greenhouses at Stanford, cloned (literally divided into parts and each part induced to root separately so that many genetically identical plants were produced), grown for six months, then transplanted to test gardens along the transect. Initially there were 11 gardens, but the number was soon reduced to three: Stanford, near sea level; Mather, in the mid-elevation Sierra Nevada; and Timberline.

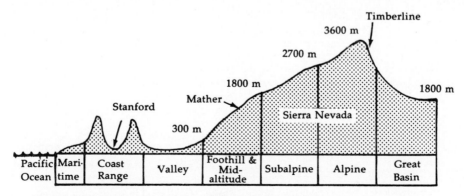

Figure 3-5 Profile, west to east (left to right), of California at approximately 37°30′N latitude, showing the location of the three principal transplant gardens used by Clausen, Keck, and Hiesey. From Clausen et al. 1940. Courtesy of the Carnegie Institution of Washington.

Table 3-3 summarizes the garden environments. About 60 species were rugged enough to survive this initial handling, and their growth, phenology, and mortality were followed for as long as 16 years.

The herbaceous perennial cinquefoil (*Potentilla glandulosa*) can serve as an example of their garden results (Table 3-4). Based on morphology, phenology, physiology, and habitat, there appeared to be four ecotypes in the species (Clausen, Keck, and Hiesey chose the more conservative taxonomic term *subspecies,* but *ecotype* is synonymous in this case). Ecotype *typica* was a lowland form. It grew best at Stanford, survived at Mather, but did not last a year at Timberline. Its herbage was relatively frost tolerant, and at Stanford it was able to grow year-round, but at Timberline it could not withstand the hard winter frosts (Figure 3-6). Ecotypes *reflexa* and *hanseni* were mid-elevation forms, distinctly taller than *typica*. One occurred on dry sites, the other on wet meadow sites, and there were subtle morphological differences between them. Both flowered late in summer, so late that when grown at Timberline they often were hit by frost before flowering or seed set could be completed. Their herbage was not frost tolerant, and cold temperatures induced them to enter dormancy. They survived at all locations, but grew best at Mather. Ecotype *nevadensis,* a timberline form, was the shortest of the four. It flow-

Table 3-3   The environments of the three principal transplant gardens used by Clausen, Keck, and Hiesey. From Clausen et al. 1940. Courtesy of the Carnegie Institution of Washington.

Garden name	Elevation (m)	Surrounding vegetation	Growing season (mo)	Annual ppt. (cm)	Ppt. as snow (cm)	Mean max. temp. (°C)	Mean min. temp. (°C)
Stanford	30	Oak woodland, chaparral	12.0	30	None	36	−3
Mather	1400	Mixed conifer forest	5.5	95	90	36	−11
Timberline	3050	Subalpine forest	2.0	73	485	23	−22

Table 3-4   Ecotypes (subspecies) of cinquefoil (*Potentilla glandulosa*) as revealed by the research of Clausen, Keck, and Hiesey. From Daubenmire. 1974. *Plants and Environment.* Copyright © 1974 by John Wiley and Sons, Inc. Reprinted by permission.

Ecotypes	Environment
nevadensis	Alpine and subalpine, 1600–3500 m
hanseni	Mid-elevation wet meadows, 250–2200 m
reflexa	Mid-elevation dry slopes, 250–2200 m
typica	Coast Ranges

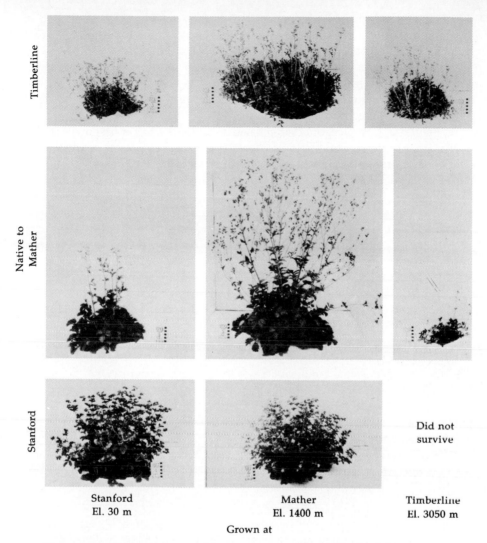

Figure 3-6 The appearance of three ecotypes of *Potentilla glandulosa* as grown at three different locations. The top row shows ecotype *nevadensis*, the middle row is ecotype *reflexa*, and the bottom row shows ecotype *typica*. Courtesy of Dr. William M. Hiesey.

ered early in the season and its herbage was very frost tolerant. Its winter dormancy may have been induced by short day length rather than temperature, for it was winter dormant even in the mild climate at Stanford. A period of winter cold may still have been physiologically necessary, however, for growth at Stanford declined in subsequent years. The herbage seemed more susceptible to disease at lower elevations than at Timberline. Crossing experiments showed that all the ecotypes were interfertile.

Clausen, Keck, and Hiesey concluded that most species are composed of an assemblage of ecotypes, each ranging in size from a single population to a regional group of many populations; the wider the species' range is, the more ecotypes there are within the species. Ecotype research by many other investigators, on many different species throughout the world, has corroborated this conclusion. The ecotype concept of Turesson may seem to give us the ecological species tool we seek, but it has limited practicality.

In Scotland, J. W. Gregor (1946) closely examined what at first appeared to be two ecotypes of the coastal plantain, *Plantago maritima.* One ecotype inhabited salt marshes regularly flooded by high tide with a soil salinity of approximately 2.5%. These plants had short leaves, small seeds, and thick, short, somewhat decumbent flowering stalks. The other ecotype inhabited nonsaline meadows farther inland, and these plants had longer leaves, larger seeds, and thinner, taller, more upright flowering stalks. Gregor collected seeds of each and sowed them in a test garden. In addition, he collected and sowed seeds from plants growing in an **ecotone,** an intermediate habitat. The resulting plants showed that field differences were genetically fixed, but more importantly, they showed a continuous gradation from one extreme, one ecotype, to the other. There were no discrete boundaries between ecotypes or even between ecotypes and plants from the ecotone (Table 3-5). Ecotone plants completely overlapped the habit grades for both salt marsh and meadow ecotype plants.

Olaf Langlet (1959), a compatriot of Turesson, brought seeds of the pine *Pinus sylvestris* from 580 sites throughout Sweden to a test garden. When he examined saplings for growth rate and morphological features, he found the extremes to be quite different, but a **cline,** a continuum of variation, connected the extremes. There were no sudden breaks in the range of variation where one could say ecotype A ended and ecotype B began.

Thus, Turesson's stairstep concept of ecotypes must be replaced with an ecocline concept. An **ecocline** is a gradation in the attributes of a species (or commu-

Table 3-5   Results of Gregor's cultivation experiments with plantain (*Plantago maritima*) ecotypes. Habit grade 1 includes the most prostrate, succulent plants, with the shortest leaves and smallest seeds; habit grade 5 is the other extreme; habit grades 2–4 are intermediate. From Briggs and Walters. 1969. *Plant Variation and Evolution.* McGraw-Hill, New York.

Habitat	Mean scape length (cm)	% of plants in habit grades 1–5				
		1	2	3	4	5
Salt marsh, soil salinity 2.5%	23.0	74.5	21.6	3.9		
Upper marsh edge, ecotone	38.6	10.8	20.6	66.7	2.0	
Nonsaline meadow above marsh, soil salinity 0.25%	48.9		2.0	61.6	35.4	1.0

nity or ecosystem) associated with an environmental gradient. Sometimes *ecocline* is used to refer to the environmental gradient itself (Hanson 1962). Turesson, and to some extent Clausen, Keck, and Hiesey, thought of ecotypes as discrete because their sampling method prejudiced their results: Plant material was selected from widely separate places, and ecotones were ignored.

Sometimes habitats are discrete, with sudden changes from one habitat to the other (usually because of steep terrain or edaphic factors), but typically habitats intergrade, and plant populations vary with the same subtlety. An ecotype, then, is only an arbitrary segment of an ecocline, which may be convenient to recognize for reference purposes.

James Quinn (1978, 1987) found the ecotype concept to be an oversimplified description of variation he found in the Australian perennial bunchgrass *Danthonia caespitosa*. He found each population to be markedly different in phenology, morphology, and physiology, even among populations located in similar habitats. He noted that if his sampling design had been less extensive, he might have concluded that a smaller number of discrete ecotypes existed. For example, the amount of caloric energy that went into annual sexual reproduction increased with increasing latitude when three particular populations were evaluated. However, when many interspersed populations were included, the change was continuous and random, not correlated with latitude at all. Quinn concluded that the ecotype concept (as meaning a *group* of populations) is misleading because in nature each population is "individualistic." If one insists on retaining the term *ecotype*, it could be merely a synonym for any one population.

The term *ecotype* is now used in Quinn's sense. For example, Nienhuis et al. (1994) specifically adopted Quinn's definition when they showed that 31 ecotypes of the mustard *Arabidopsis thaliana* differed in water use efficiency (mg dry weight gained per g of water lost in transpiration) and that the trait was genetically determined. They clearly state that what they had done was to test 31 different populations and they assumed that each was its own ecotype.

Ecophysiological Variation

One aspect of ecotype research has revealed the subtle physiological, metabolic basis of plant adaptation to the local habitat. That aspect will be illustrated by a series of investigations that take us successively closer to the ultimate control of adaptations, the genes themselves.

Harold Mooney and Dwight Billings (1961) published a classic study on the perennial herb sorrel (*Oxyria digyna*) (Figure 3-7), for which they received the 1962 George Mercer award from the Ecological Society of America for the most outstanding ecological paper published by young ecologists in the previous two years. *Oxyria* has a circumboreal distribution in the treeless arctic tundra and extends south in alpine tundra along several mountain chains. In the conterminous United States it is found at high elevations in the Sierra Nevada and Rocky Mountains.

Arctic tundra vegetation and alpine tundra vegetation share many similarities, but there are important environmental differences. Alpine areas in the tem-

Figure 3-7   Alpine sorrel (*Oxyria digyna*) above timberline in Wyoming. Note the heart-shaped leaves and the profusion of flowers, both traits of the alpine ecotype. Courtesy of W. D. Billings.

perate zone experience higher light intensity and greater extremes of temperature in summer than do arctic areas.

Just as there are two environmental extremes, there are two ecotypic extremes of *Oxyria*. Mooney and Billings showed that the arctic and alpine ecotypes differed morphologically and phenologically even when grown together from seed in controlled chambers that simulate a natural, uniform environment (the modern test garden) (Table 3-6). Mooney and Billings also showed that the metabolism of the two ecotypes differed.

By placing potted plants in a small clear acrylic chamber whose microenvironment could be controlled, then measuring the CO_2 content of air passing into and out of the chamber, the rate of photosynthesis per unit leaf area or weight could be determined. As shown in Figure 3-8a on page 48, the alpine ecotype had a significantly higher temperature optimum for photosynthesis. The amount of light required to saturate the photosynthesis system, beyond which increasing light fails to increase the rate of photosynthesis very much, was greater for the alpine ecotype (Figure 3-8b), again correlating well with the environmental differences in nature. These physiological differences most likely correlated with enzymatic and other biochemical factors, but Mooney and Billings did not pursue their investigation to that level.

One species of cattail, *Typha latifolia*, is widely distributed in the northern hemisphere. McNaughton (1966) collected dormant rhizomes from such disparate habitats as the cool, maritime, foggy Pacific coast at Point Reyes, California, and the relatively hot, arid Sacramento Valley near Red Bluff, California, more than 100 km inland. He potted the rhizomes and placed them in a greenhouse regulated at 30/25°C day/night, where they broke dormancy, produced shoots, and grew for three months. He then took samples of leaf tissue, extracted their enzymes, and

Table 3-6 Some morphological, biochemical, and phenological traits of arctic and alpine ecotypes of alpine sorrel (*Oxyria digyna*). From "Comparative physiological ecology of arctic and alpine populations of *Oxyria digyna*" by H. A. Mooney and W. D. Billings, *Ecological Monographs* 31:1–29. Copyright © 1961 by the Ecological Society of America. Reprinted by permission.

Trait	Arctic	Alpine
Rhizomes	Present	Absent
Intensity of flowering	Low	High
Leaf anthyocyanin level (red color)	Low	High
Critical photoperiod for flowering	24 hr at 70°N	15–17 hr at 38–48°N
Leaf shape	Ovate	Heart-shaped

subjected the extract to a heat stress of 50°C for periods of up to 30 minutes. This simulated the leaf temperature that Red Bluff plants might experience in nature. Point Reyes plants probably do not experience a leaf temperature in nature higher than 30°C.

After each period of heat stress, he tested the activity of three important respiratory enzymes: malate dehydrogenase, glutamate oxaloacetate transaminase, and aldolase. He reasoned that the heat tolerance of the Red Bluff ecotype must inherently lie in the heat stability of some or all of its enzymes. As shown in Figure 3-9 on page 49, malate dehydrogenase was significantly more heat stable in the Red Bluff ecotype than in the Point Reyes ecotype; the other two enzymes showed no difference. Enzymes are complex macromolecules with tertiary or quaternary structure, and the reaction site is a relatively small part of their total architecture and chemistry. Thus, malate dehydrogenase could differ in the two ecotypes in many ways that might increase stability or activity (McNaughton 1972). Why only one of the enzymes, rather than all three, showed ecotypic differentiation is not clear.

In another paper, McNaughton (1967) showed that cattail ecotypes from Point Reyes at sea level and Wyoming at 1980 m elevation differed in their photosynthetic efficiency. The high-elevation, short-growing season ecotype exhibited about twice the photosynthetic rate of the Point Reyes ecotype. He was able to trace the difference to higher reducing activity (an important process in the light reactions of photosynthesis) in the chloroplasts of the Wyoming ecotype.

Respiration is also attuned to elevation at the enzymatic level, as Klikoff (1966) showed for populations of the grass squirrel tail (*Sitanion hystrix*) from different elevations of the Sierra Nevada. Isolated mitochondria showed higher oxidative rates at lower temperatures with increasing elevation of the parent plant.

Some species are differentiated into sun ecotypes (those that germinate and develop in the open) and shade ecotypes (those that develop beneath the canopy of other plants). A European species of goldenrod (*Solidago virgaurea*), a perennial herb, has such ecotypes. Olle Björkman, in a series of papers culminating in 1968,

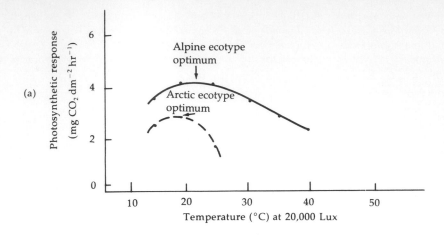

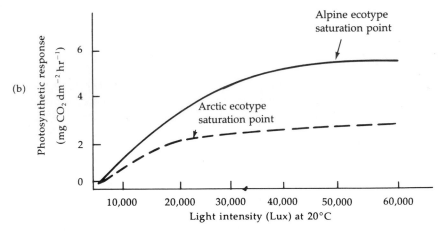

Figure 3-8   Photosynthetic response of arctic (dashed line) and alpine (solid line) ecotypes of *Oxyria digyna* (a) to varying temperatures at full sun, approximately 20,000 lux, and (b) to varying amounts of light at 20°C. From "Comparative physiological ecology of arctic and alpine populations of *Oxyria digyna*" by H. A. Mooney and W. D. Billings, *Ecological Monographs* 31:1–29. Copyright © 1961 by the Ecological Society of America. Reprinted with permission.

examined the differences between the two goldenrod ecotypes at successively more subtle levels. He collected vegetative material from plants growing in exposed heath vegetation in Norway and from plants growing beneath oak forests in Sweden, then cloned the material and grew plants in controlled environments.

Among other differences, he showed that the light saturation curve of the sun ecotype was different from the light saturation curve of the shade ecotype. The sun ecotype had a higher light saturation point and showed a higher rate of photosynthesis at that saturation point. To find the reason for the difference, he first searched

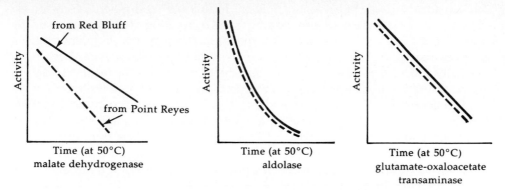

Figure 3-9   Enzyme response to incubation at 50°C. Enzymes were extracted from coastal (Point Reyes, California; dashed line) and inland (Red Bluff, California; solid line) ecotypes of cattail, *Typha latifolia*. From McNaughton, 1966. "Thermal inactivation properties of enzymes from *Typha latifolia* L. ecotypes." *Plant Physiology* 41:1736–1738. By permission of the American Society of Plant Physiologists.

at the morphological level, asking if sun ecotype leaves perhaps absorbed more light; the answer was no. He then searched at the cellular level, asking if the chlorophyll concentration was higher in sun leaves; the answer was no. Finally he examined the enzyme level. The enzyme responsible for fixation of CO_2 into the dark reaction pathway of photosynthesis is ribulose bisphosphate carboxylase (also called carboxydismutase). When he assayed for the concentration (activity) of this important enzyme, he found that it was two to five times greater in the sun ecotype, which was sufficient to account for the five times higher rate of photosynthesis of the sun ecotype at higher light intensity.

Had these ecophysiologists examined individual populations, instead of searching for ecotypes, they probably would have reported significant differences from one population to another on a much more local scale. Recent research has revealed physiological difference at this scale, and even within a single population. Genetic diversity is to be expected, of course, within a typical population because each individual is the unique product of its parental genes. What is surprising is that even small populations of rare plants or of asexually reproducing plants may contain high levels of genetic variability.

Using the techniques of molecular biology, including the multiplication of bits of nuclear DNA by an enzymatic process called polymerase chain reaction, Smith and Pham (1996) showed that populations of a rare Idaho wild onion possessed unexpectedly high genetic diversity. In another study, individual plants of the pink *Silene alba* were mapped in one 20 × 65 m area, and each had its chloroplast DNA examined (McCauley et al. 1996). The researchers found complex spatial patterns of genotypes, those most similar being found near each other. A third example of within-population research examined the rhizomatous West Coast beach strawberry (*Fragaria chiloensis*). This plant has populations whose individuals are genetically distinct, mixed with identical, clonal individuals connected by rhizomes. Alpert et al. (1993) measured genetic differences by treating drops of leaf tissue extract with gel electrophoresis, a process that separates proteins. Different proteins are presumed to represent different genes. Several genotypes coexisted at a scale as small as 10×10 m, despite asexual reproduction. Perhaps the most extreme example of within-population variation was provided by Rex Cates, who studied the leaf biochemistry of Douglas fir (*Pseudotsuga menziesii*). He reported that terpenes, acetates, sesquiterpines, and some other volatile hydrocarbons differed among individuals, and he concluded that such genetic variation is important in reducing outbreaks of herbivores and pathogens because the substrate on which they feed is too varied, from tree to tree, to allow a pest to spread widely and rapidly (Cates 1996, Cates et al. 1991).

Another way in which populations and individuals can differ is the amount of nuclear DNA in each cell. The tall prairie dominant grass, big bluestem (*Andropogon gerardii*), typically exists as a hexaploid in the wet eastern part of its range, but is octoploid or nonoploid in the arid western plains (Keeler 1992). The east-west gradient is much more complex than that, however. In one Kansas prairie, midway between the two regional extremes, the great majority of sample plots (each only 40–120 m^2 in size) contained complex mixtures of hexaploid, septoploid, octoploid, and nonoploid plants. Other North American prairie grasses that exhibit similar nuclear patterns include *Agropyron smithii, Bouteloua curtipendula, B. gracilis,* and *Panicum virgatum.*

Keeler was not able to attribute differences in nuclear DNA to ecological fitness, but Marcel Rejmanek (1995, 1996) has, among pine species. Pines have been widely transplanted in the temperate zones of the world. Some of the species have proven to be aggressive weeds in their new homes, whereas others do not spread from their plantations. Rejmanek found that the aggressive species had the following five traits in common: low seed weight, short juvenile period, short periods of time between large annual seed crops, seed dispersal by vertebrates, and small nuclear DNA mass. Apparently, a small amount of nuclear DNA correlates with a short time for each cell division cycle and, hence, with rapid growth. Similar relationships between DNA mass and aggressiveness exist for introduced *Senecio* species in Australia (Lawrence 1985), *Acacia* species in India (Hughes and Styles 1982), and *Briza* species in the United States (Murray 1975).

Ultimately, molecular ecology could have great practical importance in restoration ecology. Plants can be genetically engineered (or at least selected) for traits that enable them to grow on nutritionally poor road cuts, toxic mine spoil, coarse dredge piles, overgrazed and eroded land, and other difficult substrates. At present, we have yet to reach this level of biotechnology.

Acclimation

Acclimation (also called acclimatization) is a plastic, temporary change in an organism caused by an environment to which it has been exposed *in the past*. Matthaei (1905) may have been the first to document such a phenomenon in plants, and the effect of past temperatures on photosynthesis and respiration rates has been reviewed by Semikhatova (1960).

Billings et al. (1971) conducted an experiment that provides a good example of acclimation, and once again alpine sorrel proved to be a good study plant. Alpine sorrel (*Oxyria digyna*) seeds were collected from a range of habitats, germinated and grown in a uniform greenhouse for four months, then subdivided into three growth chamber environments: warm (32/21°C day/night), medium (21/10°C), and cold (12/4°C). After five to six months in the chambers, replicates of each collection were measured for net photosynthesis at a range of temperatures, from 10 to 43°C, and the optimum temperature for photosynthesis was noted. Table 3-7 shows that representatives of arctic and alpine ecotypes possessed different acclimation capacities. The optimum temperatures for alpine plants shifted as much as 11°C, depending on the temperatures they had been growing at before the photosynthesis measurement, but the optimum temperature for arctic plants shifted only 1°C.

Similar effects of preconditioning, or acclimation, have been shown for plants as diverse as pine tress (Rook 1969) and desert shrubs (Mooney and West 1964, Mooney and Harrison 1970, Strain and Chase 1966).

The relationship between plant and environment can be written as follows:

$$\text{Phenotype} = \text{genotype} + \frac{\text{prevailing}}{\text{environment}} + \frac{\text{past}}{\text{environment}}$$

How distant a past environment can influence phenotype? The influential past environment may reach back to parent generations, according to Rowe (1964). Seeds of groundsel (*Senecio vulgaris*) were germinated at different temperatures. The seedlings were immediately transferred to a common environment and allowed to grow for 80 days, then the shoots were weighed. The sevenfold difference in shoot weights shown in Table 3-8 is hard to explain except as a result of

Table 3-7 The effect of acclimation temperature on the optimum temperature of photosynthesis of alpine and arctic ecotypes of alpine sorrel (*Oxyria digyna*). Adapted from Billings et al. Reproduced with permission of the Regents of the University of Colorado from *Arctic and Alpine Research* 3:277–289, 1971.

| | | Optimum temperature for photosynthesis of plant from each acclimation regime (°C) | | | |
Population site	Ecotype	Warm (32/21°C)	Medium (21/10°C)	Cold (12/4°C)	Shift (warm− cold)
Sonora Pass, California	Alpine	28	21.5	17	11
Pitmegea River, Alaska	Arctic	21	20.5	20	1

Table 3-8   The effect of germination temperature on subsequent growth of groundsel (*Senecio vulgaris*). From "Environmental preconditions with special reference to forestry" by J. S. Rowe, *Ecology* 45:399–403. Copyright © 1964 by the Ecological Society of America. Reprinted by permission.

Germination temperature (°C)	Subsequent growing conditions, next 80 days	Final plant weight (mg)
10		147
14	All grown together at 17°C,	775
23	16 hr photoperiod	1078
30		390

temperature differences at the time of germination, 80 days earlier. Another study found that the nutrient content and seed size of *Abutilon theophrasti* seeds can be influenced by the nutrient level of the soil in which the mother plant grows. Seed nutrient content and size can affect the next generation by stimulating early germination and faster initial seedling growth (Parrish and Bazzaz 1985). Other environmental factors have been shown to affect seed size—and the next generation—in the annual herb baby blue eyes (*Nemophila menziesii;* Platenkamp and Shaw 1993). Several species of cacti in Baja California produce seeds that retain a "memory" of wetting and drying cycles prior to germination. Those that have experienced wetting and drying germinate several days sooner, grow faster, and experience higher survival than controls. Plants of the weedy annual prickly lettuce (*Lactuca scariola*) subjected to different day lengths or applications of growth regulators produced progeny that differed in germination, seedling growth, and time of flowering (Gutterman et al. 1975). Other examples of parental environment carrying over to germination have been reported by Baskin and Baskin (1973), Quinn and Colosi (1977), and Hume and Cavers (1981).

Highkin (1958) grew pure-line peas under two sets of conditions: 24/14°C day/night and 26°C constant. Pollen from plants in either regime was transferred to the stigmas of other pure-line peas that had been kept separate in uniform conditions. The seeds were harvested, kept separate according to the temperature regime of the pollen donor, germinated, and grown in uniform conditions. The two types of progeny differed significantly in height and number of nodes, and there was a diminishing but still measurable carryover effect for several generations.

How are these acclimation traits carried across generations? The phenotype of an organism can be influenced by maternal effects (influences of the mother plant on the offspring by non-genetic means) such as the mother plant's nutritional status having affected seed size, as previously described (Futuyma 1986). Maternal effects can also be transmitted by DNA-containing cytoplasmic organelles such as mitochondria and chloroplasts. Some of this DNA duplicates nuclear genes and the rest is unique or complements nuclear genes. According to a recent review of cytoplasmic inheritance (Mogensen 1996), organelles can also be inherited from the male parent. The use of recombinant DNA technology has allowed research in organelle inheritance to be routinely applied to population ecology studies.

Integrated Approaches to Population Research: A Case Study

An excellent example of research that utilizes a wide assortment of techniques is a study of *Dryas octopetala* ecotypes (populations) from the Alaskan tundra (McGraw and Antonovics 1983, McGraw 1985a and b). The authors combined such techniques as growth chamber studies in the Duke University phytotron, field transplants, competition trials, pollination ecology, demographic observations, environmental manipulation *in situ*, photosynthesis measurements, and determination of photosynthate allocation patterns to various plant organs. This rich mixture made their conclusions ecologically powerful and important.

Dryas octopetala is a perennial dicot herb of the rose family (Figure 3-10). It grows in two distinct habitats, which may be very near one another, and several traits distinguish the plants of each habitat (Table 3-9). Fellfield sites are on ridges or slopes that experience dry summer soils and minimal winter snow cover. Snowbed sites are in protected swales with late-melting snow; as a result, summer aridity is much less pronounced. Plant cover is dense on snowbed sites but rather open on fellfields. McGraw and Antonovics selected a study site where fellfield *Dryas* grew only 150 m away from, and 30 m above, snowbed *Dryas*.

First they demonstrated that the two forms (which taxonomists had distinguished as subspecies) were fully interfertile. Both are outcrossers, and there was no evidence that either could self-pollinate or produce seeds by apomixis. Hybrids in nature were often encountered, though they accounted for less than 1% of all plants. When grown together in a common garden (the phytotron) or when reciprocally transplanted in the field, some of the morphological differences in Table 3-9 became less pronounced, but they still remained distinct. Generally, the snowbed ecotype showed more phenotypic plasticity than did the fellfield eco-

Figure 3-10 *Dryas octopetala* subspecies *octopetala*. Courtesy of James B. McGraw.

Table 3-9 Some genetically fixed differences between *Dryas octopetala* populations. Fellfield populations are subspecies *octopetala*; snowbed populations are subspecies *alaskensis*.

Fellfield	Snowbed
Plants long-lived (>100 yr)	Plants may be shorter-lived
Rhizomatous spread modest	Rhizomatous spread extensive
Flowers early in growing season	Flowers up to 2 mo later
Leaves deciduous	Leaves evergreen
Leaves with orange scales and branched hairs	Leaves with glands
Leaves small, short	Leaves large, long
Fewer marginal teeth on leaves	More marginal teeth on leaves
Low specific leaf weight (g cm^{-2})	High specific leaf weight
Petioles short	Petioles long

type: in the common garden its petiole length, leaf area, leaf margin, and specific leaf weight approached those of the fellfield ecotype more than vice versa.

If the two are fully interfertile and are obligate outcrossers, how did this genetic differentiation occur? Observations of insect pollinator flight paths revealed, first of all, that more than 99% of all pollen was transferred within ecotypes rather than between them. Second, the onset of flowering for snowbed plants was later (up to two months later, depending on the year) than that for fellfield plants (late snow covered the snowbed plants, and wet soil warms more slowly than the drier fellfield soil).

Field manipulations of the environment revealed that the most significant microenvironmental factors for differential survival of the two ecotypes were light and nutrients. Fellfield plants beneath shade screens declined in growth more than snowbed plants. Addition of nitrogen and phosphorus stimulated snowbed plant growth, whereas it eventually had a negative effect on fellfield plant growth (mainly due to the increased growth of fellfield grasses that overtopped *Dryas*). Habitat differences in degree of wind exposure and soil moisture were not important.

Competition trials in the phytotron between the two ecotypes demonstrated that root competition was more critical than shoot competition and that the fellfield ecotype was consistently the "loser"—that is, it experienced a greater reduction of growth in the presence of competition with the other ecotype than did the snowbed ecotype. The authors concluded that the snowbed ecotype "won" because of more efficient nutrient uptake and a plasticity in leaf form that permitted less self-shading in crowded conditions.

Finally, by following reciprocal transplants started from seed for several years, they were able to determine at which life-cycle stages the ecotypes were sorted by the different microenvironments. They devised a simple scale of com-

Table 3-10   Relative mortality of *Dryas* populations at various life-cycle stages when planted from seed in two habitats and followed for several years (1.00 = complete mortality). A dash indicates maximum survival of that population relative to the other (if both populations have a dash there was no statistically significant difference between them). P&S = pollination and seed set. From McGraw and Antonovics (1983).

Population	Habitat	Germination	After first winter	After first summer	End of first yr	Adult	P&S
Fellfield	Fellfield	—	0.89	—	0.85	—	—
Snowbed		0.85	—	—	—	—	—
Fellfield	Snowbed	—	0.89	0.74	0.97	0.50	—
Snowbed		—	—	—	—	—	—

parative survival success that ranged from 0 (equivalent survival of the two ecotypes) to 1 (zero survival for one ecotype). Departures from 0 were statistically analyzed; Table 3-10 shows only the significant departures.

Note that selection has not yet perfectly matched each ecotype to its usual habitat. Only in the process of completing an entire life cycle will each ecotype ultimately be favored by the microenvironment of its usual habitat.

Summary

Plant ecologists would like to use species as deductive tools, as rather precise indicators of certain levels of environmental factors. This may not be a realistic objective for two reasons. First, plants respond to a complex of climatic, edaphic, and biotic factors, and the impact of single factors is difficult to isolate. The tolerance range of a species to factor X may be modified by factors Y or Z. Good's version of the theory of tolerance recognizes the special confounding effect of competition on tolerance ranges. Second, taxonomic species, whether recognized on morphological, biological, or statistical grounds, are partially artifacts of the human desire to classify; they are not completely natural units.

Turesson searched for "an ecological understanding of the Linnaean species." He discovered that taxonomic species were composed of ecologically important subunits, which he called ecotypes. He defined ecotype as the genetic response of a population (or group of populations) to a habitat, distinguished by morphological and/or physiological characters, yet interfertile with other ecotypes of the same species. Most wide-ranging species are now known to be made up of many ecotypes, but the practical utility of the concept was diluted when it was discovered that ecotypes are just as heterogeneous, with boundaries just as vague, as species. Turesson's ecotype concept has been replaced with the ecocline concept, recognizing each population as being subtly different from the next. Populations, rather than ecotypes, have become the accepted research focus.

Populations are known to vary in terms of plant behavior, morphology, phenology, and metabolism. Molecular ecology has now taken our level of understanding down to the genetic level. We are not yet able, however, to genetically engineer (or select) taxa for habitat restoration purposes as we can for seed production or disease resistance. A confounding factor in population ecology research is acclimation. An environment to which an organism has been exposed in the past (sometimes a generation in the past) may cause a physiological change in that organism. The capacity for acclimation is likely to be a genetic trait for most populations.

CHAPTER 4

POPULATION STRUCTURE AND PLANT DEMOGRAPHY

Describing the structure of a population and the distribution of individuals and populations is central to understanding plant dynamics. Plants are not evenly distributed in either time or space. Differences in environmental conditions, biotic neighborhoods, and site histories influence the distribution and dynamics of plant populations.

The arrangement of plants in space and time presents both unique advantages and disadvantages for the researcher. Unlike most animals, individual adult plants do not move, in general, making tracking of survivorship and mortality much easier. Plants can, however, produce new individuals asexually and can drop or add new sets of organs (flowers, leaves, roots, stems, and branches) in response to the external environment (biotic or abiotic). Thus the material of the plant population ecologist is not limited to the distribution and dynamics of individuals in a population but includes the dynamic growth of the ever-changing plant body.

To gain some perspective on plant populations, we will first examine different patterns of population density and distribution of individuals within a single species. Second, we consider plant demography as it applies to individuals and modules of plant growth. Third, we examine how immigration and emigration of propagules between populations may influence the persistence of species within an integrated landscape.

Density and Pattern

Density is the number of individuals per unit area, such as 300 sugar maples (*Acer saccharum*) per hectare in a Michigan forest, or 3000 creosote bushes (*Larrea tridentata*) per hectare in a New Mexico desert. It is not necessary, however, to count every individual in an area to arrive at a density value; one may subsample randomly with a combined area perhaps as low as 1% of the total area and still end up with a close estimate of density. The pattern of how these individuals are distributed within the study region, however, can strongly influence the method used to estimate density.

This pattern of distribution also provides information about a population. The same density of individuals can be characterized by one of three statistical distributions: regular, random, or clumped (Figure 4-1). In a **random** pattern, the location of any one plant has no bearing on the location of another of the same species. In a **regular** (hyperdispersed) pattern, the presence of one plant decreases the probability, relative to random chance alone, of finding another very nearby or very far away. A fruit tree orchard is an extreme example of a regular pattern. In a **clumped** (aggregated, underdispersed) pattern, the presence of one plant increases the probability of finding another nearby.

Although most plants tend to be clumped, there are forces that both increase and decrease this natural tendency. At a local scale, seeds or fruit tend to fall close to parent plants, and vegetative reproduction also tends to produce plants in the immediate vicinity of others of the same species. Through time this can lead to increasing patchiness (Frelich et al. 1993). Further, environmental variation tends to be spatially correlated so that sites close to one plant tend to be good for another of the same species. Forces that oppose clumping include the tendency for herbivores, disease, and other threats to find plants in clusters more easily than they find isolated individuals (Condit et al. 1992, Gilbert et al. 1994). In addition, competition among individual plants for resources may in some instances (such as water consumption in deserts) exert a force toward even or regular spacing of individuals (Milne 1992). As an aside, spacing patterns are much more difficult to detect in clonal plants, where it may be very difficult to determine the spatial extent of a single individual.

Quantitative Samples of Plant Populations

A **quadrat** is an area of any shape that can be delimited so that plant species may be listed, counted, or have their vegetation cover estimated. For herbaceous

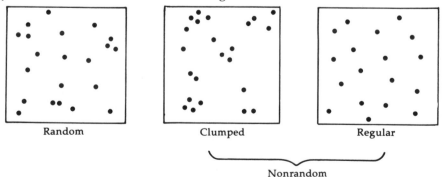

Figure 4-1   Random and nonrandom distribution patterns of individuals. Nonrandom distribution patterns may be clumped or regular. Dots represent individual plants as viewed from above.

species a quadrat that is 0.5 m to 2 m on a side is usually small enough that one person, standing at one point along its edge, can easily survey its extent. Quadrats for tree samples may be 10–50 m on a side. It is critical to understand the statistical ramifications of any plot selection procedures used to generate an estimate of population density, demographic performance, or community structure. Failure to recognize the statistical limits of the data can lead to inappropriate statistical procedures or inferences (Pielou 1977, Scheiner 1993).

In order to best characterize a large population, one would ideally choose to study a set of randomly selected portions of the entire population. Sample quadrats may be located randomly by constructing imaginary axes along adjacent edges of the stand or area being sampled, dividing the axes into units (e.g., meters) and picking pairs of random numbers that designate sampling coordinates. For the example illustrated in Figure 4-2, the random pair (5, 15) designates a sampling plot with a plot corner 5 m up and 15 m over from the origin of the axes. In this example, 12 quadrats, each 2 m^2, were placed at random in a 2400 m^2 area, resulting in a sampling density of 1%. In practice, regular sampling, such as every 10 m along a transect, often results in very similar estimates and is typically much simpler. When habitats are patchy, however, regular sampling is prone to a poor estimation of density.

Completely random plot selection is not favored in sites with a strong environmental gradient. By chance alone all samples may be clustered in one portion of the area, such as toward the wet end of a within-plot moisture gradient, resulting in an inaccurate estimate of the population if sample plots were randomly located. In such a case it would be preferable to retain certain attributes of random plot selection, yet also guarantee that the plots will sample the full range of density variation that the species displays. To accommodate nonrandom densities, the area can be subdivided into many areas of equal size. An equal number of plots are selected in random position within each subsection. This describes a **stratified random sample** experimental design.

Density estimates also can be collected using any of a variety of distance methods that do not require the establishment of plots. One simple method uses a random point within the region and a compass angle. The investigator then traverses along a straight line, following the compass angle away from the randomly selected starting point. The distance to the nearest individual is measured at periodic intervals along this transect. After many such samples are collected, density may be estimated from the average distance, based on specific geometrical and statistical assumptions. The statistical assumptions and mathematical machinations of this and other methods are detailed in works such as Greig-Smith (1964), Kershaw (1973), Mueller-Dombois and Ellenberg (1974), Pielou (1977), Cox (1985), and Engeman et al. (1994). In addition, we consider plant sampling again in Chapter 9.

Patterns in Vegetation

There are many ways to test for nonrandomness in plant populations. One method tallies the number of individuals of a species rooted within randomly located quadrats (as in Table 4-1, column A). In a randomly dispersed population,

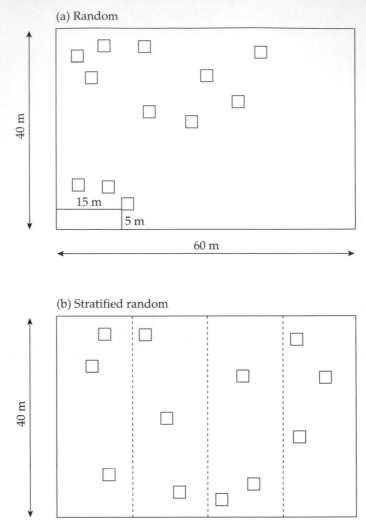

Figure 4-2   Placement of twelve quadrats by (a) a random method and (b) a stratified random method (three quadrats in each section). In each case, 1% of the total area has been included in the quadrats. One of the quadrats in (a) has been located by the random numbers 5 and 15, indicating that the investigator should locate the quadrat corner 5 m up from the baseline and 15 m down the baseline. The other quadrat corners were located using different pairs of random numbers.

the observed numbers of plants in each quadrat will approximate a Poisson distribution (column B). The discrepancies between the observed and expected values are then summed to calculate a chi-square for the test. In the example provided in Table 4-1, the chi-square value is greater than we would expect if the differences between observed and expected were generated by random chance. Thus we can

Table 4-1 Poisson analysis of quadrat data for a species with a nonrandom (regular) distribution. In the formulas, m is the average number of plants per quadrat. In this case, $m = 1.56$ plants/quadrat and $e^{-m} = 0.21$. To employ this test, each category must have an expected value of >5% of the total quadrats. To achieve this, category 5 is lumped with category 4. The χ^2 sum is based on 5 numbers, and the degree of freedom is $5 - 1 = 4$. The critical value for significance at the 99% level for 4 degrees of freedom is 13.28; any value greater than this suggests that the hypothesis that plants are distributed randomly can be rejected. With an observed χ^2 sum of 22.9, we conclude that plants are distributed nonrandomly. In this case, we observe that since more than the expected number of plots have 1 plant, plants are more regular, or evenly distributed, than expected. If plants were clumped, we would see larger than expected values in cells with many plants and no plants.

A No. of plants per quadrat (x)	B Observed no. of quadrats with x plants	C Expected no. of quadrats with x plants = $(e^{-m}) \left(\dfrac{m^x}{x!} \right)$ (100)	D $\chi^2 =$ $\dfrac{(\text{observed} - \text{expected})^2}{\text{expected}}$
0	13	21.0	3.0
1	51	32.8	10.1
2	23	25.6	0.3
3	3	13.3	8.0
4	0 } 10	5.2 } 6.8	1.5
5	10	1.6	
Totals	100	99.5	$\Sigma \chi^2 - 22.9$

conclude that it is unlikely that the members of this species are distributed at random in the sampled area. We are left to decide whether this nonrandom pattern is a result of plants being clumped or being regularly spaced. Inspection of the table shows that fewer quadrats than expected for a random distribution had either 0 or many individuals. In contrast, more quadrats than expected had a single plant in them. By deduction, then, we can say that the members of this population are distributed regularly. Distance methods (nonquadrat methods) can also be used to detect distribution patterns. For a discussion of alternative analytic methods and examples see Kershaw (1973).

Determining an appropriate plot or quadrat size is critical because frequency values are highly dependent on quadrat size (Stohlgren et al. 1997). If quadrats are too large, most species will be found in all plots; if they are too small, many species will have frequency values close to 0. Daubenmire (1968a) suggested that the appropriate quadrat size for sampling a given life form be small enough that only one or two species show 100% frequency, although Blackman (1935) suggested that the maximum frequency should be 80%.

Plant Demography

Modular Growth

Demography is the study of changes in population size and structure through time. By determining birth and death rates of individuals of each age in a population, the demographer projects how long an individual is likely to live, when it will produce offspring, how many offspring are likely to be produced, and how these variables affect population size through time. Collecting data for plant demographic study might seem to be a simple problem of monitoring individuals of known age, as one would do with animals. The morphological complexity and plasticity of plants, however, mean that plants can take different forms depending on the circumstances. Thus, counting trees may seem straightforward, but counting individuals in a poplar stand where all the tree trunks may be a single genetic individual makes demographic calculations difficult. On the other hand, the fact that plants do not move makes them easier to monitor than animals.

One approach to plant demography is to delineate the critical stages of the life history of a plant. For example, Figure 4-3 shows how John L. Harper (1977), a leader in the field of plant demography, diagrammatically represented a generalized model of the life history stages of plants. In this model, the seeds present in the soil are referred to as the **seed pool** (or seed bank, Leck et al. 1989). The researcher quantifies the number of individuals in each stage (rather than age) class. Some of these seeds germinate to become **seedlings.** The environment acts like a sieve as some seedlings become established and others remain in the seed bank or are lost through mortality. Some plants die before reaching reproductive maturity, and others form progeny by vegetative reproduction. Finally, mature individuals produce seed that enters the seed pool for the next generation.

When vegetatively formed recruits remain visibly attached to the parent it is easy to distinguish genetic individuals, called **genets.** Once this connection is lost, or if it is cryptic (underground), then deciding whether two stems are **ramets** that represent the same genetic individual or are true genets becomes very difficult. The term *ramet* is derived from the Latin word meaning "a branch" and represents a portion of a genetic individual. In contrast, a genet refers to all of the plant tissue arising from a single seed. Thus, a group of ramets no longer connected to a single parental plant remains ramets of the parental genet, never becoming new genets by virtue of their isolation. This group of ramets of the parental genet is often referred to as a **clone.** Population ecologists must decide whether, in a demographic study, they are tracking ramets or genets. A common example of extensive genets with many ramets are lawn grasses that form sod. For several years it was thought that the largest terrestrial individual on earth was an aspen grove in Montana, but a fungus that is larger has since been found.

Another complication in plant demography is that plants have **indeterminant growth,** that is, they continue to grow throughout their lives and two individuals of the same age may vary substantially in size depending on their environmental conditions. For this reason, it is appropriate to think of plant growth as the pro-

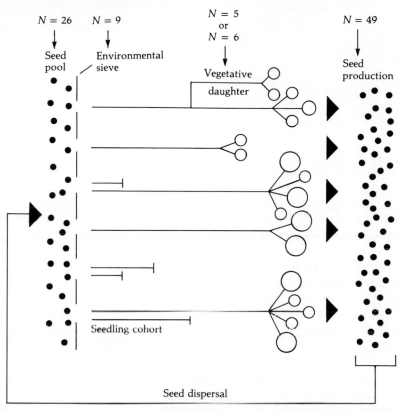

Figure 4-3 An idealized plant life history. Near the middle of the growing period, *N* is 5 or 6, depending on whether the vegetative daughter is considered an individual or part of the parent plant. Copyright © 1977 by J. L. Harper, *Population Biology of Plants.*

duction of **modules** (roots, stems, and leaves). One can conceptualize a plant as a population of modules that make up an individual (White 1979). Plants may independently allocate growth to different modules, such as exposed versus shaded leaves. Modular growth adds to the complexities of determining whether a long-lived plant is successful in its environment. For example, consider a tree whose normal allocation pattern is to produce an abundance of seed, increasing the number of genets in a location. Under stressful conditions (conditions where seedling success may be low) an individual may change its strategy and allocate more resources toward producing new ramets and increasing root biomass. When an individual changes reproductive and allocation strategies throughout its lifetime as a result of environmental variation, it becomes difficult to assess its **fitness** (lifetime reproductive success) without studying the individual for its entire lifetime. One would not want to assert, for example, that a population of individuals following the strategy described above (allocation to vegetative modules) is static because of a temporary absence of seedlings. This plasticity of plant module production in response to varying environmental conditions presents a level of population dynamics that is superimposed on population size.

Plant Ages versus Stages

Plants of a given age are rarely distributed as a normal curve, in which most individuals are of moderate size (Weiner 1995). Instead, pernnial plants typically have an L-shaped frequency distribution (Figure 4-4). One reason for this pattern may be that plants of the same species do not typically add modules at the same rate because of limitations of resources and competition between individuals. Many individuals remain small, while a few grow large and occupy a disproportionate amount of habitat space. In such a population, size is not simply proportional to age. Classic demographic theory (developed for animals) uses age as the basis for estimating population growth measures. Age, however, is often not a good predictor of reproductive status in plants. Varying environmental conditions can affect growth rates, decoupling the relationship between size and age. For example, the typical biennial produces a vegetative rosette in its first year and a flowering stalk in its second and final year of life. This oversimplifies reality, however, because vegetative rosettes that grow more slowly often remain as rosettes for an additional year or two and delay reproduction until a critical size is attained (Werner and Caswell 1977, Young 1984). Similarly, shade-tolerant trees may remain small under a dense canopy and then grow rapidly only when a forest gap appears above suppressed saplings (Henry and Swan 1974), resulting in older trees that are smaller than younger ones grown under more favorable light conditions.

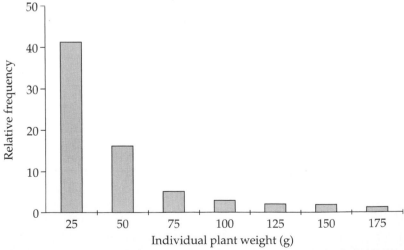

Figure 4-4   A typical frequency distribution of individual plant weight for a plant population experiencing intraspecific competition.

Population Growth Models

Although population growth models are integral to the study of plant demographics, a thorough mathematical treatment of population growth is beyond the scope of this book. (For specific reviews, see Harper 1977, Wilson and Bossert 1971, Silvertown and Lovett-Doust 1993.)

Population growth models can be described by either discrete-time or continuous models. Using a **discrete-time model,** we can predict the number (N) of plants at some time in the future ($t + 1$) by summing the number of plants (N_t) present at time (t), the number established from seed (B), and the number of seeds arriving at the site (I), and then subtracting the number that die (D) and the number of seeds that disperse out of the area (E) during the period from t to $t + 1$:

$$N_{t+1} = N_t + B - D + I - E \qquad \text{(Equation 4-1)}$$

This simple equation, however, is often not so simple to use because we are seldom able to make accurate accounts of births, deaths, immigration, and emigration for an entire population.

Using census data, we can define a measure of the rate of population growth as the ratio of future population size to past population size:

$$\lambda = N_{t+1} / N_t \qquad \text{(Equation 4-2)}$$

When the *annual rate of population increase* (λ) equals one, which is when $B + I = D + E$, the population is stable ($N_{t+1} = N_t$). By substitution we observe that

$$\lambda = 1 + [B + I - (D + E)] / N_t \qquad \text{(Equation 4-3)}$$

When $B + I > D + E$, so that $\lambda > 1$, the population will multiply by λ each year and will increase exponentially.

We can convert these state expressions to a **continuous-time model** and express birth (b) and death (d) processes in terms of rates (number per individual in the population). We will also assume, for simplicity, that immigration and emigration are approximately equal or otherwise trivial relative to births and deaths. We can then define a rate of population increase (r):

$$r = b - d \qquad \text{(Equation 4-4)}$$

From this we can calculate the instantaneous rate of population change using the differential equation

$$dN/dt = rN \qquad \text{(Equation 4-5)}$$

where N is the number of individuals in the population at time t (Lotka 1925). Equations 4-2 and 4-5 are expressing similar concepts; the relationship between λ and r is

$$\ln \lambda = r \qquad \text{(Equation 4-6)}$$

Plotting N by time for any value of $r > 0$ results in a demonstration of the exponential nature of growing populations in an unlimited environment (Figure 4-5, line A).

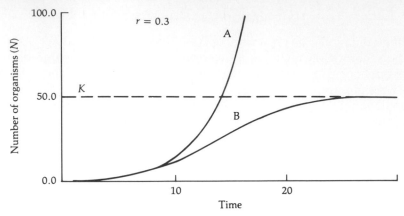

Figure 4-5   Exponential (line A) versus logistic (line B) population growth over time. *K* is the carrying capacity of the habitat for a population showing density-dependent logistic growth.

Populations in the real world, however, are constrained by limits in their environment. This environmental limitation is typically assumed to be a constant (*K*) and is termed the carrying capacity (Figure 4-5, line B). The **carrying capacity** (*K*) is defined as the maximum number of individuals that the environment can support, and is modeled to have a dampening effect on population growth. The model of population growth is modified to assume that population growth slows as it nears the carrying capacity:

$$dN/dt = rN(1 - N/K) \qquad \text{(Equation 4-7)}$$

also often expressed in a mathematically equivalent form as

$$dN/dt = rN(K - N)/K \qquad \text{(Equation 4-8)}$$

These equations, known as the Verhulst-Pearl equations, are the basis for much of theoretical ecology and serve as a good foundation for idealized population growth. Note that in each case the population growth rate (*dN/dt*) is a function of the intrinsic rate of growth (*r*) and population size (*N*) when populations are small (*N* << *K*) and decreases to zero as *N* approaches *K*. Thus the per capita population growth rate (*dN/N dt*) is equal to *r* only when *N* is a negligible portion of *K*. In reality, however, these model equations are often not directly functional because (a) population carrying capacities are not constant (they vary with environmental conditions); (b) birth and death rates are also not constant and therefore are difficult to estimate; (c) biomass in plants may have more impact on carrying capacity than the number of individuals; and (d) population boundaries are often indistinct and ill-defined. Nonetheless, this simple model at the very least provides a conceptual framework for treating population growth.

Transition Matrix Models

An alternative approach to modeling the demography of populations was developed by Leslie (1945) using matrix algebra. **Transition matrix models** use estimates of birth, growth, and death probabilities for individuals within different age classes. Using discrete time steps, populations of individuals within each age class may be projected into the future. A number of researchers have expanded this generalized model for use in plants, where the stage of life history is often a more useful representation of survivorship and reproductive performance than age (Lefkovitch 1965, Goodman 1969, Lewis 1972, Sarukhan and Gadgil 1974, Werner and Caswell 1977, Harper and Bell 1979, Silvertown 1982, Caswell 1989).

Before embarking upon a discussion of the matrix model and its applicability to plant demography, it is necessary to define basic terminology for matrix algebra. A matrix consisting of a single column is referred to as a vector or a **column matrix.** We will construct a column matrix that represents the number of individuals in each of three life-history stages of development. For a hypothetical biennial, the number of seeds (N_s) in a seed pool, the number of plants in rosette form (N_r), and the number of plants in the flowering phase (N_f) would appear in matrix form as

Column matrix   (Matrix 4-1)

$$\begin{bmatrix} N_s \\ N_r \\ N_f \end{bmatrix}$$

A **transition matrix** for three growth stages consists of a group of probability values that represent the chance that a plant in a given stage of development will arrive at a different (or remain at the same) stage during the census interval. The transition matrix for the three-stage growth example above would appear as

This census   (Matrix 4-2)

		Seed	Rosette	Flower
	Seed	a_{ss}	a_{rs}	a_{fs}
Next census	Rosette	a_{sr}	a_{rr}	a_{fr}
	Flower	a_{sf}	a_{rf}	a_{ff}

where, for example, a_{sr} represents the probability that a seed will germinate and develop into a rosette by the next census, and a_{rf} represents the probability that a rosette will mature into a flowering adult by the next census. The columns represent current status stages (seeds, rosettes, and flowering adults) from left to right. Rows represent the fates of these individuals during a future census (e.g., the next year). Values are calculated using census information of growth and survivorship of marked individuals in each size class through time. For example, by following the development of rosettes during the time period under consideration, say, one year, values for the relative number of rosettes that remain as rosettes, flower, or die become probability values. Assume for the sake of illustration that 60% of the rosettes present at this census die before the next census, leaving 40%, of which 75% flower and 25% remain rosettes. The probability of a single rosette flowering (a_{rf}) would then be $.4 \times .75 = .3$, and the probability of a single rosette remaining a rosette (a_{rr}) would be $.4 \times .25 = .1$. The values for the remainder of the matrix could be calculated from similar data on other growth stages.

This transition matrix model allows for the probability of any transition. The life histories of plants, however, often constrain certain transition probabilities to equal zero. For example, rosettes do not produce seed in species with a biennial life history, so $a_{rs} = 0$. Similarly, mature adults of long-lived plants can produce seed, but not other mature adults ($a_{ff} = 0$).

To see how the transition matrix is used to calculate future population size structure, consider the following example using a long-lived perennial. The transition matrix (**A**) is established by monitoring the probability of individuals in each stage class (s = seed, i = immature, f = flowering adult) either remaining in that stage class, or changing stages. The column matrix (**B$_1$**) is established by counting the number of individuals in each stage class. In order to predict a future population we then need to multiply the transition matrix by the column census matrix as follows:

$$
\begin{array}{cccccc}
\mathbf{A} & \times & \mathbf{B_1} & = & \mathbf{B_2} & \text{(Matrix 4-3)}
\end{array}
$$

$$
\begin{bmatrix} a_{ss} & 0 & a_{fs} \\ a_{si} & a_{ii} & a_{fi} \\ 0 & a_{if} & a_{ff} \end{bmatrix} \times \begin{bmatrix} N_s \\ N_i \\ N_f \end{bmatrix} = \begin{bmatrix} (N_s a_{ss}) & + & 0 & + & (N_f a_{fs}) \\ (N_s a_{si}) & + & (N_i a_{ii}) & + & (N_f a_{fi}) \\ 0 & + & (N_i a_{if}) & + & (N_f a_{ff}) \end{bmatrix}
$$

For the top entry of the new column matrix (**B$_2$**), elements for seeds (N_s) is the sum of three segments of the initial population (**B$_1$**) multiplied by their transition probabilities (**A**): (1) the proportion of seeds that remain as seeds ($N_s \times a_{ss}$); (2) the number of seeds produced by immature plants ($N_i \times 0$); and (3) the number of seeds produced by flowering adults ($N_f \times a_{fs}$). When its rows are summed, **B$_2$** becomes a new column matrix of N_s, N_i, and N_f for the next generation. Multiplying **B$_2$** by the transition matrix **A** will then provide an estimate of population size in the following generation. After several years (census periods) of population projections, assuming the transition matrix remains the same, the proportion of the total population found in each growth stage should become constant. These stable proportions represent the **stable stage distribution** of the population and represent a means by which the ecologist may evaluate the stability of current population structure.

The matrix model of population growth is advantageous when population units move from one identifiable stage to another at a relatively frequent rate. This allows the investigator to determine the effects of changes in transition probabilities with changing environmental conditions, or the sensitivity of the population growth model to particular transition models. As an example, Werner and Caswell (1977) conducted a study of teasel (*Dipsacus sylvestris*), a short-lived perennial that dies after flowering. After germination, teasel forms a rosette that may exist for up to five years before flowering. Vernalization is required for flowering, so no reproductive activity is possible in the first year. Teasel was studied in an open field and in an old field where much of the ground was shaded by shrubs. The transition matrices are shown in Table 4-2. The instantaneous rate of increase per individual (r) was 0.957 in the open field and -0.465 in the shrub-covered old field. For the period of time that the transition matrices do not change, teasel populations in the open field will expand geometrically ($r > 0$), and those in the shrub-covered old field will decline toward extinction ($r < 0$). It is possible, however, that immigration (I) and emigration (E) of seed may be important and maintain populations with declining numbers. Populations that produce much seed that emigrates while maintaining a positive growth rate ($r > 0$) are termed "source" populations; those that persist only as a result of immigration are termed "sink" populations. These concepts are discussed in the final section of this chapter, but the open field population could act as a source population while shrub-covered old fields may act as population sinks.

The population in the open field, however, cannot continue to increase indefinitely. The finite resources available within the environment will result in a decreasing r as the population approaches its carrying capacity. Alternatively, the environment may change, shifting transition probabilities such that r becomes negative and populations begin to shrink. From this study we might conclude that teasel is a fugitive species that invades open fields and that populations will eventually decline as the field fills in with shrub and tree growth.

The **net reproductive rate** (R_0) of a population is yet another measure of whether a population is increasing, decreasing, or stable. To determine R_0 for a population, calculate the ratio of the number of individuals in a particular growth stage (or the entire population) in two succeeding generations in a population that has reached a stable stage structure. For example, the number of rosettes in generation ($t + 1$) divided by the number of rosettes in generation (t) equals the net reproductive rate for population in a stable stage distribution. If $R_0 = 1.0$, then the population is stable. The population increases when $R_0 > 1.0$ and shrinks when $R_0 < 1.0$. If R_0 is derived from a matrix measured on an annual basis, then $\lambda = R_0$. Alternatively, we often measure population growth on the basis of generation time (τ). In this case, R_0 must be factored by the generation time to estimate annual rates of population change, and $\lambda = R_0^{1/\tau}$. To summarize, Table 4-3 provides a synopsis of variables related to population growth.

A final note on these models is in order. All of these models assume that transition probabilities for individuals are independent of previous transitions. Although this is true for age-based models, it is clearly not true for stage-based models and this may have important consequences. For example, slower growing individuals often (a) continue to grow slowly, (b) die sooner, and (c) flower at a smaller size than faster growing individuals (Young 1985). Further demographic characteristics of a species are assumed to be constant through time whenever one uses a demographic model to characterize the future fate of a species. This assumption is rarely tested, but existing research suggests that demographic characteristics can vary widely through time (e.g., Young 1985, 1994).

Chapter 4 Population Structure and Plant Demography

Table 4-2 Transition matrices for *Dipsacus sylvestris* (teasel) in open field and in shrub-covered old field. Data for stable size distribution are in percent. The number in the Flowering column is average number of seeds produced per flowering plant. Modified from P. Werner and H. Caswell, "Population growth rates and age versus stage-distribution models for teasel (*Dipsacus sylvestris* Huds.)," *Ecology* 58:1103–1111. Copyright © 1977 by the Ecological Society of America. Used with permission.

	Seeds	Seeds 1 yr	Seeds 2 yr	Rosette small	Rosette medium	Rosette large	Flowering
Open field							
Seeds	—	—	—	—	—	—	635
Seeds 1 yr	.634	—	—	—	—	—	—
Seeds 2 yr	—	.974	—	—	—	—	—
Rosette small	.013	.017	.011	.000	—	—	—
Rosette medium	.109	.004	.002	.077	.212	—	—
Rosette large	.006	.003	.000	.038	.281	.000	—
Flowering	—	—	—	.000	.063	1.000	—
Shrub-covered field							
Seeds	—	—	—	—	—	—	476
Seeds 1 yr	.423	—	—	—	—	—	—
Seeds 2 yr	—	.987	—	—	—	—	—
Rosette small	.024	.009	.006	.007	—	—	—
Rosette medium	.044	.000	.000	.050	.158	—	—
Rosette large	.001	.000	.000	.002	.008	.000	—
Flowering	—	—	—	.000	.000	.250	—
Stable size distribution for open field teasel							
	71.45	17.39	6.50	0.50	3.30	0.56	0.29

Table 4-3 A summary of values used to describe population growth and their interrelationships.

Terms

Term	Critical values for population change		
	Increasing	**Decreasing**	**Stable**
λ	>1	<1	1
r	>0	<0	0
R_0	>1.0	<1	1

Interrelationships of terms

$\ln \lambda = r \qquad r = \lambda^e$

$\lambda = R_0 \quad$ or $\quad \lambda^\tau = R_0$ (annual rate based on generation time, τ)

Population Viability Analysis

A key concern of conservation biologists is to predict whether populations of critically endangered species are increasing, decreasing, or stable. One tool that helps the conservation biologist approach the question of determining a risk of extinction for a species is called population viability analysis (Shaffer 1981). After collecting demographic data, the ecologist performs a **population viability analysis (PVA)** that predicts two vital attributes: an estimated time to extinction and the probability of extinction over a specified length of time. If a population is on a trajectory toward extinction (as is the case with *Torreya taxifolia*, an endangered conifer in Florida; Schwartz and Hermann 1993), then estimating a time to extinction would help identify the urgency for specific conservation actions. In contrast, when we are not certain that a species is heading toward extinction in the near future it is typically more useful to express population dynamics in terms of the probability of extinction over the next 50 or 100 years. Using demographic modeling we can then estimate a population size required to reduce the extinction probability within this time frame to an "acceptable" level. An appropriate minimum population size would exist, for example, if the population has less than a 10% extinction probability over the next 100 years.

PVA may also be used to test the sensitivity of a species to uncertainty in the environment. Three factors may influence the probability of species extinction: genetic, demographic, and environmental stochasticity. Each factor becomes important at a different population size. Demographic stochasticity is uncertainty that typically results from random events that may affect the reproduction of a species, such as an extremely skewed sex ratio of a dioecious plant. Genetic stochasticity results from too few breeding individuals. This may lead to inbreeding depression, reduced reproductive success, and a loss of genetic variability. Both of these factors seem to be a problem only at very low population sizes (e.g., <50 individuals). In contrast, environmental stochasticity may arise in the form of environmental changes such as drought, fire, or flood. Although these events may have devastat-

ing effects on small populations, large numbers can significantly buffer populations from the likelihood of extirpation. Populations are often able to persist
through the vagaries of environmental stochasticity only when individuals number in the hundreds to thousands. Thus, environmental stochasticity is typically
considered far more limiting a factor in population viability than either genetic or
demographic considerations. Some consider environmental catastrophes, such as
extreme drought, volcanic activity, or disease epidemics, as a separate type of stochasticity that would require significantly larger populations to prevent extirpation. Another view places catastrophes as an extreme along a continuum of
environmental stochasticity.

PVA may also be used to test what stages of a species' life history have the
largest influence on net population growth (e.g., Kalisz and McPeek 1992, Silvertown et al. 1996). For example, if a species is characterized in four stage classes
(seed, seedling, mature vegetative, and mature reproductive adult) with an estimated transition matrix, we can calculate an elasticity matrix. An **elasticity matrix**
is a matrix of partial derivatives of transition matrix elements. As such it estimates
the likely effect of changes in a matrix element on the prediction of net population
change. Thus, the elasticity matrix provides a means to determine which stages of
the species' life history are most vulnerable to changes that affect population increases or decreases. Thus, this sensitivity analysis suggests ways to focus species
management on the critical stages of the life history for species survival. For example, we might ask whether germination or competition of seedlings from other
species is most limiting to population growth in the giant sequioa (*Sequoiadendron
giganteum*) (Stohlgren 1993). Silvertown et al. (1996) reviewed the literature to construct elasticity matrices for 84 plant species and describe correlations between
elasticities and life history. In particular, they observed that the most important aspect of long-lived trees and shrubs was survivorship, while growth elements were
the most important predictors of likely success for perennial herbs of open habitats. At this time, however, the use of elasticity matrices to determine demographic
sensitivity often carries an unacceptably high rate of uncertainty. Much work remains to boost the robustness of predictions based on elasticity matrices and the
conservation recommendations derived from them.

Density Dependence

Earlier we noted that the carrying capacity of a population is not a simple measure, constant through space and time, and that biomass may be a better measure
of carrying capacity than total numbers. Size hierarchies can lead to strong dispari-
ties between the largest and smallest individuals within a single population (Nagashima and Terashima 1995, Weiner 1995). The influence of density on biomass
accumulation (yield) is exemplified by data presented in Figure 4-6. Seedling density varied from 6 to 32,500 plants m^{-2}. The total biomass of all mature plants combined at the end of the growing season was then tallied at each sowing density.
This figure demonstrates that at maturity, yields do not vary over a remarkably
wide array of planting densities. Since individual mortality was unimportant, the
differences between plots were a result of differences in individual plant sizes at
maturity. Below 1500 plants m^{-2}, interplant distance is great enough that growth

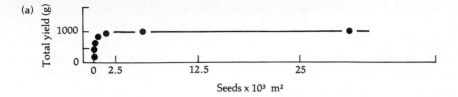

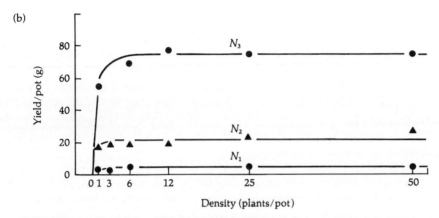

Figure 4-6 The relationship between yield (dry weight per unit area) and population density for mature populations of *Trifolium subterraneum* (a) and mature populations of *Bromus unioloides* (b) grown at three levels of nitrogen fertilization. From C. M. Donald, 1951. *Australian Journal of Agricultural Research* 2:355–376.

is not influenced by neighbors. Thus, yield depends on the number of plants at low densities. In contrast, when planting densities are high enough for intraspecific interference to be important, yield is predictable regardless of the planting density. This result, which is consistent for many plant species, has been referred to as the **"law" of constant yield** (Kira et al. 1953).

The specific magnitude of plant yield at a site, however, depends on the availability of resources. For example, the yield of *Bromus unioloides* was constant over a wide range of densities, but the maximum yield varied with nitrogen availability (Figure 4-6b). The environment sets a limit on the amount of plant biomass that can be supported at a site. Thus, carrying capacity varies with site characteristics.

Size hierarchy is used to refer to the large disparity in size frequently observed in plants in a single location. These size differences are generated by microsite differences or timing of growth. All individuals in a plant population would need to be identical and the environment completely uniform in order for each individual to be equally inhibited by its neighbors. These conditions are never met in reality. Some individuals inevitably gain more than their share of resources and grow faster than other plants of equal age. Typically, a stand of very young plants exhibits a size structure that mimics a normal distribution. As the stand matures, however, the size distributions of plant populations almost always become skewed, with many small plants and just a few large ones (see Figure 4-4). This right-skewed distribution may be a result of the exponential growth rates of plants and a few individuals growing through this curve ahead of the majority of the population. An alternative and not mutually exclusive hypothesis is that plant density has effects that do not scale linearly with size. **Asymmetric competition,** where larger plants have larger effects on small plants than small plants have on large plants, often leads to the **size hierarchies** that are commonly observed in plant populations (Harper 1977, Firbank and Watkinson 1987, Silvertown and Doust 1993, Weiner 1995).

As large plants usurp the vast majority of light, water, and soil nutrition, increasing numbers of small plants die and density is reduced. Previously we predicted that growing populations will increase in density to some carrying capacity. In particular, we predicted that as N approaches K, r decreases until it averages zero. Likewise, we know from the law of constant yield that plants respond to crowding caused not only by density but also by size of individuals. It is clear that the state of a plant population cannot be described by biomass alone, nor by density alone. It is more accurate to say that plant populations are crowding dependent rather than density dependent.

The **self-thinning rule,** proposed by Yoda et al. (1963), describes the interacting influences of density and biomass in plant populations (Figure 4-7). The self-thinning line describes a population trajectory of a single population in a single location through time under conditions where density-dependent mortality is occurring. As such, the self-thinning rule defines a boundary line toward which plant populations grow:

$$B = CN^{-1/2}$$
(Equation 4-9)

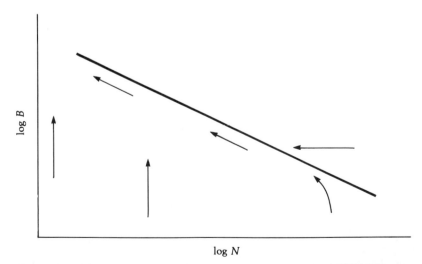

Figure 4-7   The *B-N* diagram. The line of slope $-1/2$ is the self-thinning line. Arrows indicate the trajectories that stands follow at different biomass-density combinations. From *American Naturalist* 118:581–587, M. Westoby. Copyright © 1981 by University of Chicago Press. Reprinted by permission.

where *B* is biomass per unit area (biomass density), *N* is the number of individuals per unit area, and *C* is a constant (Westoby 1981). The original formulation of Yoda et al. (1963) is $w = KD^{-3/2}$, where *w* is the average weight per individual, *K* is a constant, and *D* is density. Thus, the self-thinning rule is also known as the **−3/2 power law.** We use Westoby's formulation of self-thinning because it more readily translates to other population models.

Where stands have biomass and density values below the boundary line, biomass will increase toward the boundary condition. If a plot has biomass and density values above the self-thinning line, plant mortality will reduce density toward the self-thinning line. Once at the self-thinning line, individuals will then die at a rate related to biomass accumulation rates [i.e., the population will move up and left along the self-thinning line (Figure 4-7)]. An interesting aspect of the self-thinning law is that the slope of the line (−1/2) seems to be the same for herbs, shrubs, and trees across approximately 12 orders of magnitude in size (Figure 4-8; Harper 1977, Gorham 1979, White 1980, 1985). More recently there has been considerable debate regarding (a) the support of the rule by empirical data (Zeide

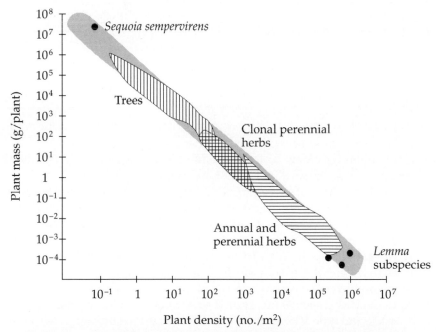

Figure 4-8   The relationship between plant density and mean plant mass for monoculture populations of plants ranging from 10^{-6} g (*Lemna* subspecies) to 10^{7} (*Sequoia sempervirens*). Extremes in mean plant size are represented by filled circles. Shaded regions represent summaries from numerous data sets of self-thinning studies. Redrawn from J. White. 1985. "The thinning rule and its application to mixtures of plant populations," pp. 291–309 in J. White, ed., *Studies on Plant Demography*. Academic Press, London.

1985, 1987; Weller 1987a,b, 1990; Lonsdale 1990), (b) the proper way to express the relationship between plant density and biomass (Westoby 1981, 1984), and (c) the mechanistic reason for the pattern.

Life Tables

Life tables were originally developed by insurance companies as a means of determining the relationship between age and the life expectancy of clients to determine sufficient insurance premiums to keep the company solvent. Insurance life tables provide basic information on survival for demographic studies, but ignore the process of birth. By expanding life tables to include information on birthrates (**fecundity**) and age, the ecologist has an effective means of organizing demographic data.

There are two basic kinds of life tables. Annual plants and short-lived perennials may be studied using a **cohort life table,** where the observer follows all of the seedlings that germinate at a particular time (a cohort) through their entire lives. Long-lived perennials, such as trees and shrubs, often live longer than the ecologist (or at least longer than the study). In these cases, the ecologist must use a time-specific static life table. In a **static life table,** the stage or age structure of a population (consisting of multiple cohorts) is used to estimate the survival and fecundity patterns of various stage or age classes.

An example of a cohort life table for the annual *Phlox drummondii* (Leverich and Levin 1979) is presented in Table 4-4. Age is used as a demographic parameter because these annual plants show nearly simultaneous germination and developmental stages. Age (x) is presented as an interval $(x - x')$ in days. By knowing the number dying during each age interval (d_x) and the number of seeds in the initial seed pool (996 in Table 4-4), we can calculate the number surviving to the beginning of the next age interval (N_x). Survivorship (l_x) is equal to the proportion of the original cohort surviving to the *beginning* of each age interval. By definition, survivorship is 1.0 for the original cohort and goes to 0 during the life span of the longest-lived individuals in a cohort. The **age-specific mortality rate** (q_x) is the proportion of the population that dies during a particular age interval (in this case, one day).

Plotting the number of survivors at each age interval against time results in a **survivorship curve.** Deevey (1947) distinguished three basic types of survivorship curves that represent the extremes of population demography (Figure 4-9). A **type I survivorship curve** is characteristic of organisms with most mortality concentrated in the later stages of life; **type II survivorship** is represented by constant mortality rates; and **type III survivorship** is characterized by high rates of juvenile mortality followed by a long period of low mortality. These are typically plotted on a log scale (Figure 4-9). Forest trees typically have a type III survivorship curve, with low probability of seed and seedling survivorship, followed by a more or less constant rate of loss with self-thinning, followed by an extended period of low mortality as adult canopy trees (Figure 4-10a). Annual plants without seed dormancy, such as *Phlox drummondii,* often exhibit a type I survivorship curve in

Table 4-4 A cohort life table for *Phlox drummondii.* From W. S. Leverich and D. A. Levin. *American Naturalist* 113:881–903. Copyright © 1979 by the University of Chicago Press.

Age interval (days) $x - x'$	No. surviving to day x N_x	Survivorship l_x	No. dying during interval d_x	Average mortality rate per day q_x
0–63	996	1.0000	328	.0052
63–124	668	.6707	373	.0092
124–184	295	.2962	105	.0059
184–215	190	.1908	14	.0024
215–231	176	.1767	2	.0007
231–247	174	.1747	1	.0004
247–264	173	.1737	1	.0003
264–271	172	.1727	2	.0017
271–278	170	.1707	3	.0025
278–285	167	.1677	2	.0017
285–292	165	.1657	6	.0052
292–299	159	.1596	1	.0009
299–306	158	.1586	4	.0036
306–313	154	.1546	3	.0028
313–320	151	.1516	4	.0038
320–327	147	.1476	11	.0107
327–334	136	.1365	31	.0325
334–341	105	.1054	31	.0422
341–348	74	.0743	52	.1004
348–355	22	.0221	22	.1428
355–362	0	.0000		

Figure 4-9 Hypothetical survivorship curves. From E. S. Deevey, Jr. 1947. "Life tables for natural populations of animals." *Quarterly Review of Biology* 22:283–314. © The Stony Brook Foundation, Inc. Used with permission.

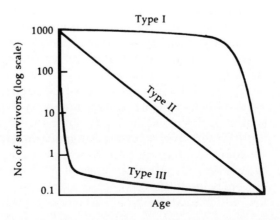

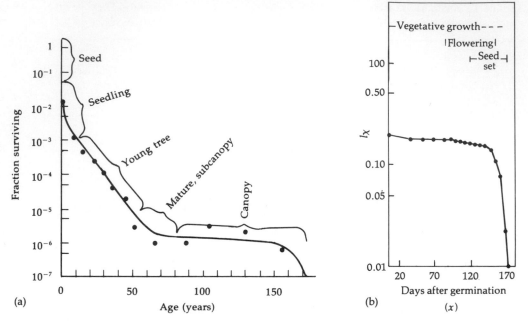

Figure 4-10 Survivorship curves for (a) the tropical palm *Euterpe globosa*. Reprinted by permission from L. Van Valen, 1975. *Biotropica* 7:260–269. (b) *Phlox drummondii*. Modified from Leverich and Levin. *American Naturalist* 113:881–903. Copyright © 1979 by University of Chicago Press. By permission.

favorable sites because most seedlings are able to survive to reproduction (Figure 4-10b).

The length of time that seeds survive in the seed pool is a life history attribute of species and is generally related to the growth form and environment of a species. Annual plants in harsh environments tend to have seeds that can remain viable in the seed pool for a long time (a seed bank), and have relatively short post-germination life spans. In contrast, seeds of long-lived perennials typically spend a relatively short time in the seed pool. In general, seed longevity is inversely proportional to adult life span and often proportional to environmental severity. Note that larger, long-lived animals have type I survivorship curves, and smaller, more rapidly reproducing species have type III curves. This is opposite to the pattern often exhibited in plants. The relationship between survivorship and life history attributes is discussed more fully in Chapter 5.

The age-specific birthrate of individuals is called **fecundity** (b_x) and is a measure of the average number of seeds produced by the individuals of a single size or age cohort during a given interval (x). If the plant is dioecious (male and female flowers being borne on separate plants), then only female plants are considered in the life table. Multiplying survivorship (l_x) by fecundity (b_x) and summing over the life span of the cohort gives an estimate of the **net reproductive rate** (R_0) of the cohort:

$$R_0 = \Sigma \, l_x b_x \qquad \text{(Equation 4-10)}$$

When a population is in a steady state, and hence $R_0 = 1$, each reproductive adult on average produces a single offspring that survives to reproduce. When $R_0 < 1$, fewer seeds are being placed into the seed pool than are necessary to replace the population. The population of *Phlox drummondii* of Table 4-4 is increasing, since $R_0 > 1$.

The success of a colonizing population, or the survival of an established population, depends on the ability of existing individuals to contribute offspring to future generations. The relative contribution that individuals of age x are likely to make to the seed pool before they die is defined as their **reproductive value** (V_x). Thus, reproductive value of an individual of age x is the sum of the average number of seeds it produces in the current year (b_x) plus the average number of seeds produced by an individual in each age class older than age x (e.g., b_{x+1}, b_{x+2}) times the probability that an individual of age x will survive to each older age category (l_{x+1}/l_x):

$$V_x = b_x + \Sigma \, (l_{x+i} \, / \, l_x) \, b_{x+i} \qquad \text{(Equation 4-11)}$$

In general, reproductive value is low for young plants because of their relatively high probability of dying before reproduction. Perennial plants that survive to maturity typically maintain a high reproductive value until they die. Reproductive values for the annual *Phlox drummondii* are low when plants are young and old, but high at intermediate age (Figure 4-11).

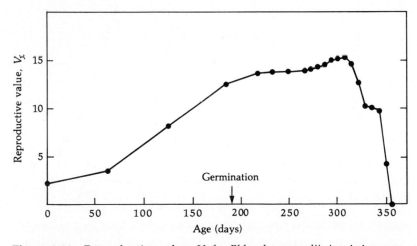

Figure 4-11 Reproductive values V_x for *Phlox drummondii*. Age is in number of days after seed dispersal. Modified from Leverich and Levin. *American Naturalist* 113:881–903. Copyright © 1979 by University of Chicago Press. By permission.

Population Dynamics

The (st)age distribution of a population can be used as a predictive tool in both population and community ecology. Figure 4-12 shows a size (dbh = diameter at breast height) distribution of longleaf pine (*Pinus palustris*) in northern Florida. Size, for this species, correlates well with age. This pine population, with a high density of adults and few seedlings or saplings, appears to be a senescent population. In contrast, hardwoods (mainly water oak, *Quercus nigra*, and wax myrtle, *Myrica cerifera*) have many saplings and seedlings but few large individuals. As a result of fire suppression, hardwoods are invading this forest and replacing the pine population. This pattern is observed for many oak forests of the eastern deciduous forests where mesophytic species such as sugar maple (*Acer saccharum*) are taking over forest dominance with the lack of fire (Abrams 1992, Shotola et al. 1992).

In contrast, Figure 4-13 shows the age distribution of red spruce (*Picea rubens*) in the White Mountains of New Hampshire. In this case we observe an L-shaped

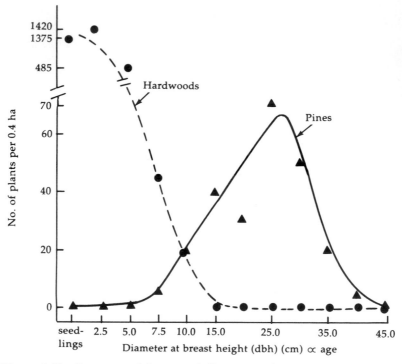

Figure 4-12 Summary of the ages of all trees in a forest near Gainesville, Florida. Succession is progressing from a pine community (solid line) to a hardwood community (mainly oak and wax myrtle) (dashed line). Based on data from "The relation of fire-to-stand composition of longleaf pine forests" by F. Heyward, *Ecology* 20:287–304. © 1939 by the Ecological Society of America, and from *Plant Communities* by R. Daubenmire © 1968 by Harper & Row, Publishers, Inc., New York.

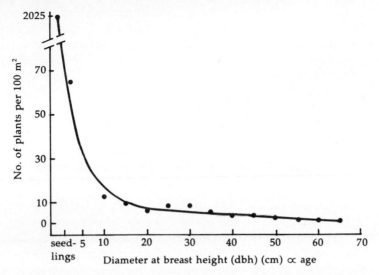

Figure 4-13   An L-shaped age distribution curve for red spruce (*Picea rubens*) in four stands in the White Mountains of New Hampshire. Based on data from "A comparison of virgin spruce-fir forest in the northern and southern Appalachian system" by H. J. Oosting and W. D. Billings, *Ecology* 32:84–103. Copyright © 1951 by the Ecological Society of America. Reprinted by permission.

curve with many seedlings and saplings and fewer larger trees. Many ages and all stages are represented in this population, which seems to experience a relatively constant rate of mortality. The high density of young trees suggests a high probability of this population maintaining itself as part of a climax community.

Not all stable climax populations show this kind of age distribution curve. Woody perennials are long-lived and so can successfully maintain themselves even if their seedlings become established only sporadically. For example, creosote bush (*Larrea tridentata*) is an evergreen shrub that dominates the warm deserts of North America. It seems to have all the attributes of a climax species (Loudermilk and Munz 1938, Vasek et al. 1975b). The species, however, has remarkably mesic requirements for good germination and seedling development (e.g., abundant moisture, moderate temperature, Barbour et al. 1977a). This combination of events happens only occasionally. In this case, we do not expect to see an L-shaped curve but an irregular curve with peaks of recruitment interspersed with intervals of virtually no recruitment (Chew and Chew 1965, Barbour et al. 1977b). In addition, age is often not correlated particularly well with size, resulting in patterns that are difficult to interpret.

Knobcone pine (*Pinus attenuata*) is a closed-cone conifer that typically occurs in single-aged (even-aged) stands dating from some destructive fire event (Vogl 1973). Reproduction in this species depends on fire melting a resin that otherwise seals cones shut and prevents seed dispersal. This life history attribute (serotiny) is found in several other conifer species (e.g., jack pine [*P. banksiana*], sand pine [*P.*

clausa]). The frequency of stand-clearing disturbances, such as fire, is sufficient to maintain populations despite an age structure that is not typical of self-replacing climax species (Loucks 1970). Similarly, demographic characteristics of a stand can generate cyclical patterns of establishment and mortality with cohorts sequentially replacing one another. This phenomenon has been observed with Hawaiian trees (Mueller-Dombois 1986) and giant *Senecio*s in Africa (Young and Peacock 1992).

Metapopulation Dynamics

So far this chapter has focused on the dynamics of individual populations. Broadening this discussion, we can envision a suite of populations that make up the distribution of a species within a region. All species are patchily distributed at some scale and populations are often semi-isolated from one another as a result of habitat heterogeneity. Species found in naturally fragmented habitats may be found at any scale from large forest patches of differing age among the boreal forests of North America, down to plants characteristic of ant mounds in British chalk grasslands (King 1977).

An explicit example of the differing scale at which species may be patchy is found in Erickson's study (1945) of a leather flower (*Clematis fremontii* var. *riehlii*) in Missouri (Figure 4-14). By mapping the distribution of *C. fremontii* var. *riehlii* on rocky outcrops (barrens), Erickson showed that the available habitat for this leather flower is distributed in discrete, isolated patches. In fact, the species sparsely occupied its distribution at nearly every scale Erickson examined.

Yet neighboring patches may be interconnected by emigration and immigration of seed. In addition, these semi-isolated populations may interchange genes through the exchange of pollen. Levins (1970) defined spatially isolated populations linked by a significant amount of individual exchange as a **metapopulation,** or a population of populations. We can characterize a metapopulation in the same way that we can measure the demographic parameters characteristic of a single population. It is likely that not all populations of a species are in equally favorable habitats. Some populations found in favorable habitats may consistently experience high seed production and hence produce potential emigrants. In contrast, some populations may be found in marginal habitats where R_0 is consistently less than one. Pulliam (1989) refers to these two extremes as **source** and **sink populations,** respectively. Examples of source-sink dynamics include *Sorghum intrans* in savanna woodlands in Australia (Watkinson et al. 1989), *Stipa capensis* in Israel (Kadmon and Shmida 1990), and teasel, described earlier in this chapter (Werner and Caswell 1977).

Using our demographic measures of population performance, we can add complexity by allowing two or more populations to interchange seed (Figure 4-15). We can then measure the vital rates of populations to understand the relationship among populations in maintaining the species as a whole. This process allows us to consider local extinction and recolonization processes.

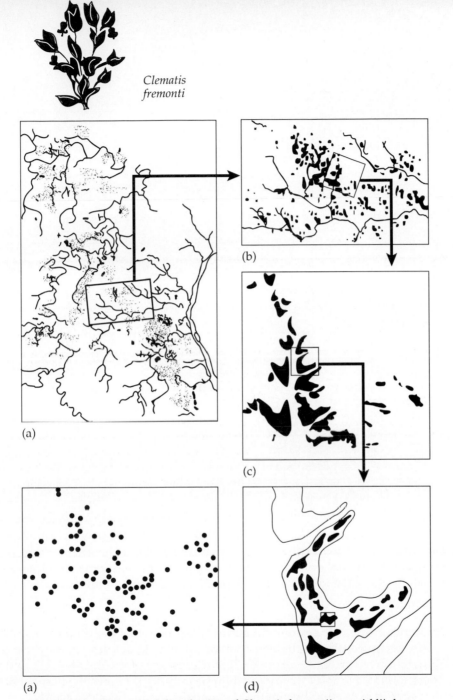

Clematis fremonti

Figure 4-14 The spatial distribution of *Clematis fremontii* var. *riehlii* showing the distribution across (a) the entire range, (b) a region, (c) a cluster of glades, (d) patches within a single glade, and (e) individuals within a single patch. From R. O. Erickson, 1945. "The *Clematis fremontii* var. *riehlii* population of the Ozarks." *Annals of the Missouri Botanical Garden* 32:413–460.

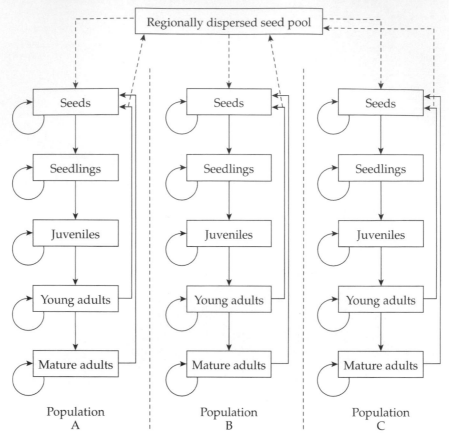

Figure 4-15 A generalized life cycle diagram for a plant species with five stage classes and seed dormancy. Metapopulation dynamics are generated by limited seed exchange between the mostly isolated populations A–C.

Critical attributes of a metapopulation may be estimated by a few simple measures. Allowing P_t to equal the number of occupied sites at a particular time (t), V_t to represent the number of vacant sites, c to equal the site colonization rate, and x to equal the site extinction rate, we can predict future site occupancy:

$$P_{t+1} = P_t + cP_tV_t - xP_t \qquad \text{(Equation 4-12)}$$

The number of sites occupied in the future is a function of the number of (1) sites currently occupied (P_t); (2) the potential for colonization of new sites, which is itself a function of current occupancy (P_t), the presence of vacant sites (V_t), and the colonization rate (c); and (3) the extinction probability (x) times the number of occupied sites (P_t). At equilibrium, the occupancy rate does not change ($P_{t+1} = P_t$):

$$cV_t = x \quad \text{or} \quad V_t = x/c \qquad \text{(Equation 4-13)}$$

Silvertown and Lovett-Doust (1993) refer to x/c as the **relative extinction rate** of populations (relative to the colonization rate). Carter and Prince (1981, 1988) postulate that a geographically expanding population may be limited by a threshold of relative extinction rates where the distribution of vacant sites is sparse. A population ceases to expand because the low colonization rate, owing to sparse occupiable patches, is matched by the extinction rate of patches. Carter and Prince use this model to suggest that gradual changes in the environment may functionally result in abrupt species limits as a result of metapopulation dynamics and the balance between colonization and extinction.

Platt and Weiss (1977, 1985) studied the distribution, abundance, and dynamics of five species of perennial plants on badger mounds along a hillside slope at Cayler Prairie (Iowa). They described patterns in the plant community that suggested differences in individual tolerances, competition, and metapopulation dynamics. Badger mound disturbances provide the habitat for metapopulations of plants that specialize in these short-lived habitats. A gradient with increasing soil moisture at lower slope positions was correlated with decreasing badger mound disturbance. Badger mound disturbances on dry upper slopes were dominated by *Mirabilis hirsuta*, which appeared to be favored by the drier habitats. At the other extreme, only *Apocynum sibiricum* occupied the sparsely distributed wetter sites, probably due to its superior dispersal capabilities. In between these extremes, varying species occupied mounds. Coexistence was attributed to a correlation between dispersal ability and extinction probability such that species that readily colonized sites did not compete well in the presence of other species.

In the same way that not all metapopulations may be net sources for seeds, not all populations are the same size. Having strong differences in population size violates the assumptions of the metapopulation model in important ways (Harrison and Hastings 1994). For many species there are observed to be one, or few, relatively large populations and many smaller ones. This condition is better labeled a **mainland-island dynamic** because predictions of stability differ somewhat from the metapopulation model in that stability of the entire population is strongly dependent on the likely persistence of the mainland population (C. Ray, personal communication, 1997). Thus, metapopulation dynamics is an emerging area of interest in ecology that uses the methodologies of population biology. Empirical tests of metapopulation theory are required in order to add empirical validity to this model.

Summary

The measurement and description of plant population structure and dynamics are central to terrestrial ecology. Factors that influence population dynamics and the distribution of plants include environmental conditions, resource availability, competitors, and disturbance. Varying external conditions have different effects on species depending on the species' life history. The outcomes, however, are expressed in a limited number of ways: by varying (a) reproductive effort, (b) growth, (c) branching patterns, and/or (d) biomass. Plant population ecologists strive to understand and predict the magnitude and pattern of responses to internal and external conditions.

The density of plants, the distribution of biomass, and the allocation to reproduction are three primary measures used to characterize plant populations. When density is combined with measures of spatial distribution, we can deduce more about habitat preference, competitive dynamics, and microhabitat distribution than with density alone. Clumping is the most common distribution pattern in plants, owing to the combined effects of limited dispersal of most seeds and a propensity for individuals within a population to aggregate in suitable micro-environments.

Plant demography is the study of changes in plant populations through time. Plant populations increase or decrease through birth and death processes as well as immigration and emigration, and by indeterminant growth of individuals and vegetative sprouting. Thus, complete knowledge of plant population dynamics requires information about the number of genetically distinct individuals (genets), the number of vegetatively reproduced individuals (ramets), and the number of growth modules on a genet (i.e., the size of a plant). Growth modules may be individual leaves, branches, tillers, or stems depending on what growth regions form repeating modules throughout the individual. Further, understanding the dynamics of seeds (e.g., movement into and out of a population as well as storage and dormancy in the soil seed pool, discussed further in Chapter 5) provides other important insights into plant demography.

Plant demography is studied using either continuous-time or discrete-time models to reveal the consequences of variations in birth and death rates for a population. Continuous models are best used for populations with continuous growth where birth, death, and size are correlated with age. Most plants have a less predictable growth rate, age at reproduction, and age of senescence than is necessary for use in continuous-time models. Matrix models deal with discrete-time periods and can be used with identifiable stages of growth, rather than age, as the basis for prediction.

The number of individuals of a species that can be supported by a unit of habitat, known as the carrying capacity, and the age of individuals at particular stages of development are complicated by the developmental plasticity of plants. This plasticity in plants is another reason that stages, rather than ages, are often used to model plant population performance. The total amount of biomass that can be supported by a site depends on the availability of resources and, once the resources are fully utilized, is constant over a wide range of plant densities (the "law" of constant yield). Biomass is a better currency for carrying capacity than density. Local variability in individual genotypes and resources typically results in large size differences among plants within a population. Population size hierarchies typically include many small individuals relative to the number of large individuals. As high-density populations mature, population density declines and plant biomass accumulates in a pattern predicted by the self-thinning rule.

Data for demographic studies can be organized into either cohort or time-specific life tables, depending on the life span of the plants. Life tables provide information on age-specific mortality rate, survivorship, fecundity, net reproductive rate, and the reproductive value for a population. Population viability analysis typically uses matrix models (but the use of continuous-time or life table models is possible) to project probabilities of local extirpation over specified time frames for the purposes of conservation management planning.

The static age and stage structure of a plant population is often used as a predictive tool to deduce trends in population replacement. Although these static views of a population must be viewed with caution, they can be a powerful tool to predict periodicity of reproductive success and to reconstruct population responses to periodic fire, climatic perturbations, or other disturbances.

CHAPTER 5

ALLOCATION AND LIFE HISTORY PATTERNS

Several unique aspects of plants create a special concern for the population biologists who study them. First, plants may repeatedly shed and regrow most of their biomass. Second, plants have indeterminant growth, that is, they grow continuously until death. Third, many plants are clonal and may, in addition to sexual reproduction, vegetatively reproduce other individuals. Finally, with no developmentally fixed separation between somatic and germ lines of tissue, plants are exceptionally flexible in their adult form. In this chapter, we describe variation in patterns of allocation and variation in life history attributes. Further, we deduce the adaptive value of several of these allocation and life history patterns. This chapter, like many others in this text, attempts to highlight the key features of a topic rather than be exhaustive. The issue of allocation, life history patterns, and trade-offs in plant strategies is rapidly developing with a steady stream of books dealing with this issue from the physiological to the population level (e.g., Willson 1983, Lovett-Doust and Lovett-Doust 1988, Wyatt 1992, Bazzaz 1997).

Allocation

The **principle of allocation** states that individuals have a limited amount of resources to spend on *growth, maintenance* (survival), and *reproduction,* the three basic functions that allow plants to succeed in their environments. Plants therefore must adopt a strategy for allocating resources. The term *strategy* has often been applied to the concept of life history variation and allocation, but it is not meant to imply that the individual plant has the ability to strategically plan resource allocation. A **strategy** is a pattern of resource allocation that has evolved over millennia through the process of natural selection. An individual's strategy is a genetically inherited pattern. Allocation and growth patterns are, to some degree, modified by the local environment. But even the degree to which a plant may be modified by its environment—its **phenotypic plasticity**—is under genetic control.

A synonym for a plant's strategy is its **life history pattern.** A life history pattern that is successful is one that enhances reproduction, survival, and/or growth in a particular environment (Figure 5-1). A successful pattern may involve remaining small and allocating most available carbohydrates and nutrients to reproduction; another may be to extend life by growing slowly and allocating resources to maximize resistance to herbivores, parasites, disease, and abiotic conditions. A variety of life history patterns are often found among the species cohabiting a site.

There may be multiple ways to succeed in a given environment, but not all strategies are equally likely to succeed in all habitats. For example, succulents, with modified stems for water storage, are not likely to be found in either tropical rain forests or arctic tundra. A successful allocation pattern for a plant is a product of both short-term success and the ability to persist for as long as possible (Slobodkin and Rapoport 1974). The environment determines which life history patterns allow a plant to successfully germinate, grow, and reproduce. For example, frequently disturbed habitats favor plants that allocate most of their resources to reproduction. Plants that reproduce just once during their lifetimes tend to occur in ephemeral or harsh habitats (Young and Augspurger 1991). In contrast, for trees in a stable habitat, devoting relatively more resources to growth than to reproduction (at least in the early stages of development) may be favored. Life history pattern is related to the type of habitat and the role a species plays in a community.

Plant growth has been considered analogous to a business operating under the principles of economic theory (Bloom et al. 1985). Individuals that have the greatest profit (grow fastest) gain an advantage (usurp more space and increase their resource base). However, allocation to growth is typically associated with trade-offs (e.g., decreased reproduction). In other words, growth, reproduction, and survival represent competing demands for a finite supply of carbohydrates, fats or waxes, and proteins produced by synthetic processes in the plant. Thus, we could view the acquisition of resources as income, with expenditures partitioned between maintenance (e.g., storage and defenses), growth, and reproduction. Allocation can then be treated by a cost/benefit analysis, allowing the prediction of optimal allocation patterns. The ultimate measure of successful allocation, however, is reproductive success.

The principle of allocation and associated hypotheses described in this chapter typically carry an assumption of trade-offs in resource acquisition (Lovett-Doust 1989). Allocating resources to the development of some modules restricts the ability of plants to develop others. Furthermore, allocation for the capture of some resources (e.g., root production for nutrient acquisition) inhibits allocation to structures to capture other resources (e.g., leaves for carbon acquisition). Gadgil and Bossert (1970) developed a theoretical model, based on cost/benefit analysis, proposing that early or large allocations of energy to reproduction lead to decreased probability of survival and/or reduced potential for future reproduction.

The cost/benefit trade-off model has been supported by several, but not all, studies. *Poa annua* completes its life cycle within one or more years, depending on

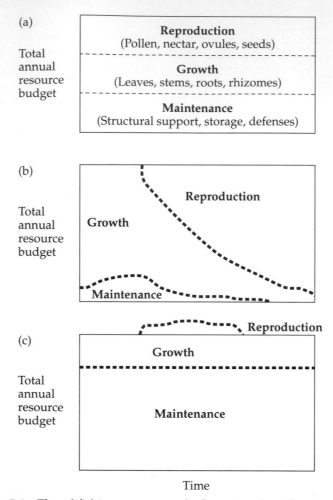

Figure 5-1   Three life history patterns of a flowering plant illustrating potential variation in allocation to basic plant functions across a single growing season. (a) Even allocation to reproduction, growth, and maintenance. (b) Annual plants allocate most resources early in the growing season and allocate most resources to reproduction late in the growing season. (c) A typical stress-tolerating tree or shrub allocates most resources to maintenance, less to growth, and only allocates resources to reproduction in growing seasons that are high in resource availability. Diagram modified from M.F. Willson. 1983. *Plant Reproduction Ecology.* John Wiley & Sons, New York.

conditions (Law 1979). High levels of reproduction in the first season are correlated with lower levels of reproduction in the second season (Figure 5-2a). Plants with high first-season reproductive output are smaller in the second season than their associates that have little or no early reproductive output (Figure 5-2b). There is also evidence that early, heavy reproduction increases the subsequent risk of

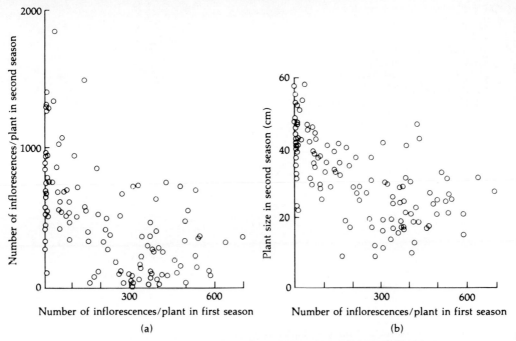

Figure 5-2   Scatter diagrams of number of inflorescences per plant (*Poa pratensis*) in the first season plotted against (a) number of inflorescences per plant in the second season, and (b) plant size in the second season. From Law. *American Naturalist* 113:3–16. Copyright © 1979 by University of Chicago Press. By permission.

mortality in plants. Pinero et al. (1982) reported that the probability of a tropical palm (*Astrocaryum mexicanum*) surviving for 15 years is essentially zero when a plant produces 40 to 50 fruits per year and increases with decreasing reproductive output (Figure 5-3). Although some studies (e.g., Horvitz and Schemske 1988, Pitelka et al. 1985) have failed to find a trade-off between reproductive effort and vegetative growth, it is likely that all reproductive effort comes at a cost.

Complications arise, however, in using the cost/benefit analogy for plant allocation. First, unlike business economics, there are three currencies rather than one in the realm of plant economics (Bloom 1986). Plants share a requirement for (a) carbon (CO_2 captured using solar energy via photosynthesis and stored in organic molecules), (b) water, and (c) different soil nutrients. Plants, though requiring different amounts of each resource, all require the various resources (currencies) in roughly similar proportions (Chapin et al. 1987). Reekie and Bazzaz (1987) suggest that carbon, because of its important role in plant energetics, integrates biomass allocation patterns of other resources and is the most appropriate common currency. Second, the net cost of biomass production is often difficult to assess. For example, flowers are considered a cost, yet some flowers have green parts that perform photosynthesis (Williams et al. 1985). Thus, Tuomi et al. (1983) caution against a simple assumption of trade-off costs in resource allocation

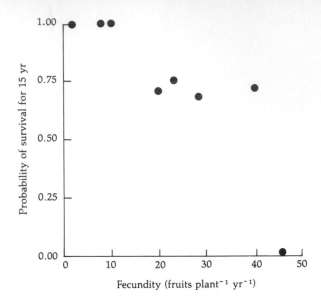

Figure 5-3 The probability of surviving for 15 years as a function of fecundity in 10-year-old plants of *Astrocaryum mexicanum* in Veracruz, Mexico. From Pinero et al. *Journal of Ecology* 70:473–481.

because various modules (reproductive and vegetative) may support their cost through resource acquisition, and plants can physiologically adapt to different costs. Ashman (1994) provides one of the few studies where the cost of reproduction is considered repeatedly within a growing season. In her work on the checker mallow (*Sidalcea oregana* ssp. *spicata*), Ashman finds that dynamic measures of nutrient investment in reproduction provide better estimators of future reproduction than do traditional measures. One reason for this is that nutrients are often resorbed from leaves and flowers before abscission, making their true costs difficult to measure. Finally, allocation patterns may be genetically fixed, or relatively flexible and dependent upon environmental conditions. The costs and benefits of remaining flexible to environmental variability are difficult to assess.

In general, it remains intuitively appealing to view plants as resource-limited and resource allocation as a combination of genetically predetermined responses and plastic allocation patterns that respond to environmental conditions. Despite the drawbacks, cost/benefit analyses remain a popular and productive way to characterize problems of how plants should optimize energy allocation toward different functions and to assess the conditions under which different strategies should prevail.

Allocation to Resource Acquisition

Even before ecology became a recognized field of study, Liebig (1855) observed that it is unlikely for an environment to provide resources in the exact proportions and amounts needed for maximum plant growth. Under such conditions, there is typically one resource that is more limiting than all others. Liebig called this the law of the minimum (also discussed in Chapter 3). The essential idea of

Figure 5-4 Curves A and B depict population growth of two species (A and B) across a range of resource availability. The two curved solid lines show the resource-dependent population growth curves for A and B. The dashed lines (m_A and m_B) represent mortality rates for A and B. The intersection of the curved solid and dashed lines represents the minimum resource level (R^*) to support an equilibrium population (indicated by the vertical solid lines). In this case, species B survives at lower resource levels than species A, and hence is regarded as the superior resource competitor. From D. Tilman. 1982. *Resource Competition and Community Structure*. Princeton University Press, Princeton, NJ.

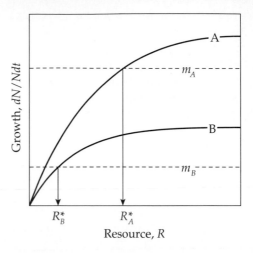

Liebig's law was formalized into the concept of a limiting resource by Armstrong and McGeehee (1980). Under this limiting resource model, R^* represents the amount of a resource that is the minimum required for an individual to maintain a positive net growth rate. Competition between individuals can then be viewed as mediated by the R^*s of different species (Figure 5-4). In this case, the better competitor is defined as the species with the lower R^* for a limiting resource because this species may continue to thrive while drawing the resource below the minimum level for the other species. Increasing the availability of the most limiting resource will increase growth until some other resource becomes limiting, according to this model.

Within this context, plants are unusual in that the structures responsible for acquiring essential resources are spatially separated (Grime 1994). Energy and CO_2 are absorbed by the leaves (and sometimes stems), but nutrients and water are normally absorbed by roots. Thus, plants are faced with balancing allocation toward developing the capacity to capture different resources through different organs (roots versus shoots). Allocation patterns of root versus shoot growth are generally more variable between species than within species. Knowing this, we can consider allocation patterns among plant species as conforming to functional types (root or shoot specialists).

The **resource-ratio hypothesis** predicts the success of species varying in root-shoot allocation patterns across environments characterized by different levels of resource supply (Huston and Smith 1987, Tilman 1988). Species with allocation patterns focusing on shoots (low root to shoot ratios) are assumed to be relatively effective competitors for light, and those allocating more heavily toward roots (high root to shoot ratios) are assumed to be good competitors for belowground resources. Further, the resource-ratio hypothesis predicts that resource supply ratios vary systematically through successional series to first favor root specialists because soil nutrition is more limiting than light in primary succession. Through

Not all plants showing the annual growth form are obligate annuals. Obligate annuals die after reproduction because all resources are committed to reproduction, leaving no potential for further vegetative growth. More rarely some annuals maintain the potential for continued growth and reproduction until some environmental factor surpasses the plant's range of tolerance. For example, *Perityle emoryi* grows as a monocarpic ephemeral in the southwestern deserts and as an iteroparous perennial in more moderate island and coastal habitats.

Situations that reduce the probability of adult survival, and therefore favor a monocarpic annual or biennial life cycle, are often related to frequent disturbance and the existence of temporary habitats. Disturbance removes competing vegetation, increases resource availability, and often favors rapid growth and reproduction. These sites, however, are often successionally ephemeral.

Disturbance is not the only phenomenon that favors the annual habit. For example, over 90% of the desert flora of Death Valley, California, is annual. Although physiological stress associated with drought and high temperature may reduce the probability of survival of perennating organs (organs capable of perennial activity, such as buds, tubers, corms, and bulbs), extreme variation in annual rainfall may be the more important factor. For example, Went and Westergaard (1949) reported that annual precipitation in Death Valley varied from 19 to 94 mm. Because about 60 mm of precipitation is required within a few hours for germination of many species, one or more years may pass between germination episodes; even then, the length of time that resources are available may be very short, and reproduction must occur quickly. *Boerrhavia repens* of the Sahara Desert goes from seed to seed in as few as 10 days (Cloudsley-Thompson and Chadwick 1964). Annual plants that occupy habitats with temporally and spatially unpredictable resources are referred to as **ephemerals,** with life spans typically less than six months.

Short-lived monocarpic perennials (biennials) typically occupy sites disturbed periodically but not every year. Postponing reproduction has the advantage that a greater store of reserves will be available for reproduction. The weakly annual sea rocket (*Cakile maritima*) outcompetes the strictly annual *C. edentula* because second-year seed production in *C. maritima* is much higher than in *C. edentula* (Boyd and Barbour 1993). For some species, a shorter growing season is associated with a delay of reproduction in order to build up reserves. For example, *Melilotus alba* (white sweet clover) is an annual at lower elevations but postpones reproduction to the second year at high elevations, where the growing season is contracted (Smith 1927). More generally, delay to a second (or third) year should be favored as long as the risk of mortality is less than the benefit of increased plant size and fecundity. Many "biennials" are found in early successional sites, habitats that are predictably short-lived but also predictably last more than a single year. Overwintering may be costly in terms of potential mortality, but the increased reproductive potential may compensate for this increased risk.

The longer an adult plant is likely to survive, the more likely it is to invest in maintenance costs. Woody perennials epitomize this relationship through investment in structural support tissues. Loehle (1988) examined patterns of growth rates, defenses, reproduction, and longevity for 159 species of North American trees, making three important observations. First, the typical age of maturity is

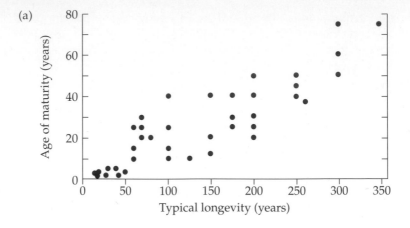

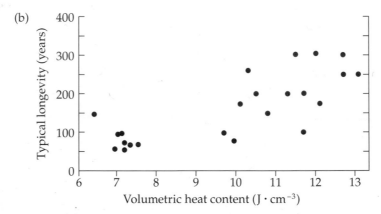

Figure 5-5   (a) Typical age of sexual maturity versus typical longevity for North American angiosperm trees and shrubs. (b) Typical longevity versus volumetric heat content of sawtimber-size stem wood of angiosperms. Two groups are evident: a low heat content group that is shade intolerant and short-lived, and a high heat content group that is longer lived and more shade tolerant. From C. Loehle. 1988. "Tree life history strategies: The role of defenses." *Canadian Journal of Forest Research* 18:209–222.

positively correlated with typical longevity, suggesting a higher early investment in growth and maintenance costs relative to reproduction (Figure 5-5b). Second, he observed a negative correlation between growth rate and longevity, suggesting that longer-lived species invest more in maintenance than growth. Finally, Loehle observed that trees with longer life spans produce wood that is more likely to be better defended against decay or disease (Figure 5-5a). Although this survey made no effort to factor out the potentially confounding effect of habitat quality on growth rates, age to maturity, and longevity, these results support the contention that trade-offs between growth, maintenance costs, and reproduction are important determinants of life history.

The cost to a plant of defending itself against herbivores or pathogens is another important area for plant allocation. Rosenthal and Kotanen (1994) split allocation to plant defenses into two categories: developing an ability to tolerate herbivory, and avoiding herbivory through chemical defenses. Tolerance mechanisms include shifts in postdamage allocation patterns toward increased nutrient uptake or photosynthesis (Chapin and McNaughton 1989, Welter 1989). Alternatively, plants may defend themselves by allocating resources to chemical deterrents to herbivores (see Chapter 7). Although it seems likely that a suite of complex secondary compounds used as a defense against herbivores and pathogens ought to result in a physiological cost to the plant, experiments conducted to measure these costs have often failed to find evidence of costs, at least in terms of reduced growth (e.g., Simms and Rausher 1989). A recent survey of 88 studies that looked for costs of defenses in terms of plant fitness found measurable costs in only about half the cases (Bergelson and Purrington 1996). Among those studies that found significant fitness costs of plant defense, most were in relation to herbicide resistance, and the fewest were found in association with herbivory. As in studies of the cost of reproduction, the difficulty in demonstrating costs of plant defenses has not prevented successful research based on the assumption that costs exist (see Chapter 7). We remain far from a general understanding of the cost of anti-herbivore and anti-pathogen defensive compounds.

Allocation to Reproduction

Assuming the plant that produces the greatest number of seeds during its lifetime has the highest fitness (i.e., maximizes the number of successfully reproducing offspring and thus gene transmission to the next generation), we can imagine certain circumstances that favor polycarpy and others that favor monocarpy. In one of the most clear-cut sets of examples of an allocation trade-off (cost), monocarpic plants have repeatedly been shown to have higher reproductive outputs in their single flowering episodes than do closely related polycarpic species in each of their flowering episodes (Young 1990, Young and Augspurger 1991).

Ecologists have tried to define circumstances under which monocarpy or polycarpy is favored (e.g., Willson 1983). One method of looking at this is through residual reproductive value (RRV) when allocating a particular fraction of the available resources to reproduction. The important parameters are seed output (b_x) and the probability of reproducing in the future (RRV). If b_x responds slowly to increases in RA (the proportion of resources allocated to reproduction), the resulting curve is concave (Figure 5-6a); if response is rapid, the curve is convex (Figure 5-6b). RRV is the product of the reproductive value of an average individual one year older (V_{x+1}; see Equation 4-11) and the probability that a plant of age x will survive to reproduce at age $x + 1$ (l_{x+1}/l_x):

$$RRV = V_{x+1}(l_{x+1}/l_x) \qquad \text{(Equation 5-1)}$$

When reproduction causes a rapid decrease in residual reproductive value, the RRV versus RA curve is concave (Figure 5-6c). When reproduction causes only a small decrease in RRV, the curve is convex (Figure 5-6d). Adding the residual

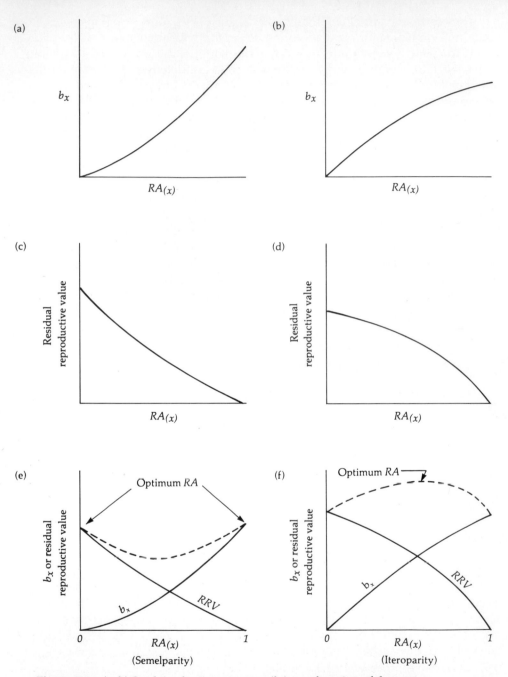

Figure 5-6 (a, b) Seed production at year x (b_x) as a function of the proportion of resources allocated to reproduction (RA). (c, d) Residual reproductive value $V_{x+1}(l_{x+1}/l_x)$ as a function of RA. (e, f) The total (dashed line)
of b_x and residual reproductive value (RRV) is highest at RA = 0 or 1
(monocarpy) when scenarios (a) and (c) are combined and at intermediate
RA values (polycarpy) when scenarios (b) and (d) are combined. See text
for details.

reproductive value and b_x for all possible RA values from 0 (no reproduction) to 1 (allocating all resources to reproduction) gives a series of numbers, the maximum of which is associated with the RA of maximum seed production (fecundity). When both the residual reproductive value and b_x are concave (Figure 5-6e), maximum reproductive output is attained either by not reproducing in year x or by committing all resources to reproduction (monocarpy). When both b_x and the residual reproductive value (RRV) are convex (Figure 5-6f), maximum seed production will result from allocating an intermediate amount of resources to reproduction (polycarpy).

In a classic paper on plant life history and demography, Cole (1954) suggested that the population growth rate (λ) of a polycarpic perennial could be matched by an annual with the addition of one single seed over that produced by the perennial. If this is the case, and annuals typically produce many more seeds than perennials, pondered Cole, then why would selection ever favor the perennial growth form? This query, termed **Cole's paradox,** triggered much research over the past several decades. Mathematically, we can express Cole's paradox as follows:

Monocarpy: $N_{t+1} = N_t \times B_m$ (Equation 5-2)

Polycarpy: $N_{t+1} = (N_t \times B_p) + N_t$ (Equation 5-3)

where B represents the birthrate expressed as the total annual seed production for monocarpic (m) and polycarpic (p) species, respectively. Rearranging this equation as per Equation 4-2:

Monocarpy: $N_{t+1}/N_t = B_m$ (Equation 5-4)

Polycarpy: $N_{t+1}/N_t = B_p + 1$ (Equation 5-5)

and recalling that $N_{t+1}/N_t = \lambda$, we find that when the rates of population increase are equal in the monocarpic and polycarpic species ($\lambda_m = \lambda_p$), then:

$B_m = B_p + 1$ (Equation 5-6)

The solution to the paradox lies in the fact that Cole did not build juvenile mortality into his model, and he made fairly simplistic assumptions regarding age of maturity and fecundity (Stearns 1992). Charnov and Schaffer (1973) include adult and juvenile mortality to derive

$B_m = B_p + P_a/P_j$ (Equation 5-7)

where P_a and P_j are adult and juvenile survival rates, respectively. In Cole's simple model P_a and P_j were equal to one, and therefore equal to each other. Charnov and Schaffer's work, later made more general by Young (1981), pointed to conditions that would favor polycarpy or monocarpy. Namely, monocarpy is favored when juvenile mortality rates are low and adult mortality rates are high. These conditions are frequently encountered at highly disturbed sites with brief but predictable periods conducive to rapid growth (ephemeral habitats). We can then generalize that environmental variation ought to select for a longer reproductive life span so that plants spread the risk of a bad year over many reproductive bouts. This is because monocarpy is a risky option when environmental uncertainty can

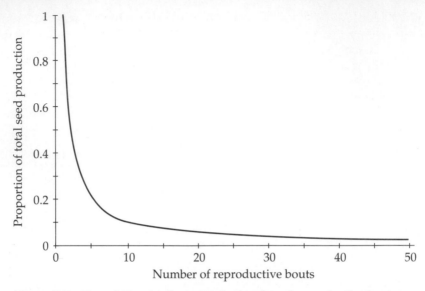

Figure 5-7   The relationship between the number of reproductive bouts and the proportion of total reproductive output in each reproductive bout (assuming each reproductive bout results in equal numbers of seeds). This graph demonstrates the challenge of annuals to achieve the same individual lifetime fitness as a perennial.

wipe out an entire cohort. Nonetheless, we should not be surprised to find a monocarpic organism in variable environments, because "in a stochastic world that should happen from time to time" (Stearns 1992).

   Monocarpy has the advantage of potentially mobilizing a single massive reproductive effort. As observed in Figure 5-7, the potential to outproduce a polycarpic species with a single reproductive bout becomes more difficult the longer the polycarpic individual lives; a long-lived polycarpic plant can produce a tiny fraction of the seeds of a monocarpic plant in any one reproductive bout and yet have an equal net lifetime production. The large reproductive effort focused in one year, however, may afford other advantages, such as allowing the individual to gain a greater share of the pollinators in a competitive system. For example, some species of *Agave, Yucca,* and *Swertia* (Figure 5-8) produce a large inflorescence with copious amounts of nectar, an energetic cost for reproduction that would be impossible in a polycarpic species. *Swertia radiata* (monument plant) populations show both temporal and spatial flowering synchrony, produce large amounts of nectar for an extended period, and have inflorescences up to 2 m tall. As a result, a small number of individuals (relative to the total number of individuals of all species competing for pollinators) in the Colorado Rocky Mountains receive up to 80% of the observed floral visits by pollinators (Beattie et al. 1973).

   Young (1990) studied the demographics of a suite of giant lobelias on Mount Kenya to test hypotheses regarding reproductive effort. These long-lived perennials (the monocarpic *Lobelia telekii* and the polycarpic *L. keniensis*) were studied for up to eight years to predict survivorship and reproductive effort. A model based

Figure 5-8   Examples of large inflorescences of the perennial semelparous *Frasera speciosa* (a) and *Yucca whipplei* (b) and, in the background, *Agave deserti*.

on the number of years between reproductive bouts and the likelihood of adult survival was constructed to predict conditions that should favor one life history strategy over another. The two polycarpic populations of *L. keniensis* studied for eight years both fell within the predicted life history, while a less well studied population of the polycarpic *L. keniensis* fell within the range of uncertain demographic fate (Figure 5-9). Thus, the observed reproductive traits for *L. keniensis* generally fit predictions.

Another advantage of monocarpic reproduction may be to swamp potential seed predators, increasing the probability of some seeds surviving. *Phyllostachys bambusoides* (a Chinese bamboo species) lives about 120 years before populations synchronously flower. Flowering is physiologically controlled and is not stimulated by external environmental cues. The seeds are vulnerable because they are not protected chemically, they contain more nutrients than rice or wheat, and they disperse close to the parent plant and often become buried deep in the soil. Granivore populations, however, are unable to predict reproduction in bamboo, and so their populations are not sufficient to consume all the seeds. Individuals that flower out of synchrony, or in areas isolated from the main group of flowering individuals, lose essentially all of their seeds to granivores (Janzen 1976).

Some polycarpic woody perennials reproduce heavily in some years, called **mast years,** followed by one or more years of little or no reproduction. Mast years

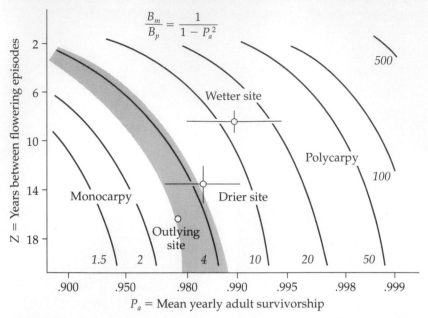

Figure 5-9   Demographic location of the three *Lobelia keniensis* populations monitored by Young (1990) on Mount Kenya relative to the putative evolutionary boundary between monocarpy and polycarpy in this system. The equation at the top represents the relative increase in fecundity that would need to be associated with a switch to monocarpy in order for that life history to be more productive than a polycarpic life history with yearly adult survivorship P and average years between reproductive episodes Z. The curved lines represent the actual reproductive advantage of monocarpy in this system. Populations experiencing values of P and Z that place them above and to the right of this threshold are better off remaining polycarpic, but those below and to the left of the line have such a low prospect for future reproduction that they do better putting all reserves into a single massive reproductive episode (monocarpy). From T. P. Young. 1990. "Evolution of semelparity in Mount Kenya lobelias." *Evolutionary Ecology* 4:157–172.

are common in temperate trees and serve to increase the chances of seeds escaping predators (Silvertown 1980). Sork et al. (1993) observed that acorn production in three midwestern oaks varied in the degree to which they exhibited mast seed production, and varied in the extent to which mast years were predicted by weather and recent reproductive efforts. Masting in western oaks is similarly variable among species and appears to depend on both abiotic conditions and recent reproductive history (Koenig et al. 1994).

Allocation within Reproductive Structures

Plant allocation to reproductive effort may be spent in many ways (Table 5-1). For example, plants vary widely in both pollen and seed dispersal vectors. Species relying on wind to disperse pollen must produce a large quantity of pollen to en-

Table 5-1 A simplified view of the predominant ways in which allocation to reproductive structures varies within and among species of terrestrial plants.

I. Sexual reproduction
    A. Investment in seed production and fertilization
        1. Pollen production vs. ovule production (male/female fitness)
        2. Pollen number vs. pollen size (e.g., wind vs. animal pollinator systems)
        3. Gamete production vs. pollinator reward (ovule number/nectar reward)
    B. Investment in potential seedling success
        1. Seed size (provisioning) vs. seed number
        2. Dispersal vs. provisioning of seed (e.g., wing loading on wind-dispersed seed; fruit quality or size)
II. Vegetative reproduction: Sprouting vs. seed production

---

sure success. Wind-pollinated plants produce tens of thousands to millions of pollen grains for every ovule that is successfully fertilized (Faegri and van der Pijl 1971). High pollination success rates can be ensured in plants pollinated by insects and vertebrates, allowing plants lower investment in pollen production. Investment in pollinator attraction, as well as larger and better provisioned pollen grains, are costs that may balance the savings from reduced pollen production. Pollen limitation, one potential outcome of failing to attract pollinator visits, reduces fitness (Dudash and Fenster 1997). Allocation to pollinator attraction versus the number of ovules produced, likewise, ought to be under strong selection because increased ovule number allows increased seed production, which should have a tangible effect on fitness. Modeling the costs and benefits of allocation to various reproductive effort components (i.e., pollen, pollinator attraction, ovules, and seeds) is complicated and remains unsatisfactory.

Allocation to seed production varies widely. Seed sizes vary by ten orders of magnitude, from 10 kg coconuts down to orchid seeds weighing less than a microgram (Haig 1996). The size of a seed within a species was classically considered a fixed trait (Harper 1977), but wide variability in seed sizes is often observed (e.g., Stanton 1984, Michaels et al. 1988). Many species maintain continuous variation in seed size (Jordano 1984, 1995) as well as seed dimorphisms that include different dispersal mechanisms and germination characteristics (Venable and Lawlor 1980). Geritz (1995) modeled the role of seed size variability in determining success within various environments. The study showed that variation in seed size is adaptive in competitive environments. Thus, variability of the biotic and abiotic environment suggests that variability in seed provisioning is beneficial.

Relative to the effort that has gone into analyzing costs and benefits of sexual reproduction, little effort has been applied to analyzing fitness consequences of vegetative reproduction. This lack of attention is curious, given the widespread occurrence of vegetative reproduction within plants. Loehle (1987) presents one of the few models of allocation to vegetative reproduction. Using a cost/benefit analysis, Loehle concludes that predictable environmental variability and habitat differences that result in differential success of sexual versus vegetative reproduction favor flexibility in allocation to different reproductive modes.

The key to the comprehensive analysis of the fitness of plant reproduction is that the simple act of reproduction is not so simple. It involves accumulation of resources, biomass protection, allocation to pollen and ovule production, aspects of pollinator attraction and seed dispersal, and sufficient seed provisioning to allow for seedling establishment. Until the energetic costs and benefits of all these can be considered simultaneously, we cannot assess relative fitness. Even were we able to simultaneously assess all these costs, we are left with problems of assigning costs to their appropriate functions. For example, when does a seedling become independent? What proportion of costs associated with protecting a plant from herbivores should be assigned to that seed crop rather than to a future seed crop the plant will produce if it survives?

Reproductive allocation within individual plants also can vary greatly. For example, allocation to seed production along a branch of the California buckeye (*Aesculus californica*) in one year has a negative effect on branch elongation during the next year (Newell 1991). This implies that production of buckeye fruits draws on storage that the branch requires to maximize growth in the subsequent year. In contrast, oaks (Marquis 1988) and hickories (McCarthy and Quinn 1992) demonstrate lower reproductive output associated with branches that grow less in a given year. Each of these studies suggests that resources for reproduction are, to some extent, compartmentalized within the individual plant.

Allocation to seed versus dispersal mechanism is another important source of variability among plants. For wind-dispersed seed, wing loading (roughly the size of the wing, or wind-resistant structure, relative to the mass of the seed) can vary by several orders of magnitude (Augspurger 1986). Wing loading determines the terminal velocity at which seeds fall through still air, so it is strongly correlated with seed dispersal distance. Variation in allocating resources to sexual reproduction for wind-dispersed seed producers is between (a) the number of seeds produced; (b) the mass of those seeds, where increased mass reduces likely dispersal distances but increases the probability of successful establishment; and (c) wing loading, which provides seeds with the ability to travel longer distances. Allocation patterns in wind-dispersed species ought to depend on whether the establishment phase for a species is very risky and is benefited by a large store of reserves, and whether the species life history is one of colonizing new habitats so that long-distance dispersal is advantageous.

Australian *Acacia*s have developed a special appendage on seeds, called an **eliaosome,** that is consumed by their ant or bird dispersers (O'Dowd and Gill 1986). Again, allocation patterns for animal dispersal vary among (a) the number of seeds produced; (b) the mass of seeds; and (c) the disperser reward provided (e.g., eliaosome size or pulp of a fruit). This disperser reward can be a very complicated issue in that the reward depends not only on the mass of the reward but also its nutritive content. *Acacia* eliaosomes vary widely in size within this genus

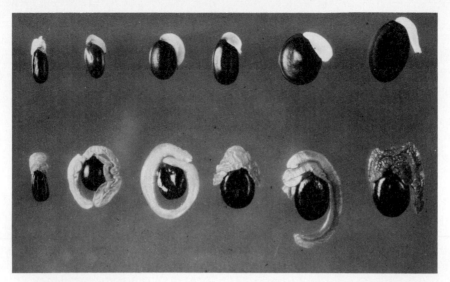

Figure 5-10 *Acacia* seeds apparently adapted for dispersal by ants (above) and birds (below). The food reward in the appendage (eliaosome) on the seed is larger and usually more colorful (red, orange, or yellow versus white) in the bird-dispersed species. From D. J. O'Dowd and A. M. Gill. 1986. "Seed dispersal syndromes in Australian *Acacia*," pp. 87–121 in D. R. Murray, ed. *Seed Dispersal*. Academic Press, San Diego.

(Figure 5-10). Examples of variability in plant strategies to attract dispersal agents are far too numerous to describe here and are left for more detailed studies (e.g., Fenner 1992).

Life History Attributes

Seed Dispersal and Dormancy

Well-timed periods of seed dormancy and suitable patterns of seed dispersal are necessary components of a successful life history pattern. **Dormancy** is a mechanism to avoid seasonally harsh conditions and, in the case of seeds, may be a means of spreading the risk of failure over long time periods.

A **seed pool** is an accumulation of living seeds in the soil. Transient seed pools are represented by seeds that will either die or germinate within a year, whereas persistent seed pools accumulate seeds over a longer period (Thompson and Grime 1979). Transient seed pools have little impact on the population, aside from their numbers and position in safe sites, and are usually formed by plants growing in predictable habitats. Persistent seed pools are characteristic of ephemeral plants in unpredictable habitats and of perennial herbs (especially short-lived mono-carpic perennials) or shrubs that germinate in response to some unpredictable event that opens habitat space within an otherwise late-successional community. For example, chaparral shrubs add seeds to the seed pool yearly, but they germi-nate only when fire removes the overstory (Keeley 1977), or when conditions allow a very long fire-free period to develop a litter layer sufficient to support seedling survival (Keeley 1992).

Freas and Kemp (1983) compared germination of *Pectis angustifolia*, a Chi-huahuan desert annual that germinates in response to predictable summer precip-itation, with *Lappula redowskii* and *Lepidium lasiocarpum*, which germinate following unpredictable winter and spring rainfall. *Pectis* had no innate dormancy (i.e., 100% potential germination), whereas *Lappula* and *Lepidium* both showed innate dor-mancy in a portion of the seeds. Thus, some seeds with innate dormancy are added to the seed pool each year in order to potentially germinate at some future time. Both the winter and summer annuals had germination rates proportional to the amount of rainfall; that is, seeds of *Pectis* would not all germinate unless there was a sufficient minimal amount (threshold) of rainfall. This fits the prediction of Ven-able and Lawlor (1980) that a sensitive, environmentally induced germination mechanism can substitute for innate dormancy in desert ephemerals.

Species differ in patterns of seed dispersal (Harper 1977, Portnoy and Willson 1993). Although most seeds fall close to the parental plants, those that fall far from their parents are more likely to produce plants (Figure 5-11; Augspurger 1984, Augspurger and Katijama 1992). Long-distance dispersal is typical of plants that occupy patchy environments. Agricultural weeds typically have effective means of dispersal that depend on wind or animal (including human) vectors. Riparian trees grow along the margins of streams and rivers and are thus adapted to widely sep-arated or patchy environments. Seeds of these trees are often dispersed by wind or water. Dispersal distance may also be related to the density of patches in the environment that contain an appropriate combination of conditions for seedling establishment.

Venable and Lawlor (1980) modeled the interplay between dispersability and seed dormancy in desert annuals. Their model predicts that plants with long-range dispersal will have few dormancy mechanisms and thus quick germination. In contrast, those without adaptations for long-distance dispersal should have de-layed germination (see also MacArthur 1972, Cohen 1966). Table 5-2 is a list of spe-cies that produce dimorphic seeds, some capable of long-distance dispersal, others not. In most cases the seeds without long-range dispersal exhibited delayed ger-mination, confirming the predictions of the Venable and Lawlor model.

Seed Dispersal and Tree Distribution Shifts: Predicting Responses to Global Warming

Global warming is projected to increase global temperatures 2–5°C during the next century (Kattenberg et al. 1996). This temperature increase is likely to cause widespread shifts in plant distributions. To put the magnitude of these shifts in

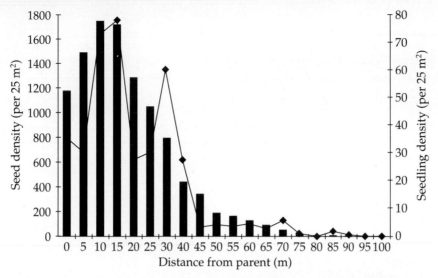

Figure 5-11 The density of *Tachigalia versicolor* seeds naturally dispersed from parental trees (bars) versus those surviving two years (solid line). Modified from C. K. Augspurger and K. Katijama. 1992. "Experimental studies of seedling recruitment from contrasting seed distributions." *Ecology* 73:1270–1284.

perspective, spruce (*Picea* spp.) shifted northward approximately 1500 km in response to the 3–4°C warming at the end of the last glaciation. Tree species in general moved north slowly, 35–50 km per century (Davis 1981). Further, there is evidence that trees were at least occasionally limited by the maximum migration rates (as determined by seed dispersal rates). Thus, we do not expect maximum migration rates to far exceed 50 km per century. There remains much to learn about dispersal. Greene and Johnson (1995), measuring seed dispersal from pines in southern Ontario, find that the wind events required to move a seed 1 km happen on the order of one minute per decade! This rate is so slow that it is difficult to reconcile how plants migrated across the landscape as fast as we know they historically did. Thus, ecologists need to better understand the potential role of secondary dispersal (e.g., by small mammals, or by wind across snow) that may increase maximum migration events. We know that populations of trees have occasionally become established hundreds of kilometers outside their distributions (e.g., Webb 1986). These observations suggest that maximal dispersal distances may be quite large.

If the global climate warms as much during the next century as is predicted by climatologists, however, trees would need to shift their distributions northward somewhere between 500 and 800 km to remain in equilibrium with climate (Davis and Zabinski 1992). Thus, ecologists expect tree migrations to lag far behind future rates of climatic change. Schwartz (1993) modeled likely future migration rates in a fragmented habitat and concluded that where habitat loss is high (most

Table 5-2 Species with two types of seeds with differences in both dispersal (N = near, F = far) and germination (D = delayed, Q = quick). Dispersal is inferred from presence versus absence of dispersal structures such as barbed or plumed pappus for animal or wind dispersal, respectively. If dispersal differences are inferred from substantial differences in size and weight of propagules, symbols are marked with an asterisk (*). Reprinted from Venable and Lawlor. 1980. *Oecologia* 46:272–282. Springer-Verlag, New York.

Species of Asteraceae	Dispersal		Germination	
	Outer	Inner	Outer	Inner
Dimorphotheca plurialis	N	F	slightD	Q
Xanthocephalum gymnospermoides (= Gutierrezia g.)	N	F	D	Q
Heterotheca latifolia (= H. lamarckii)	N	F	D	Q
Charieis heterophylla	N	F	Q	Q
Bidens bipinnata	similar		D	Q
Sanvitalia procumbens	N	F	D	Q
Verbesina enceliodes	N	F	slightD	Q
Synedrella nodiflora	N	F	D	Q
Heterospermum xanthii	N	F	D	Q
Galinsoga parviflora	N	F	Q	D
Layia platyglossa ssp. *campestris (= L. elegans)*	N	F	slightD	Q
L. platyglossa	N	F	D	Q
L. heterotricha	N	F	D	Q
Achyrachaena mollis	N	F	D	Q
Chrysanthemum segetum	N*	F*	D	Q
C. coronarium	N*	F*	D	Q
C. viscosum	N*	F*	D	Q
C. frutescens	N*	F*	D	Q
Coleostephus myconis (= Chrysanthemum m.)	N*	F*	D	Q
Chardinia xeranthemoides	N	F	D	Q
Leontodon taraxacoides	N	F	D	Q

Species of Brassicaceae	Dispersal		Germination	
	Upper	Lower	Upper	Lower
Cakile maritima	F	N	Q	D
Rapistrum rugosum				
(with capsule wall)	F	N	D	Q
(without capsule wall)	F	N	Q	D
Sinapis arcense	N	F	D	Q
S. alba	N	F	D	Q
Hirshfieldia incana	N	F	D	Q
Brassica tournefortii	N	F	D	Q

agricultural and urban areas), migration rates may be as low as 1–10 km per century. The result is that trees are unlikely to be sufficiently responsive to future warming as a result of dispersal limitations and changing land use patterns.

Classification of Life History Patterns

The immense number of plant species and the complexity of adaptations they represent have led to an effort to classify plants according to similarities of life history patterns. Several assumptions underlie the classification schemes. First, evolution leads to adaptations that maximize fitness. Second, plants growing in similar spatial and temporal microhabitats have similar life history patterns. Because spatial and temporal microhabitat differences occur within any habitat, we expect that a community will contain plants with a variety of different life history patterns and that successional changes can be characterized by predictable changes in life history patterns of the component plants. Third, the specifics of different life history patterns may lead to similar classifications because characteristics may compensate for one another (Grime 1982). For example, both increased seed production and an earlier age for first reproduction increase intrinsic rates of population growth (r). Fourth, physiological characteristics reflect life form, habitat, and reproductive patterns. Fifth, a life history classification scheme will aid in our effort to understand and communicate information about plant adaptations.

r- and K-Selected Life History Patterns

A widely used classification of life history patterns places organisms on a spectrum between the extremes of allocation to reproduction (MacArthur and Wilson 1967). Table 5-3 summarizes some important distinctions between r- and K-selected taxa.

At one extreme, selection acts to maximize the intrinsic rate of increase (r, see Equation 4-4). This extreme occurs in habitats where mortality is through density-independent factors (e.g., drought, fire). The more seeds in such an environment, the more individuals. Desert ephemerals, ephemeral plants of granitic and carbonite outcrops in the eastern United States, and roadside weeds are viewed as archetypical r-selected species (Table 5-3). This life history extreme is characterized by short-lived species, such as monocarpic annual plants, which allocate very little energy to either growth or maintenance but a great deal toward reproduction. Survivorship curves generally follow the Deevey type I curve (see Figure 4-9). Seeds of r-selected plants are typically numerous and very small, are dispersed by wind or water over large areas, and typically have a long potential seed dormancy period.

On the other extreme, selection can act to maintain populations at or near their habitat-specific carrying capacities (K, see Equation 4-7). K-selected species occur in habitats where populations are density dependent. This life history extreme is characterized by long-lived species, such as polycarpic trees and shrubs like red-

Table 5-3 Some traits correlated with typical *r*- and *K*-selection species. Modified from E. Pianka. 1970. "On r- and K-selection." *American Naturalist* 104:592–597.

Trait type	*r*-selection	*K*-selection
Climate	Variable, unpredictable	Constant or predictable
Mortality	Density independent	Density dependent
Survivorship	Type III	Type I or II
Population size	Variable	Fairly constant, near carrying capacity
Effects of competition	Often lax	Strong
Development time	Short	Long
Life span	Short, <2 years	Long, >5 years
Seed bank	Yes	No
Allocation	Reproduction focus	Survivorship, delayed reproduction
Reproductive mode	Monocarpic	Polycarpic
Overall	Productivity	Efficiency

wood or sugar maple, allocating a greater proportion of available resources to functions that increase competitive ability and survival (such as growth and maintenance) and a lesser proportion to reproductive output. Survivorship curves generally follow the Deevey type III curve (see Figure 4-9). Seeds of *K*-selected plants may often be few, large, and dispersed by animals.

This distinction between *r*- and *K*-selected species focuses on the extremes. Most plants fall somewhere between these extremes, however, and wide variability is expected even within habitats. For example, desert ephemerals typically fit the concept of *r*-selected species, but not all of them fit this definition equally well. Clark and Burk (1980) studied patterns of allocation of structural (biomass) and nonstructural (stored carbohydrates) carbon in the co-occurring desert ephemerals *Plantago insularis* and *Camissonia boothii* (Figure 5-12). These species germinate simultaneously in response to cool-season precipitation but have markedly different patterns of growth and reproduction. *Plantago* allocates significantly more structural and nonstructural carbon to reproduction early in the season, completing its life cycle in less than 60 days, whereas *Camissonia* continues both vegetative and reproductive growth for over 100 days. The extended growing season of *Camissonia* is made possible by allocating more energy to the production and maintenance of vegetative tissues and to storage. Presumably, in years when no further rainfall occurs after germination, *Plantago* will have higher reproductive output because of early, heavy commitment of resources to reproduction. In years when it rains again later in the growth period, however, *Camissonia*, by allocating resources to maintenance and survival, will realize higher overall reproduction than would *Plantago*. The two seasonal patterns of precipitation occur with sufficient frequency to maintain both life history patterns as desert ephemerals, with *Plantago* to the *r*-side of *Camissonia* on the *r-K* spectrum.

Classification by *r*- and *K*-selection has been criticized as overly simplistic (e.g., Stearns 1977, 1992; Hickman 1975; Wilbur 1976; Tilman 1987). Stearns (1992),

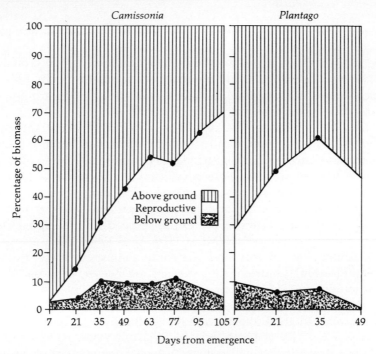

Figure 5-12 Distribution of aboveground, belowground, and reproductive biomass (as % of total biomass) throughout the growing season of the desert ephemerals *Camissonia boothii* and *Plantago insularis*. From D. D. Clark and J. H. Burk. *Oecologia* 46:86–91. Copyright © 1980 by Springer-Verlag, New York. Reprinted by permission.

for example, refutes the *r-K* continuum as potentially misleading because it equates natural selection, which works on the individual, with a process (density dependence) that regulates populations and not individuals. Further, broad categorizations of life history attributes such as those presented in Table 5-3 should be viewed as general guidelines, not hard rules. As we should expect, not all traits characteristic of any given life history pattern consistently co-vary in all species. Nonetheless, as a phenomenological description of basic life-history strategies, the *r-K* continuum affords a generality that is appealing, even if the mechanisms are not sufficiently well understood and exceptions to the rules exist.

R-, *C-*, and *S*-Selected Species

Grime (1977, 1979) expanded the *r* and *K* concept of life history classification to include patterns of species selected in various types of habitats. For example, temporary habitats with abundant resources select for rapidly colonizing (*R*—ruderal) species. Predictable habitats with abundant resources favor species that compete well for those resources (*C*—competitors). Finally, resource-stressed habitats favor species that persist under conditions of sometimes severe resource deficits (*S*—stress-tolerators). The three adaptive patterns coincide with allocation primarily to reproduction (*R*), growth (*C*), or maintenance (*S*). **Ruderals** (*R*) are typically found in temporary, often frequently disturbed habitats and are characterized by an adaptive suite identical to that described in the previous section for *r*-selected species. Plants that allocate most available resources to growth and that can capitalize on readily available resources are **competitors** (*C*). The efficiency with which these species capture resources makes it difficult for other species to capture resources in these environments.

The final category of life history patterns includes those plants that live in habitats where, because of abiotically limited resources or because physiological stress restricts resource utilization, survival depends on allocating most resources to maintenance; these are **stress-tolerant** (*S*) species. Examples of habitats dominated by stress-tolerators are deserts, arctic tundra, alpine meadows, peat bogs, and serpentine soils. Successful stress-tolerant species have slow growth rates and produce tissues that are durable. For example, needlelike evergreen leaves are common in nutrient-poor habitats (Monk 1966, Aerts 1995). Many evergreen gymnosperms exhibit slow growth rates relative to angiosperms and an inability to fully capitalize on high-nutrient sites (Keeley and Zedler, in press). Because the norm in resource-limited habitats is stress, stress-tolerator plants do not have the ability to grow rapidly even when ample resources are available. Stress-tolerant species tend to have long life spans with tissues that are resistant to stress and expensive to produce. These plants may also have a well-developed acclimation potential to variation in the environment as well as a potential for vegetative reproduction. Seedlings generally remain small for extended periods.

Table 5-4 contains an overview of the characteristics of competitive, stress-tolerant, and ruderal plants. As in the case of *r*- and *K*-selected species, attributes characterizing a life history type are general patterns and not strict rules. It is im-

Table 5-4 Some traits correlated with typical competitive, stress-tolerant, and ruderal plant species. Modified from J. P. Grime. 1979. *Plant Strategies and Vegetation Processes.* Wiley, New York.

Trait type	Competitive	Stress-tolerant	Ruderal
Life form	Variable	Variable	Herbs
Shoot morphology	Dense canopy	Variable	Small stature
Leaf form	Variable	Leathery, needles	Variable
Leaves	Deciduous	Evergreen	Deciduous
Longevity	Long or short	Very long	Very short
Flowering	Annual	Intermittent	Annual
Reproductive maturity	Late	Late	Early
Reproductive effort	Small	Small	Large
Perennation	Buds, seeds	Leaves and roots	Seeds
Growth rate	Rapid	Slow	Rapid
Stress response	Rapid	Slow	Reproduces
Litter	Persistent, copious	Persistent, sparse	Not persistent, sparse
Palatability to herbivores	Variable	Low	Often high

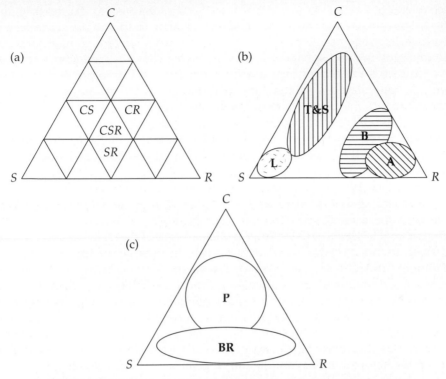

Figure 5-13   (a) Grime's model of life history variation based on percent occurrence of competition (*C*), disturbance (*R*—ruderals), and stress (*S*). General patterns of life histories with Grime's model of life history variation. Combinations (e.g., *CSR*) represent intermediate life histories as per Grime's (1979) definitions. (b) Grime's assessment of life history strategies of most trees and shrubs (T&S), lichens (L), biennial herbs (B), and annual herbs (A). (c) Grime's assessment of typical life history strategies of most bryophytes (BR) and perennial herbs (P). Modified from J. P. Grimes, et al. 1990. *Comparative Plant Ecology.* Unwin-Hyman, London.

portant to remember that plants can be adapted to any combination of disturbance, competition, and stress. This means that a classification system must be based on the relative importance of each in determining the life history pattern. To emphasize this point, Grime (1979) presents life form classification as a triangular model (Figure 5-13a) in which the competitive, ruderal, and stress-tolerant populations described in Table 5-4 occupy the corners and plants with intermediate life history patterns occupy the central area of the triangle.

These intermediate regions of Grime's plant strategies triangle represent specific combinations of life history attributes. For example, competitive ruderals (*CR*) tend to occur in resource-abundant habitats where disturbance is severe enough to prevent high-intensity competition but infrequent or moderate enough to prevent noncompetitive ruderals from dominating. Stress-tolerant ruderals (*SR*) occur in unproductive habitats with intermediate levels of disturbance. Desert ephemerals, desert geophytes, arctic/alpine annuals, and many bryophytes are examples of stress-tolerant ruderals. Where the habitat is not subjected to disturbance, and sufficient resources are available for moderate growth, the plants are characterized as stress-tolerant competitors (*CS*). Herbaceous stress-tolerant competitors have moderate growth rates, spread aggressively by vegetative reproduction, and possess evergreen leaves. Many woody plants that grow in unproductive habitats or during the later stages of succession, when many of the resources are tied up in biomass, fall into the category of stress-tolerant competitors.

Grime's model predicts a general relationship between habitat, life form, longevity, and life history pattern (Figure 5-13b). Note that, with the exception of perennial herbs and ferns, the life form groups occupy a restricted area of the Grime triangular model.

Grime et al. (1990) have characterized the establishment strategies of the British flora according to this triangular model. This attempt, not surprisingly, has resulted in some difficult classifications. For example, *Lathyrus montanus* (bitter vetch) is characterized as a *C-S-R* intermediate (center of the triangle) and an *S* species. This variation in classification may indicate the presence of different ecotypes, or different strategies during different life stages. For example, such a species may be a *C-S-R* intermediate as a seedling, whereas established adults are stress-tolerators. Because the natural world is complex and a species' adaptive possibilities are driven by numerous factors, encountering species that are difficult to classify is expected.

These classifications of life history patterns are most instructive when used as a conceptual basis for understanding the range of variety exhibited by plants. Few, if any, quantitative predictions are generated from this simple three-sided classification. Further, most species exhibit a mixed strategy that includes components of different generalized strategies. Nonetheless, categorizing species as they coexist or change temporally or spatially in a community may allow the ecologist to simplify the system, leading to a greater understanding of community organization (Whittaker and Goodman 1979). Exceptions to expected life history properties may lead the physiological ecologist to examine new adaptive modes.

Plant ecologists have divergent views on the response of plants to stress, and the role of stress in interspecific interactions (Oksanen 1990). Grime, like many others before him, stresses the relationship between a plant and its environment. Others (e.g., Cajander 1990, Tilman 1982) stress the role of interspecific interactions and competition in structuring plant communities. The Cajander-Tilman school (*sensu* Oksanen 1990) believes that strong competition characterizes virtually all communities. In this sense, Grime (1977) and Tilman (1982) view competition somewhat differently; these differences will be discussed in detail in the next chapter.

Life History Correlates of Invasiveness/Weediness

The classification of "*r*-selected" or "ruderal" species can be related to at least one real-world problem. The magnitude of the invasive weed problem worldwide is staggering. Another term for a weed is a "nonnative invasive pest plant." A species is generally considered native if it has a pre-industrial record of inhabiting a region (Webb 1985, Schwartz 1997). All other species are nonnative. Invasive plants are those that invade undisturbed or lightly disturbed habitats. Pest plants are those that interfere with agriculture or natural areas management and the maintenance of biological diversity. Thus, species can be nonnative and neither invasive nor a pest.

The flora of most regions within the United States are comprised of 25% or more nonnative species. Some ecosystems, such as grasslands of the intermountain West (cheatgrass), eastern wetlands (purple loosestrife), and the streams of Florida (water hyacinth) are often dominated in biomass and number by nonnative species. It has been estimated that undesirable nonindigenous species have directly cost the U.S. economy $97 billion (mostly through crop losses, reduced grazing potential, and weed control costs). This estimate is conservative in that it focuses on agricultural costs and ignores the damage to many native ecosystems (U.S. Office of Technology Assessment 1993). With an increasing awareness of the problem, and a recognition that harmful plant invasions continue to occur at unprecedented rates (U.S. Office of Technology Assessment 1993), it becomes increasingly important to use life history attributes to predict whether a species is likely to become a problem weed if imported. The ecological literature on invasive nonnative species is expanding quickly (e.g., Cousens and Mortimer 1995, Cronk 1995, Pysek et al. 1995, Aldrich and Kremer 1997, Luken and Thieret 1997, Radosevich et al. 1997).

Baker (1974) proposed 13 characteristics of plants that may be associated with invasiveness. Unfortunately, a successful weed may exhibit all 13 characteristics, or just one. There is wide variation in the characteristics that confer success on plant invaders. In order to improve predictions of invasiveness, several researchers have begun more focused examinations of the invasive weed problem. Rejmanek and Richardson (1996) used complex statistical methodologies (discriminant function analysis) to describe characteristics of pines that have repeatedly become invasive weeds. They found that a short juvenile period, small seed mass, and a short time interval between seed crops are the three primary predictors of invasiveness. Further, they found that their analysis correctly predicts invasive pines in 68–100% of cases tested. Similarly, Reichard and Hamilton (1997) use another method (regression tree analysis) to predict woody species invasiveness and attain overall predictive rates of over 75%. Characteristics that frequently appear in the analysis of Reichard and Hamilton are vegetative reproduction, early age of reproduction, and no required pretreatment for seed germination. Thus, within relatively constrained life histories (e.g., woody species), ecologists are beginning to isolate suites of characteristics that allow prediction of whether a species is likely to become weedy if introduced. This area of research is just beginning to show its potential for ecological application and holds promise for further development. It remains to be seen whether governmental agencies will use this information to restrict importations of potentially problematic plant species. At present, only species on a federal "dirty" list are prohibited. Yet very few plants make it on the dirty list without an on-going outbreak within the United States, when it is already too late. Many of our worst offenders (annual grasses and crop weeds) are accidental imports and less amenable to legal intervention.

Summary

Life history patterns summarize the growth, reproduction, and longevity of plants. The life history pattern of a plant is analogous to an economic balance. A finite amount of resources can be acquired by a plant, and these resources must be allocated among modules and functions promoting growth, reproduction, or survival. There are never enough resources for maximal allocation to all three, and the result is a series of trade-offs. Allocation patterns are flexible enough to allow different patterns in different environmental conditions, but they are relatively constrained within the overall adaptive suite of each plant species. In other words, an annual can adjust to different situations by changing allocation patterns, but being a successful annual precludes investing in structural support through woody growth or living a long time.

Allocation to growth is partitioned between root and shoot growth. Those species specializing in accumulating belowground biomass tend to be good competitors for water and soil nutrients, and those focusing on aboveground biomass tend to be good competitors for light. Allocation to survival (maintenance) is characteristic of long-lived polycarpic species. This allocation can be in the form of plant defensive compounds, or in structural support such as woody growth.

Allocation to reproduction is in reality a complex series of allocation patterns balancing among maximizing seed output, seed provisioning, and dispersal. Each attribute can lead to an increased likelihood of successful seedling establishment. Allocation to reproduction can also be achieved through the production of vegetative sprouts. Life span and pattern of reproduction during the plant's life are important determinants of life history. Monocarpic, or semelparous, plants reproduce only once during their lifetime and allocate all available resources to a single bout of massive reproduction. Polycarpic, or iteroparous, plants have repeated but smaller reproductive periods. Here longevity, and thus future expectation of offspring, are maximized by repeated, less intensive reproduction.

Life span is also related to environment. Annuals are monocarpic and have a short life as a means of survival in temporary, usually disturbed habitats. Biennials are monocarpic and usually live into the second season before reproducing. Postponing reproduction presumably has benefits in terms of future reproductive success that outweigh the mortality risk associated with longer life. Perennials are either monocarpic or polycarpic and are best adapted to more predictable habitats. Advantages of semelparity include a large flower display that may attract more pollinators, a higher initial seed set, and massive seed production that may satiate predators.

The successful life history pattern has well-timed periods of dormancy and seed dispersal. Dormancy is a means of avoiding harsh environmental conditions and, for plants with long-lived seeds, spreads the risk of failure over several growing seasons. Dispersal is the mechanism for efficiently utilizing a habitat and moving to other habitats. Plants in patchy environments tend to have long-distance dispersal capabilities and few dormancy mechanisms, whereas plants from uniform environments typically have shorter dispersal distance. Annuals in unpredictable environments tend to have effective dormancy mechanisms and extremely long seed life, forming persistent seed pools in the soil. In predictable environments where successful reproduction is essentially assured, plants have few dormancy mechanisms and form transient seed pools.

There may be many variants of plant allocation patterns, but because species with similar patterns tend to occupy similar habitats in space and time, they can be classified to reduce the complexity. One classification system places life history patterns along a continuum from the extreme of the transient *r*-selected species to the extreme of the perpetual competitor or *K*-selected species. The *r*-selected species have short life spans, allocate most resources to reproduction, have high dispersal distances, small size, little competitive ability, and live in temporary habitats. *K*-selected species live longer, are larger, allocate more resources to increase longevity and competitive ability, and occupy more permanent habitats. The life history classification of Grime defines three nodes of life history patterns. Ruderal (*R*) species are identical to *r*-selected species; competitive (*C*) species are those that allocate most of their energy to rapid growth and competitive ability; and stress-tolerant (*S*) species are those that allocate most of the available resources to increase longevity. Stress-tolerant species live in resource-poor habitats, competitive species in resource-rich permanent habitats, and ruderal species in resource-rich temporary habitats. Although many species can be placed in one of these categories, most tend to exhibit characteristics of more than one strategy and are considered to follow some form of intermediate, or combined, strategy.

CHAPTER 6

SPECIES INTERACTIONS: COMPETITION AND AMENSALISM

Chapters 3–5 dealt primarily with plant species and populations as though they grow in isolation. The biology of organisms grown in isolation, however, is not the same as their biology when grown in mixtures. In nature, virtually all communities consist of more than one plant population. In addition, plants within communities are influenced by nonplant populations, such as decomposers (bacteria and fungi), parasites, pathogens, pollinators, and herbivores. Many ecologists believe that the associated organisms in a community are in various ways interdependent, that they are not associated by chance, and that disturbing one organism could have consequences for many other organisms. Clements (1916) took this view to an extreme, equating climax communities with superorganisms and considering their component populations to be as interdependent as the cells, tissues, or organs of a single organism. At the other extreme, some ecologists refer to communities as completely artificial constructs of the human mind. In place of the community concept, these ecologists assert that nature consists of assemblages of species thrown together by historical accident and physiological constraints into pools of species that interact.

We do not debate the merits of either of these extreme views here. The truth is, no doubt, somewhere between these extremes. The merits of these two views have occupied the thoughts of plant ecologists for the past half century. In recent decades, however, there has been a trend toward using population studies of a few interacting species to build a mechanistic understanding of community structure from the bottom up. The objective of the next two chapters is to survey the variety of interactions among populations of different species growing in the same location. This will lead to an exploration of views on community interdependence and integrity in Chapter 8.

Simple Interactions

Ecologists have devised a simple symbolic notation, as well as a terminology, for describing the effect of one species on another in a two-species interaction. For example, species that have negative effects on one another (−,−) are competitors, and species that have positive effects on one another (+,+) are mutualists. Filling in other potential combinations of positive, negative, or null interactions creates a simple table of six types of interactions, with three others as simple mirror images of the named interactions (Figure 6-1). Note that there is no single unifying name for (+,−) interactions; the name used is often based on the trophic status of the participants (e.g., predator-prey, herbivory, parasitism).

Burkholder (1952) expanded the interaction concept to create a list of interactions that include the effects of one species on another when they are both present ("on"), as well as when the interaction is missing ("off"). For example, when an herbivore and its food plant are together, the herbivore is stimulated (its growth, reproduction, or general success is improved), and the plant is depressed (its growth, reserves, reproduction, or general success declines), resulting in a (+,−) interaction. When the two species are apart, however, the herbivore is depressed and the plant is unaffected (−,0), an altogether different effect that is not the opposite of the "on" interaction.

Bronstein (1994) further developed the interaction concept to describe conditional interactions. Conditional interactions vary in magnitude and sign under different environmental conditions. Bronstein showed that the costs and benefits of various interactions often vary in space and time. (Figure 6-2 shows an interaction realm.) For example, two species may act as mutualists under favorable environmental conditions, yet as pathogen and prey under stressful conditions.

Understanding how environmental variability can change the cost/benefit ratio of an interaction results in one of the difficulties in assessing the frequency of

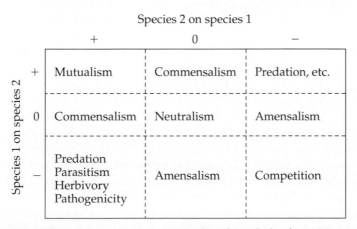

Species 2 on species 1

	+	0	−
+	Mutualism	Commensalism	Predation, etc.
0	Commensalism	Neutralism	Amensalism
−	Predation Parasitism Herbivory Pathogenicity	Amensalism	Competition

(Species 1 on species 2)

Figure 6-1 Pairwise species interactions listed in tabular form. Notice that there are many names for a mutually negative interaction.

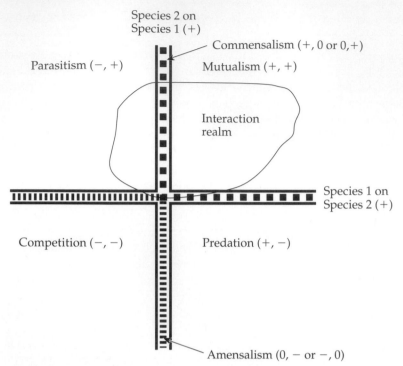

Figure 6-2   A hypothetical graph of possible interaction "space" where an interaction may fluctuate through time or space and vary in strength of the mutualism or even change sign and become a commensal or parasitic relationship.

various interactions. It is easier to label the interaction based on logic and observation than it is to measure costs and benefits (Cushman and Beattie 1991). A further complication is distinguishing among variations of basic interactions. Competition between plant species typically takes the form of two species competing for shared resources (e.g., light, water, nutrients). In this case (**exploitation competition**), the outcome of competition may depend on resource availability. Alternatively, competitors may interact directly (**interference competition**), although this is more common among animals that have the ability to fight one another over territory or resources. Notable examples of interference competition are the agonistic behavior of male red-winged blackbirds as they fight for nesting territories, or male bighorn sheep fighting for mating opportunities by butting heads. Although one might consider the shading of one plant by another as a direct form of interference competition, it is probably better described as exploitation competition where each individual strives to maximize its own capture of light resources.

Although two plant species may compete indirectly, indirect interactions, mediated through a third species, may be just as important (Connell 1990). Consider two species that share a common herbivore. Increasing the populations of plant species A results in increasing herbivore populations. Increasing the herbivore

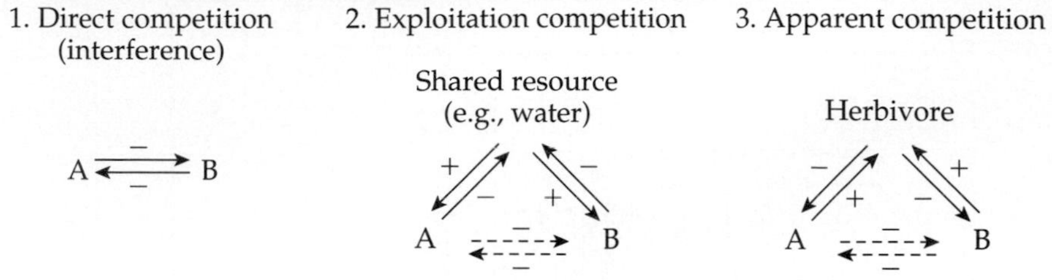

1. Direct competition (interference)

2. Exploitation competition

3. Apparent competition

Figure 6-3   Three types of pairwise species interactions. In direct interference competition, species actively confront each other. Examples include territorial fighting in animals and may include allelopathy in plants. In explotation competition, species have a negative impact on one another through a shared resource, such as water. Apparent competition is observed when species have a net negative impact on one another, but the interaction is indirectly mediated through a third species.

populations has a negative effect on plant species B (Figure 6-3). Thus, increasing A leads to a decrease in B with no direct interaction between them. **Apparent competition** is the state where two species have a negative net effect on each other via a third species. These apparent competitors may or may not also interact with each other directly.

Using Pattern to Infer an Interaction

Before an interaction is considered ecologically significant, the researcher must follow a series of steps. A frequent prerequisite for hypothesizing the importance of an interaction (positive or negative) between species is to observe a nonrandom pattern in distribution. The sampling is based on the premise that positive interactions will produce positive spatial relationships between species; where one species is found, the probability is high that the other will be found nearby. The two populations "attract" one another and exist in a nonrandom, clumped pattern (Figure 6-4a). Similarly, a negative interaction will produce a nonrandom, negative spatial association (Figure 6-4b). If there is no interaction between the populations then the location of one individual has no influence on the location of others and the two populations are randomly distributed with respect to each other (Figure 6-4c). A random pattern, however, may also arise from complex interactions. Similarly, species may cluster nonrandomly around resource concentrations and appear to be positively associated while actually interacting negatively. Thus, spatial pattern, or lack thereof, cannot necessarily be interpreted as the presence or lack of an interaction and must be interpreted with caution.

One way in which pattern can be revealed is by sampling vegetation with random quadrats of an appropriate size. In each quadrat, the presence or absence of any two (or all) species is noted, then summarized in a contingency table (Table 6-1). This constitutes the observed data. Expected data, assuming a completely random distribution of the two taxa, are compared to observed data by a

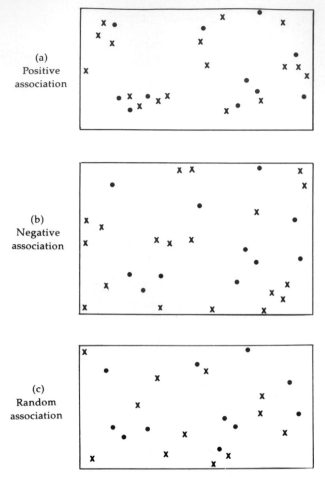

Figure 6-4 An overhead, diagrammatic view of two species (dots and x's) that are (a) positively associated, (b) negatively associated, and (c) randomly associated.

chi-square formula where the expected value is the product of the row and column sums for a cell divided by the grand total. The example in Table 6-1 yields a chi-square value higher than expected by chance, so the two species (A and B) are not distributed randomly with respect to each other. Inspection of the table shows more quadrats with only A or B than expected by chance, and fewer quadrats with both species present than expected by chance. The conclusion is that A and B are negatively associated. The statistical treatment is merely the first step; the direction of the result must be discerned from the data and the reason for the association must be determined by ecological experimentation.

Another method of determining pattern is by measuring distances between randomly chosen plants and their nearest neighbors. The dispersion index (R) of

Table 6-1 Contingency table for analysis of association between two species, A and B. In this example, 100 random quadrats were tallied for the presence or absence of each species.

Species B	Species A		Totals
	Present	**Absent**	
Present	Observed: 30 Expected: $65 \times 59/100 = 38.35$	Observed: 35 Expected: $65 \times 41/100 = 26.65$	65
Absent	Observed: 29 Expected: $35 \times 59/100 = 20.65$	Observed: 6 Expected: $35 \times 41/100 = 14.35$	35
Total	59	41	100

$$\chi^2 = \Sigma(\text{observed} - \text{expected})^2/\text{expected} = 1.81 + 2.61 + 3.38 + 4.86 = 12.58$$

Clark and Evans (1954) is then calculated as the ratio of actual mean distance and expected distance, based on a random spatial pattern:

$$R = \frac{\text{Mean distance}}{0.5 \sqrt{\text{density}}} \qquad \text{(Equation 6-1)}$$

where the mean density of plants per unit area can be estimated from quadrats. The departure of R from 1 indicates regularity ($R > 1$) or patchiness ($R < 1$). It is important in all these methods to realize that pattern may change as plants age (increase in size). A review of Sonoran and Mojave Desert plants showed that small shrubs tended to be clumped, medium-sized shrubs tended to be random, and large shrubs could sometimes be regularly dispersed (Phillips and MacMahon 1981). Although the dispersion index examines intraspecific patterns, the data can be interpreted in light of interspecific interactions. In this instance the data indicate that competition increases in intensity as plants grow larger (older). Sampling design, consequently, should take plant age into account.

The second step required to infer an interaction involves experimentation. The choice of methods to complete this step depends on the putative interaction. These may be experiments that take place in the field or in controlled environments (greenhouse or growth chamber) that attempt to imitate field conditions. The third step returns to the field to ask whether the factors discovered to be important in one's experiment operate in nature. The difficulties in carrying a suspected interaction through all three steps are numerous and well illustrated in the literature.

Armed with a classification of the types of interactions, an appreciation of the variability within these interactions, and the potential difficulties in specifying them, we now move on to discuss the ecological ramifications of the various forms of interactions.

Competition

Competition $(-,-)$, the mutually adverse effect of organisms on one another, has been a central focus of plant studies since the inception of plant ecology. Anyone who has grown a garden has personal experience observing that plants compete for shared resources and that competitive differences between species exist. Perhaps the earliest documentation of plant competition as an important ecological phenomena dates to de Candolle (1820, cited in Clements 1929). Books on plant competition have appeared regularly, from Clements (1929) to Keddy (1989) and Grace and Tilman (1990).

The extent to which competition affects the structure of natural plant assemblages remains an issue of debate (Connell 1990, Goldberg and Barton 1992, Gurevitch 1992). Plant assemblages in nature are influenced by a number of interactions. Isolating real, versus apparent, competition can be very difficult. Interactions that seem to be competitive may be artifacts of other interactions. Furthermore, varying strengths of interactions may make detection of competition difficult. Reviews of competition in field experiments found that a majority (~about 60%) of tests for competition actually revealed an amensalism $(-,0)$ (Lawton and Hassell 1981, Connell 1983). However, the failure to detect some effect may not argue very strongly for its strict absence. What these studies do show is that **asymmetric competition** is common, with one species experiencing significantly more deleterious effects of the interaction than the other.

Bengtsson et al. (1994) distinguished two basic ways of studying plant competition. Investigators can examine the *results* of competition, either through experiments or careful observation, in order to assess the role of competition in community structure and the maintenance of diversity. Alternatively, investigators can examine the *mechanisms* of competition, focusing on resource acquisition and use, to distinguish relative competitive abilities among species. During this discussion of competition, we will examine both types of studies and describe their influence on the field of plant ecology.

As previously discussed, all plants require the same basic resources (carbon, water, and nutrients) in roughly the same proportions (Chapin et al. 1992), and they acquire these through similar mechanisms (root uptake and photosynthesis). Thus, it stands to reason that plants ought to compete with one another for access to resources (above- and belowground space). Since competition involves two organisms utilizing the same resource(s) at the same time, competing organisms must have overlapping niches.

A **niche,** as defined by Hutchinson (1957), is the multidimensional description of a species with all aspects of its biotic and abiotic environmental requirements.

This definition makes it difficult to precisely define a niche for an individual species, but it captures the essence of our modern concept of a niche. A species requires a suite of environmental conditions (a niche) in order to survive, and the more similar two species are in their requirements, the more likely they are to compete for those resources. Nonetheless, we simply do not know enough about most species to fully describe their niche. Instead we make assumptions about what are likely to be important attributes (e.g., soil nutrient, water, and light requirements), describe other conditions that are correlated with a species distribution (e.g., growing season length, minimum and maximum winter and summer temperatures), and assume that this accurately characterizes a species niche (ignoring biotic interactions such as mutualist pollinators, mycorrhizae, etc.). Thus, the niche concept remains intuitively appealing, yet functionally difficult to enumerate. The condition becomes somewhat worse when one considers that the range of conditions under which it is possible for a plant to grow (its **fundamental niche**) is often broader than the conditions under which we actually see a plant grow (its **realized niche**).

The assertion that competing species must have overlapping niches gave rise to the simple statement "one niche, one species" made by the Russian microbiologist G. F. Gause (1934). Gause placed pairs of closely related species together in a homogeneous environment and noted population growth rates (dN/dt—read "the change in population size [N] with change in time [t]"). Initially, growth rates of both species were depressed compared to rates when grown in isolation, indicating that competition was occurring. Ultimately, only one species survived, with the "winner" dominating the habitat and the "loser" going extinct. Gause concluded that in order for species to coexist in nature they must evolve ecological differences. The one niche, one species concept is often called **Gause's competitive exclusion principle,** even though zoologists before him had published much the same conclusion (Krebs 1972). Laboratory work and field observations since Gause's time, using animals with complex life cycles and with heterogeneous environments, have shown that minor niche differences exist even between closely related taxa, and these differences are sufficient to permit coexistence (see, for example, Ayala 1969 and MacArthur 1958).

Animal community ecologists during the past 30 years have occupied much of their time trying to discern whether there is strong evidence of niche separations that structure animal communities. In contrast, plant ecologists have typically viewed the problem of measuring niche breadth and niche separation as secondary to focusing directly on measuring competition for critical resources, such as light, water, and nutrients, which are required by all plants (Austin 1985). As a result, testing the niche concept has not been central in the domain of the plant community ecologist as it has been among animal community ecologists. Instead it has been the population ecologists who have focused studies on the effects of interspecific competition and species niche differences in plants.

A fundamental problem ecologists have long sought to understand is: how is it that so many species persist within the same habitats? Hutchinson (1959) defined this as the paradox of diversity. Plants provide a good example of this dilemma, in that we know there are relatively few resources for which plants compete and by which their growth is limited (i.e., light, water, and nutrients). Thus, if we can discern the ways in which resource competition limits the ability of sites to maintain diversity, we might better understand how to manage habitats for biodiversity.

Measuring Competition

Recall, from Chapter 4, the logistic equation of population growth for a species in isolation:

$$dN/dt = rN\left(\frac{K - N}{K}\right)$$ (Equation 6-2)

where r is the natural, intrinsic rate of population growth, N is population size, and K is population size at saturation (the environment's carrying capacity). This equation contains an **intraspecific competition** component in the quantity ($[K - N]/K$). When population size (N) is small, this component approaches K/K or 1, meaning that the population increases at a maximal rate (rN). When N approaches the carrying capacity (K), the population growth rate is dampened because the quantity ($[K - N]/K$) approaches 0. In this case, carrying capacity is a phenomenon of intraspecific competition. With that understanding, we can now consider a joint carrying capacity in a two-species system where

$$N_1 + N_2 = K$$ (Equation 6-3)

Realistically, however, two species often do not have equal per capita effects on one another owing to differences in biomass or in resource consumption. Thus, we need to scale the effect of species 2 on species 1. This scaling is accomplished with the coefficient α as follows:

$$N_1 + \alpha N_2 = K_1$$ (Equation 6-4)

We can now model **interspecific competition** by introducing this term for a second species, one that has a negative effect on population growth, into our population growth equation from Chapter 4 (Equation 4-8). The equation for species 1 becomes

$$dN_1/dt = r_1 N_1 \frac{(K_1 - N_1 - \alpha N_2)}{K_1}$$ (Equation 6-5)

where α, once again, is the inhibiting (competitive) effect on species 1 for every individual of species 2. Similarly, the two-species competition equation for species 2 is

$$dN_2/dt = r_2 N_2 \frac{(K_2 - N_2 - \beta N_1)}{K_2}$$ (Equation 6-6)

where β is the competition coefficient for species 1 on species 2.

This pair of equations was developed independently by Lotka (1925) and Volterra (1926). Although these equations carry a number of restrictive assumptions (e.g., no spatial or temporal variability in habitats), they are a starting point for quantifying competition and discussing the nature of coexistence. For example, we can predict the outcome of competition by drawing **zero net growth isoclines** (often called "ZNGIs" but simply referred to here as "isoclines") for populations in a two-species mixture based on the Lotka-Volterra competition equations (Figure 6-5).

In Figure 6-5 we sketch an idealized depiction of the potential outcomes of two species interacting in a competitive manner. Figure 6-5a begins by showing an isocline (line) for a single species that distinguishes regions (combined abundances of species 1 and 2) where species 1 will increase (left of the line) from those where they will decrease (right of the line). The isocline represents various combinations of species 1 and 2 that result in the joint population carrying capacity. Figure 6-5b combines the previous figure with an isocline for a second species where species 1 has a higher isocline than species 2. For the species 2 isocline, population combinations that fall below the line grow (move up), and those above the line exceed carrying capacity and shrink (move down). From this diagram we can see that the intercept of a species' isocline for species 1 along the N_1 axis is simply K_1. Similarly, the intercept along the N_2 axis is defined by the effect of species 2 on K_1, which is K_1/α. A similar pattern holds for species 2. There is a simple method for reading these joint abundance diagrams. Mixed populations that fall below the isocline for a species will tend to increase for that species. Thus, a population at point A will increase in both species 1 and 2. Mixed populations that fall above the isocline for a species will decrease. Thus, both populations will decrease in size from point B. Examining the behavior of populations between the isoclines for both species (point C), we find that species 1 will increase, while species 2 will decrease. In this case, populations follow a trajectory toward the K_1 intercept to reach a **stable equilibrium point.** The important point here is that each of the diagrams has a stable equilibrium point, but not all of these stable equilibrium points include a mixture of the two competing species.

Figure 6-5c shows the isoclines for the two hypothetical species crossing. In this case, populations that fall exactly on the point of intersection will remain constant. This, however, is an **unstable equilibrium point,** as populations anywhere off this point will move toward one of two stable equilibria where only one species exists. The results of competition in this case depend on the initial conditions. Note that the carrying capacity for each species intersects its own population size axis at a point *above* that of the other species. Finally, Figure 6-5d shows stable coexistence between competitors. The point where the two isoclines cross is a **stable equilibrium point.** Notice the placement of the intercepts on this graph. Each species' isocline intercepts its own population size axis at a point below that of the other. In other words, each species limits itself more than it limits the other species. The most obvious way that this may happen is through **niche separation.**

The general prediction of the Lotka-Volterra competition models is that there is a limited set of conditions where two species may coexist. Further, these conditions (when populations that are more self- than other-limited) do not allow for the stable coexistence of more than two species. There are, however, caveats to be added to this conclusion. Natural populations may not actually come to

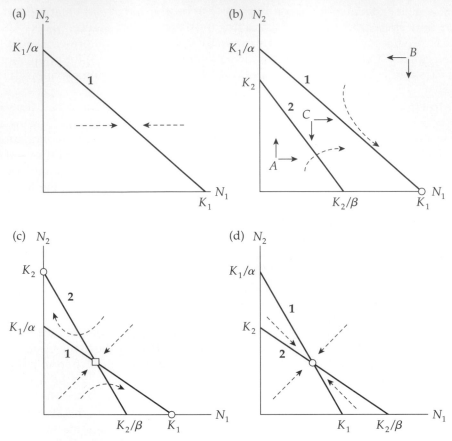

Figure 6-5   Pairwise interactions of populations as depicted by zero net growth isoclines (ZNGIs, solid lines). (a) The isocline for species 1 demarks positive population growth in combined population densities that fall to the left of the isocline and net negative growth in regions to the right of the isocline. (b) Two species isoclines, again depicting population growth under varying conditions of a two-species abundance. At point A the combined populations are below the carrying capacity of both species, hence both species grow. Similarly, at point B the combined populations of the two target species exceed carrying capacity for both species and both populations decrease. At point C the combined population exceeds (is above) the isocline for species 2, so it decreases. In contrast, the combined population falls below the joint carrying capacity for species 1, and it increases. These combined forces drive the population to a stable equilibrium at K_1, the carrying capacity for species 1 in isolation, and it competitively excludes species 2. (c) Similar to case (b), except the ZNGIs cross. Nonetheless, population trajectories are such that stable equilibria (circles) are attained at either species' carrying capacity. A two-species equilibrium (square) exists, but it is unstable and any perturbation leads to the extinction of one species or the other. (d) Similar to case (c), except that the two-species equilibrium is now stable. That is, any perturbation off this equilibrium will drive populations back toward this two-species equilibrium. In this case, species are more self-limiting than other-limiting. Modified from D. Tilman. 1982. *Resource Competition and Community Structure.* Princeton University Press, Princeton, NJ.

equilibrium very often (Connell 1983), or other interactions may inhibit the full competitive potential of a species. Mack and Harper (1977) studied the intraspecific and interspecific abilities of four dune annuals in Wales and reached several important conclusions. One species, *Vulpia fasciculata,* showed strong competitive superiority in greenhouse experiments, yet in nature it did not exclude the other species. The reason for the anomaly may be herbivory by rabbits, which preferentially graze on *Vulpia.* Another conclusion was that the typical response to competition by these annuals was a reduction in size and seed production, rather than mortality (thinning).

On theoretical grounds, however, interspecific competition should be no more important or likely than intraspecific competition. Goldberg and Werner (1983) cite three reasons to support this idea. First, most communities consist of a rich mixture of species, where the probability of contacts between many pairs of species is high, so selection pressure for minimal niche overlap between any one pair of species is not expected to be high. Second, Goldberg and Werner observe that most plants require the same basic resources in similar amounts, limiting the possibilities for resource niche separation. Finally, plant size typically confers superiority, and size depends on age or growth form rather than on species identity.

Experimental Evidence of Competition

Competition theory and experimentation were developed by zoologists during the early part of this century. De Wit (1960, 1961) pioneered the application of zoological concepts of competition to plants. One of his techniques to measure plant competition was to calculate input/output ratios for each species in a mixture, where

$$\text{Input ratio} = \frac{\text{seeds sown of A}}{\text{seeds sown of B}} \qquad \text{(Equation 6-7)}$$

$$\text{Output ratio} = \frac{\text{biomass A}}{\text{biomass B}} \qquad \text{(Equation 6-8)}$$

Variations on this model can use other output variables, such as numbers of individuals, seeds, or length of stems. The difference in the output ratio relative to the input ratio indicates the nature and strength of the interaction. The only constraint is that the variables measured must be the same for both species.

Two species are planted in a **replacement series,** where the total number of initial propagules is constant, but the ratio of the two species changes from pure species A to an even mix of A and B and on to pure species B. There are different possible outcomes of such an experiment (Figure 6-6a): (1) species A wins, in which case the input/output line is parallel to but above the equilibrium line; (2) species B wins, in which case the ratio line is parallel to but below the equilibrium line; and (3) species A and B coexist, in which case the ratio line crosses the equilibrium line. If the slope is less than 45° (Figure 6-6b), the equilibrium point is

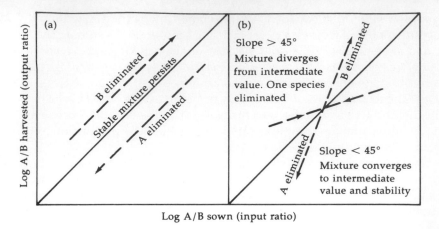

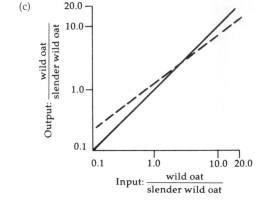

Figure 6-6 (a) and (b)   Input/output diagrams illustrating the possible outcomes of competition between two species, A and B. From J. R. Etherington. *Environment and Plant Ecology.* Copyright 1975 by John Wiley and Sons, Inc., LTD, London. Reprinted by permission. (c) Input/output diagram from an actual experiment between two species of wild oats, the two capable of coexisting at one ratio only. From Marshall and Jain. 1969. By permission of the British Ecological Society.

stable because movement away from the equilibrium point tends to be dampened by the populations' competitive qualities, and the populations drift back to equilibrium. If the slope is greater than 45°, the equilibrium is unstable because movement away from the equilibrium point tends to lead to further departure from stability, and one species will decline to extinction.

Figure 6-6c shows actual, rather than hypothetical, data of a replacement experiment using two species of wild oat (Marshall and Jain 1969). Wild oat (*Avena fatua*) and slender wild oat (*A. barbata*) were sown with a combined density of 1380 m^{-2} in a greenhouse experiment. The input/output line, with a slope of less than 45°, crosses the equilibrium line at a point corresponding to a mix of about 20% slender wild oat and 80% wild oat. One could conclude that these species should coexist in nature. Marshall and Jain, however, could not verify these greenhouse results in experimental plots in California annual grasslands. Between 90 and 100% of planted stands resulted in either pure wild oat or slender wild oat. Very few mixed stands were observed. The complexity of nature, it was concluded, exacerbates the problems of coexistence for these species. Marshall and Jain's results highlight another basic tenet of plant ecology: experimental results obtained from manipulative experiments in highly controlled greenhouse conditions are often difficult to replicate in field experiments (often for unknown reasons). The ideas and methods of de Wit, however, remain an important and essential first step toward the goal of quantifying and understanding competition.

Since de Wit, there have been numerous variations on the basic competition experiment. If one is interested primarily in the effect of one species on another, a **partial additive** design may be used (Figure 6-7a). Results from a partial additive experiment allow one to make an assertion regarding the effect of one species on the other, but not the reverse. The de Wit replacement series design is depicted in Figure 6-7b. Two additional designs, the **additive series** (Figure 6-7c) and the **complete additive** (Figure 6-7d), allow one to test the interaction of varying densities, but are more complex and require more experimental units than the simpler designs.

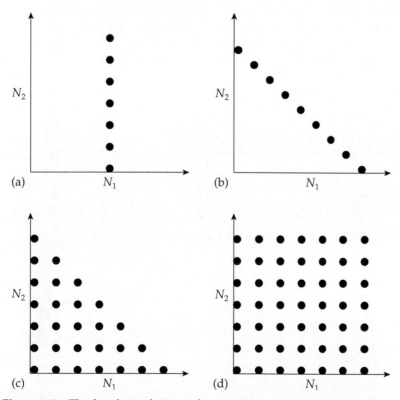

Figure 6-7 The four basic designs of competition experiments plotted on joint abundance diagrams for two species. These designs are (a) partial additive, (b) replacement series, (c) additive series, and (d) complete additive series. From J. W. Silvertown and J. Lovett Doust. 1993. *Introduction to Plant Population Biology.* Blackwell Scientific Publications, Oxford.

As previously mentioned, summaries of competition experiments frequently observe either extremely asymmetric competition or no effect of competition at all (Gurevitch 1992, Bengtsson et al. 1994). This may in part result from the way competition is measured in field experiments (Goldberg and Scheiner 1993). The experimental designs in Figure 6-7 treat competition as a function of populations. Yet resource use and competition for resources are phenomena that happen to individuals, by other individuals. Goldberg and Scheiner (1993) outline an argument in favor of using **target-neighborhood** experimental designs where the performance of single individuals of one species (the target) are tested against varying densities of a potential competitor (the neighborhood). This design is in effect the same as a partial additive design where the density of the target species is 1 per unit area of the experiment (Figure 6-7a). To assay reciprocal effects of competition, the same experiment is conducted except target and neighborhood species change identity, and the results are compared. Pacala (1997), citing unpublished data, suggests that these plant-centered experiments are much more effective at detecting the effects of competition than plot-centered experiments.

Field manipulations of plant densities are also used to measure competitive effects. For example, Fonteyn and Mahall (1978, 1981) manipulated shrub density in a uniform Mojave Desert community of *Larrea* (creosote bush) and *Ambrosia* (burrow bush). One species or the other was removed in a variety of patterns in a series of 100 m² plots (Figure 6-8). The water status of the remaining plants was monitored for several months as a measure of release from competition for soil moisture. The results indicated that the three types of competitive interactions

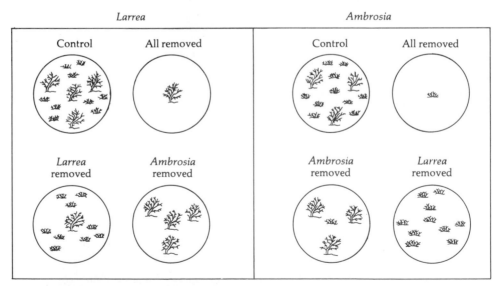

Figure 6-8   The eight treatments of plant removal in *Larrea-Ambrosia* desert scrub. Each circle represents a 100 m² test plot. Large shrubs are *Larrea,* small ones are *Ambrosia.* In the "all removed" plots, the center of the plot was occupied by one remaining plant. Controls were not manipulated. Redrawn from P. S. Fonteyn and B. E. Mahall. 1981. *Journal of Ecology* 69:883–896.

(*Larrea-Larrea, Larrea-Ambrosia, Ambrosia-Ambrosia*) were not of equal intensity. Release of *Larrea* from competition with other *Larrea* had little effect on plant water status, whereas release of *Ambrosia* from competition with surrounding *Ambrosia* did affect plant water status.

Models of Competition and Resource Limitation

It is one thing to be able to document and quantify plant competition; it is another to determine which resources are limiting or the mechanism of the interaction. Wilson (1988) surveyed the literature to identify the relative importance of root versus shoot competition. In 68% of the 23 cases reviewed, root competition had a greater effect on plant growth. Similarly, Wilson and Tilman (1991) examined the relative importance of root versus shoot competition among three coexisting grass species along a soil nitrogen gradient in Minnesota. Their results showed that when soil nitrogen is low, root competition is important and shoot competition is not. In high-nitrogen soils, however, plants exhibited signs of both above- and belowground competitive effects (Figure 6-9).They concluded that with high soil nitrogen, species compete more strongly for light, and that shoot competition should dominate the interaction.

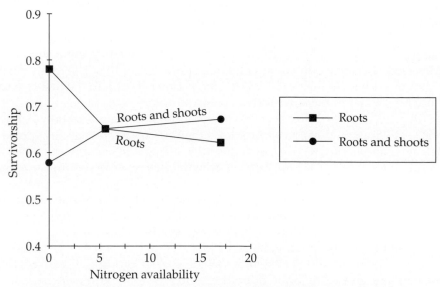

Figure 6-9 Survivorship of *Schizachyrium scoparium* in competition experiments along a gradient of nitrogen availability. Competition treatments shown here include those with just root competition (square symbols) and both root and shoot competition (round symbols). This experiment demonstrates that survivorship is lower when nitrogen availability is low and competition is with roots and shoots of competing species, but that this survivorship pattern switches at high nitrogen availability. Modified from J. B. Wilson and D. Tilman. 1991. "Components of plant communities along an experimental gradient of nitrogen availability." *Ecology* 72:1050–1065.

As the previous example implies, there is often more than one limiting resource in the real world (see Chapter 5). If more than one resource is limiting, it may be difficult to demonstrate which one determines the competitive outcome. Consider a study by Shirley (1945) in north-central Minnesota. The overstory trees in the Lake states are often the conifers white spruce (*Picea glauca*), white pine (*Pinus strobus*), red pine (*P. resinosa*), and/or jack pine (*P. banksiana*), yet these commercially valuable species do not reproduce in their own shade. Instead, hardwood seedlings and saplings grow up beneath the canopy. Why are conifer seedlings such poor competitors with the hardwoods? Shirley examined the effect of light on conifer seedlings by planting seedlings beneath screens that reduced full sun in steps from 2% to 89% (Figure 6-10a). He found that conifer survival and growth over a four-year period increased as light intensity increased. Generally, growth was not satisfactory when light intensity was reduced more than 65% (Figure 6-10a). He also examined the combined effect of shade and root competition (Figure 6-10b). Results after four years were very complex. In dry areas, shade improved conifer seedling survival, though not seedling growth. In more mesic areas, removal of the overstory stimulated conifer seedling growth, weeding of ground vegetation further improved it, and trenching around the plot improved it even further. The conclusion is that competition for shade and soil moisture strongly interact.

David Tilman (1982) developed a model for examining competition where more than one resource is limiting. Recall from Chapter 5 the idea of R^*, the minimum level of a resource where a population can sustain itself. Tilman's resource-ratio model demonstrates how two or more species may differ in R^*s, yet coexist in nature. To see how this works, we describe R^*s for two different resources and two species. Let one of our two hypothetical species (A or B) be the more effective competitor for both of the resources (R_1, R_2, Figure 6-11a,b). In the first two cases (Figure 6-11 a and b) the species that can draw both resources below the critical level (the ZNGI) of the other species wins in a competitive interaction. Allowing species A to draw R_1 down to a lower level than species B, and species B to be the superior competitor for R_2, we can define intersecting ZNGIs for both species and resources (Figure 6-11c). [These zero net growth isoclines (ZNGIs) appear as minimum resource supply thresholds for each resource for each species and are depicted in Figure 6-11 as bold lines.] If we assume that plants consume resources in proportion to the relative amount each resource is limiting, we can then define a resource consumption vector (C_A, C_B) for each species (Figure 6-11c).

In habitats where an abundance of one resource or another is found, one resource is clearly more limiting and a single species prevails (Figure 6-11c). In habitats where moderate levels of both resources are found, however, resource consumption of the two competitors draws both resources down to a point where they are equally limiting to both species, allowing coexistence. In theory, a partitioning of resource limitation trade-offs (Figure 6-11d) allows the coexistence of an infinite number of species where resources are spatially nonuniform (Tilman 1982). Similarly, Chesson (1985) has shown that temporal heterogeneity in resources also allows for competitive coexistence in what he calls the **storage effect.** Species can coexist using Chesson's storage effect by capitalizing on trade-offs in

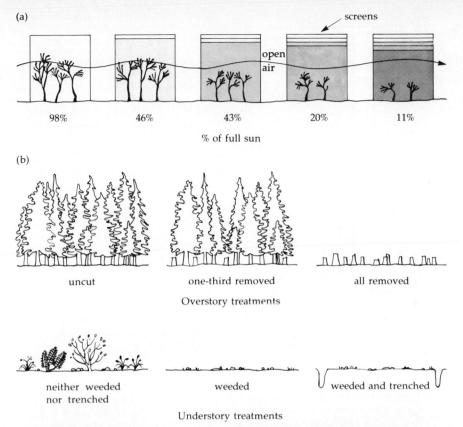

Figure 6-10 Diagrammatic summary of Shirley's experiments to determine the relative importance of competition for light and soil moisture to pine seedlings. (a) Pine seedlings were grown beneath different layers of screens to achieve different levels of sunlight. (b) Starting conditions for seedlings in another experimental set: three different overstories (closed canopy, one-third removed, all removed) and three different understories (control, all other plants removed, all other plants removed and trenches dug around plot to sever the roots of adjacent plants). From H. L. Shirley. 1945. *American Midland Naturalist* 33:537–612.

the ability to compete under different environmental conditions that vary through time. A poor resource competitor can persist in an environment by specializing in doing well under conditions that only occasionally arise (e.g., early spring rains for desert annuals) and persisting in the environment during normal conditions through a seed bank. In general, coexistence among species in each of these models requires trade-offs such that the superior competitor under the dominant environmental conditions is not the superior competitor under all conditions and variability in conditions. This variability in conditions that allows the coexistence of multiple species may be dispersed through either time or space. Coexistence can even be maintained when dispersal abilities vary such that a poorer resource

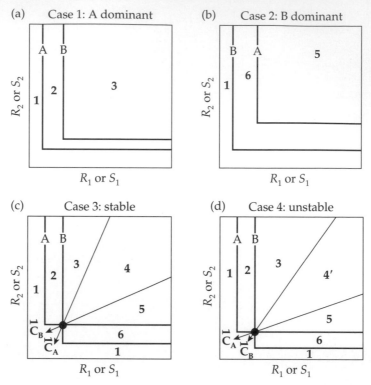

Figure 6-11   The four distinct cases of resource competition are illustrated. In case 1, the zero net growth isocline (ZNGI, solid lines) for species A is at a lower level of resource supply (R) than that of species B. As a result, A will reduce resource levels to a point below that required for survival of B and competitively exclude B in any habitat that would support its own growth (zones 2 and 3). Case 2 is the converse of case 1. In this case, B wins the competitive interaction and will displace A. In case 3, the ZNGIs cross at a stable two-species equilibrium point. This intersection is stable because each species consumes relatively more of the resource that is more limiting to its growth at equilibrium. In this case, habitats with resource supply ratios in zones 2 and 3 will result in the dominance of species A. Similarly, habitats in zones 5 and 6 will be dominated by B. Habitats with resource supply in zone 4 will result in stable coexistence of the two species. Case 4 is similar to case 3 except that the equilibrium point is unstable because each species uses more of the resource that primarily limits the other species. The outcome of this competitive interaction (except in zone 1 where neither species exists) will be dominated by either species A or B, depending on initial conditions. From D. Tilman. 1982. *Resource Competition and Community Structure.* Princeton University Press, Princeton, NJ.

competitor is a superior disperser, allowing a higher probability of new site colonization by the poorer resource competitor. Finally, coexistence of multiple species in a competitive environment is also possible if species are competitively very similar. Competitive similarity results in increased time to extinction. If species are very similar to one another such that competitive exclusion requires a long time, then disturbance cycles may disrupt competition, and reset species abundances, before any competitive exclusion can occur.

Tilman subsequently modified his original model to incorporate allocation patterns to roots and shoots (Tilman 1988). He also expanded this model to predict species life histories, diversity, and competitive effects of communities at different successional stages (Tilman 1988, Tilman and Pacala 1993). Similarly, Huston and DeAngelis (1994) examined the effects of local soil resource depletion and soil nutrient transport rates on competition and coexistence. Because nutrient transport rates in soil are low, plants have a very limited ability to affect resource availability outside their root zones. In Figure 6-12 the lightly shaded regions represent portions of the root zones where an individual plant has depleted local resources. Thus, species with similar resource requirements, but restricted rooting zones (Figure 6-12), can coexist because each can access only a small portion of the total resources available. In contrast, if soil resource depletion zones extend into the rooting zones of neighboring individuals (Figure 6-12b), then competitive effects become important.

Extreme habitats are generally observed to support relatively low biological diversity. Resource-rich and resource-poor habitats are observed to be dominated by few species relative to many intermediate habitats (Figure 6-13; Tilman and Pacala 1993). Resource competition models, such as those of Tilman, Chesson, or Huston and DeAngelis, offer several possible explanations for this commonly observed pattern. These various models typically include a mechanism that enhances coexistence through spatial or temporal escape in moderate environments. For example, in Tilman's model extremely poor soils are likely to be dominated by a few species that can compete well for a single limiting resource. At the other extreme, habitats on rich soils support high biomass production and are dominated by the few species that compete the most effectively for light. At intermediate nutrient levels there is a greater possibility of spatial and temporal variability in resource supply. This variability allows more different resource strategies to coexist in stable equilibrium.

Other mechanisms besides competition may also explain high diversity at intermediate resource availability. For example, sites of intermediate richness may also be those that are disturbed at intermediate intervals (the intermediate disturbance hypothesis predicts a similar diversity pattern). Many species with essentially equivalent competitive abilities are predicted to persist in non-equilibrium conditions for a relatively long period of time (Sale 1978). This pattern may be more common in nonextreme environments. Similarly, complex food webs may promote the maintenance of diversity and be more common in moderate environments. All told, it is very difficult to discern the specific mechanisms of multi-species persistence, but the pattern of high diversity on sites of intermediate resource supply is apparent.

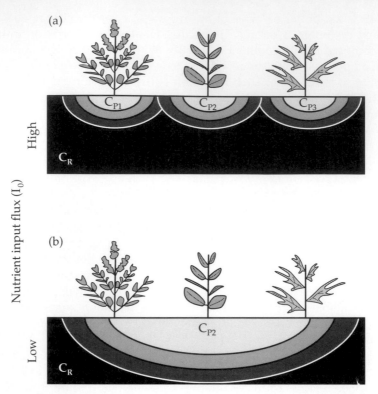

Figure 6-12   Hypothetical nutrient concentration gradients in a region with more than one plant. (a) Under high nutrient flux, plants deplete resources within a narrow region, reducing interplant competition. (b) Under low nutrient flux, plants reduce soil nutrients over a much broader range, increasing competitive effects between individuals. C_R represents the regional soil concentration of nutrients, while $C_{P\#}$ represents soil nutrient depletion zones created by individual plants extracting soil nutrients and reducing local concentrations. From M. A. Huston and D. L. DeAngelis. 1994. American Naturalist 140:539–572. University of Chicago Press, Chicago.

Environmental Tolerance as a Means to Avoid Competition: Serpentine and Saline Soils

Serpentine is a metamorphic, magnesium silicate rock, often green in color and slippery to the touch. It has a number of traits inimical to plant growth. It is low in such essential nutrients as N, Ca, K, and P; its pH may be far from neutrality (either acidic or basic); and it is high in such toxic elements as Ni and Cr. Soils derived from serpentine rock are nutrient poor; they support unusual, endemic floras; and they are covered with vegetation whose physiognomy differs from that of surrounding vegetation on adjacent nonserpentine soil. Serpentine outcrops are found throughout the world (Whittaker 1954), but are especially common in the Pacific Coast states.

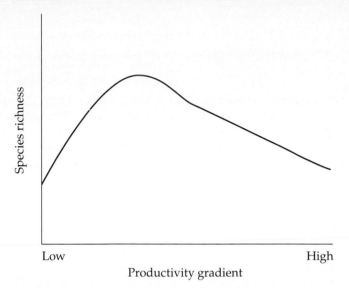

Figure 6-13   A schematic diagram showing the often observed relationship between productivity and species richness, where biological diversity is maximized under intermediate levels of productivity. Data for this schematic diagram are summarized in D. Tilman and S. Pacala. 1993. "The maintenance of species richness in plant communities," pp. 13–25 in R. E. Ricklefs and D. Schluter, eds. *Species Diversity in Ecological Communities. Historical and Geographical Perspectives.* University of Chicago Press.

Kruckeberg (1954) experimented with serpentine and nonserpentine ecotypes and species. He reported that herbaceous serpentine endemics became established from seed better and grew faster on nonserpentine soil, providing they were free from interspecific competition. When sown with typical nonserpentine species on nonserpentine soil, they became etiolated and did not survive. On serpentine soil, only the serpentine endemics survived, but there was considerable bare ground and the plants grew slowly. This is not to say that all serpentine taxa have ranges limited by competition. Tadros (1957) found that toxins released by the higher densities of certain soil microorganisms on nonserpentine soil prevent the herbaceous serpentine endemic *Emmenanthe rosea* from establishing on normal soil. McMillan (1956) showed that some serpentine taxa grow as well on serpentine soil as on nonserpentine soil, even where interspecific competition is not a factor. Thus, there is more than one way in which plants have adapted to serpentine, but tolerance as a means to avoid competition may be a major strategy.

High-salinity soils also provide a unique stress to plant growth. **Glycophytes** (literally, "sweet plants") are intolerant of salinity above that necessary to supply essential nutrients, approximately 0.1–0.2% salt. Higher salt concentrations disrupt the water balance in plant cells (through osmotic pressure) and cause cells to rupture. In contrast, **halophytes** (literally, "salt plants") are capable of growing in soil with more than 0.2–0.25% salt concentration. Many halophytes have special structures by which they secrete salts. Examples of halophytes include mangroves, coastal salt marsh herbs, beach plants that receive salt spray, or salt desert herbs and shrubs.

All halophytes are not equally tolerant of salt. Ingram (1957) used three terms to describe these plants: intolerant, facultative, and obligate. **Intolerant halophytes** show highest growth rates at low salinity and declining growth with increasing salinity. Most halophytes are intolerant, whether we look at their germination, growth, or reproduction. **Facultative halophytes** show maximum growth at intermediate salinities and lower growth rates at both low and high salinities. A few halophytes are facultative. **Obligate halophytes** show increasing growth rates with increasing salinities and are incapable of growing on low-salinity soils. There are likely to be very few obligate halophytes (Barbour 1970b), a conclusion that can be reached from both field observations and manipulative experiments in growth chambers.

Limiting Resources and Plant Strategies

The significance of competition in plant distribution and plant community composition is undisputed. The exact mechanisms of this competition and how best to model competitive systems, however, have not been agreed upon. Grime (1977, 1979) views competition as being strongest in high-productivity habitats where resources are most available. Thus, species that extract resources the fastest and apply them to plant growth (species with the highest maximal growth rate) are the best competitors in Grime's plant strategies triangle. Similarly, Keddy (1989) views competition as increasing along a productivity gradient. In contrast, Tilman's model defines competitive superiority as the ability to drive a resource down to the lowest level and still maintain a positive growth rate. Competition, in Tilman's opinion, increases as a resource becomes limiting. Grime views competitive interactions as lessening in stressful environments, whereas Tilman does not. This contrast between how Tilman and Grime view competition is among the most controversial aspects of plant ecological theory.

Thompson (1987) pointed out that much of this debate may be a semantic argument over the nature of competition. For Grime, competition is about resource capture rates when there is ready availability of resources. For Tilman, competition is about drawing limiting resources down to the lowest possible levels and tolerating these low resource levels. Grace (1990) clarifies this issue by pointing out aspects of both models that appear true. Both models cite empirical support from experimental tests of their theories (e.g., Wilson and Tilman 1991, Campbell and Grime 1992). Further understanding of the mechanisms by which plant competition structures communities will likely help sharpen our ability to predict vegetation responses to stresses such as pollutants or climate change. This area of plant ecology is likely to remain one of the most exciting and vibrant areas of study in plant ecology over the years to come.

Amensalism

Amensalism is an interaction that depresses the population of one species while the other species remains unaffected. Earlier we mentioned that more than half of the studies seeking to demonstrate competitive interactions between two species ended up demonstrating amensalism (Goldberg and Barton 1992, Gurevitch et al. 1992). These potential amensalisms are likely to be asymmetric competition where the effects experienced by one species are stronger than those experienced by another. A simple example might be the effect of a tree on an herbaceous seedling. The seedling, under the shade of the tree canopy, experiences severely negative effects of competition for light from the tree. The tree, in turn, has fewer nutrients available to it as a result of the herb seedling extracting nutrients within its root zone, but this effect may be immeasurably small. Following our example of a variable realm of interaction strength (Figure 6-2), at some point it becomes unimportant whether an interaction is strongly asymmetric competition or strictly amensal.

Another proposed example of amensalism is **allelochemic** interactions, the inhibition of one organism by another via the release of metabolic by-products (**allelochemicals**) into the environment. Allelochemicals are selectively toxic, affecting some species but not others. Since the 1940s when the word *allelochemics* was coined, a large body of literature has developed on the subject that seems to demonstrate pervasive chemical links between organisms within a community—decomposers, producers, and consumers (Muller 1965, Whittaker and Feeny 1971, Sondheimer and Simeone 1970, Rice 1984).

Allelopathy

The secretion of allelochemicals to directly inhibit the growth or reproduction of other plants is called **allelopathy.** Allelopathy is viewed by some biologists as an aggressive form of competition, but there is an important distinction to be made. In exploitative competition a plant *removes* a resource from the environment. In contrast, in allelopathy a plant *adds* a substance to the environment. Allelochemical additions suppress the growth of potential competitors and in this sense may be viewed as interference competition (Figure 6-3). Functionally, however, research has focused on the effect of the allelochemical-producing species on its neighbors and not on the effect of the neighbors on the allelopath. Given the one-sided nature of the interaction, we will classify this interaction as amensalism, understanding that these categories are somewhat artificial and may fluctuate in time and space or be immeasurably small and undetectable (Bronstein 1994).

A number of researchers have reported evidence for chemical control of plant distribution, spatial associations between species, and the course of community succession. C. H. Muller (1965, 1966, 1969) and his students brought allelopathy to the attention of plant ecologists through the 1960s and the field has subsequently grown dramatically (Williamson 1990). The ecological significance of allelopathy, however, remains unclear and has garnered both avid enthusiasts and detractors (Whittaker and Feeney 1971, Barbour 1973a, Harper 1977, Williamson 1990).

Muller's studies remain as some of the most widely cited and thoroughly documented cases of allelopathy. Muller studied the spatial relationship between coastal sage *(Salvia leucophylla)* and annual grassland in the Santa Ynez valley of southern California. A number of shrub species, including sage, dominate the foothills, whereas annual grasses and herbs dominate the valley floors. Patches of sage shrubs, however, occur in the grassland. Beneath those shrubs, and for 1–2 m beyond the shrub canopy limits, the ground is nearly bare of herbs and grasses (Figure 6-14a,b). Even 6–10 m from the canopy, annuals are stunted. Stunting is not caused by competition for water, since shrub roots do not penetrate very far into the grassland, and stunting is observed even in the wettest times of the year. Soil factors do not seem to be responsible for the negative association either, because major chemical and physical soil factors do not change across the bare zone.

Muller was able to show that *Salvia* shrubs emit a number of volatile oils from their leaves and that some of these (principally cineole and camphor) are toxic to the germination and growth of surrounding annuals. He was able to (a) detect allelochemic substances in the field; (b) demonstrate that they are adsorbed by the soil and can be retained there for months; and (c) show that they are able to enter seeds and seedlings through their waxy cuticles. He was not able to detect, however, the same amounts of oils in natural soils that were necessary to produce inhibition in the laboratory. Muller was also unable to completely eliminate other factors as contributors to the maintenance of the bare zones.

Bartholomew (1970) examined the influence of mammalian and bird herbivores. From seed predation and exclosure experiments (experiments where fencing around a plant is used to prevent access to the experimental plant(s) by the herbivores), Bartholomew was able to support the hypothesis that foraging activity increases closer to the shrubs. Thus, seed predation might be the main cause for maintenance of the bare zone. Halligan (1973) discovered a similar influence of herbivores in bare zones around California sagebrush (*Artemisia californica*).

Muller and his students presented a more convincing case for allelopathy in California hard chaparral, a dense, scrubby vegetation dominated by such taxa as chamise (*Adenostoma fasciculatum*). McPherson and Muller (1969) and Christensen and Muller (1975) concluded that substances released from leaves of chamise inhibited the germination and growth of understory herbs because other major environmental factors, such as light, moisture, and soil fertility, were successfully eliminated by experimental manipulation. Kaminsky (1981), however, argued that the story is more complex. He demonstrated several anomalies that conflicted with Muller's hypothesis. Soils beneath chamise were inhibitory to herb germination only when wetted, and the inhibition became more intense with time following wetting. The inhibitory effect could be eliminated by fumigating the soil with a bactericidal agent. Also, dry soils typically contained only 100 ppm of known inhibitors (primarily *p*-coumaric acid), whereas amounts required to induce inhibition in the laboratory were up to four times that concentration. He concluded that chamise merely releases nontoxic material in its litter, which is then converted and concentrated over time by soil microbes into allelochemic substances. The amensal

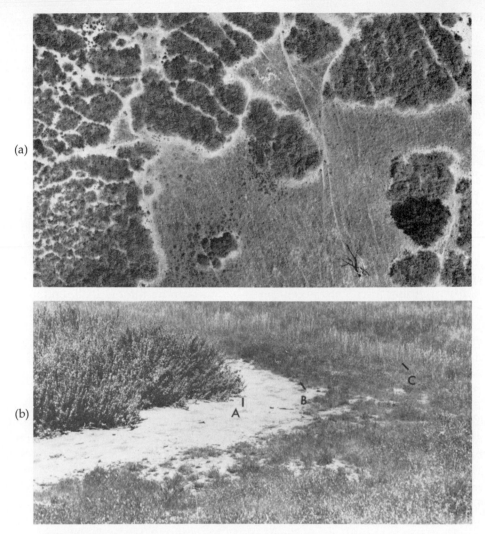

Figure 6-14   Nearly bare zones between soft chaparral and annual grassland near Santa Barbara, California. (a) Aerial photograph of *Salvia leucophylla* shrubs and adjoining grassland; the light bands beneath and next to the shrubs are devoid of all but a few species of small herbs. (b) Ground view of the same phenomenon. A indicates soft chaparral, B indicates the edge of the bare zone, and C indicates the grassland. Courtesy of C. H. Muller.

organism, then, is not just the chamise but also microbial decomposers. At some level this argument seems more one of semantics than biology.

Williamson and colleagues (reviewed in Williamson 1990) conducted a series of careful studies of potential allelopathy in sand pine scrub habitats of central Florida. Focusing on growth and establishment of species in scrub and sandhill communities, he was able to demonstrate patterns of growth and recruitment suggestive of allelopathy. Bare soil patterns, similar to those in California chaparral, were observed around shrubs (*Ceratiola ericoides*, Figure 6-15a,b) too small to harbor rodents and other herbivores. He isolated substances (e.g., ceratiolin produced by *C. ericoides*) that degrade into growth-inhibiting, toxic chemicals. Thus, although strict allelopathy has remained difficult to prove to everyone's satisfaction, it is widely accepted in its looser definition. Gardeners know through experience that many herbs will not grow beneath conifers or walnuts. Whether this effect should be labeled a strict allelopathy by the trees or a passive modification of the environment as a result of their mere presence is problematic and continues to be argued through research.

Interactions between Trophic Levels

To a large degree, decomposers in the soil and litter beneath a community are another example of amensalism because soil microbes are affected by the species of plants shedding the litter and penetrating the soil with roots. Soils beneath northern conifer forests are largely acidic because conifer litter is acidic and its decomposition influences soil pH. Many types of soil microbes, however, cannot tolerate low pH soils. As a result, fungi dominate the soil microflora, whereas bacteria dominate more neutral soil beneath deciduous forests (Eyre 1963). There are also differences even within a conifer forest: pine needles are more acidic than spruce needles, and the soil beneath most pine species has less decomposer activity and is almost devoid of earthworms compared to soil beneath spruce forests.

Understory plants can further modify the soil microenvironment. Tappeiner and Alm (1975) showed that such features as soil pH, litter decay rate, and soil nutrient status depend not only on overstory tree species but also on whether the understory is dominated by hazel shrubs (*Corylus cornuta*), a mixture of other shrubs and herbs, or is largely bare (Table 6-2). Because the understory is often a mosaic of the three types, the authors concluded that there must be considerable spatial variation in the forest soil microenvironment.

Although forest understory herbs are notably patchy, and overstory tree seedlings may be positively or negatively associated with the herbs (Maguire and Forman 1983), amensalism is not always involved. Herb pattern can be determined by (a) the physical quantity and quality of litter; (b) uneven distribution of soil nutrients; (c) microenvironmental patterns of soil drainage; (d) patterns of vegetative reproduction or seed dispersal; or (e) past land use (Sydes and Grime 1981; Rogers 1982, 1985; Dennis 1997).

Apart from pH and litter decay products, plants affect soil chemistry by passively contributing a variety of inorganic and organic compounds to the soil. Apparently, plants are very leaky systems (as are animals). Carlisle et al. (1966) analyzed the nutrient content of rainwater falling directly on the ground and of rainwater falling through the leaf canopy of sessile oak (*Quercus petraea*). The rainwater falling through the canopy contained a higher concentration of nutrients except for nitrogen (Table 6-3).

Figure 6-15   (a) *Ceratiola ericoides* growing at the scrub ecotone with a halo; and (b) *C. ericoides* colonizing a disturbed site. From G. B. Williamson. 1990. "Allelopathy, Koch's postulates and the neck riddle," pp. 143–162 in J. Grace and D. Tilman, eds. *Perspectives on Plant Competition*. Academic Press, San Diego.

Table 6-2   Effect of overstory and understory plants on soil pH, bulk density (g cm^{-3}), weights of nutrients in the top 3 cm (kg ha^{-1}), and turnover rates (in years) for all dry litter and for selected nutrients. From J. C. Tappeiner and A. A. Alm. "Undergrowth vegetation effects on the nutrient content of litterfall and soils in red pine and birch stands in northern Minnesota." *Ecology* 56:1193–1200. Copyright © 1975 by the Ecological Society of America. Reprinted by permission.

Overstory/ understory	pH	Bulk density (g cm^{-3})	Weight in soil					Turnover time (years)		
			Ca	N	K	Mg	P	Litter	Ca	K
Red pine/none	4.1	0.66	121	470	29	13	7	5.0	4.9	3.9
Red pine/hazel	4.1	0.54	111	535	29	14	7	3.2	2.5	3.1
Red pine/ herb-shrub	4.2	0.62	128	524	30	14	8	4.1	3.8	2.1
Birch/hazel	5.0	0.44	304	617	47	43	8	1.7	1.4	0.2
Birch/ herb-shrub	5.1	0.48	337	602	50	44	9	2.3	1.9	0.3

Table 6-3   Nutrient content in rainwater that falls directly to earth (rainfall) and in rainwater that falls through an oak (*Quercus petraea*) canopy (throughfall). Values are expressed in kg ha^{-1} yr^{-1}. From Carlisle et al. 1966. By permission of the British Ecological Society.

Nutrient	In rainfall (kg ha^{-1} yr^{-1})	In throughfall (kg ha^{-1} yr^{-1})
N	9.54	8.82
P	0.43	1.31
K	2.96	28.14
Ca	7.30	17.18
Mg	4.63	9.36
Na[a]	35.34	55.55
Total	60.20	120.36

[a] An essential nutrient for some plants

As a result, plants can often modify their edaphic environment in ways that adversely affect soil microbes.

Summary

Species interactions can be classified as positive, neutral, or negative in several pairwise interactions. The spatial distribution of plants often gives us a clue as to strong interactions among species. Understanding specific relationships, however, requires experimentation. Even then, it may not be easy to describe specific relationships as they may fluctuate in time and space with patterns of interactions being dependent upon environmental conditions.

Of the many combinations of interactions, competitive $(-,-)$ interactions have received the most attention. Competition can be classified into interference competition, where species directly interfere with one another, or exploitation competition, where competition is mediated through the exploitation of a shared resource. Most plant competition is the latter, although allelopathy may be an example of the former. Similarly, competition between two species may be direct, or apparent and mediated through a third species. Finally, competition may be asymmetric, with the competitive effect of one species on another being stronger than the reverse. Field experiments observing the effects of competition among species pairs suggest that competition is frequently very asymmetric. One way in which competition is often inferred is by nonrandom spatial association of individuals in the field; however, nonrandom patterns may arise for many reasons other than competition.

Gause proposed that competition should limit diversity within communities as species compete for resources. One can modify Lotka-Volterra population growth equations using a coefficient to scale the effect of individuals of one species on the carrying capacity of another to consider the negative effects of species on each other. Mathematical models show that species may coexist (have a stable two-species equilibrium) when each species has a stronger effect on its own populations than it does on its competitor's populations. Equilibrium competition models provide a context in which we may discuss the characteristics of population behavior under competition. Interactions among species and the environment in the real world, however, are often more complex than these simple equations suggest.

Both temporal and spatial variability in resources allow for coexistence of species in a competitive environment. Tilman, Chesson, and others provide models that show how variability in the environment allows for persistence of any number of species. Similarly, stable coexistence of competing species can be inferred using differences in dispersal abilities, or differences in above- and belowground allocation. All of these models, however, assume that trade-offs exist and that a species cannot be a superior competitor for all resources.

Grime models plants as following one of three basic strategies: competitor, ruderal, or stress-tolerator. Examples of stress-tolerators are those plants that thrive in stressful environments such as on serpentine soils or saline soils. The strength of competition may vary in different communities, with sites that are characterized by unusually high stress relative to surrounding habitats experiencing less plant competition in general. According to Grime's model, these stress-tolerating species persist by avoiding a competitive environment. Grime, however, uses a slightly different definition for competition, and a competitive environment, than does Tilman.

Allelopathy, one of many types of allelochemic interactions, is an example of an amensal $(-,0)$ interaction. Allelopathic interactions have been clearly demonstrated in relatively few habitats and are difficult to definitively prove given that many other interactions (e.g., increased seed predation along shrub borders) may result in similar patterns of vegetation. Observations that strongly suggest a particular amensal interaction are common, as are observations that plants produce chemicals that are toxic to other plants, but convincing evidence of actual allelopathic interactions giving rise to vegetation patterns has seldom been obtained to everyone's satisfaction.

CHAPTER 7

SPECIES INTERACTIONS: COMMENSALISM, MUTUALISM, AND HERBIVORY

I n Chapter 6, a pair of negative interactions were examined. In this chapter, we consider one other negative interaction (herbivory) and other interactions that result in positive spatial associations. Once again, we stress the fine line between a positive and a negative interaction, a line that is often crossed under the normal realm of environmental variation for many interactions (Bronstein 1994). For example, herbivory is an interaction that is traditionally considered to have exclusively negative consequences for the plant. Biologists are coming to recognize, however, that there may be complex, subtle, and unexpected ways in which some plants are stimulated by herbivores (Paige and Whitham 1987). In general, ecologists during the twentieth century can be characterized as focusing attention on interspecific interactions that are negative. More recently this has been seen as a shortcoming in need of redress. Several researchers have begun working on the possibly underappreciated role that positive interactions play in the ecology of terrestrial plants (e.g., Paige and Whitham 1987, Bertness and Callaway 1994, Callaway 1995).

Commensalism

Among positive interactions, we begin with **commensalism** $(+, 0)$. **Commensalism** is an interaction that stimulates one organism but has no effect on the other. In contrast to the vast literature on competitive interactions between plants, there is a relatively sparse literature on the effects of commensalisms within plant communities. In part this may be because positive interactions appear less pervasive and influential. Although ecologists have described a variety of important commensal relationships, commensalism has not been a focus of theoretical modeling.

Epiphytes

One example of commensalism is the growth of epiphytes on host trees (Figure 7-1). Epiphytes may be herbaceous perennials (e.g., orchids, ferns, bromeliads, and cacti) or lower plants (e.g., mosses, algae, or lichens). Although over half of all epiphytes are orchids, epiphytes are taxonomically widespread among vascular plants (Kress 1989). Over 23,000 epiphytic species are distributed in 879 genera among 84 vascular plant families representing all of the major vascular plant groups (ferns, gymnosperms, monocots, and dicots; Kress 1989). Epiphytes use their hosts for physical structure and may be autotrophic (gain no nourishment from the host) or heterotrophic (parasitic on their hosts, e.g., mistletoe). Autotrophic epiphytes may be either obligatory or facultative and live in either plant canopies or on the forest floor (Benzing 1989). Some epiphytes have expanded leaf bases or unusual root surfaces that trap, retain, and absorb rainwater. Others, such as Spanish moss (*Tillandsia usneoides*), a bromeliad common in the southeastern United States (Figure 7-1), or grandfather's beard (*Ramalina reticulata*), a lichen that festoons California oaks, absorb much of their water from humid air. *Ramalina reticulata*'s tissue moisture content fluctuates during the day, closely reflecting the trend of the relative humidity of the air, with a positive net photosynthesis rate only during the humid early morning hours (Rundel 1974). Similarly, the range limit of *Tillandsia* corresponds to areas with an average annual relative humidity of 64% or more, even though it cannot satisfy its entire water demand from the air (Garth 1964). Garth found that the distribution of *Tillandsia* correlated very well with the average routes of storms that sweep east and north from Mexico, bringing frequent and necessary rain.

Epiphytes, thus far defined as commensal or parasitic, may easily grade into other types of interactions. For example, epiphytes compete with the canopy of the host tree as well as other epiphytes for space and light. Censuses of tropical trees have shown that relatively distinct parts of the canopy are associated with different epiphyte species (Figure 7-2; Longman and Jenik 1974, Janzen 1975). This kind of fine-scale resource partitioning is considered an indication of intense competition.

A mutualism between an epiphyte and its host tree may result if the epiphyte produces nutrients that are leached by rainwater and run down the trunk to enter the soil around host roots. The upper canopy branches of trees in a Colombian rain forest (2700 m elevation, about 250 cm rain a year) are thickly covered with lichen epiphytes. Forman (1975) carefully demonstrated that 86% of these lichens contain the blue-green alga *Nostoc* as the algal symbiont, and that *Nostoc* fixed nitrogen at the rate of 1.5–8.0 kg N ha^{-1} yr^{-1}. This is equivalent to the amount of available nitrogen added to the forest from rainwater alone (N_2 is converted to NO_3^{2-} during electrical storms, and the nitrate is carried to the earth in rainwater). The nitrogen fixed by lichen epiphytes is probably widely distributed as a result of decomposition and leaching.

(a)

(b)

Figure 7-1 Examples of ephiphytes. (a) The bromeliad *Tillandsia usneoides* (Spanish "moss") in trees along the Gulf Coast. (b) A close-up of *Tillandsia*. (c) The lichen *Ramalina reticulata* (grandfather's beard) on a California oak tree.

(c)

Epiphytes also trap and retain nutrients. Several tropical tree species produce adventitious roots in their canopies that penetrate the mass of humus associated with epiphytes suspended in tree crowns (Nadkarni 1994). Aboveground adventitious roots, or canopy roots, have been observed in numerous genera of temperate and tropical trees in both the New and Old World (Nadkarni 1981, 1983; Table 7-1). These aboveground humus masses trap nutrients from rainfall and aug-

Figure 7-2   Microhabitats of epiphytes within an emergent tree of the tropical rain forest. Small epiphytes are common in zone 1, large epiphytes in zone 2, crustaceous lichen epiphytes in zone 3, and bryophytes in zones 4 and 5. From K. A. Longman and J. Jenik. *Tropical Rainforest and Its Environment.* Copyright 1974 by Longman, London.

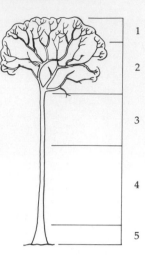

Table 7-1   Genera and species of trees that produce canopy roots. Modified from N. Nadkarni. 1994. *Oecologia* 100:94–97.

	New World	Old World
Temperate	*Acer macrophyllum*	*Coprosma* sp.
	Acer circinatum	*Grisilinia lucide*
	Alnus rubra	*Leptospermum* sp.
	Populus trichocarpa	*Meterosideros umbellata*
		Nothofagus fusca
		Podocarpus totara
		Weinmannia racemosa
Tropical	*Clusia alata*	*Acronychia* sp.
	Didymopanax pittieri	*Cinnamomum* sp.
	Nectandra sp.	*Dacrydium* sp.
	Ocotea sp.	*Podocarpus merifolium*
	Senecio cooperi	*Schizomeria* sp.
	Weinmannia pinnata	*Weinmannia* sp.
	Xylosma sp.	*Meterosideros collina*
	Ilex sp.	*Cheirodendron trigynum*
	Calyptranthes sp.	*Ceratopetalum virchowii*
	Grammadenia sp.	

ment resource uptake in trees that produce canopy roots (Nadkarni and Primack 1989). Although these studies lack the ability to analyze the net effect of epiphytes on their hosts, they at least document a positive effect of epiphytes on their host: increased plant nutrition.

Parasitism has also evolved through epiphytism when an epiphyte's roots penetrate beneath the bark into phloem and xylem and develop absorbing organs called haustoria. There seem to be many degrees of parasitism. For example, Hull and Leonard (1964; see also a review by Kuijt 1969) experimented with two genera of green mistletoe that parasitize a variety of conifers in the Sierra Nevada Mountains. They exposed host foliage to radioactive $^{14}CO_2$, then took autoradiographs of mistletoe foliage. Hull and Leonard concluded that the genus *Phoradendron* derives very little if any carbohydrate from its host. *Phoradendron* is a **hemiparasite,** a species able to live facultatively as a parasite or on its own. In contrast, the dwarf mistletoe (*Arceuthobium*) draws heavily on the photosynthate of its host and is dependent on host resources. Hull and Leonard found that the degree of parasitism closely correlated with chlorophyll content and photosynthetic rate of mistletoe tissue. *Phoradendron,* the hemiparasite, contained 0.93 mg chlorophyll per gram dry weight, about half the value of the host tissue, but *Arceuthobium* contained only 0.37 mg g^{-1}. In turn, the photosynthetic rate of *Arceuthobium* was very low compared to the rate of the conifer host. Although Hull and Leonard did not examine water utilization, undoubtedly both genera parasitized host xylem.

Epiphytes, as parasites, can also become damaging if their size or weight creates strain on the host. The strangler fig (*Ficus leprieuri*), like other species of *Ficus,* germinates in the canopy of its host and begins life as a typical epiphyte. As the strangler fig grows, aerial roots grow toward the soil. Eventually these aerial roots reach the ground and introduce a new source of nutrients to the fig. At this point, the fig is no longer an epiphyte. These roots thicken, engulfing the host trunk and preventing further growth of the host tree. At the same time, the canopy of the strangler fig enlarges to overtop the host and deprive it of light. Eventually the host dies (Figure 7-3). In this case the epiphyte parasitizes and competes with its host.

Nurse Plants

A plant that affords seedlings protection from a harsh environment while they grow large enough to withstand the travails of the environment on their own is called a **nurse plant** (Muller 1953, Niering et al. 1963, Steenbergh and Lowe 1969). Given that the seedling has no measurable effect on its nurse plant, this relationship is best considered commensal. A classic example is the saguaro cactus (*Cereus gigantea*) in the deserts of southern Arizona and Mexico (Niering et al. 1963, Steenbergh and Lowe 1969). Saguaro seedlings are nearly always found close to a shade-producing object. Occasionally the object is inanimate, but in most cases it is a perennial plant. Observations by these authors, coupled with others by Vandermeer (1980) and Turner et al. (1966), show that the positive effects of the nurse plant are (a) shading (reduces temperature and rate of soil drying); (b) hiding the young cactus from rodent herbivores; and (c) protection from frost. In the deserts of southern Arizona some 15 different species function as nurse plants. The frequency with which species appear as nurse plants is an approximation of their relative abundance. In other words, saguaro seems to have no preference for one species over another. Turner et al. (1966) showed that soils from beneath different nurse plant species differ in color, salinity, and nutrient status, but their data

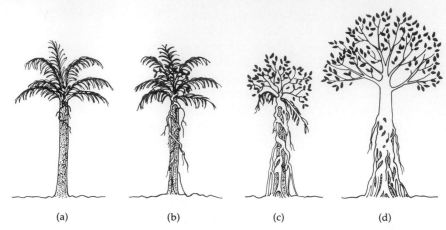

(a)   (b)   (c)   (d)

Figure 7-3   Four stages in the establishment of a strangler fig, *Ficus lep-rieuri,* on the palm *Elaeis quineensis.* (a) The young fig germinates high up in the palm and sends aerial roots down toward the soil. (b) The roots reach the soil and the fig shoot begins to expand. (c) The fig overtops the palm and the palm begins to senesce. (d) The palm has died but the fig tree remains. From K. A. Longman and J. Jenik. *Tropical Rainforest and Its Environment.* Copyright 1974 by Longman, London.

indicated that these soil differences had little effect on cactus seedling mortality. Vandermeer (1980) presented evidence to indicate that as the cactus grows it may compete for soil moisture with palo verde (*Cercidium floridum*), one of its nurse plant species. Dead palo verdes are more frequently associated with saguaros than not (Figure 7-4). As with some epiphytes, nurse plant relationships may shift from commensal to competitive over time.

Nurse plant effects also extend to desert annuals (Went 1942, Muller 1953, Muller and Muller 1956). Species of *Malacothrix* and *Chaenactis,* for example, are positively associated with the canopies of burro bush (*Ambrosia dumosa*) and turpentine-broom (*Thamnosma montana*) shrubs. Apparently the reason for the association is that the growth form of these shrubs is a suitable trap for windblown organic debris. The debris collects beneath the canopies, and this provides a better substrate for the annuals than open soil. It may be that seeds of the annuals are also trapped in abundance beneath the canopies. In any case, many desert annuals are restricted to association with a shrub. In the Great Basin desert of the western United States, bitterbrush (*Purshia tridentata*), shadscale (*Atriplex confertifolia*), and winterfat (*Eurotia lanata*) seedlings may also require nurse plants (West and Tueller 1971). Perennial bunchgrasses are positively associated with mesquite (*Prosopis juliflora*) canopies in the desert grassland of southern Arizona (Yavitt and Smith 1983).

Nurse plant effects are not limited to desert environments. Callaway and colleagues (Callaway et al. 1991, Callaway 1993) have shown that blue oak (*Quercus*

Figure 7-4 The nurse plant syndrome: saguaro cactus (*Cereus gigantea*) next to a dead palo verde (*Cercidium floridum*).

douglassii) can have positive effects on the surrounding herbaceous plants if the tree has tapped its roots into groundwater and thus uses very little soil water from near the surface. In contrast, blue oaks that do not have roots into the groundwater have strong negative impacts on the herbaceous community as a result of a massive near-surface root structure that depletes soil moisture very efficiently.

Similarly, Bertness and colleagues (Shumway and Bertness 1992, Bertness and Hacker 1993, Bertness and Shumway 1993, Bertness and Yeh 1994) provide some of the best examples of positive interactions in their studies of community dynamics in the salt marshes of Rhode Island. For example, Bertness and Yeh (1994) observed that recruitment of the marsh elder (*Iva frutescens*) is competitively inhibited by dense perennial turfs of plants in undisturbed marshes. Successful recruitment of marsh elder is typically restricted to disturbed bare patches. Yet isolated *Iva* individuals in bare patches fare poorly. High sunlight causes high evaporation rates and hence high soil salinities in bare patches with few recruits. In contrast, *Iva* that germinates under adults or in clumps on bare ground has higher success rates. The net effect of clumped plants is to increase shading, reduce evaporation, and reduce soil

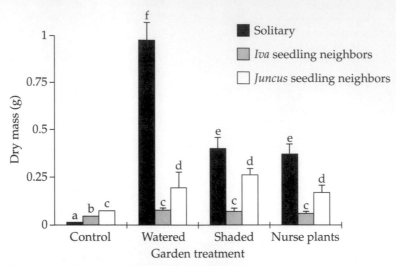

Figure 7-5   Dry mass of surviving seedlings from an experiment where *Iva* was grown solitary or crowded with other *Iva* or with *Juncus* neighbors (solid, dashed, and clear bars, respectively). Each bar (and 1 standard error) represents the mean dry mass of between 4 and 100 seedlings harvested toward the end of the growing season (August). Means with the same letter are not significantly different at the $p < .05$ level. This result shows that plants grown with nurse plants do as well as those that receive water or are shaded and much better than control plants. Thus, the nurse plants facilitate the growth of *Iva* seedlings. From M. D. Bertness and S. M. Yeh. 1994. "Cooperative and competitive interactions in the recruitment of marsh elders." *Ecology* 75:2416–2429.

salinities. In their experiments, only competitive effects remain among individual plants when soil salinities are experimentally reduced through shading or watering (Figure 7-5).

Addressing the issue of commensalism from a physiological perspective, Emerman and Dawson (1996) observed that water taken up from the groundwater by deep-rooting sugar maples (*Acer saccharum*) is passed up and out through the stomata with photosynthesis during the daytime. At night, in contrast, there is a water pressure gradient upward from the roots to the stem and then back out the near-surface roots (Figure 7-6). The process of plants moving water from deep groundwater that then "leaks" from shallow roots back into the upper soil is called **hydraulic lift** (Richards and Caldwell 1987). Emerman and Dawson (1996) observed the effect of hydraulic lift on herbaceous species performance in a sugar maple stand in New York. They found that increased soil moisture levels in the near-surface soil allowed for larger and more vigorous plants of most herbaceous species within 2 m of the base of trees. Thus, sugar maples have a positive effect on the herbaceous plant community by providing access to water that would otherwise be unavailable. The trees, however, are presumed to be unaffected by the presence of herbaceous plants.

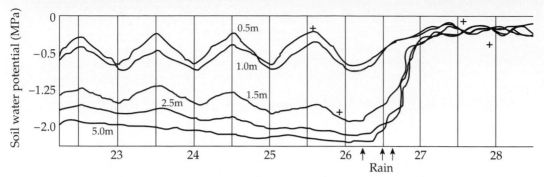

Figure 7-6 A time course of soil water potential at various distances (0.5, 1.0, 1.5, 2.5, and 5 m) away from the base of mature sugar maple (*Acer saccharum*) trees. Numbers on the horizontal axis represent midnight on each day during this nearly weeklong period. The early portion of the week was during a drought. The latter portion experienced rain. The fluctuaring soil water potential observed near the mature sugar maple during the drought was a result of higher soil water potentials during the night when the mature tree was lifting groundwater up and depositing it near the soil surface. This effect is diminished further from the mature tree and is swamped by soil saturation due to the rainfall events. From S. H. Emerman and T. E. Dawson. 1996. "Hydraulic lift and its influence on the water content of the rhizosphere—an example from sugar maple (*Acer saccharum*)." *Oecologia* 108:273–278.

Protocooperation

Protocooperation is an interaction that stimulates both partners, but it is not obligatory. Individual survivorship and growth are possible in the absence of the interaction. One example of protocooperation is root grafts, or unions, between members of the same or different species. As roots of some trees grow through a soil and come in contact with one another, a natural graft may form. This may be more common than was previously presumed. More than 160 tree species are known to form such grafts, and perhaps 20% of them can form interspecific, or intergeneric grafts (e.g., *Betula alleghaniensis* with *Ulmus americana* or *Acer saccharum;* Graham and Bormann 1966). Grafting of tree roots may occur on a wide variety of site conditions from mesic sites to dry chaparral slopes with rainfall below 50 cm per year (Saunier and Wagle 1965).

If both partners are equally successful with a mutual exchange of photosynthate, then the relationship is protocooperation (Figure 7-7a). There is evidence that hormones may also be transferred, resulting in more uniform phenology, such as synchronous spring bud break. If one individual is smaller and suppressed, however, the relationship may be primarily unidirectional and parasitic (Figure 7-7b).

Woods and Brock (1964) hypothesized that mycorrhizal hyphae may similarly link trees with one another in forest soils. They found rapid translocation of

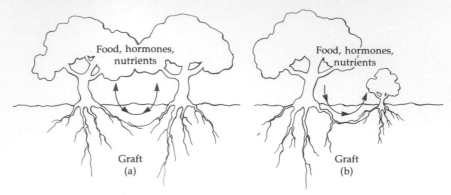

Figure 7-7   Two relationships between trees connected by a root graft, depending on the relative size of the trees. (a) The partners are both overstory trees, and the flow of nutrients is equal in both directions; the interaction is protocooperation. (b) One partner is an understory tree, and it parasitizes the larger tree.

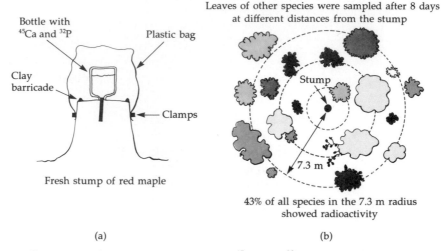

Figure 7-8   The transfer of radioactive ^{45}Ca and ^{32}P from a freshly cut stump of red maple [see detail in (a)] to the leaves of surrounding woody plants. Within eight days, 43% of all species within a radius of 7.3 m of the stump showed radioactivity, including red maple, Carolina hickory, mockernut hickory, fringe tree, persimmon, ash, holly, juniper, black oak, and blackjack oak. From F. W. Woods and R. E. Brock. "Interspecific transfer of ^{45}Ca and ^{32}P by root systems." *Ecology* 45:886–889. Copyright © 1964 by the Ecological Society of America. Reprinted by permission.

labeled (radioactive) ^{45}Ca and ^{32}P from a red maple (*Acer rubrum*) stump to the leaves of 19 other species of trees and shrubs in a North Carolina forest (Figure 7-8). Woods and Brock concluded that one should view the root mass of plants in a forest as a single functional unit.

Mutualism

Mutualisms are varied and diverse in nature. Mutual positive interactions have long been recognized, but they have at times been considered unimportant or not widespread (Williamson 1972, May 1973, Roughgarden 1979). More recent work, however, has impressed on plant ecologists the breadth and importance of mutualisms in ecological systems (Boucher et al. 1982, Boucher 1985). Indeed, mutualistic interactions are some of the most ancient that we recognize. Our current understanding of the origin of the eukaryotic cell is as an obligate mutualism between prokaryotic organisms (Margulis 1996). The term **symbiosis** is often used to describe such obligate mutualisms, although the term can be applied to any two organisms living in close contact (i.e., a symbiotic relationship need not be mutually beneficial). Conversely, mutualisms are merely defined by their mutual positive effect and may be either obligatory or facultative.

A common example of a class of mutualisms is the lichens, where an algae provides photosynthate and a fungus provides nutrients in a mutually maintained lichen body. Common mutualists with vascular plants include mycorrhizal fungi (phosphorus), symbiotic nitrogen-fixation bacteria or blue-green algae (fixed nitrogen), and insects, birds, and mammals (pollination) and zoochory—animal dispersal of plant propagules. Zoochory can be further split into different types of seed dispersal such as frugivory, where fruits are consumed and the seeds pass through the gut of the dispersal agent, or myrmechory, where ants disperse plant propagules, often while consuming a nutritious appendage (eliaosome). In each case the vascular plant provides a reward, typically in a carbon-based resource such as nectar or carbohydrates, in return for the nutrients or services provided by the mutualist.

Ants are often an underappreciated ecological force in nature. Ants are frequently mutualists with epiphytes, using them for protection and nesting sites while providing nutritional input for their epiphytic hosts (Davidson and Epstein 1989). In some cases, ants pack their feces around the rhizomes and adventitious roots of the epiphytes (Janzen 1974). In addition, epiphytes with ant mutualists may benefit from a reduction in herbivory (Rico-Grey and Thien 1986). Finally, ants are important seed dispersers in many parts of the world.

Although the theoretical notion of cooperation between species for mutual benefit is well understood, mathematical models of mutualism are historically weak. One can easily rearrange Lotka and Volterra's two-species competition equations to change a competitor into a mutualist:

$$\frac{dN}{dt} = \frac{rN(K - N_1 + \alpha N_2)}{K}$$

(Equation 7-1)

where the density of a second species (N_2) now has a positive effect on population growth. This equation, however, leads to unstable populations that exponentially increase to infinity in what May (1973) called "an orgy of mutual benefaction." Roughgarden (1975) proposed a model for the evolution of mutualisms, but the conditions were similarly constrained. Keeler (1985) developed a cost/benefit model to analyze fitness benefits of mutualism versus isolation. Keeler's was the first attempt to identify attributes (e.g., pollination success, increased seed set) that would need to be improved upon with cooperation in order for stable mutualisms to evolve.

Schwartz and Hoeksema (1998) have presented an economic model to explain why some resource exchange mutualisms, such as in the fungus-plant interactions of mycorrhizae, may be common. The model shows that specialization and re-source trade are favored by species whenever there is (a) differential difficulty in obtaining essential resources (almost always) and (b) the potential to ensure mu-tual benefits (trade is fair). Borrowing from the economic literature, they cite the theory of comparative advantage to support empirical data regarding resource exchange mutualisms. This model makes the point that it is beneficial for poor re-source competitors to establish trading relationships with superior resource com-petitors and vice versa.

A simple numeric example illustrates the Schwartz-Hoeksema model (Table 7-2). Imagine two species (A and B) that both require two resources (R_1 and R_2), but one species is more efficient than the other at acquiring these resources. We allow species A to gain R_1 and R_2 equally well and acquire 24 combined total re-source units per year. Species B can acquire 12 combined resource units per year but has three times more difficulty acquiring R_1 than R_2. Finally, both species re-quire equal amounts of each resource in order to maximize growth. In the ab-sence of resource trade, species A acquires 12 units each of R_1 and R_2 (net cost = 24 units). Species B acquires 3 units of each resource (net cost = 12 units). In a mu-

Table 7-2   A hypothetical numeric example of a resource trade mutualism demonstrating the eco-nomic principle of comparative advantage as applied to mutualisms. Species B in this example per-ceives a 3:1 ratio in the cost of R_1 compared to R_2, but species A does not differentiate the resources with respect to acquisition costs. In trade, species are allowed to trade at a 2:1 resource value, inter-mediate between each species' costs. In this case, each individual of species A trades with two indi-viduals of species B.

	Value ratio	Resource 1			Resource 2		
		Produced	Traded	Consumed	Produced	Traded	Consumed
Before trade							
Species A	1:1	12	0	12	12	0	12
Species B	3:1	3	0	3	3	0	3
After trade							
Species A	2:1	24	−8	16	0	16	16
Species B	2:1	0	4	4	12	−8	4
Gain from trade							
Species A				4 (33%)			4 (33%)
Species B				1 (33%)			1 (33%)

tualistic relationship, we allow each species to specialize in one resource and to trade resources at a 2:1 ratio, a value intermediate between each species' perceived value (1:1 and 3:1, respectively). An individual of species A trades with two individuals of species B, giving 8 units of R_1 for 16 units of R_2. The species A individual ends up with 16 units of each resource, a net gain of 4 (33%) units of each resource (Table 7-2). An individual of species B, acquires 12 units of R_2 and trades 8 of those units for just 4 units of R_1, ends up with 4 units of each resource rather than the 3 units it obtained in isolation. Both species benefit. The conditions under which this model does not work are when: (a) the poorer resource competitor is equally bad at acquiring all resources, (b) the cost of exchange is high and exceeds the benefits gained, (c) the ability to ensure fair trade is lacking, or (d) the trade ratio is not between the relative acquisition costs for the two trading partners.

This model provides for a broader array of conditions under which stable mutualisms may evolve. Another value of the model is that it shows how plants may benefit from a mutualism without the mathematical quirk of increasing population sizes infinitely. Cushman and Beattie (1991) have shown that the benefits derived by species within many supposed mutualisms are often assumed and not sufficiently documented. They conclude that more research needs to focus on careful documentation of the benefit derived by each species within a putative mutualism. The Schwartz-Hoeksema model suggests a reason for this lack of documentation: mutualisms will evolve even under conditions where participating species experience very small gains. Thus, experimental proof of net benefit may be very difficult to obtain in a variable world.

The issue of ensuring fair trade in mutualistic interactions remains a critical issue. For some mutualisms there are clear boundaries that may facilitate the maintenance of fair trade. For example, a plant and the fungus with which it forms mycorrhizae are responsive to each other in directing growth toward or away from specific roots. A well-documented example of a policing mechanism for cheating in a mutualistic relationship is the relationship between the yucca and the yucca moth (Pellmyr and Huth 1994). Yucca moths oviposit larvae in flowers of the yucca while pollinating their host plant. Pollination fails, however, when "cheating" yucca moths oviposit eggs but do not pollinate flowers. Many populations of yuccas and yucca moths are observed to contain cheaters that gain in individual fitness by not wasting time or energy on pollination of yuccas (Powell 1992). So, why don't all yucca moths cheat? Yuccas appear to have some ability to discourage cheating; they selectively abort flowers with an excess of moth larvae. These aborted flowers are likely to have been visited by cheaters as well as noncheating pollinating moths. Selective maturation of fruits with few moth larvae selects against cheaters and maintains a mutualistic relationship (Pellmyr and Huth 1994, Pellmyr et al. 1996).

Mycorrhizae

Mycorrhizae (singular: mycorrhiza) are fungal associations with the roots of higher plants. First observed in the nineteenth century (Allen 1991), these fungi were separated into types that penetrate the cell wall (**endomycorrhizae**) and those that do not (**ectomycorrhizae**). In some cases, the fungus covers the root

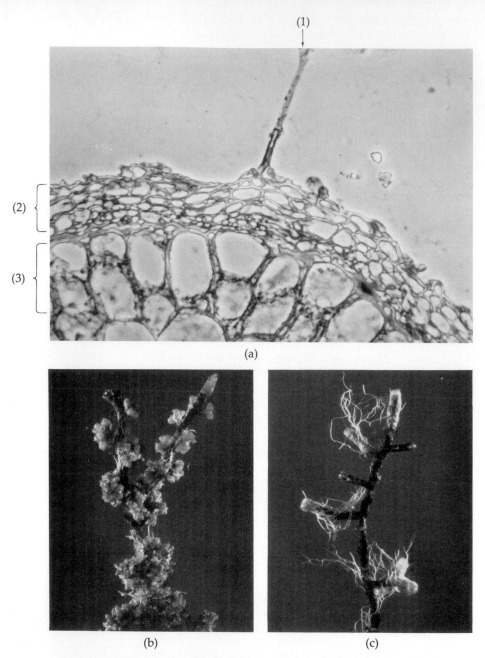

Figure 7-9   Ectomycorrhizae. (a) Cross section of a root, showing (1) hyphae branching off into the soil, (2) the fungal mantle coating the outside of the root, and (3) hyphae forming a net between root cortex cells. (b) and (c) show the various shapes that mycorrhizae may take, depending on the species involved. The fungal mantle covers short, club-shaped lateral roots. (a) Courtesy of F. H. Meyer, (b) and (c) courtesy of B. Zak in Marks and Kozlowski. 1973.

exterior near the tip of the root with a thick mantle of hyphae (haustoria), but does not penetrate the root cells. In other cases, the hyphae penetrate the host root between cortical cells to form a nutrient-absorbing network (Figure 7-9). Mycorrhizae transfer nutrients and metabolites in both directions between the vascular plant host and the fungus (Scott 1969). Radioactive $^{14}CO_2$, for example, can be fixed in the leaves of the higher plant by photosynthesis and later be detected in the fungus. Roots exude nutrients, such as amino acids, which are absorbed by the fungus. Radioactive isotopes of phosphorus, calcium, and potassium can be shown to be taken up in greater amounts by plants with mycorrhizae than by plants without them, and so the growth of the higher plant is stimulated by 25–300%. Mycorrhizal relationships are mutualistic, but not necessarily very species-specific. One tree may host as many as 100 different fungal species, and hyphae from one mycorrhiza can extend through the soil and connect plants of different species (Harley 1969, Hacskaylo 1971, Marks and Kozlowski 1973).

The biology and ecology of ectomycorrhizae have been reviewed by Harley (1969), Hacskaylo (1971), Marks and Kozlowski (1973), Smith (1980), and Allen (1991). Ectomycorrhizae (typically involving Basidiomycetes) are commonly found on northern temperate zone trees in the oak, pine, willow, walnut, maple, basswood, and birch families. Four other kinds of mycorrhizae, loosely classed under the overall heading endomycorrhizae, are ecologically important to an enormous variety of herbs, shrubs, and trees (Table 7-3). These four types typically do not form an external mantle of hyphae but penetrate cortical cells (hence the prefix *endo-*). Endomycorrhizal fungi may be Phycomycetes, Ascomycetes, or Basidiomycetes (Table 7-3). Mycorrhizal associations between plants and fungi may be as ancient as land plants themselves (Pirozynski and Malloch 1975, Pirozynski 1981, Wagner and Taylor 1981, Stubblefield et al. 1987). Very few higher plants appear to lack any of these five forms of mycorrhizal associations. These nonmycorrhizal exceptions include aquatic vascular plants and members of the families Brassicaceae, Cyperaceae, and Juncaceae. In addition, relatively few grasses (Poaceae) form mycorrhizal associations.

Symbiotic Nitrogen Fixation

Nitrogen fixation is the conversion of atmospheric nitrogen gas into organic ammonium. Only certain prokaryotic organisms are capable of this process. Some of these prokaryotes are free-living, and others live in close association with eukaryotes, receiving sugars and other energy-rich molecules from the eukaryotic symbiont. Nitrogen fixation requires energy in the form of ATP and a local anaerobic environment:

$$4\,N_2 + 6\,H_2O \xleftrightarrow{\quad ATP \rightarrow ADP \quad} 4\,NH_3 + 3\,O_2$$

the enzyme nitrogenase
no oxygen

Associations of plants with nitrogen-fixing species can contribute significantly to the nitrogen supply of the host, owing to the fact that nitrogen is often

Table 7-3   Five types of mycorrhizae and their common habitats and degree of specialization. A compilation of information from Smith (1980), Read (1983), and Allen (1991).

	Mycorrhizal type				
	Ectotrophic	Vesicular-arbuscular	Ericoid	Arbutoid	Orchidaceous
Plant associates	Temperate trees, e.g., *Quercus*, *Eucalyptus*	Mostly herbs; some deciduous trees	Ericaceae, Empetraceae, Epacridaceae	Evergreen trees and shrubs, e.g., *Arbutus*	Orchids
Habitats	Mostly temperate	Tropics and temperate, forests and grasslands	Shrublands	Mediterranean climates	Varied
Soil richness	Intermediate nitrogen, phosphorus (N, P)	High N, P	Low N, P	Intermediate N, P	Varied
Fungi	Ascomycetes, Basidiomycetes	Zygomycetes, e.g., *Glomus*	Ascomycetes, e.g., *Clavaria*	Basidiomycetes, e.g., *Armillaria*	Basidiomycetes
Host specificity	Moderate	General	Specialized	Specialized	Specialized

a limiting resource for plant growth. The host in turn provides carbon-based resources for the nitrogen fixer. The association of *Rhizobium* bacteria with the root nodules of legumes is well known to biologists, but several other nitrogen-fixing symbioses are also ecologically important to plants. Species of the blue-green algae *Nostoc* and *Anabaena* can become associated with bryophyte gametophytes, root nodules of cycads, leaf tissue of the angiosperm *Gunnera,* and leaf tissue of the aquatic fern *Azolla* (Peters 1978). The *Azolla-Anabaena* symbiosis has economic importance in addition to ecological significance. Agricultural trials in California have shown that three-fourths or more of all the nitrogen requirements of rice can be met by cultivating *Azolla* in rice paddies.

Certain soil actinomycetes (filamentous bacteria that resemble fungi, e.g., *Frankia*) are capable of invading the roots and causing elongate nodules to form in some higher plants. Within these nodules, nitrogen fixation occurs at a rate comparable to that of legume nodules. Some 285 species of woody plants, mainly of the temperate zones, have been shown to possess such nodules (Table 7-4). Many of the plants are pioneers in succession and occur in open habitats on acidic, saline, or sandy soils (Burleigh and Dawson 1994, Paschke et al. 1994).

Table 7-4 Currently identified families and genera of woody plants with actinorhizal associations. From Baker and Schwintzer 1990, compiled from data in Bond (1983) and Torrey and Berg (1988).

Family	Genus	Number of species
Betulaceae	*Alnus*	47
Casuarinaceae	*Allocasuarina*	54
	Casuarina	16
	Ceuthostoma	2
	Gymnostoma	18
Coriariaceae	*Coriaria*	16
Datiscaceae	*Datisca*	2
Elaeagnaceae	*Elaeagnus*	38
	Hippophae	2
	Shepherdia	2
Myricaceae	*Comptonia*	1
	Myrica	28
Rhamnaceae	*Ceanothus*	31
	Colletia	4
	Discaria	5
	Kentrothamnus	1
	Retanilla	2
	Talguenea	1
	Trevoa	2
Rosaceae	*Cercocarpus*	4
	Chaemaebatia	1
	Cowania	1
	Dryas	3
	Purshia	2
	Rubus	2

Pollination

Pollination is a very specialized form of mutualism that has developed in flowering plants; it may be the key to much of the variation and specialization in morphology of the angiosperms (Macior 1973). The transfer of pollen from the stamen to the stigma is essential for sexual reproduction. The attraction of pollinators, in some cases, limits the amount of seed a plant can produce (Galen 1985, Young and Young 1991, Parker 1997). In a review of recent experiments, Burd (1994) found that increasing pollination resulted in increased seed production both in terms of the quantity and quality of seed produced for 160 of 258 species (62%).

A plant species is usually morphologically adapted to the specific behavioral and morphological characteristics of its pollinator (e.g., insects, birds, bats). Adapted plants usually exhibit some or all of the following characteristics: (a) petals, sepals, or inflorescences that are attractive, either visually or olfactorily or

Table 7-5   A comparison of flowers pollinated by birds and birds that pollinate flowers. Reprinted with permission from Faegri and van der Pijl. *The Principles of Pollination Ecology.* Copyright © 1971 by Pergamon Press.

Flowers pollinated by birds	Birds that pollinate flowers
Diurnal anthesis	Diurnal activity
Vivid, highly contrasting colors, often scarlet	Visual sensitivity to red
Lip or margin absent or curved back, flower tubate and/or hanging	Too large to alight on the flower itself
Hard flower wall, filaments stiff or united, protected ovary, nectar stowed away but visisble	Hard bill
Absence of odors	Scarcely any sense of smell
Nectar very abundant	Large, requiring a large amount of food
Capillary system bringing nectar up or prevenging its flowing out	Long bill and tongue
Usually deep tube or spur, wider than in flowers pollinated by butterflies	Long bill and tongue
Distance between nectar source and sexual organs may be large	Long bill; large body
Nectar guide absent or plain	Intelligent in finding an entrance

both; (b) pollen grains often sculptured or sticky, sometimes massed together; (c) nectar, starch bodies, or pollen of nutritive value to the pollinator; and (d) attractants available at blooming time, the release of which is correlated to activity patterns of the pollinator.

In temperate regions, insects are of primary importance as pollinators, especially insects of the orders Diptera and Hymenoptera (Moldenke 1975). Solitary bees, hive bees, and the higher bumblebees (Apidae) have a greater capacity for remembering blossom characteristics, discriminating among them, and remaining faithful pollinators than any other insect group (Faegri and van der Pijl 1971). Bee-pollinated flowers are generally characterized by (a) bilateral symmetry; (b) mechanically strong flowers, often partly closed with sexual organs concealed; (c) bright blue or yellow color (bees don't see red); (d) nectar guides along a landing platform; (e) moderate quantities of nectar that is sometimes partly concealed; and (f) many ovules per ovary and few stamens. The adaptations of the pollinator include: (a) the possession of good color discrimination in the blue, yellow, and ultraviolet ranges (Jones and Buchmann 1974); (b) a relatively high degree of intelligence and a long memory; and (c) a long proboscis capable of probing for nectar.

In the tropics, birds assume a more important pollinator function. The sunbirds (Nectarinidae) of Africa and Asia, the honey creepers (Drepanididae) of

Hawaii, and the hummingbirds (Trochilidae) of both Americas are all recognized as important pollinators. Many tropical plants have flowers that are adapted to bird pollination, such as some species in the genera *Eucalyptus* and *Erythrina*. Old World bird-pollinated flowers are borne upon a stem or inflorescence branch that provides a stout landing place because Old World birds that pollinate flowers do not hover. New World hummingbirds do hover and can pollinate pendulous flowers. Table 7-5 presents a summary of the coevolved traits of birds and ornithophilous flowers that are important to their mutualistic interactions.

The obvious advantage conferred upon animal-pollinated flowers is the possibility of accurate pollen dispersal far from the host anther, allowing for outcrossing and therefore genetic variability. By providing a nutritional source for animals, flowers have influenced the evolution of pollinators and have evolved very specific traits to ensure pollination. The result is a series of mutualistic interactions, each with its own set of special adaptations. An excellent review of pollination ecology is found in a volume edited by Leslie Real (1983). Numerous more specialized texts have been published on topics such as methods for studying pollination (Kearns and Inouye 1993), often overlooked groups of pollinators (Buchmann and Nabhan 1996), and the evolution of animal pollination systems in plants (Lloyd and Barrett 1996).

Herbivory

Herbivory is the consumption of all or part of a living plant by a consumer. Taking a broad view, plant consumers include (a) parasitic and phytophagous microbes (e.g., fungi, algae), (b) phytophagous invertebrates (e.g., stem- and foliage-feeding insects, and root-feeding insects and nematodes), (c) browsing and grazing vertebrates, and (d) seed predators. Technically one would also have to consider fruit consumers (frugivores) and nectar-feeding pollinators as herbivores as well. These latter types of herbivory, however, are typically classified separately because the consumers are feeding on rewards provided by the plants for services that they perform and not consuming living plant material to the detriment of the host plant.

Herbivory is a trait that may be relatively recent compared to the development of vascular plants (Southwood 1985). Fossil evidence of herbivory on the vegetative parts of higher plants appears in the Permian period, 70 million years after the first appearance of vascular plants on the earth's surface. Likewise, herbivory is relatively restricted taxonomically (Southwood 1985). Only 27 of 97 orders of extant major taxa of animals, and 9 of 29 orders of insects, contain herbivores (Strong et al. 1984). These nine insect groups, however are very diverse (Figure 7-10). There is no denying that herbivory is an important component of the earth's ecosystems. Strong et al. (1984) estimate that 80% of the world's macroscopic species are plants, herbivores, or species that prey on herbivores.

Perhaps no other area of plant ecology has garnered as much research interest over the past two decades as herbivory. One reason for this interest is the observa-

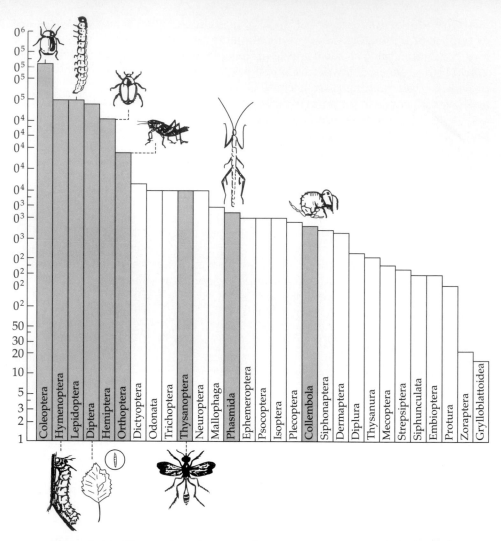

Figure 7-10   The number of species of insects in each of 29 extant orders of insects plotted on a log scale showing the relatively large numbers of species in the nine orders that contain herbivorous insects (shaded bars with picture examples). From D. R. Strong, J. H. Lawton, and T. R. E. Southwood. 1984. *Insects of Plants, Community Patterns and Mechanisms.* Blackwell Scientific Publications, Oxford.

tion that insect herbivores have the capacity for very high population growth. These capacities of herbivores raise an important question: what regulates herbivore populations such that the world remains green (Hairston et al. 1960)? Another reason that herbivory is a popular area of study is its relevance to our understanding of ecology, as so much of the world's natural energy flow is through plants and herbivory. Finally, plants and their herbivores often make for tractable systems in which to conduct ecological experiments. We address three issues related to herbivory: (a) What is the effect of herbivory on plants and plant communities? (b) What limits the intensity of herbivory? (c) What defenses do plants have and how do they work?

Although classically herbivory is considered a negative interaction, more recent evidence suggests that it can have positive effects on plant growth. We have long understood that plants can have a remarkable tolerance to herbivory and that herbivory can stimulate more efficient photosynthesis (e.g., Georgiadis et al. 1989, Frank and McNaughton 1993). Ken Paige (Paige and Whitham 1987, Paige 1992), however, has taken this one step further and experimentally demonstrated that grazing has a net positive effect on lifetime fitness of the monocarpic scarlet gilia (*Ipomopsis aggregata*) in northern Arizona. In this experiment, randomly selected plants were allowed to be grazed by ungulates (mule deer and elk). Control plants were excluded from grazing. Grazed plants responded with more growth and increased seed production over ungrazed controls. This phenomenon, called **overcompensation,** has been hotly debated in the ecological literature (Bergelson and Crawley 1992a,b, Paige 1992, Belsky et al. 1993, Paige 1994, Bergelson et al. 1996). Indeed it does seem astounding that the loss of photosynthetic tissue can have a positive effect on plant growth. Nonetheless, the debate perhaps underscores our reticence in viewing positive interactions as important in ecology. Much work remains in determining the conditions under which herbivory leads to positive effects on plant populations.

The Effect of Herbivory on Plant Communities

Gross primary productivity (GPP) is the total amount of chemical energy fixed by photosynthesis for a given unit of land surface for a given unit of time. Typically, GPP is expressed as calories per square meter per year, but it may be expressed in biomass units also. **Net primary productivity (NPP)** is gross primary productivity minus energy lost through plant respiration and is equivalent to chemical energy stored or accumulated per unit area per unit time. For most terrestrial vegetation, NPP = 30–70% of GPP. Finally, **detritus** is dead plant material and a synonym for litter. If we are concerned with an annual plant, then virtually the entire plant body can be considered litter at the end of one year; only the seeds remain living. In the case of a deciduous tree, some bark, twigs, flowers, aborted fruit, and all the leaf matter produced in spring and summer (except the fraction already eaten by herbivores) will be returned to the soil surface as litter.

About 10% of NPP is consumed by biophage herbivores for typical terrestrial vegetation. The percentage varies with vegetation: 2–3% for desert scrub or arctic/alpine tundra, 4–7% for forest, 10–15% for temperate grasslands with minimal grazing, and 30–60% for African grasslands or grasslands managed for domesticated animals. The percentage can also fluctuate from year to year. Episodic outbreaks of herbivores such as tent caterpillars, locusts, budworms, rabbits, voles, lemmings, or pathogenic microbes may raise the consumption to 50–100% of NPP. Data show that vegetation may often withstand considerable loss to herbivores without lasting damage, and may in certain instances increase in vigor as a result of herbivory (Paige 1992, 1994; Frank and McNaughton 1993). For example, tree wood growth rate is not reduced until more than 50% of the canopy's leaf surface is lost. Seedling populations, where most of the mortality for most perennial plant populations occurs, experience heavy losses to herbivory on a regular basis. Certainly, however, the death of mature plants can and does result from herbivore outbreaks, particularly if they persist for more than one growing season.

If NPP is calculated for seed production only, consumption by seed predators is typically well above 10% of NPP, and may often reach 100%. It may be that herbivores exert their principal effect on vegetation in this way. Janzen (1970) and Connell (1971) concluded that intensive seed predation in the tropics may be the single most important factor regulating tree populations. In the tropical rain forest, a high diversity of tree species coexists, and neighboring trees are usually not the same species. Janzen and Connell proposed that the driving force behind seed dispersal is escape from predators. Further, Augspurger (1983, 1988) showed that pathogen effects on seeds can be particularly high near parental trees in tropical forests. Seed predators and pathogens associate high densities of seed with the parent, and predator populations are high near the parent. Therefore, the closer the seed falls to the parent, the lower its probability of escaping predators or pathogens. The optimum dispersal distance for establishment is thought to be a compromise distance from the parent. At this intermediate distance the probability of a predator or pathogen finding the seed is moderately low, yet the probability of the seed being dispersed that far and encountering a favorable environment remains moderately high (Figure 7-11).

Limits to Herbivory

A mere 10–20% of the aboveground green biomass of terrestrial plants is estimated to be consumed each year by herbivores (Cyr and Pace 1993). This is despite the fact that on the order of 25% of the multicellular species on the planet are herbivores (Strong et al. 1984). Population outbreaks of deer, rabbits, locusts, and gypsy moths, however, provide evidence of the potential for near total defoliation by herbivores. Two primary restraints can maintain a green terrestrial world. First, predators may limit the number of herbivores, prohibiting them from attaining densities sufficient to completely defoliate plants (Hairston et al. 1960, Strong et al. 1984). Because predators of herbivores are defined to be on a higher trophic level, this type of population control is called a **top-down effect.** Although empirical evidence supports this hypothesis, predator control of herbivores is beyond the scope of plant ecology and will not be specifically addressed.

Second, many have argued that plants themselves control insect density. Plant tissue is poor food for herbivores (McNeill and Southwood 1978, Lawton and McNeill 1979, Janzen 1988, White 1993). Because plant proteins are different from animal proteins, they must be digested and resynthesized by the herbivore. Further, the protein content of plant tissue is low. Although carbohydrate content of plant tissue is high, it is typically concentrated in poorly digestible forms such as lignin and cellulose. Finally, the primary limiting nutrient for most plants and their insect herbivores is nitrogen. Herbivores often have a hard time extracting and us-

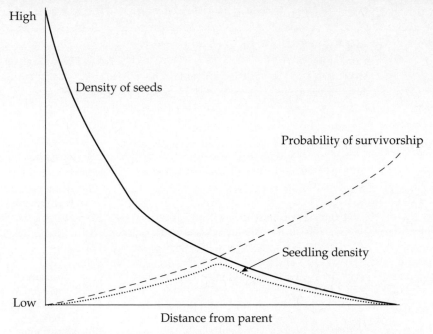

Figure 7-11 Under the escape hypothesis, maximum seedling recruitment probability (solid lines) is at an intermediate distance away from parental plants due to the combined effects of decreasing seed density with distance (dotted line) and increasing likelihood of seed/seedling survivorship with distance (dashed line).

ing plant nitrogen because it is often bound in relatively inaccessible forms, such as secondary metabolites. These **secondary metabolites** are complex organic molecules that are often toxic to herbivores.

Thus, the nutritional quality of many plant tissues is low, and they often contain anti-herbivore defensive chemicals, presenting a dietary hurdle for potential herbivores. This type of herbivore population control is called a **bottom-up effect** because plants as primary producers are on a lower trophic level than their herbivores. Ecologists currently debate the relative importance of top-down versus bottom-up effects on energy flow and population regulation.

Plant Defenses

There are several competing theories on how plants defend themselves. Plants may either *tolerate* herbivory through the production of relatively cheap plant parts and rapid growth rates, or *avoid* herbivory through some defensive system (Herms and Mattson 1992). Plant defenses, when produced, may be *physical* (e.g., spines, hairs, leaf toughness) or *chemical*. Table 7-6 describes an array of the most common defensive chemicals used by plants. These secondary metabolites may inhibit herbivory because they are toxic, inhibit growth, or repel insect attack. Many

Table 7-6 Major classes of secondary plant compounds involved in plant-animal interactions. Reprinted from *Introduction to Ecological Biochemistry* by J. B. Harborne. Academic Press, London.

Class	Approx. number of structures	Distribution	Physiological activity
Nitrogen compounds			
Alkaloids	6500	Widely in angiosperms, especially in root, leaf, and fruit	Many toxic and bitter tasting
Amines	100	Widely in angiosperms, often in flowers	Many repellent smelling; some hallucinogenic
Amino acids (nonprotein)	400	Especially in seeds of legumes but relatively widespread	Many toxic
Cyanogenic glycosides	30	Sporadic, especially in fruit and leaf	Poisonous (as HCN)
Glucosinolates	75	Cruciferae and ten other families	Acrid and bitter (as isothiocyanates)
Terpenoids			
Monoterpenes	1000	Widely, in essential oils	Pleasant smells
Sesquiterpene lactones	1500	Mainly in Compositae, but increasingly in other angiosperms	Some bitter and toxic, also allergenic
Diterpenoids	2000	Widely, especially in latex and plant resins	Some toxic
Saponins	600	In over 70 plant families	Hemolyze blood cells
Limonoids	100	Mainly in Rutaceae, Meliaceae, and Simaroubaceae	Bitter tasting
Cucurbitacins	50	Mainly in Cucurbitaceae	Bitter tasting and toxic
Cardenolides	150	Especially common in Apocynaceae, Asclepiadaceae, and Scrophulariaceae	Toxic and bitter
Carotenoids	500	Universal in leaf, often in flower and fruit	Colored
Phenolics			
Simple phenols	200	Universal in leaf, often in other tissues as well	Antimicrobial
Flavonoids	4000	Universal in angiosperms, gymnosperms, and ferns	Often colored
Quinones	800	Widely, especially Rhamnaceae	Colored
Other			
Polyacetylenes	650	Mainly in Compositae and Umbelliferae	Some toxic

secondary metabolites may have a combination of effects. Besides reducing herbivory, many of these compounds have additional functions: physiological (UV absorption, thermal protection), regulatory (expression of a growth regulation gene, signal transmission), storage (control of nutrient cycling), or other (attraction of pollinators, allelopathy).

Similarly, defensive systems may be constitutive, preformed inducible, or inducible (Karban and Baldwin 1997). **Constitutive defenses** are systems that are a fixed part of plant allocation. For example, smooth sumac experiences more phloem-sucking beetle damage than the hairy-stemmed staghorn sumac (S. Strauss, 1997 personal communication). The hairiness of stems, however, is a relatively stable characteristic and is a constitutive portion of that plant's defenses.

Preformed inducible defenses are defensive compounds, usually chemical, which only become active as anti-herbivory defenses under stimulation from attack. For example, highly localized concentrations of furanocoumarin appear within three hours of physical damage to cow parsnip (*Pastinaca sativa*) (Zangerl and Berenbaum 1995). Furanocoumarin is a highly effective deterrent to most herbivores and is quickly mobilized to the site of herbivory in the plant. As a defense against herbivory, furanocoumarins, or furanocoumarin-like compounds, are likely to be produced within plant tissue prior to damage. Thus, this defensive system is preformed, but only becomes active when induced.

Induced defenses are those that plants begin to produce in response to herbivory. For example, Baldwin (1988, 1989) has shown through a series of detailed experiments that early damage to seedlings increases nicotine production in tobacco. Similar inducible defenses have been observed in many plant herbivore systems (Karban and Baldwin 1997). We might expect to find in nature the complete continuum from fully preformed plant defenses to fully induced construction of plant defensive compounds.

Allocation to plant defenses has been explained in terms of **optimal defense theory** (Rhoades 1979). Simply stated, a plant should neither overallocate nor underallocate to its defenses. Optimal defense theory predicts levels of investment in defenses. We expect plants or plant parts that grow fast, have a high regrowth potential, and are cheap to produce to be less well defended against herbivory than those that do not fit these descriptors. The predictability of herbivore attack, and hence the allocation pattern of a plant species toward defenses, may vary by habitat and resource availability (Zangerl and Bazzaz 1992). For example, ephemeral plants with a low risk of being found and consumed by herbivores may invest little in plant defenses. Thus, the constancy, or predictability, of herbivore attack should be correlated with allocation to constitutive and inducible defenses (Figure 7-12).

Similarly, Feeny (1976) and Rhoades and Cates (1976) proposed a theory of plant defenses based on plant life history. Long-lived plants, which are often relatively large, are apparent to herbivores. **Apparency theory** predicts that these easy-to-find perennial plants require high levels of constitutive defenses throughout their green tissue because they are constantly attacked by herbivores. Apparency to herbivores is suggested to explain why many tree leaves contain high levels of tannins, resins, and lignin, as well as physical features that may make herbivory difficult (e.g., spines, tough leaves). Defense systems of apparent plants are

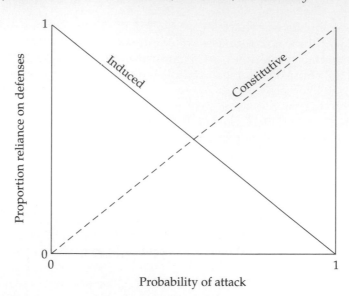

Figure 7-12   The hypothetical relationship between the type of defense and the probability of attack. From A. R. Zangerl and F. A. Bazzaz. 1992. "Theory and pattern in plant defense allocation," pp. 363–391 in R. S. Fritts and E. L. Simms, eds. *Plant Resistance to Herbivores and Pathogens: Ecology, Evolution and Genetics.* University of Chicago Press, Chicago.

expected to be expensive and have consequences in long-term growth rates. In contrast, many annual or ephemeral plants may be more difficult for herbivores to detect because they are inconspicuous and patchy in time and space. Plants that fit this description of nonapparency are more likely to invest in cheap, inducible defenses. Plant secondary chemicals act in a variety of ways against herbivores, as well having other potential functions, making it difficult to directly test the theory of apparency. Nonetheless, observed patterns of plant defenses generally seem to support this model.

Five models have been proposed to explain the types of defensive compounds a plant uses for defense (Karban and Baldwin 1997). The first three models presented here are very similar and fall under the classification of "supply-side" theories (Lerdau et al. 1994). These three theories similarly posit that plants do not regulate the amount of secondary metabolites they produce because the primary function of these compounds is to package waste material.

The **carbon-nutrient balance theory** suggests that plants vary systematically in the types of defensive compounds used depending on whether their growth is more limited by carbon or nutrients. Plants, as we observed in Chapter 5, are often limited in their growth by carbon (e.g., through light limitation or water shortage) or nitrogen (soil nutrition). Secondary metabolites that are used as plant defenses may be either carbon or nitrogen intensive. Plants that are more limited by nitrogen than carbon may invest in defenses that are primarily carbon based (e.g., phenolics, terpenoids, tannins, lignin), but species limited more by carbon than nitrogen would invest in nitrogen-based defenses (e.g., alkaloids). The **substrate-enzyme imbalance theory** is similar to the carbon-nutrient balance theory. Again, defensive compounds are primarily considered as derivatives of plant waste materials compartmentalized into secondary metabolites. In this case, either carbon-rich or nitrogen-rich secondary metabolites are produced as a result of excess metabolic activity. The **growth-differentiation balance theory** posits that plant growth and differentiation are different allocation pathways: plants do not differentiate tissues when maximizing growth and vice versa (Herms and Mattson 1992). If growth and differentiation processes are negatively correlated, then plants may produce secondary metabolites for use as anti-herbivore defenses only when not maximally allocating toward growth. Further, growth is favored when conditions are favorable to the plant. These three models make very similar predictions regarding the types of defenses produced by plants.

Although many studies have been used to support these theories (e.g., Gershenzon 1984, Waring et al. 1985, Larrson et al. 1986, Bryant et al. 1989, Price et al. 1989), support is not universal (Herms and Mattson 1992, Baldwin et al. 1993). A further problem is that the actual amount of carbon or nitrogen in a defensive compound may not accurately predict the actual cost required to manufacture that compound (Gershenzon 1994).

Alternatively, two models have been proposed that view plant defenses as part of an integrated and multifaceted strategy of plant allocation. The **generalized stress-response theory** views plants as having a centralized mechanism that allows them to simultaneously and interactively respond to diverse stresses (Chapin 1991). Thus, induced defenses are just one of many adaptable strategies that plants use to cope with a variable environment. The **active defense response theory** is similar, but suggests that the signaling system that plants use to induce the construction of defensive compounds is very specific (Chessin and Zipf 1990). Karban and Baldwin (1997) review induced chemical defenses in plants and suggest that most of the recent work on rapidly responding plant defensive systems best fits the active defense response theory. Accounting for how plants defend themselves with secondary metabolites and the degree to which plants control their allocation to defenses is an emerging and active area in plant ecological research.

Summary

Interspecific interactions may vary in both strength and sign. In this chapter we outlined a variety of types of commensal $(+,0)$, mutualistic $(+,+)$, and herbivorous $(+,-)$ interactions. Examples of commensalism are epiphytes that use a host plant as a support structure. Although our knowledge of the ecology of epiphytes has grown substantially in the past few decades, there is much yet to be learned regarding the basic ecology of plant-epiphyte interactions. Lichen epiphytes that contain nitrogen-fixing algal symbionts may contribute significant amounts of nitrogen to their host, so some epiphytic relationships can be mutualistic. Some host plants grow roots into their canopies to exploit litter accumulation from epiphytes. But many epiphytes, such as the strangler fig, can be very damaging to their hosts. Other commensal examples are the nurse plants required by several desert plants to ameliorate harsh environmental conditions during the sensitive establishment phase. Nurse plant observations have been extended to salt marsh environments, where established adults have a positive effect on seedling recruitment by providing shade that reduces evaporation and thus salinity. Advances in plant physiological ecology are unveiling previously undescribed positive interactions between species. For example, trees drawing groundwater from deep beneath the soil surface by hydraulic lift can have commensal interactions with the surrounding herbaceous vegetation.

Despite a historical bias toward the study of negative interactions, positive interactions such as mutualism are common. Plants exhibit a wide variety of mutualisms with mycorrhizal fungi, nitrogen-fixing symbionts, and pollinators. The mathematical theory of the effect of positive interactions on population dynamics is not well developed, however.

Mycorrhizae are fungal associations with vascular plants. Mycorhizal fungi gain carbon from host plants in exchange for providing soil-derived nutrients, primarily phosphorus. Mycorrhizal associations among plants are ancient and diverse. Nearly all plant families form mycorrhizal associations. There are five forms of mycorrhizae (ectomycorrhizae, vesicular-arbuscular, ericoid, arbutoid, and orchidaceous) that include three fungal groups (Phycomycetes, Ascomycetes, and Basidiomycetes).

Nitrogen-fixing symbionts are observed in relatively few vascular plants (e.g., legumes), but provide a particularly important mutualism owing to the widespread limitation of available nitrogen for plant growth. Nitrogen fixation in many habitats is a major source of available nitrogen.

Plant-pollinator mutualisms have long garnered the attention of naturalists and ecologists. Much of the diversity of vascular plants is often attributed to the specialization between plants and their pollinators. Similarly, diversification in plants may be the result of specialization between plants and their seed dispersers. For example, many ant- and bird-dispersed acacias produce seed that is specialized to a specific disperser.

Herbivory is a broadly defined and widely studied interaction that has a rich realm of theory with respect to plant defenses. Herbivores can include any type of organism that feeds on plants or plant parts, including parasites, phytophagous insects, and vertebrates. Despite the wide variety of herbivores, most green plant tissue produced each year (net primary productivity) is not consumed by herbivores. Plants are relatively well defended against herbivory through physical barriers (e.g., woody fibers, spines, and hairs) and secondary metabolites, which are plant-produced chemicals that are either toxic or unpalatable to herbivores (e.g., tannins, alkaloids, and glucosinolates). A wide variety of secondary metabolites are used by plants in defense against herbivores, although these vary somewhat systematically with plant life history. For example, apparency theory predicts that long-lived woody plants should maintain a high investment in constitutive defenses such as tannins, but ephemeral species should favor inducible defenses that are only produced in response to an herbivore stimulus. The carbon-nutrient balance theory predicts that plants limited by carbon or nutrients will specialize in defensive compounds with either a nitrogenous or carbon base, respectively. Similar theories predict that plant nutrient status and environment drive the type of defensive compounds used by plants to discourage herbivory. In contrast, the generalized stress-response and active defense response theories predict complex control systems that are used by plants to respond to specific stresses such as herbivory.

PART III

THE COMMUNITY AS AN ECOLOGICAL UNIT

In each type of habitat, certain species group together as a community. Fossil records indicate that some of these groups (or very closely related precursors) have lived together for thousands and even millions of years. During that time, it is possible that an intricate balance has been fashioned. Community members share incoming solar radiation, soil, water, and nutrients to produce a constant biomass; they recycle nutrients from the soil to living tissue and back again; and they alternate with each other in time and space. Synecologists attempt to determine what is involved in this balance between all the species of a community and their environment.

First, how is the community to be measured and how are the measurements summarized for maximum, useful information? Second, why do some communities change more rapidly over time than others? How accurately can we predict future changes and reconstruct the past? Are there community traits that transcend traits of the individual species making up the community? A synecological view asks such questions, and the answers take us to a level of complexity beyond autecology.

CHAPTER 8

COMMUNITY CONCEPTS AND ATTRIBUTES

U p to this point, we have discussed plant species as though their populations grew in isolation or, at most, alongside one other species. Most habitats, however, are simultaneously occupied by many plant species mingled together in what at first seems to be a random fashion. That is, members of one species normally grow in close spatial association with members of many other taxa. The earliest plant ecologists called these mixtures **communities** or **associations,** names that have technical meanings beyond our intuitive definitions.

The term **community** is a very general one that can be applied to vegetation types of any size or longevity. It can, for example, be applied to one stratum of plants in a very local area, such as the herbs, woody seedlings, and mosses on the floor of a streambank forest; or to a very widespread, regional vegetation type; or to a transitory plot of vegetation undergoing rapid change in the species that compose it; or to very stable vegetation that has exhibited no significant change for hundreds of years.

An **association** is a particular type of community that has been described sufficiently and repeatedly in several locations such that we can conclude that it has (a) a relatively consistent floristic composition, (b) a uniform physiognomy, and (c) a distribution that is characteristic of a particular habitat. In terms of its basic importance to plant ecology and to the classification of vegetation, the association has been compared to the species of taxonomy. Just as a species is an abstract, somewhat artificial synthesis of many individual plants, so an association is a synthesis of many local examples of vegetation called **stands.**

An observant person can measure and describe vegetation in such a way that it can be subdivided into associations (see Chapters 9 and 10). How natural are these units and how much can their existence be attributed to human bias and the need to classify? Some ecologists imply or state that associations are real, closed entities whose component species are interdependent, but other ecologists argue that there is little experimental evidence to support this notion of interdependence.

Is the Association an Integrated Unit?

The Organismic View

One of the traits of an association, first accepted by the International Botanical Congress of 1910, is a relatively consistent floristic composition. Wherever a particular habitat repeats itself in a given region, the same cluster of associated taxa is found. This does not mean that every species repeats itself, nor even that the majority of taxa are repeated. Some species are extremely widespread, with broad tolerance ranges, and these can be found in many habitats and in many associations. Other species may have narrower range limits, but nevertheless a few individuals can be found beyond the normal limits, and they are occasional members of many communities.

According to the procedures of one method of community classification, after accidental and ubiquitous taxa are removed from consideration, certain species will remain that show a large degree of association with each other and with a particular habitat. Associations are defined by the presence of such characteristic or indicator species, and a particular stand is placed in an association if it contains a significant number of such species. This method—the *table method* of Braun-Blanquet and others—is described more fully in Chapters 9 and 10.

If clusters of species do repeatedly associate together, that is indirect evidence for either positive or neutral (not negative) interactions between them. Such evidence favors a view that communities are indeed integrated units—that the whole is somehow greater than the sum of its parts, much like an organism is greater than the sum of its cells, tissues, or organs. Some of the interactions described in Chapters 6 and 7, such as mycorrhizae, are also evidence that communities are units. Clements (1916, 1920) metaphorically equated associations with organisms. The pattern of species distribution and abundance predicted by this organismic view is diagrammed in Figure 8-1. The species in an association have similar distribution limits along the horizontal axis, and many of them rise to maximum abundance at the same points (nodes). The ecotones between adjacent associations are narrow, with very little overlap of species ranges, except for a few ubiquitous taxa found in many associations.

The Clementsian view of nature is a holistic one, as described in Chapter 2, and holistic viewpoints continued into the late twentieth century under a variety of names. In Europe (and now in many other parts of the world), a standard process for locating, sampling, and defining discrete plant communities in the landscape developed under the leadership of Josias Braun-Blanquet. This work resulted in a **syntaxonomy** of associations and higher units that is every bit as formal as the parallel taxonomy of species, genera, families, orders, and classes. Braun-Blanquet's view of the plant community's cohesiveness was based more on habitat and less on species interactions than Clements's model (Noy-Meir and van der Maarel 1987). The **Gaia hypothesis,** first postulated 30 years ago by James E. Lovelock, proposes that the entire earth is one large organism whose biological and physical components continually interact to maintain an environment suitable for the maintenance of life. That is, the earth, together with oceans and atmosphere,

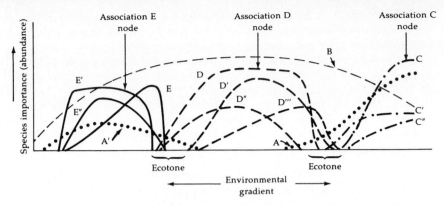

Figure 8-1 Patterns of species importance (abundance) along an environmental gradient as predicted by the association-unit view. Clusters of some species (Cs, Ds, and Es) show similar distribution limits and abundance peaks. Each cluster of associated species defines an association. Ecotones are narrow. A few species (As) have such broad ranges that they are found in adjacent associations, but are more abundant in one than another. A few other species (Bs) have very broad ranges and are ubiquitous.

is a superorganism. Some ecologists consider this hypothesis to be nonscientific because they believe it cannot be tested. Jared Diamond (1975) postulated that species, on a local scale, are limited in the ways they can be packaged with other taxa. He suggested that there are **assembly rules** that restrict the presence or abundance of some species, depending on the presence or abundance of other species. Like other animal ecologists, Diamond thought of communities as composed of **guilds**—groups of species that compete more with each other than with species in other guilds—each guild having a rather fixed composition (Pianka 1980, Fox 1989, Wilson 1989). New Zealand plant ecologist J. Bastow Wilson claims to have found evidence that assembly rules exist in several kinds of plant communities, ranging from disturbed roadsides to lawns to undisturbed salt marshes (Wilson et al. 1987 and 1996, Wilson and Roxburgh 1994).

The Continuum View

In 1926, Henry Gleason published a carefully written paper entitled, "The individualistic concept of the plant association." In an essay written 27 years later, he recounted the furor that his proposal generated and the never-never world into which his ideas were cast for some time. Gradually, his ideas did gain acceptance. The Ecological Society of America cited him as a "distinguished ecologist" in 1953 and as an "eminent ecologist" in 1959.

Gleason (1953) sampled forest vegetation along a north-south gradient in the Midwest and concluded that changes in species abundance and presence occurred so gradually that it was not practical to divide the vegetation into associations. Even within a relatively homogeneous region, there were subtle but important differences in the vegetation from one plot to another.

The sole conclusion we can draw from all the foregoing considerations is that the vegetation of an area is merely the resultant of two factors, the fluctuating and fortuitous immigration of plants and an equally fluctuating and variable environment. As a result, there is no inherent reason why any two areas of the earth's surface should bear precisely the same vegetation, nor any reason for adhering to our old ideas of the definiteness and distinctness of plant associations. . . . Again, experience has shown that it is impossible for ecologists to agree on the scope of the plant association or on the method of classifying plant communities. Furthermore, it seems that the vegetation of a region is not capable of complete segregation into definite communities, but that there is a considerable development of vegetational mixtures. . . .

Gleason's conclusions have been supported by more extensive and sophisticated studies in such widely different vegetation types as desert scrub in Nevada (Billings 1949), upland forest in Wisconsin (Curtis and McIntosh 1951), montane forest in the Great Smoky Mountains in North Carolina and Tennessee (Whittaker 1956), and conifer forests in the Siskiyou Mountains of California (Whittaker 1960). These studies show that neither dominance of single taxa nor presence and abundance of groups of species change abruptly along an environmental gradient; nodes do not exist (Figure 8-2).

It is now apparent that the method used to sample vegetation will determine whether associations appear as distinct units or as arbitrary segments along a continuum. If sampling is subjective, searching for stands that show the presence of differentiating species, then the discrete view will be supported, for only similar stands will be sampled, and intermediate stands will be ignored. It is possible that if only climax stands are sampled, many successional stands will also be ignored. If sampling is done less subjectively, ecotonal stands will be included, and all the stands will be seen to lie along a continuum. Exceptions to the continuum view do exist, of course, where there are sudden discontinuities in the environment, such as a change in soil parent material, a change in elevation or slope aspect, fire history, or the presence of landslide rubble.

The individualistic nature of species distribution is indirect evidence that communities are not more than the sum of their parts. It suggests that the level of interactions and interdependence is relatively low, or at least nonspecific. For example, forest floor herbs may be well adapted to the phenology, shade cast, and leaf chemistry of an overstory canopy, but the exact species composition of that overstory is not critical. This was shown in a large-scale, natural "experiment" 50 years ago. Between 1906 and 1930, virtually all of the chestnuts (*Castanea dentata*) in the eastern deciduous forest were killed by chestnut blight. Woods and Shanks (1959) and McCormick and Platt (1980) showed that the canopy openings were largely filled by crown expansion or sapling growth of previously associated species (mainly oaks and red maple), and there was no additional major community upheaval. What had been an oak-chestnut community became an oak community, but the understory was little changed. Studies of American elm (*Ulmus americana*) forests struck by Dutch elm disease reveal similar modest adjustments (Parker and Leopold 1983).

Studies of understory forest herbs beneath fir forests of Montana (McCune and Antos 1981), deciduous forests of the Lake states (Rogers 1981), and hemlock

(a)

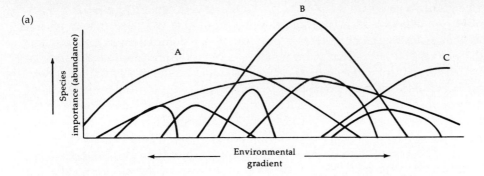

(b)

(c)

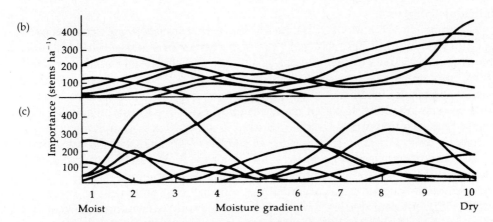

Figure 8-2   Idealized patterns (a) of species importance/abundance along an environmental gradient as predicted by the continuum view. Actual data: density of stems per hectare for all tree species encountered in the Siskiyou Mountains of Oregon (b) and the Santa Catalina Mountains of Arizona (c). Reprinted with permission of Macmillan Publishing Co., from *Communities and Ecosystems* by R. H. Whittaker. Copyright 1975 by R. H. Whittaker.

forests of eastern North America (Rogers 1980) all demonstrate that understory changes are poorly correlated with overstory composition and, further, that understory species change in abundance in an individualistic way. An elegant, two-year-long field manipulation study of tundra plants in the Brooks Range of Alaska (Chapin 1985) similarly showed that 21 associated species were individualistically affected by such factors as temperature, soil nutrients, shade, competitive interactions, combinations of all these, and evolutionary history.

A Modern Synthesis

The present American concept of the community is a synthesis of Clements's association unit hypothesis with Gleason's continuum-individualistic hypothesis, together with a recognition that there may be several different kinds of associations, each explainable according to a different model. We have come to understand that the classical theories are inadequate for completely explaining and predicting vegetation patterns, and we have come to appreciate that the scale of complexity in nature makes the existence of a single model highly unlikely. New pieces of evidence continue to be put forward that explain some communities but not others; these approaches have included an emphasis on competition as an organizing force (Tilman 1982), the role of episodic disturbance and chance recolonization (Shugart 1984), population dynamics (Bazzaz 1996), and the selective influence of herbivores (see Chapter 7). A review of community concepts by Noy-Meir and van der Maarel (1987) elegantly captured major differences among existing theories, then concluded that definitive, convincing tests of them will be difficult to construct.

One aspect of the modern synthesis view is that interdependence does exist among some, but not necessarily all, member species of a community. Previous chapters have included examples of positive interactions such as natural root grafts, nurse plants, and mutualistic relationships that have coevolved over long periods; that is, we are aware of many interdependences between pairs of species. On a larger scale, it is clear that many communities have **keystone species** that have interdependent relationships with many associated species, not just one. As defined by Mary Power and others (1996), "a keystone species is one whose impact on its community or ecosystem is large, and disproportionately large relative to its abundance." Keystone species are less abundant than dominant species, yet their organizational control of the community may be as important or more important than that of dominant species. Power and her large group of coauthors suggest that one could conceivably quantify the importance of such a species as "the change in a community or ecosystem trait per unit change in the abundance of the species."

The keystone species concept was first defined and described by the animal ecologist Paine (1966 and 1969), and typically keystone species are animals near the top of the food chain. Nevertheless, plants may be keystone species if they are an important food resource, fix nitrogen, modify soil chemistry through litter qualities, are a generalist parasite, or are an aggressive exotic woody or herbaceous weed.

We will return to concepts of the plant community in Chapter 11, when we discuss plant succession. It turns out that models of succession are tightly linked to models of the plant community.

Associations can be recognized and in some way defined in the field, but we must appreciate their subjective, arbitrary boundaries, the nature of which depends on the bias of the investigator and the sampling methods. Whether associations are real or not, stands of vegetation certainly do exist. These stands exhibit collective or emergent attributes beyond those of the individual populations that make them up. Some of these attributes (Table 8-1) will be examined in the rest of this chapter.

Table 8-1   Some characteristics of plant communities.

Physiognomy
    Architecture
    Life forms
    Cover, leaf area index (LAI)
    Phenology
Species composition
    Characteristic species
    Accidental and ubiquitous species
    Relative importance (cover,
        density, etc.)
Species patterns
    Spatial
    Niche breadth and overlap
Species diversity
    Richness
    Evenness
    Diversity (within stands and
        between stands)
	Nutrient cycling
	    Nutrient demand
	    Storage capacity
	    Rate of nutrient return to the soil
	    Nutrient retention efficiency of
	        the nutrient cycles
	Change or development over time
	    Succession
	    Stability
	    Response to climatic change
	    Evolution (?)
	Productivity
	    Biomass
	    Annual net productivity
	    Efficiency of net productivity
	    Allocation of net production
	**Creation of, and control over,
	    a microenvironment**

Some Community Attributes

Physiognomy, Species Composition, and Spatial Patterns

Physiognomy is a combination of the external appearance of vegetation, its vertical structure (its architecture or biomass structure), and the growth forms of its dominant taxa. Chapter 9 discusses methods of sampling, measuring, and numerically describing the physiognomy of communities. Physiognomy is an emergent trait of communities. It could not, for example, be accurately estimated just from a list of all the taxa present within the community.

Life form (also called growth form, see Chapter 1) includes such plant features as size, life span, degree of woodiness, degree of independence, general morphology, leaf traits, the location of perennating buds, and phenology. Vertical structure refers to the height and canopy coverage of each layer within the community.

Canopy coverage can be expressed as the percentage of ground covered by the canopy, when the edges of the canopy are mentally projected down to the surface. Canopy cover gives some indication of the amount of ground that is shaded by canopy, but it does not convey the deepness of that shade. Two forests may each have 100% canopy cover, but one canopy may reduce solar radiation only slightly while the other reduces it considerably. If one forest is composed of trees with an architecture that spreads leaves thinly through the canopy, the fraction of sunlight intercepted will be modest. If the other forest has trees that spread many overlapping layers of leaves through the canopy, the fraction of sunlight intercepted will be greater. To quantify the amount of leaf overlap, one must measure the community's **leaf area index (LAI).** All leaves are collected that project into a column of space that lies above an area of ground (say, 1 m^2), and their cumulative surface area is measured. Then,

$$\text{LAI} = \frac{\text{total leaf area, one surface only}}{\text{ground area}}$$

Many crops, such as corn, have an LAI of about 4, meaning that for every square meter of ground, 4 m^2 of leaves lie above it. Some examples of LAI for natural vegetation types are given in Table 8-5 on page 199.

The species composition of a community is also extremely important because communities are also defined on a floristic basis. Several communities may have similar physiognomies yet differ in the identity of dominants or other species. The abundance, importance, or dominance of each species can be expressed numerically, so that different communities can be mathematically compared on the basis of species similarities and differences (see Chapter 9).

The relative spatial arrangement of species within a community is another community trait. As described in earlier chapters, individuals within a species or of different species may be distributed at random with respect to each other, clumped (positive or neutral interactions), or overdispersed (negative interactions).

The importance of species interactions and interdependence to a discrete view of the community has already been discussed in this chapter, and that assumption will be touched on again in Chapter 11. One theory of community stability discussed in Chapter 11 suggests that stable, long-lasting communities exhibit more species interactions and more species than transient, seral communities. If this is so, then the niches of species in stable communities must be narrower, with less overlap in niche boundaries, than those of less stable communities. Some ecologists have tried to measure niche breadth and overlap to test such hypotheses (Huey and Pianka 1977, Parrish and Bazzaz 1976, Pickett and Bazzaz 1976), but the data are so sparse that generalities cannot yet be made.

Part of a species' niche may include its unique metabolism. For example, flowering plants exhibit three different metabolic pathways of photosynthesis, C$_3$, C$_4$, and CAM, each of which is best fitted to different environmental conditions or to different species' phenologies (see Chapter 14). There have been attempts to include such metabolic traits in desert and chaparral community descriptions (Mooney and Dunn 1970b, Johnson 1976).

Community descriptions based on physiognomy, life form, niche overlap, and other functional traits (as opposed to taxonomic traits such as the identity of species in the communities) are useful because they permit comparison of widely disjunct stands that have little or no floristic similarity. These comparisons often show a convergence of vegetation types, given a similar macroenvironment. Perhaps the most comprehensive recent scheme for classifying plants into functional types (PFTs) is that of Elgene Box (1996). He combines structural traits (e.g., tree, palm-like, shrub, forb, grass), phenology (evergreen, summergreen, raingreen, wintergreen), and leaf traits (hardness, shade tolerance) to generate 15 major plant types sufficient to describe all the world's main terrestrial vegetation types (Table 8-2). Sandra Diaz (1998) has attempted to do the same with 18 traits.

Biodiversity

Biodiversity is a term so widely used that it has lost its original, technical meaning. In the popular press and even within the general conservation literature, "biodiversity" (often shortened to "diversity") means the number of something present in an area. Most often **biodiversity** is the number of species in a given area, but sometimes it is the number of habitats or communities or ecosystems. Conservation efforts are often aimed at maintaining some degree of biodiversity in the face of development threats, pollutants, or natural resource extraction. The success of conservation management is also often measured by the degree to which biodiversity declined over time.

More technically, diversity of species is called **species richness.** It is simply the number of species per unit area. Different communities can be compared in their species richness, but only when the procedure of counting has been standardized. For example, if one spent 30 minutes counting species in community A and three hours counting species in community B, should the data be considered equally accurate and will the comparison be fair? To remove this bias, **equal effort sampling** (also called species effort sampling) can be a standardizing method, whereby the same amount of time is spent counting species in each community to be compared. Another method of standardizing is to count a preselected number of individuals and then to express species richness as a ratio of species encountered per individuals sampled; the ratio is called a **species proportion** (Hayek and Buzas 1997, Krebs 1989, Ludwig and Reynolds 1988). Empirically, field-workers have determined that 200–500 individuals must be encountered to obtain a reasonable species proportion.

An incidental benefit of measuring species richness by species proportion is that the data can be used to predict the number of species in either a smaller area or a larger area. Several formulas have been developed for predicting species number in a smaller area (a procedure called **rarefaction**) and for predicting species number in a larger area (a procedure called **abundification**), but they should be used with caution (Hayek and Buzas 1997, Krebs 1989).

One other approach to standardizing counts of species richness emphasizes the relationship of increasing species number with increasing sample area. This technique was developed by Robert Whittaker (Shmida 1984), and it is described in Chapter 9.

Each species in a community is not likely to have the same number of individuals. One species may be represented by 1000 plants, another by 200, and a third by only a single plant. The distribution of individuals among the species is called **species evenness,** or species equitability. Evenness is maximum when all species have the same number of individuals. **Species diversity**, technically, is a combination of richness and evenness; it is species richness weighted by species evenness, and there are formulas that permit the diversity of a community to be expressed in a single index number.

It is important to emphasize that richness and diversity are quite different. Although richness and diversity are often positively correlated, environmental gradients do exist along which a decrease in richness is accompanied by an increase in diversity (Hurlbert 1971). Community A, with five species but uneven numbers

Table 8-2   Major plant functional types (PFTs). Determinate branching = long laterals; monopodial branching = short laterals; malacophyll = soft-textured, light-demanding leaf; coriaceous = leathery, light-demanding leaf; sclerophyll = hard or brittle, light-demanding leaf; laurophyll = shade-tolerant leaf of any texture; sarcophyll = succulent. Modified from Box (1996).

Plant functional type	Vegetation	Traits
1. Tropical evergreen broad-leaved trees	Tropical rain forest	Tall, determinate branching, laurophyll
2. Tropical deciduous broad-leaved trees	Raingreen forest, woodland, and scrub	Various, determinate, malacophyll or laurophyll
3. Extra-tropical evergreen broad-leaved trees	Evergreen broad-leaved forest, temperate rain forest	Tall, determinate, malacophyll or laurophyll
4. Temperate deciduous broad-leaved trees	Summergreen broad-leaved forest and woodland	Various, determinate, winter-dormant, malacophyll
5. Temperate/boreal needle-leaved evergreen trees	Needle-leaved forest and woodland	Various, monopodial, winter-dormant, coriaceous
6. Boreal/cool temperate deciduous needle-leaved trees	Deciduous boreal needle-leaved forest and woodlan	Various, monopodial, winter-dormant, malacophyll
7. Sclerophyllous trees	Subhumid woodland and scrub	Short, determinate, evergreen, sclerophyll, light-demanding
8. Sclerophyll/coriaceous shrubs and subshrubs	Scrubland (chaparral, semidesert, heath)	Short, basally determinate, evergreen, sclerophyll or coriaceous, light-demanding
9. Deciduous shrubs and dwarf shrubs	Scrubland, kurmmholz, semidesert	Short, basally determinate, seasonally dormant, malacophyll or coriaceous
10. Short-season broad-leaved dwarf shrubs	Tundra	Short, basally ramifying, winter-dormant, evergreen or deciduous, malacophyll or coriaceous or sclerophyll
11. Diurnally active tufted arborescents and forbs	Tropical alpine	Various, monopodial rosettes, evergreen, coriaceous or sclerophyll
12. Grasses and graminoids	Grassland	Short herbs, usually perennial, malacophyll
13. Stress-tolerant succulents	Semidesert scrub	Various, perennial, evergreen, sarcophyll
14. Ephemeral herbs	Semidesert scrub	Short, annual to perennial but short growing season, malacophyll
15. Stress-tolerant mosses and lichens	Tundra, cold desert	Short, winter-dormant

of individuals in each species, has a lower diversity than community B, with four species that have a very similar number of individuals in each. Community A has a higher species richness, however.

Many simplifications are made in every calculation of species diversity (Peet 1974). In the equations, all individuals of any species are equal. This may not be true, especially in regard to animals of different sex or of different developmental stages, or to plants of different phenological stages (dormant, full-leaf, flowering, juvenile, senescent adult), different sizes (seedling, sapling, suppressed adult, over-story adult), or different ecotypes. For this reason, diversity is often calculated for each stratum in a community rather than for the entire community. It is also assumed that all species are equally different, whether the difference is in morphology or niche breadth. This assumption is generally made even when it is unlikely to be valid, simply because we have no way to quantify species differences. Finally, diversity is sometimes expressed as numbers of individuals, but other times as biomass, productivity (see Whittaker 1975), canopy cover, or in other ways. Obviously, plant diversity from community to community can be compared only if the same units are used. Some units are more appropriate to plants, and some to animals; possibly there is no one unit suitable for calculation of an ecosystem-wide diversity index (Hurlbert 1971).

A further assumption, and one that is made in any statistical procedure, is that the sample of the community was large enough to represent the community adequately. The data on numbers of species and the relative abundance of each are dependent on sample size. Given a large enough sample, there will be some species with few individuals, some species with many, and many species with an intermediate number. The data will form a bell-shaped curve on a log-normal plot (Figure 8-3). A small sample, however, will probably not include the full spectrum of rare species. Notice, in Figure 8-4, how the relative abundance of butterfly species shifted and began to approach a bell-shaped curve as the sample collecting periods increased from less than 1 year to 4 years. The truncated appearance of small sample size curves is an artifact, then, and the truncated end is a "veil line" that hides additional categories of rarity (Preston 1948).

Diversity Indices   Several indices have been proposed over the past six decades (Auclair and Goff 1971, Hill 1973, Peet 1974, DeJong 1975, Magurran 1988; see Table 8-3), but we shall emphasize only two here: **Simpson's index** (Simpson 1949) and **Shannon-Wiener's index** (Shannon and Weaver 1949; sometimes erroneously called the *Shannon-Weaver index*).

Simpson's index reflects dominance because it weights (is more sensitive to) the most abundant species more heavily than the rare species. The advantage of this attribute is that index values are unlikely to vary much from sample to sample, because it is the rare species that would vary from place to place more than the common ones. Its formula is

$$C = \sum_{i=1}^{s} (p_i)^2$$

Figure 8-3 (a) The log-normal distribution model of species in a community. Species are ranked according to the number of individuals in each. Note that the horizontal axis is on a log scale. Most species have a moderate number of individuals, about 1–32, in this idealized example. From "The commonness and rarity of species" by F. W. Preston, *Ecology* 29:254–283. Copyright © 1948 by the Ecological Society of America. Reprinted by permission. (b) A large sample of British birds, which approaches Preston's log-normal curve. From Williams 1964 (*Patterns in the Balance of Nature*, by permission of Academic Press) and Krebs 1972 (*Ecology: The Experimental Analysis of Distribution and Abundance*, Harper and Row, Publishers, New York).

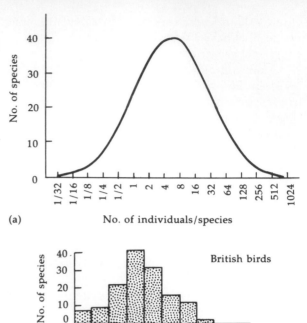

(a)

(b)

Figure 8-4 Log-normal distributions of butterfly species abundances as determined from sampling at Rothamsted Station, England, for periods of (a) 1/8 year, (b) 1 year, and (c) 4 years. As sample size increased, the distribution approached Preston's idealized log-normal curve. From Williams 1964 (*Patterns in the Balance of Nature,* by permission of Academic Press) and Krebs 1972 (*Ecology: The Experimental Analysis of Distribution and Abundance,* Harper and Row, Publishers, New York).

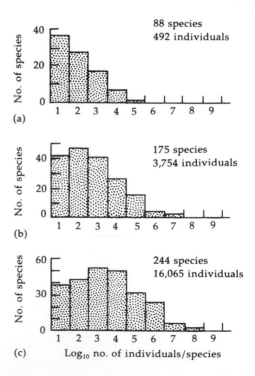

(a)

(b)

(c)

Table 8-3 Some dominance and diversity indices. For additional details, see Auclair and Goff (1971). N_o = total number of species in the sample; X_o = total number of individuals in the sample; X_i = number of individuals in species i; ln = natural logarithm.

Index name or description	Formula
Species per unit area	N_o
Species/(individuals/species)	$N_o/(X_o/N_o)$
Species/individuals	N_o/X_o
Species/ln individuals	$N_o/\ln X_o$
ln species/ln individuals	$\ln N_o/\ln X_o$
Species/sq. rt. individuals	$N_o/\sqrt{X_o}$
Simpson dominance index	$\Sigma(X_i/X_o)^2$
Shannon-Wiener information diversity	$-\Sigma(X_i/X_o)\ln(X_i/X_o)$
McIntosh density diversity	$1 - \sqrt{\Sigma(X_i/X_o)^2}$

where C is the index number, s is the total number of species in the sample, and p_i is the proportion of all individuals in the sample that belong to species i. Common variations, which measure diversity rather than dominance, include

$$D = 1 - C$$

$$D' = \frac{1}{C}$$

The Shannon-Wiener index formula is

$$H' = - \sum_{i=1}^{s} (p_i)(\ln p_i)$$

and a common variation is

$$H'' = 2^{H'}$$

If proportions are expressed on base 10 ($\log p_i$) instead of 2 ($\ln p_i$), one can simply convert the summation value to base 2 by multiplying it by 3.32.

H' is thought to represent the "uncertainty" or "information" of a community. The more variable its composition, the more variable (the more uncertain or unpredictable) each sample of it would be. The index units have been called *bits, bels, decits,* or *digits* per individual, referring to the binomial units of information used in computers. H' varies from 0, for a community of one species only, to values of 7 or more in rich forests such as those of the Siskiyou Mountains of Oregon and California (DeJong 1975). The deciduous forest of the eastern United States is of intermediate diversity, with H' values for tree species ranging only from >3.0 for the mixed mesophytic forest in the Cumberland and Allegheny Mountains to values <2.0 at the western and northern edges (Monk 1966).

Because the formulae are different, the absolute values of Simpson's and Shannon-Wiener's indices will differ for the same community. Table 8-4 and Figure 8-5 compare variations of the indices for two herb communities.

What Does Diversity Signify?   A lot of attention has been paid to species diversity in the ecological literature: first, in the development and comparison of several formulae; second, in searching for general trends in diversity along environmental gradients; and third, in trying to attribute functional explanations for diversity gradients and some ecological value that diversity might convey about communities. Despite all this attention, we think it best to treat an index of diversity as simply one descriptive attribute of a community, on a par (for example) with a list of all species found in the community, or an estimate of tons of biomass above the ground, or the leaf area index.

Generally, there is a gradient of increasing species diversity (and richness) from the poles to the equator, and from high elevations to low elevations. These gradients follow complex environmental gradients of increasing warmth, among other factors. It has been stated that diversity increases as any particular stress lessens (Krebs 1972), but this is not true for all stress gradients. An aridity stress gradient may show the opposite trend, with greater diversity in semiarid grassland and desert than in savanna, woodland, or forest (Hurlbert 1971). Diversity has been equated with productivity and stability, but some very diverse semiarid grasslands and deserts have low productivity and stability, and there are many other exceptions (van Dobben and Lowe-McConnell 1972).

Large-scale, long-term field experiments in southern England (Naeem et al. 1994 and 1995) and in Minnesota (Tilman and Downing 1994, Tilman 1996) have supposedly shown that increasing species richness or diversity increases the temporal stability of communities, as measured by productivity and biomass. Subsequent analyses by Huston (1997) and Grime (1997), however, concluded that the

Table 8-4   Diversity indices calculated for community samples A and B in Figure 8-5. Note that only the Simpson index (C) shows community A to be the more diverse, reflecting that index's high sensitivity to dominance by one species and insensitivity to low numbers of other species.

Index	A	B
Simpson (C)	0.38	0.20
$D = 1 - C$	0.62	0.80
$D' = 1/C$	2.66	4.92
Shannon-Wiener (H')	1.78	2.31
$H'' = 2^{H'}$	3.43	4.96

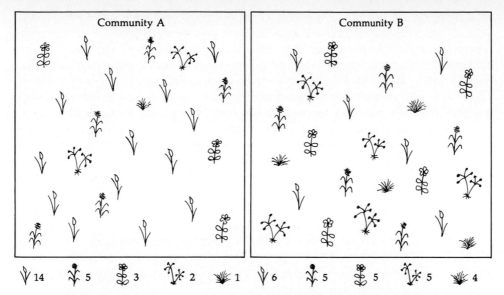

Figure 8-5 Diagrammatic representation of sample plots from two different herbaceous communities, each with the same five species but in different abundances. Community A is heavily dominated by one species, whereas community B has a more even distribution of species abundances. The number of individuals of each species is summarized below the diagrams. Diversity indices are shown in Table 8-4. Normally, diversity increases as evenness increases. Courtesy of M. Rejmanek, University of California, Davis.

experimental designs were not appropriate to reach such conclusions. Huston and Grime concluded that the identity of the species was as important to biomass and productivity as the number of species, because all the species were not uniform in their biology and behavior. It was not possible to distinguish the effect of the number of species from the effect of particular species.

Diversity has been taken as a reflection of the many interactions that supposedly characterize complex communities. However, one model of community structure that assumes no species interactions shows that species diversity "may be maintained in spite of, rather than because of, such interactions" (Caswell 1976). It is fair to say that the significance of species diversity is not well understood at this time.

Models of Species Diversity Some ecologists have asked the question, "Are there any fixed rules or patterns in nature that regulate the number of common, intermediate, and rare species packed into any one community?" The late Robert Whittaker, in his 1975 plant ecology textbook, summarized three theoretical answers to that question.

(1) In 1957, the mathematical ecologist Robert MacArthur proposed that the number of rare, intermediate, and common species was random. His **broken stick**

model suggested that we let a long stick represent all the species and all the individuals in a community. If we break the stick into random pieces, let each piece represent a different species, and let each piece's length represent the number of individuals in that species, we wind up with a random number of common, intermediate, and rare species. Publication of his model stimulated fieldwork to test it. The results suggest that few communities in nature have a random number of rare, intermediate, and common species. We can reject the broken stick model.

(2) A **geometric model,** not closely connected to any one researcher, proposes that each community is populated by a series of increasingly rare species, the abundance of each declining in a geometric, rather than arithmetic, progression. This model predicts that most species in a community are uncommon, and it does correspond to certain communities in severe environments, such as boreal forest.

(3) Another mathematical ecologist, Robert Preston, proposed a **log-normal model** in 1948 (refer back to Figures 8-3 and 8-4). According to this model, there are few rare and few common species, but many intermediate species. This model does conform to natural communities, providing that sampling has been intensive enough to include the rarest species. The log-normal model remains our best empirical description of species diversity.

These three models presume that communities being sampled have been undisturbed for long enough periods to achieve equilibrium composition. Connell (1978), Pickett (1980), and Roberts and Gilliam (1995a), however, have reminded us that it is more realistic to expect periodic random disturbance, which will prevent a community from ever reaching equilibrium. Huston (1979) proposed that at some intermediate frequency of disturbance, species diversity will be greater than at equilibrium (Figure 8-6a). This is because some r-selected taxa will still be present in locally favorable microsites, as well as some representatives of equilibrium K-selected taxa. If disturbance becomes more frequent, we lose all K-selected taxa, and if disturbance becomes too infrequent, we lose all r-selected taxa. Huston's explanation of natural diversity has since been called the **intermediate disturbance hypothesis.**

Maintenance of high diversity appears to require episodic, random (**stochastic**) disturbance. Very stable, regionally extensive, and homogeneous communities exhibit lower species diversity than communities composed of a mosaic of patches disturbed at various times in the past by wind throw, fire, disease, etc. Following disturbance, diversity increases with time up to a point where dominance by a few long-lived, large-sized species reverses the trend, and diversity falls thereafter. We can see from a computer simulation (Figure 8-6b,c) of spruce forest in the Krkonoše Mountains of the Czech Republic that episodic avalanche disturbance maintains three diverse life forms and groups of species indefinitely, whereas absence of disturbance leads to loss of two-thirds of the groups within 40 years.

Nutrient Cycles and Allocation Patterns

Sixteen elements are known to be required for normal growth and development of all higher plants: carbon, hydrogen, oxygen, phosphorus, potassium, nitrogen, sulfur, calcium, magnesium, iron, boron, manganese, copper, zinc, chlorine, and molybdenum. A few other elements are required by certain plant groups,

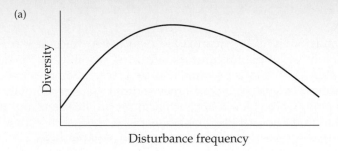

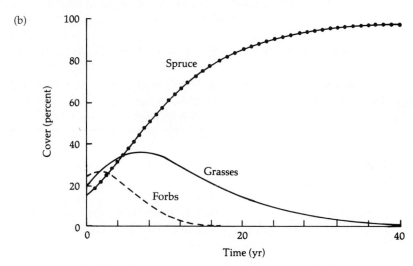

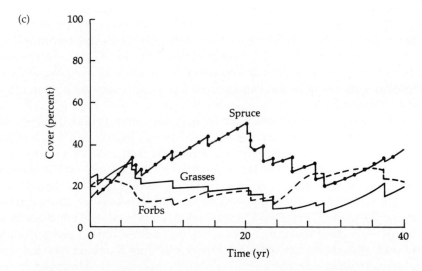

Figure 8-6 The intermediate disturbance hypothesis. (a) Idealized relationship between species diversity and frequency of disturbance. From Roberts and Gilliam (1995a). (b) Computer simulation of canopy cover over time following a single, major disturbance at time 0 and a subsequent period of 40 years without disturbance. (c) Same vegetation, but disturbed by a small-scale avalanche every fifth year. (b) and (c) Courtesy of M. Rejmanek, University of California, Davis.

but not by most plants. Sodium, for example, is required in trace amounts by plants with the C_4 pathway of photosynthesis; silicon is required by horsetails (*Equisetum* spp.; Epstein 1972) and is accumulated by grasses. Some elements— heavy metals such as lead or gold, and sometimes sodium—are accumulated by certain plants to the point where their foliage is poisonous to livestock, but these elements are not required for normal plant metabolism.

Communities differ in their utilization of certain essential nutrients (that is, how much of each element is required for normal growth, or at least how much is absorbed from the soil solution and translocated to leaves and growing points). They also differ in the rate at which the nutrients are returned to the soil in litter fall and in the efficiency of the plant-soil-plant cycle. Early successional communities, for example, may require little soil nitrogen, accumulate very little of any nutrient in their tissues, and return nutrients rapidly to the soil, but erosion removes a large fraction of the returned nutrients because of their low cover or seasonal absence. Climax communities may require greater quantities of some nutrients, store large quantities of nutrients in biomass, and return only a small fraction to the soil in leaf litter and decaying roots, but prevent erosive losses by shielding the soil with a permanent, closed canopy or a well-developed root system (see Chapter 11). Thus, climax communities have fewer leaks in nutrient cycles and more efficiently hold the nutrients in the plant-soil-plant cycle.

A few remarks about the carbon cycle will illustrate major community differences. If vegetation types are ranked according to the amount of **standing** (aboveground) **biomass** per hectare, they range from desert communities of only 100 kg ha^{-1} to tropical rain forests of 500,000 kg ha^{-1} (Figure 8-7). Generally, greater biomass indicates greater leaf area, which means that more radiant energy can be trapped each year and a faster growth rate will result, with greater net productivity. **Net productivity** ranges from 10 kg ha^{-1} yr^{-1} for desert communities to 40,000 kg ha^{-1} yr^{-1} for tidal zone, mangrove, marsh, and swamp communities (Figure 8-7 and see also Chapter 12). **Efficiency,** the fraction of radiant energy converted into kilocalories of tissue, ranges from 0.04% in deserts to 1.5% in tropical rain forests (Table 8-5).

The low efficiency and net productivity of desert communities does not mean that the component species are themselves inefficient or incapable of attaining high photosynthetic rates. Net productivity is a *community* trait and to some extent a climatic one, for it is affected by the leaf area index, temperatures during the growing season, distribution of rainfall, soil moisture storage, and the length of the growing season. If net productivity or efficiency is expressed on a 12-month basis, it is clear that the tropical rain forest, with a leaf area index of 10–11 and warm temperatures and adequate moisture all year, should indeed have a greater net productivity and efficiency than a desert, with a leaf area index of 1 or less and a growing season (based on water supply) of only a few months. However, the photosynthetic rates of individual desert shrubs and shrubs of mesic forests are very similar (Barbour 1973b).

The **allocation** of net productivity to different organs also differs from community to community. Grassland communities channel much of their energy to belowground biomass, scrub communities less so, and forests even less so. If we

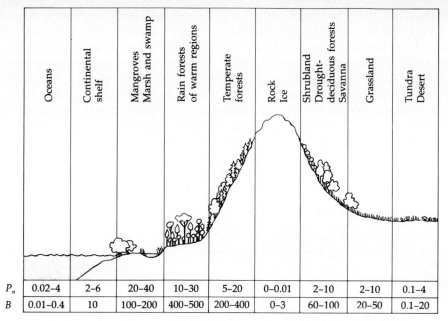

	Oceans	Continental shelf	Mangroves Marsh and swamp	Rain forests of warm regions	Temperate forests	Rock Ice	Shrubland Drought-deciduous forests Savanna	Grassland	Tundra Desert
P_n	0.02–4	2–6	20–40	10–30	5–20	0–0.01	2–10	2–10	0.1–4
B	0.01–0.4	10	100–200	400–500	200–400	0–3	60–100	20–50	0.1–20

Figure 8-7 Net primary productivity (P_n) and standing biomass (B) of various vegetation types. Data are in metric tons per hectare for B and metric tons per hectare per year for P_n. From W. Larcher 1995. *Physiological Plant Ecology.* By permission of Springer-Verlag.

Table 8-5 The leaf area index (LAI, square meters of leaf surface per square meter of ground) and efficiency of net productivity for major vegetation types. Efficiency is calculated by converting grams of net production per unit area to kilocalories per unit area (there are approximately 4.2 kcal in each gram dry weight of plant tissue), then dividing by the total radiation received in a year. Only radiation between 400 and 700 nm is entered into the calculation because these are the only wavelengths usable for photosynthesis. From W. Larcher 1995. *Physiological Plant Ecology.* By permission of Springer-Verlag.

Vegetation type	LAI	Efficiency (%)
Tropical rain forest	10–11	1.50
Deciduous forest	5–8	1.00
Boreal conifer forest	9–11 (7–38)	0.75
Grassland	5–8	0.50
Tundra	1–2	0.25
Semiarid desert	1	0.04
Agricultural	3–5	0.60

compare the distribution of carbon in a subalpine conifer forest, a temperate broadleaf forest, and a tropical rain forest (Figure 8-8), there are significant differences. Most carbon in the subalpine forest is in the form of humus atop or in the surface soil. This is because the acidic, cold soil is not favorable to decomposers. Litter half-life is 10 or more years (Whittaker 1975). Most carbon in the tropical rain forest, in contrast, is locked up as inert wood. This means that leaf litter has a major function in the tropical rain forest, for it represents the only mobile, cyclable part of the nutrient bank. This is one reason why tropical soils cleared of forest vegetation soon become infertile. The annual litter rain has been stopped, and the limited soil reserves are soon depleted by crops or leached from the soil. The litter decay rate is very rapid beneath the rain forest canopy; litter half-life is a fraction of a year. Nutrient and productivity relationships of plant communities are discussed further in Chapters 12 and 13.

Change Over Time

1–500 Years: Succession All communities are dynamic, changing entities. Change, however, is a relative term, and the time framework must be stated.

Plant communities that exhibit no directional change for several centuries are considered to be in equilibrium with their environment, and are called **climax** communities (see Chapter 11). Other communities may exhibit significant changes in such a time period. Some species decline in abundance and may disappear from the site; invasive species may increase in abundance; and the vegetation type itself may change, for example, from a meadow to a forest, or from a pine forest to a hardwood forest. Such transient communities are called **successional,** or **seral,** communities. It may be possible to recognize and describe an entire sequence of successional communities that replace each other on one site, finally culminating in a climax community. This sequence of communities is called a **succession** or a **sere.**

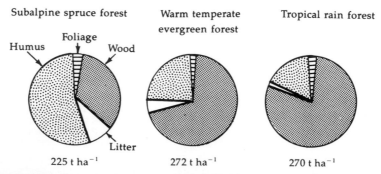

Figure 8-8   Content of organic carbon in the foliage, wood, litter, and humus for three forest types. From W. Larcher 1995. *Physiological Plant Ecology.* By permission of Springer-Verlag.

Stability   Communities differ in their response to disturbance or stress. Intuitively, one can use terms such as *stable* and *fragile* to describe how easily communities are perturbed by stress. More technically, however, *stability* is a complex term that includes several distinctly different qualities (Leps et al. 1982, Pimm 1984).

One component of stability is **resistance,** which is the ability of a community to remain unchanged during a period of stress. Resistance appears to be characteristic of vegetation dominated by long-lived perennials with moderately high species diversity and many paths of interdependence (links or connections) between the component species. Climax communities generally fit this definition. Such communities have considerable inertia, changing slowly even in the face of regional climatic change. Evidence from the fossil record, for example, indicates that vegetation change lagged 1000–3000 years behind a major warming at the Pleistocene-Holocene boundary 13,000–16,000 years ago (Cole 1985).

Resilience is a second component of stability; this is the ability of a community to return to normal, or the rate at which this occurs, following a period of stress or disturbance. Resilience appears to be characteristic of vegetation dominated by short-lived, rapidly maturing species of low diversity and with few interdependence links among them. Early seral communities fit this definition. Climax communities require long times to recover from destructive disturbance—thus they have lower resilience—but they have greater resistance to stress.

A third component is **persistence,** which is the ability to remain relatively unchanged over time. Some persistent communities are neither resistant nor resilient but owe their continued existence to a protected, buffered environment. The redwood forest along the northern California coast is a good example; it exists in a very narrow, foggy, temperate belt.

Thousands of Years: Climatic Change   Succession is thought to be driven by biological interactions, such as competition, which occur in the microenvironment created by the plants themselves; it is not driven by macroclimatic change. Climate is assumed to be constant when a succession is investigated and described.

Climate, however, has not been a constant throughout time and it has always undergone significant fluctuations. Weather records, such as those taken by the U.S. Weather Bureau, only exist for the past 100 or so years. Nevertheless, even these records show statistically significant (though minor) changes. For example, rainfall has been declining and temperatures rising in the southwestern United States (Hastings and Turner 1965).

The climate before the nineteenth century can be inferred by several methods (see, for example, Flint 1957 and Strahler and Strahler 1974). One method, applicable to the southwestern United States, is the analysis of tree rings (**dendrochronology**). Climate of the past 8200 years has been estimated by examining sequences of growth ring widths in the wood of living and dead trunks of bristlecone pine (*Pinus longaeva = P. aristata*). The data indicate cycles of warm, arid periods followed by cool, wet periods (Figure 8-9; LaMarche 1974, Ferguson 1968). The period of time shown in Figure 8-9 begins with an exceptionally arid period called the **Xerothermic period,** which started approximately 7000 years ago and ended approximately 5000 years ago. This period is thought to be responsible for the present distribution limits of many southwestern vegetation types and communities (Axelrod 1977). See Figure 8-10 for average temperatures during geologic times.

Relatively small changes in global temperatures have created striking changes in climate and vegetation. Most careful analyses suggest an average global surface temperature difference between full glacial eras and the present of only 4–6°C (Bryson 1974). This temperature difference, however, has been correlated with changes in cloudiness, rainfall, length of the frost-free season, and other factors that have a large impact on vegetation (Figure 8-11).

The nature of vegetational change since the Ice Age can be documented in several other ways. In the arid southwestern United States, the dried remains of plants cached in underground middens by wood rats (*Neotoma*) have in some cases remained intact and identifiable for thousands of years. Since the foraging activity of wood rats is restricted to a rather limited radius around the nest, the composition of the plant material gives some indication of the nearby vegetation at the time the midden was formed. The plant material can be carbon dated. This kind of evidence permitted Wells and Berger (1967), among others, to determine the changing elevation of vegetation zones following the retreat of the last glaciation in what is now the Mojave Desert.

A more widely used method of documentation uses pollen grains, which accumulate at the bottom of slowly filling lakes or ponds. As pollen is shed and transported by wind, some falls onto a lake surface, sinks to the bottom, and becomes incorporated with silt and organic matter into sediment. Pollen of many species is resistant to decay in the anaerobic, cold sediment and may remain intact for thou-

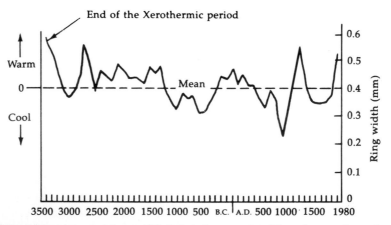

Figure 8-9   Average ring width in bristlecone pine (*Pinus longaeva*) trunks dated back to 3500 B.C., near the end of the Xerothermic period. Positive departures from mean ring width of 0.4 mm indicate temperatures warmer than average during April–October. From V. C. LaMarche, Jr. *Science* 18:1043–1048. Copyright © 1974 by the American Association for the Advancement of Science. By permission.

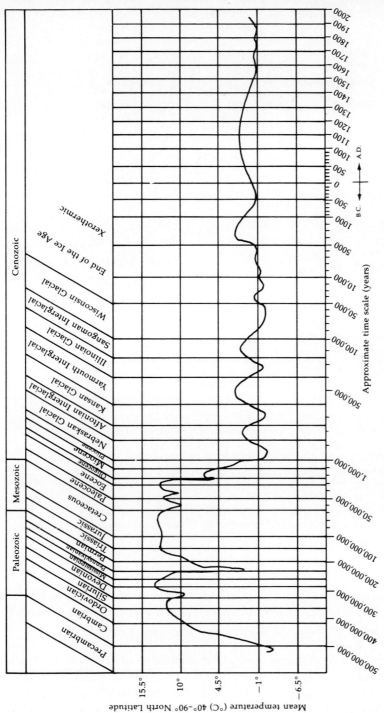

Figure 8.10 Average temperatures in north temperate and polar latitudes during geologic time. The time scale is distorted to show more detail for the last 1 million years. Notice the four glacial advances, separated by warm interglacial retreats, and also the Xerothermic period at about 5000–3000 yr B.C. From Dorf 1960. Reprinted by permission of *American Scientist*, Journal of Sigma Xi, the Scientific Research Society.

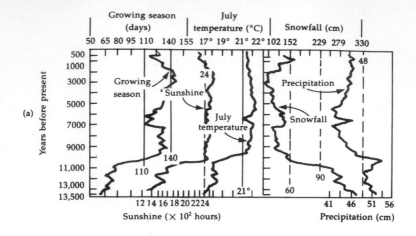

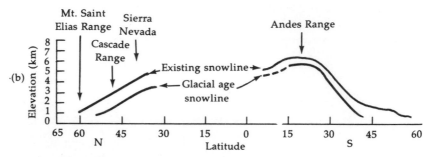

Figure 8-11   (a) A 13,500-year record of growing season length, hours of bright sunshine annually, mean July temperature, annual snowfall, and precipitation during the growing season at Kirchner Marsh, Minnesota. From R. A. Bryson. *Science* 18:753–759. Copyright © by the American Association for the Advancement of Science. By permission. (b) The effect of climate change over the past 13,000 years on the elevation of permanent snow (the upper limit of alpine vegetation) in western mountain ranges of the Americas. From Flint 1957. *Glacial and Pleistocene Geology,* copyright © 1957 John Wiley and Sons, reprinted by permission and Strahler and Strahler 1974. *Introduction to Environmental Science,* Hamilton Publishing Co., Santa Barbara, CA.

sands or millions of years. The family, genus, or even species of the pollen can be determined under the microscope. A core of sediment, then, reveals a chronological sequence of surrounding vegetation; the deeper the pollen occurs in the sediments, the older it is.

Within limits, ecologists assume that the abundance of pollen in the core is related to the abundance of species in the surrounding vegetation. This assumption, of course, is applied only to wind-pollinated plants such as sedges, grasses, most trees, many shrubs, and some forbs (herbaceous plants other than grasses). The assumption is upheld by measurements of modern pollen "rain." Griffin (1975), for example, found that modern pollen rain in Minnesota plant communities correlated very well with the nearest community type. Pollen is not carried in significant quantities further than 50 km in forested regions (Livingstone 1968). If we consider a pond surrounded by vegetation in a radius of 50 km, then the pond can receive pollen from a total area of 7850 km^2. Most vegetation is not uniform over such an area, but it is likely that the pollen profile will give a good general picture of regional vegetation.

A **pollen profile** constructed from sediment beneath a Nova Scotia lake is shown in Figure 8-12. The present vegetation consists of a deciduous forest on the hilltops with sugar maple (*Acer saccharum*), beech (*Fagus grandifolia*), and yellow birch (*Betula lutea*), and a coniferous forest in the valleys with balsam fir (*Abies balsamea*) and white spruce (*Picea glauca*). Figure 8-12 shows that this pattern of vegetation is relatively recent. Sediments at a depth of 5–6 m, corresponding to an age of about 9000 years before the present, reveal pollen mostly from herbs and shrubs characteristic of tundra vegetation near permanent ice. The nearest such vegetation today occurs 500 km to the north of the lake. Shallower depths, corresponding to an age of about 6000 years before the present, show peaks in spruce and fir, indicating the dominance of typical northern taiga forest, widespread today farther to the north. In yet shallower depths, the pollen of deciduous species (birch and maple) increases in abundance and the pollen of coniferous species declines, indicating a continual warming trend in climate.

Millions of Years: Evolutionary Change   **Microfossils,** such as pollen grains, are used to document vegetational changes over the course of thousands of years. **Macrofossils,** such as leaf impressions, are more useful to document changes over

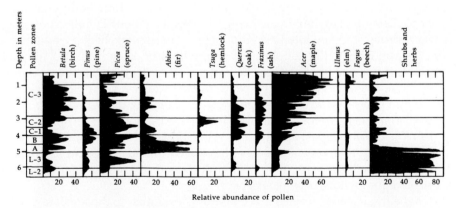

Figure 8-12   Pollen profile of sediment from a lake in Nova Scotia. The numbers along the horizontal axis refer to relative abundance of each type of pollen. Depth corresponds to age: layer L-2 has been carbon dated to 9000 yr B.C. and the top of layer C-3 represents the present. From "Some interstadial and postglacial pollen diagrams from eastern Canada" by D. A. Livingstone, *Ecological Monographs* 38:87–125. Copyright © 1968 by the Ecological Society of America. Reprinted by permission.

millions of years. As with pollen, it is assumed that the abundance of fossils represents the abundance of species in past vegetation. Care must be taken in interpreting site geology to determine whether plant material was deposited in place or carried by water for some distance and then deposited.

By examining general leaf shape and the pattern of leaf venation, paleoecologists have been able to identify fossil species and relate them to their nearest living species. The fossil species are usually extinct, but in many cases they are so close to living species that they can be written as, for example, *Pseudotsuga (menziesii)*, which means a fossil very closely related to the modern species of Douglas fir, *P. menziesii*. It is possible that fossil species were physiologically different from modern taxa, much as modern ecotypes of the same species differ from each other (Axelrod 1977). Nevertheless, the assumption is made that the present is the key to the past, and that modern relatives of fossil plants are in a climate similar to the climate that existed at the time and place the fossil material was deposited. In this way, past climates as well as past vegetation may be reconstructed.

One of the most complete records of fossil plants for western North America during the Cenozoic era (the past 65 million years) is in the John Day Basin of eastern Oregon (Chaney 1948). The record shows a cooling and drying trend. About 60 million years ago the prevailing community contained cinnamon, palms, figs, cycads, avocados, and tropical ferns, which are now found in cool mountain forests of Central America with an annual rainfall >1500 mm and no frost. The leaves of these plants were large, with entire margins. About 40 million years ago there was a change to a mixed conifer-hardwood forest with birch, alder, oak, dawn redwood, elm, sycamore, beech, maple, chestnut, sweet gum, and others. The leaves of these plants were smaller, with dentate or convoluted margins, indicating a drier climate, and some trees were deciduous. This exact mixture does not appear anywhere today, but close approximations exist along the cool, wet California coast and on the Cumberland Plateau of Tennessee. Annual rainfall was still high, about 1250 mm, but the climate had grown cooler. About 25 million years ago there was a strong shift to winter-deciduous trees, such as oak, hickory, and maple. This indicates a climate like modern Indiana, with 1000 mm of rainfall annually and prolonged freezing temperatures in winter. Today, the John Day Basin is dominated by sagebrush. Trees are absent except along waterways, and precipitation is about 250 mm per year, including some snow in a cold winter period.

Fossil assemblages for other localities in North America have permitted paleoecologists to reconstruct past climate and vegetation zones on a continental basis (Figure 8-13). In general, vegetation zones have been shifted south and compressed over the past 40 million years; that is, environmental and vegetational gradients from pole to equator have become steeper.

Since the fossil record reveals that communities similar to those existing today have a history extending back millions of years, it is reasonable to ask whether communities evolve. That question has not yet been resolved. Whittaker and Woodwell (1972) have suggested that evolution does occur at the community level, developing their argument as follows: (a) All species evolve in communities rather than in isolation; thus, (b) the evolution of a community occurs as a process of coevolution of the associated species, making the whole community an interactive assemblage; therefore, (c) communities must change in structure and func-

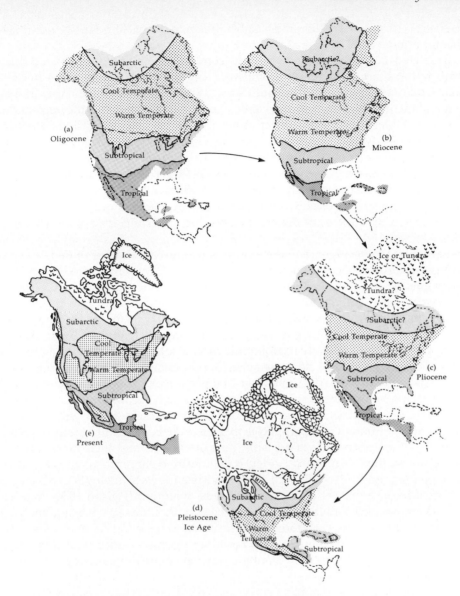

Figure 8-13 Climatic zones in North America during the past 40 million years. (a) Oligocene, about 35 million years ago; (b) Miocene, 20 million years ago; (c) Pliocene, 5 million years ago; (d) peak of the Pleistocene Ice Age, 60,000 years ago; and (e) present. From Dorf 1960. Reprinted by permission of *American Scientist*, journal of Sigma Xi, the Scientific Research Society.

tion from ancestral, simpler communities to modern, complex communities, as the component species become more interdependent; and finally, (d) communities have emergent characteristics corresponding to those of organisms, such as growth and maturity, structural differentiation, energy flow, material turnover, homeostasis (the tendency to remain stable or regain stability), adaptive optimization, and organization.

This view is supported by data on the convergence of widely separated communities that share a similar environment, such as chaparral vegetation in California, Chile, southern Australia, South Africa, and the Mediterranean region. The climate of the five regions is similar, and so is the physiognomy of the vegetation. "Similar," of course, is a very subjective term. An extensive International Biological Program study of Chilean and southern Californian chaparral did illustrate parallelisms, but comparative studies of all five chaparral areas have revealed significant differences in physiognomy, growth forms, biomass, phenology, species richness, and response to fire (Barbour and Minnich 1990). Environmental differences in climate, land use history, incidence of fire, and soil nutrient levels have also been identified, and some nonconvergent chaparral differences may be related to those factors.

Examples of convergence in the morphology and/or behavior of widely separated and unrelated species, such as cacti in the southwestern United States and similar-looking euphorbs in Africa, have long served as examples of natural selection at work. Can we accept community convergence in the same light? Is there an optimal solution for community structure, given a certain climate, and does natural selection drive community development toward that solution? So far, the answer appears to be no, because if communities were integrated units and a product of evolution, they would be closed systems like organisms, with sharp boundaries. In fact, however, as we have seen earlier in this chapter, groups of species do not parallel each other in their distribution curves, and when communities are recognized they are relatively arbitrary units, sharing broad ecotones with adjacent communities. Although pairs of species may have coevolved and become interdependent (pollinators and certain plants, parasitic plants and their hosts, mycorrhizal unions), there is no hard evidence that entire communities are integrated, interdependent units. Nevertheless, there have been several attempts to develop a theoretical basis or model of community-wide evolution (Wilson 1980, Aarssen and Turkington 1983, Salthe 1985). We can be sure that the topic has not yet been resolved. It is ironic that such recent models bring us back full circle to Clementsian ideas of the community as some kind of superorganism—ideas that stimulated the very beginnings of American plant ecology. The more things change, the more they stay the same.

Summary

The community concept is of general importance to synecology, just as the ecotype concept is central to autecology and the species concept is central to taxonomy. The precise nature of the community is ambiguous because of the biases of individual ecologists and their sampling methods. This does not make the concept useless, however; we must simply appreciate its subjectivity. For the purposes of classification, stands can be grouped into associations that have a fixed floristic composition, physiognomy, and habitat range.

The discrete view of associations assumes that they are closed systems, with interdependent species that synchronously peak in abundance and with narrow ecotones. The discrete view assumes that the whole is greater than the sum of its parts. The individualistic view assumes that associations are open systems, with independent species that happen to associate together wherever their range limits and the chance arrival of propagules overlap; consequently, associations are at most arbitrary units along a continuum.

Whether or not associations exist in the abstract sense, real stands do exist, and it is useful to consider attributes exhibited by communities beyond the attributes of the component species. These emergent attributes include physiognomy, species importance, spatial and niche patterns, species richness, species evenness, species diversity, the rate of nutrient cycling, the pattern of nutrient allocation to above- and belowground parts, and community change over time.

The significance of species diversity to community stability, productivity, interdependence, and environmental stress is still not clear. This may be because we have not yet developed a reasonable method to measure diversity. Simpson's index may provide the best general index of diversity, if for no other reason than the fact that it is relatively insensitive to sampling error. Diversity is at some intermediate frequency of disturbance.

Communities differ in their demand for essential nutrient elements, in the efficiency with which radiant energy is converted into net productivity, in the fraction of the nutrient pool that is stored, in the rate at which nutrients are returned to the soil in litter fall, and in the efficiency of the plant-soil-plant cycle.

Succession is community change in a period of up to 500 years, with climate and plant genomes assumed to be constant. In fact, however, climate has not been constant. There have been changes in temperature, rainfall, length of the growing season, and duration of sunshine that have accompanied glacial retreat in the temperate zone over the past 10,000 years. The resulting vegetation changes can be shown by pollen profiles in lake sediment. Vegetation changes over millions of years are documented in macrofossil deposits. These long-term changes, however, include genetic (evolutionary) change in the floras.

There is some argument and indirect evidence to suggest that communities evolve, directed by natural selection toward optimal solutions of environmental problems. However, the conservative opinion is that the most complex level at which natural selection has been shown to operate is with pairs of species, such as a parasite and a host.

CHAPTER 9

METHODS OF SAMPLING
THE PLANT COMMUNITY

Vegetation scientists would like to understand the degree of species inter-dependence within communities, how the distribution of communities depends upon past and present environmental factors, and what the role of com-munities is in such ecosystem activities as energy transfer, nutrient cycling, and succession. However, communities must first be measured and summarized in some effective way before these questions can be addressed. The ongoing attempt to inventory the world's vegetation is based on a small sample of the total vege-tation cover because of limitations in people, time, and resources. These samples must be taken very carefully to ensure that the resulting estimates will be accurate and useful.

In this chapter, we will summarize only a few sampling methods; not a very complete review but enough to get a taste of the diversity among the meth-ods. In Chapter 10, we will discuss data analysis methods that can be used to con-vey the essence of a vegetation type to others who may be half a world away and have never seen the type, yet wish to compare it to types with which they are familiar.

Overview of Sampling Methods

For sampling to be done rationally and efficiently, the continuum of vegeta-tion that covers the earth must be divided into discrete, describable community or vegetation types, just as the taxonomic continuum of individual plants has been divided into species. Most of the world's vegetation is known to science, but at a relatively simplistic level. It has been broadly classified, mapped for large areas, and photographed from the air or from space. The dominant species and physiognomy of major vegetation types are generally known, but more detailed information about all the component species, the relative importance of each species, and relationships among the species is frequently lacking.

Even to describe a particular plant community in a relatively circumscribed region, scientists usually do not make a complete census of the community but instead take measurements on perhaps only 1% of the total land on which the community exists. If the samples are chosen carefully, the investigators feel confident in extrapolating from their sample data to estimate the true values or the **parameters** for the entire community. If the samples are not chosen carefully, the samples will not be representative of the true community parameters and are said to be **biased.**

There are five approaches to locating representative samples. One approach is complete **subjectivity.** A seasoned field-worker who has traveled extensively in a region first formulates a concept of a particular community type. Representative stands of that type are found in the field, and one or more sample quadrats are placed so that each quadrat encloses the essence of that stand. This is the **relevé method.**

The other approaches involve either random or regular placement of samples. A completely random approach would randomly choose stands of vegetation and then randomly locate replicate subsamples (quadrats) within each stand. Such a **complete random design** is statistically desirable because it removes all bias. In practice, however, there are several limitations to this approach: (a) Random selection of stands may place some in difficult-to-reach locations and the small amount of information they provide would not compensate for the investment in time it takes to reach them. In any event, the field time required to visit randomly located samples is large and will probably be inappropriately large for surveys of more than several thousand hectares. (b) Random selection of either stands or subsamples within stands may—by chance—result in the location of samples in clumps, leaving large sections of the survey area unsampled and failing to include all possible variation. (c) Completely random sampling will inevitably undersample rare but ecologically informative patches of vegetation, which *would* be sampled in a subjective design.

A **stratified random design** allows the field-worker to subdivide the survey area—or any given stand—into several homogeneous regions, then to locate the stands or samples randomly within each homogeneous region. This design ensures that samples will be dispersed throughout the entire survey area and throughout each stand (Figure 9-1b), and it does not compromise the concept of random sampling (Avery 1964).

A **complete systematic design** locates stands and subsamples within stands by reference to a grid superimposed over the survey area (Figure 9-1c). Intersections of north-south and east-west lines at one scale define stands to be sampled and intersections at a finer scale define sample points within stands. This design has some of the same limitations that a complete random design has. It is rarely used, although Sequoia–Kings Canyon National Park in California has been using this method to document its vegetation for nearly a decade. Ecologists have argued with statisticians for half a century about the wisdom of applying paramet-

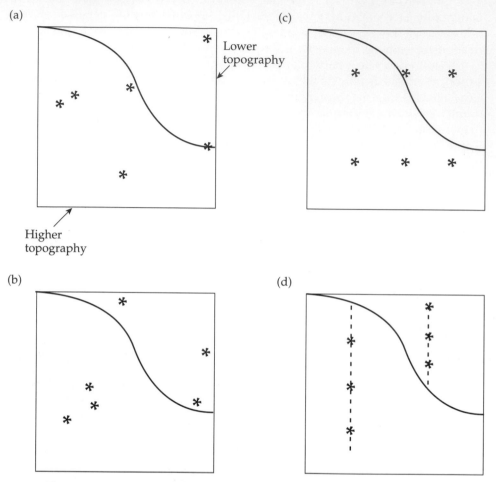

Figure 9-1   Basic sampling designs. (a) Completely random, (b) stratified random, (c) completely systematic, (d) random-systematic. (* = sample point)

ric statistics to data collected systematically, rather than randomly. Most ecologists would argue that statistical analyses require independence of data points and that randomization neither guarantees this nor is the only way to achieve it. That is, regularly placed samples along a transect line should be appropriate for statistical analysis unless that placement just happens to coincide with some underlying environmental periodicity, such as some topographic pattern.

The most frequently used design employs both random (or stratified random) and systematic aspects. We will call this compromise the **random-systematic design.** We may first select stands by either a random or a stratified random plan. We then locate the starting point—and direction—of a transect within a stand by some random or stratified random plan (Figure 9-1d). Finally, along that starting point we will take samples according to some systematic plan. This compromise design is probably the most frequently chosen design by field ecologists. It incorporates some degree of randomness, yet it is efficient of field time because the ultimate stand samples lie along convenient transects. Later in the chapter we will discuss the gradsect method, which is a variant of this random-systematic design.

The Relevé Method

The relevé method was largely codified, if not developed, by Josias Braun-Blanquet, an energetic Swiss ecologist who helped classify much of Europe's vegetation, wrote an impressive text on plant ecology in 1928, founded and directed a center of synecology at Montpellier, France, called SIGMA (Station Internationale de Géobotanique Mediterranéene et Alpine), and was an active editor of the technical journal *Vegetatio* until he was 90. His methods of sampling and classification are sometimes called the **relevé, SIGMA, Braun-Blanquet,** or **Zurich-Montpellier (Z-M) school.**

Our description of the relevé method will be brief. Expanded discussions can be found in van der Maarel et al. (1980), Mueller-Dombois and Ellenberg (1974), Shimwell (1971), Becking (1957), Küchler (1967), and Poore (1955a,b). The Braun-Blanquet method is extensively used throughout the world—with the exception of North America. Until the end of the twentieth century, the relevé method was seldom applied to North American vegetation. Henry Conard (1935)—who helped translate Braun-Blanquet's book into English—was one of the first Americans to use the method. Vera Komarkova (1978, 1979) was another early user, applying the relevé technique to Colorado alpine vegetation. Professor Jack Major at the University of California supervised several master's and doctoral candidates in the use of Braun-Blanquet's method for various California vegetation types in the 1960s through the 1980s, but their results went unpublished except as dissertations with limited distribution.

During the 1990s, the relevé approach found wide use with government agencies, nonprofit conservation organizations, and academic vegetation scientists. The Nature Conservancy adopted the relevé approach in its project to survey, classify, and map all national park vegetation in the United States. The California Department of Fish and Game regularly employs the relevé approach to document threatened vegetation types ranging from riparian forest to desert scrub. Plant ecologists in the U.S. Fish and Wildlife Service have used the method on vast acreages of Alaskan tundra (e.g., Talbot and Talbot 1994). Academic ecologists have classified North American deserts (Peinado et al. 1995), western woody vegetation (Peinado 1997, Rivas-Martinez 1997), eastern deciduous forests (Miyawaki et al. 1994), and Pacific Northwest conifer forests (Krajina 1965, Strong et al. 1990, Klinka et al. 1996).

In the relevé method, an investigator familiar with the vegetation of a region begins to develop concepts about the existence of certain community types that appear to repeat themselves in similar habitats. A number of stands that represent a given community are subjectively chosen. The investigator walks through as much of each stand as possible, compiling a list of all species encountered. Next, an area that best represents the community is located. It is then necessary to determine the **minimal area**—the smallest area within which the species of the community are adequately represented. The minimal area may be determined by a species-area curve. The resulting sample quadrat, based on the concept of minimal area, is called a relevé.

A **species-area curve** is compiled by placing larger and larger quadrats on the ground in such a way that each larger quadrat encompasses all the smaller ones, an arrangement called **nested quadrats** (Figure 9-2a). As each larger quadrat is located, a list is kept of additional species encountered. A point of diminishing return is eventually reached, beyond which increasing the quadrat area results in the addition of only a very few more species. The point on the curve where the slope most rapidly approaches the horizontal is called the *minimal area* (Figure 9-2b–d). Because this definition of minimal area is subjective, some define it instead as that area which contains some standard fraction of the total flora of the stand, for example, 95%. Problems in defining minimal area have been discussed by Rice and Kelting (1955). The most recently proposed solution (Dietvorst et al. 1982) is to plot the mean similarity among all samples as the cumulative area sampled increases. Beyond some critical area, similarity remains relatively constant, and this critical area is the minimal area. Experienced phytosociologists do not usually conduct a minimal area exercise every time they take a relevé, instead employing widely agreed-upon rules of thumb that suggest an adequate relevé size for herbaceous vegetation would be <10 m^2, for a shrubland <25 m^2, and for forests <1000 m^2 (though another rule of thumb for forests is to use an area equivalent to the square of the height of the overstory canopy).

Whatever the method, the actual relevé area is somewhat larger than that which gives the graphical minimal area, for the sake of being conservative. For this reason, it might be more appropriate to call the relevé area "appropriate" rather than "minimal." The appropriate/minimal area is thought by some ecologists to be an important community trait, just as characteristic of a community type as the species that make it up. Table 9-1 shows minimal areas associated with various vegetation types.

In the relevé method, each species' cover is recorded. Cover is not measured precisely but is placed in one of seven categories by a visual estimate (Table 9-2). Braun-Blanquet and others, such as Rexford Daubenmire (1968a), recognize that plant cover is very heterogeneous from point to point and from time to time even within a small stand. They believe that an exact estimate at one place gives an aura of precision to community description that is not warranted. As another ecologist said, "The ecological world is a sloppy place" (Slobodkin 1974).

Another argument against overly precise cover estimates is that there is a differential bias from one individual to another; it is unlikely that any two estimates would agree closely (Sykes et al. 1983). Schultz et al. (1961) dramatically demonstrated this bias by bringing an artificial quadrat (a 1 m^2 board with plants represented by discs of different size and color) to a national meeting of professional range management people and asking 100 of them to estimate the total cover of all

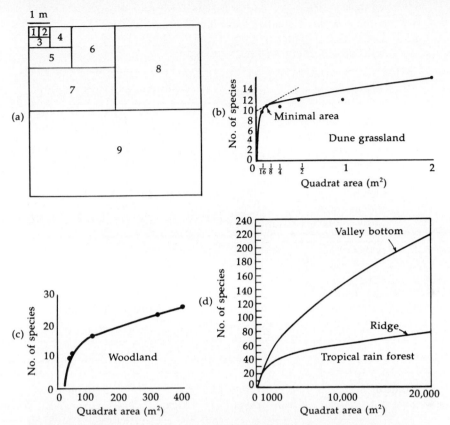

Figure 9-2 The species-area curve. (a) Nested plots for determining minimal area. (b) Minimal area for dune grassland in North Carolina is about 0.13 m². (c) Minimal area for an English woodland is about 100 m². (d) Minimal area for two stands of tropical rain forest in Brunei are 1000 m² on a dry ridge and >20,000 m² in a mesic valley bottom. (a) and (d) From *Aims and Methods of Vegetation Ecology.* Mueller-Dombois and Ellenberg. Copyright © 1974 John Wiley and Sons, Inc. Reprinted by permission. (b) From "A discussion of the application of a climatological diagram, the hythergraph, to the distribution of natural vegetation types," by A. D. Smith, *Ecology* 21:184–191. Copyright © 1940 by the Ecological Society of America. Reprinted by permission. (c) From Hopkins. 1957. By permission of the British Ecological Society.

the "plants" on the board. The resulting range of estimates was impressive: from 6 to 62%. Just as impressive was the fact that the average estimate (27% cover) was 33% in error of the true cover (20%). If each percentage estimate is converted to a class, such as those in Table 9-2, however, more than half of the percentage estimates will fall into the correct class. Seven classes do not provide as much precision as 100 percentage points, but using classes results in greater agreement among investigators. The range of percentage points within each class allows for each observer's deviance from the correct cover percentage.

Table 9-1   Minimal areas for various vegetation types. From *Aims and Methods of Vegetation Ecology*. Mueller-Dombois and Ellenberg. Copyright © 1974 John Wiley and Sons, Inc. Reprinted by permission.

Type	Minimal area (m^2)
Tropical rain forest	1000–50,000
Temperate forest:	
Overstory	200–500
Undergrowth	50–200
Dry temperate grassland	50–100
Heath	10–25
Wet meadow	5–10
Moss and lichen communities	0.1–4

Table 9-2   Cover classes of Braun-Blanquet, Domin-Krajina, and Daubenmire. From *Aims and Methods of Vegetation Ecology*. Mueller-Dombois and Ellenberg. Copyright © 1974 by John Wiley and Sons, Inc. Reprinted by permission.

Braun-Blanquet			Domin-Krajina			Daubenmire		
Class	Range of cover (%)	Mean	Class	Range of cover (%)	Mean	Class	Range of cover (%)	Mean
5	75–100	87.5	10	100	100.0	6	95–100	97.5
4	50–75	62.5	9	75–99	87.0	5	75–95	85.0
3	25–50	37.5	8	50–75	62.5	4	50–75	62.5
2	5–25	15.0	7	33–50	41.5	3	25–50	37.5
1	1–5	2.5	6	25–33	29.0	2	5–25	15.0
†	<1	0.1	5	10–25	17.5	1	0–5	2.5
r	<<1	*	4	5–10	7.5			
			3	1–5	2.5			
			2	<1	0.5			
			1	<<1	*			
			†	<<<1	*			

*Individuals occurring seldom or only once; cover ignored and assumed to be insignificant.

When all stands have been visited, a summary table of species and stands is prepared; this table will be discussed in Chapter 10. The summary table reveals **synthetic traits,** which are traits of a community rather than of a single stand. Two synthetic traits are presence and constance. **Presence** is the percentage of all stands that contain a given species. If species A occurs in 8 of 10 stands, the species has

80% presence. Presence is calculated from the presence lists that were generated as the investigator walked through the stands. **Constance,** in contrast, is based on species encountered in *relevés.* One relevé, recall, is placed in each stand, and those relevés are all of equal area (though not necessarily of equal shape). Generally, presence is higher than constance. Species A may have been present in 8 stands, but in only 6 of the 10 relevés, thus having 60% constance (sometimes called *constancy*).

Quadrat Methods

Care must be taken in selecting the shape, size, and number of quadrats. A considerable body of literature developed on these subjects in the 1940s and 1950s. Some of that research was done on scale drawings of plots of real vegetation as seen from above, with miniature quadrats of various sizes and shapes placed randomly over the drawings. Some of these maps have been published (e.g., Curtis and Cottam 1962). Other research was done on maplike models of artificial vegetation, generally by placing discs of different size and color in random or other patterns, and then sampling these models with miniature quadrats.

In any case, whether the maps represent real or artificial vegetation, the point is that one knows the true number of plants and the true cover. Sample estimates of these parameters can then be compared for accuracy. The best sampling method will be both accurate and precise. **Accuracy** is close agreement of sample means with actual parameter means. In Figure 9-3, method A gives values that are very accurate, within 10–20% of the true mean. **Precision** is close agreement of sample

Figure 9-3   Accuracy and precision of sampling. The center of the target represents the true cover for a stand of vegetation, and the radiating circles represent departures from accuracy of up to 40% error. Means estimated by sampling methods A, B, and C are shown. Methods A and B are both precise (the points are in tight clusters), but method A is the more accurate. Method C has both poor precision and accuracy.

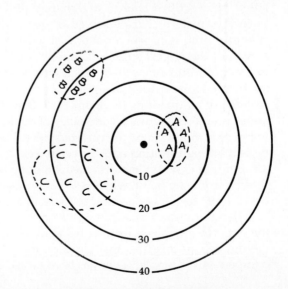

means to each other, without reference to the true mean. In Figure 9-3, methods A and B are equally precise because their sample means cluster equally tightly. Method B, however, is much less accurate, sample means being within 30% of the true mean. Method C is neither precise nor accurate.

Precision can be measured without knowing the true mean; it is equal to 1 divided by the variance of sample means. Accuracy, however, can only be measured when the actual parameters are known, and in vegetation sampling this is never the case.

Quadrat Shape, Size, and Number

By sampling a map of forest vegetation in North Carolina, Bourdeau (1953) and Bormann (1953) concluded that precision is best when quadrats are long, narrow rectangles that tend to cross contour lines. Square and round quadrats are often less precise because each one encompasses less heterogeneity within it than a long, narrow plot placed parallel to the major environmental gradient (see also Clapham 1932 and Lindsey et al. 1958).

Accuracy, however, may decline as the plot lengthens because of the **edge effect.** The more perimeter there is to a quadrat, the more often an investigator will have to make subjective decisions as to whether a plant near the edge is "in" or "out," and these decisions are likely to be biased by the taxonomic knowledge of the investigator, how alert the investigator is that day, and how close it is to dinnertime. In this respect, round quadrats are the most accurate because they have the smallest perimeter for a given area. They are also easier to define in the field with a tape measure and center stake, and so the perimeter of a large quadrat need not be marked first. Obviously, compromise choices on quadrat shape are often made.

The best quadrat size to use depends on the items to be measured. If cover alone is important, then size is not a factor. In fact, the quadrat may be shrunk to a line of one dimension or to a point of no dimension and cover can still be measured, as described later in this chapter. But if plant numbers per unit area, frequency, or pattern of dispersal are to be measured, then quadrat size is critical. One rule of thumb is to use a quadrat at least twice as large as the average canopy spread of the largest species (Greig-Smith 1964); another is to use a quadrat size that permits only one or two species to occur in all quadrats (Daubenmire 1968a); another is to use a quadrat size that permits the most common species to occur in 63–86% of all quadrats (Bonham 1989, Blackman 1935). Quadrats should also be large enough to capture a wide range of density, so that the data can be considered to be continuous. There are no fixed rules, however, and the choice is often made by combining intuition and convenience. One person working alone in a desert scrub might choose a quadrat area 2 m long on each side because the area can all be seen from one point, whereas a team of two to three people in the same vegetation might choose a quadrat area twice that size.

The number of quadrats to use can be determined empirically by plotting the data for any given feature, using different numbers of quadrats, and picking the number of quadrats that corresponds to a point where fluctuations become damped. For example, cover has been tallied in this way in Figure 9-4. Some spe-

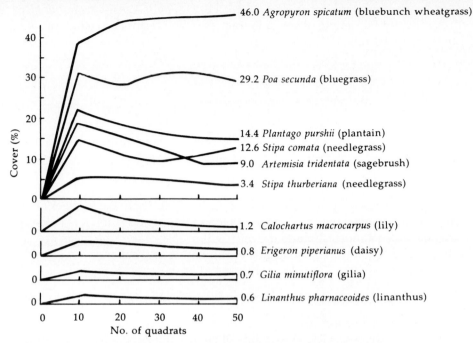

Figure 9-4   Fluctuations in cover are damped out as the number of quadrats increases. In this example from a sagebrush/grassland in the eastern Washington area, 40 quadrats might be sufficient to conduct the study, with diminishing reward for added effort beyond that point. Reprinted by permission from Daubenmire (1959).

cies, such as bluebunch wheatgrass (*Agropyron spicatum*), were underrepresented by the first 10 quadrats, while others, such as sagebrush (*Artemisia tridentata*), were overrepresented. By 30 to 40 quadrats, values had leveled out, and 50 quadrats did not give additional information, with the exception of one species, needlegrass (*Stipa comata*). Therefore, one assumes that 40 quadrats will give a fairly accurate estimate for all additional sampling in this community type.

Alternatively, one can sample until the standard error of the quadrat data is within some previously decided, acceptable bounds. Some field-workers suggest that the standard error be ±15–20% of the mean. The number of quadrats necessary to reach even this modest accuracy can be enormous if variance among quadrats is high. Standard statistical formulae are available to predict quadrat numbers required for selected degrees of accuracy (Bonham 1989, Hayek and Buzas 1997, Krebs 1989). Warning: these formulae presume a normal, continuous density distribution.

Suppose, for example, we sample a grassland with five quadrats, and we count the number of individuals of species A in each quadrat. Our objective is to determine how many quadrats will be necessary, based on this initial sample, to come within 10% of the real average density of species A with 95% probability. Our five density measures for species A are 2, 3, 6, 8, and 11. The mean density for A, from this initial sample of five quadrats, is 6. The formula for the number of desired quadrats (Q) is

$$Q = \frac{(t \text{ value for 95\% confidence and 4 degrees of freedom})^2 \times (\text{coefficient of variation})^2}{(\text{percentage accuracy desired})^2}$$

The coefficient of variation (CV) in the formula is (100 × standard deviation)/(mean). Solving for CV first, (100 × 3.67)/6.0 = 61. Substituting this value in the Q formula, and substituting 2.78 for the t value, we have

$$Q = 2.78^2 \times 61^2/10^2 = 288 \text{ quadrats}$$

Note that as the percent accuracy gets smaller by half, the number of quadrats desired will quadruple. In the face of such sampling intensity, it is not surprising that most vegetation estimates are only within 25% of the true mean because only about 1% of a stand is included in the cumulative quadrat area.

Cover, Density, Frequency, Dominance, and Importance

Cover (also called **coverage**) is the percentage of quadrat area beneath the canopy of a given species. The canopy of an overstory species creates a microenvironment that smaller, associated species must contend with. The overstory canopy, therefore, exerts a biotic control over the microclimate of the site. No doubt the root system of the overstory species extends beneath the ground out to a perimeter corresponding with the canopy edge or even further, so the soil microenvironment is also under the biotic influence of the overstory species. It is assumed that a comparison of cover for each species in a given canopy layer will reveal the relative control or dominance that each species exerts on the community as a whole, such as the relative amount of nutrients or other resources each species commands.

For the practical measurement of cover, holes in the canopy may be viewed as nonexistent, and the canopy edge can be mentally "rounded out," the rationale being that such space is still under the root or shoot influence of the plant in question. The canopy of a plant rooted outside the quadrat is tallied to the extent that the canopy projects into the quadrat space. Thus, in Figure 9-5 and Table 9-3, shrub E does extend into the quadrat space when its canopy is rounded out (dashed lines), and it contributes 7.9% cover. Similarly, the radiating, basal leaves of B that project into the canopy space are tallied by rounding out the edges and estimating each plant separately, but the overlap is not counted twice. The D plants, some small herbs, are mentally grouped together and their cover is estimated separately from the overtopping B plants. In some vegetation, with many overlapping canopies, total cover could exceed 100% and there could still be bare ground. For this reason, bare ground cannot be estimated by subtracting total plant cover from 100%. Some ecologists, however, do not round out the canopies to the extent shown in Figure 9-5; they would award 0% cover for shrub E. Also, some ecologists do count overlapping canopy areas in the same stratum twice.

Relative cover is the cover of a particular species as a percentage of total plant

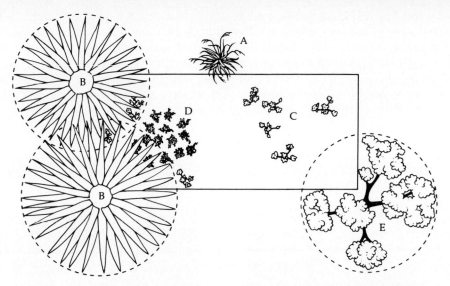

Figure 9-5 Estimation of cover. As seen from above, two members of species B contribute the most cover within the quadrat. The leaf tips describe the perimeter of a circle, and cover is estimated as though the circle were completely covered with leaves. Shrub E has a canopy that does not actually penetrate into the quadrat, yet if one fills in the canopy holes with an imaginary circle, then there is cover. Cover values are shown in Table 9-3. Reprinted by permission from Daubenmire (1959).

Table 9-3 Absolute and relative cover and density, based on the 0.1 m^2 quadrat shown in Figure 9-5.

Species	Absolute cover (%)	Relative cover (%)	Absolute density Per quadrat	Absolute density Per ha	Relative density (%)
A	0.2	0.4	0	<1	<1
B	33.2	63.6	1	100,000	4.2
C	4.7	9.0	9	900,000	37.5
D	6.2	11.9	14	1,400,000	58.3
E	7.9	15.1	0	<1	<1
Total	52.2	100.0	24	2,400,000	100.0
Overlap	5.3	—	—	—	—
Bare ground	53.1	—	—	—	—

cover. Thus, relative cover will always total 100%, even when total absolute cover is quite low, as in the case of Figure 9-5 and Table 9-3.

Cover of tree canopies can be difficult to estimate, as well as painful after a few hours of neck bending. One solution is to use a "moosehorn" crown-closure estimator (Garrison 1949). A periscope-like device is attached to a staff so that the eyepiece is easy to use for viewing while standing; the view of the canopy is seen superimposed on a template of dots (Figure 9-6). The percentage of dots "covered" by canopies is equivalent to percent cover. These readings could be taken at one location in each quadrat. Other versions of the same approach include densiometers and densitometers. A **densitometer** is an L-shaped tube that reflects one's vision

(a)

(b)

(c)

Figure 9-6   Several devices for estimated overstory tree cover. (a) Crown-closure estimator (moosehorn) with a grid of dots in the view. (b) Densitometer with a single point created by two crosshairs in the field of view. (c) Spherical densiometer with 24 squares etched into the reflective surface.

directly upward (Figure 9-6b). A bull's-eye type level—visible in the field of view—directs the person holding the densitometer to hold it in an unbiased way. Cover is then recorded as either yes or no, depending on whether tree canopy covers the intersecting point of two hairs that cross in the center of the field of view. If readings are taken at 100 positions along a transect route and "yes" is noted 85 times, then there is an average of 85% cover along the transect for the canopy of all species.

A **densiometer** is a hemispherical mirror about 6 cm in diameter and mounted on a wooden platform held in one's palm parallel with the ground and about 20 cm in front of one's chest. The mirror reflects a large segment of overstory canopy, much like what would be seen through a fish-eye lens. The mirror is marked off into 24 squares (Figure 9-6c) and the recorder tallies cover in each. Readings are repeated at regular intervals along a transect to gain an estimate of cover throughout the vegetation type. A more elegant method is to take a picture of the canopy with a fish-eye lens from one location in each quadrat, then to analyze the photographs later for percent cover.

Typically, however, canopy cover of trees is assumed to correlate with trunk cross-sectional area (**basal area, BA**) or with trunk **diameter at breast height (dbh).** To obtain the basal area, the tree is usually measured with a special diameter tape that converts circumference to diameter units. Sometimes **relative dominance** is used as a synonym for relative basal area or relative cover. In this book, we prefer not to equate dominance with BA or any single measure, but to use it as a sum of several measures (see below).

Cover of shrubs and herbs is usually estimated to the nearest whole number or put into cover categories, but if greater detail is required the quadrats may be photographed from above or a scale drawing can be made. Cover of grasses—especially bunchgrasses—is often assessed by estimating the area that a group of tillers occupy at the ground surface, rather than tediously trying to estimate cover of all the grass blades. One can quickly capture the perimeter's shape and its enclosed area by bending a piece of wire around the cluster of tillers. The wire forms can be saved for repeated sampling over time, to determine shape as well as area changes.

Density is the number of plants rooted within each quadrat. The average density per quadrat of each species can be extrapolated to any convenient unit area. For example, Figure 9-5 and Table 9-3 show that herb D had a density of 14 plants per 0.1 m^2 quadrat, which converts to 1.4 million plants per hectare. **Relative density** is the density of one species as a percent of total plant density. **Mean area** is plot area/density; it is the area per plant. Density may not be proportional to cover. For example, many young, slender trees may have a higher density but a lower cover than a few older, branching trees. **Abundance** is a rather nebulous term, but often it is used as a synonym for density.

Frequency is the percentage of total quadrats that contains at least one rooted individual of a given species. Rarely, frequency is expressed on a cover basis: any plant, whether rooted in the quadrat or not, which contributes cover for species A is tallied as "present," and frequency becomes the percent of all quadrats in which the canopy of A was "present." **Relative frequency** is the frequency of one species as a percentage of total plant frequency. Frequency is highly dependent on quadrat size, and in this respect it is a more artificial statistic than cover or density.

Frequency and density are often independent of each other. A clumped species may have a high density but a low frequency, whereas a much less abundant species distributed singly and regularly throughout a stand will have a low density but a high frequency.

It is unfortunate that such an important ecological term as **dominance** is still ambiguously defined by many ecologists. Generally, the dominant species of a community is that overstory species that contributes the most cover or basal area to the community, compared to other overstory species. This definition is based on physiognomy. If oak has the highest relative cover in an eastern deciduous forest with oak, hickory, and elm in the overstory, then oak is said to be the *dominant* species. If all three species contribute about the same amount of cover, or if the balance shifts from one to the other depending on the stand, then the three species are *codominants.* In a savanna or a semidesert woodland, where tree canopies may contribute only 10–30% cover and understory plants such as grasses or shrubs contribute more, then some grass or shrub species will usually be considered the dominant species.

Another view of dominance is **sociologic dominance** (Kershaw 1973). Sociologic dominants control the reproduction and continued existence of a community, and they may be understory species. For example, the regeneration of ponderosa pine saplings in some ponderosa pine forests is inhibited by root competition for moisture by the understory grass *Festuca arizonica*, but not by another grass, *Muhlenbergia montana* (Pearson 1942). The grasses are the sociologic dominants of the ponderosa pine community, even though ponderosa pine is the physiognomic dominant. For this reason, some methods of community description name communities by both overstory and understory species and separate the two species by a slash, for example, the *Pinus ponderosa/Muhlenbergia montana* community, or the *P. ponderosa/Purshia tridentata* community (Daubenmire 1952).

Foresters and tropical ecologists call any individual overtopping tree whose canopy is more than half exposed to full sun a dominant, even though it may not be a member of a species that is a physiognomic or sociologic dominant of that community. In this sense, *dominant* is a synonym for **emergent,** and the latter term should be used. In some other forest studies, dominance is equivalent to trunk basal area. The species with the most basal area per hectare is called the dominant. This use of the term dominant is similar to our definition.

Finally, the term **aspect dominance** is applied to species that are very noticeable and at first glance appear to dominate a community by cover. Careful sampling would reveal, however, that other, less conspicuous species in the same canopy layer contribute more cover and are the actual dominants. Aspect dominance is most common in herbaceous communities, such as grasslands or meadows, where all members of one species will flower synchronously and in this way stand out from the rest of the vegetation at certain times of the year.

Throughout this text, we will use the term *dominant* in the physiognomic sense.

Importance refers to the relative contribution of a species to the entire community. It can be used in a very nebulous, almost intuitive, informal sense, or it can be calculated in a precise way. At the investigator's pleasure, importance may be synonymous with any one measure—for example, density. Originally, however, importance was defined as the sum of relative cover, relative density, and relative frequency (Curtis and McIntosh 1951). In the latter case, the **importance value (IV)** of any species in a community ranges between 0 and 300. Table 9-4 illustrates the calculation of IVs for all overstory trees in a Hawaiian rain forest. Notice that two species with similar IVs could have entirely different values for relative cover, density, and frequency; any differences are submerged in the addition process, and the one number that results is a synthetic index of importance. Other formulas for IV calculation have been developed; they may sum only two relative values rather than three (Bray and Curtis 1957, Ayyad and Dix 1964), or sum more than three values (Lindsey 1956). To avoid confusion, all IVs should be made relative to a 0–100 unit scale and called an importance percentage (Bonham 1992).

Biomass and Productivity

Biomass is the weight of vegetation per unit area; synonyms are *standing crop* and *phytomass.* The dominance or importance of any species can be expressed as its percentage of total biomass. For small quadrats in herbaceous vegetation, biomass may be measured by clipping all aboveground matter, drying it in an oven, and weighing it. Ideally, roots are also excavated, but they are often ignored; consequently, most biomass data represent only aboveground plant matter.

The clearing of large plots in woody vegetation is not practical. Instead, relatively few individuals of different age or size classes are harvested, and a regres-

Table 9-4 Calculation of importance value (IV) for an open tropical rain forest at 450 m elevation near Honolulu, Hawaii. Only the four most abundant overstory trees are summarized below. "Cover" here is the total basal area of all stems >3 cm dbh. From *Aims and Methods of Vegetation Ecology.* Mueller-Dombois and Ellenberg. Copyright © 1974 John Wiley and Sons, Inc. Reprinted by permission.

Species	Relative density	Relative cover	Relative frequency	IV	IV rank
Koa tree					
(*Acacia koa*)	30.0	78.4	30.8	139.2	1
Ohia lehua					
(*Metrosideros collina*)	20.0	13.9	23.1	57.0	3
Ohia					
(*M. tremuloides*)	5.0	5.8	7.7	18.5	4
Guava					
(*Psidium guajava*)	45.0	1.9	38.5	85.4	2

sion line is developed between size and biomass. Sampling for biomass over larger areas can then proceed by measuring plant size, and the data can be converted to biomass (see Bonham 1989 for sample formulas).

Productivity is the rate of change in biomass per unit area over the course of a growing season or a year. Productivity and biomass may not be related. A mature forest has a large biomass but may exhibit a small productivity; a grassland has a smaller biomass but may exhibit a larger productivity. Productivity and biomass data may serve to characterize a particular vegetation type, as summarized by Rodin and Basilevic (1967) and Whittaker (1975). Methods of biomass and productivity sampling have been thoroughly reviewed by Chapman (1976) and Whittaker and Marks (1975).

Species Richness and Diversity

Avi Shmida (1984) has proposed that a quadrat technique developed and widely used by Robert Whittaker until his death in 1980 be adopted worldwide as a standard method to take data for species richness and the calculation of species diversity.

In this method, a 20×50 m plot (0.1 ha) of homogeneous vegetation is subjectively chosen as representative of a community by the investigator. A 50 m tape is stretched across the plot to serve as a central axis (Figure 9-7a). A central 10 m portion is selected, and ten contiguous 1 m^2 quadrats are marked along one side of the tape. Species presence is noted for each of these ten quadrats. Two 1×5 m quadrats on the other side of the tape are then searched for species not already noted in the ten small quadrats, and these are added separately to the list. A single 10×10 m quadrat is then searched for new species, and so, finally, is the entire 20×50 m plot. Canopy cover for each species can be noted, and additional measures or notes on growth forms, vertical foliage profile, tree and shrub density, and distribution patterns can be taken. Two individuals can complete one plot in as little as one hour (desert vegetation) or as many as four hours (tropical vegetation).

Two important comparative values can then be calculated. If cumulative number of species is plotted against increasing area on semilog paper, a best-fit linear relationship can be formulated. The slope of that line is a convenient measure of habitat and biotic diversity and can readily be compared to other habitats and communities. Species diversity can also be calculated for the 0.1 ha plot, weighted by cover rather than density.

Tom Stohlgren and others (1995) have pointed out that the Whittaker method has two design flaws: the nested quadrats do not all have the same general shape, and the dispersal of the plots overemphasizes the central portion of the 0.1 ha area. They then modified the design, making all plots rectangular and scattering them throughout the 0.1 ha sample (Figure 9-7b). Tests demonstrated that the new design was strikingly more accurate in capturing species richness in the 1–100 m^2 range. The Whittaker design consistently underestimated species richness by about one-third compared to the modified design.

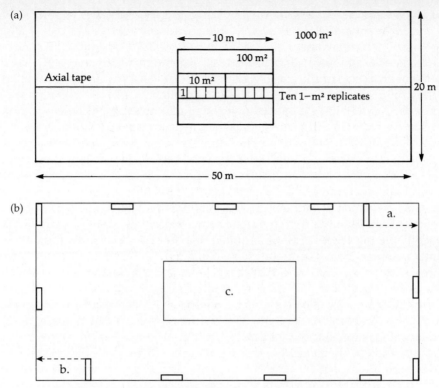

Figure 9-7 Nested quadrats for estimating species richness. (a) Whittaker's plot, as summarized by A. Shmida. *Israel Journal of Botany* 33: 41–46, published in 1984 by the Weizmann Science Press of Israel. (b) Modified Whittaker plot developed by Stohlgren et al. (1995). Diagram courtesy of Thomas Stohlgren.

The Gradsect Method

Gradient-oriented transect (gradsect) sampling was developed in the 1980s to solve the problem of sampling large areas (>10,000 ha) within narrow constraints of limited time, staff, and budgets (Gillison and Anderson 1981, Gillison and Brewer 1985, Austin and Heyligers 1989). Transects that follow the steepest gradients of environment and vegetation are subjectively selected to maximize sampling efficiency and minimize travel time. Once a transect is chosen, homogeneous pieces (vegetation types or topographic positions) within it are sampled with quadrats or relevés in a stratified random design. Sampling intensity for each vegetation type is proportional to the fraction of the landscape that it occupies. Thus, if 10,000 ha of land is to be sampled and aerial photographs indicate that 6000 ha are covered by many patches of relatively open savanna, 3000 ha are cumulatively covered by patches of denser forest, and the remaining 1000 ha are covered by patches of grassland, then 60% of the quadrats will be randomly scattered within the 6000 ha of savanna, 30% within the forest, and 10% within the grassland. Field trials and computer simulations (e.g., Austin and Adomeit 1991) have revealed that gradsect sampling detected more species and vegetation units at lower cost than more complete restricted random sampling.

Gradsect sampling is a natural outgrowth of a quadrat placement technique first used by Robert Whittaker (1956) when he sampled vegetation in the Great Smoky Mountains. He subjectively chose transects that progressed up in elevation along major slope faces (e.g., north-facing, south-facing, west-facing), beginning in mesic valleys or coves and ending on xeric ridges. At regular intervals of elevation change along each transect he placed quadrats to sample the vegetation. Thus there was no element of random quadrat placement at all. Whittaker applied this technique later to mountain slopes in the Pacific Northwest (1960) and in the warm deserts (1975). In each location he assumed that the environmental gradients were complex combinations of temperature, precipitation, soil moisture storage, and exposure to drying winds, but he did not take microenvironmental measurements to quantify those gradients.

Whittaker's nonrandom sampling technique has been called **direct gradient analysis** (see Chapter 10). Mohler (1983) later analyzed the effectiveness of Whittaker's regular placement of quadrats in comparison with various other patterns, including random. He concluded that regular placement was very effective but it could be even more so if the intensity of sampling was greater toward the ends of the transects. That is, if quadrats were placed every 100 m change in elevation along most of the transect, they should be placed every 50–25 m near the ends.

Line Intercept, Strip Transect, and Bisect Methods

R. H. Canfield (1941) and H. L. Bauer (1943) developed the **line intercept** method for dense, shrub-dominated vegetation. They found it to be as accurate as traditional quadrat methods but less time-consuming. If a quadrat is reduced to a single dimension, it becomes a line. The line may be thought of as representing one edge of a vertical plane that is perpendicular to the ground; all plant canopies projecting through that plane, over the line, are tallied. The total decimal fraction of the line covered by each species, multiplied by 100, is equal to its percent cover. Just as with quadrats, total cover can be more than 100%. Disadvantages of the method are the loss of density and frequency measures because there is no area involved (although frequency can be expressed on a cover basis if the line is broken up into segments).

Often, a lengthy line intercept is combined with quadrats that run alongside it. Cover is measured along the line, and density or frequency is noted in the quadrats. If the quadrats run continuously along the line, the method is called the **belt transect, strip transect,** or **line strip method.** These methods have been most often applied to forest vegetation (Lindsey 1955).

Bisects are scale drawings of the vegetation within line strips. The idea was originally applied to tropical forests (David and Richards 1933, Richards 1936, Beard 1946); Figure 9-8 is an example from the British West Indies. All plants in a strip approximately 60 m long and 8 m wide are shown, drawn as accurately as possible. For those who are not good artists, bisects can be drawn in highly diagrammatic fashion using symbols (see Dansereau 1951). Other bisect studies have used areas of size 10 × 10 m, 10 × 50 m, and 8 × 40 m (Ashton and Hall 1992, Iremonger 1990, Peters and Ohkubo 1990, Swaine 1992).

These three methods can record cover as a function of height above the ground, if the sampling is done carefully enough. When the data are summarized in bar graphs, such as the ones in Figure 9-9, striking differences between vegetation types become apparent. The eastern deciduous forest is seen to be composed of four canopy layers, with the most cover being contributed by the overstory tree layer. In contrast, the boreal forest has three canopy layers, with the trees and ground (herbaceous) layers providing nearly continuous cover.

The Point Method

If a quadrat is reduced to no dimension, it becomes an infinitely small point. In practice, metal pins with sharp tips serve as the points, and cover is equal to the fraction of total pins that touch any plant part as the pin is lowered. Typically the pins are arranged in frames that rigidly limit the pin to a vertical path perpendicular to the ground (Figure 9-10). The frame may be located at several random places in a stand, and one to ten pins can be lowered at each place. For the best precision of estimated cover, lowering only one pin at each place is better than lowering several (Goodall 1957). To measure cover up to 1.5 m tall, modifications in frame construction permit rapid placement of as many as 100 pins and projection of pins upward as well as downward (Baker and Thomas 1983, Taha et al. 1983).

As the pin is lowered, the first plant it touches is recorded, then the pin is lowered more until it touches another leaf (of the same or a different plant than the first touch), and so on until bare ground is reached. If no plant is hit, then the point is tallied as bare ground. These data permit two calculations. One is percent cover:

$$\% \text{ cover} = \frac{\text{no. of pins that hit species A at least once}}{\text{total no. of pins}} \times 100$$

The other calculation is percent of sward, which weights each species by its canopy thickness, or cover repetition, at each point:

$$\% \text{ sward} = \frac{\text{no. of contacts with species A}}{\text{total no. of contacts}} \times 100$$

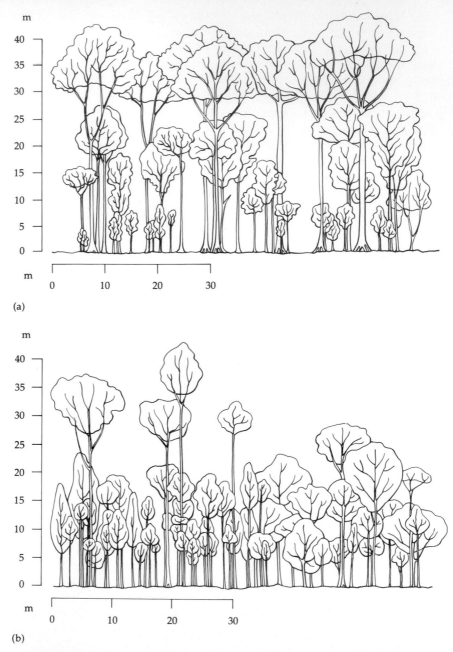

Figure 9-8   Bisects of tropical rain forest. (a) Trinidad, British West Indies. (b) Borneo. Both bisects represent all woody vegetation within a strip 61 m long and 7.6 m wide. (a) From Beard (1946). By permission of the British Ecological Society. (b) From Richards (1936). By permission of the British Ecological Society.

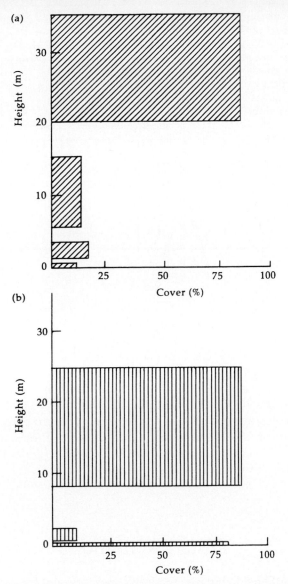

Figure 9-9 Canopy profiles. The thickness of the horizontal bars represents the span of canopy height and the length of each bar represents the total cover of that canopy. (a) Typical eastern deciduous forest. (b) Typical conifer forest in the Canadian taiga.

Disadvantages of the point method are that density cannot be measured (although cover frequency can), and the method is limited to low vegetation, such as grassland, for obvious reasons. But for measuring cover of low vegetation, it may be the most trustworthy and objective method available (Goodall 1957). Modifica-

Figure 9-10   A point frame suitable for one to ten pins. Cover is expressed as the percent of pins that touch a given species as the pins are allowed to drop vertically to the ground. Such frames can be modified to permit the pins to move up instead of down to sample overstory shrub cover. From *Aims and Methods of Vegetation Ecology.* Mueller-Dombois and Ellenberg. Copyright © 1974 by John Wiley and Sons, Inc. Reprinted by permission.

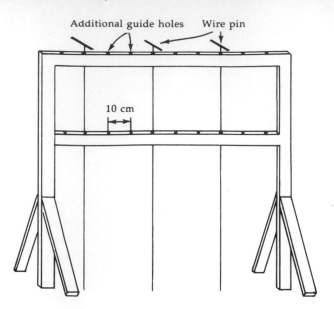

tions of the method have included decisions to drop the pins at some angle, rather than perpendicular to the ground (vertical pins tend to weight broad-leaved herbs more than grasses); to nest the pins in a circular arrangement rather than a linear one (Baker and Thomas 1983); to insert pins upside down so that they can be used to touch shrub cover above the point frame, in addition to herb cover beneath it (Floyd and Anderson 1982); and to refine the diameter of the sample point by replacing the pins with sighting devices that use crosshairs (Bonham 1989) or a laser beam deflected by a mirror (Eek and Zobel 1997).

Distance Methods

Distance methods do not use quadrats, lines, or point frames. Only distances (from a random point to the nearest plant, or from plant to plant) are tallied. Average distance, multiplied by an empirically determined correction factor, becomes density. The basic distance methods were developed by Grant Cottam and John Curtis at the University of Wisconsin in the 1950s and were tested and refined on maps of real and artificial forest vegetation. The five methods briefly described here have been best summarized and compared by Cottam and Curtis (1956), Lindsey et al. (1958), Mueller-Dombois and Ellenberg (1974), and Bonham (1989). Four methods are illustrated in Figure 9-11. These methods have been used with many different types of plants, but most often with trees.

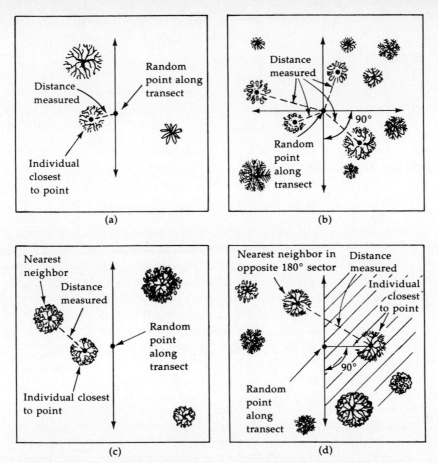

Figure 9-11 Four distance methods. (a) Nearest individual, (b) point-centered quarter, (c) nearest neighbor, and (d) random pairs. The shaded area of (d) is an excluded area that contains the first plant but cannot contain the second.

Nearest Individual Method

Random or regular points are located in a stand. At each point, the distance to the nearest tree of any species is recorded, the species is identified, and its basal area is measured. Only one measurement is made from each random point. All distances for all species are summed and divided to yield one average distance. Density per hectare (10,000 m^2) for all trees is then

$$\text{Density} = \frac{10,000}{2 \, (\text{average distance, in meters})^2}$$

(If distance is measured in feet, then the numerator is 43,560, the number of square feet in an acre.) The 2 in the denominator is a constant correction factor. Absolute density of each species is

$$\begin{matrix} \text{Absolute density} \\ \text{of species A} \end{matrix} = \frac{\begin{matrix} \text{no. of trees of} \\ \text{species A encountered} \end{matrix}}{\begin{matrix} \text{no. of all} \\ \text{trees encountered} \end{matrix}} \times \begin{matrix} \text{density for} \\ \text{all trees} \end{matrix}$$

Cover or dominance of each species can then be calculated as its relative density times its average basal area.

Point-Centered Quarter Method

Again, random or regular points are located. The area around each point is divided into four 90° quarters of the compass, and the nearest tree in each quarter is sought. Each tree is identified, its basal area is measured, and its distance from the random point is measured. Again, average distance for all trees taken together is computed, and this is converted to total density by the formula given for the nearest individual method, except that the correction factor vanishes. (The correction factor is 1, not 2.) Because more information is gained at each point than in the nearest individual method, the point-centered quarter method is more efficient, requiring one-fourth the number of points to achieve the same level of accuracy and precision. This method has also been used in grasslands (Dix 1961, Risser and Zedler 1968).

Problems with accuracy for the point-centered quarter (PCQ) method revolve around steepness of the terrain and nonrandom dispersion of trees. The distance measured from point to tree must be along a horizontal surface; if instead the distance is measured parallel to steeply sloping ground, it will be greater than the true (horizontal) distance and tree density will be underestimated. If trees are clumped rather than random, the PCQ method underestimates density; if trees are regularly distributed, PCQ will overestimate density (Bonham 1989, Mueller-Dombois and Ellenberg 1974). Consequently, a test for random tree dispersion should be conducted wherever the PCQ method is used (Ludwig and Reynolds 1968).

An alternative distance method specifically for use with clumped populations is the **wandering quarter method** (Figure 9-12; Catana 1963, Bonham 1989). A transect is taken along a compass line, and at the first point the distance to the nearest tree within a 90° quadrant (bisected by the compass line) is measured. From that tree a new line parallel to the original is erected, a 90° quadrant is established, and again the distance to the nearest tree is measured, and so on. Several transects per forest stand are recommended. Ultimately, the distance data can be analyzed to reveal mean clump radius, mean distance between clumps, and overall tree density.

A thorough comparison of these and other distance methods was conducted by Richard Engeman and others (1994). They concluded that the most accurate, efficient, and computationally straightforward method is a modification of the

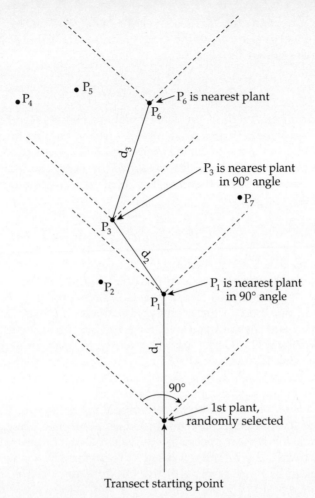

Figure 9-12 The wandering quarter method, a distance method for estimating density for species that have a clumped distribution. Redrawn from Bonham (1989).

nearest individual method, where up to three individuals are measured: the nearest, next nearest, and second next nearest. They reported that the PCQ method required too much time in the field determining the boundaries of each quarter.

Nearest Neighbor Method

Random or regular points are located in a stand, and the nearest plant is located. The distance measured in this case is from that plant to its nearest neighbor (of any species). Total density is calculated as for the nearest individual method, except that the correction factor is 1.67. This nearest neighbor method has also been used to determine whether trees of the same species are distributed at random, are clumped, or are regular (Clark and Evans 1954, Pielou 1961).

Random Pairs Method

In the random pairs method, the nearest plant to a point is located. A line from point to plant is imagined. Perpendicular to the line and passing through the point is an exclusion line. In Figure 9-11d, the exclusion line happens to correspond with the transect. A nearest neighbor is now searched for, but it cannot be on the same side of this exclusion line as the first tree. The constant correction factor in the density formula is 0.8.

Bitterlich Variable Plot Method

A final distance method can be used to calculate basal area only, but basal area is important in calculating board feet of lumber, and the method is extremely fast and has been widely adopted by foresters. It yields more reliable data for less field time than quadrat methods or other distance methods (Lindsey et al. 1958). The method is named after its German inventor (see Grosenbaugh 1952), who originally used a sighting stick 100 cm long with a crosspiece at one end that was 1.4 cm across (Figure 9-13a). The stick was held horizontally with the plain end at one eye, and the viewer would slowly turn in a complete circle. Every tree whose trunk was seen in the line of sight on the circuit was tallied and identified as to species if its trunk appeared to exceed the width of the crosspiece; all other trees were ignored. Using a stick of these dimensions, and based on geometric principles, the total basal area in $m^2\,ha^{-1}$ for any species is equivalent to the number of trees of that species tallied, divided by 2. (If English units are preferred, then a stick 33 inches long with a crosspiece 1 inch across will give basal area in $ft^2\,acre^{-1}$ if the trees tallied are multiplied by 10.)

As shown in Figure 9-13b, an angle is being projected—an angle whose size depends on the relative lengths of the stick and the crosspiece. For example, the 33 inches × 1 inch arrangement produces an angle of 1°45′ and an English units basal area factor (BAF) of 10. BAF is the number that is multiplied by the number of tallies to obtain basal area (in $ft^2\,acre^{-1}$). If the angle becomes smaller, more trees will be tallied and the BAF will become smaller; for example, an angle of 0°33′ has an English units BAF of 1. If the angle becomes larger, then fewer trees are tallied and the BAF will become larger; this is useful in a dense forest to avoid miscounting. In the United States, angles that give BAFs of 5–20 are commonly used.

More recently, small hand held prisms or sophisticated viewing scopes have replaced sighting sticks (Figure 9-13c). Looking both through and over the top of a prism, the lower trunk will appear to be offset partially or completely from the upper trunk; if it is not completely offset, the tree is tallied.

Note that plot size is variable in this method. Plot radius is not fixed but extends as far as the largest tree with an apparent diameter big enough to be tallied.

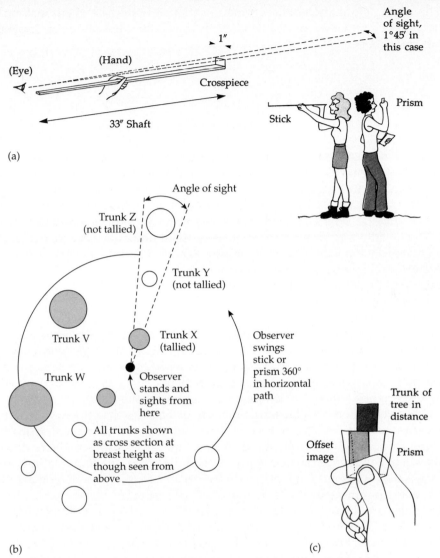

(a)

(b)

(c)

Figure 9-13 The Bitterlich variable plot method for calculation of basal area. (a) The original Bitterlich stick sighting device. This particular one will give total basal area in square feet per acre if the number of trees tallied is multiplied by 10. (b) A bird's-eye view of trees that would be tallied as an observer at the center point turns in a complete circle. Trunk X (shaded) is tallied because its trunk diameter exceeds the angle projected by the sighting device, but trees Y and Z (not shaded) are not tallied. In one complete circle, the observer would tally four trees (all shaded). (c) The prism type of sighting device. In this case, the lower trunk is not completely displaced from the upper trunk, therefore the tree would be tallied.

A rather complete analysis of the geometric rationale at the base of this method has been compiled by Dilworth and Bell (1978). The method has also been applied to shrubland (Cooper 1957).

Summary

A major objective of vegetation scientists is to complete an inventory of the earth's plant resources—not an inventory of individual species, but an inventory of communities and vegetation types. The results of such an inventory have application to applied and basic science and to autecology as well, for the environment of a plant includes adjacent organisms as well as the physical factors of climate and soil. The inventory is far from complete because of limitations in researchers, scientific interest, and accessibility of some areas. Conclusions will have to be based on samples representing 1% or less of the earth's surface vegetation. If the estimates are to be accurate and precise, the samples must be chosen carefully. This chapter described procedures for choosing and measuring those samples.

The relevé method uses a more subjective choice of sample locations than any other mentioned, but the process of recording data is relatively rapid and non-mathematical, and its widespread use in the non-English-speaking world makes it an attractive sampling method. Each stand is represented by one large quadrat, whose size must meet minimal area requirements. Data are recorded by cover class. Important synthetic values are presence and constance.

Quadrat methods involve fewer subjective decisions than the relevé method. Questions that have to do with the size, shape, number, and placement of quadrats are time-consuming to answer, and there is no one best solution to them. A statistical level of confidence can be applied only to data from random quadrats. Such data can include absolute and relative cover, density, frequency, and biomass. Generally, these four measurements are unrelated. (If plants are distributed randomly, density and frequency are related, but plants are usually not distributed randomly.) Cover may be the most meaningful of the four. It can be estimated as actual crown cover, but for trees it is assumed to correlate with trunk basal area. Cover and basal area are both aspects of size. Frequency is the most artificial of the four, being highly dependent on quadrat size.

Two synthetic values from quadrat data are dominance and importance. A single definition of dominance has not yet been accepted. Most often, however, dominance is related to the cover of overstory species, but in some cases density is also considered. Typically, importance is a summation of relative cover or basal area, density, and frequency, but the summation may involve one more or one less component, and it can be expressed as a phytograph.

Restricted random quadrat placement is generally preferred over complete random. Gradsect sampling is a combination of subjective and restricted random quadrat placement.

The line intercept sampling method reduces quadrats to a single dimension;

consequently, only cover can be measured. Line intercepts may be combined with quadrats, as in strip transect and bisect methods. The point method reduces quadrats to no dimension; it is most useful for estimating cover in short vegetation types.

Distance methods measure distances from random points to nearest plants or distances between plants. The data can be converted to plant density, and in some cases the data can reveal whether plants are randomly or nonrandomly distributed. The most efficient distance method is the nearest individual method. The Bitterlich variable plot method was treated as a distance method, even though distances are not measured, because the geometric assumptions behind it relate to distance from observer to tree and to tree diameter.

CHAPTER 10

CLASSIFICATION AND ORDINATION OF PLANT COMMUNITIES

Virtually all sampling methods—relevés, quadrats, line intercepts, points, or distances—generate data too voluminous and disorganized in their raw form to be meaningful, whether to the investigator or to others with whom the investigator would like to communicate. This chapter describes some common methods of summarizing sampled data in order to relate one stand to another.

Sampling and data analysis, of course, are not independent of each other. Some methods of analysis require certain methods of sampling. For example, association analysis requires the use of random quadrats, and the Braun-Blanquet table method requires relevé information.

The biases and objectives of the investigator also influence the choice of a data analysis method. If the aim is to draw lines and describe discrete entities, then the table method and vegetation mapping are good choices. If the aim is to let the reader draw the lines between entities and to emphasize the continuum of vegetation, then gradient analysis and ordination are good choices.

A word of caution: The statistical tests and the mathematical rigor behind them are only superficially presented in Chapter 9 and in this chapter. The old saying, "A little knowledge can be a dangerous thing," fits the area of sampling and analysis very well. Although we will continue to cite the original literature in this chapter, these citations may not be helpful unless you possess some mathematical background, for their subject matter is technical and the reading sometimes difficult. Our objective is to survey some commonly used procedures so that you are reasonably familiar with the options available.

Many analytical methods that once required a mainframe computer are now available for today's personal computers with large data storage and high-speed microprocessors. We will highlight these where appropriate.

Classification

Classification methods attempt to place similar stands together in discrete entities (associations or other units), cleanly separated from all other stands and units. In North America, particularly in the West, classification has emphasized dominant species. In Europe and some other parts of the world, classification has emphasized indicator or character species that seldom are dominants. They are often understory species of low cover but exhibit strong fidelity to certain units; typically they are present in one unit and absent in all others.

Classification Based on Dominance

Clements classified North American vegetation on the basis of only one or two dominant species in each unit. Sometimes the dominant taxon was listed only to genus, implying that the species might vary locally within the vegetation unit. This is an extreme emphasis on overstory dominants, and the results can mask important variations. In some regions, the overstory dominant may remain constant while ecologically important understory changes occur. Daubenmire (1952) modified Clements's simple scheme by listing two species per unit: an overstory and an understory species, separated by a slash. In this way, a *Pinus ponderosa/Purshia tridentata* tree—shrub community could be classified separately from a *Pinus ponderosa/Agropyron spicatum* woodland—grass community. In a simpler scheme, they would be lumped together as a *Pinus ponderosa* unit.

In order to classify vegetation on the basis of dominance, any quadrat, transect, point, or distance sampling method can be used if it yields data that can be relativized. Dominance (as described in earlier chapters) can be based on relative canopy cover, density, basal area, biomass, or on importance values.

Classification Based on the Entire Flora

The Braun-Blanquet relevé technique (Chapter 9) lends itself to floristic classification of vegetation. Indeed, this method involves several stages that begin with the field sampling described in Chapter 9 and may continue toward the development of classification of vegetation at many levels ("orders"). Because of its historical significance and widespread use among some plant ecologists, the Braun-Blanquet method is described here. Once a sufficient number of stands have been sampled to represent variation within an area, the results are summarized in a **primary data table** (matrix) such as Table 10-1, which gives data for English pastures. In such a primary matrix, the stands and species are listed in the order in which they were sampled and encountered. The data for the primary matrix (e.g. Table 10-1) include cover, using cover classes as shown in Table 9-2 and sociability.

Sociability is an estimate of the dispersion of members of a species, which does not necessarily have any relationship to cover. Two species may have the same cover, for example, but one could be restricted to a few, dense clumps of individuals while the other may be uniformly scattered throughout the quadrat or

Table 10-1 The raw data table, listing species (down) chronologically or phylogenetically, and listing relevés (across) in numerical sequence. The units digit represents cover, and the tenths digit represents sociability. The cross (†) indicates <1% cover and then sociability is understood to be 1 unless noted. Reprinted from *The Description and Classification of Vegetation* by D. W. Shimwell. Copyright © 1971 by University of Washington Press, Seattle.

Relevé number	1	2	3	4	5	6	7	8	9	10	11	12	13	14	15	16	17	18	19	20
Seed plants																				
Hippophae rhamnoides	5·1	5·1	5·1	5·1	5·1	3·2	2·2	4·3	2·2	5·1	3·2	5·1	3·3	4·3	3·2	5·1	5·1	5·1	5·1	5·1
Senecio jacobaea	1·1	+	1·1	+	+	+	+		+		+	+	+	+	1·1	+		+	1·1	+
Solanum dulcamara	2·1	2·1	+	+	+	+	+			1·1	+	+		+		1·1		+	+	1·1
Rubus fruticosus	+	1·1		+	1·1											+	+		+	1·1
Urtica dioica	3·3	1·3	1·3																†3	2·3
Rumex crispus	+	+	+		+	+		+	+	+	+		+	+		+		+	1·1	+
Montia perfoliata				3·4	4·4					4·5		2·3				4·5	2·3	1·3		
Stellaria media				1·2	+							3·4				+	2·3	1·2		
Festuca rubra					+	1·1		1·1	+		2·3		3·3	1·3	1·3					
Agropyron repens						+	2·3	+	3·3	+	+		+	2·3	†2					
Ammophila arenaria						2·3	4·3	2·3	+		3·3		1·3	†3	4·3					
Sonchus arvensis						+		+	+	+				+	+					
Ononis repens						1·1					1·1									
Galium verum									+					+						
Calystegia soldanella							+	1·1	+	+		+		1·1				+		
Poa pratensis									+		1·1		+	+	+					
Agrostis stolonifera								+	†2						†2		+			
Ranunculus bulbosus									+		+		+	+						
Plantago lanceolata									+		+		1·1		1·1					
Veronica chamaedrys									+	+				+					+	
Chamaenerion angustifolium										+		1·1		+		+	1·1		2·3	
Cerastium vulgatum										+		+				+	1·1			+

(continued on next page)

Table 10-1 (*continued*)

Relevé number	1	2	3	4	5	6	7	8	9	10	11	12	13	14	15	16	17	18	19	20
Seed plants (*continued*)																				
Sambucus nigra																+				†2
Cirsium vulgare																+	+			
Heracleum sphondylium																	+			
Inula conyza																	+	1·1		+
Cardamine hirsuta																	+	+		
Hypochaeris radicata																			+	
Arrhenatherum elatius																			1·1	
Sonchus asper																				+
Seedless plants																				
Eurynchium praelongum	1·3	†3	1·3																1·3	1·3
Hypnum cupressiforme	+	†2	†2																	
Brachythecium rutabulum		+	1·3	+	+												+	†3		
Geastrum fornicatum				+	+					†3		+					+	+		
Brachythecium albicans						†3			1·3					+						
Bryum inclinatum						†3					†3			+						
Tortula ruraliformis								†3						†3						
Cladonia rangiformis													+							
Bovista nigrescens														+						
Lophocolea heterophylla																				+
Number of species	8	9	8	8	9	11	7	10	14	12	12	10	10	22	10	11	12	12	12	13

243

stand. Sociability is recorded on a scale of 1 growing singly to 5 growing in large, nearly pure stands and is written as a decimal addition to the cover value. Thus, species A on a data sheet may be represented by a number such as 2·1, which translates as 5–25% cover, with plants occurring singly.

The objective now becomes to generate a second, **differentiated table,** where similar stands lie near each other and species with similar distributions also lie near each other. As the rows (species) and columns are shifted about, it soon becomes apparent that some species are of little use in differentiating groups of stands. Some species occur so rarely that only one or two stands show them; others occur so often that nearly every stand shows them; still others do not form species groups. Such species will be eliminated from the differentiated table. In the case of Table 10-1, 29 of the total 40 taxa were in this unusable category, and they are not included in Table 10-2.

The remaining species are called *differential* or *characteristic* species. Their **fidelity** (faithfulness) to a given association can be expressed on the basis of how few stands outside the association contain them. The exact level of fidelity demanded varies from investigator to investigator, but a general rule is that a species that helps to define an association cannot occur in more than 20% of the stands outside that association.

Fidelity is not related to constance. A species may be restricted to association X, but it may occur rarely even there and have a constance of only 10%. Useful species must have both moderately high fidelity and moderately high constance. Again, the required level of constance varies, but in general it must be above 50%.

As characteristic species are searched for, a reciprocal problem must be solved simultaneously: How many characteristic species must be shared by any two stands before they are considered part of the same association? A common answer is that more than 50% of the total list of characteristic species for association X must be present in any of the stands belonging to it.

The Braun-Blanquet method of classification has not been without criticism and scrutiny in the literature. One of the more notable critiques of this method was offered by Egler (1954), who pointed to several perceived shortcomings but was especially critical of the subjective nature of both sampling and analysis. Some of the subjectivity associated with the method has been reduced (i.e., the method has been made more objective) by computer programs (Wildi and Orlóci 1990).

Classification at Higher Levels

Once there has been sufficient sampling of associations in a region, it may be possible to group similar associations together into a higher level of classification, analogous to the way many species can be grouped into a genus. Recently there has been an attempt in the United States to standardize vegetation classification at the national level, an effort that has combined the expertise of several federal agencies (such as the USDA Forest Service) and nongovernment organizations (such as The Nature Conservancy and the Ecological Society of America) (Federal Geographic Data Committee 1996). One of the main purposes of such an effort is to develop the National Vegetation Classification System (NVCS). The NVCS establishes

Table 10-2 The differentiated table. Species and relevés shown in Table 10-1 have been rearranged so that similar groups (associations or groups A, B, C, boxed) stand together. A large group of species that did not contribute to defining associations has been left out of the table. Also, relevé 14 has been omitted because it was a disturbed site. Reprinted from *The Description and Classification of Vegetation* by D. W. Shimwell. Copyright © 1971 by University of Washington Press, Seattle.

Revised Relevé order	1	2	3	19	20	4	5	10	12	16	17	18	7	6	8	9	11	13	15
Group A																			
Urtica dioica	3·3	1·3	3·3	†3	2·3														
Eurynchium praelongum	1·3	†3	1·3	1·3	1·3														
Group B																			
Montia perfoliata						3·4	4·4	4·5	2·3	4·5	2·3	1·3							
Stellaria media						1·2	+	+	3·4	+	2·3	1·2							
Geastrum fornicatum						+	+	+	+	+	+	+							
Cerastium vulgatum								+	+	+	1·1	1·1							
Group C																			
Festuca rubra							+						+	1·1	1·1	+	2·3	3·3	1·3
Agropyron repens													2·3	+	+	3·3	+	+	†2
Ammophila arenaria													4·3	2·3	2·3	+	3·3	1·3	4·3
Poa pratensis														1·1	+	+	1·1	+	+
Plantago lanceolata																+	+	1·1	1·1

245

a hierarchy of organization for vegetation classification that is quite similar to that used for taxonomic classification of species, as follows:

Division
Order
Physiognomic class
Physiognomic subclass
Physiognomic group
Subgroup
Formation
Alliance

See Table 10-3 for examples of the classification of two types of bogs, cultivated cranberry and sphagnum. Note that this approach allows for equal handling of both natural and human-dominated landscapes.

Cluster Analysis

The objective of **cluster analysis** is to simplify the data and to present them in graphical, rather than table, form. The first step is to express the similarity between any two stands in a single number, called the **community coefficient (CC).** There

Table 10-3   Examples of bog community types as established by the National Vegetation Classification System (Federal Geographic Data Committee 1996). Note that only the Alliance level contains floristic (taxonomic) information.

Cultivated cranberry bog

Division	Vegetated (>1% vegetation cover)
Order	Shrub dominated
Physiognomic class	Shrubland
Physiognomic subclass	Deciduous
Physiognomic group	Cold-deciduous
Subgroup	Planted/cultivated
Formation	Fruit/leaf/nut shrub crop
Alliance	Cultivated cranberry (*Vaccinium macrocarpon*) bog

Sphagnum bog

Division	Vegetated
Order	Herbaceous/nonvascular dominated
Physiognomic class	Nonvascular
Physiognomic subclass	Bryophyte vegetation
Physiognomic group	Temperate
Subgroup	Natural/seminatural
Formation	Saturated
Alliance	*Sphagnum cuspidatum*

Table 10-4   Four methods of calculating the community coefficient (CC) from presence and % cover data.

	Presence and % cover data	
Species	**Stand (quadrat) A**	**Stand B**
No. 1	10	20
No. 2	4	12
No. 3	—	7
No. 4	—	15
No. 5	32	15
No. 6	15	—
No. 7	2	—
No. 8	1	1
Total % cover	64	70

Calculation methods

Formula	**Calculation**	**Community coefficient**
Jaccard, presence only	$\dfrac{C}{A + B - C} \times 100 = \dfrac{4}{6 + 6 - 4} \times 100 =$	50
Jaccard, weighted by cover	$\dfrac{MC}{MA + MB} \times 100 = \dfrac{(10 + 4 + 15 + 1)}{64 + 70} \times 100 =$	22
Sorensen, presence only	$\dfrac{2C}{A + B} \times 100 = \dfrac{2(4)}{6 + 6} \times 100 =$	75
Sorensen, weighted by cover	$\dfrac{2MC}{MA + MB} \times 100 = \dfrac{2(10 + 4 + 15 + 1)}{64 + 70} \times 100 =$	45

A = total number of species in stand A
B = total number of species in stand B
C = total number of species in both stand A and stand B
MA = total % cover of species in stand A
MB = total % cover of species in stand B
MC = total % cover of species in both stand A and stand B, using the lower % cover figure for each species

are several ways to calculate CC, as Table 10-4 shows (see also Goodall 1973 and Mueller-Dombois and Ellenberg 1974). Basically all the formulae indicate the relative number of species shared by any two stands or quadrats. A CC of 100 represents identity (all species shared between two stands), whereas a CC of 0 represents complete difference (no species shared). A generally accepted criterion for acceptance is that any two plots with a CC of more than 50 represent the same association. The result of this type of analysis is a dendrogram such as the one in Figure 10-1.

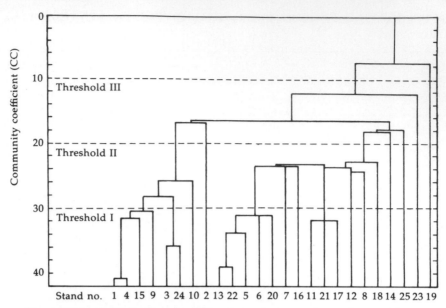

Figure 10-1   A dendrogram resulting from cluster analysis of 25 stands. If one decides that members of an association must exhibit a community coefficient (CC) of 30 or more, then this dendrogram indicates that 15 associations exist, because 15 lines project up through the dashed line called Threshold I. If a CC of 10 were instead selected, there would be only two associations (two lines project through the dashed line called Threshold III). Simplified from *Aims and Methods of Vegetation Ecology.* Mueller-Dombois and Ellenberg. Copyright © 1974 by John Wiley and Sons, Inc. Reprinted by permission of John Wiley and Sons, Inc.

After CC values are computed for every pair of stands, the two stands with the highest CC (closest similarity) are plotted on a graph as vertical lines, which are joined by a horizontal line at that CC value. In Figure 10-1, stands 1 and 4 have the highest CC, 41, and they are joined at that level. Now these two stands are lumped into a new, artificial, second-level stand, and CC values are computed all over again. In this case, it turned out that the highest CC value in the second calculation was for stands 13 and 22. They are joined, CC values are recomputed, and so on until all stands are joined at some, generally low, CC value (a CC of 8 in Figure 10-1).

A cluster analysis dendrogram can be used for classification if the investigator selects some threshold value at which to define associations. As already mentioned, stands of one association are often expected to share a CC of above 50. Using that criterion, every one of the 25 stands in Figure 10-1 represents a different association. If a threshold value of 30 is chosen instead of 50, then 15 associations would result, as represented by the 15 vertical lines projecting up through the dashed Threshold I line in Figure 10-1. If a threshold of 20 is selected instead, then 7 associations would result.

The advantage of cluster analysis over the table method is that the classification process can be quantified when some threshold value is chosen as the lower limit to an association. Also, the relatedness of different associations can be quantified. A computer program (CLUSTAN) has been developed specifically for this technique (Wishart 1987). Kent and Coker (1992) review this and several other clustering techniques.

Other computer programs have been developed to produce cluster diagrams, one of the more popular among vegetation scientists being TWINSPAN (Two-Way INdicator SPecies ANalysis). TWINSPAN is a polythetic, divisive classification procedure (Hill 1979b, Gauch and Whittaker 1981, Gauch 1982). *Polythetic* means that classification and assignment of groups is based on all the species data (as opposed to *monothetic*—see below). A *divisive* method is one that starts with the total population of individuals, then progressively divides them into smaller and smaller groups (Kent and Coker 1992). Indeed, TWINSPAN begins with all samples together in a single cluster and then divides the samples successively and hierarchically into smaller and smaller clusters. The final result is a dendrogram of clusters with each cluster containing one to a few samples. TWINSPAN can now be performed on a personal computer with software from the Cornell Ecology Programs package. It has been widely used on a variety of spatial scales. For example, Monk et al. (1989) used TWINSPAN to provide a classification of deciduous forests of eastern North America. Young and Peacock (1992) used TWINSPAN to test the validity of subjectively drawn vegetation maps of Mount Kenya in East Africa. A modification of the method, called COINSPAN, has been developed recently by Carleton et al. (1996).

As widely used as TWINSPAN has been in recent years among vegetation scientists, it has not been without its detractors. For example, a critique by Groenewoud (1992) concluded that TWINSPAN should clearly not be used for vegetation analyses.

Association Analysis

Association analysis builds a diagram from the top down, rather than from the bottom up, as does cluster analysis. Although association analysis, like TWINSPAN, is a divisive technique, it differs sharply from TWINSPAN and other polythetic techniques in being monothetic. A *monothetic* technique is one that assigns individual samples to groups (associations) based on the presence or absence of one species. Associations are divided on the basis of differential species just as in the table method, but the selection of such species is based on probabilistic, statistical equations rather than on fidelity and constance.

Recall that a positive or negative association between two species can be revealed with a contingency table and a chi-square calculation, and the higher the chi-square value, the stronger the positive or negative association. Use of the chi-square formula requires the placement of random quadrats, so this method of analysis cannot be applied to relevé data. The first step is to compute chi-square values for every pair of species. The species having the highest sum of chi-square values with all other species is selected as the first differential species. In the salt marsh example shown in Figure 10-2, with 77 species, *Puccinellia maritima* (species

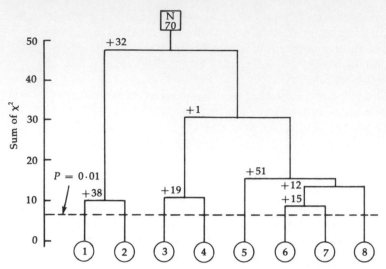

Figure 10-2   A dendrogram resulting from normal association analysis of 77 species in 70 quadrats in a salt marsh. The +32, upper left, represents a line of quadrats that all contain species no. 32, *Puccinellia maritima*. The +38 represents *Spergularia media*, and so on. Each of the seven species that are numbered (*Puccinellia, Spergularia,* etc.) serves to separate quadrat groups, and each species exhibited positive and/or negative associations with other species, as indicated by the high chi-square values along the left axis. A chi-square value below 7 was considered nonsignificant in this example. The result was a division of the quadrats into eight groups, each group representing an association. From Ivimey-Cook and Proctor (1966). By permission of the British Ecological Society.

no. 32) is the primary differential species, with a chi-square sum of 48. The 77 species encountered in the 70 quadrats are represented as a vertical line that splits at a chi-square sum of 48; at the left end of the horizontal line are those quadrats that contain *Puccinellia,* and at the right end are those quadrats without this species.

In those quadrats with *Puccinellia,* the chi-square procedure is repeated to find a second differential species, which is *Spergularia media* (species no. 38) in this example. A horizontal line at a chi-square value of 10 separates the quadrats once again. At the left end are quadrats with both *Puccinellia* and *Spergularia,* and at the right end are quadrats with *Puccinellia* but not *Spergularia.* Each of these groups is analyzed by the chi-square procedure, but in this example no species with a significantly high value was found (a 99% level of significance was used, which is represented by the dashed line in Figure 10-2). The chi-square procedure is repeated for those quadrats without *Puccinellia,* resulting in a total of 8 groups of quadrats. Each group is relatively homogeneous and differs from other groups by the inclusion or exclusion of one of the 7 differential species. Each group may be considered an association.

This method was first described in a slightly different form by Goodall (1953) in the days before widespread computer use. It was elaborated and given the name *association analysis* by Williams and Lambert (1959); their approach requires computer facilities. Still later (1961), they referred to this method as **normal association analysis,** and they presented a complementary procedure called **inverse association analysis.** Inverse association analysis results in groupings of species rather than quadrats (see also Ivimey-Cook and Proctor 1966, and Goldsmith and Harrison 1976). Together, normal association analysis and inverse association analysis do the same thing that the table method does—that is, species and stands, or quadrats, are rearranged so that similar ones lie near each other in the summary picture—but these methods do at least give a numerical value to the degree of similarity and they invite the reader to decide what an association is.

Vegetation Mapping

The most generally useful vegetation maps show both physiognomic and floristic information. In such maps, both the life form and the identity of the dominants of each type are indicated. Needleleaf forests are separated from broadleaf forests, and pine-dominated needleleaf forests are separated from fir-dominated needleleaf forests. If physiognomy alone is mapped, then considerable information is lost.

An early map that proved useful to plant ecologists was developed by E. Lucy Braun as part of her classic book, *Deciduous Forests of Eastern North America* (1950). In this map, Braun used a somewhat hierarchical system of mapping: sections within regions within formations. The sections were often based on geomorphic characteristics, whereas the regions were based on dominant vegetation. For example, most of south-central Kentucky and central Tennessee was mapped as the Mississippian Plateau Section of the Western Mesophytic Region of the Eastern Deciduous Forest Formation.

At even finer spatial scales, maps of individual plants (e.g., trees) of a community can be constructed to indicate specific sizes and locations of these plants. Such maps have been particularly useful in characterizing old-growth forests (e.g., Platt et al. 1988, McCarthy and Bailey 1996), as well as managed forests (McCarthy and Bailey 1994).

The most sophisticated maps, such as those of Henri Gaussen, use symbols and colors to provide environmental as well as vegetational information. A temperate zone deciduous forest with moderately high humidity and moderately warm summer temperatures, for example, would be colored light blue (for moderately high humidity) applied in a flat tint (for a forest), plus yellow (for moderate temperatures), giving an overall effect of light green. An overlay symbol of $\female$ indicates that the dominant species are broadleaf, deciduous trees. A warm desert scrub would be colored in red lines (red for very hot, lines for scrub), plus orange (very dry), with an overlay symbol of $\forall$ to indicate xeromorphic shrubs. UNESCO adopted this method and published a booklet that describes the symbols in some detail (UNESCO 1973). Most vegetation maps, however, use color only to help the reader separate one vegetation type from another; Küchler's maps of the conterminous United States (1964), of Kansas (1974), and of California (1977) are good examples.

Mapping generally begins by examining aerial photographs, such as that shown in Figure 10-3. These pictures are taken from directly above with black-and-white, color, or infrared-sensitive film, and they are generally timed so that there is considerable overlap from one frame to the next. When prints from adjacent frames are placed together correctly and examined with a stereoscopic viewer, the vegetation and topography take on a three-dimensional quality that makes mapping easier. From the photographs, the investigator tries to distinguish as many different communities or vegetation types as possible, outlining the boundaries of each on acetate sheets placed on top of the prints.

The next step is to go back to the field and examine representatives of each tentatively mapped type in order to identify the dominants and check the reliability of the interpretation of the photographs. This phase is often called *ground truth*. A short description of each mapped type is prepared, the boundary lines may be amended, and then the acetate overlays are transferred to some political, topographic, or other kind of base map.

The map being constructed by the USDA Forest Service of ecological units of the eastern United States is an example of current vegetation map construction. Each map unit is composed of physical and biological elements that significantly influence ecological relationships and processes. Principal land-use classes are included to describe human utilization of land and water resources. Each map unit is identified by symbol and name and the following information is included: geomorphology, Quaternary geology, soil taxa, climate (including mean precipitation, temperature, and growing season), potential vegetation, surface water (e.g., rivers, lakes, wetlands), and human use (Keys et al. 1995).

Figure 10-3 Aerial photograph showing several contiguous vegetation types or communities. A synecologist has marked the boundaries of each type with ink lines in preparation for publishing a vegetation map. From *Vegetation Mapping* by A. W. Küchler (1967). By permission of Ronald Press, New York.

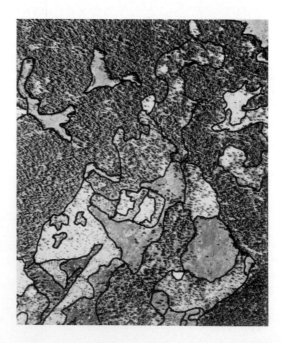

Limitations Imposed by Scale and Objectives   The level of detail that can be shown on a map is limited by its scale. The smallest area that can be easily seen by a reader is a circle 1 mm in diameter. At a scale of 1:1,000,000, this corresponds to a real area of 247 acres (100 ha). Obviously, such a scale is inappropriate to show the location of associations that might have a mean area of 100 acres or less; for that level of detail, a larger scale would have to be used, perhaps 1:100,000.

Many of the boundary lines drawn on vegetation maps are not so abrupt when seen in nature. One community or vegetation type will typically change gradually into an adjacent one through a broad ecotone. The line on the map may represent the midpoint of that ecotone. If the ecotone is unusually large, or if several types coexist in some complex mosaic, then the ecotone or mosaic may be mapped as a separate unit in its own right.

A final word of caution: Some vegetation maps represent **actual vegetation,** existing at the time the map is made; other maps represent **virgin, prehuman vegetation;** and still others represent **potential vegetation,** which could return if human activities ceased in the area. Depending on the objective, the results may be quite different. Küchler's 1977 map of California vegetation is a blend of all three types of maps. For example, the central California valley is mapped as bunchgrass prairie, which is a primeval, virgin type. Today the area is either cultivated land or an annual grassland dominated by aggressive, introduced species; it is highly unlikely that the original grassland could return even if humans and domesticated animals were to vanish. Sierran montane brush fields that today cover thousands of hectares are mapped instead as forest. Fire and logging have removed the original forest cover, but young conifers are slowly growing up through the brush, and forest is the potential vegetation if given enough time. In contrast, many other montane areas are mapped according to existing vegetation. It is likely that most vegetation maps show a bit of all three approaches.

Remote Sensing

Remote sensing literally means observation and measurement of an object without contact. In vegetation classification and analysis, remote sensing generally refers to the use of sensors carried above the earth's surface either by aircraft or, more commonly, by satellites to detect electromagnetic radiation that is either reflected (visible, near infrared, shortwave infrared, and microwave radiation) or emitted (thermal infrared radiation) by vegetation and soil (Short 1982, Curran 1985). Satellite sensors enhance photographic data because they acquire data in nonvisible spectral bands, including infrared and microwave spectra. Recent developments in satellite-based remote sensing is quite timely considering the large spatial scale in which environmental research is done.

Satellite remote sensing systems generally comprise four main components: (1) the satellite, with a platform containing a sensor and an associated computer-based transmitter to receive and transmit information; (2) a ground-based receiving station to receive transmitted information; (3) a computer to do initial processing, translating received information into a computer format and storing it on tape, disk, or CD-ROM; (4) the user's computer with the necessary software to read stored information, translate it into visual images, and perform appropriate statistical analyses (Aber and Melillo 1991).

Some of the more commonly used satellite remote sensing systems are the SPOT, GOES, Landsat, and AVHRR systems. SPOT (*Satellite Pour l'Observation de la Terra*) is a European earth observation program developed by France, Sweden, and Belgium. It has provided a permanent source of geographic data since its launch in February 1986. GOES (Geostationary Operational Environmental Satellites) is actually a series of satellites (GOES I through M) launched by NASA, but owned and managed by the National Oceanic and Atmospheric Administration (NOAA). Begun in the mid-1990s, the GOES I-M mission is scheduled to operate into the 2010s. Although the current Landsat satellites (Landsats 4 and 5) are still in orbit around the earth, the first three are not. The first Landsat satellite was launched in July 1972 to provide repetitive global coverage of the earth's land masses. In 1985, the Landsat system was commercialized, but the Landsat data archive is managed by the U.S. Geological Survey, as it has been since the beginning of the program. Beginning in October 1978, NOAA has operated a series of satellites carrying the AVHRR (Advanced Very-High Resolution Radiometer). This instrument has been especially valuable for several earth observation applications, including regional and global vegetation monitoring.

The satellite systems vary with respect to what part of the electromagnetic spectrum they can sense. The sensor on each system contains one to several bands, with each band detecting radiation at a particular range of wavelength. Most systems contain at least one band in the visible portion of the spectrum and some, including the Thematic Mapper (TM) of the Landsats 4 and 5, have bands at higher wavelengths (Figure 10-4).

Other remote sensing systems keep the detection process closer to the earth's surface through the use of aircraft, including airplanes and helicopters. One of the more commonly used aircraft-based instruments is the Airborne Visible Infrared Imaging Spectrophotometer (AVIRIS), which flies onboard a NASA ER-2 airplane (which is actually a modified U2 spy plane). The uniqueness of the AVIRIS lies in its optical sensor, which senses in 224 contiguous spectral bands from the ultraviolet to the infrared. Compared to satellites, aircraft-based remote sensing has the disadvantage of being limited to smaller spatial scales, but the advantage of enhanced accuracy and precision in characterizing vegetation. Whatever their nature, remote sensing systems provide data that can be applied to a wide variety of uses.

Although photointerpretation for environmental classification and mapping remains a widely used application of remote sensing data (Colwell 1983), other applications are becoming quite common as well. Treitz and Howarth (1996) reviewed several practical applications of remote sensing for Canadian forest ecosystems, including forest resource management. Lachowski (1995) established guidelines for use of remotely sensed digital imagery to map vegetation in the United States. These guidelines describe its application for range allotment mapping, forest plan monitoring, mapping of ecological units, and even the development of a wild and scenic river. Remote sensing data have even been used to quantify seasonal and spatial variation in biogenic emissions of organic com-

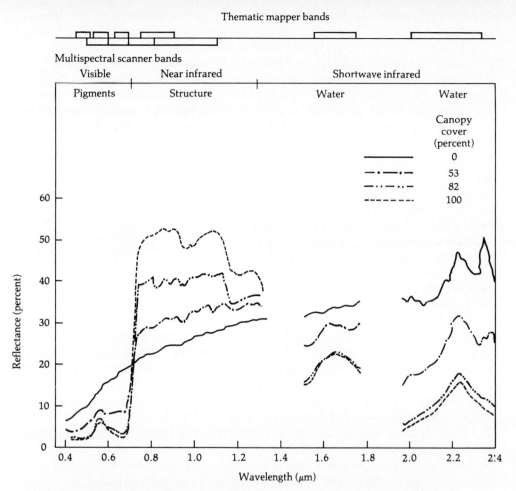

Figure 10-4   Reflectance properties of bare soil (0 canopy cover) and of different amounts of vegetation cover (53, 82, and 100% cover by alfalfa). Reflectance for two wavelength regions, about 1.4 and 1.9 μm, are omitted from the figure because they contain "noise" from nonvegetation sources. Redrawn from Short (1982).

pounds from vegetation (Guenther 1997, Kinnee et al. 1997). A major effort to integrate remote sensing (using satellites and aircraft) with field monitoring of vegetation and micrometeorological measurements is the FIFE project.

The FIFE Project   Although it was not designed specifically for vegetation mapping, the First ISLSCP (International Satellite Land Surface Climatology Project) Field Experiment (FIFE) is an excellent example of how data from a number of remote sensing systems may be combined to provide useful information about natural vegetation. One of the main objectives of FIFE was to understand the biophysical processes that control energy, moisture, and CO_2 fluxes between the land surface and the atmosphere. Because these fluxes are so closely linked to vegetation, FIFE was able to yield important insights into the use of remote sensing data to characterize spatial and temporal patterns of vegetation.

This project can be divided into two separate research efforts, which together provided some of the more useful information on land surface–climate interactions. First, the monitoring effort of FIFE acquired data from many of the satellite systems we have just discussed: AVHRR, Landsat, SPOT, and GOES. These were taken for the Konza Prairie Research Natural Area in north-central Kansas. Second, the Intensive Field Campaigns took observations on the ground ("ground truthing"), including different phases of vegetative development of dominant prairie species. From quantitative comparisons between remotely sensed data and ground-truthed data (essentially a calibration), plant ecologists were able to use the landscape-level information from satellite and aircraft systems to provide an accurate understanding of vegetation patterns. Sellers et al. (1992) provide an excellent overview of the FIFE project.

Spectral Characteristics of Leaves and Canopies   Reflectance spectra for leaves are different from those for bare soil, and they differ from species to species. Even within a species, the spectra differ with leaf area index (LAI) or biomass, as shown for alfalfa in Figure 10-4. Relatively little energy is reflected by leaves in the visible portion of the spectrum (0.4–0.7 μm), most is reflected in the near infrared (0.7–1.3 μm), and less is reflected in the shortwave infrared (1.5–2.5 μm). This is principally due to the absorption of visible light by photosynthetic pigments, reflection of near infrared by cell wall–air interfaces, and absorption of shortwave infrared by leaf water (Gates et al. 1965; Gausman et al. 1977, 1978; Gates 1980).

These spectral features are the basis for obtaining information about two basic physiological processes in plants, photosynthesis and transpiration, which are dependent on solar radiation. When chlorophyll content and leaf moisture change, such as during development, senescence, or under environmental stress, reflectances change in predictable ways. This makes it possible to infer and model the physiological condition of vegetation. The development of such models is an active area of research that will be useful for many ecological purposes, including estimating regional and global primary productivity, energy budgets, and evapotranspiration (Goetz et al. 1985b, Jackson 1985, Tucker et al. 1985).

Although the leaves of most species are spectrally similar to each other, canopy features such as the orientation of leaves and branches, LAI, moisture content, and surface texture and chemistry of species alter the reflectance spectra and provide a basis for the identification of individual species (Gates 1970, 1980; Barrett and Curtis 1982; Bauer 1985; Vanderbilt 1985; Vanderbilt et al. 1985). Fundamental canopy parameters, such as LAI, leaf angle distribution, and green biomass, have been measured for a number of crop types (Bauer 1985, Goel and Thompson 1985). These estimates have led to the routine prediction of the worldwide regional wheat yield one to two months before harvest with better than 90% accuracy (Barrett and Curtis 1982) and a procedure for the yearly estimation of crop inventory and acreage for the state of California (Thomas et al. 1984).

Once thought to be inherently more difficult to interpret because of greater environmental heterogeneity, spectral characteristics of natural vegetation are now interpreted with the same exactness once confined to crops. Many of the current remote sensing studies of natural vegetation determine the **normalized difference vegetation index (NDVI)** to estimate plant biomass. The NDVI is calculated by the following equation: NDVI $= (R_{NIR} - R_{RED})/(R_{NIR} + R_{RED})$; RED$_{NIR}$ and R_{RED} are reflectances of near-infrared and red light, respectively, and are bands used by Landsat satellites. Tieszen et al. (1997) used data from the NOAA AVHRR sensor to derive NDVIs for biomass estimates of Great Plains grasslands. They were further able to use this information to determine the relative contribution of C_3 and C_4 plants to biomass and net primary productivity throughout the Great Plains landscape. Wessman et al. (1997) found that the NDVI was able to differentiate between fire frequencies in Kansas tallgrass prairie, but concluded that it was a poor indicator of canopy biomass. Instead, they found that the multispectral data from the AVIRIS, as opposed to the two-wavelength nature of the NDVI, did a better job of estimating biomass and other vegetation measures such as vertical structure, cover, and greenness.

Data obtained by the AVIRIS, especially those from the higher wavelength end of the spectrum, can also be used to estimate foliar chemistry of natural vegetation. Because near-infrared reflectance spectroscopy (NIRS) detects signals from bending and stretching vibrations in chemical bonds between C, N, H, and O, it can measure concentrations of prominent chemical compounds in organic matter (Wessman 1990). NIRS has been shown to provide a rapid, reliable alternative to wet chemistry methods for determining protein, lignin, and cellulose concentrations in foliage of forest trees (McLellan et al. 1991). Bolster et al. (1996) collected over 550 samples of fresh foliage from hardwood and conifer species at three sites (Massachusetts, Wisconsin, and Maine) in conjunction with overflights of the AVIRIS. They found that NIR data from the AVIRIS predicted quite closely foliar concentrations of N, lignin, and cellulose that were determined in the laboratory.

Ordination

Derived from the German *Ordnung* (order, arrangement, classification), **ordination** means "setting in order." In the context of vegetation analysis, it refers to the arrangement of samples (i.e., stands, plots, quadrats) of vegetation relative to each other based on similarities in species composition and, in some instances, in environmental factors. Ordination summarizes sampled data in a simpler, more space-efficient fashion than do table methods. For example, the differentiated table in Table 10-2 is relatively small, but still contains 285 cells (bits) of data (19 stands × 15 differential species). Ordination of these data might yield a single Cartesian graph with 19 points in what is called **ordination space.** Each point represents a single stand, quadrat, or plot, with the closeness of points representing the degree of similarity in species composition and/or environmental conditions. The location of points in the ordination space can show patterns of more continuous relationships than table methods can.

Discussion of the use of ordination in vegetation analysis brings up the trade-off of information versus meaning. Because ordination is a form of data reduction, some of the information contained in the original data is lost in the ordination diagram. However, as we will see in the following discussion, ordination provides a largely objective analysis of plant communities that uses all the sampled species together. Consequently, ordinations convey a great deal of meaning with regard to patterns of species and environmental factors in plant communities. Because of this, and because of the development of ordination software for personal computers, ordinations have become some of the more common techniques in plant community analysis. Indeed, an informal survey of volume 7 (1996) of the *Journal of Vegetation Science* revealed that at least 25% of the papers published used at least one type of ordination technique.

Some of the more popular techniques are discussed here in an approximately chronological order of development and use. It is not our intent to provide an in-depth mathematical explanation of each technique, although such a background will be important to anyone wishing to pursue these techniques for research. See Pielou (1984) and Manly (1986) for more complete descriptions of the mathematics behind ordinations. Excellent overviews of all these techniques (and others) can be found in Causton (1988) and Kent and Coker (1992).

Polar Ordination

The simplest, earliest methods of ordination involve the least complicated computations and the most involvement by the investigator. The method described here, called **polar ordination** to emphasize that the reference points of each axis (the "poles") are chosen by the investigator, was developed by Bray and Curtis (1957). It was the first ordination technique to receive widespread application (Cottam et al. 1978). Although it is not as commonly used as it once was, it provides interpretation that is often similar to that of more sophisticated methods (Beals 1984), in addition to being easy to learn and a good introduction to the concept of ordination and its objectives.

The first step is to create a matrix of community coefficient (CC) values between all possible pairs of stands. For *n* stands, there will be $(n)(n-1)/2$ different CC calculations. Figure 10-5a shows hypothetical data for seven stands, A–G; the CC values range from 90 (stands A and G) to 20 (stands A and B). The second step is simple calculation of a matrix of dissimilarity rather than similarity. Each pair of stands has an **index of dissimilarity (ID),** wherein $ID = 100 - CC$ (Figure 10-5b). The third step is a transfer of the ID values to a graph. In polar ordination, the two most dissimilar stands are selected as endpoints (poles) on a horizontal axis. In Figure 10-5c, stands A and B are the most dissimilar with an ID of 80, which becomes 80 graph units on axis 1. All other stands are now placed on the same axis by plotting their IDs with reference to A and B. For example, stand C has an ID of 67 with A and an ID of 63 with B. A compass with a radius of 67 units is swung from A, and a compass with a radius of 63 units is swung from B, forming two arcs.

The intersections of the two arcs define a line perpendicular to the axis. Where that line crosses the axis is the position of stand C on that axis.

To spread the points in two dimensions, another axis must be generated, with its own reference stands. One reference stand may be the stand with the largest *e* value, as shown in Figure 10-5, which would be stand C. However, the objective of creating a second axis is to spread out stands that are close together on axis 1, so by agreement the search for the other reference stand is limited to ±0.1*L* units

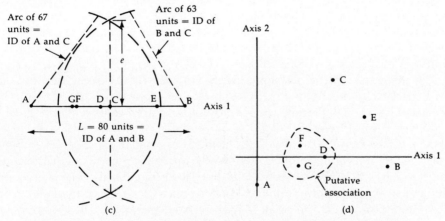

	A	B	C	D	E	F	G
A	—	20	33	65	30	86	90
B		—	37	60	70	52	50
C			—	55	68	49	45
D				—	70	70	75
E					—	45	47
F						—	40

CC values
(a)

	A	B	C	D	E	F	G
A		80	67	35	70	14	10
B			63	40	30	48	50
C				45	32	51	55
D					30	30	25
E						55	53
F							60

ID values
(b)

Figure 10-5   The location of stands in an ordination figure. (a) Community similarity matrix for seven hypothetical stands, A–G, showing community coefficients (CC). (b) Community dissimilarity matrix for the seven stands, showing indices of dissimilarity, or difference (ID). (c) After the endpoint stands A and B have been located on axis 1, all other stands are located between them by swinging arcs that correspond to their ID values with stands A and B. Stand C, for example, has an ID of 63 with B and an ID of 67 with A. The crossing arcs define a line perpendicular to the axis; where that line crosses the axis is the location of stand C. The poorness of fit is the distance *e* from the crossing arcs to the axis. (d) Creation of a second axis serves to pull the stands apart much better than did the first axis alone. Stands F, G, and D (within the dashed line) appear to cluster together and they represent an approximately homogeneous unit, perhaps an association.

away from C on axis 1. In this case, *L* is 80 units long, so stands ±8 units away from C are examined. Stand D then becomes the other reference point of axis 2.

A third axis also may be constructed, generally at right angles to the first two axes, but often two dimensions serve to spread the stands out well enough to account for the greatest variability among stands. Step-by-step procedures for this method are well illustrated in an ecology laboratory manual by Cox (1990).

The end result of polar ordination is a physical arrangement of stands, such as that shown in Figure 10-5d, based on similarity/dissimilarity of species composition. Although this graphical representation is useful in many respects, it has numerous limitations, the correction of which served as the basis for development of other ordination techniques.

The first limitation is the large role of subjectivity on the part of the investigator in determining which stands will serve as the "poles." Second, because the stands are plotted on the basis of ID values, information on spatial variation of individual species is lost. Finally, selecting a different CC formula may change the array of stands.

Principal Components Analysis

Second- and third-generation ordination techniques have been designed to minimize these limitations. Such techniques are true multivariate analyses because they use the original data matrix, wherein each species or environmental factor represents a single variable. Consequently, (1) the investigator is not required to provide subjective endpoints during the computation process, and (2) information on individual species is conserved.

The method of **principal components analysis (PCA)** came into wide use among plant ecologists following the publication of a paper by Orlóci (1966). PCA is still used occasionally today, and it has numerous applications in ecological literature (e.g., McCarthy et al. 1987).

PCA is based on the degree of duplication or correlation in species and environmental factors among samples (i.e., stands, quadrats, plots). The mathematical computations of PCA reduce this duplication by finding highly correlated combinations of all species or environmental factors, with each new combination called a **component,** or more appropriately, a **principal component.** These components are usually expressed two at a time as axes for graphing the samples. For example, Figure 10-6a shows the results of PCA that reduced a 60-plot by 30-species data matrix to an ordination of 60 points (plots) by 2 principal components. Furthermore, by including additional information, such as stand age, we can confer certain meaning on these axes. In this example we can see that principal component axis I (PCA-1, Figure 10-6a) suggests a possible stand age gradient.

The physical arrangement of points (plots, stands, quadrats) in ordination space is analogous to that described for polar ordination—the closer the points, the more similar they are in species composition. PCA goes much further, however, in its utility. By using environmental variables (such as soil moisture, pH, organic matter, and texture) instead of species data, PCA also can show how plots or stands compare to each other with respect to such environmental factors. Also, because data for PCA are input from the original matrix, patterns of occurrence of individual species can be determined. In Figure 10-6a, sugar maple and northern red oak are similar to one another, but quite different from yellow-poplar and, especially, black cherry.

Although PCA represented a vast improvement over polar ordination, it was not without its limitations. First, the main assumption on which PCA is based (that correlations among species or environmental variables are linear) is rarely met. Second, the components (axes) are calculated to be orthogonal (i.e., assumed to be completely noncorrelated with each other), a potentially serious problem considering that important environmental gradients are often themselves intercorrelated (Kent and Coker 1992).

Detrended Correspondence Analysis

Limitations noted for PCA were corrected in large part by the development of another ordination technique, called **correspondence analysis (CA),** or **reciprocal averaging (RA).** Although we will not describe this technique in detail here, it gained great popularity in its applications in vegetation analysis beginning in the early 1970s (Gauch 1982). One of the problems of CA, however, which is shared by PCA, is referred to as the "arch" or "horseshoe" effect. This problem, seen as an arching pattern of points in ordination space, arises because the second axis in these techniques is a curvilinear (quadratic) function of the first axis (Gauch 1982). The second problem of CA is that it tends to compress points toward the ends of the first axis, thereby leading to the spurious conclusion of similarities in species composition (Gauch 1982).

These problems were corrected by the development of yet another ordination technique that removes the "arch" effect by a mathematical process known as **detrending,** hence its name, **detrended correspondence analysis (DCA)** (Hill 1979a, Hill and Gauch 1980, Peet et al. 1988). Mathematically, DCA ensures that similar floristic or ecological differences between stands will be expressed as similar distances in ordination space. The improvements that DCA made with respect to earlier techniques were so great that it is now one of the more commonly used ordination methods (Palmer 1993).

It is also possible to assess the relative importance of environmental factors in influencing species' patterns, an approach used by Greer et al. (1997) to determine factors that determined the composition of fern communities in hardwood forests of southeastern Ohio. It should be noted, however, that such an approach assesses the importance of environmental factors indirectly through correlation (see below). Greer et al. (1997) were able to examine direct gradient effects through another, nonordination analytical technique.

Nonmetric Multidimensional Scaling

Another ordination technique that removes the "arch effect" of earlier methods is **nonmetric multidimensional scaling (NMDS).** Indeed, it has been shown to be superior to DCA for displaying known species' gradients when used on

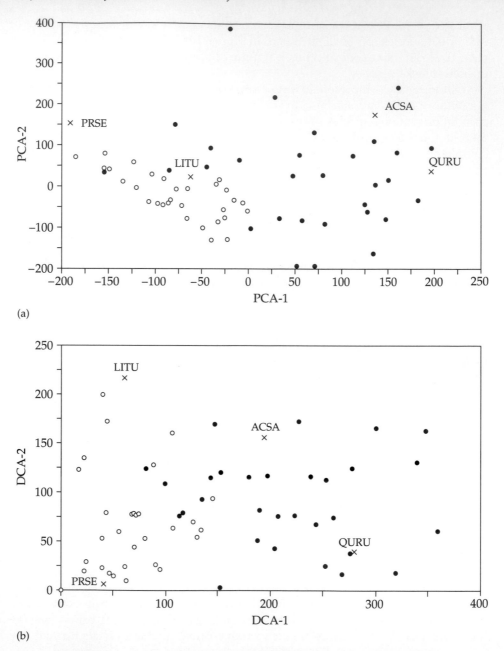

(a)

(b)

Figure 10-6   Comparison of data from Gilliam et al. (1995) using (a) principal components analysis (PCA); (b) detrended correspondence analysis (DCA); (c) nonmetric multidimensional scaling (NMDS); and (d) canonical correspondence analysis (CCA). Open circles represent young stands, and filled circles represent mature stands. Important species' locations are

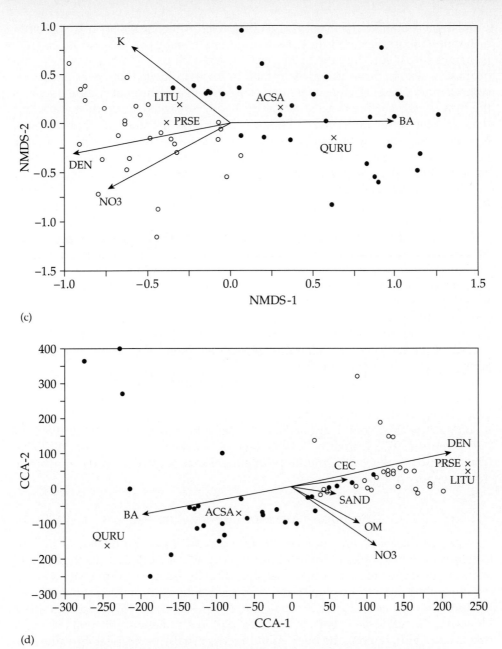

(c)

(d)

shown as x with the following codes: ACSA = sugar maple, LITU = yellow poplar, QURU = northern red oak, PRSE = black cherry. Environmental variables for NMDS and CCA are as follows: BA = tree basal area; DEN = tree density; NO3 = soil nitrate; K = soil potassium; OM = soil organic matter; SAND = sand content of soil; CEC = cation exchange capacity.

simulated vegetation data (Kenkel and Orlóci 1986, Minchin 1987). NMDS finds an ordination of sampling units (e.g., plots, quadrats, stands) in a given number of dimensions (axes) such that the distances between each pair of sample units in the ordination are, as far as possible, in rank order with their differences in species composition. These differences are measured by some dissimilarity index, such as the Bray-Curtis coefficient used in Figure 10-5b, computed from the community data. It is the term "rank order" that gives NMDS its "nonmetric" characteristic, for rather than deal with specific measured distances between sample points (a parametric measure), NMDS prepares a rank of pairs of sample points from closest to farthest apart (a nonparametric measure). It then correlates these orders of rank with the dissimilarity values. This is comparable to the distinction between Pearson product-moment correlation (parametric) and Spearman rank correlation (nonparametric) (Zar 1996).

An optimum NMDS ordination is one that finds the best possible rank order fit between the dissimilarities in species composition of samples and the respective distances between samples in ordination space. To achieve this, it is necessary to use a successive improvement approach. Starting with some initial ordination, the sample points are gradually moved around in an iterative (repeated) fashion to improve the fit between distances and dissimilarities. The main objective of these iterations is to minimize "stress," which is a measure of how badly the distances are in rank order agreement with the dissimilarities. It often takes from 30 to 100 iterations to reach the minimum stress.

Although NMDS has not had the same widespread use as DCA, as mentioned previously it has been shown to be superior to DCA in many respects (Minchin 1987). Its use in nonecological applications actually predates that of DCA for vegetation data, with ecological application of NMDS beginning in the 1970s (Fasham 1977, Prentice 1977). Currently, NMDS software is available through the DECODA (Database for Ecological COmmunity DAta) package written by Peter Minchin at the University of Melbourne.

Canonical Correspondence Analysis

One of the main limitations of DCA and NMDS may be related to the objectives of the investigator. For example, the "units" of DCA express the average standard deviation of species turnover, thus directly assessing β-diversity (i.e., diversity among stands or communities) and gradient lengths among stands. DCA, therefore, would be adequate for such simple assessments of plant communities. However, the relationship between species' patterns and environmental factors, if that is one's aim, can only be inferred indirectly, usually by correlating values for environmental variables with ordination axis scores (e.g., Roberts and Christensen 1988, Gilliam et al. 1993, Greer et al. 1997, McCarthy et al. 1987). This limitation is not unique to DCA but is shared by all techniques described thus far. Because such correlations determine the importance of environmental factors in influencing species' patterns (gradients) indirectly, these are all forms of **indirect gradient analysis,** and all suffer the common limitation of assuming that ordination axis scores and environmental variables are related in an inherently linear fashion.

The latest development in ordination techniques was designed to compensate for this most troubling limitation of all, as well as to alleviate some of the problems associated with DCA, such as an occasional inability to handle complex sampling designs properly (Palmer 1993). **Canonical correspondence analysis (CCA)** is a technique developed by ter Braak (1986) to include environmental variables as part of the mathematical computations. Palmer (1993) assessed the advantages of CCA and found, as did Minchin (1987) for NMDS, that it "outperformed" DCA in displaying species' gradients under a variety of conditions with simulated data. Because CCA performs a least-squares regression of plot scores (species' weighted averages) as dependent variables onto environmental variables as independent variables, CCA is an example of **direct gradient analysis** (Palmer 1993). Thus, CCA is able to produce plot and species locations in ordination space and to generate environmental vectors (lines) originating from the center. Furthermore, because the length of these vectors represents the gradient length of an environmental variable, vector length is proportional to the importance of the gradient; that is, important environmental factors will have long vectors and unimportant factors will have short vectors. The result is a useful, meaningful way to look at species' patterns in communities and how they relate to, and are influenced by, their environment. Indeed, CCA is replacing DCA as the most widely used ordination technique. It is also quite common to see CCA used in combination with other techniques, such as cluster analysis, to characterize vegetation (e.g., Young and Peacock 1992).

Brief Comparison of Methods

Most of the comparative studies that assess the analytical performance of ordination techniques (e.g., Kenkel and Orlóci 1986, Minchin 1987, Groenewoud 1992, Palmer 1993) have used simulated data to allow ultimate knowledge about the degree of variation in species composition and distribution. Here we provide a comparison of four ordination techniques using actual data from the field.

Gilliam et al. (1995) compared tree species composition among stands of two contrasting ages in a central Appalachian forest: young, even-aged stands following clear-cutting (20 years old) versus mature, uneven-aged stands (>80 years old). They sampled about 3500 trees in sixty 0.04 ha circular sample plots, fifteen plots per watershed with two watersheds combined for each stand age. The data set from this study was subjected simultaneously to PCA, DCA, NMDS, and CCA, and the results are compared in Figure 10-6. Gilliam et al. (1995) concluded that the relative importance of species changed appreciably over time in these stands, with shade-intolerant species dominating early in succession and shade-tolerant species replacing them later in succession. Let us now ask, can we arrive at this conclusion with all four techniques?

PCA did a poor job in separating stands on the basis of age. Although the first axis (PCA-1) suggests a stand age gradient, there was an extremely high degree of overlap of young and mature stand plots (Figure 10-6a). PCA did a reasonable (but not good) job of distinguishing between early- and late-successional species (black cherry/yellow-poplar versus sugar maple/red oak).

DCA did a much better job in depicting the age gradient on axis 1 (DCA-1), with much less overlap than was found with PCA. DCA appeared to do well in comparing species. Early successional species are actually close together along the age gradient (axis 1), but are quite distinct from the late successional species. Red oak is separated from sugar maple toward the mature end of the gradient (Figure 10-6b), though we can get no further information than this from the ordination. NMDS and CCA will explain why this is so.

Let us first interpret Figure 10-6c. NMDS calculates environmental vectors in a manner that appears to be similar to CCA. The fundamental difference, however, is that environmental data in NMDS are analyzed independently of species data. Thus, this is indeed another form of indirect gradient analysis. However, the combination of both environmental and species data is quite informative, showing that the mature plots had high tree basal area and low tree density, whereas young plots had low basal area and high density. NMDS improved upon DCA's ability to compare yellow-poplar and black cherry, placing them closer together in ordination space.

Ordination axes (e.g., CCA-1 and CCA-2) are less important in CCA than are the environmental vectors, shown as lines originating from the center. The arrows indicate the direction of positive values of a given factor in explaining species' patterns (negative values project symmetrically in the opposite direction, but this is not shown). These vectors offer essentially the same types of interpretations as those in NMDA, but it must be kept in mind that they are calculated in CCA along with the species data and thus are considered direct gradient analyses. As with DCA and NMDS, CCA effectively depicted the age gradient, and like NMDS it was more effective than DCA in showing early successional species to be very similar to each other and dissimilar to late successional species (Figure 10-6d). Although CCA also separated red oak from sugar maple toward the mature end of the age gradient, it further showed which factors best explain that gradient. Plots aligned closely along two environmental gradients, themselves opposites of each other, stem density and basal area, indicating (1) high numbers (density) of stems in young stands and low numbers in mature stands; and (2) large trees (basal area) in mature stands and small trees in young stands. The red oak–sugar maple difference indicates failure of red oak to reproduce in these stands, that is, there are more large oak trees than small oak trees. Some soil variables (SAND, CEC) are relatively unimportant factors, whereas others (OM, NO_3) are of intermediate importance to explain species' patterns not explained by the primary gradient of age, density, and basal area.

In conclusion, all techniques except PCA were able to discern effectively between young and mature stands. Furthermore, NMDS and CCA were able to provide useful information on underlying environmental gradients that may have determined the observed distribution of plots, and did a better job than DCA in comparing species. Finally, note that the objective of this comparison is not to prescribe one ordination technique over another but to present the various methods that are available and the advantages and disadvantages of each.

Summary

The investigator's biases and objectives affect the choice of method for data analysis. Further, some methods of analysis require certain sampling procedures. In the table method, associations are defined on the basis of differential or characteristic species that have high fidelity and constance values. Associations are presented in a large differentiated table, which manages to preserve most of the original sampling data of species and stands.

In contrast, ordination reduces the sampled data to a graph showing stands as points in space. The closeness of stands on such a graph represents their degree of similarity. For indirect gradient analyses, axes of ordination graphs may correspond (i.e., be correlated) to gradients of environmental factors.

Direct gradient analysis is a form of ordination that includes environmental data in the determination of stand locations on the graph. Direct gradient graphs display stand location along with environmental vectors, the length of a vector being proportional to the importance of the environmental gradient in explaining species' patterns and stand locations.

Cluster analysis uses community coefficient (CC) values of stand pairs to construct dendrograms that show the relatedness of stands. In contrast to the table method, the result invites readers to choose their own criteria for defining associations, and it quantifies the amount of relatedness for stands of the same association and between associations. However, the table method and the cluster analysis method are equally subjective.

Association analysis also builds a dendrogram of stand-to-stand relationships, but its construction is based on differential species rather than on CC values. The differential species chosen are those that show the greatest degree of nonrandom association with other species; they are not chosen on the basis of fidelity and constance, as in the table method. Normal and inverse association analyses, however, accomplish the same effect as the table method.

Sampling results can also be summarized in the form of vegetation maps. Some mapping schemes permit environmental data to be shown in addition to vegetational data. The sources of mapping data are aerial photographs and ground truth. The level of detail is dependent upon map scale and the skill and biases of the cartographer. Some vegetation maps show only existing vegetation, but others may include potential vegetation and virgin (prehuman, climax) vegetation. Recent developments in remote sensing techniques not only improve vegetation mapping accuracy and extent but can describe the growth and vigor of plant biomass over large areas.

CHAPTER 11

SUCCESSION

Plant succession is a directional change in the species composition or structure of a community over time. For such a simple definition, succession is a deceptively complex concept and process. The source of the complexity lies in the temporal (time-related) and spatial (space- or area-related) scales over which succession occurs, coupled with the logistical difficulties of studying natural communities across these scales.

The complexity surrounding plant succession results in a lack of consensus regarding the underlying mechanisms that direct successional change in plant communities. Indeed, there has been considerable and often heated debate on this topic, occasionally leading to some degree of rancor among plant ecologists. Mostly, however, the debates have been beneficial by inspiring new ideas, new insight into old ideas, and perhaps most important to the science of plant ecology, research efforts to test predictive hypotheses. We will discuss many of these hypotheses and highlight some of the controversies in this chapter.

Succession has been the subject of a great deal of research by many naturalists and plant ecologists for the past three centuries. The term itself may have originated with Thoreau (1860). There have been papers published on vegetation change associated with bogs, rocks, and dunes since as early as 1685, although it was the pioneering work of Johannes Warming and Henry Chandler Cowles (see Chapter 2) that galvanized the study of succession as a new subdiscipline of plant ecology (Olson 1958). Frederick Clements published a book devoted entirely to plant succession in 1916, a time when the topic was gaining prominence among ecological studies.

It is notable that these last two plant ecologists—Cowles and Clements—set the stage for the long-standing debates on the nature of succession by expressing diametrically opposed views on the subject. Clements, with his "superorganism" view of the plant community, saw succession as a highly predictable process. In contrast, Cowles, with his dynamic view of vegetation, described succession as "a variable approaching a variable, rather than a constant."

The Direction and Time Scale of Succession

Of central importance in our definition of succession is the term *directional.* It refers to progressive change in the dominance of the plant community such that once-dominant species will not return to dominance unless some exogenous disturbance occurs that reinitiates succession. Accordingly, our definition of succession implies a particular time scale. It excludes seasonal and short-lived changes in plant communities, regardless of how spectacular they may be. For example, in most years in tallgrass prairie, C_3 plants (including grasses and perennial dicotyledonous herbs—see Chapter 15) are dominant in early spring when heat and moisture stress are minimal. These decline rapidly in dominance later in the spring when C_4 grasses take over at a time when heat and moisture stress are high (Archibold 1995). This repeated seasonal pattern is not considered succession.

Similarly, in ephemeral herb communities of warm deserts, many annual plants are present in large number as seeds in the soil, being maintained in the dormant state by a water-soluble germination inhibitor (Vankat 1979). A large rainfall washes off the inhibitor, not only facilitating germination but also providing sufficient soil moisture to allow each plant to complete its life cycle—growth, production of flowers, and release of more seed to replenish the desert soil seed bank. However, because these are all annual plants, they die at the end of their short life cycle, as brief as eight days for a species of the Sahara Desert (Cloudsley-Thompson 1977), creating little change for the next year's community and beyond. In fact, succession is virtually absent in deserts (Vankat 1979).

At the other extreme of temporal scales—from tens of thousands to even millions of years—other processes, such as climate change, evolution of new species, and even major landscape-scale changes (e.g., glaciation and sea-level fluctuation) exert an overriding influence on the species composition of plants in a given area. The successional processes we will describe in this chapter, and the mechanisms that best explain them, do not operate on such broad temporal scales. Thus, most studies of plant succession have dealt with time periods of less than 500–1000 years.

Types of Succession

A "unit" of succession, from the beginning of the temporal sequence to a state of minimal species change, is called a **sere,** and each distinct community type within the sere is a **seral stage.** The initial seral stage is referred to as the **pioneer community** (composed of pioneer species), whereas the final seral stage is the **climax community** (composed of climax species). A community exhibiting change toward the climax stage is a **seral community.** Species change does not cease once a community has come to a climax stage, but this change is small and nondirectional (random). At this point the community is in a state of **dynamic equilibrium** (Figure 11-1).

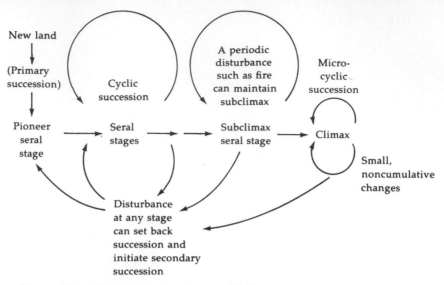

Figure 11-1   Diagrammatic pathway of different types of succession: primary, secondary, and cyclic. The climax stage is in a state of dynamic equilibrium.

Primary versus Secondary

New land is continuously being exposed to colonization by plants, such as when volcanic explosions create new islands or cover a pre-existing landscape with a new surface of ash or lava, when a landslide buries a swath of vegetation beneath coarse rubble, or when an advancing sand dune smothers a coastal forest. New land is also exposed by mountain building, by the filling of ponds, by the changing of river courses, and by the expansion of tropical deltas under the influence of silt-trapping mangroves.

The establishment of plants on land not previously vegetated is called **primary succession.** If the pioneer community becomes established on a wet substrate, such as the edge of a gradually filling pond, then the invasion is a **hydrarch** primary succession. If the pioneer community becomes established on a dry substrate, such as exposed granite, then the invasion is a **xerarch** primary succession. In either case, the direction of succession is usually toward the most mesic site and the most mesophytic community possible, given the limitations of regional climate, topography, and soil parent material.

Secondary succession is the invasion of land that has been previously vegetated, the pre-existing vegetation having been destroyed by natural or human disturbances such as windthrow, fire, logging, or cultivation. However, the barren surface is not as severe as the surface in primary succession because much of the soil remains (although it is sometimes depleted of nitrogen or other nutrients), and many plant propagules (seeds, bits of rhizomes, etc.) are already present in the soil.

Consequently, rates of secondary succession are generally much higher than those of primary succession (Huston 1994). In some cases, however, primary succession can be nearly as rapid as secondary succession. As we will discuss later in this chapter, the approximately 200-year period for primary succession in areas of glacial retreat in Alaska (Chapin et al. 1994) is about the same as that required for secondary succession from abandoned cropland (known as **old-field succession**) in North Carolina (Christensen and Peet 1984).

Other types of disturbance create a mosaic of primary and secondary successional sequences. Consider the eruption of Mount Saint Helens in Washington State in 1980. Areas adjacent to the eruption are undergoing primary succession as a result of the force and heat of the blast and the deposition of pyroclastic material. Interestingly, Edwards and Sugg (1993) have found that pioneer microorganisms and plants in this primary succession are helped by arthropod (insect) fallout as a source of organic matter and nutrients. Farther away (20–40 km) from the eruption site, Zobel and Antos (1997) found that tephra (airborne volcanic ejecta) deposition 5–10 cm thick greatly modified structure and composition of forest understory communities, although it neither totally eliminated existing species nor favored establishment of new ones.

Autogenic versus Allogenic

Successional change in community composition responds to two general types of environmental changes: (1) those caused by the organisms themselves are **autogenic factors,** and this type of succession is called **autogenic succession;** and (2) those caused by external forces not affected by organisms are **allogenic factors,** and this type of succession is called **allogenic succession** (Huston 1994).

Environmental change caused by organisms in autogenic succession include light capture (shade) by canopy leaves, production of detritus, uptake of water and nutrients from the soil by roots, and additions of nitrogen to the soil via symbiotic and nonsymbiotic fixation. Thus, some autogenic factors (e.g., nitrogen fixation) can be facilitative, allowing for the establishment of other species. Other autogenic factors (e.g., light capture and root uptake) can be inhibitory, preventing establishment of plants that might otherwise occupy the site. By contrast, regional climate change (such as drying, warming, or cooling trends) can alter species composition, but it is allogenic because it is not under the control of the plants themselves. Also, river meanders that fluctuate in time (i.e., periodic flooding) may serve to move succession back to a meadow or tundra stage (Walker 1997). Another example would be the tephra deposition described previously (Zobel and Antos 1997).

Although we will emphasize autogenic succession in this chapter, it is important to remember that autogenic and allogenic processes invariably interact throughout the course of any successional sequence. For example, Gilliam and Turrill (1993) concluded that herbaceous layer development is influenced greatly by allogenic factors early in secondary succession of hardwood forests, but is influenced more by autogenic factors (such as canopy closure) as the stand matures and becomes more stratified.

Progressive versus Retrogressive

Successional communities usually increase in diversity and biomass over time, with the habitat becoming more mesic (moist); this is called **progressive succession** and will be the primary focus of discussion in this chapter.

By contrast, **retrogressive succession** results in a decrease in diversity and biomass, often with the habitat becoming either more hydric (wet) or xeric (dry). Walker et al. (1981) studied vegetation development on coastal sand dunes of Australia and found lower plant height and biomass on the oldest dunes. This was allogenic succession, being driven by soil weathering and nutrient depletion over long periods. Other retrogressive successions are autogenic. Drury (1956) noted that succession in Alaskan flood plains may at first be progressive, leading from a sedge meadow to a white spruce forest with low shrubs of cranberry and blueberry. The dense shade, however, encourages the growth of a dense moss carpet and the encroachment of a shallow permafrost (frozen soil and water) layer. As the soil moisture rises, sphagnum moss invades, white spruce is replaced by black spruce, and ultimately retrogression to a sedge meadow can result (Figure 11-2).

Another example of retrogressive succession comes from the Mendocino

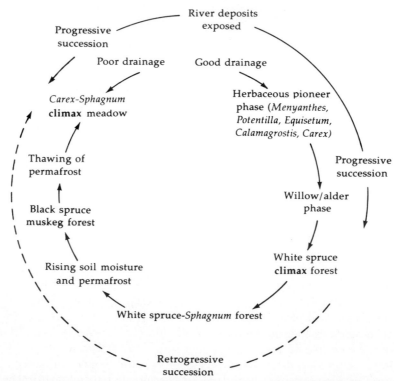

Figure 11-2 Progressive and retrogressive succession on Alaskan flood plains. From Drury, W. H., Jr. 1956. "Bog flats and physiographic processes in the upper Kuskowin River region, Alaska." *Contributions from the Gray Herbarium* 178:1–30. By permission of the Gray Herbarium.

coast of California. A series of coastal terraces, elevated and exposed to plant in-vasion for 500,000 years, shows degeneration in vegetation and soil environment over time, leading from a rich, mesic forest of redwood, spruce, fir, and hemlock to an open, dwarf forest of pygmy pine and cypress. The dwarf forest is underlain by a very acidic, leached podzolic soil with a shallow hardpan that creates flooding in winter and drought in summer (Figure 11-3; Jenny et al. 1969).

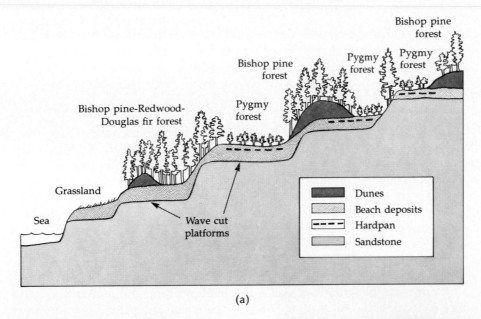

(a)

Figure 11-3 (a) The sequence of older terraces along part of the Mendocino coast of California. (b) On the oldest (highest) terraces, soil development has led to the formation of a shallow hardpan; above such soil is the climax commu-nity—an open, dwarf forest of pine and cypress, much more depauperate than earlier seral stages of spruce, fir, hemlock, and redwood. (a) Reprinted by permis-sion from Jenny et al. (1969). (b) Courtesy of H. Jenny.

(b)

Our discussion thus far has been on **directional succession,** characterized by accumulating changes leading to community-wide changes. The climax community is generally considered to be at a state of dynamic equilibrium (Huston 1994), wherein **cyclic successional** changes may occur at finer spatial and temporal scales. These changes occur because the life span of overstory plants is finite, and their disappearance from the canopy may open the site to an invasion by new species. In some climax communities, the juvenile forms of overstory plants are well adapted to life beneath the parent; when the parent dies, they will replace it in the overstory. In such cases, there will be no local, cyclic succession. In other communities, however, the overstory may inhibit the growth of juveniles beneath it—juveniles of its own kind or those of any species—and in such cases, local, cyclic succession will occur when the overstory plant dies.

Shrub-dominated communities often exhibit cyclic succession. Open areas within desert scrub in Texas, for example, appear to go through a short cycle of invasion by creosote bush (*Larrea tridentata*) followed by invasion by a cactus, Christmas tree cholla (*Opuntia leptocaulis*), followed by a reversion back to bare ground (Figure 11-4; Yeaton 1978). Bare sites may be invaded by *Larrea* seedlings because the small seeds and fruits are abundant and may be widely wind-dispersed. Once a shrub is established, it may attract birds and rodents that scatter the seeds and fruits of *Opuntia*. As the cactus grows, its roots may compete for soil moisture with *Larrea,* leading to *Larrea* mortality. Now removed from the protective influence of a shrub canopy, the shallow root system of the cactus may be subject to erosive forces. Large *Opuntia* plants also attract burrowing rodents that further weaken the root system, and the plant dies. Now open space is available for *Larrea* seedlings to invade.

A northern hardwoods climax forest in New Hampshire offers another example of local, cyclic succession. Seedlings of the sugar maple (*Acer saccharum*), American beech (*Fagus grandifola*), and yellow birch (*Betula alleghaniensis*) are not positively associated with overstory parents (Forcier 1975). Beech seedlings and saplings, for example, are positively associated with overstory sugar maple but are negatively associated with overstory beech. This means that when a beech tree dies, its space in the canopy will not immediately be filled by another beech. Forcier concluded that the following cyclic microsuccession would occur (Figure 11-5, p. 276). Yellow birch, whose seedlings were the most widespread of all three species, would most likely be the first species to fill the gap. It would grow rapidly and a sugar maple understory would develop beneath it (sugar maple seedlings are positively associated with birch overstory). When the relatively short-lived birch died, it would be supplanted by a sugar maple tree. However, beech seedlings are positively associated with maple overstory trees, and beech would ultimately succeed to the canopy when the maple later died.

Chronosequence versus Toposequence

Typically, many plant communities coexist in a complex mosaic pattern. That is, one climax community does not cover an entire region. Sometimes the mosaic reflects a periodic, local disturbance, such as fire, or it might reflect the progressive

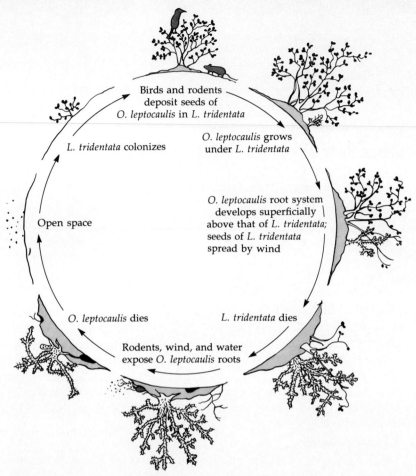

Figure 11-4 Cyclic succession in a desert in Texas (*Opuntia leptocaulis* and *Larrea tridentata*). From Yeaton (1978). By permission of the British Ecological Society.

exposure of new land, as behind a retreating glacier. In the boreal forest of Canada, for example, a spruce-fir forest forms a matrix within which localized patches of meadow, aspen, and aspen with an understory of spruce-fir occur. These patches represent different stages of recovery (seral stages) from fire, windthrow, or other disturbances to the matrix type. The mosaic expresses a successional relationship and is called a **chronosequence.**

In other cases, the mosaic reflects topographic differences, such as south-facing versus north-facing slopes, basins with poor drainage and fine-textured soil versus upland slopes with good drainage and coarser soil, or different distances from a stress such as salt spray. In such cases, the communities within the mosaic do not bear a successional relationship to one another; they constitute a **topo-sequence.**

Figure 11-5   Cyclic microsuccession in a climax hardwood forest in New Hampshire. From "Reproductive strategies and the co-occurrence of climax tree species," by L. K. Forcier, *Science* Vol. 189, pp. 808–810, 5 September 1975. Copyright © 1975 by the American Association for the Advancement of Science.

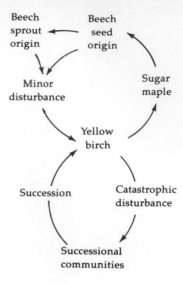

It is possible that the various communities within a toposequence represent different climax communities. As we will see later in the chapter, Clements's view of the climax was on a broad spatial scale, controlled ultimately by climate—hence the term **climatic climax.** However, because of potentially great topographic discontinuities (such as different parent materials, poor drainage, or abrupt changes in soil texture), some communities represent discrete **edaphic climaxes.** It was the awareness of this type of spatial variability that led Whittaker (1953, 1973a) to reject Clements's climatic climax in favor of a **pattern climax.**

Methods of Documenting Succession

One of the serious challenges to plant ecologists in studying succession is the great temporal scale on which it occurs. Whereas an active career for an ecologist might be, at the longest, about 60 years, most seres require at least 100 years to reach a climax stage. Thus, it is nearly impossible for a single ecologist to follow species changes at a disturbed site for the entire period of successional development.

Two approaches usually allow plant ecologists to address this logistical difficulty: (1) use of historical, long-term data from single plots; (2) observations taken from closely adjacent plots of different stand ages (i.e., chronosequences). Each approach has its own set of advantages and disadvantages.

The most direct, unambiguous way to document succession is to make repeated observations of the same area over time. Permanent plots can be established and a variety of vegetation measurements (see Chapter 8) can be made. These can be repeated (e.g., every year, decade, or longer time period) and the data archived for later use by other ecologists. Assuming that the sampling techniques are consistent from one investigator to the next, advantages of this method include accuracy in documenting vegetation change through time. Disadvantages include the lack of adequate assessment of spatial variability.

Most successional schemes have been determined intuitively from observations on nearby plots of different successional ages. The objective is to find a series of plots that have been exposed to primary succession or disturbed and opened to secondary succession at different, known times. In contrast to the repeated measures approach, this method has the advantage of accounting for spatial variability, but the disadvantage of a lower level of accuracy.

Mark Roberts combined both of these approaches in an extensive study of secondary successional aspen forests in northern lower Michigan (Roberts and Richardson 1995, Roberts and Christensen 1988, Roberts and Gilliam 1995b). Here we highlight the results and conclusions of these studies in the context of the two sampling approaches.

Repeated Measures

In 1938 Frank Gates and W. F. Ramsdell initiated a long-term study at the University of Michigan Biological Station in the northern tip of the lower peninsula of Michigan. They established a single 0.04 ha square plot on each of seven different soil types. Four of these were chosen for Roberts's study to represent forested upland sites of the region and a gradient from dry-mesic to wet-mesic sites and vegetation types. In 1979, the year of Roberts's study, stands were all aspen dominated, but were 60–70 years old—close to the life span for the fast-growing, shade-intolerant aspens. Archived data on tree species composition, basal area, and density were available for six sample years prior to Roberts's study: 1938, 1945, 1951, 1955, 1968, and 1973. Thus, the entire study was based on continuous data for 40 years with seven repeated measures (Roberts and Richardson 1985).

This extensive data set allowed Roberts and Richardson (1985) to identify five pattern changes in individual tree species that influence successional trends: (1) early dominance, (2) delayed dominance, (3) persistence, (4) progressive recruitment, and (5) late recruitment. Furthermore, they concluded that the importance of these patterns varied greatly with site type. For example, patterns 1, 3, and 4 predominated on dry-mesic soils, whereas all five occurred on mesic soil. The long-term, detailed nature of the data even allowed them to construct thinning curves (i.e., thinning lines, see Chapter 4), wherein they were able to detect the effects of insect defoliation on changes in stand biomass and density (Figure 11-6b). Finally, and more germane to their study of successional patterns, the authors were able to provide accurate records of change in stem density and basal area for individual tree species over a long period (Figure 11-6c,d).

Observations of Nearby Plots of Different Successional Ages

Roberts and Christensen (1988) established a single 0.1 ha (20 m × 50 m) plot in each of 70 successional aspen stands ranging in age from 1 to 90 years old. In addition to collecting the vegetation data described above, they also measured cover

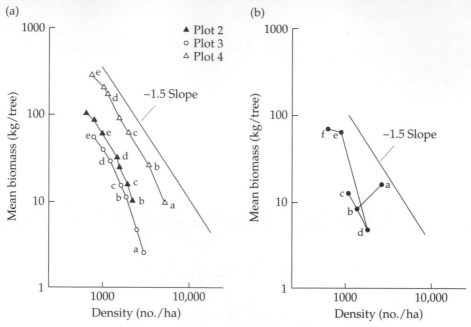

(a)

(b)

Figure 11-6   (a) Logarithmic graphs of changing largetooth aspen biomass versus tree density for three long-term plots in successional aspen forests of northern lower Michigan (Roberts and Richardson 1985). Approximate stand ages (with sample dates) are as follows: *a,* 20 (1938); *b,* 30 (1945); *c,* 35 (1951); *d,* 49 (1955); *e,* 60 (1973); *f,* 67 (1979). A line representing the theoretical −3/2 law of thinning (Yoda et al. 1963) is included for comparison. Self-thinning (see Chapter 4) is evident for healthy stands beginning with a stand age of 30–35 years. (b) Graph similar to (a) for a plot that experienced insect defoliation beginning about 1940 and continuing until 1955. Initial defoliation caused extensive mortality (a decline in both density and

of herbaceous layer species and several environmental variables, including soil physical and chemical characteristics. These data allowed them to examine several aspects of successional change that could not be determined from the four-plot, long-term study. Using multivariate analyses (DCA—see Chapter 10), they determined that vegetation responded significantly to soil nutrient gradients in mesic but not dry-mesic stands. A subsequent study using the same data set (Roberts and Gilliam 1995b) found that the degree of successional change was linked to the degree of change in soil fertility; both successional and soil change were greater on mesic sites than on dry-mesic sites (Figure 11-7, p. 280).

The two approaches to documenting succession are complementary, rather than redundant, in studying successional aspen forests. That is, the long-term plot data of Roberts and Richardson (1985) led to results not found in the one-time sampling of plots of varying age of Roberts and Christensen (1988) and Roberts and Gilliam (1995b), and vice versa. Overall, however, the most important conclusion

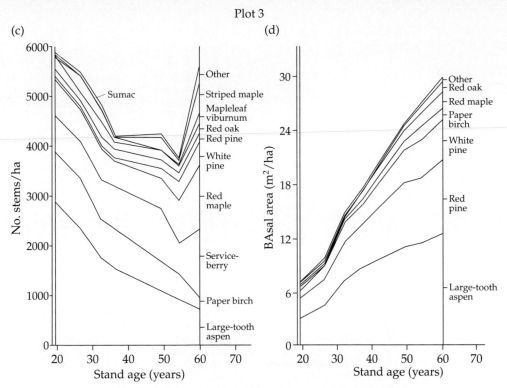

Plot 3

biomass from 1938 to 1945) followed by sprouting of aspen. (c) Changes in stem density over a 40-year period for tree species at plot 3 of the study of Roberts and Richardson (1985). (d) Changes in basal area for tree species over a 40-year period at the same plot. The repeated measures approach taken by this study allowed for such accurate records of change over time. Reprinted with permission from the *Canadian Journal of Botany*.

(that is, the site-dependent nature of succession in aspen forests) is shared by both studies. By employing both approaches a clearer picture of succession emerges.

Examples of Succession

Primary Succession

Glacier Bay, Alaska, has provided the opportunity for numerous excellent studies on changes in soil and vegetation as new terrain is exposed by the retreating glacier. These studies, beginning with the pioneering work of William Cooper (1923, 1931, 1939) and followed by Crocker and Major (1955) and several others,

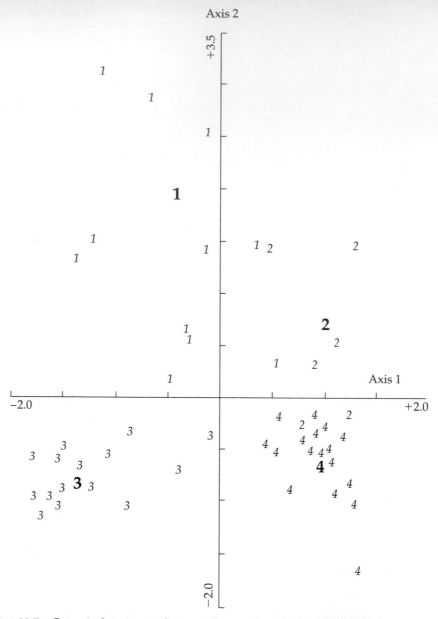

Figure 11-7 Canonical correspondence analysis ordination (see Chapter 10) of 0.1 ha sample plots of successional aspen forest on two site types (mesic, fertile versus dry-mesic, infertile) at two successional ages: young (13–14 years) versus mature (55–82 years). Numbers represent site type and age classes as follows: 1 = young, mesic; 2 = young, dry-mesic; 3 = mature, mesic; 4 = mature, dry-mesic. Small numbers represent individual plots of a given type; large numbers represent the positions of the age–site class centroids (i.e., the center of each cluster). This figure depicts the site-dependent nature of these aspen forests. The closeness of centroids **2** and **4** (and the overlap of plots in ordination space) indicates minimal successional change in species composition on dry-mesic sites, whereas the greater distance between **1** and **3** (and nonoverlap) indicates substantial species change during succession on mesic sites. Figure is taken from Roberts, M. R., and F. S. Gilliam. 1995b. "Disturbance effects on herbaceous layer vegetation and soil nutrients in *Populus* forests of northern lower Michigan." *Journal of Vegetation Science* 6:903–912. With permission from International Association for Vegetation Science.

describe a fascinating 200-year sequence of community development from pioneer algal crusts and mosses on bare rock glacial debris to an intermediate cushion-plant stage and finally to a spruce-hemlock climax forest. More recently, Chapin and colleagues (Chapin et al. 1994, Fastie 1995) have performed the most extensive investigation into the mechanisms that direct the Glacier Bay prisere (a **prisere** is a primary successional sequence).

Glacial retreat in Glacier Bay, well documented since 1760 (Crocker and Major 1955), has been extremely rapid. Even John Muir's time there provides indirect evidence of such rapid retreat. His cabin site, located near the ice front in the 1890s, is now nearly 30 km from the ice front. Overall, since its initiation about 230 years ago, glacial retreat at Glacier Bay has averaged 0.4 km a year, or 15 times more rapid than the rate of any other tidewater glaciers (Chapin et al. 1994).

The time span of just over 200 years for the Glacier Bay prisere is exceptionally short for primary succession, likely the result of (a) glacial till parent materials (metamorphosed sandstone and limestone) that weather easily, (b) a wet, moderate climate, and (c) nearly constant presence of N-fixing organisms (beginning with blue-green algae, shifting to *Dryas* and *Alnus*). This time period is the same as that required for our other example of ecological succession—old-field secondary succession—and many times shorter than many other priseres reported in the literature. Consider the following: rain forest from fresh lava in Hawaii—300 years (Vitousek et al. 1995); pine scrub from bare granite outcrops—700 years (Burbanck and Phillips 1983); spruce-hemlock forest from a Hoh Valley river terrace in Washington—750 years (Luken and Fonda 1983); deciduous forest from Lake Michigan sand dunes—1000 years (Olson 1958); moss-birch-tussock grass from glacial debris in Alaska—5000 years (Viereck 1966).

Seral stages of vegetation and temporal changes in several soil variables of the Glacier Bay prisere, based on Chapin et al. (1994), are summarized in Table 11-1. Pioneer organisms that colonized immediately following exposure of bare glacial till included blue-green algae (which are N-fixers and form "black crusts"), gametophytes of horsetail (*Equisetum variegatum*), lichens, and liverworts. There were also sparsely scattered vascular plants, such as river-beauty (*Epilobium latifolium*), *Dryas drummondii* (an N-fixing, mat-forming dwarf shrub of the rose family; see Figure 11-8, p. 283), willows (*Salix* spp.), cottonwood (*Populus trichocarpa*), and spruce (*Picea sitchensis*). Following the pioneer stage, there was a transition into a *Dryas* stage 35–45 years following glacial retreat. This was characterized by a continuous mat of *Dryas* covering the ground, along with individuals of willow, cottonwood, alder (*Alnus sinuata*, an N-fixing tree), and spruce. By 60–70 years following deglaciation, alder became established as dense thickets, and *Dryas* mats declined. Finally, about 100 years after glacial retreat, spruce began to outgrow alder, leading to spruce dominance 200–225 years after the initiation of the Glacier Bay prisere.

Numerous changes also occurred in the soil (Table 11-1). Much of this change was the result of the variety of plants that occupy the sites. It is important to note, however, that some aspects of soil development also may determine the rate and direction of species change. For example, it is unlikely that a spruce forest will grow on soil that is only 5 cm deep.

Soil depth increased threefold as a result of rapid weathering of glacial till. As

Table 11-1 Primary xerarch succession at Glacier Bay, Alaska. The table includes dominant species, soil depth, pools of carbon (C), nitrogen (N), and phosphorus (P), N accumulation rate, litterfall rate, and pH. Data are means ±1 SE, $n = 10$ sites per stage. From Chapin et al. (1994). By permission of the Ecological Society of America.

Time since deglaciation (yr)	Seral stage and dominant species	Soil depth (cm)	C pool (km m⁻²)	N pool (g m⁻²)	P pool (mg g)	N accumulation (g m⁻² yr⁻¹)	Litterfall (g m⁻² yr⁻¹)	pH
5–10	Pioneer: blue-green algae, *Equisetum*, lichens, liverworts	5.2±0.5[a]	1.3±0.4[a]	3.8±0.8[a]	0.57[a]	—	1.5[a]	7.22[a]
35–45	*Dryas: Dryas drummondii*, scattered individuals of willow (*Salix* spp.), cottonwood (*Populus trichocarpa*), alder (*Alnus*), and spruce (*Picea sitchensis*)	7.0±0.7[a]	1.3±0.3[a]	5.3±1.1[a]	0.51[b]	0.07	2.8[a]	7.28[a]
60–70	Alder: *Alnus sinuata*	8.8±0.2[a]	2.9±0.2[a]	21.8±3.4[a]	0.55[a]	0.83	277.5[b]	6.77[b]
200–225	Spruce: *Picea sitchensis*	15.1±2.5[b]	9.7±2.6[b]	53.3±14.0[b]	0.27[c]	0.18	261.0[b]	3.57[c]

[a,b,c] Values followed by the same superscript within each row are not significantly different at $P < .05$.

Figure 11-8   (a) *Dryas drummondii* (× 1/2).
(b) Scattered *D. drummondii* discs and small
thickets of *Alnus crispa sinuata* on a 15-year-old
surface near Glacier Bay, Alaska. *Dryas* is a
dominant of the pioneer community and contains symbiotic nitrogen-fixing actinomycetes in
root nodules. Photo courtesy of D. B. Lawrence,
Professor Emeritus, Dept. of Botany, University
of Minnesota, St. Paul, MN.

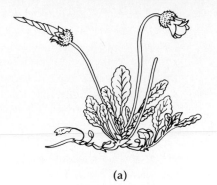

(a)

(b)

vascular plants, with their extensive root systems, continued to occupy these sites,
soil C increased from a combination of litterfall and root turnover. Because most N
in soil is in an organic form (see Chapter 13), it is not surprising that increases in
soil N followed those for C. However, the rate of increase for N exceeded that for
C, resulting in C-to-N ratios decreasing from 342 (highly N-limited) in the pioneer
stage to 182 (less N-limited) in the climax forest. In fact, lower soil P in the spruce
stage suggests a shift from N limitation to P limitation. Much of the change in soil
N may have been from the prevalence of N-fixers at each of the first three seral
stages, with a maximum rate of N accumulation occurring with the dominance of
the N-fixing alders. Finally, soil pH declined rapidly—from 7.22 in the pioneer
stage to 3.57 at climax, a 4000-fold increase in acidity in the 200-year period. Initially circumneutral from the high calcium (Ca) content of glacial till, the soil be-

came acidic first from weathering and leaching of Ca, and then from inputs and decomposition of acidic spruce litter in the climax forest.

A final note should be added concerning the representativeness of the Glacier Bay prisere. Walker performed work (Walker et al. 1986, Walker and Chapin 1986) in a floodplain prisere of the Tanana River in interior Alaska, work that was similar enough to the Glacier Bay studies to make some direct comparisons. Although there were large differences between the sites in soil physical properties, general patterns of N accumulation were remarkably similar, likely the result of a prominent successional stage dominated by N-fixing alders, *Alnus sinuata* at Glacier Bay and *A. tenuifolia* at Tanana River (Walker 1995). The two sites also differ in which spruce species dominates the final stage of succession—*Picea sitchensis* at Glacier Bay and *P. glauca* at Tanana River. Overall, primary succession is remarkably similar between the two sites. The specific mechanisms (to be discussed later in this chapter) that drive these priseres, however, may differ greatly (Walker 1995).

Secondary Succession—Old-Field Succession

The Piedmont (*pied* = "foot," *mont* = "mountain"; literally, "foothill") is the physiographic region between the Appalachian Mountains and the southeastern Coastal Plain from Georgia to northern New Jersey. It has a gently rolling terrain and supports a patchy landscape of hardwood and pine forests, along with recently abandoned and currently cultivated agricultural fields. Unlike the expansive farms of the Midwest, farms of the Piedmont are usually only a few hectares in size and are often planted in corn, beans, and tobacco. Following abandonment of Piedmont farms, secondary succession proceeds toward a climax forest of hardwood species in a sequence that has been well documented by numerous early studies. This has especially been the case for old fields of North Carolina (Crafton and Wells 1934, Billings 1938, Oosting 1942, Keever 1950). The clearest picture of old-field succession was presented by Henry Oosting and by his student, Catherine Keever, who began investigating old fields in North Carolina as a doctoral student at Duke University in 1947 (Keever 1983).

Oosting (1942) sampled abandoned fields of varying age in North Carolina, trying to replicate each age with fields as similar as possible in soil, slope, aspect, and other physical features. On older, forested plots, he compared sapling density with tree density as an additional source of indirect evidence for succession. Keever (1950) did careful experimental studies to understand what caused early shifts in species dominance. Their sere is summarized in Table 11-2, and is discussed below.

One-year-old fields, sampled in the early summer one year after abandonment, had a total of 35 species, all annual and perennial herbaceous species. Although not every field had all 35 species, two species consistently had the highest density and frequency on all fields: crabgrass (*Digitaria sanguinalis*) and horseweed (*Conyza canadensis*). Accordingly, these were considered the pioneer dominants. Two-year-old fields maintained virtually all first-year species (including crabgrass and horseweed), but exhibited a shift in dominance to aster (*Aster ericoides*, absent from first-year fields) and ragweed (*Ambrosia artemisifolia*, a minor component of

Table 11-2 Secondary (old-field) succession in the Piedmont region. Only the common names of the dominant species are listed; many other taxa are associated with each seral stage. From Oosting 1942 (reprinted by permission of *American Midland Naturalist*), Bard 1952 (by permission of the Ecological Society of America), and Richardson 1977 (*Dimensions of Ecology,* by permission of Williams and Wilkins, Baltimore).

Years after abandonment	North Carolina Piedmont
0	Cropland ↓
1	Crabgrass, horseweed ↓
2	White aster, ragweed ↓
3	Broomsedge ↓
5	Broomsedge, pine seedlings ↓
10	Young pines, broomsedge
	Shortleaf pine (drier sites) ← → Loblolly pine (moister sites)
20	
30	
40	
60	Shortleaf pine, hardwood understory	Loblolly pine, hardwood understory
100	
150	White oak, post oak, hickory, dogwood, etc.	White oak, many hickories, dogwood, sourwood, etc.

first-year fields). Although 26 new species were added in the second year, the Sorensen community coefficient (CC, based on presence—see Chapter 10) for first-versus second-year fields was quite high—0.63.

Species richness dropped sharply in third-year fields because of a rapid increase in dominance of broomsedge (*Andropogon virginicus*), a perennial grass (Figure 11-9). *Andropogon* maintained dominance for several years, during which time seeds of pines (particularly loblolly pine, *Pinus taeda,* but also shortleaf pine, *P. echinata*) and some hardwood species, such as white ash (*Fraxinus americana*), winged elm (*Ulmus alata*), sweetgum (*Liquidambar styraciflua*), yellow-poplar (*Liriodendron tulipifera*), and red maple (*Acer rubrum*), arrived at these sites via wind dispersal.

Figure 11-9   Old fields dominated by broomsedge (*Andropogon virginicus*), typical of secondary succession three to six years after abandonment and prior to dominance by pine. From *Introduction to Ecology* by P. A. Colinvaux. Copyright © 1973 John Wiley and Sons, Inc. Reprinted by permission.

Of these species, only the pines became well established by years 3–5 following abandonment, and they formed a closed canopy by year 10. Germination and survivorship of pine seedlings is quite low, however, under the low-irradiance conditions of even-age pine canopies; thus the understory was composed largely of mixed-age hardwood saplings by year 20.

Using an elaborate field experiment, De Steven (1991a,b) investigated possible mechanisms to explain the transition from pine to hardwoods. She was particularly interested in assessing the importance of two factors that had received little attention in studies of old-field sequences—seed predation and old-field herbaceous cover—in influencing seedling emergence and growth of loblolly pine and five early successional hardwood species common in the North Carolina Piedmont: yellow-poplar, white ash, sweetgum, winged elm, and red maple. She found that both factors had significant effects on seedling emergence and growth of many (but not all) of these species. Regardless of treatment combinations, loblolly pine exhibited the highest levels of seedling emergence, seedling survival, and seedling height growth. Thus, her data are useful in explaining why, although all these species produce wind-dispersed seeds (often called *seed rain*) that arrive at old-field sites in potentially large numbers, it is loblolly pine that initiates the woody species stage of old-field succession. Which hardwood species become established initially with the pines may be as dependent on their amount of seed rain as on factors affecting seedling performance (De Steven 1991a,b).

Oosting further found that some hardwood species, such as dogwood (*Cornus florida*) and redbud (*Cercis canadensis*), were confined genetically to the understory, and were found in that stratum in great number in older sites. Other hardwoods, notably hickories (*Carya* spp.) and oaks (*Quercus* spp.), with greater growth capacities and longer life spans eventually replaced overmature pines. Pine stands 50–75 years old were over 25 m tall on productive sites; beneath them were oaks and hickories, up to 10 m tall, and beneath them were scattered understory hardwoods (Figure 11-10). Stands 100 years old had as many hardwoods as pines in the overstory; this balance shifted greatly toward hardwoods thereafter (Figure 11-11). Climax hardwood stands >200 years old had scattered pines in the overstory and a few pine seedlings and saplings, but the overstory, understory, and sapling strata were clearly and completely dominated by hardwoods.

Old-field succession can be maintained at the pine seral stage (i.e., in subclimax state) by fire, which can occur at a natural frequency of once every 5 to 7 years (Christensen 1987, Gilliam 1991; see also Chapter 16). Hardwood seedlings and saplings are susceptible to damage (mortality) from typical ground fires, whereas pines >10 years old are quite fire-resistant because of their thick bark and greater height than the hardwoods.

In a retrospective of studies of old-field successional sequences since her ear-

Figure 11-10   A loblolly pine stand with a hardwoods understory, typical of secondary succession in the Piedmont 50 years after abandonment.

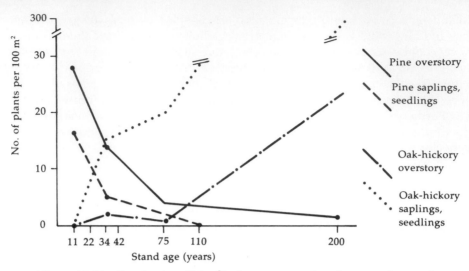

Figure 11-11   Density (per 100 m^2) of overstory and understory pines and oaks and hickories throughout secondary succession in the Piedmont of North Carolina. From Oosting (1942). Reprinted by permission of *American Midland Naturalist*.

lier work in the 1940s, Keever (1983) concluded that old-field succession was quite distinct from other types of secondary succession:

> The sequence of species and the timing of these changes in old-field succession in the Piedmont of the Southeast are not typical of such succession elsewhere. Nowhere else is there such a fast and distinct change in species dominance. In most places there is a gradual overlapping of species dominance often extended over a much longer time.

Evidence certainly supports her conclusion. Furthermore, not only do Piedmont old-field seres differ from other secondary seres, but they also differ from old-field seres in other physiographic regions. Work by Vankat and Snyder (1991) in southwestern Ohio described an old-field sere that (a) lacked the initial buried seed component of the Piedmont, (b) had very few tree seedlings as late as 10 years following abandonment, (c) exhibited herbaceous dominance by year 50, and (d) lacked a conifer (pine) stage.

General Trends during Succession

For nearly a century, ecologists have been describing successional pathways for many types of vegetation. More recently there has been an attempt to synthesize such information into general theories on the process of succession. One of the first ecologists to truly embrace the ecosystem concept, Odum (1969) made the transition from the taxonomic (descriptive) level of the community to the functional level of the ecosystem. He presented an impressive table listing 24 ecosystem traits that he thought would change significantly through succession, carefully explaining that data supporting these trends were, in most cases, few. His was essentially a table listing hypotheses of successional change that invited testing.

Indeed, Odum's table had the immediate impact of stimulating further research and development of ideas on succession (see Drury and Nisbet 1973, Vitousek and Reiners 1975). These hypotheses have met with various fates—some supported, some rejected, and others amended. We have chosen 14 community and ecosystem attributes illustrating successional trends (Table 11-3). Each trait, some paraphrased from Odum's original table, will be discussed briefly. Several things should be kept in mind in this discussion. First, trends are only for progressive, directional succession. Second, early seral stages are being compared to late stages in a relative sense (i.e., not necessarily pioneer to climax stages, because some trends peak in mid-to-late succession, not at climax). Third, there are exceptions to virtually every item. Fourth, rates of change are not uniform throughout succession (Major 1974). Finally, it should be pointed out that Odum's conclusions were based largely on Georgia old-field successions; thus, his conclusions may be influenced by his geographic region of study.

Table 11-3 Some vegetation and ecosystem traits that often change during progressive succession. The status of each trait is shown for early and late stages of succession (*not* for pioneer and climax stages necessarily, because some trends peak at intermediate seral stages). Each trend is briefly discussed in the text.

Trait	Early stages	Late stages
Biomass	Small	Large
Physiognomy	Simple	Complex
Leaf orientation	Multilayered	Monolayered
Major site of nutrient storage	Soil	Biomass
Role of detritus	Minor	Important
Mineral cycles	Open (leaky), rapid transfer	Closed (tight), slow transfer
Net primary production	High	Low
Site quality	Extreme	Mesic
Importance of the macroenvironment	Great	Moderated and dampened; less
Stability (absence or slowness of change)	Low	High
Plant species diversity	Low	High
Species life history character	*r*	*K*
Propagule dispersal vector	Wind	Animals
Propagule longevity	Long	Short

Vegetation and Site Quality

Biomass increases during succession. Plant cover, the density of foliage above the ground (the leaf area index; see Chapter 8), and the height of the plants also increase. If succession leads from herbs or shrubs to trees, the percentage of total biomass that is below the ground may decline, but the total biomass still increases.

Physiognomy increases in complexity because the variety of growth forms increases as succession proceeds. If successional stages involve trees, the structure and leaf orientation of those trees may change (Horn 1971, 1975). Tree species characteristic of early successional stages, such as aspens or pines, tend to be tall, thin, conical, and capable of rapid growth; their leaves are small, numerous, and randomly oriented and are borne in such a way that many leaves are shaded by others above them (multilayered) (Figure 11-12a). Tree species characteristic of the climax community may have a similar profile but grow more slowly, have fewer and larger leaves, and bear those leaves in a planar fashion such that self-shading is minimized (monolayered) (Figure 11-12b).

Vitousek and Reiners (1975) proposed a hypothesis to explain changes in nutrient dynamics through ecosystem succession, particularly secondary succession in forests. The disturbance that initiates succession disrupts the biotic control of most nutrient cycles (e.g., the nitrogen cycle), resulting in their leaky nature early in succession, as indicated in Table 11-3. However, they also demonstrated that mature forests are more leaky with respect to nitrogen than are intermediate-aged forests, a finding inconsistent with Odum's prediction. Gilliam and Adams (1995) found only slight differences in soil fertility between 20-year-old and mature (>85-year-old) stands of central Appalachian hardwood forests. They concluded, however, that the importance of soil organic matter as a source of plant nutrients increases dramatically during succession.

Net primary production (photosynthesis minus respiration) may decline with succession, for a variety of reasons. (a) There is a great deal of supporting but non-photosynthetic tissue whose maintenance (respiration) reduces net production. (b) Nutrients may be limiting because of their storage in inert tissue. (c) Many plants in the overstory may be senescent, with lower photosynthetic rates than young plants. (d) The leaf orientation of climax trees may be inappropriate for high photosynthetic rates (Horn 1974).

The environment becomes more mesic during succession, and the effect of the macroclimate is dampened. As the canopy closes, diurnal fluctuations in temperature and humidity are moderated. Increased humus, increased soil depth, and a finer soil texture in later successional stages lead to greater soil moisture retention and a buffering of seasonal changes in precipitation.

Stability and Diversity

Stability can be defined in a wide variety of ways, and depending on the definition, stability may increase or decrease with succession (Huston 1994). The simplest definition—the one used in Table 11-3—is that stability is equivalent to

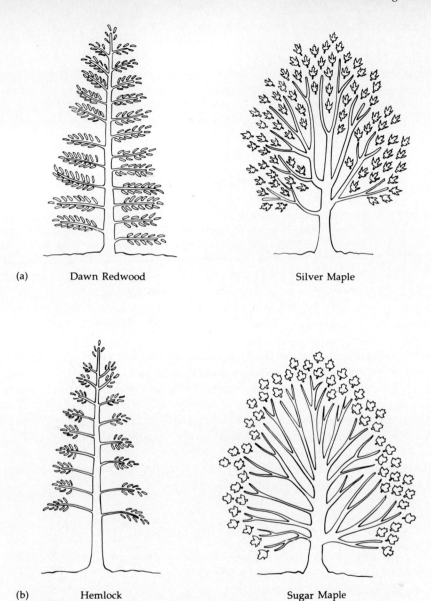

(a) Dawn Redwood   Silver Maple

(b) Hemlock   Sugar Maple

Figure 11-12 Distribution of leaves in multilayered trees (inner leaves shaded) and monolayered trees (self-shading minimized). (a) Dawn redwood and silver maple, which are multilayered; and (b) hemlock and sugar maple, which are monolayered. Reprinted by permission from "Forest succession" by H. S. Horn. Copyright © 1975 by Scientific American, Inc. All rights reserved.

lack of change. Stability, so defined, increases with succession. Climax species are long-lived, thus changes in the undisturbed climax community are merely random, minor fluctuations around a long-term mean. Stability defined as resistance to minor changes in the macroenvironment also increases with succession because plant cover dampens macroenvironmental fluctuations and extremes. However, if stability is defined as the ability to return rapidly to **homeostasis** (equilibrium) following major recurring disturbances, then preclimax communities are much more stable than climax communities, with stability declining over time. This view was taken by Horn (1974), who concluded that climax communities are fragile:

> Conservationists . . . have often cited the conventional generalization that diversity conveys stability, arguing that diverse [climax] natural communities should be conserved for their stabilizing influence. . . . One could equally argue that if complex [climax] systems were inherently stable, they should need no protection. The opposite view, that . . . climax communities are inherently fragile, is a much more powerful reason for their requiring protection.

Plant species diversity (see Chapter 8) increases throughout early succession, but it decreases in the temperate zone in late succession as the canopy closes and a few species become major dominants. Thus, in the temperate zone, a periodic local disturbance that sets succession back to earlier seral stages is necessary to maintain maximum diversity (Loucks 1970). There are exceptions to this trend of increasing diversity, just as there are exceptions to all the trends listed in Table 11-3. Furthermore, there are inconsistencies in these trends even within a single community type. For example, working in successional aspen (*Populus grandidentata* and *P. tremuloides*) forests of Michigan, Roberts and Gilliam (1995b) found that nutrient availability changed (declined) significantly with time on mesic, fertile soils, but not on adjacent drier, less-fertile soils. Similarly, species diversity of the herbaceous layer did not change from young to mature stands on the drier soils, but declined significantly on mesic soils.

Life History Traits

Pioneer species are typically *r*-selected (*R* species of Grime 1979), exhibiting rapid growth rates, short life spans, large allocations of energy to sexual reproduction, high photosynthetic rates, and small, wind-dispersed seeds. By contrast, climax species are typically *K*-selected (*S* species of Grime 1979), exhibiting slow growth rates, long life spans, small allocations of energy to sexual reproduction, low photosynthetic rates, and large, animal-dispersed seeds. Both groups of species, however, may coexist in climax communities. For example, *r*-selected species may occupy sites beneath openings in the canopy and have seeds dormant in the soil (**seed pool**) that will germinate following creation of such openings.

These patterns of change in dominant life history traits among plant species are important in explaining successional pathways. As we will see in the next section, differences in life history traits may be one of the more crucial mechanisms that drive succession.

Mechanisms Driving Succession

That succession occurs in plant communities has always received widespread acceptance among plant ecologists. However, in addition to describing general pathways and stages of succession, ecologists also have been interested in determining the various mechanisms that drive successional change. On this subject, ecologists have virtually never been in agreement. Indeed, the heated debates on succession discussed at the beginning of this chapter have been largely about which biotic and abiotic factors are most important in directing primary and secondary succession. These mechanisms, often suggested in the form of successional theories and hypotheses by the ecologists who champion them, appear in this section mostly in the chronological order of their proposal.

This is not intended to be an exhaustive list of published studies; there are several excellent reviews on causal factors to explain succession (e.g., Pickett et al. 1987, McCook 1994, Walker in press) that offer a more complete summary. Given the recent interest in succession as a process that results in changes in species diversity (see Chapter 20), some of these reviews focus on mechanisms that determine changes in diversity (e.g., Huston 1994, Tilman 1994, Roberts and Gilliam 1995a). Although there will be a brief synthesis at its conclusion, this section is not intended to present the "final word" regarding succession. Indeed, this summary will take the approach of McCook (1994) that a single, universal cause of succession is unlikely. Instead, what should be gained from these various ideas and viewpoints is that such debate is beneficial from a scientific perspective—requiring inductive and deductive reasoning to make observations and formulate predictive hypotheses, hypotheses that then inspire further research in the form of experimentation and field studies to test their validity and generate new hypotheses.

Much of the debate on the causes of succession is based on which view of the nature of community is taken by the ecologist. In Chapter 8 we mentioned that there are two sharply contrasting views: (1) the organismic view (also called "holistic"), which considers the biotic community as a single, integrated, interacting entity; and (2) the continuum view (also called "individualistic" or "reductionist"), which considers the community as simply composed of species with similar enough ecological requirements to bring them together in the same habitat. Certainly, how one views the biotic community will have much to do with how one views change in the community.

Early Debates

In his classic paper wherein he coined the term *ecosystem*, Tansley (1935) described the early work by Cowles (1899, 1901, 1911) on dune succession as "the first thorough working out of a strikingly complete and beautiful successional series." At the time, Cowles's ideas on the dynamic nature of vegetation, based on his dune prisere, were quite new to the developing field of plant ecology. Although he did not focus specifically on the mechanisms that best explained his observed successional changes, he emphasized several abiotic factors, such as light, heat, wind, soil, water, topography, and fire, to explain the variable vegetation patterns he found, downplaying biotic interactions. It was Cowles's awareness of the dynamic nature of the plant community along with the variability of the abiotic environment that led him to conclude that succession is "a variable approaching a variable, not a constant" (Olson 1958).

Ecological succession as envisioned by Frederic Clements was in sharp contrast to that described by Cowles. Clements championed the holistic view of the plant community, going so far as to call it a "superorganism," with each species of the community being analogous to the anatomical parts composing an organism. Thus, it is small wonder that Clements saw succession as an entirely predictable process, for it was merely "reproduction" of the superorganism.

Clements relied heavily on jargon in his writings (e.g., Clements 1916, 1936) and used specific terms to describe a six-step mechanism for succession: (1) **nudation** (disturbance initiating succession), (2) **migration** (movement of plant propagules to the disturbed area), (3) **ecesis** (establishment, growth, and development of plants), (4) **competition** (among established plants), (5) **reaction** (alteration of environmental conditions by competitive plants), and (6) **stabilization** (attainment of the climax stage).

Clementsian succession is composed of discrete seral communities, such that each stage undergoes steps 2–5 (migration through reaction). A given seral stage is replaced by the succeeding stage when new species are more competitive under the conditions created by the previous stage ("reaction"). This process is repeated until the climax stage is reached, a time when the establishment of new taxa is prevented by the self-propagating climax community. Such a view of succession, diagrammed in Figure 11-13, has been called **relay floristics** by Egler (1954) and **facilitation** by Connell and Slatyer (1977).

Two central tenets of Clementsian succession were (1) the community as superorganism and (2) the **climatic climax** as a predictable endpoint, somewhat uniform in a region and determined by the selection of the regional climate for characteristic plant assemblages (Golley 1993). Clements focused on the importance of biotic processes driving succession, de-emphasizing the importance of abiotic factors. He was a strong advocate for his own views of the nature of plant communities and of succession; these views were accepted widely among plant ecologists of his time.

In a controversial paper, Gleason (1926) seemed to take pains to avoid mentioning Clements and the term "superorganism" (and rarely addressed the concept of climax). However, it is clear in his writing that he was rejecting virtually all of Clements's ideas on succession and especially on the nature of plant communities. Consider Gleason's opinion (1926) of the concept of plant association (similar in this case to Clements's climax stage):

> It seems we are treading upon rather dangerous ground when we define an association as an area of uniform vegetation, or, in fact, when we attempt any definition of it.

Thus, Gleason established (not without invective from the ecological mainstream, mentioned in Chapters 2 and 8) "the individualistic concept of the plant association."

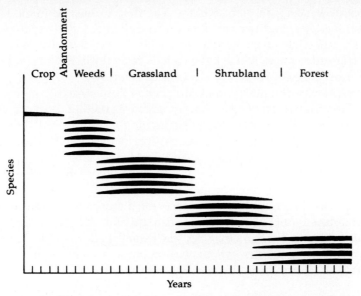

Figure 11-13 Diagrammatic summary of the relay floristics model of succession. The thicker the line, the more important the species at a given time. From "Vegetational science concepts. I. Initial floristic composition, a factor in old-field vegetation development" by F. E. Egler. *Vegetatio* 4:412–417. Reproduced by permission of Dr. W. Junk BV.

Gleason saw the plant association as being too complex and spatially and temporally variable to have the simplistic label of "superorganism" attached to it. Without direct reference to Cowles's work, Gleason (1926) echoed the essential themes brought out by Cowles—that plant processes and environmental variability created a dynamic system in plant communities. Gleason reasoned that the principal mechanisms driving succession were migration (rates and modes of which varied greatly among species) and environmental selection (the action of a variable set of abiotic conditions, coupled with a response to those conditions that was unique to each species present).

Later Debates

As has been suggested, Clementsian succession was firmly in place in prevailing ecological thought at the time of Gleason's challenging paper of 1926. Gleason's individualistic hypothesis gradually gained wider acceptance as it became more and more supported by published studies on a variety of vegetation types, such as Curtis and McIntosh (1951) working in prairie-forest ecotones and Whittaker (1956) working in the Great Smoky Mountains. Whittaker (1953) went further to reject the Clementsian climatic climax concept in favor of a **pattern climax** concept, wherein the climax condition of a broad region is a pattern of species abundances that is spatially quite variable, though possibly constant locally (Horn 1974).

In 1954, Frank Egler published two important papers that also challenged widely accepted aspects in plant ecology. The first, published in the journal *Castanea*, questioned the validity of the Braun-Blanquet method of vegetation analysis (see Chapter 10), whereas the second, published in *Vegetatio*, was a contradiction of Clementsian succession, which Egler called "relay floristics" to contrast with his own initial floristic composition model of succession.

Working in experimental plots at Aton Forest (his estate in Connecticut), Egler found that the progression of seral communities was neither fixed nor predictable, as would be expected from the relay floristics model. He concluded that at least in the early stages, mechanisms driving succession included chance (of migration of initial propagules to the disturbed site) and differential longevity of plants. As depicted hypothetically in Figure 11-14, according to Egler's **initial floristic composition (IFC)** model, all the pioneer species, many of the seral species, and some of the climax species are present initially following disturbance. Some germinate and become established quickly, others germinate quickly but grow more slowly and for a longer period, and others may become established still later. Larger, longer-lived, slower-growing species eventually outcompete small pioneer species; thus community dominance shifts and succession proceeds.

Recent Models

Since the replacement in ecological thought of Clements's superorganism and climatic climax with Gleason's individualistic hypothesis and Whittaker's pattern climax, hypotheses of successional mechanisms have focused on the contribution

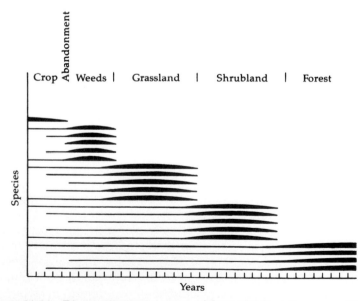

Figure 11-14   Diagrammatic summary of the initial floristic composition model of succession. The thicker the line, the more important the species at a given time. From "Vegetational science concepts. I. Initial floristic composition, a factor in old-field vegetation development" by F. E. Egler. *Vegetatio* 4:412–417. Reproduced by permission of Dr. W. Junk BV.

of the population dynamics of individual plant species in explaining successional change (Peet and Christensen 1980). Drury and Nisbet (1973) suggested an explanation, similar to that of Egler, which emphasized the importance of life history traits of successional species:

> Most of the phenomena of succession can be understood as consequences of differential growth, differential survival (and perhaps different colonizing ability) of species adapted to growth at different points on environmental gradients.

Such a view was further elaborated on by Pickett (1976), who saw succession as a dynamic, regional process, as opposed to a single pattern for a given site.

In a very readable, thoughtful review and synthesis of successional hypotheses, Connell and Slatyer (1977) stated that succession is driven by one of three overriding mechanisms, which they proposed as models. The **facilitation** model was essentially a rewording of Clementsian relay floristics, but without the superorganism and climatic climax. The **tolerance** model was a paraphrase of Egler's IFC model, suggesting that climax species are merely better competitors, longer-lived, and bigger than seral species; all may be present early in succession. The **inhibition** model de-emphasized biotic interactions, such as competition, and suggested that all species resist invasions by competitors by preempting space and continue to inhibit invasion until they die or suffer damage.

Walker and Chapin (1987) questioned the Connell-Slatyer models, seeing them as mutually exclusive and overly simplistic, and suggested that succession is an inherently complex phenomenon driven by interactions of numerous mechanisms. Their criticism was based in large part on experimental work on primary succession in an Alaskan floodplain (Walker and Chapin 1986, Walker et al. 1986). In contrast to the Connell-Slatyer facilitation model, which Connell and Slatyer (1977) stated would be important in primary succession, Walker and Chapin found that facilitation alone was not essential in explaining successional change, but instead concluded that life history traits, competition, and stochastic (random) processes (such as flooding) interacted as a complex of mechanisms to drive primary succession.

Grime (1977, 1979) expanded his theory of life history traits of plant species (see Chapter 5) to explain mechanisms driving succession. This theory emphasizes disturbance, stress, and competitive interactions as selective forces that change over the course of plant succession. Figure 11-15 depicts hypothetical secondary successional trajectories under varying levels of productivity (determined largely by resource availability—see Chapter 12). Following any disturbance that initiates secondary succession, essential resources (e.g., light, nutrients, moisture) are abundant, selecting for ruderal (R) species, regardless of level of potential productivity. For more productive sites, competition will be keen following establishment of the ruderal community, selecting for competitive (C) species, which are eventually replaced by stress-tolerant (S) species when resources are depleted to a stress-inducing level. The contribution of C species in this successional pathway decreases with potential productivity, according to Figure 11-15. Consequently, Grime's theory predicts that the magnitude of species change through succession is directly (positively) proportional to productivity.

Tilman (1985, 1988) proposed a theory, which he called the **resource-ratio**

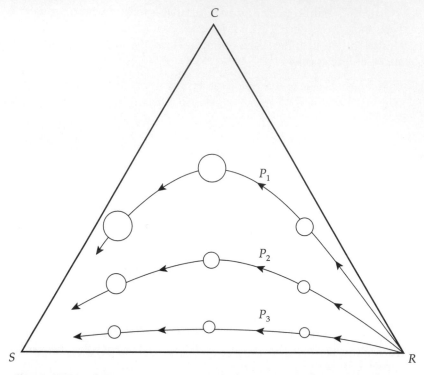

Figure 11-15   Grime's triangular model showing the path of secondary succession under conditions of high (P_1), moderate (P_2), and low (P_3) potential productivity. The circles indicate the amount of plant biomass at each stage of succession. Letters represent predominance of plant life history traits, as discussed in Chapter 5: R = ruderal, C = competitive, S = stress-tolerant. The trajectory for P_1 indicates the greater importance of competitive species when productivity is high. Such species would be absent when productivity is low (P_3). Thus, Grime's model predicts succession with higher species change on fertile, productive soils and lower species change on infertile, less productive soils. Modified from J. P. Grime. *American Naturalist* 111:1169–1194. Copyright © 1977 by University of Chicago Press. By permission.

hypothesis, which predicts an outcome opposite that of Grime. This hypothesis assumes that each species is a superior competitor at a given ratio (balance) of resources (most commonly, light and nitrogen—Figure 11-16). There are two key elements in his model: (1) interspecific competition for resources and (2) long-term patterns of supply of limiting resources (which Tilman called "resource-supply trajectories").

Tilman's hypothesis is based on principles of allocation (see Chapter 5) and, not unlike Grime's theory, predicts that resource-supply trajectories (usually seen as decreases in resource availability) change through succession, a change that results in successional change in the plant community. However, unlike that of Grime, Tilman's hypothesis predicts that the magnitude of species change in succession is inversely related to site productivity (Figure 11-16).

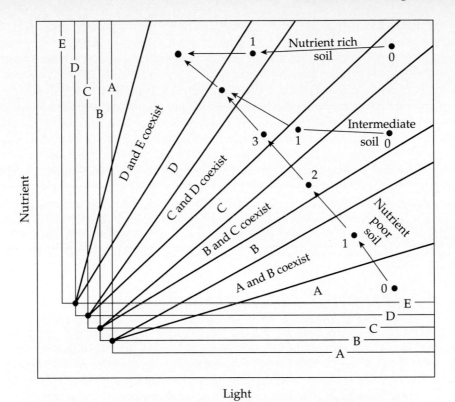

Figure 11-16 Tilman's resource-ratio hypothesis of secondary succession on sites of different productivity (soil fertility). The *x*-axis is light availability and the *y*-axis is nutrient (usually nitrogen) availability. Lines A through E parallel to the axes represent zero growth rate lines for species A through species E, i.e., minimum combinations of nutrient and light availability to support existence (but not growth) of a population. Diagonal lines represent ranges of nutrient and light availability to allow for either dominance of a single species or coexistence of more than one species. Numbers represent time intervals. Thus, Tilman's model predicts rapid succession with minimal species change on fertile, productive soils and slow succession with greater species change on infertile, less productive soils. From D. Tilman. *American Naturalist* 125:827–852. Copyright © 1985 by University of Chicago Press. By permission.

The controversial ideas of both Grime and Tilman have been fruitful in inviting further research and thought in testing their hypotheses (e.g., Wilson and Keddy 1986, Loehle 1988, Grime 1988, Shipley and Peters 1990, Olff 1992). Though now less vehement than it once was, the Grime-Tilman debate was a prominent component of successional thought among plant ecologists in the 1980s. Grace (1991) provides an excellent review of this debate, summarizing the two theories and discussing points of dispute.

With recent concerns regarding loss of biodiversity on a global scale (see Chapter 20), many current successional models emphasize succession as a process integrating several factors that determine the diversity of a biotic community. Using the intermediate disturbance concept (Huston 1979—see Chapter 8), Huston and Smith (1987) took the life history–based explanation of succession as a population process (e.g., Peet and Christensen 1980) and extended it down to the level of the individual plant. Their **dynamic equilibrium model** of succession stresses the importance of competition between individual plants and is based on life history and physiological traits, such as shade tolerance and growth rates, along with the following premises: (a) Competition among individual plants occurs in communities, although pertinent resources and competitive intensity may vary temporally and spatially. (b) Plants change relative resource availability, altering outcomes of competition. (c) Physiological and energetic constraints prevent any one species from being competitive under all environmental conditions, producing an inverse correlation between certain groups of traits and resulting in relative competitive abilities that vary over a range of environmental conditions. This model has received support from both computer simulations and field studies (Huston 1994).

Future Directions

What mechanisms have been found to operate in our examples of primary succession in Alaska and old-field succession in North Carolina? Clearly, no single model discussed here is sufficient. For example, Chapin et al. (1994) supported earlier conclusions based on work in an Alaskan floodplain (Walker 1986, Walker et al. 1986) that alder both inhibited and facilitated later successional stages. Furthermore, Fastie (1995) warned that Glacier Bay communities of varying age do not necessarily constitute a single chronosequence, and inferences about long-term successional trends should be made with caution.

Initial change in North Carolina old fields is likely an interaction between stochastic factors, such as seed dispersal, and competition. Allelopathy (Chapter 6) has been shown to be an insignificant factor (Keever 1983). The surprisingly consistent outcome of old-field succession toward hardwood dominance (Keever 1983), a phenomenon known as **convergence,** arises out of an interaction of life history traits of woody species and the decreasing importance of stochastic factors with successional age (Christensen and Peet 1984).

Plant ecologists will continue to be preoccupied with the challenge of studying succession. Because it is unlikely that a single general model will ever emerge to explain all successional sequences, two questions might be asked: What is the value of past studies of succession? and What is the necessity of future studies? These may be answered from both pure (basic) and practical (applied) contexts. The basic science of successional studies has taught us much about the essential nature of biotic communities, as well as the interaction of biotic and abiotic factors that influence them.

This knowledge of plant communities can be applied toward their responsible management (Luken 1990). For example, forest management (e.g., harvesting) practices invariably initiate secondary forest succession (Halpern and Spies 1995, Gilliam et al. 1995). Our knowledge of successional change in a variety of forest types will provide guidelines for management techniques that are best suited for a particular forest type (Gilliam and Roberts 1995).

Applying knowledge of succession to plant invasions should also be valuable. Invasions of non-indigenous plant species continue to pose a serious threat to the structure and stability of natural communities (Luken and Thieret 1997). For example, McCarthy (1997) found that garlic mustard (*Allaria petiolata*), a biennial weed invading many forests of eastern North America, has had negative impacts on the composition and structure of herbaceous understories of these forests. He further found, however, that simple removal of this individual species is not enough to alleviate the problem. The question then arises: What is the impact of exotic invader species on the process of succession? As these and other studies suggest, it is anticipated that future research on succession will be applicable on a variety of levels.

Summary

Succession is directional change in species composition and/or structure of a community over time. Seral (successional) communities exhibit measurable, significant change during succession, whereas climax communities do not.

Primary succession begins with plant colonization of new land; secondary succession occurs on land that still supports some residual soil and plant propagules. A secondary sere may be completed in 5–300 years, but a primary sere requires 200–1000+ years because of the greater severity of the pioneer conditions. Progressive succession leads to more mesic sites and communities of greater complexity and biomass. However, there are many well-documented cases of retrogressive succession (both allogenic and autogenic), which result in a more extreme habitat and simpler communities. There is also a type of local, nondirectional succession called *cyclic succession*. And in some severe habitats there is no succession, the pioneer community being the climax community as well. Some seres do not reach a climax because the periodicity of some disturbance, such as fire, is shorter than the time required for the progression of a complete sere. Most regions are characterized by a mosaic of climax types and seral stages leading to them.

Succession may be documented by repeated measures over time on a single plot or by reference to historical records for that plot, but most seres have been inferred from indirect evidence. One indirect method is to sample vegetation on many separated plots that differ in age. Also, the species composition in seedling and sapling strata can be compared to the overstory stratum.

Many changes in soil properties, microclimate, and vegetation occur during progressive succession. Some trends peak in the climax and others peak in mid-to-late succession; most do not have a constant rate of change. The following attributes increase during succession: biomass, complexity of the physiognomy, nutrient storage in biomass, the importance of detritus in mineral cycles, the mesic and equable nature of the site, stability (in the sense of longer-lived species and slower organism turnover times), species diversity, and the proportion of species that are *K*-selected. Net productivity may decline.

The mechanisms that drive succession have been debated for nearly a century, and will likely continue to be debated in the future. Clements envisioned succession as a highly predictable process aimed at replicating the community "superorganism" and emphasized the importance of biotic effects on the community and the environment. Clementsian succession is a series of discrete seral stages that relay the site eventually to the climax community. Gleason's individualistic view of the community emphasized migration of plant propagules to the disturbed area along with environmental selection. With his initial floristic composition model, Egler elaborated on Gleason's ideas, saying that the chance of migration and the differential longevity of plant species were the predominant mechanisms. More recent ideas concerning successional mechanisms see succession as a population process, emphasizing the interaction of each species' life history traits with environmental change. Thus, after all the debates through the intervening years, it is Cowles who was correct nearly 100 years ago with his first assessment of succession as "a variable approaching a variable, not a constant."

CHAPTER 12

PRODUCTIVITY

For terrestrial ecosystems **productivity** reflects the ability of plants, whether trees and herbs of a forest, grasses of a prairie, or annuals and shrubs of a desert, to capture energy in the form of solar radiation and convert it to biomass. More specifically this is referred to as **primary productivity** to indicate that it represents the initial entry of energy into an ecosystem, forming the most basic level of the trophic (food) pyramid and representing the amount of energy potentially available for heterotrophs. Because plants require energy to satisfy their metabolic demands, however, not all of the energy captured by plants via photosynthesis is available to the heterotrophic community. Accordingly, we distinguish between **gross primary productivity** (**GPP**—total amount of energy captured by autotrophs) and **net primary productivity** (**NPP**—total amount of energy captured minus energy used in respiration).

In this chapter, we will discuss the interaction of plants and the global carbon cycle, consider the function of plant communities in the context of energy flow, and contrast the abilities of different communities to convert solar radiation into plant biomass. Understanding the factors that influence productivity is becoming increasingly important in light of the relationship between the global carbon cycle and global climate change and in the context of global sustainability of ecosystems (Vitousek et al. 1986).

Energy Flow Model

To begin our discussion of terrestrial plant productivity, let us return to some of the terminology used to introduce the chapter. Energy enters the ecosystem as solar radiation and is fixed into chemical energy via photosynthesis; this is GPP (see Figure 12-1). Again, because a significant amount of GPP is released by respiration to supply energy for plant metabolic activities, we are generally more con-

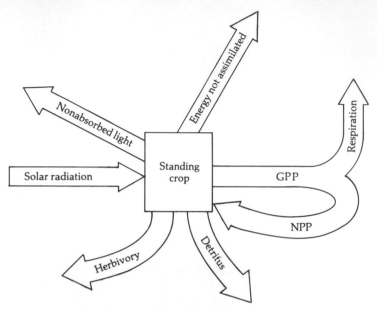

Figure 12-1   Potential pathways of energy at the primary trophic level.

cerned with NPP, which is equal to GPP minus respiration. Because solar energy is ultimately fixed in the form of carbon-carbon bonds and because carbon is the most prevalent element in living biomass, NPP can be expressed in either energy, carbon, or biomass units, but always as a rate adjusted to area (i.e., per unit area per unit time). So, plant **biomass** generally refers to the dry weight of plant material present at any one point in time (also usually on a per area basis), also called **standing crop** and **phytomass.** Existing biomass is rarely a useful measure of NPP because other factors (e.g., herbivory and decomposition) can affect biomass somewhat independently of NPP.

Indeed, part of NPP is the food source for herbivores and decomposers of detritus; the remainder accumulates within the community as standing crop biomass, or is consumed by fire. The distribution and rate of turnover is closely related to abiotic factors such as temperature and moisture, to the density and types of herbivores present, and to the relative importance of decomposers in the system. These factors are discussed in more detail later in this chapter.

It is useful to compare the efficiency with which a species or an ecosystem transmits energy from one form or physical state to another. Efficiency values are the ratio of output to input of energy at various points along the pathways of energy flow within a plant or an ecosystem (Perry 1994, Ricklefs 1997). Efficiency values can be calculated for any energy transfer. Note that the second law of thermodynamics demands that even the most efficient plants or ecosystems will have values less than 100%.

Three of the measures of efficiency that are important in understanding differences in ecosystem productivity are as follows: (1) **Exploitation efficiency**

Table 12-1   Formulae defining common efficiency values as applied to primary producers. Modified from Ricklefs. 1997. *Ecology*, 4th ed., by permission of J. H. Freeman & Co., New York.

Exploitation efficiency (%)	=	$\dfrac{\text{gross primary productivity}}{\text{solar radiation}} \times 100$
Assimilation efficiency (%)	=	$\dfrac{\text{gross primary productivity}}{\text{absorbed radiation}} \times 100$
Net production efficiency (%)	=	$\dfrac{\text{net primary productivity}}{\text{gross primary productivity}} \times 100$

(Table 12-1) is related to the ability of plants to intercept light. Characteristics important in this regard include latitude, topographic location, leaf area index (LAI; see Chapter 8), and leaf orientation. (2) **Assimilation efficiency** (or **quantum yield**) refers to the ability of plants to convert absorbed radiation into photosynthate. Factors modifying assimilation efficiency are those governing the photosynthetic process (see Chapter 15 for details), such as resistance to CO_2 assimilation, water and light availability, evaporative demands of the atmosphere, and temperature. (3) **Net production efficiency** is a measure of the capacity to convert photosynthate into growth and reproductive biomass rather than utilizing it for maintenance respiration. The amount of energy used for maintenance depends on such factors as temperature (because of the direct effect on rates of chemical reactions) and the amount of nonphotosynthetic biomass that must be supported.

Methods of Measuring Productivity

The most accurate means of measuring net primary productivity would be to measure the net photosynthetic rates of photosynthetic tissues, then subtract the respiration rates of nonphotosynthetic tissues, and finally extrapolate to the community level, using the net production per gram of biomass of each species in the community. This assessment of NPP is not possible on a large scale because we do not have photosynthesis and respiration measurements for all the species in any community, nor are we prepared to predict physiological responses to the wide range of conditions experienced by plants in the natural environment. Consequently, we usually use methods that depend on the accumulation and disappearance of biomass through time.

Net primary productivity (NPP) is frequently measured by calculating the change in biomass through time:

$$\text{NPP} = (W_{t+1} - W_t) + D + H \qquad \text{(Equation 12-1)}$$

where $W_{t+1} - W_t$ is the difference in standing crop biomass between two harvest times, D is the biomass lost to decomposition, and H is the biomass consumed by herbivores during the period between harvests. Productivity may be expressed as $g \, m^{-2} \, yr^{-1}$ or, if the caloric content of the material is known, as $cal \, m^{-2} \, yr^{-1}$. The latter units, in calories rather than grams, are more meaningful when efficiency of light conversion or respiration is important.

Aboveground herbaceous vegetation can be measured with reasonable accuracy by harvesting replicate, random quadrats within a grid. Although this technique is most effective with annual vegetation, where little biomass is lost to decomposition during the growing season, it can also be used for low-growing perennial vegetation. In grasslands, Knapp et al. (1998) concluded that harvesting a minimum of twenty $0.1 \, m^2$ quadrats at peak biomass can provide a highly accurate estimate of NPP (Briggs and Knapp 1991, Biondini et al. 1991). In forest herbaceous layers, Gilliam (1991, Gilliam and Turrill 1993) found that combining harvested biomass measurements with estimated cover from $1 \, m^2$ quadrats yielded accurate models to allow for biomass estimates based on simple, nondestructive cover measurements. Arthur and Fahey (1990, 1992) employed a similar approach for the herb layer, but used plant height instead of cover.

Dimension analysis is an alternate way of estimating productivity in ecosystems dominated by vegetation too large or too slowly growing to allow for direct harvest methods. This is especially the case for forests. The technique assumes that some easily measured parameter, for example, plant height or diameter at breast height (dbh), can be directly correlated with standing crop. A few individuals must be harvested to determine the slope (and intercept) of a regression line, which may then be used to predict (estimate) plant biomass from the easily measured parameter. Harcombe et al. (1993) used such an approach with the following equation:

$$\log_{10} Y = A + B \log_{10} X \qquad \text{(Equation 12-2)}$$

where Y is estimated biomass, X is dbh, A is the y-axis intercept, and B is the slope of the regression line. Initially developed using $\log_e$ by Kittredge (1944), this approach is commonly used in studies of forest NPP (Prescott et al. 1989, Arthur and Fahey 1992).

In cases where herbivore activity is significant, these approaches must be used in combination with exclosures to eliminate herbivory (McInnes et al. 1992). These are especially effective for larger vertebrate herbivores (e.g., ungulates), whereas insect herbivores (e.g., gypsy moth and grasshoppers) present a more formidable challenge.

Production studies must also account for loss of leaves, branches, etc., during the period between harvests. Several types of **litter traps** have been devised to quantify the litter production of shrubs and trees (Figure 12-2). The most effective trap design will vary depending on local conditions and the nature of the material being sampled.

Root production is generally estimated to exceed aboveground production, but technical limitations make it difficult to assess the accuracy of the estimates. The usual approach is comparable to the methods described for measuring aboveground productivity—that is, to estimate the relative biomass of roots extracted

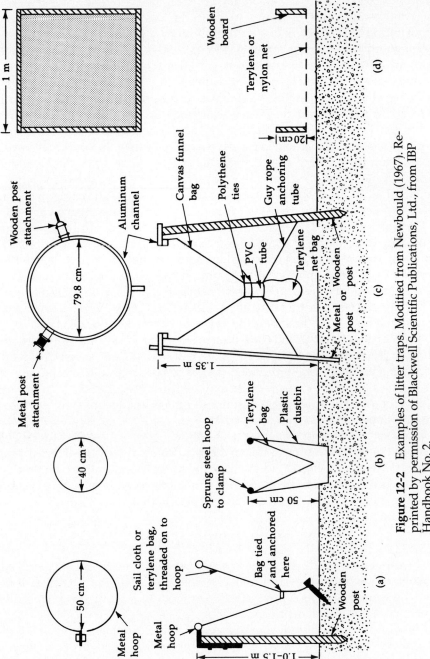

Figure 12-2 Examples of litter traps. Modified from Newbould (1967). Reprinted by permission of Blackwell Scientific Publications, Ltd., from IBP Handbook No. 2.

from soil cores taken periodically over a growing season (Vogt et al. 1982). Accuracy is problematic because we cannot know when to sample to obtain estimates that represent minimum and maximum root biomass levels during the growing season and because there is no way to gauge the loss of roots to herbivory and decomposition. The problem is further complicated by a high degree of variability in biomass estimates from soil cores (Singh et al. 1984). The problem of determining maximum and minimum biomass levels can be reduced by frequent sampling, and variability can be decreased by taking as many core samples as can be processed. The values obtained by harvest techniques may be conservative because production balanced by death and herbivory occurring between sampling dates is not accounted for. For example, estimates taken at consecutive harvests may show no increase in root biomass when, in reality, a large portion of the roots may have been shed and replaced during the period. Others (e.g., Lieth 1968, Newbould 1968) have monitored root growth by direct observation in glass-sided root pits. This technique has all the problems of sequential harvests for estimating productivity but is very effective in determining the temporal aspects of root growth.

No single method is best for estimating belowground production (Hansson and Andrén 1986, Jongen et al. 1995). Although the soil core method described here is the most widely used, an attractive alternative is the use of **root ingrowth bags** (Owensby et al. 1993). In this method, a mesh bag is filled with soil and inserted back into the soil for a given period, after which the amount of root tissue growing into the bag can be measured directly. Neill (1992) compared soil core versus root ingrowth methods and found that they yielded comparable results.

Another alternative to soil cores is an *in situ* method called the **mini-rhizotron technique** (Hendrick and Pregitzer 1992). A clear plastic tube is inserted in the soil along with a miniaturized color video camera to record root images. Root parameters (e.g., length, planar area, elongation) are thus measured directly and can be related allometrically to biomass (similar to dimension analysis) to estimate productivity (Snook and Day 1995).

Belowground productivity has been estimated to be 10–75% of total ecosystem productivity, with much of this in the form of fine roots (<3–5 mm diameter). Whereas earlier production studies either did not account for fine root production or did so with great error, the more recent methods provide us with a clearer picture of the dynamics of belowground plant productivity.

Productivity Models

For estimating productivity across still broader spatial and temporal scales, computer modeling is the best, and perhaps only, method. Although it is not within the scope of this book to describe these models in detail, several recently developed models deserve mention.

Many of the earlier models used to predict and estimate regional and global productivity would best be described as regression-based because they used known regressions between system variables, such as NPP, and climate variables, such as evapotranspiration (Meentemeyer et al. 1982). The more recent models are considered process-based because they integrate responses of individual processes, such as photosynthesis, respiration, decomposition, and nutrient cycling (Melillo et al. 1993). Ågren et al. (1991) argued that regression-based models are less likely than process-based models to make accurate predictions of productivity on a global scale because regression models only predict steady-state distributions of vegetation and carbon. The following is a brief mention of some currently used productivity models. The associated references should be consulted for further details on how the models are constructed.

The TEM (Terrestrial Ecosystem Model) predicts major carbon and nitrogen fluxes and pool sizes at continental to global scales, and was used by Raich et al. (1991) to estimate patterns of NPP for South America. Melillo et al. (1993) used TEM to predict the response of vegetation to rising concentrations of CO_2 and different climate change scenarios. Warnant et al. (1994) developed CARAIB (CARbon Assimilation In the Biosphere) as a model to estimate NPP for global ecosystems and found values similar to those presented later in this chapter. Like the TEM, the CENTURY model simulates carbon and nitrogen dynamics, and was used by Seastedt et al. (1994) to determine factors controlling productivity of a semihumid temperate grassland. Williams et al. (1997) used the MBL/SPA (Marine Biological Laboratory/Soil-Plant-Atmosphere) model to predict with great accuracy GPP in several terrestrial ecosystems.

Patterns of Productivity and the Distribution of Biomass

Production and Biomass in Shoots and Roots

The relative productivity and biomass of shoots and roots have important adaptive significance in plants. For example, colonizing species with annual life histories should show greater aboveground productivity because the following generation depends on seed output. On the other hand, a perennial plant growing in a nutrient- or water-limiting environment might have a greater belowground productivity and root biomass because uptake of limiting nutrients and water is increased by having a large and rapidly growing root system (Grime 1977, Chapin 1980, Waring 1983). Root production is costly because it requires transport of carbohydrate from the shoot. Therefore, plants growing in xeric or nutrient-poor habitats should maintain a larger active root system and have a lower rate of root tissue turnover than plants of resource-abundant habitats (Orians and Solbrig 1977, Chapin 1980). This explains why we find larger root-shoot ratios in plants of resource-stressed habitats. Do roots of plants in resource-limiting environments therefore have higher productivity? The answer is apparently no, because of a rapid fine-root turnover in plants of resource-abundant habitats. For example, Chapin and Van Cleve (1981) found that the fine roots of boreal forest trees growing in resource-limiting situations were maintained for a full year, whereas those from less limiting habitats had few fine roots that survived the winter. It appears, then, that plants of resource-limiting situations have larger root-shoot ratios but lower root productivity than plants of resource-abundant habitats. Productivity of fine roots and associated mycorrhizal fungi represents as much as 75% of the total net primary productivity in mature Pacific silver fir stands in western Washington (Vogt et al. 1982).

Litter Production and Decomposition

Biomass may remain living, serve a support function, be consumed directly by herbivores, or become detritus. The latter possibility is most likely; more than half of annual net productivity is deposited as litter. Litter is the food source for decomposers and detritivores and the means by which nutrients are returned to the cycling pool. It is thus important to consider the rate of accumulation and decomposition of dead plant parts. Olson (1963) calculated the ratio of litter production to litter accumulation as an expression of the rate of decomposition. The ratio is high in tropical environments, reflecting the high rate of leaf production and rapid decomposition, and decreases on a gradient toward the poles. Litterfall is generally related to LAI, causing the ratio to decrease with latitude in a fashion similar to that of NPP. Within forests, litterfall decreases with latitude and/or altitude. However, Jordan (1971) pointed out that litterfall does not show the same relationship along a moisture gradient. No differences in litter production are found when grassland, old-field, and tundra values are compared with forests at similar latitudes (Table 12-2).

Plant decomposition rates have been compiled by Singh and Gupta (1977), showing that decomposition varies with vegetation type and environment (Table 12-3, p. 312). Broadleaf temperate forests require about 1 year for complete litter decomposition, whereas coniferous forests typically require 3–5 years. The extremes reported by Singh and Gupta are represented by high mountain pine forests of California, which need over 30 years for complete degradation, and tropical forests, where complete breakdown may occur in less than 2 months. These turnover rates depend primarily on the chemical makeup of the litter, the temperature, and the moisture conditions of the habitat.

Moisture and temperature together exert a significant influence on rates of decomposition (Wiegert and Evans 1964, Rochow 1974). It is not possible to separate the effects of temperature and moisture because they are not environmentally independent, and extreme levels of either will exceed the tolerance of decomposers. Douglas and Tedrow (1959) found decomposition rates of only about 1.0 t ha^{-1} yr^{-1} in arctic tundra communities. Such low rates are probably a result of the very short growing season. Contrast the eight times greater rate in tropical rain forests with continuous growing seasons (Wanner 1970). Bleak (1970) measured the disappearance of litter under winter snow in central Utah and found losses of up to 50% despite temperatures of $+1.2°$ to $-2.5°C$. Fungi and bacteria were active at these temperatures and were the primary agents for breakdown. Winn (1977) examined decomposition rates in chaparral stands in California where temperatures

Table 12-2 Rate of litter production in various forest, perennial herb, and grass ecosystems. From data compiled by Jordan 1971. Reprinted by permission of *American Scientist,* magazine of Sigma Xi, The Scientific Research Society.

Community	Location	Litterfall (g m^{-2} yr^{-1})	Source
Tropical rain forest	Thailand	2322	Kira et al. 1967
Tropical rain forest	Average of several	1600	Rodin and Basilevic 1968
Subtropical forests	Average of several	1200	Rodin and Basilevic 1968
Dry savanna	Russia	290	Rodin and Basilevic 1968
Oak forest	Russia	350	Mina, cited in Rodin and Basilevic 1968
Fir-taiga	Russia	250–300	Rodin and Basilevic 1968
Oak-pine forest	New York	406	Whittaker and Woodwell 1969
Pine forest	Virginia	490	Madgwick 1968
Tropical seasonal forest	Ivory Coast	440	Muller and Nielsen, cited in Kira et al. 1967
10 Angiosperm forests	Europe	280	Bray and Gorhman 1964
10 Angiosperm forests	Tennessee	320	Whittaker 1966
13 Gymnosperm forests	Tennessee	267	Whittaker 1966
Old-field upland	Michigan	312	Wiegert and Evans 1964
Old-field swale	Michigan	1003	Wiegert and Evans 1964
Perennial herbs	Japan	1484	Iwaki et al. 1966
Tallgrass prairie	Missouri	520	Kucera et al. 1967, Dahlman and Kucera 1965
Mesic alpine tundra	Wyoming	162	Scott and Billings 1964

rarely limit decomposer activity. Microbes responded quickly to small fluctuations in moisture conditions, and decomposition is apparently controlled by the level of hydration of litter.

Numerous other factors, such as secondary metabolites leached into the litter, soil characteristics, the kinds of organisms in the detritus food chain, and herbivore activity influence decomposition on a local basis. Many of these factors can be related to water, temperature, and the chemical nature of the litter. Meentemeyer (1978) studied actual evapotranspiration (AET), as a measure of energy and water in the environment, and lignin content, as a measure of litter quality, to develop a model predicting decomposition in temperate and boreal forests. More than half (52%) of the variation in the data could be explained by AET. Litter quality added little predictive power to the model.

Melillo et al. (1982) expanded on the Meentemeyer model, arguing that it was not just the content of lignin in litter that should be measured but the balance be-

Table 12-3   Representative rates of plant litter decomposition.

Biome and location	Decomposition rate (% per day)	Source
Tropical climate		
Rain forest:		
Trinidad	0.45	Cornforth 1970
Grassland:		
India	0.30	Gupta and Singh 1977
Other	0.17–1.5	
Temperate climate		
Oak forest:		
Minnesota	0.018	Reiners and Reiners 1970
Missouri	0.095	Rochow 1974
New Jersey	0.018	Lang 1974
England	0.30	Edwards and Heath 1963
Pine forest:		
California	0.0027–0.0082	Jenny et al. 1949
Missouri	0.036	Crosby 1961
Southeastern U.S.	0.07	Olson 1963
Other	0.0027–0.12	
Deciduous forest:		
Eastern U.S.	0.057	Shanks and Olson 1961
England	0.043–0.06	Anderson 1973
Australia	0.04–0.15	Ashton 1975
Grassland:		
North Dakota	0.082–0.11	Redmann 1975
Missouri	0.14	Koelling and Kucera 1965
Utah	0.082–0.14	Bleak 1970

tween lignin content and nitrogen in the litter; that ratio would have more effect on how rapidly decomposition occurred. They found extremely high levels of correlation (nearly 90%) between decay rates and the ratio of initial lignin (%) to initial nitrogen (%) in forest litter across a variety of forest types.

Global Patterns of Productivity and Biomass

Terrestrial vegetation occupies approximately 30% of the globe's surface and provides 62% of the total world primary productivity. Also, most of the world's biomass consists of terrestrial vegetation (Lieth 1973). Vitousek et al. (1986) estimated that total global NPP was 224.5 Pg yr^{-1} (Pg = petagram = 10^{15} g = 10^9 metric tons). Of that, 132.2 Pg yr^{-1} was from terrestrial ecosystems, or about 60% of global NPP. Table 12-4 reveals the dramatic differences in productivity and biomass

Table 12-4 Net primary productivity and related characteristics of terrestrial biomes. From R. H. Whittaker and G. E. Likens. 1975. "The biosphere and man." In *Primary Productivity of the Biosphere*, Lieth and Whittaker, eds. By permission of Springer-Verlag, New York.

Ecosystem type	Net primary productivity (dry matter)				Biomass (dry matter)			Leaf surface area	
	Area (10^6 km^2)	Normal range (g m^{-2} yr^{-1})	Mean (g m^{-2} yr^{-1})	Total (10^9 t yr^{-1})	Normal range (kg m^{-2})	Mean (kg m^{-2})	Total (10^9 t)	Mean (m^2 m^{-2})	Total (10^6 km^2)
Tropical rain forest	17.0	1000–3500	2200	37.4	6–80	45	765	8	136
Tropical seasonal forest	7.5	1000–2500	1600	12.0	6–60	35	260	5	38
Temperate forest:									
Evergreen	5.0	600–2500	1300	6.5	6–200	35	175	12	60
Deciduous	7.0	600–2500	1200	8.4	6–60	30	210	5	35
Boreal forest	12.0	400–2000	800	9.6	6–40	20	240	12	144
Woodland and shrubland	8.5	250–1200	700	6.0	2–20	6	50	4	34
Savanna	15.0	200–2000	900	13.5	0.2–15	4	60	4	60
Temperate grassland	9.0	200–1500	600	5.4	0.2–5	1.6	14	3.6	32
Tundra and alpine	8.0	10–400	140	1.1	0.1–3	0.6	5	2	16
Desert and semidesert scrub	18.0	10–250	90	1.6	0.1–4	0.7	13	1	18
Extreme desert: rock, sand, ice	24.0	0–10	3	0.07	0–0.2	0.02	0.5	0.05	1.2
Cultivated land	14.0	100–4000	650	9.1	0.4–12	1	14	4	56
Swamp and marsh	2.0	800–6000	3000	6.0	3–50	15	30	7	14
Lake and stream	2.0	100–1500	400	0.8	0–0.1	0.02	0.05	—	—
Total	149		782	117.5		12.2	1837	4.3	644

values reported for the major terrestrial ecosystems. Productivity estimates are lowest in deserts and highest in tropical rain forests, spanning the entire range between 0 and approximately 3000 g m^{-2} yr^{-1}. Whittaker suggested that 3000 to 3500 g m^{-2} yr^{-1} is a maximum productivity value for terrestrial systems. Biomass estimates range from 0.1 kg m^{-2} in desert and tundra ecosystems to 200 kg m^{-2} in some temperate rain forests. The usual range, however, is between 1.0 and 60 kg m^{-2}. More detailed summaries can be found in Röhrig (1991), Holmén (1992), and Monserud et al. (1995). A careful evaluation and comparison of these values later in this section will further refine our understanding of the distribution of productivity and biomass.

The relationship of biomass to productivity is often expressed as the **biomass accumulation ratio** (**BAR**) (Whittaker 1975). BAR is the ratio of dry weight biomass to annual net primary productivity. BAR values are a measure of the accumulation of primarily woody material, a characteristic related to environmental harshness and the potential age of dominant species. The BAR represents the average residence time in years of organic matter in the community. There is a broad overlap in BAR values between ecosystems. Representative values are 1 for communities of annuals, 2–10 for deserts, 1.3–5 for grasslands, 3–12 for shrublands, 10–30 in woodlands, and 20–50 for mature forests. Whittaker and Niering (1975) calculated BAR values for an elevational gradient from desert to subalpine forest in the Santa Catalina Mountains of Arizona and found that in forest and woodland zones, BAR decreased as a function of biomass. However, once the low-elevation desert shrublands were encountered, BAR values and biomass did not change significantly with further drops in elevation.

Jordan (1971) also recognized the relationship between major environmental gradients and the biomass allocation patterns of the dominant plants. He calculated the ratio of wood production to litter production and found high correlations between the ratio and total light energy available during the growing season (Figure 12-3) and between the ratio and annual precipitation. This suggested that rapid wood production is an advantage in areas of low light such as the boreal forests. Perhaps larger plants have more energy reserves and can thus resist greater environmental stress. Greater energy storage capacity of larger plants increases their chances of survival in stressful environments, where they may frequently suffer defoliation or very short growing seasons.

In general, trends in world productivity are loosely related to biomass, in that K-selected plants in less harsh or more predictable environments are able to accumulate extensive root and branch systems. These form the bases for efficient light interception, nutrient uptake, and, in turn, productivity. It should be remembered, however, that much of the accumulation of biomass serves only a support function, which explains the different rates of change in productivity and biomass values observed from the poles toward the tropics.

When we consider the information on forested ecosystems in Table 12-4, several generalizations are apparent. Even though the productivity of tropical rain forests has been estimated to exceed 3000 g m^{-2} yr^{-1}, the range of productivity values overlaps all other forest and woodland types and may even include values for some temperate grassland systems. Some evidence suggests that temperate

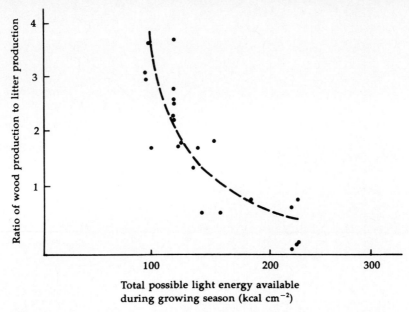

Figure 12-3   Ratio of wood production to litter production in forest communities as a function of light energy available during the growing season. From Jordan 1971. Reprinted by permission of *American Scientist,* magazine of Sigma Xi, The Scientific Research Society.

evergreen forests are more productive than their deciduous counterparts (e.g., Schlesinger 1997). At first thought, this seems improbable because evergreen leaves have 30–60% lower maximum photosynthetic rates per unit surface area than deciduous leaves (Mooney 1972); however, this is offset by high leaf area index (LAI) for evergreen forests. Kira (1975) explained that deciduous plants have very low LAI during periods of leaf development and may miss periods of favorable moisture and temperature following photoperiodically induced leaf drop. Kira also suggests that the high net productivity of tropical evergreen forests is related to increased height and greater solar radiation, which allows increased LAI and therefore higher exploitation efficiency. Needle-leaved evergreens frequently have nearly twice the LAI of deciduous trees of similar biomass and height, thus allowing plants with a lower photosynthetic capacity per unit leaf area to have higher potential stand production. This very close relationship between LAI, evergreenness, and gross productivity is plotted in Figure 12-4. A high LAI is made possible by large size, necessitating larger amounts of nonproductive structural tissue, thereby reducing the net production efficiency of forest trees.

Waring (1983) has developed a conceptual approach of expressing growth efficiency in trees as the ratio of wood production to leaf area. Because the area of sapwood involved in water conduction is directly related to canopy leaf area, the growth efficiency and potential aboveground productivity of trees can be calculated simply by regression. The product of growth efficiency and canopy leaf area

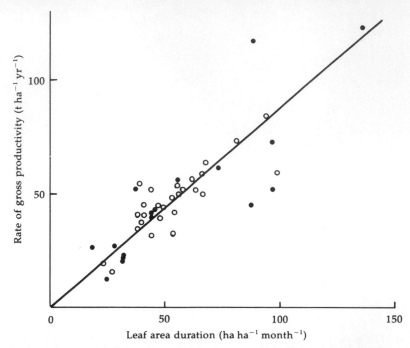

Figure 12-4 The relationship between leaf area index, evergreenness, and gross productivity. Leaf area duration values are calculated as LAI times length of growing season in months. Filled circles are for broad-leaved forests and open circles for needle-leaved forests. From Kira (1975) in *Photosynthesis and Productivity in Different Environments,* IBP Handbook No. 3, J. D. Cooper, ed. Reprinted by permission of Cambridge University Press.

gives an estimate of production and can thus be used to assess the degree to which water stress, nutrient stress, overcrowding, and so on influence a particular forest stand. Such general indices of vigor may be useful in predicting the susceptibility of a forest stand to disease or insect attack. (See "Herbivores" in this chapter.)

Grassland productivity is typically lower than in communities with trees because of the LAI relationships considered above. Ovington et al. (1963) compared prairie productivity with the productivity of adjacent savanna and oakwood communities and found grassland productivity to be 10–20% of the productivity of communities with woody plants. The absence of woody tissues results in low BAR values (1.3–5) and short-lived aboveground parts in grasslands.

Arctic and alpine communities include life forms that range from shrubs to cryptogams and have LAIs comparable to other communities in severe climates. Net primary productivity is typically 100–150 g m^{-2} yr^{-1}, with biomass as low or lower than in desert communities. Miller and Tieszen (1972) modeled production processes and noted the same increases in production in response to removal of dead material in arctic and alpine communities as in grasslands. Lemmings remove large quantities of biomass in the natural system (Dennis and Johnson 1970) and may cause increases in production similar to the increases caused by fire in grasslands. The main restrictions on production in the tundra are low LAI, low temperatures, and low angles of incident radiation.

Primary production in arid lands varies from 10 to over 200 g m^{-2} yr^{-1}, depending on the amount and pattern of rainfall. Uptake of CO_2 is severely limited by stomatal closure during dry periods, and variations occur seasonally in response to water availability (Chew and Chew 1965, Burk and Dick-Peddie 1973).

Successional Patterns of Productivity and Biomass

Productivity follows a pattern of change during succession similar to the pattern of change noted in Chapter 11 for species diversity. Figure 12-5 depicts this trend of gradually increasing productivity during the pioneer and early tree stages, followed by decreasing productivity as the self-perpetuating climax trees reach maturity (Loucks 1970). Botkin et al. (1972) constructed a model of forest growth predicting that standing crop reaches a peak within about 200 years, then drops 30–40% during the following 200 years. This biomass reduction could, in itself, account for some declining productivity. Other trends, such as a decline in the photosynthetic efficiency of older trees (Waring and Schlesinger 1985), the allocation of a greater proportion of net productivity to nonphotosynthetic structural biomass, limitations imposed upon LAI by canopy form and leaf orientation (Horn 1974), and the binding of nutrients into structural biomass (Connell and Slatyer

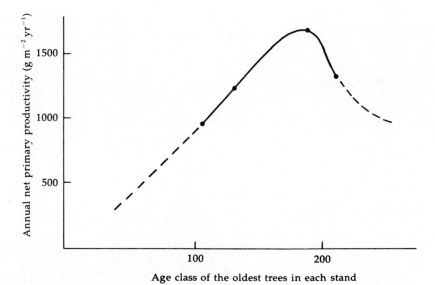

Figure 12-5   Annual net primary productivity, as estimated by tree stem measurements, plotted by age class of the oldest trees in each stand. Recalculated from Loucks. 1970. "Evolution of diversity, efficiency, and community stability." *American Zoologist* 10:17–25.

1977), also lead to reduced productivity in mature communities. Murty et al. (1995) tested several of these explanations of decline in forest productivity as a stand approaches maturity. Using an ecosystem model called G'DAY (Generic Decomposition And Yield—it might be noted that the first two authors were from Australia), they concluded that explanations that attempt to relate increases in sapwood respiration to the temporal pattern of declining NPP are unsatisfactory. They instead suggested that lowered efficiencies of stomatal closure and photosynthesis offer a better explanation. These changes in biomass allocation and physiological response lead to greater susceptibility of many communities to fire, insect attack, and windthrow. Loucks (1970) suggested that these characteristics are adaptations to repeating patterns of environmental change that stimulate periodic returns to states of high net primary productivity.

The biomass accumulation ratio mentioned earlier is often used in connection with succession studies (Whittaker and Likens 1975). The BAR increases from 1 in the pioneer stage of succession to between 30 and 50 in mature forest communities (Figure 12-6). Attiwill (1979) suggests that the growth of a forest may be considered a sequence of three stages, defined by changing relationships between aboveground net primary production and aboveground biomass. The first stage consists of rapid productivity without significant increases in biomass; thus the BAR remains low. This stage lasts up to about 20 years. After that, continued increases in

Figure 12-6 Relationship between biomass and NPP for temperate forests of the United States, and the trajectory of the relationship (biomass accumulation ratio, BAR) with age in years for *E. obliqua* (the heavier dashed line). A range of values of BAR in years is shown as a family of curves. From "Nutrient cycling in a *Eucalyptus obliqua* (L'Herit.) forest. III. Growth, biomass, and net primary production" by P. M. Attiwill, *Australian Journal of Botany* 27:439–458, 1979.

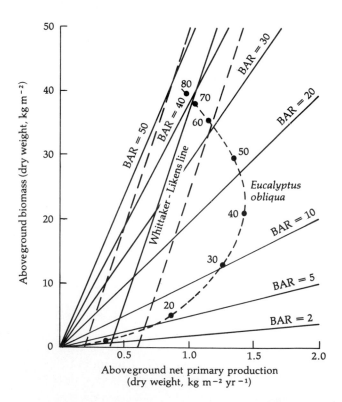

biomass depend on attaining larger size. The second stage is characterized by a period of rapid heartwood formation and rapidly increasing values of BAR. As the forest trees reach their genetic or environmental size limit, the rate of increase in BAR values tapers off. This final stage is characterized by an average residence time of organic matter in the biomass of 40–50 years.

Environmental Factors and Productivity

Light and Temperature

Global radiation varies with atmospheric conditions, latitude, and altitude. However, the effect of radiation on community productivity is indirect and is seen primarily through differences in growing season (Whittaker 1975) and temperature (Figure 12-7). Productivity increases along the mean annual temperature gradient from the poles toward the equator. Optimum temperatures for productivity coincide with the 15–25°C optimum range of photosynthesis. This close correlation between the photosynthetic temperature optimum and productivity may be coincidental; length of growing season may be a more important determinant than temperature itself. The wide spread of points in Figure 12-7 is indicative of the

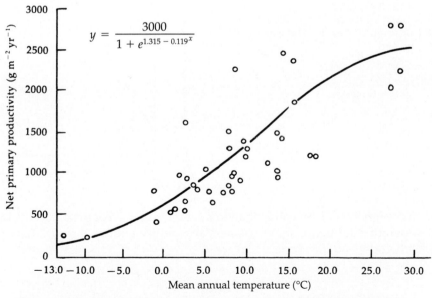

$$y = \frac{3000}{1 + e^{1.315 - 0.119x}}$$

Figure 12-7 Net primary productivity, including both above- and below-ground productivity, in relation to mean annual temperature. From "Primary production: Terrestrial ecosystems" by H. Lieth, *Human Ecology* 1:303–332, 1973. By permission of Plenum Publishing Corporation.

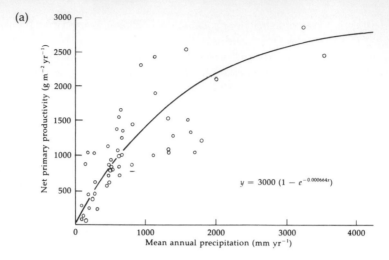

(a)

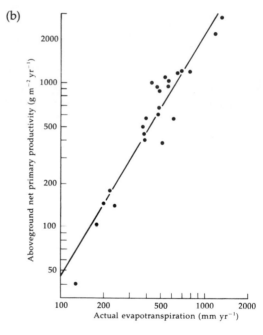

(b)

Figure 12-8   Patterns of terrestrial net primary productivity as predicted (a) by Lieth (1973), using mean annual precipitation, and (b) by Rosenzweig (1968), using actual evapotranspiration. (a) From "Primary production: Terrestrial ecosystems" by H. Lieth, *Human Ecology* 1:303–332, 1973. By permission of Plenum Publishing Corporation. (b) From Rosenzweig, M. L. 1968. "Net primary productivity of terrestrial communities: Prediction from climatological data." *American Naturalist* 102: 67–74. Copyright © 1968 by The University of Chicago.

many environmental factors other than temperature that affect productivity. Light and temperature influence water use and availability and, as a result, have few completely independent effects.

Water

The environmental factor most directly correlated with productivity is water. Lieth (1973) showed the correlation between productivity and mean annual precipitation (Figure 12-8), and Rosenzweig (1968) compared productivity with actual evapotranspiration (AET; Figure 12-8b). Evapotranspiration is a measure of the total amount of water lost by transpiration and evaporation. Evapotranspiration has high predictive value in productivity studies because it includes the influence not only of water but also of light and temperature. Precipitation and evapotranspiration approach equality in arid environments, so the relationship between productivity and annual rainfall is nearly linear at values below 500 mm. At higher levels of precipitation, more water is lost as runoff or drains below the root zone, where it no longer influences production. A fourfold increase in precipitation and associated cooler temperatures resulted in a species-specific 100–600% increase in aboveground productivity for Mojave Desert shrubs (Bamberg et al. 1976). Cable (1975) studied the response of perennial grasses near Tucson, Arizona, and developed a model in which the product of current precipitation and the past summer's precipitation was 203 times more accurate as a predictor of growth than the current precipitation alone. The reason we see such close correlations between annual means and overall productivity (annual means are usually poor predictors of plant response) may be this delayed response, which tends to reduce the impact of unusually wet or dry years.

Direct evidence for productivity trends can be obtained by studying adjacent communities along a gradient of temperature, moisture, and evaporation. Whittaker and Niering (1975) conducted such a study in the Santa Catalina Mountains near Tucson, Arizona, along an uninterrupted vegetational gradient from subalpine forest through woodlands and grasslands to desert. Figure 12-9 compares the Santa Catalina Mountains results with those predicted by Rosenzweig (1968) and Lieth (1973) for the relationship between AET or mean annual precipitation and net primary productivity. The actual measurements show a more complex relationship than predicted. There is an abrupt change of slope (line c in Figure 12-9) at 400–500 g m^{-2} yr^{-1}. If one were to guess at what point such a change might occur, the margin of the desert would seem a reasonable place because of the dramatic change in life form and the nature of precipitation extremes in that ecotone. Interestingly, however, the transition does not occur at the desert margin, but at the transition from open to dense woodland, where trees become the dominant producing life form. Net productivity of about 1300 g m^{-2} yr^{-1} represents the upper level measured. This must indicate another abrupt change in slope, because estimates of maximum productivity for climax temperate forests are approximately 1500 g m^{-2} yr^{-1}. (Note that the dashed portion of line c is hypothetical.)

Assuming that this sort of nonlinear relationship is typical, Whittaker and Niering offer some hypotheses. They suggest that the concave lower part of curve c resembles the theoretical curve a of Rosenzweig because in these arid environments evapotranspiration is essentially the same as precipitation. Productivity falls lower because a leaf area index small enough to minimize water loss necessarily limits productivity. The steep central portion of curve c represents evergreen forests in environments humid enough to support a less limiting leaf area index, thereby allowing a more rapid increase in productivity per unit of available moisture than in more arid habitats. Further increases in available moisture (above 800–900 mm) are associated with distinctly less rapid increases in net productivity.

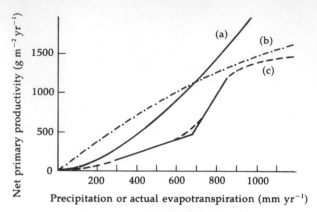

Figure 12-9   Three interpretations of the relation of net primary productivity (dry g m^{-2} yr^{-1}) to precipitation and actual evapotranspiration. (a) The curve fitted by Rosenzweig (1968) for aboveground net primary productivity of forests and shrublands in relation to actual evapotranspiration (mm yr^{-1}) using the formula NPP = 1.66 log$_{10}$ AE − 1.66. (b) The curve fitted by Lieth (1973) for total net primary productivity in relation to mean annual precipitation (mm yr^{-1}) using the formula NPP = 3000/(1 − $e^{1.315 - 0.119\,\text{MAP}}$). (c) Curve relating the two slopes of the Santa Catalina Mountains total net primary productivity estimates to probable mean annual precipitation, and adding to these a third, upper slope for limitation of climax temperate forest productivity at around 1500 g m^{-2} yr^{-1}. The dashed portion of the curve is hypothetical. (a) From Rosenzweig, M. L. 1968. "Net primary productivity of terrestrial communities: Prediction from climatological data." *American Naturalist* 102:67–74. Copyright © 1968 by The University of Chicago. (b) From H. Lieth, "Primary production: Terrestrial ecosystems," *Human Ecology* 1:303–332, 1973. (c) Modified from Whittaker and Niering 1975. Copyright © 1975 by the Ecological Society of America.

Here, factors such as nutrient turnover, balance of supporting and photosynthetic tissue, and light absorption may severely limit increases in productivity in these temperate climax forests. Webb et al. (1978, 1983) suggest that the lack of continued increase in productivity at high levels of water availability is typical of ecosystems where water is not the most critical limiting factor. The lower slope for desert and open woodland communities may be due to the fact that because of unpredictable conditions or more restricted water availability, the plants are not capable of high levels of productivity even when water is available. Those communities at intermediate elevations in the Whittaker and Niering (1975) data have sufficient aboveground biomass and metabolic adaptations to respond more dramatically to increases in precipitation. The forest communities represented by the dashed portion of curve c may represent communities that are not water-stressed. In non-water-stressed forests, Webb and his colleagues suggest that such factors as light, temperature, leaf biomass, and aboveground biomass are better correlated with productivity.

Carbon Dioxide

As we will see in Chapter 15, CO_2 is generally a limiting factor for C_3 plants. The predominance of C_3 species in the world flora suggests that global production might dramatically increase in response to increasing atmospheric CO_2. (We shall look at global patterns of atmospheric CO_2 at the end of the chapter.) The relationship is not that simple, however, because CO_2 is not evenly distributed on smaller spatial and temporal scales (Figure 12-10). Also, the distribution of CO_2 depends on other factors, such as light and wind. In a windless daytime condition, CO_2 concentrations are lowest around midcanopy—usually the portion of the canopy with the highest LAI. At this time, the gradient of ambient CO_2 is greatest, varying about 20% between the plant canopy and the forest floor, where we find the high-

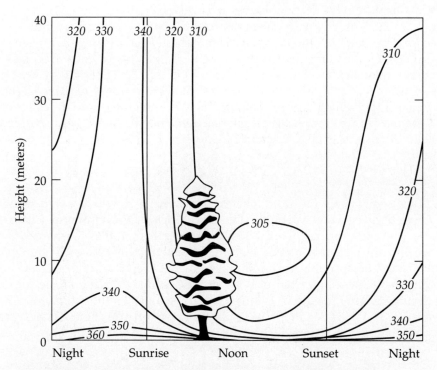

Figure 12-10 Typical profiles of concentrations of CO_2 for a forest canopy at different times of day. Isopleth lines for CO_2 are in parts per million (ppm). From Holmén, K. 1992. "The global carbon cycle." In *Global Biogeochemical Cycles*, pp. 239–262. London: Academic Press. Used with permission.

est CO_2 concentrations from (a) fine-root respiration, (b) respiratory activity of soil heterotrophs, and (c) low diffusivity of CO_2 near the ground. Because most plants emit CO_2 at night (Holmén 1992), concentrations inside the canopy rise quickly following sunset (Figure 12-10). It is notable in Figure 12-10 that the decline in mid-canopy CO_2 actually precedes sunrise. This is the result of an interesting phenomenon known as **predawn stomatal opening,** a characteristic of many plants wherein stomates actually open before the sun rises (Zeiger et al. 1981, Gilliam and Turrill 1995).

Our ability to predict the response of productivity on a global scale to increases in CO_2 has been increased substantially by the development of, and continued improvements in, the process-based models discussed earlier in this chapter. A model of particular promise, the TEM, was used by Melillo et al. (1993) to predict the response of the ecosystem types shown in Table 12-4 to a doubling of the current concentration of CO_2, both with and without subsequent increases in global temperatures. They found that doubling CO_2 concentrations without a temperature increase would increase global NPP by just over 16%. This response varied greatly among ecosystem types. Although increases in global temperature without a change in CO_2 resulted in little change in NPP, a combination of temperature increase and CO_2 doubling increased global NPP by 20–26%, depending slightly on the global climate (change) model (GCM) used. Northern and temperate ecosystems responded more than other ecosystems did to increasing temperature under CO_2 doubling, however, because of temperature-mediated increases in nitrogen availability (Melillo et al. 1993). McGuire and Joyce (1995) used the TEM for temperate forests throughout the United States and found variable responses of NPP to CO_2 doubling under different GCMs. As suggested earlier, results of the TEM emphasize the complexity of interactions among factors that can influence productivity in terrestrial ecosystems.

Soil Nutrients

It has been known for quite some time that plant growth can be determined (i.e., limited) by nutrient availability (see Liebig's law of the minimum, Chapter 3). Factors controlling nutrient availability are themselves complex (Barber 1995), and the specific nutrients that potentially limit productivity vary greatly among ecosystem types. Because there is an intimate link between the movement of carbon and nitrogen (Tateno and Chapin 1997) and the resultant high demand for nitrogen, productivity in most ecosystems tends to be nitrogen limited (Vitousek and Howarth 1991). However, in highly weathered or poorly developed (sandy) soils, phosphorus and base cations such as potassium may limit productivity also (Stone and Kszystyniak 1977, Gilliam and Richter 1991).

Other edaphic characteristics, such as soil texture and depth, can influence productivity, although it is often difficult to separate which characteristics are exerting direct influences. For example, Pastor et al. (1984) found a highly significant positive correlation between NPP and nitrogen availability (measured as net nitrogen mineralization) among different forest stand types of Blackhawk Island, Wisconsin. However, they found a similar highly significant positive correlation between NPP and soil texture, measured as percent silt plus clay in the subsoil (Figure 12-11). Although it is tempting at first to claim that these forests are nitrogen limited, it may be that both ecosystem processes (i.e., NPP and nitrogen mineralization) are controlled by the same factor (water availability), which itself is influenced by soil texture (Pastor and Post 1986).

Herbivores

In contrast to aquatic ecosystems, where most primary productivity is consumed by herbivory, terrestrial ecosystems typically lose <10% of annual NPP to herbivores (McNaughton 1985). Thus, most of the NPP of terrestrial ecosystems directly enters the detrital food chain, becoming the energy source for decomposers. Exceptions to this generally fall into two categories: (1) chronic herbivory by vertebrates in grazing-maintained ecosystems; and (2) episodic herbivory by invertebrates, usually insect larvae in forest ecosystems. The two categories of herbivory have contrasting effects; each will be treated here briefly.

A large body of evidence indicates that species of the grass family and grazing mammals have undergone long-term coevolution, from their mutual first appearance in the Eocene to their simultaneous adaptive radiation in the Miocene to Pleistocene epochs (Clayton 1981). In fact, the words *grass* and *graze* share the same Old English word origin, *græs*. Some of the better-known grazing ecosystems include the large ungulates (especially bison) of North American prairies (Knapp et al. 1998), and an impressive diversity of large vertebrate herbivores of the Serengeti Plains of Africa (McNaughton and Georgiadis 1986).

Given such a long period of coevolution, it is not surprising that although the degree of herbivory occurring in grazing ecosystems is quite high, grazers consume NPP in a way that is incorporated as part of the ecosystem. McNaughton (1985) found that herbivores of the Serengeti consume plant biomass in amounts that ranged from 17% to 94% of NPP. It should be noted that the low value, 17%, is well above the level generally associated with terrestrial ecosystems. However, grazing can actually increase NPP of remaining plant material (i.e., grass foliage) by preventing accumulation of detritus (Knapp and Seastedt 1986) and through compensatory growth (McNaughton 1983). Clearly, grazing can be viewed as a regulator of NPP in grazing ecosystems (McNaughton 1985).

Although forest ecosystems constantly experience some degree of herbivory by both vertebrates, such as deer, and invertebrates, such as insects, this "background," or endemic, herbivory typically consumes a very small amount of NPP, often as low as 1% (Perry 1994). Measurable impacts on NPP occur, however, when herbivory reaches epidemic levels, usually in the form of outbreaks of insect larvae. It is not surprising that many, if not most, reported outbreaks are associated with insect species exotic (i.e., nonnative) to the impacted forest. The U.S. Forest Service recognizes at least 17 introduced insect species that together consume enough NPP to create potential health problems for forests of the northeastern United States (USDA Forest Service 1993). Larvae of some insects can defoliate entire forests, that is, they can consume 100% NPP as leaf material. In 1993, for example, the gypsy moth, introduced from Europe to the United States in 1869,

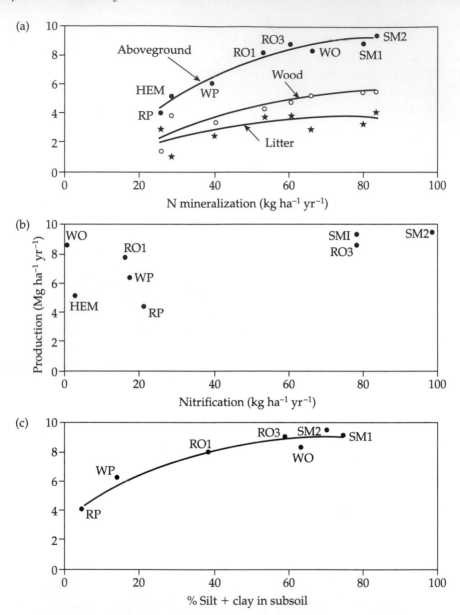

Figure 12-11   Relationship of productivity in different ecosystems of Blackhawk Island, Wisconsin, to (a) N mineralization, (b) nitrification, and (c) subsoil silt + clay content. Ecosystems are as follows: RP = red pine; HEM = hemlock; WP = white pine; RO = northern red oak; WO = white oak; SM = sugar maple. Curves are second-order polynomials fitted to data, all significant at $P < 0.001$. From Pastor, J., J. D. Aber, C. A. McClaugherty, and J. M. Melillo. 1984. "Aboveground production and N and P cycling along a nitrogen mineralization gradient on Blackhawk Island, Wisconsin." *Ecology* 65:256–268. Used with permission.

defoliated >490,000 ha of hardwood forests throughout an eleven-state area of the northeastern United States including 161,600 ha in Michigan, 135,300 in Pennsylvania, and 82,000 in West Virginia (USDA Forest Service 1993).

Herbivore activity influences primary productivity directly, by reducing photosynthetic area, and indirectly, by modifying other environmental factors. For example, nutrients may leach more readily from the damaged foliage, and litter may have higher nutrient content when removed by herbivore activity. This is because natural abscission is preceded by remobilization and reabsorption of certain nutrients in the expendable part. Herbivore feces and urine make nutrients more readily available for decomposition, thus increasing the turnover rate. Availability of partially broken-down plant material may stimulate microbial activity, thereby increasing the turnover of nutrients. Removal of living biomass may change species composition, increase light penetration into the canopy, or reduce competition for light, water, and nutrients, any of which will affect primary production.

Mattson and Addy (1975) considered the importance of plant-eating insects in regulating primary productivity of forest communities. They concluded that insects help maintain consistently high primary productivity in natural systems by consuming biomass from less vigorous plants, thus opening the canopy for younger, more vigorous plants to establish themselves. Such insect activity may occur on a large scale, as in the case of the Douglas fir tussock moth, the gypsy moth, the spruce budworm, and the southern pine beetle, which periodically cause widespread defoliation and destruction of trees (Figure 12-12). Insect outbreaks

Figure 12-12   Spruce forest in southern Colorado killed by insect herbivory. Photo compliments of Harold Bradford.

may be related to increased nutritional value of the insect food and lower host resistance brought about by tree age, marginally adequate habitat conditions such as drought, or low levels of soil nutrients.

Limitation of Productivity by Multiple Resources

Although we can examine separately the importance of the various factors that influence productivity in terrestrial ecosystems, in reality they exert their respective influences simultaneously and interactively. That is, it may be greatly shortsighted to consider any one factor as acting alone. This kind of complexity represents a serious challenge in attempting to understand the controls of terrestrial ecosystem productivity. Several attempts have been made toward such an understanding, and one of the more promising models considers multiple factors that limit productivity in ways that shift temporally and spatially—the transient maxima hypothesis (TMH) (Seastedt and Knapp 1993).

Although the TMH is intended to explain the patterns of productivity of any nonequilibrium ecosystem, it was initially developed for tallgrass prairie, based largely on work done at the Konza Prairie Research Natural Area in north-central Kansas. The term *nonequilibrium* refers to ecosystems, such as tallgrass prairie, which experience frequent temporal shifts in limiting factors. Ultimately, these factors include most of the ones discussed here—light, water, and nutrients—all mediated in tallgrass prairie by fire and herbivory (grazing by bison). Burning and grazing alter the presence of persistent detritus, which remains largely in place at the end of the tallgrass prairie growing season. Standing detritus has a high C:N ratio from efficient resorption of nitrogen at the end of the growing season, and thus is slow to decompose (Seastedt 1988). Standing detritus serves as an effective filter for light (Knapp and Seastedt 1986, Knapp and Gilliam 1985), water (Gilliam et al. 1987), and atmospheric deposition of nutrients, especially nitrogen (Seastedt 1985, Gilliam 1987), thus influencing the availability of the essential resources that determine NPP.

Although fertilizer experiments have shown that annually burned prairie is nitrogen limited (Turner et al. 1997), productivity is higher in annually burned prairie than in unburned prairie (Briggs and Knapp 1995, Turner et al. 1997). Highest NPP occurs, however, in a transient fashion following fire in long-unburned prairie (Figure 12-13a). The TMH explains this response best: under equilibrium conditions, NPP is limited by light in unburned prairie, where we find accumulations of available soil nitrogen not used by the light-limited plants of unburned prairie. Fire removes the standing detritus that has accumulated under no-fire conditions. The plants are no longer light-limited and can take advantage of the high nitrogen availability in the soil, resulting in a nonequilibrium "pulse" of NPP.

All the above assumes that moisture, another potentially limiting factor, is adequate. Quite often, however, moisture is not adequate—periodic drought is very much a part of the ecology of most grassland ecosystems (Knapp et al. 1998). Long-

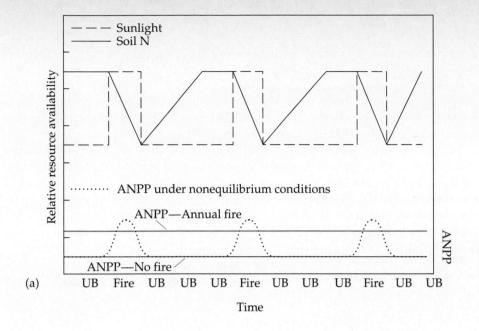

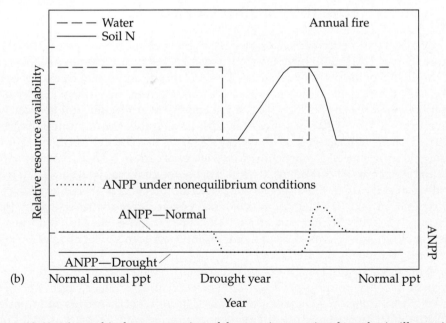

Figure 12-13 A graphical representation of the transient maxima hypothesis, illustrating how multiple limiting resources—light, nutrients (nitrogen), and water—potentially interact under nonequilibrium conditions to produce transient maximum responses in net primary productivity in tallgrass prairie in two scenarios: (a) Periodic fire occurring in long-unburned prairie. The *x*-axis represents environmental conditions (UB = unburned); *y*-axes represent relative resource (light and nitrogen) availability in unburned prairie and relative productivity (annual net primary productivity, ANPP) in annual fire or no-fire conditions. (b) Drought-caused decreases in productivity via water limitation. From *Grassland dynamics: long-term ecological research in tallgrass prairie*, edited by A. K. Knapp et al. Copyright © 1998 by Oxford University Press, Inc. Used by permission of Oxford University Press, Inc.

term precipitation data show declines in rainfall from June though August, but there is great interannual variability (Gilliam et al. 1987). During drought, "normal" NPP is greatly reduced, being limited by water. During this time available soil N again accumulates beneath the water-limited plants. When normal precipitation patterns return, there is once again a pulse of NPP with plants no longer being water limited and having adequate nitrogen (Figure 12-13b).

Thus the TMH, which states that NPP in ecosystems responds to simultaneous and transient (temporary) releases from multiple resource limitations, furthers our understanding of how NPP is controlled in terrestrial ecosystems. It has withstood the rigors of experimental testing (e.g., Blair 1997) and may apply to a variety of ecosystem types outside the realm of the tallgrass prairie. That is, depending on temporal and spatial scale, most (if not all) ecosystems can be considered nonequilibrium.

Terrestrial Vegetation and the Global Carbon Cycle

One of the more dynamic components of the global carbon cycle is the atmosphere (Figure 12-14), where fluxes of carbon (150–160 metric gigatons [Gt] C yr^{-1}) are about 20% of its total carbon store (750 Gt C). In the absence of human activity, CO_2, a relatively stable form of carbon, is removed from the atmosphere by photosynthesis and released to it by respiration and fire in an equilibrium fashion. Human combustion of fossil fuels, however, adds measurably to the amount of carbon returned to the atmosphere, as do changes in land use (Houghton 1995). For example, clearing a forest for agriculture can decrease carbon held in live biomass by 90% and decrease soil carbon by 20–50%. Part of the growing awareness of the importance of soil carbon dynamics in the global carbon cycle is a viewing of the soil component as both a potential source and sink for carbon (Kirschbaum 1995, Turner et al. 1995, Michaelson et al. 1996). Furthermore, debate continues on whether forest ecosystems as a whole represent sources (e.g., Cohen et al. 1996) or sinks (e.g., Kauppi et al. 1992) for carbon.

Widespread use of fossil fuels and clearing of forests has contributed to global increases in concentrations of CO_2, which now occur at a rate of 1 ppm yr^{-1} (Figure 12-15). Exchange of CO_2 with the oceans (Figure 12-14) serves as an important buffer for such increases. Daily ambient concentrations of CO_2 at local scales may exceed 400 ppm in areas with a high level of fossil fuel combustion and topographic conditions that reduce atmospheric mixing. The pre-industrial concentration of CO_2 is generally assumed to have been 280 ppm (Figure 12-15), with variations of about 10 ppm occurring from the years 1000–1700 (Barnola et al. 1995). It is expected that continued fossil fuel combustion and forest clearing will maintain the current pattern of increasing CO_2 concentrations well into the future (Amthor 1995).

The importance of vegetation in global patterns of CO_2 is illustrated in Fig-

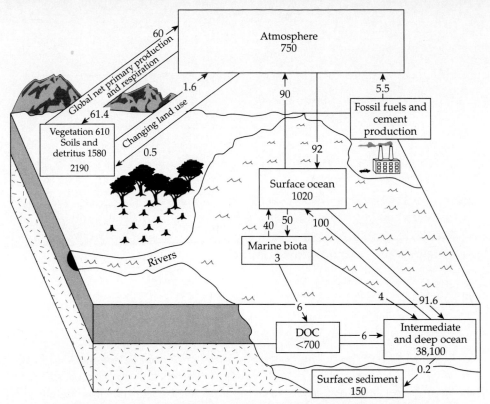

Figure 12-14 Major pools (boxes—in Gt carbon) and fluxes (arrows—in Gt carbon/yr) of the global carbon cycle. From Schimel, D. S. 1995. "Terrestrial ecosystems and the carbon cycle." *Global Change Biology* 1:77–91. Blackwell Science Ltd.

ure 12-15. The annual oscillations of CO_2 can be attributed almost entirely to annual cycles of net productivity of forests of the temperate zone. The highest levels of CO_2 at Mauna Loa are found in winter, when temperate forests are essentially dormant, and lowest in summer, when photosynthetic rates are highest. Similar shifts are found in the appropriate season in the Southern Hemisphere (Machta 1983). The magnitude of shifts in the Southern Hemisphere, however, is not as great as that in the Northern Hemisphere, likely the result of the greater land mass (and thus greater biomass of plants in temperate zones) of the Northern Hemisphere.

The global ecological implications of the temporal patterns shown in Figure 12-15 are as serious as they are complex. For some time environmental scientists have debated whether annual increases in atmospheric CO_2 would lead to an enhancement of the "greenhouse effect," based on the permeability of CO_2 to high-energy, low-wavelength radiation from the sun and its impermeability to low-energy, high-wavelength radiation (i.e., heat). It has been further theorized that this enhancement would lead to increases in global temperatures (i.e., global

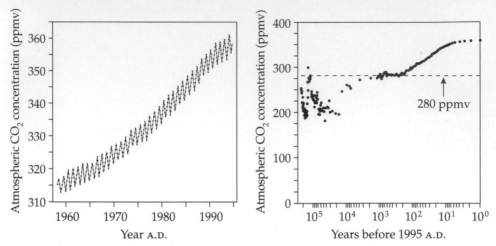

Figure 12-15   Temporal changes in global concentrations of CO_2 (ppmv is parts per million based on volume). From Amthor, J. S. 1995. "Terrestrial higher-plant response to increasing atmospheric [CO_2] in relation to the global carbon cycle." *Global Change Biology* 1:243–274. Blackwell Science Ltd.

warming). Although the debate continues, more and more atmospheric scientists and climatologists are finding data that suggest that global temperatures have indeed been increasing measurably over the past several decades (Schneider 1989, Roodman 1994). It is anticipated the changes in global climate will bring about measurable changes in global patterns of productivity (Long and Hutchin 1991).

Increasing ambient concentrations of CO_2 generally enhance photosynthesis by C_3 plants, the metabolic group that includes an overwhelming majority of plant species (see Chapter 15). It is possible that the competitive interactions of some species will be altered due to CO_2-induced changes in their photosynthetic rates or to temperature-related responses (Bazzaz and Fajer 1992). We might expect subtle changes in physiological response and vegetation structure and composition (Amthor 1995). At this point, one of the few certainties is that predicting and studying biosphere-scale responses to climate change will remain an important, but difficult, challenge to ecologists and other environmental scientists well into the new millennium.

Summary

Plants use part of the total energy fixed by photosynthesis (gross primary productivity) as maintenance energy. The remainder (net primary productivity) is used for new plant biomass and/or reproduction, and is the food for herbivores. The amount, distribution, and rate of turnover of biomass determine many important characteristics of the community: physiognomy, diversity of herbivores, and activity of decomposers. These relationships are constrained by the efficiency with which plants absorb light (exploitation efficiency), incorporate absorbed energy into photosynthate (assimilation efficiency), and convert photosynthate into biomass (net production efficiency).

Productivity is measured by several means: (a) determining the rate of photosynthesis and respiration, (b) estimating changes in biomass over time, (c) correlating biomass to an easily measured variable (e.g., plant height) to estimate standing crop (dimension analysis), and (d) use of computer models to determine productivity over broad spatial scales and predict responses to changing environmental conditions.

Terrestrial vegetation supports about 62% of the total world primary productivity, totaling about 100×10^9 t yr^{-1}. Productivity varies from a theoretical maximum of 3000–3500 g m^{-2} yr^{-1} in tropical rain forests to near zero in some deserts. These trends in productivity are related to severity of habitat and are paralleled by changes in biomass. High levels of productivity occur in communities with maximum leaf area indices (LAI). LAI is maximized in trees, but the large amounts of energy necessary to maintain nonproductive support biomass reduces the potential advantage of the tree life form.

Litter production and decomposition rates have an important influence on mineral cycling and the composition and abundance of detritivores. Rates of decomposition vary with species, the chemical composition of litter, the availability of nitrogen and water, and temperature. Litter production and decomposition rates are highest in the tropics and generally decrease toward the poles.

Productivity in terrestrial ecosystems is influenced by numerous factors: carbon dioxide, light, temperature, moisture, nutrients, soil texture, herbivory. The most productive ecosystems are those with optimal levels of such factors, ultimately to maximize photosynthesis. These factors exert their influence simultaneously and interactively, often doing so in a nonequilibrium manner.

Future studies on the productivity of terrestrial ecosystems will be challenged by the complexities brought on by the likelihood of global climate changes. These would include changes in patterns of temperature and precipitation, superimposed on a pattern of increasing concentrations of CO_2 in the atmosphere.

CHAPTER 13

MINERAL CYCLES

Nutrients move through living systems in pulses and floods, not in smooth, even-flowing transitions. The cycling of matter is inherent in the functioning of ecosystems and is integral to their structure. Matter moves in cyclic fashion, whereas energy flows one way and is noncyclic: it may be stored for a time in biomass, but it is continually replaced by the sun and is lost from the system as heat or exported as energy-rich plant and animal parts.

The goal of nutrient cycling research is to quantify (a) the sum total of nutrients and other elements present within a system, (b) the cycling times, turnover rates, and residence times of nutrients and energy within the system, and (c) the nutrient-use efficiency of organisms and systems. Researchers also work to discover how nutrient cycles regulate productivity and other ecosystem processes, how they interact with atmospheric and aquatic systems, and how biotic and abiotic factors govern nutrient cycles and cycling times. How ecosystems regulate nutrient loss following disturbance is of great interest, as are anthropogenic influences on nutrient cycling. Our approach in this chapter will be to define the roles of various physical and biotic forces with respect to nutrient cycling in several ecosystems and to briefly examine nutrient-use efficiency and ecosystem mechanisms that promote nutrient retention.

Plant Nutrients

Mineral nutrients, that is, essential elements and inorganic compounds, are categorized by plant physiologists according to relative quantities required by plants for adequate nutrition. Major nutrients are those required in large amounts (macronutrients). Those required in only small or trace amounts are minor nutrients (micronutrients). Table 13-1 lists mineral nutrients essential to plants, the form in which the mineral nutrient is obtained, and its content in dry tissue. Note the division into macronutrients and micronutrients.

Table 13-1 Essential plant nutrients, the form available to plants, mg/kg, or percent of dry weight. Modified from *Biology of Plants,* 6th ed. By Peter Raven, et al. © 1971, 1976, 1981, 1986, 1992 by Worth Publishers, Inc. © 1998 by W. H. Freeman and Company. Used with permission of W. H. Freeman and Company.

Element	Principal form in which element is absorbed	Usual concentration in healthy plants (% or ppm of dry weight)	Important functions
Macronutrients			
Carbon	CO_2	~44%	Component of organic compounds
Oxygen	H_2O or O_2	~44%	Component of organic compounds
Hydrogen	H_2O	~6%	Component of organic compounds
Nitrogen	NO_3^- or NH_4^+	1–4%	Component of amino acids, proteins, nucleotides, nucleic acids, chlorophylls, and coenzymes
Potassium	K^+	0.5–6%	Involved in osmosis and ionic balance and in opening and closing of stomata; activator of many enzymes
Calcium	Ca^{2+}	0.2–3.5%	Component of cell walls; enzyme cofactor; involved in cellular membrane permeability; component of calmodulin, a regular of membrane and enzyme activities
Phosphorus	$H_2PO_4^-$ or HPO_4^{2-}	0.1–0.8%	Component of energy-carrying phosphate compounds (ATP and ADP), nucleic acids, several essential coenzymes, phospholipids
Magnesium	Mg^{2+}	0.1–0.8%	Part of the chlorophyll molecule; activator of many enzymes
Sulfur	SO_4^{2-}	0.05–1%	Component of some amino acids and proteins and of coenzyme A
Micronutrients			
Iron	Fe^{2+} or FE^{3+}	25–300 ppm	Required for chlorophyll synthesis; component of cytochromes and nitrogenase
Chlorine	Cl^-	100–10,000 ppm	Involved in osmosis and ionic balance; probably essential in photosynthetic reactions that produce oxygen
Copper	Cu^{2+}	4–30 ppm	Activator or component of some enzymes
Manganese	Mn^{2+}	15–800 ppm	Activator of some enzymes; required for integrity of chloroplast membrane and for oxygen release in photosynthesis
Zinc	Zn^{2+}	15–100 ppm	Activator of component of many enzymes
Molybdenum	MoO_4^{2-}	0.1–5.0 ppm	Required for nitrogen fixation and nitrate reduction
Boron	$B(OH)_3$ or $B(OH)_4^-$	5–75 ppm	Influences Ca^{2+} utilization, nucleic acid synthesis, and membrane integrity
Elements essential to some plants or organisms			
Sodium	Na^+	Trace	Involved in osmotic and ionic balance; probably not essential for many plants; required by some desert and salt-marsh species and may be required by all plants that utilize C_4 pathway of photosynthesis
Cobalt	Co^{2+}	Trace	Required by nitrogen-fixing microorganisms

The nutritional state of a plant for any nutrient may be deficient, adequate, or excessive (toxic). Deficiencies may result in stunted growth or premature sexual maturity and senescence. Specific deficiencies, caused by lack of only one or a few of the essential elements, are often revealed by characteristic symptoms. If nutrient concentrations exceed the limits of tolerance, nutrients may be toxic. Examples of excess include high salinity in estuarine habitats, which limits the distribution of glycophytes, and, at extremes, even halophytes such as cord grass (*Spartina alterniflora*).

The availability and metabolism of carbon and nitrogen are tightly linked (Figure 13-1). The incorporation of nitrogen requires energy and a molecular framework supplied by carbon metabolism. In turn, the increase in biomass is often limited by available nitrogen (refer to Chapter 15). Plant biosynthesis—the production of new tissue—requires nitrogen. Nitrogen is a constituent of proteins and nucleic acids and is structurally involved in most catalytic molecules. It accumulates in young tissues, seeds, and storage organs.

Phosphorus is involved in the structure of many vital molecules, such as nucleic acids and phospholipids. The energy that is used in the cell is released largely

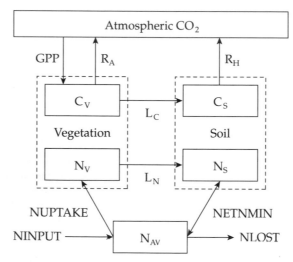

Figure 13-1   Carbon and nutrient cycles in terrestrial ecosystems. Carbon enters the vegetation pool (C_V) as gross primary production (GPP) and transfers either to the atmosphere as autotrophic (plant) respiration (R_A) or to the soil pool (C_S) as litter production (L_C); it leaves the soil pool as heterotrophic respiration (R_H). Nitrogen enters the vegetation pool (N_V) from the inorganic nitrogen pool of the soil (N_{AV}) as 3NUPTAKE. It transfers from the vegetation to the organic soil pool (N_S) in litter production as the flux L_N. Net nitrogen mineralization (NETNMIN) accounts for nitrogen exchanged between the organic and inorganic nitrogen pools of the soil. Nitrogen inputs from outside the ecosystem (NINPUT) enter the inorganic nitrogen pool; losses leave this pool as the flux NLOST. From "The role of nitrogen in the response of forest net primary production to elevated atmospheric carbon dioxide" by A. D. McGuire, J. M. Melillo, and L. A. Joyce. *Annual Review of Ecology and Systematics* 26:437–503. Copyright © 1995. With permission, from the *Annual Review of Ecology and Systematics*, Volume 26 © 1995 by Annual Reviews.

by hydrolysis of phosphate bonds in ATP (adenosine triphosphate). Sulfur forms part of some amino acids and is crucial to the stability of the tertiary structure of enzymes and other proteins.

Potassium is a major electrolyte, and it functions in enzyme activation in virtually every life form, from prokaryotes to higher plants. It has no apparent structural role, but in its catalytic role it is used in meristematic zones in young tissues and is important in regulating guard cell turgor. It is very soluble and is easily leached from light or sandy soils, but because it is so abundant in soil minerals it is a common cation.

Magnesium is a structural part of the chlorophyll molecule and serves to activate many of the enzymatic reactions that transfer phosphates. It tends to accumulate in young leaves; some will be translocated out before leaf abscission. Calcium is essential for the formation and metabolism of the mitochondria and nucleus. Cell membranes lose their integrity when there is a calcium deficiency. Apparently Ca^{++} also alters the permeability of cell membranes to other minerals. Calcium is virtually immobile in the plant, due to its structural role, and is therefore usually released only by decomposition.

Iron is essential for the synthesis of chlorophyll, though it is not part of the structure of the molecule. It is the center of the porphyrin ring of the cytochromes and so is involved both in the transformation of radiant energy and in the utilization of energy within the cell.

Deficiencies of copper and chlorine in the soil are rare, but deficiencies of boron, manganese, and molybdenum occur often. Chloride, which probably is absorbed in ionic form and remains so, plays a vital role in photosynthesis. Molybdenum is essential for nitrogen fixation and nitrate reduction. Manganese and copper are enzyme catalysts, and boron may exert influence on the activity of various enzymes. Zinc is essential to the synthesis of the important plant hormone indoleacetic acid (IAA) and may be involved in protein synthesis.

Certain elements behave as analogs to nutrients and will replace them to some extent. For example, strontium mimics calcium and will relieve deficiency symptoms for a time. Radioactive strontium released by aboveground testing of nuclear devices during the middle of this century tended to be concentrated in the Northern Hemisphere. It was transferred to humans in both the Old World and New World arctics as the lichens took up the strontium as if it were calcium, and reindeer and caribou fed upon the lichens. Consequently, humans who were nourished by those animals displayed high concentrations of the harmful isotope in their tissues, a source of worldwide concern. After the meltdown at Chernobyl in April 1986 the reindeer herds in that area were destroyed to prevent a recurrence.

Biogeochemical Cycles

We refer to the movement of chemical elements within the environment as **biogeochemical cycles.** *Bio* refers to living systems, and *geo* to the rocks, water, and air of the earth. Geochemistry deals with the exchange of elements among the physical components of the earth. Thus, *biogeochemistry* refers to the transfer or flux of materials back and forth between the living and nonliving components of the biosphere.

Nutrient cycling is the movement of those materials that are essential to organisms. Cycling is a homeostatic mechanism whereby plants build up an available (that is, exchangeable) pool of nutrients, to be used in new biomass. The movement of nutrients within an ecosystem (intraecosystem cycling) is often an order of magnitude more rapid than the movement into and out of that system (interecosystem cycling) from the surrounding global ecosystem. The retention of nutrients, through time, accompanies succession (see Chapter 11 for a complete discussion). Conversely, a drastic reduction in the nutrient pool accompanies the degradation or disturbance of an ecosystem.

Gaseous and Sedimentary Cycles

The movement of materials is largely by one of two basic avenues. (1) Those elements that have a major gaseous phase are involved in gaseous cycles and participate in regional and/or global cycles. (2) Those elements that lack a major gaseous phase move in sedimentary cycles. For gaseous nutrients, perturbation is quickly compensated for and equilibrium rapidly reestablished. Some nutrients, such as S, may have both a gaseous and a sedimentary phase; however, they may be placed in a specific category depending on what phase is incorporated into living tissues.

An example of a gaseous cycle, the nitrogen cycle, is diagrammed in Figure 13-2. Contrast this cycle with the phosphorus cycle, shown in Figure 13-3 on p. 340. Note that the reservoir for nitrogen is the atmosphere. The shallow and deep sediments and deposits of the earth's crust are the reservoir for phosphorus. Further, phosphorus is rare compared to nitrogen.

Those elements that do not have a prominent gaseous phase, such as phosphorus and iron, share the earth's crust as their reservoir. All of these elements tend to be ultimately deposited in the sea, or in a pond or lake, and only returned to the nutrient pool by geological activity—mountain building or some other tectonic phenomenon. Exceptions to this limitation include the short-term recycling that occurs when predatory birds return to land after feeding on ocean fishes. Their rookeries become covered with gleaming white deposits of guano, which is rich in phosphorus and other materials. Anadromous fishes (which have matured in the sea) return to spawn and die inland. This migration also recycles nutrients quickly, but the contribution of such transfer is minimal except in the immediate area.

Early research on nutrient cycling focused on cycling of calcium, magnesium, and potassium (Bormann and Likens 1967; Likens et al. 1970, 1977; Siccama et al. 1970). More recent literature (Aber et al. 1989, Matson and Vitousek 1990, McGuire et al. 1995, Vose et al. 1995, Neill et al. 1996, O'Lear et al. 1996) and this chapter reflect the modern focus on nitrogen cycling and on the increasing concentration of greenhouse gases in the atmosphere. Our focus is appropriate for several reasons:

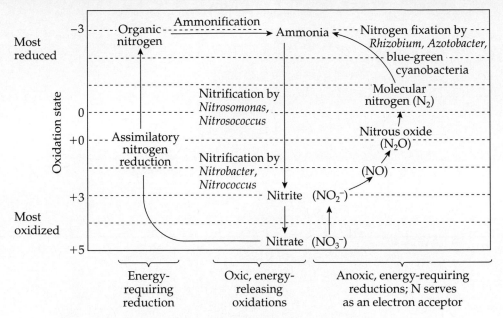

Figure 13-2   Transformations and oxidation states of compounds in the nitrogen cycle. The most reduced state of the atom, having an electric charge of -3, has the highest chemical energy potential. From *The Economy of Nature,* 4th ed., by R. E. Ricklefs. © 1976, 1983 by Chiron Press; © 1993 and 1997 by W. H. Freeman and Co. Used by permission.

1. Anthropogenic influence has increased both N and C to the level of atmospheric pollutants (Aber et al. 1989, Calloway et al. 1994).

2. Nitrogen is the primary growth-limiting nutrient in many communities, including grassland and forest biomes, and it may also limit recovery following disturbance (Vitousek et al. 1982, Gilliam 1987, Nadelhoffer et al. 1995).

3. A nitrogen-saturated forest could become a source, not a sink, of nitrogen, leading to increased nitrous oxide flux to the atmosphere and increased nitrate in water systems (Aber 1989).

4. Increasing levels of carbon dioxide in the atmosphere could lead to global climate change (Sampson et al. 1993).

Nutrient Pools and Flux

Figure 13-4 gives a diagrammatic view of nutrient storage and fluxes of potassium in a temperate hardwood forest ecosystem in New Hampshire. The nutrients are viewed as pools occurring in mineral soil (see Chapter 17), in the forest floor, in the atmosphere, and in living or dead biomass, both aboveground and belowground. As early as 1971, Odum pointed out that the flux of nutrients is more

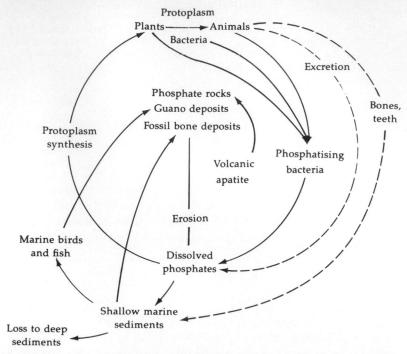

Figure 13-3   The phosphorus cycle. Phosphorus is a rare element compared with nitrogen. Its ratio to nitrogen in natural waters is about 1 to 23. Chemical erosion in the United States has been estimated at 34 t km^{-2} yr^{-1}. Fifty-year cultivation of virgin soils of the Midwest reduced the P_2O_5 content by 36%. As shown in the diagram, the evidence indicates that return of phosphorus to the land has not been keeping up with the loss to the ocean. From *Fundamentals of Ecology*, 2nd ed. by Eugene P. Odum and Howard T. Odum © 1959 by Saunders College Publishing and renewed 1987 by Eugene P. Odum and Howard T. Odum. Reproduced by permission of the publisher.

significant for the functioning of ecosystems than the absolute amount present in a given compartment in a specified time. Cycling may occur via plant uptake and assimilation of nutrients, leaching of nutrients from plants and plant parts, and biological and physical decomposition.

The factors involved in intersystem cycling, which transport nutrients to and from the surrounding ecosystem, include (a) meteorological conditions, such as rain bringing particulates and dissolved sulfates and carbonates, and wind bringing **aerosols** (suspensions of fine particles, solid or liquid, in a gas); (b) geologic forces, such as surface or subsurface water, which move nutrients such as phosphates and calcium into or out of a system and facilitate weathering of primary and secondary minerals; (c) biologic flux, which includes symbiotic and free-living prokaryotes that fix atmospheric nitrogen, as well as animal transport of nutrients into or out of the system. Each of these factors is influenced by soil temperature and water content, soil solution chemistry, and the phenology of living things (Yin et al. 1993).

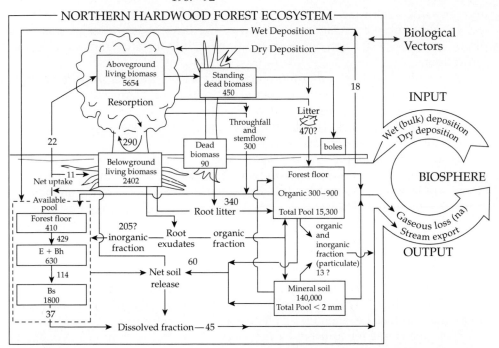

Potassium
1987–92

Figure 13-4 Ecosystem pools (boxes) and fluxes (arrows) of potassium during six years at a temperate hardwood forest study site. A nutrient element such as potassium may be found within the ecosystem in the organic pool, where it is a part of living and dead biomass, or in the available pool, where it occurs as a cation on exchange surfaces or is dissolved in the soil solution. Shown also are extrasystem vectors, representing input and output (atmospheric, geologic, and biologic). Potassium may be added to the system by weathering (geological input) or by atmospheric deposition (meteorologic input). Nutrients are moved within the system (intrasystem cycling) along such pathways as throughfall, stemflow, nutrient uptake, plant assimilation and use, resorption, and biological decomposition. Nutrients depart by the formation of new minerals, by leaching from the system, or as gaseous loss. The ecosystem is thus connected to the larger biogeochemical cycles by meteorologic, geologic, and biologic vectors that move nutrients across ecosystem boundaries. Note: data are from Watershed 6 of the Hubbard Brook Experimental Forest in New Hampshire, described later in this chapter. Some data have been rounded from original. E, Bh, and Bs are soil horizon designations. Modified from Likens et al. 1977. *Biogeochemistry of a Forested Ecosystem*, by permission of Springer-Verlag, and from "The biogeochemistry of potassium at Hubbard Brook." *Biogeochemistry* 25:61–125. Used by permission from Kluwer Academic Publishers.

Does a nutrient storage compartment ever become a vault, locked against internal cycling? Yes, large, old trees become repositories of enormous quantities of organic matter. Yet wood is notoriously low in nutrient concentration, so that the lockup, although not trivial, is not disastrous for the ecosystem. In a boreal jack pine (*Pinus banksiana*) forest, for example, the vault is in the litter and humus of the forest floor.

A jack pine stand up to about 20 years old is very productive, and organic matter accumulates on the forest floor at about 800 kg ha^{-1} yr^{-1}. After that, the annual increments of organic matter added to the litter decrease to about half as the growth of the stand slows. The decline in productivity becomes especially apparent in stands between 30 and 65 years old.

Foster and Morrison (1976) pointed out that very early in the life of the stand, phosphorus concentration stabilizes at about 12–15 kg ha^{-1} in the aboveground biomass of jack pine. At age 20–30 years, N and P are immobilized to an increasing degree in microbial biomass. Incorporation of elements into the forest stand is a measure of growth, and there is little or no incorporation in stands 30–65 years old. The flow of minerals from soil to roots to vegetation and back to soil is maintained, but the percentage of nutrients retained in plant biomass is very small. The lack of decomposition locks up N, P, and Mg during this stage in the life of the stand, and so plant growth falters. Return to high productivity may be accomplished by wildfire, which reduces litter to soluble mineral ash—available nutrients—that can then be taken up by plants (see also Chapter 16).

The time frame for nutrient cycling depends upon the compartment. Metabolic processes may turn over elements in minutes. For example, in photorespiration (a process that consumes oxygen and releases CO_2 in the light), carbon is moved from the atmosphere into the plant and back in a matter of minutes. Nutrients in the leaf litter in a forest may turn over within an ecosystem annually or require several years or decades to turn around. In some temperate deciduous forests, potassium within the forest floor and in the mineral soil recycles rapidly, with residence times of 0.4 and 3.4 years, respectively (Likens et al. 1994). Some biogeochemical processes (e.g., deposition, tectonic movement that exposes sediments, and subsequent retrieval by some root system) may require hundreds of millions of years to turn, but the consequences of that turning are of incredible ecological importance.

Factors in Nutrient Cycling

Every physical force and organism impedes, hastens, or otherwise affects nutrient cycling. Our knowledge of general pathways of nutrient movement has preceded our understanding of how much is moved and in what time frame the transfer occurs. However, in the past two decades, quantitative studies have become the norm (e.g., Aber et al. 1993b, Bowden et al. 1993a, Sampson et al. 1993, Hedin et al. 1994, Likens et al. 1994, Peterjohn 1994, Killingbeck 1996, and Luo et al. 1996).

Abiotic Factors: The Hydrologic Cycle

The mnemonic device $P = E + T + R + I$ reminds us of the balance that must exist in the hydrologic cycle, such that P (precipitation) is equal to E (evaporation) plus T (transpiration) plus R (runoff) plus I (infiltration, the downward entry of water into the soil, and, ultimately, deep drainage and movement out of that region). Both intrasystem and intersystem cycling are associated with the hydrologic cycle.

Precipitation carries nutrients in solution, runoff and infiltration remove nutrients from a system or move them down the soil column, and evapotranspiration of water concentrates and conserves nutrients. In Europe and North America, the contribution of nutrients from the atmosphere (Ca, H, K, Mg, N, Na, and S) is significant, and is elevated by human activity, especially in northern Europe.

Groundwater and water rising by capillary action transport nutrients into the root zone. Although transfer by runoff is often only to a region of lower elevation, stream-flow nutrients must be viewed as loss on a local scale. This section will give several illustrations of the role of water in nutrient cycling.

Deposition and Canopy Interaction   Ecoystems get a significant nutrient contribution from the atmosphere. Clouds process trace gases and atmospheric aerosols (Collett et al. 1993, Stahelin et al. 1993) that bring not only rain and snow but also cloud droplets, particles, and gases. Precipitation may provide the largest nutrient source to many forest systems (Parker 1983). To accurately estimate total deposition to an area requires monitoring bulk deposition—everything deposited in a continuously open collector, which includes all materials, wet and dry, heavy enough to have fallen from the atmosphere (Lovett and Lindberg 1993). In the eastern deciduous forest, atmospheric inputs include Ca^{2+}, Mg^{2+}, Na^+, and K^+, as well as NH_4^+ and H^+. Wet and dry deposition delivers nutrients such as Ca^{2+}, Mg^{2+}, Na^+, K^+, NH_4^+ , H^+, NO_3^-, and SO_4^{2-} to the tallgrass prairie ecosystem (Gilliam 1987, Likens et al. 1994).

Rainfall is often acidic. In southern Appalachian hardwood forests the precipitation may have a pH of <4.3, with cloud water as low as 3.7 (Table 13-2; Gilliam 1989). Acid precipitation may contribute significant nutrients, but may also enhance foliar leaching and necrosis (Lin and Saxena 1992a,b). We will discuss acid deposition and human influence later in this chapter. When acid precipitation infiltrates the soil, cations such as Ca, K, and Mg may be displaced by the hydrogen ions of the acid and will then be leached from the root zone.

Gilliam and others have studied atmospheric deposition of water and mineral nutrients in tallgrass prairie in the Konza Prairie Research Natural Area of Kansas. The prairie canopy may intercept as much as 48% of incident precipitation: the researchers' data suggest that the interception of precipitation by grass canopies greatly influences both quality and quantity of nutrients actually available for plant use (Gilliam 1987, Gilliam et al. 1987, Seastedt 1985). Two types of interception are **stemflow,** the fraction of precipitation that flows down the stem or trunk of a plant and enters the soil, and **throughfall,** rainwater that falls through a canopy. Though monitoring stemflow and throughfall nutrients is difficult, it is essential to include those data (along with bulk deposition data) to obtain a true pic-

Table 13-2   Cloud/fog chemistry for various sites. Note the patterns of regional cloud chemistry. Those data from New Hampshire, New York, and Virginia show greatly elevated concentrations of hydrogen ion, nitrate, and sulfate relative to western United States sites. Blank spaces indicate no data reported. Units for ions are microequivalents per liter. From "Atmospheric deposition and its potential significance in southern Appalachian hardwood forest ecosystems" by F. S. Gilliam. In: *Environment in Appalachia. Proceedings, 4th Annual Conference on Appalachia.* J. W. Bagby, ed. Lexington, KY © 1989. Reprinted by permission of the University of Kentucky Appalachian Center.

Site	H^+	Na^+	K^+	Ca^{2+}	Mg^{2+}	NH_4^+	NO_3^-	SO_4^{2-}	Cl^-
					μEq/L				
Alaska[a]	32		4	5	9	8	6	46	48
Oregon[a]	21		3	5	11	13	10	36	44
California[b]		33	13	15	17	211	42		
Sequoia N.P., CA[c]	11					237	132	96	
New Hampshire[d]	288	30	10			108	195	342	
Whiteface Mt., NY[e]	280					89	110	140	
Southwestern VA[f]							144	327	
Northwestern VA[g]	204	13	3	9	4	84	154	275	26

[a] Bormann et al. (1989)
[b] Azevedo and Morgan (1974)
[c] Collett et al. (1989)
[d] Lovett et al. (1982)
[e] Castillo and Jiusto (1983)
[f] Mueller and Weatherford (1988)
[g] Present study

ture of atmospheric contribution of nutrients (see Ollinger et al. 1993 for additional information). In prairies and forests in North America and in Europe, leaf uptake of nitrogen and immobilization by microbes cause lower inorganic nitrogen and higher organic nitrogen in throughfall and stemflow relative to wet deposition (Gilliam 1987, Lovett and Lindberg 1993). Epiphytic lichens also affect the magnitude of nutrient deposition due to their contribution to canopy biomass (Knops et al. 1996).

Studies of a mixed hardwood forest have demonstrated the role of canopy interception in forest nutrition. Lindberg et al. (1986) reported that in a typical eastern U.S. forest, the major mechanism of atmospheric input was dry deposition to the canopy, supplying sulfur, nitrogen, calcium, and potassium. They estimated that such deposition supplied calcium and nitrogen for 40% of the annual needs of the forest, and sulfur up to 100% of the forest requirement. Standard bulk deposition collectors would underestimate these contributions by a significant amount.

It is important to note that cloud processes and deposition are not uniform over a given area. In mountains, forest edges may receive both more nutrients and more pollutants than the interior of that same forest. For example, in a spruce forest in the Catskill Mountains, the exposed edges received 1.5 to 3 times more deposition than did trees 28 m into the forest. Individual trees, because of size or exposure differences, are subject to differential deposition (Weathers et al. 1995). Furthermore, cloud droplets that impact mature edge trees differ in ion concentration from those encountered by young saplings in forest gaps (Lindberg and Owens 1993).

Cloud drops scavenge aerosols and trace gases, thus removing pollutants from the cloud through a process called **riming,** which occurs when falling ice crystals capture cloud drops. (However, if ice crystals form via water vapor deposition, without significant riming, the cloud drops instead isolate atmospheric pollutants from precipitation and thereby inhibit their deposition.) At high elevations there is orographic enhancement of riming, meaning that near the summit of mountains the deposition of nutrients and pollutants will be enhanced (Collett et al. 1993).

Leaching   Leaching refers to the removal of soluble constituents from the soil or litter by percolating water. Leaching is a primary cause of nutrient loss from a system, the process being positively correlated with increasing temperature and precipitation. The most easily leached macronutrients are Cl and K. Hunt (1977) calculated that 45–65% of the **labile**—the form of the nutrient that readily undergoes chemical change—component may be leached from grassland soils and litter in as little as four hours.

Leaching from attached leaves, on the other hand, adds nutrients to the soil (see Chapter 6). As rain runs over the foliage, it passively absorbs nutrients. The nutrients are then carried to the ground in throughfall and stemflow. Potassium, sodium, and sulfur are very leachable and therefore cycled predominantly in this fashion. In work at Hubbard Brook Experimental Forest in New Hampshire, described later in this chapter, Likens and his colleagues (1994) measured the K flux in both throughfall and in bulk deposition during the growing season, finding the former some 40 times higher than the latter. In other words, throughfall is washing off deposited dry materials but also leaching K from the leaves in the canopy.

Canopy interception and subsequent foliar leaching is the major process controlling nutrient enhancement in throughfall and stemflow. At ground level, mosses gain nutrients primarily from throughfall and stemflow. In those systems with continuous moss cover on the ground, nutrients derived from throughfall must pass through the mosses before they are available to trees (Foster and Morrison 1976). Such a regime generates a very sterile layer for seedlings to deal with, especially if the moss is *Sphagnum.*

Evapotranspiration, Runoff, and Infiltration   In a region with an impermeable geologic substrate, water loss by evapotranspiration may be inferred from the difference between precipitation and stream flow. Chemical concentrations in precipitation are a measure of nutrient input, and concentrations in stream water are a direct measure of nutrient output in such a region. Stream-gauging stations, anchored to bedrock, can be used to meter nutrients in stream water flowing from a forested watershed underlain by a relatively impermeable substrate (Figure 13-5). Nutrient and evapotranspiration losses are relatively stable from year to year in an undisturbed forest. In fact, transpiration actually conserves mineral elements because excess water, which would promote loss through leaching, is removed without harm to the system, leaving the nutrients in the system.

Lysimetry   In a region where the substrate is permeable, evapotranspirational losses and water movement through the soil may be estimated by the use of a **lysimeter,** a device for measuring the percolation of water through soils and the soluble constituents removed in the drainage. A weighing lysimeter consists of a container holding a mass of soil mounted flush with the landscape surface and arranged so that it can move up and down on a weighing device. A scale may be used to follow changes, but electronic devices are also used. Soil solution chemistry may be analyzed by means of **minilysimeters,** made of Pyrex tubes with discs attached to Erlenmeyer collection flasks (Arthur and Fahey 1993, Johnson et al. 1995).

Water collected by lysimeters can be used to determine the concentration of the nutrients that are leached. The product of the volume of water and the concentration of each nutrient is the loss rate for that nutrient. Evapotranspirational losses are inferred from the difference between precipitation input and percolation output. One drawback of lysimetry is that the hydrologic regime is disturbed, sometimes greatly so, by manipulation of the soil to put a lysimeter in place (Haines et al. 1982, Arthur and Fahey 1993, Johnson et al. 1995).

Salt Spray   The nutrients that are received in a dune community from salt spray are essential to the continued functioning of that system. Seawater contains all the

Figure 13-5   A weir showing the V-notch recording house and ponding basin. Courtesy of the Northeastern Forest Experiment Station, Forest Service, U.S. Department of Agriculture.

mineral ions that are necessary for plant growth, except nitrates and phosphates (Boyce 1954). Young dune soils may be composed of up to 99% sand, with very poor capacity for cation storage because they lack clay and organic matter (see Chapter 17). Dune soils are known to have a **cation exchange capacity** [CEC—the sum of positive (+) charges of the cations that a soil can absorb at a specific pH] of only about 10–15 cmol kg^{-1} of soil. For comparison, one may expect a CEC of 40–50 cmol kg^{-1} of clay.

The dune system at Cape Hatteras, North Carolina, may receive the following amounts of nutrients (in kg ha^{-1} yr^{-1}) from salt spray: Na, 25–1300; Mg, 37–120; C, 19–120; and K, 13–77. By contrast, rainfall contributes only 1.0–1.3, 21–26, 2.3–2.4, and 3.0–3.9 kg ha^{-1} yr^{-1} of each nutrient, respectively (van der Valk 1974). In spite of the salty input, the habitat is not saline, due to the extreme leachability of the porous soils. Dune soils are said to be depauperate because of N and P deficiencies.

Other Abiotic Factors

Fire   Wildfire plays an important role in the maintenance and functioning of many communities. The reproduction of giant sequoia (*Sequoiadendron giganteum*), jack pine (*Pinus banksiana*), Bishop pine (*P. muricata*), and various species of California lilac (*Ceanothus*) is enhanced by fire of a certain intensity. Fire is also important to cycling of nutrients in certain regions. In the boreal forest, low temperatures greatly slow decomposition, and the result is an acid soil covered by raw humus. In Norway, for example, burning actually improves forest site quality by raising the soil pH, thus favoring the growth of populations of symbiotic nitrogen-fixing bacteria (*Azotobacter, Rhizobium*). In chaparral regions, fire may serve to (a) break down sclerophyllous litter, which is resistant to biotic decomposition; (b) remove inhibitors of microbial decomposition; and (c) alter wettability of the soil.

In the tallgrass prairie ecosystem, fire affects the availability of nitrogen, water, and energy. Litter is a sieve for inorganic nitrogen, water, and energy, and depresses productivity overall. Fire removes standing dead and surface detritus, allowing sunlight to reach mineral soil and making nutrients available (Seastedt 1985, Knapp and Seastedt 1986, Gilliam 1987, Gilliam et al. 1987).

Nutrients are released in the form of soluble mineral ash by slash-and-burn agriculture. Thus, in a tropical rain forest, slash-and-burn clearing temporarily enhances the nutrient regime. Crop yields are good at first but diminish quickly. Kauffman et al. (1995) measured some of the highest total losses of nutrients ever reported in pristine tropical evergreen forests in the wake of slashing and burning. Consider these facts: (a) The nutrients are released in a single large pulse; their availability probably exceeds the exchange capacity of the soil, and thus nutrients are quickly leached out of the root zone. (b) The fire may destroy such recycling mechanisms as ectomycorrhizae within the soil. (c) Nitrogen is volatilized (removed as a gas) by fire. (d) Weeds and other pests can accumulate and overwhelm the agricultural system (Vitousek 1997).

Earth Movement   Mountain building, movement along a lateral fault, volcanism, and other types of tectonic activity exert major influence on mineral cycling. The impact may be sudden or require eons. On February 20, 1943, Paricutín volcano in Mexico erupted violently, pouring an estimated one billion tons of ash, cinders, and bombs (molten material thrown from volcanoes that solidifies and falls as an igneous rock with a bomblike shape) on the land in its first year. The eruption, almost continuous, lasted more than nine years, until March 4, 1952, and covered 24.8 sq km. The total rock material erupted must have occupied about 1.4 cubic km in the magma chamber. The early flows were found to contain some 7.40% calcium oxide—an almost instant contribution to the nutrient requirements of that region (*Geological Survey Bulletin* 1956).

Such events, rare in a human lifetime, are common in geological time and even in ecological time. In addition, the magnitude of the event counterbalances its infrequent occurrence. Geophysical phenomena rank as primary physical agents of biogeochemical cycling.

Biotic Factors

The importance of biotic factors in the cycling of mineral nutrients could hardly be overstated. The transfer of biomass, with its nutrient content, from producer to herbivore to carnivore to decomposer to minerals in the soil is well known. We recognize that respiration, excretion, defecation (in animals), leaf fall (in plants), and death are integral components of nutrient cycling (see Figures 13-3 and 13-4). In this section, we will begin with litter production, then discuss microbial decomposition, ingestion, and digestion, following the pathways by which organically bound nutrients are returned to the soil. We will also discuss some roles of forest trees.

Litter   In spring, tree buds overcome dormancy and grow. Fine branchlets, leaves, and flowers are produced. Shrubs put out new green materials, bulbs push up leaves and flower stalks. Within the soil, fine roots grow and flourish. The portion of a plant root actively taking up water and nutrients, commonly called a "feeder root," occupies the upper meter of soil. The surface area of fine roots found in a few liters of soil could easily measure hundreds of square meters. As the roots age, they become suberized—their cell walls become corky from suberin—so they are protected from the bath of enzymes, fungi, and parasites that reside in the soil; they also stop absorbing water and nutrients.

Eventually, both above and below the ground, biomass becomes litter. Two factors determine the amount of litter in the system: the total litter produced in a unit of time, and the rate at which it is decomposed. Litter production is governed by the type of vegetation, soil quality, and, in some cases, environmental factors that vary with latitude (Berg et al. 1995).

Aboveground litterfall mass in annually burned tallgrass prairie may range from 370 to 620 kg ha^{-1} yr^{-1}, and in unburned tallgrass prairie from 970 to 1800 kg ha^{-1} yr^{-1}. Rates of decomposition there range from 25 to 40% yr^{-1} for foliage down to as little as 10% yr^{-1} for flowering stems, so significant material accumulates in the absence of fire. Decay rates for roots may range as high as 50% yr^{-1} (Seastedt 1988).

Litter production in a tropical rain forest can be very high, ranging from 16,000 to nearly 45,000 kg ha^{-1} yr^{-1} (Walter 1979). Decomposition in a tropical rain forest, however, can be very rapid due to elevated temperatures and high rainfall.

At higher latitudes, productivity is less, especially in less fertile soils, than at lower latitudes. However, decomposition is slower in higher latitudes, so even with lower productivity there is buildup of soil organic matter. In a Scots pine forest in Sweden, average litterfall may be 1620 kg ha^{-1} yr^{-1}, with rapid mass loss to decay the first year, slowing dramatically in the second year (Berg et al. 1995). With incomplete decomposition and the absence of fire, organic matter builds up in the soil at any latitude.

Microbial Decomposition   The action of microorganisms in the cycling of nutrients has been well studied. The transformation and cycling of plant nutrients and organic matter in the soil depends heavily on the microbial community (Gallardo and Schlesinger 1992). In the nitrogen cycle, for instance, microbial action can **immobilize** nitrogen (convert inorganic ions to organic forms) or can **mineralize*** it. The ratio of carbon to nitrogen in the substrate to be decomposed largely determines which of these processes will occur. At ratios of C:N greater than 20–30:1, some immobilization will occur, as microorganisms will then use soil nitrogen (as NO_3 or NH_4) to build their own proteins. At lower ratios, some mineralization will occur, as decomposition proceeds. In early spring, soil temperatures increase and mineralization of nitrogen can precede leaf production by deciduous trees. Thus, a pool of available N prior to uptake of water and nutrients by the trees is subject to loss by leaching or to the atmosphere (Gilliam 1987, Groffman et al. 1993).

Soil Microbes   Soil microbes serve as both source and sink for nutrients. Microbial biomass consists mainly of fungi and bacteria; the remainder is composed of algae and microfauna. Levels of nitrogen and organic carbon determine microbial biomass levels in the soil. In general, grassland and forest soils harbor greater microbial biomass than do arable soils, with levels lower in temperate forests than for most other natural ecosystems due to nitrogen limitation. In coniferous forest soils, more than 40% of the annual cycling of nitrogen relies on mycorrhizal fungi, with the fungal biomass containing a significant proportion of the total N within the system (Schulze and Mooney 1993). Recent work reveals close links between decay rates and microbial biomass (Wardle 1992). In forests, decomposition may provide 69–87% of the nutrients needed each year for growth (Sinsabaugh et al. 1993).

Field ecologists are studying decomposition in temperate forest ecosystems, grasslands, deserts, and shrub communities. Consequently, quantitative and qualitative decomposition studies are published in journals around the world with increasing frequency (e.g., Seastedt 1988; Aber et al. 1990; Bowden et al. 1993a,b; Sinsabaugh et al. 1993; Berg et al. 1995; Berg et al. 1996).

Models of Decomposition   Global cooperation among scientists allows for more and longer-term studies covering a variety of ecosystems. Rates of decomposition

---

*Mineralization is that portion of the nitrogen cycle in which organic matter is decomposed and inorganic ions are released.

vary from one ecosystem to another. The rates are governed by many factors, notably climate (moisture and heat availability), and soil microflora and fauna.

To quantify litter loss, foliar and stem materials, roots, and other litter may be placed in polypropylene or nylon bags with mesh size openings that range from 0.1 mm to 2 mm or more. The litter is sometimes air-dried and placed in the bags, which may be tethered among standing materials, simply placed on the soil surface, or placed within the A horizon of mineral soil. A large quantity of the bags will be randomly placed, and subsequently individual bags will be harvested at certain intervals (Table 13-3; Seastedt 1988).

The soluble substances—the more easily metabolized components—tend to disappear first. Nitrogen concentrations increase linearly, concomitant with cumulative mass loss during the early periods (Table 13-3). Finally, the disappearance of more **recalcitrant** substances, the materials highly resistant to decay, dominates the loss of litter mass, but this latter phase may take so long that the annual disappearance is almost imperceptible. With the availability of long-term data on decomposition, researchers have designed models to account for the disappearance of litter mass as well as for the persistence of the recalcitrant fraction.

Even though a given model may not fit every situation, if it is carefully developed it will provide generalizations that are valid in many ecosystems. Aber and his colleagues (1990) worked with hardwood and conifer data from Harvard

Table 13-3 Mass and nitrogen dynamics of decaying *Andropogon gerardii* (big bluestem) stems. "Bagged" refers to grass litter harvested and bagged in 5 × 5 cm litterbags and placed on the soil; "tethered" refers to senescent grass litter harvested and tethered by wrapping a wire and tag around a 10-cm-long bundle of litter. Data are means with standard deviations in parentheses. Modified from "Mass, nitrogen, and phosphorus dynamics in foliage and root detritus of tallgrass prairie" by Seastedt. *Ecology* 69:59–65. © 1988. Used by permission.

Days in field	Treatment	% of initial mass	Nitrogen % of mass	Nitrogen % of initial amount
0	—	100	0.18 (0.04)	100
203	Bagged	69.7 (12.8)		
	Tethered	72.9 (6.7)	—	
668	Bagged	47.5 (14.5)	0.87; $n = 1$	230
	Tethered	59.5* (8.2)	0.66; $n = 1$	218
899	Bagged	41.7 (8.0)	1.37* (0.31)	317
	Tethered	45.3 (18.1)	0.51 (0.04)	128
1254	Bagged	36.2 (8.5)	1.56* (0.39)	313
	Tethered*	44.2* (11.0)	1.02 (0.10)	250

*Significantly greater than paired value.

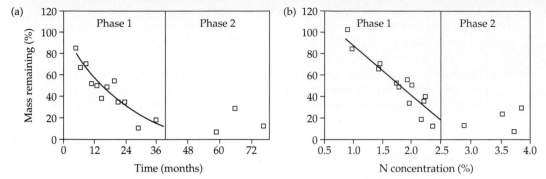

Figure 13-6   Example of separation of decomposition data into phase 1 and phase 2 according to changes in decay rates and nitrogen dynamics. Decomposition data are for paper birch foliage litter in a mixed hardwood stand at Harvard Forest. (a) Mass remaining through time. (b) Mass remaining versus nitrogen concentration (the inverse-linear function). For both panels, phase 1 continues until mass remaining equals the mean mass remaining value for all points in phase 2. From "Predicting long-term patterns of mass loss, nitrogen dynamics, and soil organic matter formation from initial fine litter chemistry in temperate forest ecosystems" by J. Aber et al. *Canadian Journal of Botany* 68:2201–2208. ©1990. Used by permission.

Forest, Massachusetts, and from Blackhawk Island, Wisconsin. They proposed a two-phase, carbon-based model for litter decay. Phase 1 of Figure 13-6a, in which percent mass remaining is plotted against time, shows constant fractional loss, but mass loss in phase 2 is almost impossible to detect.

In phase 1 of Figure 13-6b, in which percent mass remaining is plotted against nitrogen concentration, the decay function shows an inverse-linear relationship. Again in phase 2 the loss is minimal. Aber et al. (1990) estimated that the first phase included about 80% of mass loss.

The decomposition of the extractive portion, or easily metabolized components, is strongly and inversely related to the ratio of lignin to lignin + cellulose (LCI), an index of litter carbon quality. As the extractives disappear, the remaining mass is higher in lignin and ligninlike substances, which are recalcitrant substances (Aber et al. 1990). Only a few types of microorganisms are able to decompose lignin, among them the white-rot fungi. In the presence of N-rich compounds such as amino acids and ammonium, white-rot fungi may not synthesize the necessary lignin-degrading enzymes (Berg et al. 1996), and decomposition slows or stops.

Remaining biomass is often highly correlated with the initial ratio of lignin to nitrogen (Melillo et al. 1982), but sometimes the initial decomposition rate is increased by a good supply of nitrogen (Berg et al. 1996). Later on, that higher N supply may enhance the production of ligninlike materials and other recalcitrant substances (Figure 13-6b), and may inhibit the activities of lignin-degrading white-rot fungi.

Therefore, initial litter chemistry data, or at least litter decay data over a short term, are needed to predict the course of decomposition (Aber et al. 1990). With the development of near-infrared reflectance spectroscopy as a tool to determine nitrogen, lignin, and cellulose content of detritus, researchers may be able to predict decay rates from those data (McLellan et al. 1991a).

Soil Organic Matter   What is the mechanism for the buildup of organic matter in the soil? Obviously, litter is formed faster than it is decomposed. The instantaneous rate of decay decreases over time, until the disappearance of organic matter is almost imperceptible. Initial decay is rapid, and may even be exponential, but accumulated mass loss shows an asymptotic* function (Berg et al. 1996). As has been said, litter quality influences rates of decay.

Most of the foliage produced in a natural system is not consumed by herbivores but rather falls to earth to become part of the detritus. Much of what falls has already been damaged by herbivores, physical stress, and atmospheric pollutants. Findlay et al. (1996) examined the impact on leaf chemistry of mite herbivory and ozone damage. The resulting cellular damage led to increases in complex phenolic materials in the leaves, which in turn caused qualitative changes in litter, and those changes reduced the rates of litter decomposition. Cellular damage leads to predictable changes in litter quality, and many agents and events cause damage to living leaves, including acid deposition and ozone pollution.

Fire frequency can also influence the buildup of soil organic matter. Wooden dowels, 1.25 cm in diameter and 61 cm long, were placed in tallgrass prairie soil: after three years, an average of only 15% of initial wood mass remained on annually burned watersheds, but 34% remained on unburned watersheds. Burn treatment also affected nitrogen concentration: dowels decomposing in burned watersheds had a higher percentage (0.5% after three years exposure) than dowels in unburned watersheds (0.43%). The effect of fire on belowground decomposition in tallgrass prairie is a complex story (O'Lear et al. 1996).

Ingestion and Digestion   Perhaps less widely recognized than microbial decomposition is the nutrient regeneration in soils due to ingestion and digestion of bacteria and fungi by other organisms. Within the soil, multicellular animals account for only 10% of the total metabolism, meaning that microorganisms make up the other 90%. The digestion and subsequent mineralization of bacteria by microscopic protozoa is a vital part of nutrient cycling. An added benefit: protozoan predation removes cells from overcrowded bacterial populations, and the removal stimulates active bacterial growth, enhancing bacterial nitrogen fixation.

Decaying plant material becomes the focus of microbial and faunal activity. Furthermore, the placement of litter can influence both the species composition of the decomposer community and its total biomass. For example, in agricultural systems, biomass and abundance of all microbial and faunal groups is greater on buried litter than on surface litter. Bacteria are important in regulating both rates of decomposition of buried litter and sizes of populations of bacterivorous fauna (Beare et al. 1992).

The cycling of nutrients, especially of nitrogen, in a salt marsh or a boreal for-

---

*An asymptotic curve very nearly approaches some level but would never meet it even if infinitely extended.

est, for instance, is enhanced by grazing and subsequent deposition of fecal material. In a positive feedback situation, foraging can enhance sustained growth of plants when the grazers need it most. Organisms ranging from lesser snow geese to moose influence the nutrient cycling regime by selective foraging (Ruess et al. 1989, Holland and Detling 1990, Pastor et al. 1993).

Some Roles of Forest Trees   In general, trees not only have a very extensive root network but often have longer and deeper roots than do shrubs and herbs. Therefore, the portion of the soil column confronted by tree roots is greater, making possible a unique contribution to the ecosystem. Deep-lying minerals are extracted by tree roots, raised into the plant and incorporated into plant biomass. Nutrients such as potassium may be leached from the leaves or bark, and other nutrients, such as calcium, may be returned as litter. Then, decomposition and mineralization add these nutrients to upper soil layers; making these formerly deep-lying nutrients accessible to the herbaceous and shrub strata of the forest.

A similar situation exists in the salt-marsh ecosystem. Phosphorus and other nutrients from sediments more than a meter deep are transferred to the water in several steps: (a) cordgrass root systems take in P; (b) bacteria degrade the cordgrass; (c) detritus feeders ingest bacteria and return P to the marsh waters. This "pump" role of cordgrass is analogous to the action of deep roots of forest trees, bringing nutrients up from the depths and depositing them in surface layers.

Substances released into the soil from roots may provide significant amounts of carbohydrates and the like to the rhizosphere, the zone in the immediate vicinity of the plant root. Smith (1976) analyzed root exudates from sugar maple (*Acer saccharum*), yellow birch (*Betula alleghaniensis*), and beech (*Fagus grandifolia*), all of a northern hardwood forest. The beech released the most amino acids and organic acids per hectare, and the birch released the most carbohydrates. The cations released were mainly Na, K, and Ca; the anions were chiefly sulfate and chloride. Smith pointed out that the significance of root exudation is twofold. First, root exudates enhance the growth of the microbial saprophytes and parasites in the rhizosphere. Second, although root exudates are beneficial to microbial saprophytes and parasites, they may be either beneficial or harmful to individual plants and to the ecosystem.

Nutrient Cycling in Different Vegetation Types

Maritime Ecosystems

How does nutrient cycling differ from one maritime ecosystem to another? Have conservative mechanisms for retention and internal recycling evolved as a means of obtaining elements in short supply? A coastal salt-marsh* ecosystem ex-

---

*A coastal salt marsh is an herb-dominated ecosystem in which the rooting medium is inundated by tidal water for long periods, if not continuously. The substrate is chiefly mineral, typically with high humus content.

hibits a very tight, conservative mineral nutrient regime, in contrast to the loose, open regime of the dune system at Cape Hatteras, North Carolina, for instance.

In general, the annual output of nutrients from a terrestrial ecosystem exceeds the meteorological input, and that input is small compared to the nutrient reservoir of the system. This is true of the salt marsh. The marsh ecosystem is very stable, although it has little biological structure, is not very diverse, and is often subjected to a stressful environment. There is efficient internal cycling. The contributions of the large storage compartment in the clay sediments and the internal biological processes that hold and recycle nutrients provide a base for a highly productive system. However, in dune ecosystems the atmospheric contribution of cations slightly exceeds the calculated output through leaching and greatly exceeds the nutrients in storage in the system.

The dune systems at Cape Hatteras National Seashore, North Carolina (which includes the dunes of Bodie, Ocracoke, and Hatteras Islands), were studied by van der Valk (1974), with special emphasis on their nutrient cycles. Bodie Island, in the northern part of the national seashore, is oriented such that it receives the full fury of winter storm winds, which blow mostly from the northeast. The coastline of Ocracoke Island to the south runs in an east-northeast direction and so is parallel to the winds' path. Also, it is partly protected by Hatteras Island. Bodie Island receives greater cation input than Ocracoke Island, in salt spray and rainfall, probably due to its orientation. The foredune (the most seaward dune) at Bodie is higher than that at Ocracoke Island, and this may also influence salt spray deposition.

The vegetation of the dunes, primarily American beachgrass (*Ammophila breviligulata*) and sea oats (*Uniola paniculata*), forms a sparse cover; coverage is 3–4% for each of the grasses. There are few or no forbs (herbaceous plants other than grasses) on the front of the foredune, and perhaps 20 forb species on the back of the foredune. Some contribute as much or more cover than the grasses. The principal forb species are horseweed (*Conyza canadensis*), cudweed (*Gnaphalium obtusifolium*), and goldenrod (*Solidago sempervirens*).

Nutrient Supply   The primary nutrient source in this community is salt spray; there is almost no internal reservoir of nutrients. At Bodie Island, the salt spray contribution of mineral ions exceeds that of rainfall as much as eightfold (Table 13-4). Notice, for instance, that the atmospheric input of cations at Bodie Island in bulk precipitation (rainfall plus salt spray) ranges from 14 kg ha^{-1} yr^{-1} for K, to 433 kg ha^{-1} yr^{-1} for Na, with annual inputs of 35 kg ha^{-1} for Ca and 64 kg ha^{-1} for Mg. By comparison, Ca input from bulk precipitation for an inland Tennessee *Liriodendron* forest was estimated at 10.5 kg ha^{-1} yr^{-1}.

What happens to the large supply of nutrients that arrive on the wind? Recall that the foredune is not a truly saline habitat, and that the exchange capacity in soils with a very high sand to clay ratio is poor. Excess cations are quickly leached away.

Output of each cation for a certain depth of soil may be calculated, because

$$S_2 = S_1 + \text{IN} - \text{OUT}$$

Table 13-4 Average concentration of cations (ppm) and annual input of cations (kg ha^{-1} yr^{-1}) in the bulk precipitation (rainfall plus salt spray) and in the rainwater at Bodie Island and Ocracoke Island. From "Mineral cycling in coastal foredune plant communities at Cape Hatteras National Seashore" by A. G. van der Valk. *Ecology* 55:1349–1358. Copyright © 1974 by the Ecological Society of America. Reprinted by permission.

	K	Na	Ca	Mg
Bodie Island				
Bulk precipitation (ppm)	1.30	26.00	2.30	3.90
Rainfall (ppm)[a]	0.15	7.16	1.02	1.30
Bulk precipitation (kg ha^{-1} yr^{-1})[b]	14.40	432.60	34.80	64.20
Rainfall (kg ha^{-1} yr^{-1})	2.30	118.00	16.00	21.00
Ocracoke Island				
Bulk precipitation (ppm)	1.00	21.00	3.40	3.00
Rainfall (ppm)[a]	0.15	4.36	0.82	0.59
Bulk precipitation (kg ha^{-1} yr^{-1})[b]	15.60	319.80	51.60	46.20
Rainfall (kg ha^{-1} yr^{-1})	2.20	65.00	12.00	8.90

[a]Data from Gambell and Fisher (1966).
[b]Estimated from data for May 1971 to January 1973.

where S_2 is the quantity of the cation present at a certain depth at sampling time T_2, S_1 is the quantity of the cation present at the same depth at the previous sampling time T_1, IN is the input of the cation in salt spray or in salt spray leachate during the time period from T_1 to T_2, and OUT is the output of the cation due to leaching by rainwater during the period from T_1 to T_2. About 160 kg ha^{-1} yr^{-1} of Ca is exported from the 60 cm depth, which is significantly more than the amount of Ca that entered as salt spray. The difference is probably due to leaching of calcium carbonate from shell fragments in the sand. The turnover times for nutrients in dune systems are very short compared to other terrestrial ecosystems: 11–37 days at Bodie Island for K, Na, and Mg. For Ca, a longer time of 32–206 days reflects input from dissolution of shell fragments. Total annual meteorological inputs are always higher than exports for the cations K, Na, and Mg.

Limits to Growth Many of the dune perennials have extensive root systems, or rhizome networks, which form clones over large areas. These may pull nutrients from the front of the dune, which receives an ample supply of K, to the back, where K may be limiting to vegetation growth. As shown in Figure 13-7, the annual input of Na and Mg at Bodie Island greatly exceeds that present in vegetation.

The sparse vegetation provides almost all the nutrient storage of the system. There is no soil reservoir. The dune system shows virtually no conservative mechanisms for retention and internal cycling of nutrients. Thus, mineral cycles in

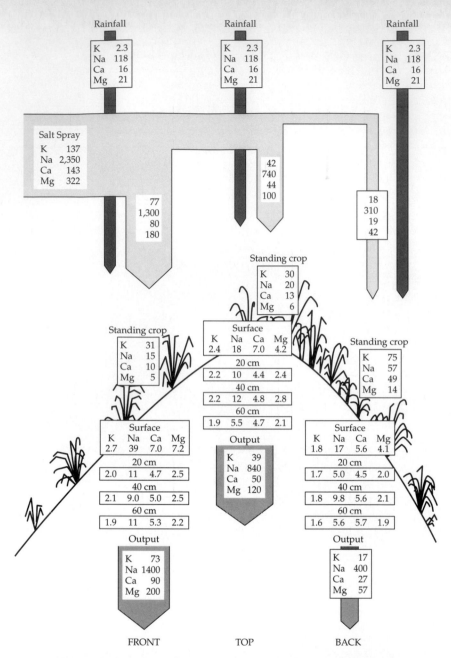

Figure 13-7 Estimated or calculated annual inputs (kg ha^{-1} yr^{-1}) of cations in salt spray and rainwater, and output (kg ha^{-1} yr^{-1}) in drainage water from 60 cm below the surface on the front, top, and back of the foredune at Bodie Island. The cation content (kg ha^{-1}) of vegetation at maximum standing crop and average concentrations (kg ha^{-1}) of cations in soil at the surface and 20, 40, and 60 cm below the surface are given in the boxes. Outputs equal salt spray leachates plus rainfall. From "Mineral cycling in coastal foredune plant communities at Cape Hatteras National Seashore" by A. G. van der Valk. *Ecology* 55:1349–1358. Copyright © 1974 by the Ecological Society of America. Reprinted by permission.

seashore dunes are less stable than in other terrestrial ecosystems (Jordan and Kline 1972).

Grasslands

Tallgrass prairie is the largest vegetative province in North America. It is possible that 99.9% of the native prairie has been converted into agro-ecosystems, and true prairie is found only as isolated remnants (Samson and Knopf 1994). Those escapees from the plow provide valuable information to science; they include the tallgrass prairie of the Flint Hills, which occupy some 10,000 km^2 and stretch from northeastern Kansas south into northeastern Oklahoma, and the Konza Prairie Research Natural Area, 10 km south of Kansas State University. The information in this section comes in part from studies that have been conducted in those areas.

In grasslands, primary productivity is limited by available energy, nitrogen, and water. Although loss of nitrogen is unlikely during normal functioning, both grasses and decomposers use up soil nitrate and ammonium ions so rapidly that their concentration remains low. Nitrogen is supplied in bulk precipitation and throughfall, but nitrogen fixation rates, both by symbiotic and by free-living organisms, are low in grasslands compared to forests.

The amount of inorganic nitrogen available to plants is largely governed by the amount of detritus present. In unburned tallgrass prairie, both the amount of water and the relative amounts of inorganic and organic nitrogen reaching the ground are reduced by litter. Furthermore, litter shades young grass shoots, reducing the amount of photosynthetically active radiation by as much as 50% early in the growing season (Knapp 1984, Seastedt 1985).

The flux of nitrogen in a terrestrial ecosystem was presented back in Figure 13-1. Plants take up nitrogen (NUPTAKE) through living roots as ammonium ion or nitrate ion. Translocation and distribution within the plant vary seasonally, with the ratio of live shoot N concentration to live root N concentration changing with plant phenology. The ratio of shoot to root N varies from 3.0 to 1.0, remaining high during the first third of the growing season, then declining.

Nitrogen is retranslocated into storage organs as the grass leaves senesce. In this year's standing dead material, C:N ratios can reach 110:1, well above the critical level (20–30:1) for net N mineralization discussed earlier, and immobilization proceeds. The retranslocation pulls vital N back to living, persisting roots late in the season, as aboveground biomass dies back, and is an important intrasystem nutrient cycling mechanism.

Eventually, decomposition and mineralization transfer N back to the various compartments (L_N to N_S, N_{AV}, as NETNMIN or NLOST). The rate of those processes is governed by the concentration of organic nitrogen and by abiotic factors such as temperature and soil water status (McGuire et al. 1995).

Fire in Grasslands   Fire has always been vital to the continued health of the tallgrass prairie ecosystem. In years of normal rainfall, unburned prairie is less productive than those prairies that are periodically burned. Burning increases both water supply and inorganic nitrogen, and favors C_4 grasses over forbs and C_3

grasses. However, prairie fires may result in significant losses of nitrogen through particulate updraft and volatilization, releasing more nitrogen than is received in bulk precipitation. Consequently, annually burned sites may experience serious nitrogen limitation (Seastedt 1985, Gilliam 1987, Gilliam et al. 1987, Seastedt 1988, Seastedt et al. 1991). It is likely that burning every few years, rather than burning every year, will result in the highest productivity in many grasslands.

Hubbard Brook Experimental Forest

Our understanding of ecosystem function has been enhanced through the use of models, which are essentially of two types: (a) a microcosm, utilizing a closed system with quantifiable inputs and outputs of energy and materials—for instance, a balanced aquarium—and (b) systems analysis, using computers to simulate the functioning of a real ecosystem (e.g., Aber et al. 1982, Aber et al. 1993a, Ollinger et al. 1993, Berg et al. 1996).

However, generalizations that derive from a microcosm study may not be valid at the ecosystem level. Simulation also has its drawbacks. Data collection to provide an informational base must be painstakingly accurate. The simplification that is essential to modeling necessarily eliminates consideration of short-term fluctuation, or of separate categories of producers and consumers, and thus certain data are lost. Analysis of real ecosystems is needed.

Ecosystems Analysis   To understand biogeochemical cycling fully, researchers must accomplish long-term monitoring and analysis of real ecosystems. The study of an entire ecosystem requires many workers of related and widely divergent disciplines, many hours, and an excellent integration system. The "small watershed approach" for ecosystem analysis has been used at Hubbard Brook Experimental Forest (HBEF; Bormann and Likens 1967, Siccama et al. 1970, Likens and Bormann 1972, Bormann et al. 1977, Reiners 1992, Likens et al. 1994, Likens et al. 1996). HBEF is situated in the lower elevations of the White Mountains of New Hampshire, just north of Plymouth. There, work began with deforestation of Watershed 2 during the winter of 1965–1966.

The watersheds studied contain cool-temperate, mesophytic, broadleaf deciduous trees, with some conifers at higher elevations. Although the area was logged extensively in the early part of this century, some forest stands are believed to represent old-growth composition. These are characteristics of the HBEF study site: (a) The watershed ecosystem is part of a larger, homogeneous biotic and geologic unit. (b) The basement unit of nearly impermeable gneiss is overlain by glacial till. (c) The site is characterized by cool, relatively humid, continental conditions. (d) The input/output budget for nongaseous nutrients can be determined from the difference between the meteorologic input (dissolved substances and particulate matter in rain and snow) and the geologic output (dissolved substances and particulate matter in drainage waters).

From data compiled over the last three decades, general principles and concepts about nutrient cycling in the eastern deciduous forest have been derived. Experimental manipulation has provided additional information, which allows predictions about ecosystem behavior in response to perturbation, whether natural or anthropogenic (Likens et al. 1977, Reiners 1992, Likens et al. 1994). We will discuss deforestation and subsequent recovery of several watersheds, nutrient budgets at HBEF, and the biogeochemistry of potassium.

Deforestation   Watershed 6 (W6), the standard reference watershed, was logged in 1910–1917. Watershed 2 (W2) was clear-cut in 1965 but the timber was not removed, nor was the soil disturbed by road-cutting or vehicular traffic. Regrowth was suppressed for three summers by herbicide application. Watershed 4 (W4) was divided into 25-m-wide strips, and every third strip was harvested in years 1970, 1972, and 1974. Watershed 5 (W5) was harvested in 1983–1984. Consequently, by 1995 there existed a chronosequence of clear-cut watersheds: W6, W2, W4, and W5 (Likens et al. 1977, Reiners 1992, Pardo et al. 1995). In that year stands were 78, 26, 21, and 11 years old, respectively. Vegetation research in those stands has had various objectives, primarily to discover the relationship between the hydrology and biogeochemistry of the stand and vegetation recovery.

Watershed 2 has been carefully analyzed in terms of maturity: the degree of development relative to an accepted standard for old-age, steady-state forests in the area (Reiners 1992, Likens et al. 1994). At W2, herbicides were used to test the resilience of the system, that is, to determine the consequences of severing the uptake of nutrients by vegetation: if one major flux or process is removed from the system, how will the others function? (Refer back to Figure 13-4.)

The deforestation and herbicide application had a great effect on the hydrologic cycle, and nutrient budgets are inextricably connected to that cycle. Brief, intense runoff from the deforested region carried an increased load of particulates. From 1966–1970, the average annual particulate loss of potassium from W2, both in organic and inorganic form, was about 6.8 times that from the reference (or control) site (W6).

Immediate cessation of transpiration resulted in increased stream flow, from about 40% above what would be expected in an undisturbed site the first year to 26% above expected in the third year after deforestation. Increased stream flow due to lack of transpiration is not surprising: stream flow regularly increases on undisturbed sites in autumn, before the rains begin, in response to synchronous leaf fall.

Recovery   Once suppression ceased, W2 displayed a vigorous capacity for regeneration (Reiners 1992). Calcium, nitrogen, sulfur, potassium, magnesium, and phosphorus all increased during the recovery along with increased biomass (Figure 13-8a). Calcium, after a slow start, increased linearly, resulting in the greatest absolute increase of any element. That pattern reflects the structural nature of calcium in plants (Table 13-1). By contrast, nitrogen and potassium increases were very high the first 11 years, but then the rates declined, yielding a sigmoid curve. Slow, asymptotic patterns of accumulation were exhibited by phosphorus, magnesium, and sulfur.

It is instructive to compare the changes in aboveground vegetation nutrient pools in W2 with those in the reference watershed, W6 (Figure 13-8b). Comparison

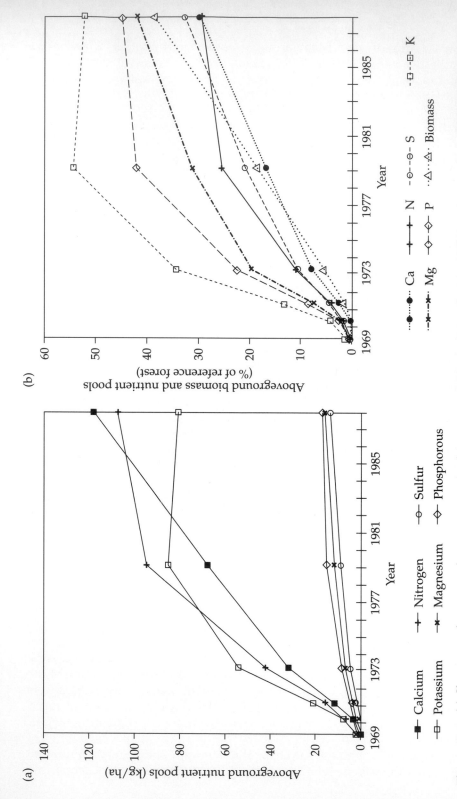

Figure 13-8 (a) Changes in aboveground vegetation pools of six macronutrients at Watershed 2 of HBEF, following experimental deforestation and herbicide treatment. (b) Changes in biomass and in aboveground vegetation pools of six macronutrients as percentages of biomass pools and of a 55-year-old forest on a comparable site. From "Twenty years of ecosystem reorganization following experimental reforestation and re-growth suppression" by W. A. Reiners. *Ecological Monographs* 62: (4)503–523. © 1992 by the Ecological Society of America. Used by permission.

of absolute and relative rates yields a dramatic difference: the greater accumulation of potassium, phosphorus, and magnesium probably is due to the use of these nutrients in regenerating canopies, which regrow faster than do the tree stems. The lower percentages (relative to the reference) of nitrogen, sulfur, and calcium indicate that these elements are being used in the heavier branches and boles of a growing forest. The percentages will increase as the forest matures (Reiners 1992).

Nutrient Input, Nutrient Output, and Balance   Nutrient input and nutrient output—geological, biological, and atmospheric—can be estimated for an ecosystem. Nutrients in bulk precipitation, throughfall, and stemflow can be gauged as the product of the concentration per unit of time and the total precipitation. Dry deposition provides a substantial fraction, perhaps as much as 63%, of bulk potassium deposition, but some portion of that potassium may come from within the forest itself. Potassium flux increases dramatically in throughfall and stemflow, carrying potassium to the forest floor. These processes represent intrasystem cycling, about which we will have more to say later.

Biological output is assumed to balance biological input because the area holds no special attraction for migratory animals. Unlike a system such as the Serengeti grasslands in Tanzania, for instance, biological transport of nutrients (e.g., potassium) across watershed/ecosystem boundaries is probably trivial. That situation may change as the moose population there grows (Likens et al. 1994). In addition, a mature forest is unlikely to lose many nutrients as windborne particulates or aerosols.

The streams and soil water at HBEF were monitored carefully, and lysimeters were put into place in the fall of 1983 to analyze soil solution chemistry at HBEF. The assumption is that the streams carry most of the water leaving the system (remember that the substrate is relatively impermeable).

Streams carry or deposit all of the particulates and dissolved nutrients leaving the watershed. A concrete ponding basin behind a V-notch gauging weir was anchored to bedrock at the base of each watershed (Figure 13-5). Flow was monitored, as were chemical concentrations and particulate load. As the stream flows over the weir, periodic samples are taken to measure solution nutrient loss from the system. Forest harvest dramatically increased the release of potassium and other nutrients to stream water. The enhanced rates of loss caused by clear-cutting may continue for years after the disturbance.

Biogeochemistry of Potassium   Potassium is the single most abundant cation in throughfall at HBEF, representing 33% of the total cations during the growing season. It is very soluble and easily lost to leaching in humid areas. On an annual basis, stream-water outputs of potassium exceed atmospheric inputs at all of HBEF, and the losses are significant. Inputs of strong acid anions may facilitate the leaching of potassium from the system, but foliar leaching is evidently more strongly correlated to precipitation flux than to acidity of that precipitation: when the canopy is wet, potassium is leached. During the growing months, June, July, August, and September, abiotic and biotic storage exceeds inputs plus soil release of K (Likens et al. 1994).

At HBEF and at other forested regions, weathering from parent material is an important source of potassium. One can derive an estimate of chemical weathering as the difference in input-output budgets. For soluble potassium, the mass balance is expressed as

$$P_K + W_K = S_K + \Delta B_K + \Delta O_K + \Delta X_K + \Delta M_K \qquad \text{(Equation 13-1)}$$

in which P represents the atmospheric input, W is the weathering release (primary minerals), S is the stream-water loss, ΔB is the uptake into biomass, ΔO is the change in the soil organic matter pool, ΔX is the change in exchangeable pool, and ΔM is the change in the secondary mineral pool. One could rearrange equation 13-1 to express the rate of net soil release of potassium as the sum of W_K, ΔM_K, ΔX_K, and ΔO_K, which is also equal to the sum of S_K and ΔB_K minus P_K. In other words, the value of soil release corresponds to plant assimilation. The equation is important because it allows one to discern relationships among the various components and compartments of the ecosystem.

Root litter is a significant source of potassium to the forest floor (refer back to Figure 13-4). Leaching and mineralization from fine, dead roots can be very rapid. The root litter may lose up to 80% of its original potassium in 60 days or less (Likens et al. 1994). Intrasystem cycling mechanisms are needed to capture that potassium before it is leached out of the ecosystem.

During the years 1965–1977, annual plant assimilation of potassium at W6 was at its highest (Table 13-5). From 1982 to 1992 that assimilation was less than 17% of the maximum. Concomitant with the decline in rates of biomass accumulation, potassium in throughfall and biomass storage of potassium decreased, and net soil release of potassium and resorption (the withdrawal of nutrients from leaves as they senesce) increased in the later period compared to 1964–1969 (Table 13-6). Potassium, a very mobile ion, moves in the phloem to overwintering plant parts. Resorption is an important, conservative, cycling mechanism.

Table 13-5   Rate of storage of potassium in living and dead biomass (B_K) and net soil release rate of potassium (estimated from $S_K + \Delta B_K - P_K$) in W6 at Hubbard Brook Experimental Forest, New Hampshire, 1965–1992. Units are in mol ha^{-1} yr^{-1}. From "The biogeochemistry of potassium at Hubbard Brook" by Likens et al. *Biogeochemistry* 25:61–125. © 1994 by Kluwer Academic Publishers. Used with kind permission from Kluwer Academic Publishers.

Period	Storage	Net soil release
1965–77	200	230
1977–82	130	160
1982–87	−30	1
1987–92	33	60

Table 13-6 Estimate of resorption of potassium for W6 at the Hubbard Brook Experimental Forest, New Hampshire. From "The biogeochemistry of potassium at Hubbard Brook" by Likens et al. *Biogeochemistry* 25:61–125. © 1994 by Kluwer Academic Publishers. Used with kind permission from Kluwer Academic Publishers.

Process	K (mol ha^{-1})	
	1964–69	1987–92
Leaf content before senescence = F_1	699	896
Leaf content after senescence = F_2	350	449
Leaching = L	276	160
Resorption = $(F_1 - F_2) - L$	73	287

Greenhouse Gases

Meteorological conditions influence intersystem cycling. For many decades we have known that acid rain is detrimental to living things, and we are learning that acids are deposited in fog and cloud water, and in gases and aerosols as dry deposition. Scientists became aware of acid rain in the northeastern United States sometime between 1950 and 1955 as subtle but significant clues began to emerge (Likens et al. 1996). Once-healthy streams and lakes became clearer as changes in pH caused the demise of phytoplankton and the zooplankton that fed on the producers. Nutrient cations, including calcium, magnesium, sodium, and potassium, were leached from forest systems as atmospheric acid concentrations increased. We now need to consider the negative consequences of other contributions from the atmosphere, including the increasing levels of carbon dioxide and of other greenhouse gases (Table 13-7).

Gas Flux   The increasing concentration of carbon dioxide in the atmosphere is of global concern. The concentration of carbon dioxide (CO_2) in the atmosphere is increasing steadily at about 1.8 μbar or 1.78 ppm per year, and currently stands at about 353 μbar, or 348 ppm, compared to 280 μbar, 276 ppm, in pre-industrial times (Sampson et al. 1993, Calloway et al. 1994). In 1993, 60 scientists from 13 nations gathered in Germany to assess global carbon dioxide flux and to estimate the potential for reducing carbon dioxide emissions and enhancing long-term carbon sinks (Table 13-7 columns C and D).

Methane (CH_4) affects the water budget of the stratosphere as well as the oxidation capacity of the troposphere and is increasing at an annual rate of 0.9%. During the past century, its concentration in the atmosphere has nearly doubled, from 0.9 to 1.72 parts per million by volume (ppmv; Castro et al. 1993, Castro et al. 1995). Nitrous oxide participates in the destruction of stratospheric ozone, a serious

Table 13-7   Current and future carbon fluxes in the terrrestrial biosphere, with atmospheric carbon dioxide doubling, under different management scenarios. (Quantities are in petagrams [Pg, 10^9 metric tons] of carbon per year.) From "Workshop summary statement: Terrestrial biospheric carbon fluxes— quantification of sinks and sources of carbon dioxide" by R. N. Sampson et al. *Water, Air, and Soil Pollution* 70:3–15. © 1993 by Kluwer Academic Publishers. Used with kind permission from Kluwer Academic Publishers.

Global biotic system	Current C flux (A)	Future C flux (doubled CO_2 climate) (B)	Future C flux with optimum vegetation management (doubled CO_2 climate) (C)	Fossil C offset potential from biomass-energy management (doubled CO_2 climate) (D)
			(Pg C yr^{-1})	
Tundra/boreal forests	+0.5 to +0.7	−1.0 to −0.5[a]	0	
Temperate forests	+0.2 to +0.5	−2.0 to +2.0	+0.3 to +2.0	+0.1 to +0.9
Tropical forests[b]	−2.2 to −1.2	−1.0 to −0.5	−0.5 to 0	0 to +0.2
Grasslands, savannas, and deserts	0 to +0.6	−0.3 to +0.1	+0.1 to +0.5	0 to +0.3
Agro-ecosystems	−0.1 to +0.1	0.0 to +0.1	0.0 to +0.3	+0.4 to +2.4
Wetlands	+0.2	+0.1	+0.2	
Total	−1.4 to +0.9	−4.2 to +1.3	−0.1 to +3.0	+0.5 to +3.8

Sink = (+); Source = (−).
[a] During transient (50–100 yr) response. In the long term (200–1000 yr), may revert to sink if climate stabilizes.
[b] From land use changes only.

environmental problem. Its atmospheric concentration is increasing about 2.5% decade^{-1} (Castro et al. 1993).

These three are all **greenhouse gases,** called such because they trap long-wave (that is, heat) radiation emitted from the surface of the earth, just as do panes of glass in a greenhouse. Our global blanket of increasing greenhouse gases retains energy that would otherwise be re-radiated to space, serving to influence the global heat budget and potentially altering global climate (Bowden et al. 1993b). This section will present information relating to greenhouse gases.

Gases flow both into and out of an ecosystem. Soil respiration in a temperate forest results in the return of carbon dioxide to the atmosphere; nitrous oxide is emitted from pasture and forest; and methane from both natural and human sources is pumped into the air. Though scientists know that the atmospheric concentrations of these gases are increasing, the global impact of the increases is imperfectly known (Matson et al. 1991, Bowden et al. 1993a, Reiners et al. 1994, Kauffman et al. 1995, Winkler et al. 1996).

In natural systems, carbon is released to the atmosphere by the respiration of

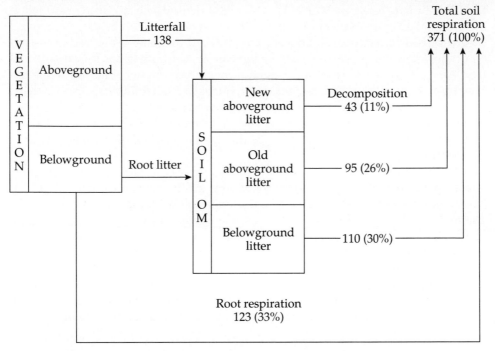

Figure 13-9   Soil respiration budget for mixed hardwood forest at the Harvard Forest, Massachusetts. Numbers are flux rates for C (g m^{-2} yr^{-1}) and percentages (in parentheses) of total soil respiration for each component. OM = organic matter. From "Contributions of aboveground litter, belowground litter, and root respiration to total soil respiration in a temperate mixed hardwood forest" by Bowden et al. *Canadian Journal of Forest Research* 23:1402–1407. © 1993. Used by permission.

living roots, and by decomposition of both aboveground and belowground litter: autotrophic and heterotrophic respiration, respectively (Figure 13-9). Soil nitrogen content, organic matter, moisture, temperature, forest development, and forest management practices control the relative contributions to total soil respiration in various forests. At Harvard Forest in Massachusetts, a temperate mixed hardwood forest, total soil respiration amounts to 371 g C m^{-2} yr^{-1}. Of that contribution, nearly 2/3 is from root activity (Bowden et al. 1993a).

The Harvard Forest research site is in that part of New England that is prone to autumn hurricanes. Bowden and his colleagues (1993b) analyzed a simulated hurricane blowdown for its impact on the flux of the greenhouse gases CO_2, CH_4, and N_2O. To measure gas exchanges, removable chambers were installed in the blowdown area on the forest floor and sampled each month. They found no increase in N_2O emissions and no significant differences in CO_2 or CH_4 fluxes compared to a control plot, though net nitrification was greater in the blowdown than in the control.

By contrast, earlier research in subtropical wet forest by Steudler et al. (1991)

found that disturbance by Hurricane Hugo and by forest clear-cutting resulted in increased N_2O emissions, short-term increases in net N mineralization, decreased carbon dioxide emissions, and decreased uptake of methane compared to the reference forest. Most impressive was the increase in N_2O emissions four months after clear-cutting: levels rose two orders of magnitude higher than the reference. Hurricane Hugo also caused an increase in nitrogen availability in South Carolina forest soils. The authors point out that forest clearing in a variety of tropical forests often causes elevated N_2O emissions, and that human disturbance such as clear-cutting has a more dramatic effect on gas flux than do natural disturbances such as hurricanes.

Tropical forests circulate more nitrogen at higher concentrations than do most boreal or temperate forests, and consequently exhibit high nitrous oxide fluxes (Garcia-Mendez et al. 1991). Table 13-8 presents a global budget for nitrous oxide. The major pathway or sink for N_2O is reaction with activated oxygen in the stratosphere: 10×10^{12} g (10 Tg) are consumed in this way each year. It is clear that tropical and subtropical forests and woodlands are the major natural sources for N_2O (Matson and Vitousek 1990).

What is causing the increases in greenhouse gases? We know that deforestation and increasing use of fossil fuels are major factors in the increase of atmospheric carbon dioxide. Deforestation in the Brazilian Amazon had cleared at least

Table 13-8   Global budget of N_2O—N (in Tg showing an estimated range). The estimated annual increase in nitrous oxide in the atmosphere is around 3.5×10^{12} g (or 3.5 Tg). Note the apparent discrepancy between "total sinks and accumulation" and "total sources," which arises in part from the relatively large uncertainty factor. From "Ecosystem approach to a global nitrous oxide budget" by P. A. Matson and P. M. Vitousek. *BioScience* 40:667–672. © 1990. American Institute of Biological Sciences. Used by permission.

Sinks and accumulation	
Stratospheric photolysis and reaction	10.5 ± 3
Accumulation in the atmosphere	3.5 ± 0.5
Total	14 ± 3.5
Sources	
Ocean	2 ± 1
Combustion	
coal and oil	4 ± 1
biomass	0.7 ± 0.2
Fertilized agricultural land	0.8 ± 0.4
Temperate grassland	0.1
Boreal and temperate forests	0.1 ± 0.5
Tropical and subtropical forests and woodlands	7.4 ± 4
Total	15.2 ± 6.7

230,000 km^2 of tropical forest, and perhaps as much as 415,000 km^2, by 1990 (Kauffman et al. 1995). It is clear that the increase in greenhouse gases is bringing changes that may be detrimental to natural and other ecosystems. It is also clear that more research is needed to understand gas flux. We will address the problem of increasing nitrogen deposition in the next section.

Nitrogen Saturation   The biogeochemical cycles of earth have been dramatically altered by humans. The damage caused by sulfur-laden pollutants producing acid rain in Europe, in the northeastern United States, and increasingly in the western United States, has been well studied. We scientists and students must now broaden our view to include nitrogen deposition as a factor affecting air, water, and ecosystem quality (refer back to Table 13-2). Many forest ecosystems that have been considered nitrogen-limited are receiving added nitrogen from fertilization and from precipitation and deposition. The additions range from <2 kg N ha^{-1} yr^{-1} in the most pristine areas to >40 kg N ha^{-1} yr^{-1} in regions subject to industrial pollution (Aber et al. 1989, 1993a,b, 1995; Gilliam et al. 1996).

In agricultural lands, the latter figure may go to 100–400 kg N ha^{-1} yr^{-1} due to fertilizer application, but that contribution is often a one-time phenomenon, whereas the nitrogen delivered in wet and dry deposition is a continuing contribution. Chronic nitrogen fertilization in tallgrass prairie may encourage forbs over grasses, which is not a desirable result (Seastedt et al. 1991).

Eventually, supply exceeds biological demand, a situation that represents **nitrogen saturation:** "the availability of ammonium and nitrate in excess of total combined plant and microbial nutritional demand" (Aber et al. 1989). Nitrogen saturation should be considered as a series of continuing changes that gradually reduce the nitrogen-retaining capacity of the ecosystem (Peterjohn et al. 1996). One implication of nitrogen saturation is that other resources will become limiting, such as water for plants or carbon for microbes.

Nitrogen in wet deposition over the northeastern United States may be from 5–20 times global, ambient background levels. Such deposition will cause greater soil acidification in that region. Even with the Clean Air Bill, which is intended to improve air quality, the total N deposited in the Northeast will remain high for years, exceeding outputs for the forests there and resulting in nitrogen saturation (Aber et al. 1993b). There are now indications of eutrophication—increases in nutrients, including nitrogen—in northeastern streams flowing from forested regions, whereas before this time forests there were limited by nitrogen availability.

Figure 13-10 presents a summary of nitrogen biogeochemistry and forest production in response to nitrogen saturation at four study site–stand combinations (two in Harvard Forest, one at Bear Brook, and one at Mt. Ascutney), and one transect in spruce-fir forest from New York to Maine. Potential consequences of nitrogen saturation are as follows: (a) large increases in net N mineralization, from 1.3 to 4.0 times the control; (b) increases in foliar N concentration, accompanied by a decrease in Ca:Al ratios (Ca:Al imbalances are associated with forest decline); (c) increased cation leaching losses; (d) increased emissions of N$_2$O; (e) declining tree growth and vigor; (e) potential loss of frost-hardiness in certain conifers (Aber et al. 1989, 1993b, 1995).

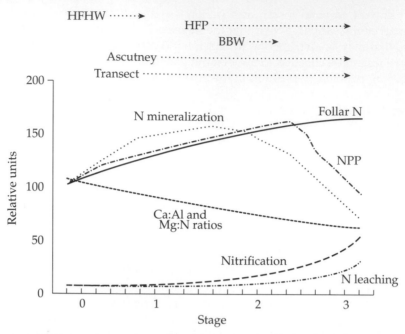

Figure 13-10   Summary of changes in nitrogen cycling and forest production with long-term chronic nitrogen additions for temperate forest ecosystems. Letters represent the initial position of each study site along this synthetic nitrogen saturation gradient. HFHW and HFP are at Harvard Forest hardwood and pine stands, respectively. BBW is Bear Brook stand. Ascutney is the Mt. Ascutney site and Transect is for the set of 161 native spruce-fir stands sampled across the northeastern United States. From "Forest biogeochemistry and primary production altered by nitrogen saturation" by J. D. Aber et al. *Water, Air, and Soil Pollution* 85:1665–1670. © 1995. Used with kind permission from Kluwer Academic Publishers.

In the Fernow Experimental Forest, Peterjohn et al. (1996) found several of the symptoms noted above. Their synthesis of more than 20 years of research in central Appalachian hardwood forest ecosystems included data that "support current conceptual models of nitrogen saturation and provide a strong, and perhaps the best, example of nitrogen saturation in the United States" (Peterjohn et al. 1996).

In that same forest, Gilliam and Adams (1996) found that precipitation is more acidic during the growing season than at other seasons, and more acidic at higher elevations than at lower sites. For hardwoods, the timing of that increased acidity is concomitant with a highly vulnerable stage, namely spring canopy development. Therefore, acid precipitation can affect productivity in hardwood forests.

Nitrogen Saturation at HBEF   Reflecting the growing concern over nitrogen saturation, Pardo et al. (1995) investigated changes in nitrogen capital and nitrogen retention, and seasonal patterns of NO_3 in stream water in the four watersheds at

HBEF, using data collected since 1963. They found that nitrate concentrations in stream water peaked in March (during snowmelt), declined during the summer growing season, and increased again in late fall in all four watersheds. The three younger stands evidenced increased losses of the nutrient cations Ca^{2+}, Mg^{2+}, and K^+, as well as H^+ and NO_3^- during the years immediately following clear-cutting.

Though atmospheric deposition of NH_4 and NO_3 was strongly retained in young, regrowing stands, that retention has declined. In fact, there was increased nitrate export at two watersheds (W2 and W4) in 1989 and 1990. In those years, W6 (the oldest stand) experienced a 10-year high in annual nitrate loss. Scientists will need to assess land-use history as well as climatic patterns and other year-to-year variability in order to predict and determine soil and water acidification, critical atmospheric contributions, and nitrogen saturation (Pardo et al. 1995).

These problems are serious and demand attention if we want the forests to remain healthy. A rapid assessment technique is needed to signal workers that a system is approaching nitrogen saturation. One promising technique is remote sensing, which could detect early stages of saturation by means of near-infrared spectroscopy (Aber et al. 1989; McLellan et al. 1991a,b; Bolster et al. 1996). Estimates of nitrogen, lignin, and cellulose concentrations in both green foliage and in litter are possible with the use of near-infrared reflectance.

Matson and her colleagues (1994) determined canopy chemicals along an Oregon transect using high-resolution spectral data in the near-infrared region. The remote assessment of canopy chemicals with these tools, in combination with detailed up-close leaf-level work, is very promising (see Chapter 10). Any temperate forest ecosystem has a limited capacity to assimilate nitrogen brought on the wind. Once that capacity is exceeded, the forest becomes a source of nitrogen, no longer a sink, leading to the consequences outlined earlier.

Stability

The concept of stability can be applied both to plant communities (see Chapter 8) and to ecosystems to describe a response to exogenous disturbance. For communities, stability generally refers to the response of species composition and diversity to disturbance (Coleman et al. 1983, Roberts and Gilliam 1995a). For ecosystems, **stability** generally refers to the response of nutrient dynamics to disturbance. In either case, stability is determined by two innate characteristics of the community or ecosystem: resilience and resistance.

Resilience refers to the ability of an ecosystem to return to the predisturbance condition. Although it may experience numerous changes through time, a resilient system rapidly returns to the predisturbance steady state following perturbation. **Resistance** refers to the ability of an ecosystem to resist changes in response to disturbance; a resistant system will change very little following perturbation and thereafter. It should be kept in mind, however, that the overall response to disturbance is determined by an interaction between resilience/resistance of an ecosystem and the nature of the disturbance, including frequency and intensity of disturbance.

Ecosystem stability involves nutrient pools that are small and quickly revolving and that rely upon the soil microflora and fauna. Also involved, in an ecological time frame of perhaps many hundreds of years, are pools that are very large and slowly turning: recalcitrant litter or marine sediments, for instance (Berg et al. 1996).

What factors are involved in the stability of ecosystem nutrient cycles? Efficiency of nutrient use and resorption, size of storage pools, and immobilization of nutrients are properties of ecosystems that serve to conserve and recycle nutrients.

Resorption

In an earlier section we discussed the resorption of potassium (see Table 13-6). Killingbeck (1996) defined these aspects of resorption: the percentage reduction of nutrients between green and senesced leaves he calls resorption efficiency; the level to which nutrients in green leaves have been reduced in senesced leaves he labels resorption proficiency; and the maximal withdrawal of nutrients that can be achieved he terms ultimate potential resorption.

Killingbeck analyzed resorption data on 89 species of evergreen and deciduous woody perennials. He found that concentrations of 0.3% nitrogen represent the ultimate potential resorption for that nutrient in senesced leaves. Plants incapable of symbiotic nitrogen fixation are much more proficient at resorbing nitrogen than are plants with symbiotic nitrogen fixation. In general, conifers retrieve more nitrogen from senescing leaves than do flowering plants.

Deciduous species are less proficient than evergreens in recovering phosphorus: 0.067% and 0.045% in senesced leaves, respectively. For phosphorus, 0.01% in senesced leaves represents ultimate potential resorption for that nutrient.

Resorption proficiency varies not only from species to species but also within the same species from one year to the next. For example, *Fouquieria splendens,* a drought-deciduous shrub of the Sonoran Desert, exhibited resorption of nitrogen of 11% in 1986 and 72% in 1989 (Killingbeck 1996).

Proficiency of retrieving nitrogen is often significantly and positively correlated with proficiency of resorbing phosphorus among woody perennials, though the two processes are not always closely coupled. For example, six species of rain forest trees, not congeners, reduced phosphorus in senesced leaves to essentially the same low levels (0.04–0.05%) yet retained >1.0 % nitrogen, indicating incomplete resorption of that nutrient (Killingbeck 1996).

Ecosystem stability hinges in part on the fitness of its members, and plant fitness is enhanced by resorption (May and Killingbeck 1992). Experimental defoliation of scrub oak (*Quercus ilicifolia*) to prevent resorption over a three-year period caused (a) declines in acorn production, (b) smaller radial stem growth increments, and (c) an increase in foliar biomass. Fitness is a function of success in survival, growth, and reproduction. Resorption of nutrients from senescing leaves, clearly an important intrasystem cycling mechanism, influences those three parameters in complex ways. The degree of resorption proficiency is positively correlated with the potential for success. The movement of phloem-mobile nutrients into perennial tissues conserves the nutrients for use in next season's new growth. The stability of the ecosystem—any ecosystem—is enhanced by highly proficient nutrient resorption (May and Killingbeck 1992, Ashman 1994).

Nutrient Use Efficiency

With regard to long-lived perennials, Vitousek (1982) defined **nutrient use efficiency (NUE)** as "the grams of organic matter lost from plants or permanently stored within plants per unit of nutrient lost or permanently stored." NUE could also be viewed as grams of tissue produced per gram of nutrient. In other words, nutrient use efficiency is the inverse of nutrient concentration in the organic matter increment and in litterfall and root turnover.

What are the ecological and evolutionary origins of nutrient use efficiency? Net mineralization of nitrogen is low in forest floors that have low nitrogen flux. That is, there is a positive biofeedback system wherein litterfall with a high C:N ratio would favor nitrogen retention by decomposers (immobilization) and would reduce available soil nitrogen. This condition leads to greater efficiency of nitrogen use, enhancing the C:N ratio further. A long-term ecological (and ultimately evolutionary) advantage in nitrogen-poor sites would go to those species with high nitrogen use efficiency, which provides litter with high C:N ratios (Vitousek 1982).

Storage

The coupling of metabolism and mineral cycling is a stabilizing force within an ecosystem. One bridge between metabolism and mineral cycling is the availability of large storage pools within the system. Large storage capability tends to buffer nutrient cycles against disturbance. A large standing crop (e.g., trees in a tropical rain forest) provides a spacious storage pool; so do ample exchange sites for cations (e.g., young volcanic soils or deep, fertile grassland soils). The magnitude of an element's potential for replenishment on soil surfaces is one measure of the resistance of that nutrient cycle to loss.

Some apparently open systems are resistant to change. Salt marshes, for example, seem to have a superabundance of energy supply. Nutrients are rarely limiting, and storage pools are never empty. Although diversity is low, stability remains high. This stability is probably attributable to a remarkably high metabolic level appropriate to the abundance of nutrients available in the system.

Immobilization of Nutrients

The largest source of biologically available nitrogen is the ammonium released from decaying organic matter; the sink is uptake of ammonium and nitrate by microorganisms and plants. If biological uptake is effective, nitrogen pools are small and turnover is rapid (Vitousek and Matson 1985a).

There is, however, potential for significant loss from fertile sites, directly related to the amounts of nitrogen circulating in a successional, undisturbed system. In the face of disturbance, the ability of plants to take up nitrogen may be reduced, but uptake by plants is only one of many processes that prevent or delay solution losses of nitrogen following disturbance. Ammonium immobilization, nitrate immobilization, lags in nitrification, clay fixation, low net nitrogen mineralization, and/or lack of water for nitrate transport could be involved in maximizing retention. Most of these mechanisms to prevent loss of nutrients are mediated by microbial activity (Vitousek and Matson 1985b).

Summary

Nutrients that are essential for plant growth include C, N, P, S, K, Mg, Ca, Fe, Mn, Zn, Cu, Mo, B, and Cl. Materials, including mineral nutrients, cycle within the biosphere, whereas energy flows only one way. Biogeochemical cycling includes nutrient cycling, which may be one of two types: gaseous, which tends to be regional or global (e.g., CO_2); or sedimentary, which operates primarily within one ecosystem (e.g., Ca). Nutrients may accumulate with successive stages. Nutrients are compartmentalized, occurring in soil and rocks, in soil solution, in the atmosphere, and in biomass. Transfer into or out of an ecosystem may be meteorologic, geologic, or biologic. Cycles may require minutes, decades, or millennia to revolve.

Abiotic factors involved in nutrient cycling include fire and earth movement and the various aspects of the hydrologic cycle: rainfall and leaching, atmospheric deposition and canopy interception, evapotranspiration, runoff, infiltration, and salt spray.

Biotic factors include litter production and organic matter accumulation, decomposition, ingestion and digestion, gas flux, root exudation, and retrieval of nutrients from deep soil layers.

Productivity in grasslands is limited by the availability of energy, nitrogen, and water. Nitrogen is provided by atmospheric inputs and by fixation, both free-living and symbiotic, and lost through volatilization. Fire is a significant factor in nutrient cycling in grasslands through its effect on detritus. Detritus influences the amount of water, nitrogen, and solar radiation reaching young grass shoots.

Experimental deforestation of a watershed in the Hubbard Brook Experimental Forest resulted in greatly increased runoff and enhanced export of particulates. The watershed has exhibited a vigorous recovery. Accumulation of biomass and nutrient pools reveal a maturing forest. Potassium is the most abundant cation in throughfall. Some stands at HBEF are experiencing increased nitrate export. The undisturbed system is stable and conservative, with efficient intrasystem cycling of nutrients.

The increase in the atmospheric concentration of greenhouse gases is of global concern. Concentrations of carbon dioxide, nitrogen oxides, and methane in the atmosphere have increased and are continuing to increase. Nitrogen saturation is becoming a serious problem in certain ecosystems.

Stability relies upon resilience—the ability of an ecosystem to return to the predisturbance condition—and resistance—the ability of an ecosystem to resist changes in response to disturbance. Proficiency of resorption, efficiency of nutrient use, replenishment power and retention of nutrients, effective storage, and immobilization of nutrients are essential homeostatic mechanisms for ecosystems.

PART IV

ENVIRONMENTAL
FACTORS

We have seen in earlier chapters that the population and community attributes of plant species are determined by the response of individuals to the external environment. It is possible that autecological information may, when enough taxa have been characterized, lead to predictability at the community level. In this part, we will consider the environmental factors of light, temperature, fire, soil, and water, including their effect on physiological and community processes.

It should be kept in mind that separate discussions of various factors is artificial; the interaction of all environmental factors determines the response of the plant. Chapter 20 attempts to show how the distribution of major vegetation types in North America is a response to interacting environmental factors.

CHAPTER 14

LIGHT AND TEMPERATURE

R adiant energy is one of the most important environmental qualities influencing plants. Visible light (radiation with wavelengths of 380–750 nm) is
the ultimate source of energy for most organisms, so the variations of light in time
and space and the photosynthetic responses to those variations are basic to our
understanding of plant ecology. Higher-energy ultraviolet radiation (1–380 nm) is
potentially harmful to organisms, and infrared radiation (756–100,000 nm) determines the temperature of organisms and their environment. In this chapter we will
consider the physical nature of radiant energy, look at how it varies spatially and
temporally, and survey related adaptations.

Heat is the aggregate internal energy of motion of atoms and molecules of
a body. It can be transferred by means of radiation, convection, or conduction.
Convection refers to heat exchange by circulation; conduction refers to the trans-
ference of heat through solids. Radiation refers to energy propagated as an elec-
tromagnetic wave, which passes through a vacuum with the speed of light (3×10^{10} cm sec^{-1}) but also travels through a medium such as air or water. Physicists
have shown that radiant energy is transported in discrete particle units called pho-
tons. Therefore, radiant energy has both a particulate and a wave form. In actual
use, the term *radiation* refers both to the energy transferred and to the process of
transferring energy. The solar energy that supports the life of this planet comes to
us as radiation, transmitted in a broad spectrum that includes mostly visible light
but also high-energy, short-wavelength ultraviolet radiation at one extreme and
low-energy, long-wavelength infrared radiation at the other extreme.

Energy is continually exchanged between objects and the environment. Direct
sunlight is absorbed by plants, rocks, soil, and the air around them. Infrared radiation (heat) is transferred from plants to the air, from soil to plants, and so forth.
When the energy gained (absorbed) by a body equals that lost (dissipated), the
temperature of that body remains constant. When the energy gained by a body is
more or less than that lost, its temperature increases or decreases respectively.

Because our eyes do not perceive the longer wavelengths, the radiant energy
given off by rocks, trees, and the air itself is invisible to us. On a warm spring day

(at 300 K, or 27°C), trees, rocks, open soils, herbaceous cover, and the various woodland animals radiate energy across a rather broad spectrum but at a minimum wavelength of about 10 microns (10,000 nm), invisible to human eyes.

Solar Energy Budget

Let us develop an energy budget and envision the allocation of solar energy in equation form:

$$S = R + C + G + Ps + LE \qquad \text{(Equation 14-1)}$$

where S = incoming solar radiation; R = reflected (shortwave, diurnal) plus reradiated (long wave, both day and night); C = convective heating of air; G = heating of solid objects by conduction (soil, for instance); Ps = energy used in photosynthesis; and LE = latent heat of evaporation.

The amount of energy reaching any given spot on the earth varies seasonally and diurnally, but on the average there are 1.94 gram-calories (or cal) cm^{-2} min^{-1} received as direct solar radiation measured perpendicular to the sun's rays just outside the earth's atmosphere at the mean distance of the earth to the sun. (A gram-calorie is that quantity of energy that is required to raise the temperature of 1 gram of water from 14.5°C to 15.5°C.) This 1.94 cal cm^{-2} min^{-1} is known as the **solar constant,** and it represents one of the most fundamental and vital components of our environment.

Total solar radiation (S) is usually expressed in cal cm^{-2} min^{-1}, W m^{-2}, or J m^{-2} s^{-1}, all units of **radiant flux,** which is the amount of energy received on a unit surface in a certain time. Net radiation at the earth's surface (R_n) is equal to solar radiation absorbed (S_{abs}) plus infrared radiation absorbed (IR$_{abs}$) minus infrared radiation emitted (IR$_{emit}$):

$$R_n = S_{abs} + IR_{abs} - IR_{emit} \qquad \text{(Equation 14-2)}$$

This can be diurnally negative because at night S_{abs} is zero, and there is a net loss of heat. During the day, S_{abs} and IR$_{abs}$ exceed IR$_{emit}$, and there will be a positive energy balance. During the long winter of the polar regions, R_n is seasonally negative; it may be so even in more temperate regions. In Siberia, the annual net radiation is negative (Figure 14-1a). Or there may be virtually no monthly period that shows a negative balance, with a resulting large net positive balance; such is the case at Aswan, Egypt (Figure 14-1b). In the tropics, there is little seasonal change. Because the angle of the sun does not vary enough to create seasonal deficits, R_n is positive throughout the year. In fact, diurnal variation can exceed the seasonal variation of daily temperature means (Figure 14-1c).

The distribution of solar energy (Equation 14-1) varies over time and from one community to another. For instance, the maximum S incident upon a tropical dry deciduous forest during the growing season is +1.3 cal cm^{-2} min^{-1}. The vegetation has an albedo of 0.16 during the growing season and 0.24 during the leafless

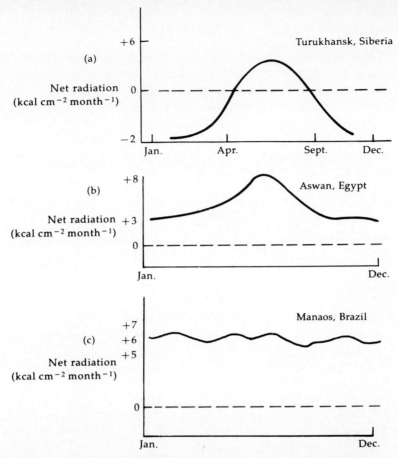

Figure 14-1 (a) Net radiation for a weather station in Siberia. Note that the annual average is negative. (b) Net radiation for a weather station at Aswan, Egypt. Note that there is virtually no period of negative radiation balance. (c) Net radiation for a tropical weather station. Note that there is very little seasonal flux in the radiation balance. These are monthly averages and do not necessarily reflect diurnal changes.

period (Barradas and Adem 1992). **Albedo** refers to the reflectivity of a material summed over all wavelengths for a sunlit surface. S times albedo equals R, so in this example $(0.16)(1.3 \text{ cal cm}^{-2}\text{ min}^{-1}) = 0.78 \text{ cal cm}^{-2}\text{ min}^{-1}$ during the growing season and $0.31 \text{ cal cm}^{-2}\text{ min}^{-1}$ during the dry season. The rest of the solar radiation is apportioned among G, C, Ps, and LE.

Mountain meadows are often densely covered with herbaceous vegetation, and at 100% cover, conduction (G) will be rather low. For various physical reasons, convective heat exchange (C) is also low. Only 1–2% of incident light is usually converted into chemical energy, and photosynthesis in mountain meadows may be lower than in other environments due to low temperatures. Although photosynthesis (Ps) is low, heat loss due to evaporation (LE) will be high, due to the wetness of this mountain meadow system.

A contrasting situation is seen in an arid area such as the Mojave Desert. Again, S_{max} is 1.3 cal cm^{-2} min^{-1}. In very arid regions plants are widely spaced, and the open expanses of exposed soil absorb a relatively high fraction of S. Thus, G will be much higher than in communities with a higher leaf area index (LAI) such as tropical forests, and photosynthesis will be very low, less than 0.1% of S. R will also be higher than in the tropical forest. C will be high, as the denser air absorbs radiant energy from sunlight and rises. There is very little liquid water available, and LE may be correspondingly low.

Leaf Energy Budgets

Radiation

In the rare situation where a leaf is in a steady state, the energy absorbed is equal to the energy emitted from the leaf ($R_n = 0$) and its temperature remains constant. Usually, however, the leaf is either gaining or losing energy. Leaves gain energy from shortwave solar radiation (250–4000 nm) and from long-wave radiation (4000–80,000 nm). Shortwave radiation (S_o) is referred to as radiant flux or solar irradiation and amounts to 1.3 cal cm^{-2} min^{-1} at the earth's surface. A narrower wave band of shortwave radiation (400–700 nm) is also of key importance because it is the energy used in photosynthesis. In full sun **photosynthetic photon flux density (PPFD)** is about 2000 μmol m^{-2} s^{-1}.

Several forms of radiant energy are important influences on an exposed leaf (Figure 14-2). Solar radiation may arrive at the surface of a leaf directly or indirectly, as diffuse radiation scattered by the atmosphere (skylight, S^{sky}) and by clouds (cloud light, S^{cloud}), or as radiation reflected from the soil or other objects in the habitat (rS^{direct}).

The spectral qualities and intensity of radiation vary depending on the distance the radiation travels through the atmosphere and the features of the habitat that absorb, reflect, or transmit the light (Figure 14-3, p. 382). The atmosphere effectively removes a portion of direct solar radiation, reducing the intensity from the extraterrestrial level of about 1.94 cal cm^{-2} min^{-1} (the solar constant) to about 1.3 cal cm^{-2} min^{-1} at sea level on a clear summer day (Gates 1965b). Most radiation received is direct, with only a small amount that is rich in the blue wavelengths arriving as skylight. The energy of solar radiation can be measured with devices called pyranometers, which are discussed later in this chapter.

As light passes through the atmosphere, not only is the amount of energy reduced but the balance of wavelengths changes. Caldwell and Robberecht (1980) have shown that maximum daily total shortwave irradiance received in the arctic alpine life zone of the Northern Hemisphere varies by a factor of only 1.6 along a

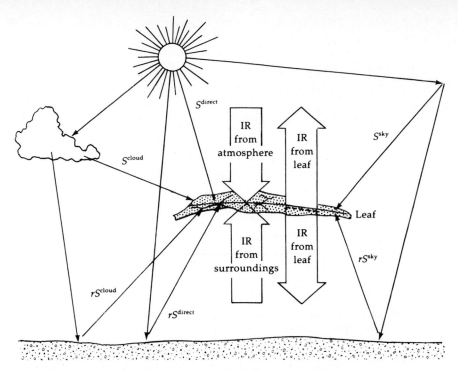

Figure 14-2 Schematic illustration of eight forms of radiant energy incident on an exposed leaf, and the infrared radiation emitted from its two surfaces. From P. S. Nobel. *Introduction to Biophysical Plant Physiology*. W. H. Freeman and Company. Copyright © 1983.

latitudinal gradient from polar regions at low elevations to alpine regions at high elevations in low latitudes. In contrast, along the same gradient, the maximum integrated, effective UV-B irradiance (280–320 nm) varies by a full order of magnitude—sevenfold for total daily effective radiation.

The light relationships of an idealized green leaf are diagrammed in Figure 14-4 on page 383. Reflection is measured by placing an appropriate light sensor above the leaf surface so that it reads only light reflected by the leaf. Placing the light sensor in the shadow of the leaf, one can measure the amount of light transmitted through the leaf. Total incident light minus reflectance and transmission is equal to absorbance. The amount of radiation absorbed is relatively high at wavelengths shorter than 0.7 μm (700 nm), which includes the visible and ultraviolet wavelengths. Brown (1994) measured only a 1–2% transmittance of UV-B radiation by closed canopies of mixed deciduous forests of Maryland. The very high energy ultraviolet light (wavelengths shorter than 0.4 μm [400 nm]) can be damaging to biological material, but water present in leaf cells is efficient in absorbing these wavelengths. High rates of absorption in the visible portion of the spectrum are caused by the presence of chlorophylls, carotenes, and xanthophylls—the pigments in greatest abundance in plant cells. Light transmitted through vegetation is dramati-

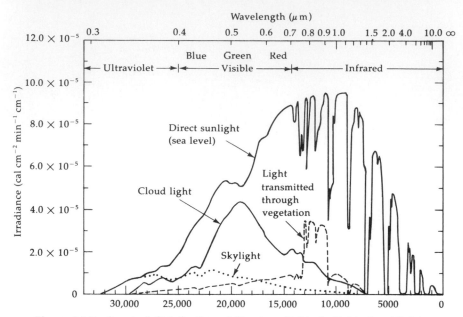

Figure 14-3   Spectral distribution of direct sunlight, skylight, cloud light, and sunlight penetrating a stand of vegetation. Each curve represents the energy incident on a horizontal surface. From Gates et al. 1965. By permission of the American Meteorological Society.

cally reduced in the visible region (Figure 14-3) because the chlorophylls absorb much of the violet, blue, and red light while the carotenoids (including xanthophylls) absorb heavily in the blue to green range. This leaves little visible light under the canopy of a dense forest.

The amount of shortwave radiation (S_o) that a leaf intercepts depends on the angle of the leaf relative to the sun. The solar radiation incident on a surface (S_j) is highest when the leaf is held perpendicular to the direct beam radiation. S_j declines as a function of the cosine of the angle of incidence (j). The angle of incidence is the angle between a perpendicular from the leaf surface and the incident angle of direct beam radiation.

$$S_j = S_o \cos j \hspace{3cm} \text{(Equation 14-3)}$$

Radiation incident on the leaf surface depends not only on leaf angle but on time of day and season. We can summarize leaf energy balance related to shortwave radiation as

$$S_{abs} = \cos j\, S_o\, [a(1 + r)] = S_j\, [a(1 + r)] = S_j a = S_j ar \hspace{1cm} \text{(Equation 14-4)}$$

where S_{abs} = solar radiation absorbed, j = angle of incidence to the direct beam radiation, S_o = direct beam radiation, a = the absorptance of the leaf to that wave band, and r = reflectance of shortwave radiation from the surroundings.

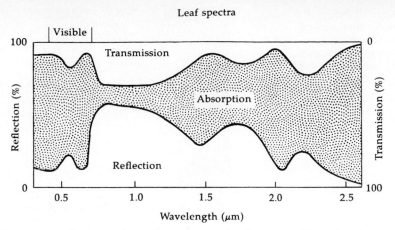

Figure 14-4   Transmission, absorption, and reflection with respect to wavelength for an idealized green leaf. After J. L. Monteith. 1973. *Principles of Environmental Physics.* Edward Arnold, London.

Leaf energy balance also depends on the gain and loss of heat energy (long-wave radiation from 4000–80,000 nm). Absorbed infrared radiation (IR_{abs}) is heat radiated from the surroundings, such as air, soil, rocks, other plants, etc. (Figure 14-4). The amount of heat radiated or reradiated by an object is a function of its temperature. Leaves are objects, so they emit long-wave radiation (IR_{emit}) as a function of their temperature. Equation 14-2 summarizes leaf energy balance. When leaf R_n is negative the leaf is cooling and when it is positive the leaf is warming. Transpirational cooling often modifies leaf R_n. We will consider that aspect of leaf energy balance in Chapter 18.

Measuring Radiation

Through a global system of weather stations, we have a reasonably accurate picture of the world's climate. We can also measure radiation, the source of heat, but measurement of energy flux has rarely been done, and only recently has the instrumentation been developed to support global-scale quantification of the energy environment.

Radiation can be measured by devices called radiometers, which integrate all incoming wavelengths and measure in energy units. Pyranometers are used to measure direct and diffuse sunlight. They are sensitive to 360–2500 nm, which includes some ultraviolet light (<450 nm), all visible light, and considerable infrared light (>700 nm). Filters can be added to a pyranometer, restricting its sensitivity to only the visible light, or all but a portion of the visible spectrum can be filtered out if that is needed.

An Eppley black-and-white pyranometer (Figure 14-5) uses the temperature differential between hot and cold areas to monitor radiation intensity. However, it does not separate direct rays from indirect rays. On a clear day, direct sunlight ac-

Figure 14-5   The Eppley pyranometer. The sensing element under the glass dome is a thermopile with hot (black) and cold (white) areas.

counts for approximately 85% of the energy received. On an overcast day, up to 100% of the incoming energy is due to diffuse, indirect sunlight. If we wish to monitor only diffuse sunlight, a shading device called an occulting ring, or shadow hand, would be used with the pyranometer. The Eppley Precision Spectral Pyranometer (Figure 14-6) is probably the most accurate instrument produced commercially for measurement of global sun and sky radiation. It may also be used with a shading device to screen out either the sun or the diffuse sky component.

An instrument called a **pyrheliometer** measures only direct solar radiation. It consists of a sort of tube and a sensor to track the sun such that only solar radiation of normal incidence enters the monitoring device, the tube. Without the use of the pyrheliometer, the intensity of direct beam (direct sunlight) may be estimated by comparing data obtained with the pyranometer alone and the pyranometer combined with the occulting ring:

Direct beam radiation = total global radiation
− diffuse radiation   (Equation 14-5)

The monitoring of net radiation is of special interest to plant ecologists. Ideally, the net radiometer absorbs all incoming (downward) radiation (I_R), and all outgoing (upward) radiation (O_R). The difference between I_R and O_R gives the net radiation, which is the energy available in the community for work.

In a forested area, estimates of light received, length of sunflecks (moving spots of sunlight on an otherwise shaded forest floor), and annual variations have relied upon continuously recording instruments, requiring considerable time and effort, and have produced results that are imprecise at best. Patterns of direct beam irradiance may be analyzed by hemispherical canopy photography, using a 180° equidistant fish-eye lens. The camera is placed at the desired position and the appropriate level beneath the canopy, a standard is placed to indicate true north, and a slide or series of slides is taken. Later, the slides are projected onto graphs of solar tracks enlarged from Smithsonian Institution tables for the appropriate latitude (List 1951). The researcher can then measure minutes of potential incident radia-

Figure 14-6 The Eppley Precision Spectral Pyranometer. This instrument uses a plated, wire-wound thermopile that is temperature compensated to be virtually unaffected by ambient temperature and is able to withstand severe mechanical vibration and shock.

tion and number and length of potential sunflecks for an entire year (Evans and Coombe 1959, Ustin et al. 1985, Selter et al. 1986). Growth of saplings of species in the understory of a Hawaiian forest was highly correlated with estimates of minutes of sunflecks received by that stratum. Photosynthetic photon flux density sensors and hemispheric fish-eye photographs were employed to estimate sunfleck duration and photon flux densities (Pearcy 1983).

The wavelengths absorbed by chlorophyll, and therefore active in the photosynthetic process, are between 400 and 700 nm. Light in this band has been labeled **photosynthetically active radiation (PAR).** A mole of photons (6.02×10^{23} photons) is referred to as a mole photon or einstein (E), and the unit of photosynthetic photon flux density (PPFD) is often expressed as μmol m^{-2} s^{-1}. The intensity of PPFD is expressed as the photon (quantum) flux per unit area, which, for example, would be about 2000 μmol m^{-2} s^{-1} near midday on a clear summer day.

Other units of light measurement were used in plant ecological studies prior to the advent of instrumentation to measure only PPFD. Common units and their equivalents are listed in Table 14-1. It is not possible to convert light values reported in conventional units directly to PPFD because lux and foot-candle measurements report light in the broad spectral range perceived by the human eye, which is not identical to PAR. Portable light meters fitted with quantum sensors that measure only PPFD are available commercially.

Temperature Measurement

Temperature may be sensed by a thermocouple, a mercury or alcohol thermometer, or various other devices. Integration may then be used if temperature data over a long period of time are needed. An inexpensive chemical integrator of average temperature over time is sucrose. The rate at which a sucrose solution is hydrolyzed (inverted) to glucose and fructose is a function of temperature. About 10–15 ml of buffered sucrose solution is sealed in a tube and left at a desired height aboveground or belowground at a field site. The amount of sucrose that was in-

Table 14-1 Common units for the measurement of light energy.

Unit name	Units or equivalents	Value at sea level in full sun
langley (ly)	1 cal cm^{-1}	1.3 ly
foot-candle	10.76 lux	10,000 ft-c
lux	1 lumen m^{-2}	107,600 lux
watt	1 joule s^{-1} or 10^7 ergs	$1000 \text{ watts m}^{-2}$
mol photon (einstein)	6.02×10^{23} photons	$2000 \text{ μmol m}^{-2}\text{ s}^{-1}$
		$200 \text{ nmol cm}^{-2}\text{ s}^{-1}$

verted during the field exposure period can be determined by measuring the optical rotation of the solution with a polarimeter. The method has been described in detail by Berthet (1960, in French) and by Lee (1969).

Workman (1980) has described another chemical method for the measurement of mean integrated temperature, with an accuracy of ±0.5°C. A spectrophotometer, which is more commonly available than a polarimeter, is employed to trace the course of a chemical reaction. The employment of such chemical devices has economic and logistic advantages over the use of climatological data or of automatic temperature recorders (Workman 1980).

Variations in Light and Temperature

Average global air temperature increased by about 0.5°C during the twentieth century, probably as a result of the effects of increased atmospheric carbon dioxide on the global energy budget. In some areas the increased average is due to minimum temperatures increasing at a faster rate than maximum temperatures, thus decreasing the daily temperature range. In much of the United States daily maximum temperatures have not increased even though increases are seen in average temperatures. This is in contrast to New Zealand (Salinger 1995) and the high mountains of central Europe (Weber et al. 1994) where both maximum and minimum temperatures have shown similar increases. The global reduction in daily temperature range appears to be related to changes in atmospheric circulation in the Northern Hemisphere where one of the contributing factors to reduction in the daily temperature range may be increased cloudiness (Easterling et al. 1997).

Insolation (exposure to solar radiation) varies temporally and spatially. The primary factors that influence spatial variations in temperature are latitude, altitude, and proximity to water. Topography including slope aspect, cloud cover, and vegetation are contributing factors to temperature variation. The obvious and immediate temporal differences derive from the earth's rotation. During the day, radiant energy from sunlight usually replaces what was lost during the night, so there is an energy balance ($R_n = 0$).

Because the earth moves about the sun in an elliptical orbit, there is some variation in the earth-sun distance. However, it is the tilt of the earth that gives us our seasons. There are periods of equability, when sunlight shines from pole to pole, and day and night are virtually equal in length. These periods, when the earth's polar axis is perpendicular to the radius between the earth and the sun, are called vernal and autumnal (spring and fall) equinoxes. Alternatively, when the plane of earth's polar axis is furthest from perpendicular to the earth-sun radius, one pole will be in 24-hour darkness, the other bathed both day and night in sunlight. In the Northern Hemisphere, our longest night, or winter solstice, is on 22 December; our longest day, or summer solstice, is on 22 June.

Latitude

The general distribution of total yearly radiation incident upon the surface of the earth is plotted in Figure 14-7. There is a gradual decrease in global radiation over the land areas from latitudes of about 30°N and S toward the poles. This decline is due primarily to the reduced number of daylight hours during the winter and to the greater distance solar radiation must pass through the atmosphere in temperate regions and poleward. From the equator to 30°N and S, radiation is generally high, with limited areas of very high radiation or very low radiation. Radiation incident on the Amazon Basin is low because of frequent heavy cloud cover and heavy rainfall. Areas of very high solar radiation occur on all the major continents where high-pressure systems prevail and cloud cover is very low.

A comparison of solar radiation received throughout the year at latitudes ranging from 0° to 80° shows the greatest equability at the lowest latitudes, and the greatest range and the highest energy received in a day at the highest latitudes (Figure 14-8).

Altitude

At higher elevations, light rays pass through less atmosphere. As a result, the ultraviolet light is a larger fraction of the total incoming radiation.

Atmospheric pressure is lower at the higher elevations, so air there expands and loses energy. Conversely, air coming downslope is more compressed and therefore warmer. Thus, in addition to latitudinal differences, elevational differences in temperature occur (see Figure 20-12). This inverse correlation of altitude with temperature creates a gradient or **lapse rate,** usually given as about 10°C per 1000 m of elevation. There are several lapse rates; the one given here is a **dry adiabatic** lapse rate, which means the cooling of air as it rises without condensation and without cloud formation. The wet adiabatic lapse rate is about 5°C per 1000 m in elevation. Lowry (1969) discusses lapse rates and their significance.

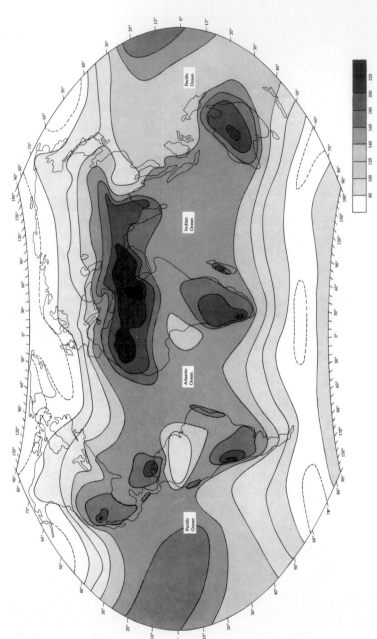

Figure 14-7 World map showing the distribution of solar radiation over the earth's surface in kcal cm^{-1} yr^{-1}. From H. E. Landsberg et al. 1966. *World Maps of Climatology.* By permission of Springer-Verlag.

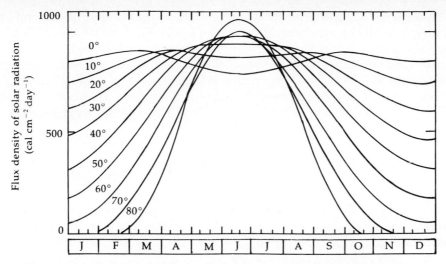

Figure 14-8 Daily total of the undepleted solar radiation received on a horizontal surface as a function of latitude and time of year. From D. M. Gates. *Energy Exchange in the Biosphere* (fig. 2, p. 8). Copyright © 1962 by Harper & Row, Publishers, Inc. Reprinted by permission.

Topography

The decline in air temperature due to increased elevation is not constant, especially in dissected topography. The situation is complicated by various physical phenomena, including temperature inversion. During the night, the soil gives up heat to the air, sometimes so quickly that the soil cools below the temperature of the overlying layers of air. These layers return heat by conduction to the soil and may become cooler than the overlying strata, resulting in a **temperature inversion.** The normal vertical temperature gradient, that is, temperature decreasing with increasing elevation, is inverted. There is thus an upslope thermal belt warmer than the lower-lying regions. The warm air capping the cold eventually is cooled (usually by 300 m in elevation) and the normal lapse rate is resumed.

Another means by which the temperature gradient can become inverted is by cold air drainage. Cold air is denser than warm air and drains into ravines and valleys at night. These low-lying canyon bottoms and river valley floors are thus colder than the slopes above them, augmenting the thermal belt effect. The flow of air may be reversed in the daytime as warmed air moves back upslope.

There are often tiny basins or frost pockets in which cold air accumulates and remains due to feeble air movement. A further complication of temperature variation in mountainous regions is the width of ravines and valleys. Narrow mountain valleys or canyons receive proportionately less sunlight. The walls therefore reflect less heat into the canyon, and it will be colder and damper than the surrounding area. By contrast, broad, open mountain valleys build up high temperatures due to the accumulation of heat reflected from the wide sides, and they are consequently hotter and drier during the day than surrounding regions.

Due to the inclination of the sun, a south-facing slope in northern temperate latitudes will always experience greater total insolation than will the north-facing slope of the same region. Thus, there will be warmer air and soil, less moisture, and sparser vegetation, in general, on the south-facing slope. North-south slope aspect difference is often manifested in very different plant cover. In central California, for instance, a north-facing slope might be oak woodland (blue oak, *Quercus douglasii*) and the south-facing slope of the same ravine will be covered by chaparral (chamise, *Adenostoma fasciculatum*). See Figure 14-9.

Figure 14-10 illustrates the variation in energy reception on sloping surfaces as a function of the time of day and across season at 40° latitude. Figure 14-10a depicts energy received at winter solstice, 22 December. Note that north-facing slopes steeper than 22.5° receive no direct insolation at that date. Figure 14-10b depicts energy reception at the vernal equinox. The north vertical receives no direct radiation, but the south slope at 45° intercepts almost as much as it receives at the summer solstice, which is depicted in Figure 14-10c. As we would predict, the total solar energy received is the greatest for a community at the summer solstice, 22 June (see also Figure 20-11).

The study of the relationship between climatic factors and seasonal biological phenomena is referred to as **phenology.** A significant climatic factor is temperature. Phenological events concerned with reproductive activity in plants may be retarded by several days on a north-facing slope with respect to plants of the same species on an adjacent south-facing slope. Jackson (1966) found significant correlation between air temperature sums and flowering dates. The increased insolation

Figure 14-9   Slope aspect contrasts in a California community mosaic of grasslands, chaparral, and oak woodland. The oak woodland–covered slope in the foreground on the right faces north and the chaparral-covered slope in the foreground on the left faces south.

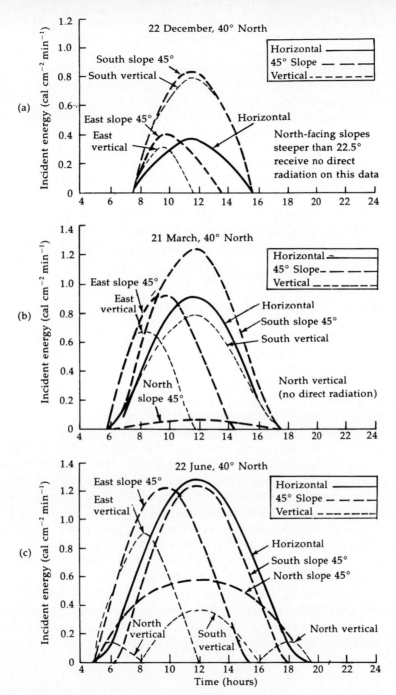

Figure 14-10   The amount of solar radiation incident upon sloping surfaces as a function of time of day at latitude 40°N for (a) the winter solstice, (b) the vernal equinox, and (c) the summer solstice. From D. M. Gates. *Man and His Environment: Climate* (p. 36). Copyright © 1972 by David M. Gates. Reprinted by permission of Harper & Row, Publishers, Inc.

received by south-facing slopes, as compared to nearby north-facing slopes, results in warmer temperatures earlier in spring and therefore earlier flowering.

We might relate time range of flowering dates to latitude (northward) and to elevation (upward). For instance, one day of retardation or advance of flowering might be equivalent to more than 27 km in distance northward, or more than 30 m in elevation. Jackson found flowering variation of an average of 6.0 days between a north-facing slope of a large gorge and the south-facing slope less than 50 m away. Across a small east-west gorge, plants of the same species blooming less than 8 m apart but on opposite sides evidenced a delay of 2.8 days on the north-facing slope as compared to the south-facing one. This could be assigned a difference comparable to 80 km (northward) or 85 m (upward). He found a mean range of flowering dates of 7.2 days for all species studied. This may be interpreted as equivalent to a geographic distance of 200 km northward and 220 m upward (Hopkins 1938), or as 220 km northward and 170 m upward (Jeffree 1960).

Proximity to Water

The temperature of a body of water, especially a large lake or ocean, is much more stable than that of a comparable land mass. This is due to the high specific heat of water, its ability to lose energy by evaporation, its high reflectivity, and the contribution of vertical mixing to stability of temperature. Therefore, proximity to a large lake or ocean buffers the climate for adjacent land areas. The water gives up energy during the winter, warming the nearby area, whose average winter temperature is therefore higher than a comparable area at the same latitude further inland. Conversely, summer temperature will also be more moderate in a marine climatic region as water absorbs energy from the surrounding air. The inland region is said to have a continental climate and experiences a greater amplitude of temperature oscillations than that of a maritime region. The effect of a maritime climatic regime may be considered with regard to dates of last and first killing frost. These data are especially important to agriculturists and horticulturists, as well as plant ecologists. Miller and Thompson (1975) have shown in map form the average dates of the last killing frost in spring and the first killing frost in autumn for the continental United States (Figure 14-11). The effects of a continental versus a maritime climate can be seen, for example, by comparing frost patterns of northern California with Nebraska at about the same latitude.

We can also combine considerations of diurnal variation with the effect of proximity to water. Note the narrow range of temperatures and the minor seasonal differences at a maritime station, San Francisco, California, compared with a more continental station, El Paso, Texas, shown in Figure 14-12.

Cloud Cover

The influence of cloud cover is twofold. Water vapor is opaque to certain wavelengths of solar radiation. A place frequently covered by clouds will not be warmed during the day nearly as much as an area under clear skies. There is, how-

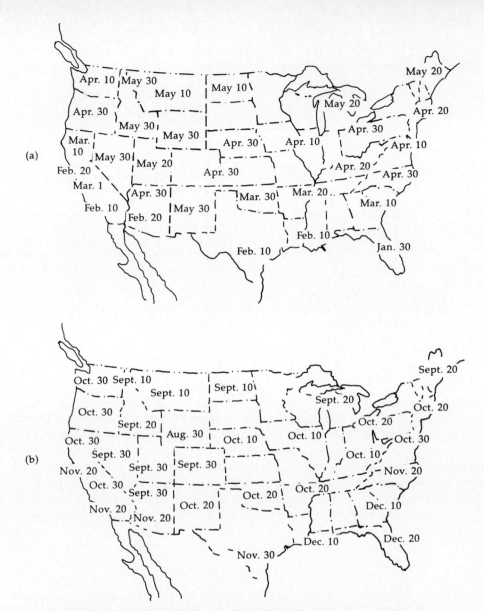

Figure 14-11 (a) Average dates of last killing frost in spring and (b) average dates of first killing frost in autumn. From Miller and Thompson. 1975. *Elements of Meteorology,* 2nd ed. Reprinted by permission of Charles Merrill Publishing Co.

ever, some reflection of sunlight from clouds to the ground and some reradiation heat from vegetation, soil surface, and rocks, so that cloud cover may augment direct sunlight reflected from clouds and may simultaneously warm an alpine slope such that radiation values are as high as 2.2 cal cm^{-2} min^{-1}, which is in excess of

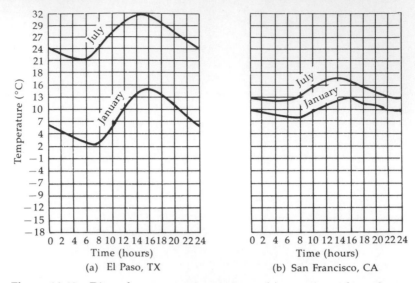

Figure 14-12   Diurnal temperature variation at (a) a continental weather station and (b) a maritime station. From Miller and Thompson. 1975. *Elements of Meteorology*, 2nd ed. Reprinted by permission of Charles Merrill Publishing Co.

the solar constant. The considerations mentioned regarding the moderating influence of liquid water apply also to water vapor. We cannot overemphasize the important role of moisture in the atmosphere in holding and radiating energy. Temperature variations in the tropics vary as little as 2°C on cloudy days but can vary as much as 9°C on sunny days. Without the moderating influence of cloud cover, deserts experience great diurnal extremes of temperature. Also, frosts are more likely on clear nights because of the absence of that moderating influence. In general, moisture in the atmosphere has a stabilizing influence on temperature variations.

Vegetation's Influence on Temperature

The modification of temperature by plant cover is both significant and complex. Shaded ground is cooler during the day than open area, if there is little exchange of air. Vegetation interrupts the laminar flow of air, impeding heat exchange by convection. The sensual impact of a forest is usually one of moist coolness. This is attributable first to the reduction of soil heating due to shading: The soil absorbs energy from the warmer air above it, cooling the forest air. Second, because the forest vegetation transpires as photosynthesis proceeds, the forest microclimate will be more humid than adjacent clearings, and the quantity of energy needed to raise the air temperature is increased. The reradiation of heat is impeded by plant cover, so that the nocturnal forest air and soil are warmer

Table 14-2 Comparison of certain habitat factors in a virgin pine forest and in an adjacent clearing in northern Idaho. Data for the month of August. From J. A. Larsen. "Effect of removal of the virgin white pine stand upon the physical factors of site." *Ecology* 3:302–305. Copyright © 1922 by the Ecological Society of America. Reprinted by permission.

Factor	Forest	Clearing
Air temperature (°C)		
Maximum	25.9	30.0
Minimum	7.4	3.9
Range	18.5	26.1
Mean relative humidity at 5 P.M. (%)	38.8	35.2
Mean daily evaporation[a] (ml)	14.1	36.1
Mean soil temperature at 15 cm (°C)	12.8	17.0
Mean soil moisture at 15 cm (%)	32.0	43.2

[a]Livingston atmometer mounted 15 cm above the ground.

than those of nearby clearings. Temperature oscillations are thus damped both day and night by heavy, continuous plant cover (Table 14-2) just as they are by cloud cover.

Vegetation's Influence on the Light Regime

Vegetation modifies the light environment in different ways, depending on the geometry of the plants, the distribution of biomass, and the spectral properties and orientation of leaves. The modifications due to these factors are variable because of seasonal differences in vegetation and in the elevation of the sun.

Hutchison and Matt (1977) studied the distribution of solar radiation in a tulip poplar (*Liriodendron tulipifera*) forest in Tennessee (Figure 14-13a). The year was divided into **phenoseasons,** each of which represents a period when the solar elevation and phenological state (the state of organism properties that change seasonally, e.g., leaf formation, bud formation, and flowering) of the forest canopy create a unique set of influences on the radiation regime within the forest. Variations above the canopy were due to the phenological state of the canopy and fluctuations in solar radiation (Figure 14-13b). The highest radiant flux reaches the forest floor between the spring leafless phenoseason and early summer, when the canopy is in full leaf (Figure 14-13a). It is during this period that sufficient light and adequate temperature allow the spring bloom of herbaceous plants on the forest floor. The gradient in light reduction during the leafless phenoseasons was much less than during the fully leafed phenoseasons at equivalent solar angle (Figure 14-13b). Other less obvious variations in leaf orientation and canopy structure modify the extinction rate of light as it passes through the forest canopy.

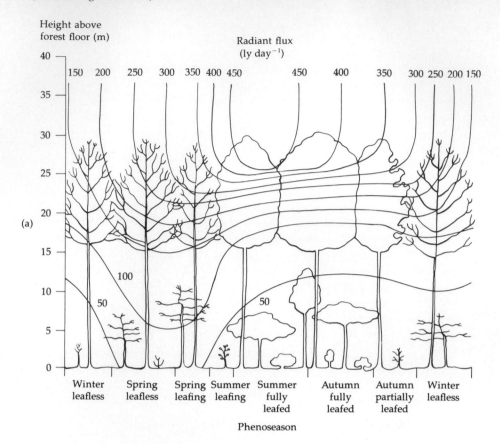

Height above
forest floor (m)

Radiant flux
(ly day^{-1})

Phenoseason

(a)

Winter leafless · Spring leafless · Spring leafing · Summer leafing · Summer fully leafed · Autumn fully leafed · Autumn partially leafed · Winter leafless

Full sunlight radiant flux (1y day^{-1})

| 400 | 475 | 610 | 800 | 905 | 975 | 980 | 940 | 820 | 675 | 530 |

(b)

Jan.   Mar.   May   July   Sept.   Nov.

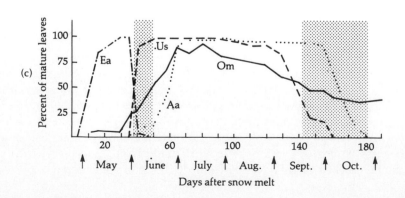

(c)

Percent of mature leaves

Days after snow melt

◀ **Figure 14-13**   Distribution of solar radiation. (a) Light and phenology in deciduous forests: synthesized annual course of average daily total solar radiation received within and above a tulip poplar forest. (b) Approximate daily total of the undepleted solar radiation received on a horizontal surface at 35°N. (c) A comparison of the percentage of leaves in the mature leaf stages of four taxa at various times of the year. Stippled bands indicate expanding or falling leaves in the canopy. (a) From B. A. Hutchison and D. R. Matt. "The distribution of solar radiation within a deciduous forest." *Ecological Monographs* 47:185–207. Copyright © 1977 by the Ecological Society of America. Reprinted by permission. (c) From B. Mahall and F. H. Bormann. *Botanical Gazette* 139:467–481. Copyright © 1978 by University of Chicago Press. By permission.

Mahall and Bormann (1978), in a detailed study of forest herb phenology in New Hampshire, found that the plants could be divided into four groups, depending on phenoseason activity. The spring annual photosynthetic species (Ea in Figure 14-13c) developed and died synchronously, taking advantage of the spring leafless and leafing phenoseasons. Summer green species (Us in Figure 14-13c) developed together in early spring along with the vernal species but remained active through most of the summer. Late-summer species (Aa in Figure 14-13c) developed gradually in summer and died as a group after taking advantage of additional light on the forest floor in late autumn. The last group, which is green from early spring through autumn (Om in Figure 14-13c), can photosynthesize on the low light of the forest floor in summer, enhanced by the periods of more light in spring and autumn. Clearly, light is a significant limiting factor for these forest floor herbs.

The extinction of light as it travels through a vegetation canopy depends on the total leaf area index—the area of leaves projected on a unit area of ground surface. Campbell (1977) considered the theoretical leaf area that would receive direct sunlight as a function of leaf area index (LAI) and leaf orientation (Figure 14-14a). Very little additional sunlight penetrates a canopy of horizontal leaves with a leaf area index greater than 3, but when more leaves have a vertical orientation, more light penetrates canopies of higher LAI. Not all leaves within a canopy or even on an individual plant are at the same inclination. Forest communities have LAIs of 8 or more. Campbell has simulated gross photosynthesis in model canopies assuming that light was the only limiting factor. Figure 14-14b shows the results of LAI values of 1, 3, and 5. Notice that high photosynthesis is possible with large LAI values when leaf orientations are random, but at low LAI, leaf orientation has little effect.

Light is far from uniform either on the forest floor or in the open. Chen and Klinka (1997) measured light extinction from a PPFD of 0.7 mol m^{-2} s^{-1} above the canopy to only 0.05 mol m^{-2} s^{-1} in the understory of a Douglas-fir forest in British Columbia. The average sunfleck durations of PPFD > 0.5 mol m^{-2} s^{-1}, > 0.2 mol m^{-2} s^{-1}, and > 0.05 mol m^{-2} s^{-1} were 8.5, 31.5, and 270.3 min in the understory. In the open the durations were 559.1, 700.7, and 803.3 min, respectively. See Chapter 15 for a discussion of the photosynthetic importance of sunflecks on the forest floor.

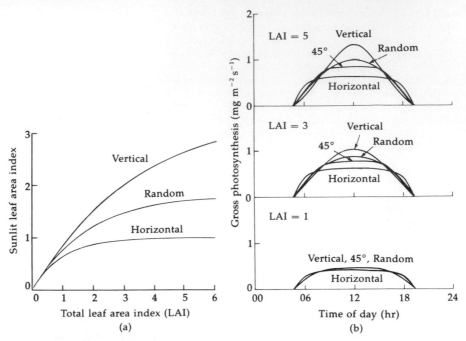

Figure 14-14   (a) Light interception versus leaf area for vertically, randomly, and horizontally oriented leaves. Sunlit leaf area index is found by subtracting the sunlit area below the canopy from 1. (b) Gross photosynthetic rate in model canopies having various leaf areas and leaf distributions as a function of time of day (48°N on 1 June). From G. S. Campbell. 1977. *An Introduction to Environmental Biophysics.* By permission of Springer-Verlag.

Light- and Temperature-Mediated Plant Responses

We have seen how energy is allocated within the community. We have recognized that a balance is needed if a plant, or even a leaf, is to live. Outgoing energy must equal incoming energy, except for radiant energy that is transformed into chemical energy and stored. How is the energy that reaches the plant dealt with so that balance is achieved? We will discuss the concept of energy allocation by the plant in broad terms and then examine some of the phenological events that are governed in part by changes in temperature: events associated with thermoperiodism, dormancy, stratification, vernalization, and sumorization.

If a plant is to survive and reproduce, the timing of various phenologic events must be in response to environmental cues: light duration, light quality, light intensity, moisture availability, various chemicals in solution, gravity, touch, and temperature. Plant responses to many of these stimuli are discussed elsewhere in this book and in general botany and plant physiology books.

Energy Balance Adaptations

How is the right side of the solar energy budget in Equation 14-1 dealt with by plants? Recall that energy is continually being exchanged between a plant and its environment (Figure 14-4). Incident radiation from a variety of sources will be absorbed by a leaf: direct sunlight, scattered skylight, reflected sunlight, and infrared wavelengths. If the air is cooler than a leaf, heat moves away from the leaf. If the air is warmer, the leaf will take on heat.

Evaporation of water (LE) is one of the ways that heat is dissipated from leaves. The evaporation of a gram of water at 28°C requires about 580 cal of energy, a process that cools the leaf in the same way an evaporative cooler cools a room. But the water supply may be limited, and desert plants, for example, cannot depend upon unlimited transpiration for balancing the energy budget. Laminar flow of air removes heat from the leaf if its temperature is warmer than that of the air. The transfer of heat is directly proportional to the square root of the wind velocity and inversely proportional to the square root of the leaf width perpendicular to air flow. In other words, the larger the leaf, the less effective the cooling. Given the same air temperature and wind velocity, a small leaf will lose heat more effectively than a larger leaf. Small leaves, dissected leaves, and lobed leaves are characteristic of plants of arid regions.

Either a very simple system or a stressed one, such as a desert plant community, can provide a model system of heat transfer. Daytime temperatures of desert soils and the air immediately above them are often a great deal higher than that of the air at the standard height above the ground for measuring temperature (1.5 m). For instance, the highest standard air temperature ever measured was 57.8°C (Azizia, Tunisia, on 13 September 1922, and San Luis, Mexico, on 11 August 1933). However, air next to soil surfaces (dark soil, in full summer sun) may exceed 70°C (Gates 1972). As the spring season gives way to early summer, some small desert annuals assume a basket shape, as the rosetted leaves are lifted into the cooler air just a few cm above the soil surface. This adaptation of leaves of desert plants increases chances for heat loss and reduces absorption of heat from the soil (G). Raising leaves may also change the angle of the leaf relative to the sun and thus reduce S.

Desert plants must deal with strong insolation, and often still air. Insolation (S) often raises leaf temperature above that of the surrounding air. If convection (C) and evaporation (LE) cannot dissipate energy as rapidly as it is absorbed, the leaf will store heat and may then have a temperature 10°C, or even 15–20°C, above the ambient temperature. Heat exchange is enhanced by winds, which may remove air to within a few millimeters of the leaf surface. Conversely, heat exchange is hampered in still air, although the layers of warm air above the leaf will tend to rise above cooler air, generating a turbulence that also hastens cooling. Reduction of leaf area, and therefore reduction of leaf exposure to insolation (*S*), is achieved in a variety of ways. Most members of the Cactaceae, for instance, have greatly reduced leaves. A different method for avoiding heat from the ground is seen in members of the Fouquieriaceae, a family with only one genus and with several species found in Mexico and the southwestern United States. Ocotillo (*Fouquieria splendens*) (Figure 20-43) loses its leaves seasonally in response to drought stress, replacing its leaves as many as five times in a single year. Vertical leaf orientation, an adaptation often exhibited by desert shrubs, is another way to avoid excessive *S*.

Several winter and summer desert ephemerals change leaf orientation to track the sun during the day, a phenomenon called heliotropic leaf movement (Ehleringer and Forseth 1980). These annuals also have the capacity to use high levels of solar radiation (*S*) in photosynthesis (Ps). Some species utilize diaheliotropic leaf movements (orientation of the leaf blade perpendicular to the sun's rays) to enhance reception of solar radiation. Paraheliotropic leaf movements (orientation of leaf blade parallel to the sun's rays) are used to reduce transpiration rates and leaf temperature, both adaptive strategies to deal with drought stress (Ehleringer and Forseth 1980).

Two species of winter annuals in Imperial County and Death Valley, California, *Malvastrum rotundifolium* and *Lupinus arizonicus*, have contrasting patterns of heliotropic movements that may relate to the origins of these taxa (Forseth and Ehleringer 1982). *M. rotundifolium* maintains tracking movements over a wider range of leaf water potential, namely, to the wilting point, −4.0 MPa (megapascal), than does *L. arizonicus*. The latter uses paraheliotropic leaf movements at −1.8 MPa and exhibits complete stomatal closure at a higher water potential than does *M. rotundifolium*. The behavior of *M. rotundifolium*, which includes alteration of leaf osmotic potential to maintain turgor, is characterized as *drought tolerance*. This winter annual probably invaded the California flora from desert ancestors to the south. *L. arizonicus* exhibits *drought avoidance* and may be derived from the northern, Arcto-tertiary flora (see also Raven and Axelrod 1978, Mooney and Ehleringer 1978).

Potential enhancement of solar radiation (*S*) received by diaheliotropic leaves and possible drought avoidance by paraheliotropic leaves can be inferred from Figure 14-15.

Reflectivity (*r*) is enhanced by a surface with high albedo. Possession of a white or light-colored leaf surface increases *R*. Desert holly (*Atriplex hymenelytra*) in the Chenopodiceae provides a good example of how one desert shrub species is adapted to increase reflectivity by different means and thereby reduce heat load and drought stress. Unlike the cacti, which have virtually no leaves that are photosynthetic, or ocotillo, which is drought deciduous, desert holly is an evergreen perennial of the hot desert. Mooney et al. (1977) found that this plant changes the characteristics of the leaves during the year. Those leaves produced during the cool season are nearly twice the size of the summer leaves. Furthermore, young leaves are covered with expanded, hydrated salt bladders that collapse as the summer progresses. The salt solution contained in the bladders crystallizes, giving the leaf a white appearance and increasing the reflectance (*r*) of the leaf surface. Leaf reflectance (*r*, measured at 550 nm) is inversely correlated with leaf water content.

Figure 14-15 Photosynthetic photon flux density (PPFD) incident on three leaf types over the course of a midsummer day: a diaheliotropic leaf (cosine of incidence = 1.0) with a fixed leaf angle of 0°, a horizontal leaf, and a paraheliotropic leaf (cosine of incidence = 0.1.) Modified from J. Ehleringer and I. Forseth. *Science* 210:1094–1098. Copyright © 1980 by the American Association for the Advancement of Science.

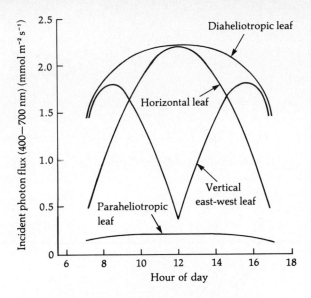

Leaves with 60% leaf moisture exhibit a reflectance of about 75% at some wavelengths. At all wavelengths, their reflectance is greater than that of leaves with 89% leaf moisture.

The leaves of desert holly are held at a steep angle. Field measurements of 50 leaves yielded a mean angle of 70° from the horizontal, with random orientation to the azimuth. Although the angle would seem to be more adaptive in summer than during the very active spring period, it does not change seasonally. This tough shrub is a C_4 plant, but it photosaturates at a rather low light level. Thus, photosynthesis is not reduced, even in spring, by the steep angle. The angle does provide protection by reducing heat load and by allowing greater interception of light during the early morning and late afternoon, which are the times when relative humidity is most favorable (Figure 14-16). Angle of leaf, reduced leaf size, and high reflectance from the leaf surface enhance the capacity of this desert shrub to remain photosynthetically active even during the hottest summer months, adaptations that allow evergreenness.

The most immediate and significant response to high temperature stress in plants is seen in photosynthetic functions (Weis and Berry 1988). Like desert holly, large-leaved evergreen species in summer-dry Mediterranean climates are particularly susceptible to overheating. *Heteromeles arbutifolia,* an evergreen California chaparral shrub, has very steep leaf angles in sun-acclimated plants. Valladares and Pearcy (1997) reported that reorienting sun-acclimated leaves from vertical to horizontal positions caused photoinhibition even in well-watered plants. They observed long-term physiological acclimation to high leaf temperatures and found increased temperature resistance in the photosynthetic processes of water-stressed

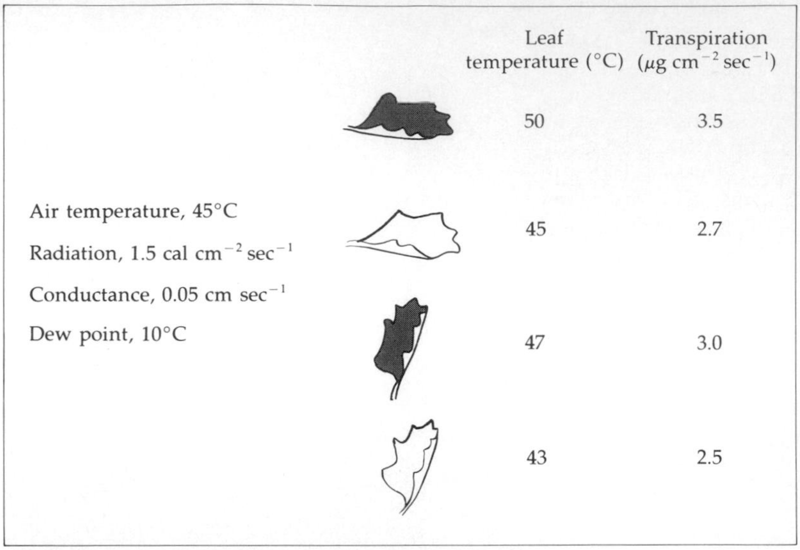

	Leaf temperature (°C)	Transpiration (μg cm^{-2} sec^{-1})
	50	3.5
	45	2.7
	47	3.0
	43	2.5

Air temperature, 45°C

Radiation, 1.5 cal cm^{-2} sec^{-1}

Conductance, 0.05 cm sec^{-1}

Dew point, 10°C

Figure 14-16   Leaf temperature and transpiration of leaves of differing absorptances and angles as determined by the energy equation. The wavelength absorptance (ratio of absorbed to incident radiation) is 0.5 for a typical desert holly (*Atriplex hymenelytra*) leaf (gray) and 0.25 for a summer desert holly leaf (white). Leaf conductance to water vapor transfer was measured on summer leaves. Environmental conditions are those characteristic of Death Valley at midday during the summer. Leaf orientation is horizontal or at 70° as indicated. Leaf size is 3.9 cm^2. From H. Mooney, J. Ehleringer, and O. Björkman. 1977. "The energy balance of leaves of the evergreen desert shrub *Atriplex hymenelytra*." *Oecologia* 29:301–310.

plants. In the cases of *Vitis californica* and *Quercus cerris*, water stress makes the photosynthetic apparatus more susceptible to high temperature damage (Gamon and Pearcy 1990, Valentini et al. 1995). Clearly, the interaction of PPFD, leaf temperature, water stress, and photoinhibition are quite complex and require further study (Mooney 1989).

Success of any taxon, measured by distribution or abundance or both, may be as greatly influenced by lack of heat as by high heat. Nobel (1980a,b, 1982) investigated the relationship between northernmost limits and low-temperature tolerance in certain cacti, particularly as that tolerance relates to morphology, tissue cold sensitivity, and cold hardening. In columnar cacti, minimum surface temperatures evidently occur at the apex. Increasing stem diameter, increasing shading by apical spines, and increasing apical pubescence (a surface with hairs) increase the surface temperature minimum and thereby provide protection against cold. Morphological differences such as these explain the presence of *Ferocactus acanthodes* in colder regions, as compared to *F. wislizenii*, which is restricted to warmer areas. Similarly, *Carnegiea gigantea* extends farther north than does its southern counterpart, *Stenocereus gummosus,* due to the greater protection against cold afforded by morphological differences.

Coryphantha vivipara var. *rosea* has an elevational limit about 600 m higher than does its counterpart, var. *deserti*. Nobel attributes this to differences in tissue cold sensitivity. He also found that certain cacti respond to decreasing day or night air temperatures by cold hardening. The ability to cold harden (and to withstand sub-zero temperatures) allows cacti to be successful in regions of considerable winter-time freezing (Nobel 1982).

Pubescence also helps maintain favorable flowering temperatures in tropical alpine arborescent rosette plants by insulating against radiant heat loss (IR_{emit}). The proportion of bract dry weight contributed by hairs in four alpine species of *Puya* is correlated with elevation and, therefore, decreasing temperatures (Figure 14-17). Miller (1994) removed the hairs from *Puya* inflorescences, causing a

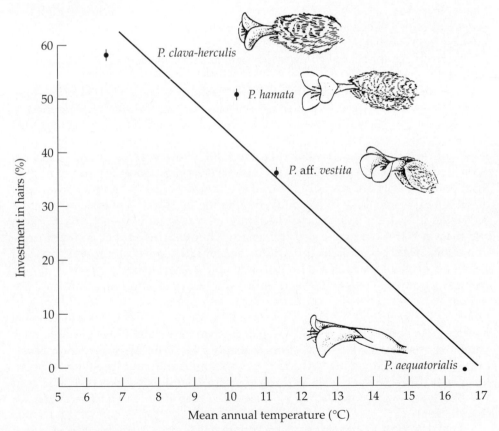

Figure 14-17   Investment (% of dry weight accounted for by hairs) in insulating pubescence of four *Puya* species. Values are the means ± 1 standard deviation. Equation for line: $y = -6.09x + 104.3$; $r = -0.96$. Modified from Miller (1994).

significant increase in IR_{emit}. Inflorescence shape and pubescence orientation may also direct reflected light to increase heating (S_{abs}) during the day.

Response to Ultraviolet-B Radiation

Reports of the Antarctic ozone hole in 1985 (Farman et al. 1985) initiated a flurry of research to determine the consequences of stratospheric ozone depletion on plants. Satellite measurements of stratospheric ozone over Antarctica indicated that levels dropped from >300 DU (Dobson units) in the late 1950s to <100 DU in late 1993 (Kerr 1994). Ozone is the only major atmospheric gas that absorbs significant quantities of ultraviolet radiation at wavelengths of <300 nm. A 10% reduction in ozone results in an approximately 20% increase in UV-B radiation. Until anthropogenic pollutants such as chorofluorocarbons, methane, and nitrous oxide generated these changes, global solar radiation in the UV-B range (280–320 nm) constituted a minor percentage of the total shortwave flux on the surface of the planet. Increases in UV-B radiation are already significant at middle and high latitudes but only small changes are seen in the tropics (Madronich et al. 1995). The increase in UV radiation is of major concern because the short wavelengths are capable of causing deleterious changes in organisms (Dahlback et al. 1989, Middleton and Teramura 1994).

The highest prepollutant level of UV-B was in the tropics. Therefore we might expect to see an increase in protective response in tropical plants, especially in young, sensitive leaves. To the contrary, young leaves on six tropical rain forest species exhibit lower reflectance in the UV-B region of the spectrum than mature leaves of the same species (Lee and Lowry 1980). However, the young leaves are protected in other ways: they contain high levels of anthocyanins and total phenols, both of which are strongly absorptive of deleterious UV radiation. Searles et al. (1995) experimentally reduced UV-B radiation on selected plants and found a reduction in leaf UV-B absorbing chemicals, increased plant height, reduction in leaf mass per area, and increased leaf length. They concluded that even preozone depletion levels of UV-B were harmful to tropical plants. Singh and Agrawal (1996) experimentally increased UV-B radiation and measured responses of four tropical legumes. Photosynthesis declined due especially to reductions in carotenoid pigments. Stomatal conductance was reduced because of loss of turgor in guard cells related to UV-B exposure (see Negash 1987 for details). The tropical legumes also had reduced protective epidermal flavonoids and lower growth rates when exposed to high UV-B radiation. Enhanced UV-B had less impact on development and developmental chemistry of sweetgum trees and seedlings (Dillenburg et al. 1995, Sullivan et al. 1994). Long-term accumulated effects of increased UV-B may be even more damaging than those seen in these short-term studies (Musil 1996).

Leaf UV optical properties were examined along a latitudinal gradient of UV-B radiation (Robberecht and Caldwell 1980). At latitudes with low UV-B radiation, mean epidermal transmittance of UV-B was >5%; in regions with high UV-B radiation, transmittance was <2%. Various pigments, including flavonoids and anthocyanins, were apparently responsible for the lower transmittance, as they absorb strongly in UV-B ranges, transmitting in the visible portion of the spectrum. In all these studies (Lee and Lowry 1980, Robberecht and Caldwell 1980, Teramura 1983), there is evidence that UV-B-absorbing pigments and phenols have adaptive value in plants that grow in regions of high UV-B radiation flux.

Not all plants respond in the same way or in the same magnitude to changes in ultraviolet radiation. We can therefore expect that, just as global warming will have dramatic community effects, one long-term and difficult-to-predict outcome of increasing UV-B radiation will be a shift in the competitive relationships of taxa and a shift in community structure (Yakimchuk and Hoddinott 1994, Musil 1995).

Thermoperiodism

Many plants require a day-night temperature difference for optimal growth. For instance, red fir (*Abies magnifica*) and Jeffrey pine (*Pinus jeffreyi*) respond positively to a thermoperiod of 13°C. A positive response to such a temperature regime is termed thermoperiodism. Hellmers (1966) investigated thermoperiodism in coast redwood (*Sequoia sempervirens*) seedlings and determined that a thermoperiod of 4°C elicits a response. However, that response was not significantly different from the growth of the plant under a constant temperature regime.

The coast redwood does not form dormant buds but can continue to increase in height in response to favorable light and temperature regime. This is in contrast to trees that occupy inland, high-elevation sites, such as Jeffrey pine and red fir. These species do set terminal buds, and this restricts upward growth even though the plant may continue to acquire new biomass. A thermoperiod requirement is adaptive in the pines and firs, since they live in areas of great diurnal temperature fluctuation. The coast redwood does not require a thermoperiod; it does not live in an area with a large diurnal temperature difference.

Thermoperiodism in woody plants has a counterpart in seed response to a fluctuating diurnal temperature regime. Figure 14-18 shows the effect of varying temperature on seed germination for three herbaceous species (Grime and Thompson 1976). *Carex otrubae*, a sedge, requires no temperature fluctuation for germination, but the other two species show a positive response to temperature depression during the dark period. Sorrel (*Rumex sanguineus*) achieved 50% germination at a thermoperiod of 2.5°C, and yellow cress (*Rorippa islandica*) did so at 9°C.

Dormancy

Dormancy is a period of inactivity for seeds or plant organs; it can be broken only when certain environmental requirements have been met. A seed that is supplied with moisture, oxygen, and the proper photoperiod and temperature but still does not germinate is said to be *dormant*. A key feature in the maintenance of seed dormancy is the need for oxygen. Plant tissues are predominantly aerobic, and the exclusion of a large molecule such as oxygen can therefore help to maintain dormancy in seeds. However, the morphology of a dormant bud does not provide for oxygen exclusion. Thus, the correlation between the maintenance of an anaerobic environment and the maintenance of dormancy may not apply to buds.

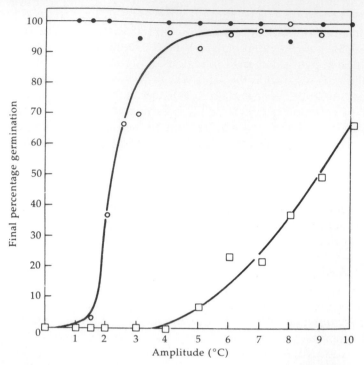

Figure 14-18   The germination response to various amplitudes of diurnal temperature fluctuations in seeds of *Carex otrubae* (●), *Rumex sanguineus* (○), and *Rorippa islandica* (□). The depression in temperature coincided with a dark period of 6 hours. During the photoperiod of 18 hours, all seeds experienced a constant temperature of 22°C. From Grime and Thompson. 1976. *Annals of Botany* 40:795–799.

Dormancy is an adaptive mechanism whereby plants avoid periods of unfavorable weather or periods of high competition. During times of a short photoperiod and a lack of heat or sunlight or liquid water, growth—whether of an established plant or a new one—is very expensive, if not impossible. The protection afforded by the dormant state, especially in extratropical regions, is very significant to the plant, and so dormancy is an adaptive feature. The dormancy of apical buds of many woody plants in temperate climates is initiated by short days interacting with cold temperature. This is a photoperiodic response mediated hormonally.

Environmental cues, such as lengthening photoperiod and increasing temperatures, trigger the resumption of activity in more favorable times. The correct interpretation of these cues is vital to the dormant plant or seed. If a dormant beech, for instance, began to leaf out in response to unseasonably warm weather in early November, subsequent frost would kill or injure the new soft tissues, at great cost to the plant. The environmental stimulus that enables the plant to perceive the passage of winter is cold temperature. Degradation of inhibitors and concomitant biosynthesis of growth-stimulating substances take place during the cold period and early spring, and the seed or plant is then prepared to respond to more favorable weather in an appropriate manner.

Stratification

Stratification is the layering of seeds in a moist medium that is then kept at a low temperature to enhance germination. Most plants growing at higher latitudes and/or altitudes possess seeds that require cold temperatures for germination. That is, the seeds are dormant prior to exposure to cold. Actual germination following release of dormancy may be in response to a variety of environmental cues, including changing photoperiod, increasing heat and water availability, and a changing hormonal regime within the seed. Germination may also rely upon a physical or chemical disruption of the seed coat, a process called scarification.

Vernalization

The flowering of winter rye (*Secale cereale*) and of other cereal plants is influenced by exposure to cold temperature during the time of germination. The time required for flowering of winter rye may be reduced by half, from 14 weeks to 7 weeks, if the seeds are planted in autumn and exposed to winter cold rather than being planted in spring. This postgermination cold exposure, copied from nature and used commercially, is called **vernalization,** from the Latin word for spring, *vernus.* Seeds of winter strains of certain cereal grasses, if kept near freezing temperature during germination, will flower the same summer even though not planted until late spring. We repeat an often-stated but very important qualifier: Even with vernalization, the appropriate photoperiod is necessary for flowering to be initiated.

Sumorization

Heat cracking is necessary to rupture the seed coat of some fire-adapted plants that possess resistant seeds. But what is the influence of temperatures below those of a fire? We mentioned the high temperatures of desert soils. The seeds of most desert annuals are subject to this heat soon after dissemination, and it probably promotes seed maturation in these ephemerals. Capon and Van Asdall (1966) found that germination of eight species of desert annuals native to the Mojave and Sonoran Deserts was enhanced by heat pretreatment. This treatment is called **sumorization,** after the Anglo-Saxon word *sumor,* for "summer."

Maximum germination for the species manipulated by Capon and Van Asdall was reached by the fifth week of storage at 50°C (Table 14-3). Untreated seeds and those stored at 4°C showed poor germination, especially the three Sonoran species tested. Of these, none showed more than 12% germination without heat pretreatment. Seeds of all species failed to germinate after storage at 75°C.

Table 14-3 Percentage germination of seeds of several species of desert annuals in response to temperature pretreatment. From B. Capon and W. Van Asdall. "Heat pretreatment as a means of increasing germination of desert annual seeds." *Ecology* 48:305–306. Copyright © 1966 by the Ecological Society of America. Reprinted by permission.

Species	Un-treated	Storage at 20°C for		Storage at 50°C for					
		4 weeks	8 weeks	1 week	2 weeks	3 weeks	4 weeks	5 weeks	10 weeks
Mojave Desert spp.									
Coreopsis bigelovii	9	14	30	26	16	24	20	32	2
Eriophyllum wallacei	5	16	24	32	48	30	18	36	2
Euphorbia polycarpa	1	1	2	2	10	12	4	2	0
Geraea canescens	14	14	16	60	74	32	34	48	4
Salvia columbariae	6	4	14	80	26	28	14	16	10
Sonoran Desert spp.									
Lepidium lasiocarpum	0	2	4	53	90	72	61	54	40
Sisymbrium altissimum	10	10	12	12	13	92	73	71	48
Streptanthus arizonicus	0	3	8	21	23	29	50	50	30
Plantago insularis (not leached)	0	0	0	0	0	0	0	0	0
Plantago insularis (leached)	100	—	—	100	—	—	—	—	—

Summary

Solar radiation is one of the most important aspects of the plant's environment. Energy that fuels this planet comes to us as solar radiation, including ultraviolet light, visible light, and infrared light. Earth is shielded from most of the UV and IR wavelengths by its atmosphere. Recent pollutant accumulations have damaged stratospheric ozone and increased penetration of UV light.

All objects absorb and transmit or reflect radiation. Energy gains and losses must balance within a leaf, an organism, and the biosphere itself. Heat may be transferred by radiation, convection, or conduction.

A solar energy budget may be symbolized as

$$S = R + C + G + Ps + LE \qquad \text{(Equation 14-1)}$$

where S = solar radiation; R = reflected plus reradiated; C = convection; G = heating of solid objects; Ps = photosynthesis; and LE = latent heat of evaporation. The solar constant (energy reaching the earth's outer atmosphere) is 1.94 cal cm^{-2} min^{-1}. Locally, radiation balance may be diurnally or seasonally negative, or may be always positive, as in some tropical regions. Energy distribution within a tropical deciduous forest differs dramatically from that of, for instance, a desert region or a wet mountain meadow, due to plant cover, amount of open soil, and standing water.

The forms of radiant energy that are important to plants include direct and indirect and radiation reflected from other objects. Most radiation received is direct. Leaves absorb incident solar radiation primarily in the visible and UV wavelengths. Visible light is absorbed by chlorophyll and carotenes, leaving little visible light under the canopy of a dense forest. There is a latitudinal gradient for UV-B irradiance in the arctic alpine zone. Wavelengths active in photosynthesis are between 400 and 700 nm (PPFD).

Radiation is monitored by means of chemical integrators, pyranometers, pyrheliometers, and photographic devices.

Factors that influence variation in surface temperature include latitude, altitude, topography and slope aspect, proximity to water, cloud cover, and vegetation. Mean annual temperatures decline poleward, but diurnal and seasonal fluxes increase. Air cools at a certain lapse rate as it moves upslope. Plant cover and cloud cover dampen temperature oscillations locally. South-facing slopes will be warmer, in general, than nearby north-facing slopes in northern temperate latitudes, the situation being reversed in southern temperate latitudes. Phenological events may be retarded on the cooler slope with respect to the warmer.

A plant allocates energy to maintain a balance, losing heat by convection, reradiation, reflectivity, and evaporation of water. Desert ephemerals exhibit heliotropic leaf movements for better irradiance or for drought avoidance. Small leaves aid in loss by convection; white or light-colored leaves enhance reflectivity. Plants in cold regions exhibit morphological and physiological protection against cold. Thermoperiodism, a day-night temperature differential required for optimal growth, characterizes certain plants. Some seeds also exhibit a need for temperature fluctuation for germination. Onset of dormancy is triggered, in part, by cold temperature, and cold is used as an environmental cue by plants to perceive the passage of winter. Exposure to cold enhances germination, as well as early flowering, in some seeds and plants. Heat pretreatment enhances germination of some desert annuals.

CHAPTER 15

PHOTOSYNTHESIS

Photosynthetic systems are extraordinarily unique: no other basic biochemical pathways in higher organisms demonstrate such dramatic diversity, and the relationship between photosynthetic pathway and ecological conditions is perhaps the most elegant story in all of physiological ecology.

Recent predictions of global warming related to accumulating greenhouse gases have brought photosynthetic research to the forefront: plants are major players in global carbon dioxide fluctuations. Global simulation models depend heavily on accurate projections of photosynthesis and respiration rates of plants in the face of increasing global temperature and CO_2. Primary productivity is basically ecosystem-level photosynthesis. Many of the productivity patterns discussed in Chapter 12 have parallels in the physiological ecology of individuals discussed in this chapter. Accurate predictions of plant response to global changes are still elusive. Exciting areas of research are opening up in the ecophysiology of photosynthesis. Understanding photosynthesis depends on a better comprehension of (a) metabolic processes at the molecular level, (b) organismic-level control and communication between energy-related processes and plant hormones, and (c) extrapolation of physiological processes to landscape and global levels (Schulze and Caldwell 1995).

Photosynthetic Pathways

Our understanding of the photosynthetic process and the ecological ramifications of various photosynthetic adaptations are essential aspects of plant ecology. Most higher plants perform photosynthesis only by the C_3 pathway. However, a few use the C_4 photosynthetic pathway discovered by Kortshak et al. (1965) in sugarcane and later described in detail by Hatch and Slack (1966). This discovery stimulated a flurry of research on aspects of the C_4 pathway, from biochemistry to the population and community levels. The photosynthetic properties of many plants have been characterized but many more remain to be studied. It is possible that additional pathways will be discovered. Certainly, as research proceeds, we will further modify our concepts of the ecological importance of photosynthetic variations.

Our goal here is not to describe the detailed aspects of the various photosynthetic processes but simply to focus attention on the characteristics of photosynthesis that are most likely to be of ecological significance. If you wish to have a more complete review of these processes, most recent textbooks in plant physiology and biochemistry have detailed information.

The photosynthetic process is divided into light-dependent reactions and light-independent (dark) reactions. Light-dependent reactions convert light energy into chemical energy and are common to photosynthetic higher land plants. Light-independent reactions convert carbon dioxide into sugars and starches. In the light-dependent reactions, 8–12 photons of light are absorbed by pigment systems within the chloroplasts (Figure 15-1). There the light energy is converted into

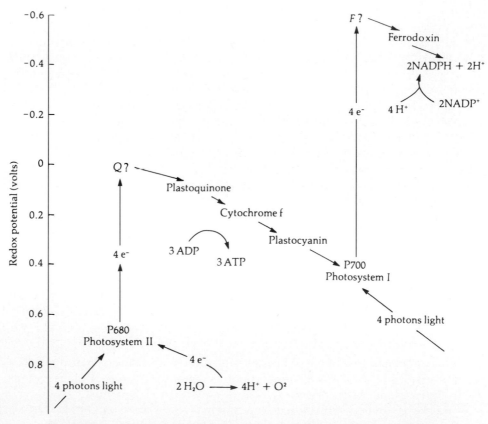

Figure 15-1   Overview of the light reactions of photosynthesis. *Q* and *F* are unidentified electron carriers.

chemical energy in the form of ATP and NADPH (reduced NADP), which are the sources of energy for the light-independent reactions. Briefly, the process is as follows. Water is oxidized and is a source of hydrogen to reduce NADP as well as the source of oxygen released during photosynthesis. ATP is also formed in this process from ADP and inorganic phosphate.

$$2H_2O + 2NADP^+ + 3ADP^{-2} + 3H_2PO_4 \xrightarrow{8-12 \text{ photons}} 2NADPH + O_2 + 2H^+ +$$
$$3ATP^{-3} + 3H_2O)$$

C3 Photosynthesis and Photorespiration

Prior to the discoveries of Hatch and Slack and Kortshak et al., it was thought that plants fixed CO_2 only by the **photosynthetic carbon reduction (PCR) cycle** or Calvin cycle. This type of photosynthesis is often referred to as C_3 *photosynthesis* because the first stable product formed is phosphoglyceric acid (PGA), which has a skeleton of 3 atoms of carbon. A series of reactions follow (Figure 15-2a) in which sugars and starch are formed as the products of photosynthesis. The energy necessary for the incorporation of atmospheric CO_2 into photosynthate is obtained from the ATP and reduced NADP formed in the light reactions. Additional ATP from the light reactions provides energy and phosphate to regenerate ribulose bisphosphate (RuBP, a 5-carbon sugar), which then can combine with CO_2 to form PGA. This process takes place during the day within mesophyll cells and in cortical cells of stems that contain chloroplasts. Anatomically the leaves of C_3 plants are typically divided into palisade and spongy layers, with chloroplasts present in both (Figure 15-3a).

Coupled with the PCR cycle is a series of reactions leading to the light-stimulated release of CO_2 known as **photorespiration.** Photorespiration is not the same as true cellular respiration, and both processes may go on independently of each other in the same cell at the same time. True cellular respiration takes place exclusively in mitochondria, oxidizing carbohydrates (such as glucose) and generating energy in the form of ATP and NADH. Photorespiration oxidizes organic acids, using up oxygen and releasing carbon dioxide, but it requires more energy than it makes available. From a photosynthetic standpoint, photorespiration is wasteful of the plant's resources; it reduces the rate of net photosynthesis by 30–50% (Goldsworthy 1976, Downes and Hesketh 1968). However, there may be advantages in other aspects of the plant's physiology that offset the reduction in photosynthesis. For example, in transgenic tobacco, photorespiration levels correlate with the plant's ability to avoid photosynthetic inhibition caused by excess light energy (Kozaki and Takeba 1996). The ability to avoid damage in high light environments may be a significant benefit of photorespiration. However, because photorespiration continues when excess energy does not need to be dissipated, it is difficult to interpret photorespiration as an adaptation for energy dissipation (Bjorkman and Demmig-Adams 1995).

The level of photorespiration in plants depends on the ratio of carbon dioxide to oxygen at the site of the PCR reactions. In the presence of today's atmospheric levels of oxygen (21%) and carbon dioxide (0.036%), the enzyme responsible for

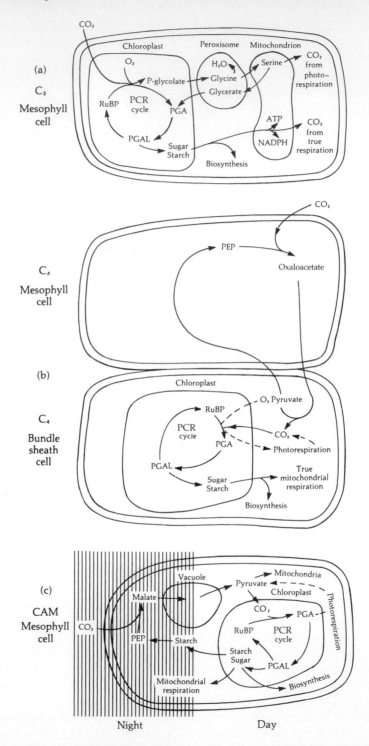

Figure 15-2 Basic light-independent reactions and product fate of (a) C_3, (b) C_4, and (c) CAM photosynthesis.

Figure 15-3 Cross sections of (a) C_3, (b) C_4 (Kranz anatomy), and (c) CAM photosynthetic tissues. P = palisades mesophyll; S = spongy mesophyll; M = mesophyll; B = bundle sheath.

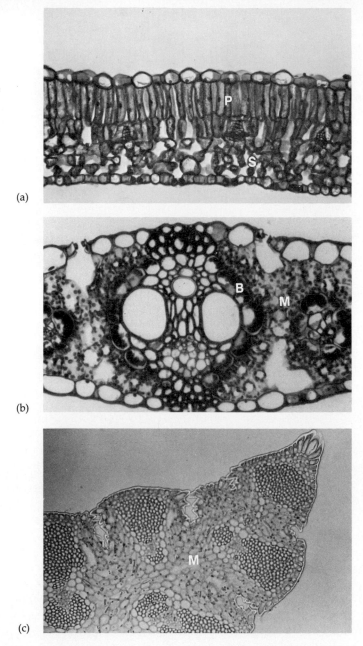

(a)

(b)

(c)

fixation of carbon dioxide in photosynthesis (RuBP carboxylase) may fix oxygen instead of CO_2 to the receptor RuBP molecule (Figure 15-4). Thus, oxygen and carbon dioxide compete for the same enzyme but the enzyme has a greater affinity for CO_2. For convenience, the enzyme is called RuBP carboxylase when it fixes carbon dioxide and RuBP oxygenase when it fixes oxygen, but it is the same enzyme. When oxygen is fixed instead of CO_2, only half as much phosphoglycerate (PGA) is ini-

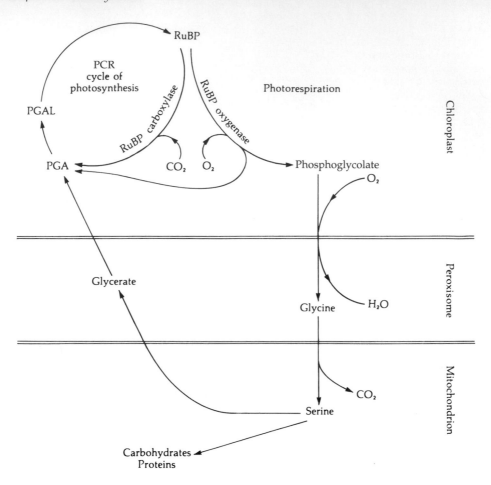

Figure 15-4   The relationship between photosynthesis and photorespiration.

tially produced; the remaining carbohydrate is converted in steps to phosphoglycolate. Through a series of steps in the peroxisome that require energy and oxygen,
phosphoglycolate is converted to the amino acid glycine. Glycine is then transported to the mitochondria where two molecules of glycine are combined into one
molecule of serine, and carbon dioxide is released. Serine can, with additional consumption of energy, be chemically altered back to PGA and transported back to the
chloroplast. Thus, some glucose is produced from the products of photorespiration, but the energy gained does not balance that lost through photorespiration.
The inhibition of photosynthesis by the fixation of O_2 is related to two factors: first,
competition between O_2 and CO_2 reduces the amount of carbon fixed; second, CO_2
is lost during photorespiration.

C4 Photosynthesis

In the C_4 **pathway,** carbon dioxide is fixed by the enzyme PEP carboxylase to form oxaloacetate (a four-carbon acid) in the mesophyll cells (Figure 15-2b). Oxaloacetate then diffuses via plasmodesmata from the mesophyll cell into a bundle sheath cell, where CO_2 is released to enter the PCR cycle (Hatch and Osmond 1976). The three-carbon carrier molecule (pyruvate) returns to the mesophyll cell, where it is converted to PEP to receive another CO_2 molecule. This mesophyll–bundle sheath shuttle thus concentrates CO_2 at the C_3 fixation site. Photosynthesis proceeds within the bundle sheath cells just as it does in the mesophyll cells of C_3 plants. The special anatomical characters associated with these processes are referred to as **Kranz anatomy** (Figure 15-3b). Kranz (wreathlike) anatomy is characterized by the presence of a dense layer of large, thick-walled, chloroplast-containing cells that form a sheath around the vascular bundles and by a lack of differentiation of the mesophyll into recognizable palisade and spongy layers. Mesophyll cells are often arranged cylindrically around the bundle sheath. Kranz anatomy is easily seen with a hand lens in sections of fresh material. Plants with C_4 photosynthesis can be identified in the field by this means.

The oxygen inhibition and associated photorespiration noted in C_3 plants is not a problem in plants with C_4 photosynthesis. PEP has a far greater affinity for CO_2 than RuBP. Also, the lack of oxygen inhibition is apparently due to the high concentration of CO_2 at the site of RuBP carboxylase—up to 10 times higher than atmospheric concentration. This higher CO_2/O_2 ratio almost eliminates RuBP oxygenase activity and allows C_4 plants to reduce the concentration of CO_2 to nearly zero. CO_2 released by photorespiration is normally refixed by PEP carboxylase in the mesophyll cells. The CO_2 compensation point (the point at which the concentration of CO_2 is sufficient to support net photosynthesis) has been used to distinguish between C_3 and C_4 plants (Downton and Tregunna 1968).

Crassulacean Acid Metabolism (CAM)

Many succulent plants in some 20 families (e.g., Agavaceae, Orchidaceae, Crassulaceae, and Cactaceae) exhibit a carbon fixation scheme referred to as **crassulacean acid metabolism (CAM).** All CAM plants have succulent photosynthetic tissues but lack the specialized bundle sheath cells of C_4 species (Figure 15-2c and 15-3c). At night, when the stomates of CAM plants are open, carbon dioxide is fixed by PEP carboxylase, forming malic or isocitric acid. In contrast, the stomates of C_3 and C_4 species are open during the day. In CAM photosynthesis, acids are accumulated at night within the extraordinarily large cell vacuoles. This causes a pattern of low nighttime pH followed by increasing pH during the day when photosynthesis proceeds by the PCR cycle. During the day when stomates are closed, CO_2 from the organic acids is released, then fixed by RuBP carboxylase to enter the PCR cycle. The elevated CO_2 concentrations within the cell probably counteract the inhibitory effect of oxygen by photorespiration (Osmond 1976). Thus temporal separation of CO_2 uptake and fixation by RuBP carboxylase during the light and dark periods results in the same benefits as the spatial separation of CO_2 uptake and fixation by RuBP in C_4 plants. Not all plants that have these characteristics are obligate CAM plants.

There is considerable variation in the dependence of plants on CAM photosynthesis. Some species with CAM rely on the CAM machinery for essentially all CO_2 uptake. Even these so-called obligate CAM plants have periods of C_3-type CO_2 fixation in early morning and in late afternoon, especially when environmental conditions are optimal. Most cacti, for example, are considered obligate CAM plants. Under extreme stress, CAM plants can enter a state where there is no external CO_2 exchange but they continue to recycle internal CO_2 and show diurnal fluctuations in acidity. Others are facultative CAM plants that may shift from the C_3 mode of CO_2 uptake to the CAM mode as conditions become more stressful. Age is sometimes a determinant of photosynthetic mode in plants. For example, *Mesembryanthemum crystallinum* is a facultative CAM plant in younger phases but becomes an obligate CAM plant later in the life cycle. These species apparently take advantage of moderate environmental conditions by shifting to the C_3 mode of photosynthesis and maintain nighttime CO_2 fixation in more stressful conditions. Examples of photosynthetic response to environmental conditions are covered later in this chapter.

Welwitschia mirabilis is a unique succulent gymnosperm that grows in the deserts of southern Africa. The literature is not clear concerning the mode of photosynthesis of *Welwitschia;* some evidence suggests CAM and other suggests C_3 (Dittrich and Huber 1974, Schulze et al. 1976, von Willert et al. 1982). Experimental studies suggest that *Welwitschia* has the capacity for CAM but apparently does not fix atmospheric CO_2 at night. Even under stress, plants grown in a greenhouse or growth chamber do not shift to nighttime CO_2 uptake but do show significant diurnal fluctuation in organic acid content of the leaf tissue. This phenomenon has been referred to as **cycling.** Carbon dioxide released from respiration during the night is fixed by PEP carboxylase into primarily malic acid and stored in vacuoles where it is released during the day to enter the PCR cycle. Thus *Welwitschia* has the biochemical capacity for CAM but does not open stomates at night; in the strict sense, it is not a CAM plant (Ting and Burk 1983). It is possible that cycling was an early step in the evolution of true CAM.

Carbon Isotope Discrimination

The ^{13}C and ^{12}C isotopes of carbon occur in a constant ratio as carbon dioxide in unpolluted air. One would expect that plants would fix carbon isotopes in the same ratio as they occur naturally. However, both carboxylating enzymes (PEP carboxylase and RuBP carboxylase) discriminate against $^{13}CO_2$, resulting in a fractionation of atmospheric carbon in the tissues (Smith and Epstein 1971).

Measurements of tissue or air ^{13}C composition are standardized by expressing the ratio $^{13}C/^{12}C$ relative to a fossilized carbonate skeleton of the cephalopod *Belemnitella* as the ^{13}C index ($\delta^{13}C\text{‰}$):

$$\delta^{13}C\text{\textperthousand} = \left(\frac{^{13}C/^{12}C \text{ sample}}{^{13}C/^{12}C \text{ standard}} - 1 \right) \cdot 1000 \qquad \text{(Equation 15-1)}$$

The $^{13}C/^{12}C$ ratio can be obtained by burning a tissue sample and subjecting the captured gas to analysis by mass spectrometer.

The carboxylating enzymes (RuBP carboxylase and PEP carboxylase) differ in the degree of discrimination against ^{13}C. This provides a means of identifying the active enzyme by monitoring $\delta^{13}C\text{\textperthousand}$ in plant tissues. The ^{13}C composition of unpolluted air is $\delta^{13}C\text{\textperthousand} = -7\text{\textperthousand}$ (Keeling et al. 1979). On the average, C_3 plants, which fix CO_2 with RuBP carboxylase, have a $\delta^{13}C\text{\textperthousand}$ of $-27\text{\textperthousand}$. C_3 plants, therefore, have a ^{13}C composition of $-20\text{\textperthousand}$ (2%) less than air. C_4 plants are all close to $-13\text{\textperthousand}$. The range for C_3 plants ($-22\text{\textperthousand}$ to $-35\text{\textperthousand}$) does not overlap with the range for C_4 plants; therefore, we can distinguish plants with C_3 photosynthesis from those with C_4 by establishing the ^{13}C-index ratio. Details concerning the mechanism of discrimination can be found in Farquhar et al. (1989), Farquhar (1983), Farquhar et al. (1982), Berry and Farquhar (1978).

The ^{13}C index for CAM plants varies into the range of both C_3 and C_4 depending on whether a plant fixes more CO_2 in the C_3 mode or in the C_4 mode (Ting and Gibbs 1982). CAM plants fix CO_2 with PEP carboxylase when stomates are open at night and with RuBP carboxylase when stomates are open during the day (O'Leary and Osmond 1980). *Kalanchoe fedtschenkoi* changed its ^{13}C-index ratio from $-16\text{\textperthousand}$ to $-33.3\text{\textperthousand}$ depending on conditions (Bender et al. 1973).

Evolution of Photosynthetic Pathways

Throughout much of the earth's history, C_3 photosynthesis has been the predominant photosynthetic pathway. C_4 and CAM photosynthesis evolved from C_3 plants many times and in several major groups of plants (Ehleringer and Monson 1993). Over a dozen families of angiosperms fix carbon dioxide by the C_4 dicarboxylic acid pathway. Lists of C_4 species have been published in several places (e.g., Downton 1975, Krenzer et al. 1975, Smith and Epstein 1971, Teeri and Stowe 1976, Mulroy and Rundel 1977, Ehleringer et al. 1997), and the lists keep growing as more C_4 species are discovered.

Paleoecologists have determined the ^{13}C index for more than 100 fossils of different taxa from the mid-Triassic (about 230 million years ago) to late Tertiary (about 2 million years ago). The index for all identified taxa was in the range of C_3 plants. Two unidentifiable fragments of fossil tissue showing values characteristic of C_4 or CAM plants were found in Cretaceous fossils that lived when angiosperm expansion was under way (Bocherens et al. 1993).

C_4 plants are responsible for about 18% of total global productivity. Most of the C_4 production is by monocots in grasslands. The importance and productivity of C_4 monocots is highest in warm climates and declines poleward from about 43°N in the Great Plains (Tieszen et al. 1997). C_4 dicots are rare, reaching greatest importance in arid climates with summer rainfall such as the Sonoran Desert.

Holm et al. (1977) observed that 14 of the 18 most noxious and aggressive of the earth's summer weeds are C_4 dicots. The ecological, geographical, and paleoecological distributions of C_4 plants are reviewed by Ehleringer et al. (1997).

A major selective advantage of C_4 and CAM photosynthesis is the ability to gain CO_2 in CO_2-limited environments. Atmospheric carbon dioxide was very low around 300 million years ago during the mid-Paleozoic (Figure 15-5; Berner 1997). Even though there is no direct evidence that C_4 or CAM evolved that early, this was the beginning of low atmospheric carbon dioxide conditions where the C_4 and CAM plants would have higher photosynthetic efficiency than C_3 species. Carbon dioxide limitations occur daily in aquatic systems dominated by C_3 plants. Keeley and Busch (1984) and Keeley et al. (1994) suggest the aquatic primitive vascular plants in the genus *Isoetes* evolved CAM characteristics as adaptations to low daytime CO_2 concentrations in its aquatic habitat. Similar daytime CO_2 limitation occurs in plants of hot, arid environments that reduce internal CO_2 concentrations to avoid daytime water loss. Avoiding daytime CO_2 limitation by adopting CAM or C_4 would have also given the plants the secondary benefit of more efficient water use.

Table 15-1 summarizes and compares 17 traits of C_3, C_4, and CAM photosynthetic pathways.

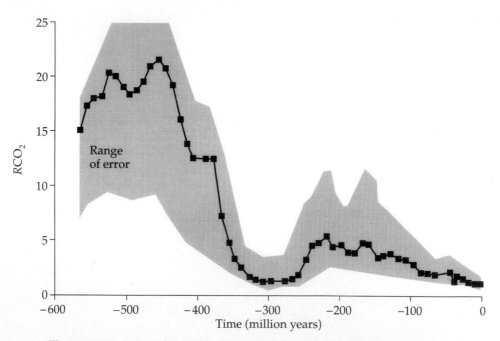

Figure 15-5   Atmospheric CO_2 versus time for the past 550 million years. RCO_2 is the ratio of the mass of CO_2 in the atmosphere at some time in the past to that at present. The shaded area encloses the approximate range of error of the modeling. Modified from Berner, R. A. The rise of plants and their effect on weathering and atmospheric CO_2. Science 276:544–546. Copyright 1997 American Association for the Advancement of Science.

Table 15-1 Comparison of photosynthetic pathways. Modified with permission from the *Annual Reviews of Plant Physiology,* vols. 24 and 27. Copyright © 1973 and 1976 by Annual Reviews, Inc.

Trait	C_3 heliophyte (adapted to a high-light environment)	C_4	CAM
Taxonomic diversity	Very wide: algae to higher plants	No algae, lower vascular plants, or conifers; wide among flowering plants	Some species in about 20 families of flowering plants + *Welwitschia*
Typical habitat	No pattern	Open, warm, saline (some exceptions)	Open, warm, saline (sometimes cool)
Leaf anatomy	Palisade + spongy parenchyma	No mesophyll differentiation; large bundle sheath; Kranz	No mesophyll differentiation; big cells with large vacuoles
Light saturation point ($mmol\ m^{-2}\ s^{-1}$)	0.6–1.2	1.6–2.0	Like C_3 (?)
Light use efficiency (mol CO_2 fixed per mol photon absorbed)	Higher than C_4 at leaf temperatures < 25–$30°C$	Higher than C_3 at leaf temperatures > 25–$30°C$	?
Optimum temperature	20–30°C (lower in tundra)	30–45°C (as for light saturation, can be lower for C_4 species in different habitats)	30–35°C for CAM mode; lower for C_3 mode
Maximum photosynthetic rate:			3 (maximum reported = 13)
mg $dm^{-2}\ hr^{-1}$	30	60	
mg $g^{-1}\ hr^{-1}$	55	100	1 or less
Maximum growth rate (g $dm^{-2}\ day^{-1}$)	1	4	0.02
Water use efficiency (g $CO_2\ kg^{-1}\ H_2O$)	1–3	2–5	10–40
Photorespiration	High	Low	Low
Na required?	No	Yes	No (but salts stimulate CAM mode)
Fixation path and enzyme	$CO_2 + 5\text{-}C \rightarrow 3\text{-}C$ PGA; carboxydismutase or also called ribulose bisphosphate carboxylase	$CO_2 + 3\text{-}C \rightarrow 4\text{-}C$ acids; PEP carboxylase	Still some debate; possibly just as C_4 but enzyme is light-inhibited, thus may be structurally different

(continued)

Table 15-1   (*Continued*)

Trait	C₃ heliophyte (adapted to a high-light environment)	C₄	CAM
Stomate behavior	Open in day, closed at night	Open in day, closed at night	Closed in day, open at night (unless environment shifts plant to C_3-like mode)
Space-time relations	Entire PCR cycle in any mesophyll cell	Initial fixation in mesophyll, then transfer of acid to bundle sheath for PCR cycle	Initial fixation at night in any mesophyll cell; storage of acid in vacuole; PCR cycle during day
Effect of environment on pathway	None	None	Moist, warm night temperature and long day length put plant in C_3 mode
CO_2 compensation point	50 ppm	5 ppm	2 ppm (in dark)
$\delta^{13}C$	−24 to −34‰	−10 to −20‰	Possibly intermediate, although mainly like C_4

Environmental Factors and Photosynthetic Response

Gas Diffusion

The maximum potential rate of CO_2 uptake and the efficiency with which available CO_2 can be fixed are qualities of great ecological importance. Photosynthesis of plants is sometimes limited by the concentration of CO_2 at the site of fixation. Therefore, the rate of diffusion of CO_2 through the stomates and into the plant is a critical factor. Because water leaves the plant when the stomates are open, the balance between water loss and CO_2 uptake is important in determining the relative success of terrestrial plants. The pathway of CO_2 and water diffusion into and out of a leaf may be visualized as a series of steps, any of which can be limiting. The application of this concept to water diffusion is considered in Chapter 18; here we consider the process of CO_2 diffusion.

The steps in the movement of CO_2 can be visualized as a series of resistances and potential gradients defined by a restatement of Fick's law for gas diffusion:

$$J_{CO_2} = \frac{\Delta c}{\Sigma r}$$

(Equation 15-2)

where CO_2 flux (or the flux of any diffusing substance $[J_x]$) depends on the change in concentration from the source to the reaction site (Δc) and the sum of the resistances (Σr) to diffusion. Figure 15-6 depicts the pathway along which CO_2 diffuses from external air to the reaction site within a chloroplast-containing cell. The concentrations and resistances dictate the rate of CO_2 assimilation. Therefore, the consideration of relative values is important in understanding photosynthetic adaptations.

Boundary layer resistance (r^{bl}) is encountered near the leaf surface where a zone of still air may become depleted of CO_2 and through which CO_2 must move by diffusion rather than by turbulent mass transfer as it does further from the surface. **Stomatal resistance** (r^{st}) restricts CO_2 diffusion at the leaf surface. **Cuticular resistance** (resistance imposed by the waxy cuticle covering stems and leaves) has been shown to be so great in most mature leaves that we may assume that essentially all of the CO_2 that enters a leaf diffuses through the stomates. **Intercellular**

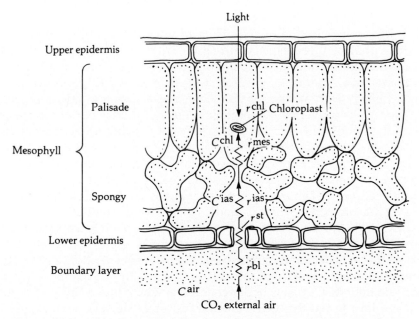

Figure 15-6   Transport pathway, transport resistances, and concentration gradients in a photosynthesizing leaf. See text for definition of terms.

air space resistance (r^{ias}) is resistance to diffusion in the gaseous state inside the leaf. The sum of r^{bl}, r^{st}, and r^{ias} is called **leaf resistance.** Resistance encountered from where CO_2 leaves the gas phase in air and is dissolved in the cytoplasm of the photosynthesizing cell and moves to the chloroplast is referred to as **mesophyll resistance** (r^{mes}). Limitations placed on CO_2 fixation by the photosynthetic process itself represent the final resistance to CO_2 transport. These limitations collectively are called **chloroplast resistance** (r^{chl}). Mesophyll and chloroplast resistance can be equal to or greater than stomatal resistance, depending on the plant and environmental conditions.

The concentration of CO_2 in the atmosphere (c^{air}), in intercellular spaces (c^{ias}), and at the chloroplast surface within the mesophyll cells (c^{chl}), in conjunction with the various resistances, determines CO_2 flux. In a rapidly photosynthesizing cell, CO_2 concentrations at the chloroplast (c^{chl}) are quite low relative to c^{air}, and so a gradient exists. This gradient becomes greater toward the outside of the leaf where average CO_2 concentrations are approximately 350 ppm. Carbon dioxide released by photorespiration and true mitochondrial respiration adds to the CO_2 concentration within the leaf. The concentration becomes high enough at night to reverse the gradient (except in CAM plants), and CO_2 passes into the atmosphere.

Stomatal aperture (and therefore stomatal resistance) appears to be responsive to light, intercellular CO_2 concentration, temperature, atmospheric humidity, and leaf water status. Stomatal aperture responds to these factors by active transport of ions into and out of the guard cells, thereby modifying turgidity by changing osmotic concentrations within the cell. For a detailed account of the mechanisms and causes of stomatal response see Kappen et al. (1995), Jarvis and Mansfield (1981), Hall et al. (1976), or Raschke (1976).

The CO_2 compensation point is reached when CO_2 concentration is at the level where photosynthesis equals respiration. Different plants have different abilities for utilizing CO_2 at low concentrations and therefore different CO_2 compensation points. In C_4 plants, for example, CO_2 is fixed by PEP carboxylase, which has an extremely high affinity for CO_2, and thus a very low CO_2 compensation point. The CO_2 is converted into organic acids that diffuse out of the mesophyll cells into the bundle sheath cells. This creates a CO_2 sink in the mesophyll cells that establishes a steeper concentration gradient from atmosphere to mesophyll cell, allowing C_4 plants to reduce the concentration of CO_2 in mesophyll to near zero, whereas C_3 plants reach their CO_2 compensation point at much higher levels (Table 15-1). Figure 15-7 shows the results of studies conducted by the Carnegie Institution of Washington in which C_3 and C_4 plants were subjected to various intercellular CO_2 concentrations (c^{ias}). The ability of the C_4 plant to maintain a positive photosynthetic rate at near-zero CO_2 concentration is apparent. It is also interesting to note that at high (about 700 ppm) intercellular concentrations of CO_2, C_3 and C_4 plants had nearly identical photosynthetic rates. Thus, C_4 plants have a higher fixation rate than do C_3 plants under conditions where intercellular CO_2 concentrations are low. Recall that high stomatal resistance decreases water loss while also reducing intercellular concentrations of CO_2. Because C_4 plants have a lower CO_2 compensation point, they can continue positive net photosynthesis at high

Figure 15-7 Comparison of the photosynthetic reactions of *Atriplex glabriuscula*, a C_3 plant, and *Atriplex sabulosa*, a C_4 plant, to various intercellular CO_2 levels. Modified from Björkman et al. 1975. Courtesy of the Carnegie Institution of Washington.

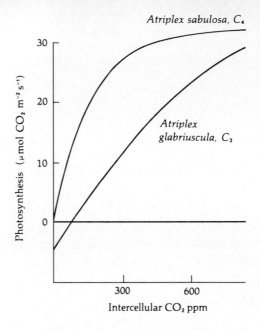

stomatal resistances. This ability to simultaneously conserve water and maintain positive net photosynthesis gives C_4 plants an advantage in arid, warm environments (Nobel 1991).

Light Utilization

It has long been recognized that some plants are adapted to high light environments (**heliophytes**) and some to low light environments (**sciophytes**). Individuals or leaves of the same genotype grown in contrasting light conditions develop morphological and physiological differences. Plants growing in the shade tend to have larger, thinner leaves with larger, less dense mesophyll cells, longer internodes, and less overall pubescence. These traits increase the efficiency of light utilization, increase the area for light interception, reduce reflection, and are associated with a variety of biochemical adaptations to low light environments (Taylor and Pearcy 1976). Figure 15-8 shows a typical response of sun-acclimated leaves and shade-acclimated leaves to increasing incident radiation in the 400–700 nm range. Note that shade leaves require less light to reach their CO_2 compensation point and they saturate at lower light levels than do sun leaves.

The ability of sciophytes to maintain a positive carbon balance in the shade is not necessarily due to an ability to photosynthesize more rapidly than heliophytes in low light. For example, Loach (1967) found that shade-tolerant and shade-intolerant species had essentially equivalent rates of photosynthesis when grown in low light. One key to differential success in shade was a difference in dark

Figure 15-8   Typical light saturation curves for sun-adapted and shade-adapted leaves of C_3 plants. Arrows indicate saturation intensity: about 1/3 full sun for sun leaves, about 1/6 full sun for shade leaves.

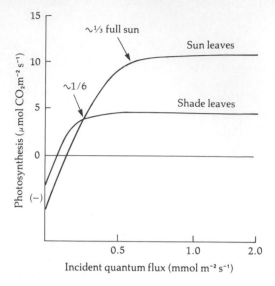

respiration rate. Shade-tolerant species had a lower dark respiration rate and therefore a lower light compensation point, allowing them to maintain a positive carbon balance (positive net photosynthesis) even at very low gross photosynthetic rates. Support for the idea that the dark respiration rate is one key to survival in the shade has been developed by several other authors (e.g., Logan 1970, Willmot and Moore 1973).

The photosynthetic apparatus of arctic vascular plants is seldom light saturated because of the low angle of the sun in the far north. This is in spite of a lower **light compensation point** (PPFD where photosynthesis and respiration are equal) and a lower light saturation point for arctic as compared with temperate species (Tieszen 1978). The low light compensation point allows continuous positive photosynthesis during the constant light period of midsummer. Photosynthetic characteristics of arctic plants are similar to those of shade plants (Chapin and Shaver 1985). Light is the common limiting factor in arctic plants. By contrast, alpine species have higher light compensation and saturation values, but because of the high light environment they show a daily course of photosynthesis that responds more directly to temperature. Consequently, temperature is the common limiting factor in alpine plants.

C_4 plants typically have higher light saturation levels than C_3 species (Figure 15-9). However, it is now apparent that at least some C_3 and C_4 plants do not saturate even at full sunlight. *Camissonia claviformis,* a C_3 desert annual, has a very high photosynthetic rate for a C_3 plant (Mooney et al. 1976). The reason for this very rapid CO_2 fixation rate lies in low stomatal resistance and high levels of RuBP carboxylase. As a result, *Camissonia* responds almost linearly to increasing light up to full sunlight. *Encelia farinosa* and *Larrea tridentata* are other C_3 desert species reported to have a similar capability (Ehleringer and Björkman 1978, Ehleringer et al. 1976, Cunningham and Strain 1969). *Amaranthus palmeri* (C_4) has the highest

Figure 15-9 Response of photosynthesis in warm desert plants to changes in quantum flux (400–700 nm) for *Amaranthus palmeri* (a C_4 summer annual), *Camissonia claviformis* (a C_3 winter annual), *Encelia farinosa* (a C_3 drought-deciduous perennial), and *Larrea tridentata* (a C_3 evergreen perennial). Measurements were made under normal atmospheric conditions and at a leaf temperature of 30°C, except for *Amaranthus palmeri*, which was measured at 40°C. Modified from J. Ehleringer. 1985. *Physiological Ecology of North American Plant Communities:* with kind permission from Kluwer Academic Publishers.

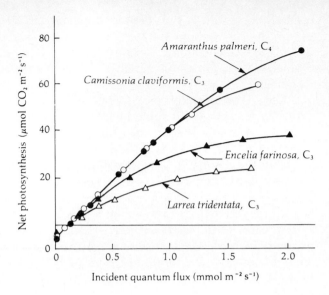

photosynthetic rate yet measured (82 μmol m^{-2} s^{-1}) in a terrestrial plant (Ehleringer 1983). The C_4 desert species *Tidestromia oblongifolia*, with maximum photosynthetic rates of 56 μmol m^{-2} s^{-1}, is comparable to the C_3 annual, *Camissonia claviformis*. It is clear that the C_4 pathway does not, in itself, give plants the ability to have high photosynthetic rates in very high light environments. It is not clear, however, why so few C_3 plants have evolved this capacity.

Light use efficiency (also called quantum yield) is the amount of CO_2 fixed per unit of light absorbed. We mentioned earlier that C_4 monocots are more successful than C_4 dicots. One possible explanation for the greater success of C_4 monocots than of C_4 dicots is the higher light use efficiency of the dicots. The average light use efficiency at 30°C for C_3 plants is 52 μmol mol^{-1}, for C_4 dicots it is 56 μmol mol ^{1}, and for C_4 monocots it is 62 μmol mol^{-1}. The causes of these differences are not fully understood (Ehleringer et al. 1997).

PPFD is frequently limiting to desert plants with CAM (Nobel 1982a). Consequently, cladodes (the flattened stem joints of platyopuntias) of desert cacti tend to grow oriented in a way that maximizes light interception during the season when other environmental factors are optimal for photosynthesis (Nobel 1982b). Thus, even in the high light desert environment, some species are light limited.

Leaves in the overstory canopy of a plant community not only receive full intensity sunlight, they receive it all day. Farther down in the canopy leaves receive a complex mosaic of light: short exposures to nearly full sunlight separated by long periods of very dim light. Chazdon and Pearcy (1991) measured hundred-fold changes in PPFD within a few seconds on the forest floor. The pulses of intense sunlight are caused by holes in the canopy through which direct sunlight passes only when the sun is directly overhead. If there is wind, the pulses become shorter and less predicable in time because branches and leaves are first blown into, then out of, the canopy gap. As the seasons progress, the sun's path changes; consequently a sunfleck area in spring is not usually a sunfleck area in late summer. Sunflecks, then, are highly variable. Can plants make use of such almost random resources?

Recent research in tropical forests strongly indicates the answer is yes and that sunflecks may even contribute a majority of the light energy utilized in growth by understory saplings, herbs, vines, shrubs, and trees. In Hawaii, Pearcy and Calkin (1983) found that 40–60% of the daily carbon gain by seedlings of the trees *Euphorbia forbesii* and *Claoxylon sandwicense* came from sunflecks. Growth rates, as well as photosynthetic rates of those species, were highly correlated with daily accumulative duration of sunflecks: Plants receiving 60 minutes of sunflecks a day had a threefold to fivefold more rapid growth rate than plants receiving 20 minutes a day (Pearcy 1983). Brief sunflecks lasting 5 to 10 seconds can be utilized with surprisingly high efficiency; further, the rate of net photosynthesis and its efficiency are both stimulated several times over as the leaf experiences a sequence of sunflecks (Chazdon and Pearcy 1986a,b). There is some evidence to show that early successional species in these tropical forests, which require full sunlight for optimal growth, do not exhibit the efficiencies and acclimation potentials described above for shade-tolerant trees.

Water Stress

Several authors (e.g., Lösch and Schulze 1995, Kappen et al. 1995, Boyer 1976, Hsiao 1973) have reviewed the impact of water stress (plant water status when evaporative demands exceed water supply) on the photosynthetic process. There is general agreement that the most immediate effect is an increase in resistance due to stomatal closure (r^{st}), which restricts the movement of CO_2 and water. Stomatal closure may be due to a decrease in leaf water status or a response to soil water depletion not yet associated with leaf water status (Bates and Hall 1981, Lösch and Schulze 1995). There is good evidence that nonstomatal inhibition of photosynthesis does occur in many species (Palta 1983). Johnson and Caldwell (1975) measured CO_2 resistance in arctic and alpine species and found that only a small part of the increased resistance measured in drying plants was due to stomatal closure. Bunce (1977) examined species from habitats with varying levels of available water and reported that increases in mesophyll resistance to CO_2 uptake occurred as the leaves dried. A range of coadaptations are found in *Quercus velutina* that are apparently related to its high tolerance of light and drought in the forests of New England. When compared to *Q. rubra*, which is shade adapted and drought intolerant, *Q. velutina* has greater plasticity in leaf anatomy, higher net photosynthesis over a wide range of light conditions, and a lower stomatal density (Ashton and Berlyn 1994). Further evidence of the effect of water stress on photosynthesis comes from studies of arctic and desert lichens conducted by Lange and Kappen (1972). Photosynthetic rates in these plants vary with changes in water availability even though there are no stomates. The actual biochemical relationship between water stress and photosynthetic rate remains to be determined.

C_4 and CAM species have an enhanced ability to utilize light for photosynthesis even while restricting water loss. That is, less water is lost while fixing a molecule of CO_2 in C_4 and CAM plants than in C_3 species. The ratio of water loss to CO_2 fixed is called water use efficiency (WUE). CAM plants conserve water by closing stomates during periods of high temperature and low humidity when the evaporative demands of the air are greatest. This is also the time when the greatest amount of water is at the evaporative surface of the mesophyll cell wall (Nobel 1983). The mesophyll cell surface is always saturated with water, but more water is present at higher temperatures. For example, with leaf temperatures of 25°C or 5°C in *Agave deserti* the saturation water vapor concentrations would be 23 g m^{-3} or 6.8 g m^{-3}, respectively. With an ambient water vapor concentration of 4 g m^{-3} the water loss would be seven times higher at 25°C than at 5°C. Stomates are open in CAM plants when both internal and external conditions are conducive to water conservation.

WUE for C_4 plants is nearly double that for C_3 plants. This is due in part to the ability of C_4 plants to dramatically reduce internal concentrations of CO_2 (c^{ias}), as discussed earlier, which steepens the concentration gradient of CO_2 between the outside air (c^{air}) and the internal air space (c^{ias}). The steep concentration gradient offsets increases in resistance caused by reduced stomatal opening; thus CO_2 flux is maintained even while H_2O loss is reduced. Water flux is reduced more than CO_2 flux because r^{st} is a greater percentage of the total resistance for H_2O than for CO_2. That is, r^{bl}, r^{st}, and r^{ias} are the only resistances important for water, whereas CO_2 also encounters r^{mes} and r^{chl} (Nobel 1983).

High WUE is one factor that leads to increased numbers of C_4 and CAM species in hot, dry environments. This does not mean that C_4 plants have a significant competitive advantage over C_3 species in arid environments (Syvertsen et al. 1976). Cover and biomass of C_4 and CAM species seldom exceed those of C_3 species even in arid habitats. However, the relative contribution of C_4 and CAM species to the flora and to cover and biomass increases in arid habitats (Wentworth 1983). The adaptability of C_4 dicots to arid regions is evidenced by the overwhelming predominance of C_4 species in the summer annual populations in the hot desert areas of the southwestern United States. Caldwell et al. (1977) considered the growth and gas exchange characteristics of a C_3 and a C_4 shrub in cold desert regions of Utah and found no advantage for the C_4 species. Even though the C_4 shrub was able to photosynthesize during the summer, it had a lower overall rate of CO_2 fixation during the more moderate springtime, so the annual total CO_2 accumulation was no greater than for the C_3 shrub.

Some CAM plants respond to increased water availability by shifting to the C_3 pathway and opening stomates during the day. Hartsock and Nobel (1976) heavily watered transplanted *Agave deserti* for 12 weeks under laboratory conditions, causing a shift in stomatal movements so that 97% of the CO_2 uptake occurred in the daytime. Artificial watering in the field, however, did not cause a similar change. However, when water is not limiting, desert *Agave* has a postdawn period of reduced stomatal resistance and a concurrent period of rapid CO_2 fixation similar to that in C_3 plants (Nobel 1976).

Certain CAM plants will revert to complete and continuous stomatal closure when exposed to water stress (Hanscom and Ting 1978). These plants still have nocturnal increases in acidity because internal CO_2 from respiration is recycled. Thus the loss of carbon by respiration is at least partially counteracted by the phenomenon of cycling.

Water stress caused by salinity can induce photosynthetic responses. Salinity of 200 mM NaCl causes the facultative CAM plant *Mesembryanthemum crystallinum* (ice plant) to switch from daytime to nighttime CO_2 fixation (Figure 15-10). Also, there is an apparent switch from C_3 to CAM as the leaves get older (Winter and Luttge 1976). Ice plant also has nocturnal CO_2 fixation in naturally saline environments in Israel.

Nutrient Responses

Carbon fixation is closely tied to the availability of nitrogen compounds that are used in the photosynthetic process (Figure 15-11). Maximum photosynthetic rates depend on nitrogen availability (Field and Mooney 1986, Evans 1989). Therefore, increased CO_2 concentrations in the atmosphere will only lead to higher long-term photosynthetic rates, as expected for C_3 plants, if sufficient N is available.

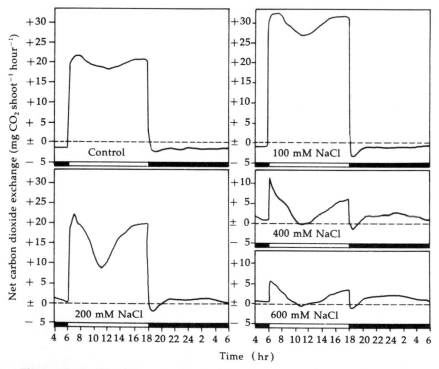

Figure 15-10   Net CO_2 exchange for *Mesembryanthemum crystallinum* at various levels of salinity. From Winter (1975) as redrawn by Winter and Luttge. 1976. "Balance between C_3 and CAM pathway of photosynthesis." *Ecological Studies* 19:323–334. Springer-Verlag, Heidelberg.

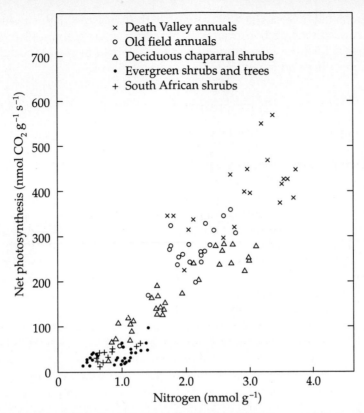

Figure 15-11 Maximum photosynthesis and nitrogen per unit leaf weight for 21 species grown under natural conditions. From Field and Mooney (1986).

Tateno and Chapin (1997) believe that higher initial photosynthetic rates will lead to high nitrogen use efficiency (the ratio of nitrogen used to CO_2 fixed) and to tissues with lower N concentrations. Decomposing plant tissues will have less N; therefore N will cycle more slowly and, by negative feedback, counter the increases due to atmospheric CO_2 increases.

Defoliation by herbivores often results in compensatory photosynthetic rates per unit leaf area in new and old leaves left on partially defoliated plants. One explanation for higher photosynthetic rates is the increased soil N available from herbivore-assisted breakdown. Compensatory photosynthesis was observed in artificially defoliated red oak (*Q. rubra*) grown with both high and limiting amounts of nitrogen. However, high photosynthetic rate was sustained only in plants with high N (Lovett and Tobiessen 1993).

Seastedt and Knapp (1993) refer to responses including compensatory photosynthesis as "transient maxima" and suggest that they cannot be maintained because of multiple limiting factors. By contrast, Drake and Leadly (1991) have measured long-term increases in photosynthetic rates of 88% and 40% in C_3 and C_4

plants, respectively, in brackish salt marshes of the Chesapeake Bay in response to increased atmospheric carbon dioxide. They argue that global enhancement of photosynthesis due to CO_2 enrichment will be sustained over time and that nutrient limitations will be adjusted for at the ecosystem level.

Temperature Responses

Temperature influences photosynthetic responses in various ways. We have already mentioned the ability of C_4 plants to carry on positive photosynthesis at higher temperatures than do C_3 plants. The most apparent influence of temperature is the limitation of all enzymatically catalyzed reactions. We would expect that reaction rates would increase gradually to optimum temperatures and then decrease sharply at higher temperatures when enzymes become denatured. However, membrane phase changes apparently reduce photosynthetic rates at temperatures well below the point of enzyme denaturation (Raison et al. 1980).

Photorespiration is much more responsive to changes in temperature than is photosynthesis. A major reason for the different temperature response of C_3 and C_4 species is that photorespiration rates increase as temperature increases. Thus a C_3 plant may have higher light use efficiency at lower temperatures, but will lose that advantage because of greater photorespiration at higher temperatures (Figure 15-12). There is, however, much more involved in understanding temperature responses of photosynthesis because plants are able to acclimate to previous temperature regimes. In addition, photosynthesis has ranges of temperature tolerance that are related both to the geographic location of the population and to the

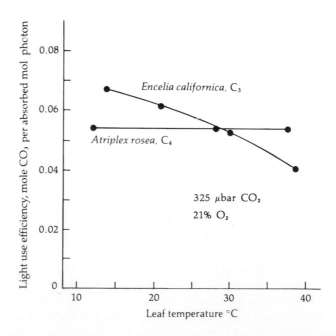

Figure 15-12 Light use efficiency (quantum yield) for CO_2 uptake in C_3 species *Encelia californica* and C_4 species *Atriplex rosea*, as a function of leaf temperature. Quantum yield was measured in air of 325 ppm CO_2 and 21% O_2. From J. Ehleringer. *Oecologia* 31:255–267. Published by Springer-Verlag, Heidelberg.

age of the tissue. Berry and Björkman (1980) and Nilsen and Orcutt (1996) reviewed photosynthetic temperature responses.

Fryer and Ledig (1972) illustrate the very sensitive responses of plants to changes in temperature in a study of the photosynthetic temperature optimum of balsam fir seedlings. The seedlings were grown from seeds collected along a 730–1460 m elevational gradient of the White Mountains of New Hampshire. The photosynthetic temperature optimum changed 4.3°C in 500 m of elevation change, which is very close to the change in mean air temperature of 3.9°C in 500 m, as recorded by weather stations along the gradient. This rather precise relationship between photosynthesis and temperature is characteristic of habitats with a short growing season. The restricted time available for growth makes the highest possible photosynthetic rates critical for success.

Where the time for acclimation is greater, we observe temperature acclimation of the photosynthetic apparatus over time. For example, McNaughton (1973) compared Quebec and California populations of *Typha latifolia* and found that the Quebec populations had a narrower range of tolerance for temperature and a much more restricted capability for acclimating to temperature changes than did the California plants. The Quebec population also had an increase in photosynthetic temperature optimum with age, which was not present in the California population. The California plants could shift the temperature optimum depending upon the season and maintain this ability throughout the life of the leaf, whereas the Quebec plants had lower optima in the younger stages and a more precisely fixed response in mature leaves. Pearcy (1976) reported that desert populations of *Atriplex lentiformis* have a much greater capacity for temperature acclimation than the population from a more constant coastal environment. The mechanism of temperature acclimation in desert plants such as *Atriplex lentiformis* (Pearcy and Harrison 1974, Osmond et al. 1980), *Simmondsia chinensis* (Al-Ani et al. 1972), *Atriplex polycarpa* (Chatterton 1970), and *Larrea tridentata* (Strain and Chase 1966) appears to involve, in part, an adjustment in respiration rate. Instead of continuing to increase with temperature, dark respiration rates tend to be constant with increasing temperature. In stress situations that restrict photosynthesis, a positive carbon balance can be maintained in some species by reducing respiration rates at temperatures in excess of 45°C. Recall that similar adjustments were the mechanism for survival in sciophytes for which gross photosynthesis was limited by light. We can generalize that plants from areas with either shorter growing seasons or predictable habitats will have a reduced ability to acclimate to temperature changes and will tend to have a narrower range of tolerance to temperature.

Temperature is also a critical factor regulating CAM photosynthesis because nighttime CO_2 uptake requires low temperature. For example, Neales (1973) found that the normal night-day stomatal rhythm of *Agave americana*, a CAM plant, was inverted when night temperatures reached 36°C or above (Figure 15-13). Kluge (1974) measured no night fixation of CO_2 when plants were subjected to 25°C day and 30°C night temperatures, and Nobel (1976) reported that stomatal resistance increased fivefold for *Agave deserti* when leaf temperatures increased from 5°C to 20°C. There is, therefore, strong evidence that the success of CAM plants in water-limiting environments is dependent on low night temperatures. This conclusion is

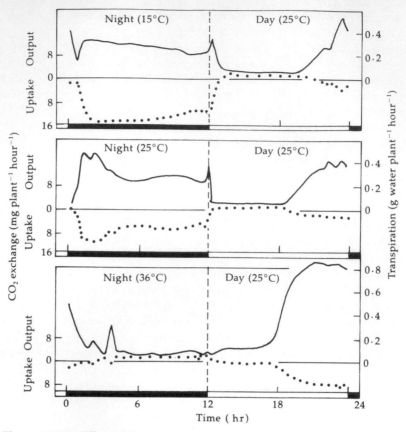

Figure 15-13   Effect of three night temperatures on the patterns of CO_2 exchange (dotted line) and water vapor exchange (solid line) of *Agave americana*, by day and night. From T. F. Neales. 1973. "The effect of night temperature on CO_2 assimilation, transpiration, and water use in *Agave americana* L." *Australian Journal of Biological Sciences* 26:705–714.

further supported by the actual geographic distribution of CAM plants; CAM plants usually do not live in water-limiting areas with high night temperatures. The biochemical reason for low night temperature requirements in CAM plants remains a mystery.

Methods of Photosynthetic Measurement

Most current ecological studies of photosynthesis measure either gas exchange or chlorophyll fluorescence. Fluorescence measurements are useful for comparing experimental groups subjected to different environmental conditions, and gas exchange measurements have wide utility in determining photosynthetic and transpiration rates.

The most direct way of determining photosynthetic rates of land plants is to measure the rate of CO_2 exchange. There are two principal ways that gas exchange rates are determined in terrestrial studies. The most widely used method incorporates an infrared gas analyzer (IRGA) to measure the flux of CO_2 to or from a plant or part of a plant sealed in an environmentally controlled chamber. Photosynthesis and dark respiration can be continuously monitored through time and the environmental conditions within the chamber can be regulated. This allows the measurement of plant responses to individual or multiple environmental factors. The other way of measuring CO_2 uptake is to briefly expose photosynthetic tissues to an atmosphere containing radioactively labeled CO_2 and measuring the amount of $^{14}CO_2$ fixed per unit time. Later in this section we will discuss the methods of determining photosynthetic gas exchange rates by IRGA. Detailed descriptions of the processes and theoretical considerations are available in Sestak et al. (1971) and Pearcy et al. (1989).

The most suitable method of measuring photosynthetic gas exchange depends on the questions the data will be used to answer. Both gas exchange methods have limitations. The IRGA measures net photosynthesis whereas ^{14}C techniques measure gross photosynthesis. If an investigator wants to have a continuous record of plant response and be able to artificially alter environmental conditions, the gas analyzer is more versatile and more accurate than repeated exposures to $^{14}CO_2$. The most severe limitations to laboratory IRGA systems are the small number of different plants that can be measured and the need to use plants growing in artificial conditions. Rapid field measurements of photosynthesis are possible using portable IRGA systems such as the one pictured in Figure 15-14. When the field measurement of many plants growing *in situ* is important, these portable photosynthesis meters are most satisfactory. In addition, there are errors inherent in the $^{14}CO_2$ measurement process, limiting its application to comparative studies or to surveys where absolute rates of photosynthesis are not critical.

Figure 15-14   Photograph of Licor LI-6400 portable photosynthetic system. Courtesy of Licor, Inc.

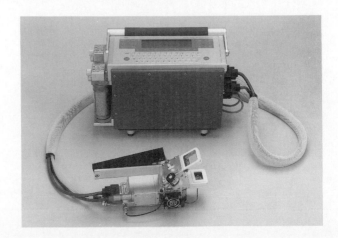

A major source of error in the $^{14}CO_2$ method is due to enzymatic discrimination between different isotopes of carbon and to the isotopes' different rates of diffusion (Yemm and Bidwell 1969). Whatever the reason, one cannot be sure that $^{14}CO_2$ is absorbed at the same rate as $^{12}CO_2$ under normal conditions. There are additional errors caused by the dilution of $^{14}CO_2$ by $^{12}CO_2$ released by respiration, which modifies the ratio of labeled and unlabeled carbon at the photosynthetic site. Therefore, factors that influence respiration rates will alter measurements obtained by this method. Respiration measurements can only be inferred by measurements of $^{14}CO_2$ dilution by $^{12}CO_2$; such measurements are not as dependable as those obtained by infrared gas analysis. Contemporary field photosynthetic gas exchange research is conducted primarily with portable infrared gas exchange systems (Field et al. 1989).

IRGA Gas Exchange Systems

Gas exchange systems used to monitor CO_2 exchange rates vary widely according to application and, more often, according to the financial resources available. We will discuss a basic open flow system that can be modified for specific purposes (Figure 15-15). The gas (air) is pumped through a series of tubes, usually stainless steel, in which flow rates are carefully monitored. Three flow paths are incorporated so that conditioned gas flows directly to the analyzer, and either through the plant chamber or through a bypass and then to the analyzer. The pur-

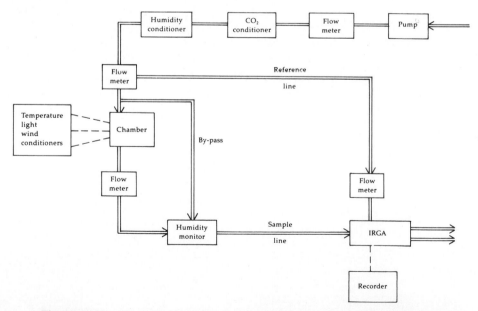

Figure 15-15   Flow diagram showing the major components of an open flow gas exchange system.

pose of the bypass is to ensure that all modifications being made to the sample gas are due to plant activity. It is also convenient to have bypass valves on environmental conditioning components so that any malfunction can be easily identified and corrected.

The plant or leaf cuvette is simply a sealed chamber into which monitoring probes are inserted to constantly measure environmental conditions in the chamber. More sophisticated systems have electronic circuitry that feeds information to the conditioning systems where rates are adjusted to maintain constant conditions. Some field systems are made so that chamber conditions track changes in ambient conditions. The components that condition the air vary widely in sophistication. If a study is being conducted where CO_2 concentration of the atmosphere is relatively constant, CO_2 conditioning may not be necessary. It is important in all systems, however, to be able to control and measure gas flow rate, photosynthetic photon flux density (PPFD), temperature, wind speed, and humidity within the chamber. Humidity changes that occur as the air moves through the chamber, when accurately measured, give reliable measurements of transpiration rates.

Additional values such as dark respiration, intercellular CO_2, stomatal and mesophyll resistances, etc. can be measured or calculated from data obtained by manipulating the conditioning components and monitoring plant response.

Chlorophyll Fluorescence Techniques

Major recent technological advances have led to the production of portable devices that measure photosynthetic capacity and response to environmental stress of plants in the field. These **chlorophyll fluorescence** monitoring systems allow nonintrusive estimates of the rate of electron transport from photosystem II (photosystem I shows little or no fluorescence). The light energy absorbed by photosystem II ends up as heat, drives photosynthetic electron movement, or is fluoresced (Figure 15-16). The optimal quantum yield is measured by relating fluorescence when the potential for photosynthetic electron movement is maxi-

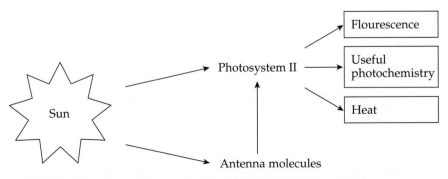

Figure 15-16 Fate of solar radiation absorbed by photosystem II in the thylakoid membrane.

mum (F_o = minimal fluorescence) with fluorescence when the photosynthetic electron transport chain is saturated (F_m = maximal fluorescence):

$$\text{Optimal quantum yield} = \frac{F_m - F_o}{F_m} \qquad \text{(Equation 15-3)}$$

Björkman and Demmig (1987) looked at many species and found a mean optimal quantum yield of 0.83. This "model leaf" optimal quantum yield is often used as a standard by which measurements can be compared. However, optimal quantum yield depends on light quality, leaf reflectance, and chlorophyll content, thus limiting the general application of "model leaf" values to comparative studies.

Light response curves of plants growing in natural or experimental conditions are constructed to determine photosynthetic responses to environmental conditions (Figure 15-17). Multiplying light intensity (PPFD) and quantum yield from fluorescence measurements gives an estimate of relative electron transport rate. This allows comparison of control plants (solid circles in Figure 15-17) with the predicted "model leaf" response and with the response of experimental treatments (open triangles in Figure 15-17). The amount of photosynthetic limitation imposed by the environment on the control is taken as the ratio of the length of line segment AB to that of AB'. That result can be compared to the photosynthetic limitation im-

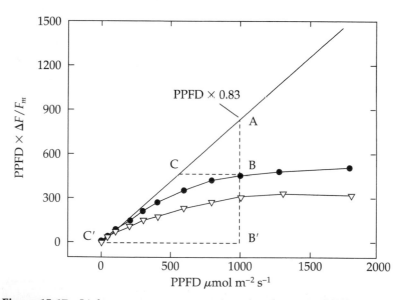

Figure 15-17   Light response curves compared to the optimal light use efficiency line of a "model leaf" (PPFD × 0.83). Solid circles are the control and open triangles are for a leaf exposed to 43°C for 5 min. See the text for explanation of dotted line segments. Modified from Schreiber et al. (1995) with permission of Springer Verlag, Berlin.

posed by the experimental treatment (AC/AB′). Alternatively, control and experimental plants can be compared as BC/BB′ (Schreiber et al. 1995).

The development of fluorescence techniques greatly expanded our understanding of photosynthetic responses. Fluorescence meters are used to determine plant response to a wide range of natural and imposed environmental circumstances (e.g., Larcher 1994, Bilger et al. 1984). They are also valuable for comparing photosynthetic rates of plants of different genotypes (Schreiber et al. 1995).

Summary

There are three photosynthetic pathways in plants, designated as C_3, C_4, and CAM. C_3 plants are the most successful and widespread of all plants, being common in all major climates. C_4 and CAM photosynthesis evolved from C_3 ancestors as global atmospheric CO_2 concentrations declined. Efficiency of C_3 photosynthesis is lowered by photorespiration, the process where O_2 substitutes for CO_2, causing the formation of glycolic acid. C_4 plants are most successful in hot, arid environments because of their low mesophyll resistance and their low CO_2 compensation point, which permits them to fix CO_2 even when high stomatal resistance reduces transpiration. This results in a more efficient use of water by C_4 plants. CAM plants are succulents that have adapted to arid environments by absorbing CO_2 at night when transpirational water loss is lower. CAM plants have the highest water use efficiency but are restricted by very low rates of photosynthesis and growth, and a requirement for cool nighttime temperatures. C_4 species have intermediate adaptations to arid environments; in fact, they are most important in hot, dry habitats. C_3 species, in general, have a lower water use efficiency but many species are successful in habitats supporting CAM and C_4 species. It is important to keep in mind that these are generalizations and that success in any given environment depends on a wide variety of adaptations other than the photosynthetic pathway. Alternate photosynthetic pathways may be a mechanism to partition resources between species and reduce competition by dividing the limited resources in time or space.

Primary limitations are imposed on photosynthesis by resistances to CO_2 uptake, water stress, light, and temperature. The rate of CO_2 uptake is determined by the boundary layer resistance, stomatal resistance, resistance as CO_2 dissolves into the liquid phase within the mesophyll cell, and the rate of fixation by photosynthesis. Different plants have different abilities to use CO_2 in low concentrations. For example, C_4 plants can reduce intercellular CO_2 concentrations to near zero whereas C_3 plants reach CO_2 compensation at much higher concentrations. Low intensities of light limit photosynthesis and high intensities may surpass the capacity of the photosynthetic apparatus. However, some desert plants have been found in which the photosynthetic apparatus does not saturate even at full light. The most immediate effect of water stress is stomatal closure, which increases resistance to CO_2 uptake. The diurnal pattern of stomatal movements in CAM plants can, under certain conditions, switch between CAM and C_3 pathways. Plants are able to acclimate to a variety of temperatures and, as a result, show different temperature optima at different seasons of the year.

The two most common ways of measuring photosynthesis in plants are systems incorporating an infrared gas analyzer or a fluorescence meter. The IRGA can be used, with proper conditioners, to measure plant responses continuously over numerous combinations of environmental factors. Chlorophyll fluorescence measurements are a sensitive means of measuring photosynthetic responses to natural or imposed environmental circumstances.

CHAPTER 16

FIRE

Forest fires attract beetles of the genus *Melanophila*. They fly from as far away as 50 km to lay eggs in the freshly burned wood, the only substrate in which their larvae can successfully develop. It is not the crackle of flames nor the smell of smoke that draws the beetles: they respond to the infrared emissions of the fire (Schmitz et al. 1997).

Powerful, even awesome to humans, fire must be viewed as an important evolutionary and ecological force in the development of most communities, particularly those subject to lightning. Many species have evolved intimate linkage to fire, as illustrated by the beetle example above. Lightning is the major cause of natural fires. Even in northern circumpolar ecosystems, natural fires caused by lightning are common (Kuhry 1994). There are an average of 100 lightning strikes to the earth every second, 24 hours a day, 365 days a year (Komarek 1964). Astronauts in space have commented on the spectacular light show on this blue planet provided by lightning striking the earth: over 3 billion strikes per year. The energy of these strikes is impressive; some range into the hundreds of thousands of volts, with a current of as much as 340,000 amperes.

In nature, a fire is classified as a **disturbance,** but it is neither good nor bad; it is simply the consequence of natural conditions. Fire has influenced the evolution of the various species of the forests and grasslands, as well as the xeric shrub communities (chaparral, fynbos, maqui, matorral) of the mediterranean climate regions of the world (Table 16-1). Fire, in varying severity, affects even the soils of those regions. With the possible exceptions of the wettest or coldest or driest regions of the earth, great tracts of land have been subject to periodic fires for millennia (Beyers and Wirtz 1995, Whelan 1995, Agee 1996, Bond and van Wilgen 1996, Pyne et al. 1996, and others).

There are three requirements for a fire: there must be fuel, weather conditions such that the fuel is dry, and a source of ignition. Fire behavior and fire intensity are affected by season, meteorological conditions (e.g., wind and temperature), terrain, fuel loading, and fuel moisture and soil moisture. Fires may be caused by lightning, by sparks from falling rocks, by volcanic activity, by spontaneous com-

Table 16-1 Examples of fire regimes for important vegetation types worldwide; the examples are drawn from nonmodified systems. From W. J. Bond and B. S. van Wilgen. 1996. *Fire and Plants.* Chapman and Hall. Used with permission.

Vegetation type	Frequency of fire	Season of fires	Intensity of fires
Grasslands	Annual or longer, depending on grazing pressure and rainfall, which determine fuel load	Fires occur in the dry season when grasses are dormant	Fire intensities range from <100 to >5000 kW m^{-1}
Californian chaparral shrublands	Fires occur at intervals of between 25 and 100 years	Fires concentrated in dry summer periods	Fire intensities can be high (>50,000 kW m^{-1})
South African fynbos shrublands	Fires occur at intervals of between 5 and 40 years	Fires concentrated in dry summer periods	Fire intensities range from 500 to 30,000 kW m^{-1}
African savannas	Frequencies range from annual to once every 30 years or more, depending on rainfall and grazing pressure	Fires occur in dry seasons when grasses are dormant	As for grasslands
Brazilian cerrado	Frequencies of 1–3 years	Fires occur in dry winters	No data
Australian eucalyptus woodlands	Some areas burn frequently, even annually (surface fires); stand-replacement crown fires can occur every 100–300 years in wet sclerophyll forests	Dry season	Usually low or moderate intensity (500–3000 kW m^{-1}); high-intensity crown fires (7000–70,000 kW m^{-1}) are rare
North American coniferous forests	Surface fires at frequent intervals (1–10 years); crown fires at intervals of 100–1000 years	Dry summer periods	Low intensity (200–800 kW m^{-1}) for surface fires, but very high (>50,000 kW m^{-1}) for crown fires (stand-replacement fires)
South American rain forests	Fires very infrequent and restricted to single trees or small patches	Only possible after several rainless days	Very low intensities (<20 kW m^{-1}).

bustion, and by human activity. Of these, only the first and the last may develop any periodicity and thus act as a consistent evolutionary force in the community. In both the Old World and the New World, periodic burning by humans has played a part in the evolution of the biota. Accelerated deliberate burning in the tropical rain forest in recent years has even affected global climate and air quality.

Human History

Humans have used fire as a manipulative tool for at least 0.5 million years (Stewart 1963). Even before people learned to start fire, humans used and carried fire from place to place. We can infer that grassland and forest fires may have been inadvertently caused by those whose practice was to leave a fire banked at their home base, to avoid calamity in case the fire being carried went out. Primitive humans living today are keen observers of minutiae, as undoubtedly our human and prehuman ancestors were. They observed the consequences of fire: Game is driven, though casually and without panic; the capture of insects, rodents, and reptiles is facilitated; visibility is enhanced; travel is easier; and forage is renewed. What had been a serendipitous accident surely became deliberate, purposeful arson. In the tropics, grasslands can be maintained by fire at the expense of trees if there is a periodic dry season. Thick forests held little of use for Stone Age peoples; grasslands and savanna were of greater value.

In the forests of the northeastern United States, Native Americans probably made deliberate use of fire near camps or villages. There is little evidence that they systematically burned large tracts of forest lands. Their burning was instead confined to the local area. However, their presence and their use of fire for whatever reasons increased fire frequency above the levels of lightning-caused fires and therefore selected for fire-adapted plants (Russell 1983).

Native American women were ethnobotanists; their role in the economy of their culture cannot be overstated. In western North America, fire was used by Native American women as a principal form of care and culture to increase the density, abundance, and diversity of desirable plant species and to reduce the competition from other plants. They tested and tended native plants for basketry materials, but also for food, medicine, dyes, and even games.

The use of fire to increase the supply of harvestable bulbs, fruits and seeds, and even rhizomes and leaves along with stimulating epicormic branching for basketry supplies was common practice. Ecological effects of that pyroculture also included the recycling of nutrients and the reduction of detritus, thus maintaining desirable habitat and decreasing the possibility of undesirable wildfire (Anderson 1998).

The Native American practice of burning forested areas was recorded by various early explorers and naturalists, such as Galen Clark (Guardian of Yosemite), Dr. L. H. Bunnell (a member of the 1851 Yosemite discovery party), and Joaquin Miller (Biswell 1974, van Wagtendonk 1995).

The role of fire is known in virtually every country and certainly in those countries in temperate and tropical latitudes. There has been, however, surprisingly little communication within the global scientific community about fire ecology (Bond and van Wilgen 1996). Whether to maintain a wilderness, to increase forest production, to decrease the devastation of wildland fire, to enhance the habitat for game, to maintain grasslands, or to increase grassland productivity, the naturalist, scientist, and manager seek to understand fire behavior and the vital role that fire plays in the ecosystem.

In this chapter we will discuss the role of fire in several communities, including southeastern forests, grasslands, western and boreal coniferous forests, and the shrublands of mediterranean climate regions. We will also consider the impact of catastrophic wildland fire and the possible role of fire as a management tool in various communities. The occurrence of fire results in changes in the environment. Our goal should be understanding, as best we can, why such changes come about, and in what manner fire is responsible for them. Inherent in that goal is gaining the ability to predict fire behavior.

Characteristics of Fire

Classes of Fire

The various types of fires are divided into three main classes, based on stratum and intensity: (a) ground, (b) surface, and (c) crown fires.

Ground Fires   Ground fires, which burn and smolder at and below the surface, are known to be important only in Histosols (organic soils), such as the Okefenokee Swamp in the southeastern United States, though they do occur elsewhere. They are termed "retrogressive agents" by Vogl (1974). In grasslands, ground fires consume soils down to the mineral substrate, creating depressions that then sometimes become ponds. Although of infrequent occurrence, ground fires can be very destructive, not only of roots, tubers, and rhizomes but of the organic matter in the soil itself. Thus, recovery of the plant community may take tens to thousands of years following such a catastrophe.

Surface Fires   Surface fires burn fuels in contact with the ground, and are generally cool, fast-moving fires. They do not build up high temperatures at the plant and ground levels because they are usually fed by lightweight fuels that are quickly converted to ash. Consequently, the basal portions, root stocks, and tubers are not harmed in grasslands and shrublands, and accumulated detritus is removed. In forested regions, cones may be opened, bark and needles scorched, and seedlings and saplings killed, but few mature trees will be severely damaged.

Crown Fires   Crown fires burn well above the ground surface and so, by definition, take place only in forest and tall shrub situations. They may result from lightning storms or from intense surface fires fueled by heavy accumulations of litter and debris. This type of fire may occur with surface fire. Crown fires often kill and consume mature trees, dropping boles and branches to ignite and further spread the fire at lower levels (Phillips 1974, Romme and Despain 1989).

In some forests, a clean burn is necessary to open the canopy and thus allow seedling survival in full sunlight. For example, in Canada and the Lake States, nearly pure stands of jack pine (*Pinus banksiana*) often burn during hot dry periods, spectacularly crowning, generating fire storms, and killing all aboveground vegetation. Periodic fires of such intensity maintain jack pine vigor (Gauthier et al. 1996). Certain populations of lodgepole pine (*P. contorta*) and black spruce (*Picea mariana*) bear cones that remain closed until a crown fire opens them to release the seeds (Johnson and Fryer 1996, Despain et al. 1996, Viereck and Johnston 1996).

Fuel and Fire Intensity

Fuel   The primary determinants of fire behavior are fuel, weather, local landforms, and the ignition pattern (Figure 16-1). The chemical constitution and various physical characteristics of the fuel influence its heat load. For example, the pine needle litter has a high surface-to-volume ratio and is not compacted, so an abundance of oxygen is available. This type of fuel carries a fire readily but will produce a relatively cool fire. The arrangement of plant parts in space and the presence of secondary compounds also influence the ability of fuel to burn (Bond and van Wilgen 1996).

Fire intensity refers to the relative rate of energy release of a fire, that is, high, moderate, or low intensity. The intensity depends on the heat yield, the availability of fuel, and the rate of spread of the fire. This relationship has been symbolized as

$$I = HWR$$

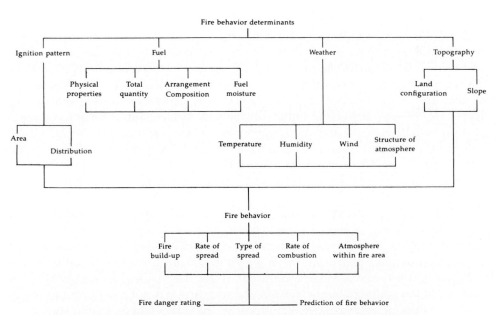

Figure 16-1   Determinants of fire behavior. In years of normal rainfall, fuel may be the major determinant, but in big fire years weather may be the critical factor in fire severity. In the Yellowstone fires of 1988, weather played the key role. Modified from R. H. Wakimoto (1977). Courtesy of R. H. Wakimoto.

where *I* is energy per length of fire front (kilowatts m^{-1}), *H* is the heat yield per unit mass of fuel (joules g^{-1}), *W* is the mass of fuel consumed by the flaming front (g m^{-2}) and *R* is the rate of spread of the fire (m s^{-1}; Borchert and Odion 1995). Frontal fire intensity refers to the energy output (kilowatts) from a strip 1 m wide of the active combustion area (Alexander 1982).

In part productivity determines the available fuel. Deserts rarely burn, because litter does not accumulate rapidly (Bond and van Wilgen 1996), but western coniferous forests may produce as much litter as 30 kg ha^{-1} day^{-1} for short periods (Hartesveldt and Harvey 1967). Rates of decomposition of litter and fire frequency are the other determinants of fuel availability. Management of a fire-prone ecosystem requires some estimate of available fuel. Though quantitative assessment of standing crop and litter in certain ecosystems is costly, such estimates are easily obtained in grasslands, and indeed are being used in fire management in many grasslands.

Fire Effects on Soil

In general, fire has the following effects on soil: soil temperatures are higher, detritus and standing dead will be ashed, some soil nutrients and organic matter may be lost, water-holding capacity and soil wettability are changed, and populations of soil microbes are altered. The impact of fire on soil will be considered as it affects the chemical and physical aspects of soil and its biotic components.

Fire Effects on Organic Matter   The effect that fire has on the soil and on soil organisms depends on the intensity and duration of the fire. A very intense burn will consume all the organic matter above and at the soil surface. Some destructive distillations will occur at levels within the soil column. At temperatures of 200–300°C, 85% of the organic substances will be destroyed. Nitrogen compounds are subject to volatilization and will be released and lost by distillation at temperatures of 100–200°C (Knapp et al. 1998). Other nutrients, such as calcium, sodium, and magnesium, will be released and deposited on the soil even at moderate temperatures and rapidly recycled into the biota (DeBano et al. 1977). One consequence of the loss of organic matter by intense burning is the concomitant loss of the high cation exchange capacity that characterizes organic matter, thus removing a holding substrate for nutrients.

Organic matter decreases bulk density (the weight of the soil solids per unit volume of total soil) of the soil because it is lighter than a corresponding volume of mineral soil, and because it gives increased aggregate stability to soil. The heat of an intense fire may break down these aggregates, leading to a loss of soil structure and to a lowered rate of water infiltration. It has been shown that, in prescribed burning in the southern Sierra Nevada, the soil carbon is not lost during burning, nor is the bulk density changed significantly. The carbon content is inversely correlated with the bulk density of the soil: the higher the carbon, the lower the bulk density. However, a ground fire, in which almost all of the organic matter is destroyed, would change both carbon content and bulk density (Agee 1973). There is a reduction in the thickness of the humus layer due to compaction after a fire.

Chemical Considerations   In general, soil pH will be higher after a fire than before. The degree of change is based on the nature of the cations released by burning, and on the acidity of the soil before the fire. For example, conifer forest soils often have a low pH (3–6) due to the high acidity of the litter (Gilliam 1991b, Gilliam and Christensen 1986). Thus, the magnitude of change when soluble basic cations are released by burning will be greater in those forest soils than in deciduous forest, tropical forest, or chaparral soils (DeBano et al. 1977). Following fire, there is often significant increase in potassium and phosphorus levels in soils of forests in the southeastern U.S. Coastal Plain (Gilliam 1991a). The activity of nitrogen-fixing bacteria may be enhanced by fire, both by the addition of nutrients by fire and by the higher pH due to release of mineral bases in the soluble ash (Christensen 1987).

Physical Effects of Fire on Soil   Soil acts as an effective insulator. Even the ash layer that results from the combustion of the litter can insulate the soil. Heat energy may be transferred downward by conduction, convection, and vapor flux (DeBano et al. 1977). The rate of this transfer is affected most by the level of soil moisture, although there are many other factors involved. Moisture is important because temperatures will not rise above 100°C in a given layer until all the water has been evaporated; this evaporation requires a great deal of energy, slowing the transfer of heat downward.

Other factors that affect soil heating include the nature of the litter. For example, two different points in the same chaparral fire may exhibit very different time-temperature curves (Figure 16-2). The curve that peaks and quickly drops to a low level (solid line in Figure 16-2) probably represents fine flashy fuels. The line representing the temperature at the soil surface is generated by the smoldering of less aerated or larger fuels burning after the flame front has passed (Borchert and Odion 1995).

The very high temperature lasts but a short time, and even at shallow depth (5 cm) the temperature does not rise to levels greater than 60°C for 1 minute, a situation that is lethal for plants. In other words, a light surface fire would heat only the top few centimeters of mineral soil to near the boiling point.

More than 30 years ago, Hartesveldt and Harvey (1967) working in Redwood Mountain Grove, Kings Canyon National Park, used the product Tempilaq to determine temperature in burn piles. Tempilaq consists of strips of a special paint that fuses at a predetermined temperature. Today, thermocouples attached to a data logger can be used to determine maximum temperatures. Temperature-sensitive crayons, Tempil heat-sensitive tablets, and even evaporative water loss from cans painted black have been used for such determinations, but the method of choice is still Tempilaq because of its low cost and the lack of a better product (Borchert and Odion 1995, Glitzenstein et al. 1995, Borchert 1997).

Soil Moisture and Fuel Moisture   Soil moisture levels are affected by fire, but not in a simple fashion. In certain communities, moss and humus can absorb large amounts of water, effectively preventing rainwater from reaching tree roots. In grasslands, the detritus serves the same function (Knapp and Seastedt 1986). The

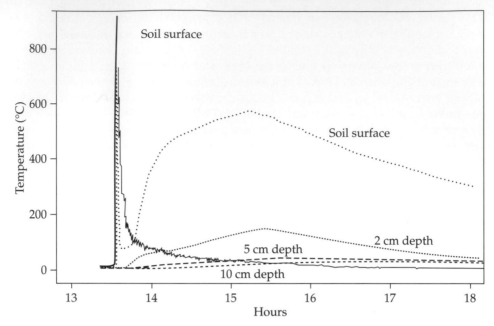

Figure 16-2   Time-temperature curves for two different areas in a fire in maritime chaparral. Temperatures at the soil surface are shown for one area by a solid line, which represents fine, flashy fuels. For the second area, temperature curves are represented for the soil surface and at depths of 2 cm, 5 cm, and 10 cm (dotted and dashed lines), representing heat transfer from larger and less aerated fuels. Modified from M. I. Borchert and D. C. Odion. 1995. "Fire intensity and vegetation recovery in chaparral: A review." In: J. E. Keeley and T. Scott, eds. *Brushfires in California: Ecology and Resource Management.* International Association of Wildland Fire. Used with permission.

removal of that layer by fire should allow infiltration of more water. However, the layer also prevents evaporation, so that water-holding capacity is lower and evaporation is higher on burned plots than on unburned ones (Viro 1974). Burned sites also have higher temperatures than unburned sites, due to the increased absorption of sunlight by blackened soil surfaces. This is especially true in grasslands. The higher temperature initially increases evaporation, but quick drying breaks capillary connections and thus reduces evaporation.

The influence of fire on water runoff and infiltration is temperature dependent. Following extremely hot fires that completely volatilize hydrophobic substances and thus increase wettability (see next section), there may be greater infiltration of water into the soil. However, after less intense fires, runoff is increased, due to lowered infiltration, and there may be an increase in the sediment yield (Booker et al. 1995). Such increases may correlate with, for instance, changes in soil structure, increased nonwettability, or both of these. Increased water runoff and sediment yields are significantly correlated with the decrease in the forest floor litter and duff observed following a forest fire. Although heavy surface fuels and aerial fuels (attached branches, epiphytes, etc.) are often unaffected, the weight, depth, and water-holding capacity of fine surface fuels are reduced by fire (Agee 1973).

Fuel moisture is estimated by a number of means. Very fine fuel moisture can be directly estimated from relative humidity. Moisture in fuels the diameter of a human finger is measured by means of **fuel sticks.** These are wooden rods that weigh exactly 100 g when oven-dry. After exposure to ambient conditions, they are reweighed in the field, and the assumption is made that additional weight (over the original weight, expressed as a percentage) is an approximation of the fuel moisture levels in available fuels at the site. Larger fuel moistures are usually modeled from longer-term weather data.

Soil Nonwettability   In certain communities, hydrophobic substances prevent the movement of water into the soil. Repellency is a function of oily resiny vegetation, hot fires, and/or coarse-textured soils. This condition is called **nonwettability** and is characteristic of ponderosa pine (*Pinus ponderosa*)–incense cedar (*Calocedrus decurrens*) forest, white fir (*Abies concolor*)–giant sequoia (*Sequioadendron giganteum*) forests, and the chaparral community of California. Hydrophobic substances that reside in decomposing plant parts accumulate on the soil surface and in the upper part of the soil column in the years between burns. Temperature gradients established during fires may alter the translocation of these substances into the lower soil layers. The nonwettable layer is thus moved lower in the soil column.

Water can move into the soil above the nonwettable layer and cause sheet erosion. Following the devastating Oakland hills, California, fire of 20 October 1991, the USDA Soil Conservation Service identified hydrophobic (water repellent) soils in the area, and predictions were for heavy erosion: a "fire-flood" event, meaning the erosional response of severely burned lands to winter rains. Mitigation efforts included reseeding, straw-bale check dams, and hydromulching. The feared fireflood never materialized, and sediment loss was minimal (Booker et al. 1995).

Effects of Fire on Soil Biota   Ahlgren (1974) pointed out that the ecology of soil organisms following a fire is not well understood. Even so, numerous studies have been made of soil microflora and fauna in reference to fire (Dunn and DeBano 1977, Ahlgren 1974, Parmeter 1977, Wright and Tarrant 1957, Wright and Bailey 1982). The effect of fire is usually to reduce fungi but to increase populations of soil bacteria and actinomycetes. Any changes in the microbiota are most obvious in the upper soil levels.

One may infer that pathogens will be destroyed by fire if these pathogens sporulate on litter and if that litter is consumed by fire. Damping-off fungi, seedling root rot fungi, and organisms that decay seeds will be removed from thoroughly heated seedbeds. The brown needle disease pathogen (*Scirrhia acicola*) of longleaf pine (*Pinus palustris*) is controlled by burning. Thus, the success of seedlings in a burned-over area may be attributable in part to the removal of soil pathogens or their substrates (Parmeter 1977).

There are, however, a number of early postfire species of ascomycetes, mostly pyrophilous (fire-loving) fungi such as *Pyronema*. The succession of fungi on burned lands has been studied and is thought to be analogous to the succession of higher plants on disturbed lands (Ahlgren 1974).

Nitrogen losses due to volatilization of nitrates, ammonia, and amino acids (as well as through alterations in forms of both organic and inorganic nitrogen) may be recovered, both on grasslands and in the chaparral, by early dominance of the postburn community by herbaceous legumes and nitrogen-fixing shrubs. Examples of these plants are such nodulated plants as *Trifolium, Lotus, Lupinus,* and *Ceanothus,* which dominate postburn regions in southern California. Nitrifying bacteria, such as *Nitrosomonas* and *Nitrobacter,* are very sensitive to fire. Their populations are completely destroyed at 140°C and evidence great mortality even at 100°C. Following a burn, the recovery of such nitrifying bacteria is slow (Dunn and DeBano 1977).

Fire-Prone Ecosystems

A General Fire Hypothesis   "For those ecosystems in which fire occurs predictably and frequently enough to result in a degree of fire-dependence, fire serves to increase and/or maintain the availability of essential resources which would without fire be growth-limiting for the organisms in the system" (Gilliam 1991a). Fire enhances resources that are limiting in a community, and flammable species are literally enhanced by fire. Thus a positive feedback system exists in which the interaction between fire and vegetation is such that fire may be as important as climate in determining the physiognomy of the vegetation.

Over time, if burning is frequent, there may be a sequential replacement of physiognomic type: taller trees replaced by shorter trees, woodlands giving way to shrublands, and finally woody species yielding to grasses (Bond and van Wilgen 1996).

It has been thought—and taught—that the similarities in the vegetation of those regions of the world with a "mediterranean climate" are generated by the climate. It is becoming clear that fire is also of paramount importance in the development of those ecosystems. Darwin said, "The change of climate being conspicuous, we are tempted to attribute the whole effect to its direct action . . . but this is a false view. Climate acts in the main part indirectly" (Bond and van Wilgen 1996).

Some General Properties of Fire-Maintained Systems   Properties of plants that make them susceptible to burning include the arrangement of plant parts in space, as well as the size and shape of those parts. Fine, xeromorphic leaves and twigs have great surface-to-volume ratio, dry out quickly, and allow a favorable fuel-to-air ratio. If litter decomposition is slow, fuel with low moisture content will accumulate: dead plant parts typically carry 5–15% moisture in dry weather.

Table 16-2 Fuel characteristics and resultant types of fire in major vegetation groups of the world. From W. J. Bond and B. S. van Wilgen. 1996. *Fire and Plants.* Chapman and Hall. Used with permission.

Vegetation type	Fuel characteristics	Types of fire
Semidesert	Sparse plants	Fires rare or absent due to sparse fuel
Grassland	Fine grass litter and live grass	Surface fires
Savanna	Fine grass litter and live grass. Trees not normally considered part of the fuel complex	Surface fires. Tree layer affected by intensity of surface fire
Heathlands and shrublands	Fine-leaved shrub fuels with varying amount of dead material. In some cases (e.g., fynbos) herbaceous fuels co-occur	Fires in crowns of shrubs
Tundra	Sparse plants	Fires rare or absent due to sparse fuel
Coniferous forest	Stratified fuel beds consisting of needle litter, twigs, cones, and understory shrubs below a canopy of live foliage	Surface fires under mild conditions. Under hot, dry conditions, dependent crown fires develop. Under severe conditions, running crown fires will occur. Huge differences in fire intensity between these types
Tropical rain forests	Stratified fuel beds with leaf litter layer, understory shrubs and trees, and canopy trees	Fires rare or absent due to moist conditions, transient litter layers, and high moisture content of leaves in canopy trees

Table 16-2 shows fuel characteristics and types of fire in major vegetation groups of the world.

Adaptations of Plants to Fire   Adaptations of plants to fire include the following: (a) in-soil seed storage and fire-stimulated germination; (b) fire-induced flowering; (c) bud protection and sprouting subsequent to fire; (d) on-the-plant storage and fire-stimulated seed dispersal. In addition, thick bark, evanescent branches, rapid growth, and early maturity help to ensure survival.

Southern Pine Savannas and Forests

History of the Southeast

A brief history will help us to understand this fire-prone ecosystem. The region of the United States that stretches from the Appalachian Mountains to the Atlantic is geologically very old. When the northern parts of the continent were glaciated, this southeastern portion became a refuge for many species of plants and animals (Delacourt and Delacourt 1987, Webb 1990). The geological history of the region and the complex patterns of weather influenced by topography and by proximity to the ocean have created a banded mosaic of communities in the southeastern United States ranging from warm temperate to subtropical savannas and forests (Figure 16-3). Komarek (1968) divided the Southeast into two bioclimatic regions, defined by variations in fire weather and other meteorological phenomena and by forest types: inner Coastal Plain forests and outer Coastal Plain savannas and prairies.

Fire has been an integral player in the evolution of many communities in the southeastern United States. The Coastal Plain is characterized by soils that are oligotrophic (nutrient poor) due to strongly acidic litter from the pines. For vigorous growth in the herb layer, and indeed for growth and reproduction of all the plants including the pines, fire is necessary to convert the acidic detritus to nutrients that enhance productivity and reproduction (Gilliam 1988, 1991a,b). The as-

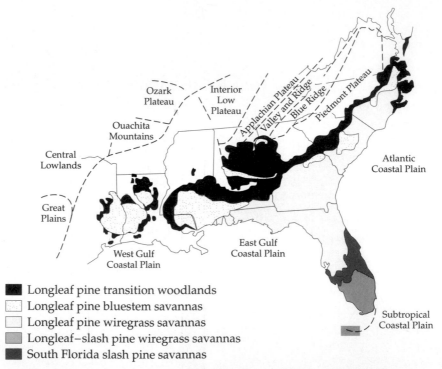

■ Longleaf pine transition woodlands
▫ Longleaf pine bluestem savannas
□ Longleaf pine wiregrass savannas
▨ Longleaf–slash pine wiregrass savannas
■ South Florida slash pine savannas

Figure 16-3   The range and distribution of pine savannas in the southeastern United States. Pine savannas constitute the dominant upland vegetation in most of the Atlantic, East and West Gulf, and Subtropical Regions of the Coastal Plain Physiographic Province and extend northward into the Appalachian, Valley and Ridge, and Piedmont Plateau Physiographic Provinces. From Anderson, R. C., J. S. Fralish, and J. Baskin, eds. 1998. *The Savanna, Barrens, and Rock Outcrop Communities of North America.* Cambridge University Press. Cambridge, England. Used with permission.

sociation of pines and depauperate soils has no doubt come down through family lines. Competition from angiosperms that evolved to deal with high nutrient content in soil has strengthened that association (Keeley and Zedler 1997).

Though the outer Coastal Plain is incredibly flat, the local species richness in the pine savannas is the highest reported for North American savannas, rivaling even that of tropical forests; as many as 15 plant species have been recorded in areas as small as 100 cm^{-2}, and several hundred species occur in areas of a hectare in size (Platt 1998).

Dominant species of pines in the southeastern pine savannas include longleaf pine (*Pinus palustris*) and south Florida slash pine (*P. elliottii* var. *densa*). The very diverse ground cover contains many endemic species, and is dominated by one or more species of grasses including wiregrasses (*Aristida, Sporobolus*), bluestem grasses (*Andropogon, Schizachyrium*), Indian grasses (*Sorghastrum*), and panic grasses (*Dichanthelium, Panicum*), along with many herbaceous forbs. More than 95% of the ground cover species are perennials.

When the first Europeans arrived in the Southeast, they found pine savannas to be the most prominent plant communities within the 2,000,000 km^2 Coastal Plain. Many of the savannas have since become degraded and fragmented into little more than pine plantations, but some excellent examples survive: the Boyd Tract in North Carolina, the largest old-growth stand of longleaf pine in the state, unburned for more than 80 years (Gilliam et al. 1993); the Wade Tract in southern Georgia (Platt and Rathbun 1993, Platt 1998); the Patterson Natural Area in the Eglin Air Force Base in the panhandle of Florida (Platt 1998); and Lostman's Pines in Big Cypress National Preserve in southern Florida (Doren et al. 1993). High-quality second-growth habitats with the ground cover fairly intact occur in many national forests and wildlife refuges, as well as state and national parks.

In the pine savannas, the frequency of adiabatic thunderstorms during the wet summer season is one of the highest in the world. Savanna pines, such as longleaf pine and south Florida slash pine, are probably struck by lightning more often than other tree species (Platt et al. 1988, Myers 1990). Longleaf pines, by virtue of their needles, reduce lightning strikes to low-intensity, widespread surface fires. The needles of longleaf savanna pines are the longest and most resin filled in North America, and they stay on the tree a maximum of two years (Platt 1998).

Resistance to fire is universal in the pine savannas. The pines exhibit rapid increases in height and bark thickness and **evanescent** lower branches, meaning that the branches are shaded by the forest canopy, drop off, and are burned up in the frequent fires. Plants of every stratum are adapted to environmental conditions produced by fire, and most ground cover species flower and set seeds after growing-season fires. Both south Florida slash pines and longleaf pines are famous for their adaptation to fire (Platt 1998). We will examine the role of fire in the life history of the latter.

Longleaf Pine

Longleaf pine (*Pinus palustris*) and the role of fire in its ecology have been well studied. The pines of the Southeast, including longleaf pines, require mineral soil as seedbeds, and most of the seedlings are shade intolerant. Favorable conditions for seedling recruitment occur often as a result of the frequent fires. Following germination, longleaf pines grow rather uniformly, reaching a height of several centimeters in just a few weeks, but then upward growth ceases.

By the time the seedlings are six months old, they sprout a pompon of long, slender, drooping secondary needles that help them survive fire (Figure 16-4). The needles that surround the terminal bud contain resins that both slow combustion and lower the heat produced if there is a fire during this "grass" stage. Thus longleaf pine seedlings form a "juvenile bank" to survive fire. If the young tree can survive a full year before fire returns, it is then very tolerant of fire for several years to more than a decade. Then the young seedling resumes upward growth (terminating the grass stage). The mature pines possess the thick layered bark that protects the cambium from frequent fire, and even young saplings are rather well insulated with corky bark (Platt et al. 1988, Platt 1998).

Longleaf pine is susceptible to a fungal disease called brown spot disease or brown needle blight. The causal agent, brown spot fungus (*Scirrhia acicola,* Dothideaceae), matures during the summer months. Multitudes of spores are released to the autumn rains, and they are splashed on low-growing plants, including seedlings of various kinds. The frequent fires control the fungus by burning the infected needles.

Persistent communities, the pine savannas are maintained by a combination of lightning and pyrogenic vegetation. **Pyrogenicity** refers to those characteristics of plants that favor fire. The highest fire frequencies are known to favor longleaf pine over all other tree species (Figure 16-5). The pine savannas are not succeeded by hardwoods, though there are occasional evergreen broad-leaved hardwood trees and shrubs growing in stands dominated by pines. Instead, the frequent fires kill young oak and other competing hardwoods, but the grass stage protects the perennating bud of the pine from heat damage, and the young seedling lives (Platt et al. 1988, Glitzenstein et al. 1995, Biondo 1997).

Figure 16-4   The grass stage of longleaf pine. The long, slender needles protect the perennating bud against fire.

(a)

(b)

Figure 16-5   (a) This prescription burn was set in May on an experimental plot in a xeric sandhill pine savanna in St. Marks National Wildlife Refuge, Wakulla County, Florida. (b) Cluster of recruits of juvenile (grass stage) longleaf pine about one year after a May fire in an old-growth stand in the Wade Tract, Thomas County, Georgia. Notice the patch of recruits at the far right. The juveniles are about nine years old, and are vigorous despite the recent fire. Photos courtesy of William J. Platt, Department of Biological Sciences, Louisiana State University, Louisiana.

Grasslands

Both grassland climate and grassland vegetation favor fire. These arid and semiarid regions may be subject to prolonged drought. The periods of lowest moisture may correspond with those of the highest temperatures. The prairie of North America, which begins south of 55°N latitude, is a prime example of grassland (see also Chapter 20). The abundance of certain grassland species relies upon "episodic fire," and that reliance is pronounced in time of drought (Bock et al. 1995). Summer and early fall lightning fires favor grasses over woody species. Within the prairie region of North America, the only forests are riparian forests, those sinuous stretches of trees along perennial waterways, where conditions are too wet for fire to spread (Collins and Wallace 1990, Knight et al. 1994).

Before the prairies were subject to overgrazing, fences, roads, and the like, there were miles of continuous flammable vegetation on the prairie, which allowed fires to sweep unimpeded across large areas. A series of thunderstorms can travel great distances, igniting multiple fires in its path. Under primeval conditions, these fires simply spread until they burned out because the fuel was gone, the fires reached a barrier such as a watercourse, or the weather changed (Komarek 1965).

The intentional burning of grasslands, probably the first land-management tool, would have been rapidly rewarded. After a few days of good regrowth, practically any burned grassland will act as a powerful attractant for animal life. Actually, deliberate burning in many ecosystems serves a second function, very meaningful to us today: reducing the danger of wildfire. Wildfire may spread quickly and destructively over large areas that have been protected from fire for a long period (Botti 1995, Pyne et al. 1996): witness the conflagrations in Yellowstone National Park in 1988 and Point Reyes National Seashore in 1995. Toward the end of this chapter we will discuss wildland fire and management of fire-prone regions.

Interaction of Fire and Vegetation

There are complex, ecosystem-wide consequences to grassland fires. Fire attracts herbivores, thereby increasing herbivory, but the silica content of the surviving grass stalks may increase, enhancing resistance to herbivory and decay and thereby providing fuel for the next fire (Vogl 1974). High silica content may have served to select mammals with unique teeth, who have dominated the grassland.

Nutrients that are required for rapid decay (e.g., nitrogen) may be removed by fire. Decay-resistant foliage immobilizes nutrients in plant tissue. Frequent fire is the agent of soil nutrient renewal, releasing the nutrients other than N and S from plant tissue and recycling them into new biomass (Gilliam 1987). Climate, fire, grazing, and the cycles of production and decay generated and maintain the great grasslands of the world.

Fire provides specific benefits for the dominant grasses in the tallgrass prairie (*Andropogon* spp.). The C_4 grasses are well adapted to frequent fire, and their ability to function effectively under reduced nitrogen regimes allows them to outcompete forbs (Knapp and Seastedt 1986, Seastedt et al. 1991). In addition, a late spring fire prevents early annuals from reseeding and removes the green tops of certain perennials, thus reducing competition.

Research at the Konza Prairie Research Natural Area showed that certain perennial forbs on frequently burned sites exhibit lower productivity and significantly lower reproductive effort than they do on unburned sites. Prairie coneflower (*Ratibida columnifera*) plants growing in sites that had not burned in several years were both significantly larger and produced more stems and more seeds per plant than did those in frequently burned sites. The responses are probably due to the influence of fire on the relative competitive abilities of grasses and forbs (Hartnett 1991).

Though one might be tempted to think of grasslands as simple systems, the truth is that tallgrass prairies comprise a complex mosaic of herbaceous vegetation ranging from wetland plants to drought-adapted plants in drier locations that resemble shortgrass prairie. Furthermore, grasses are among the most fire-tolerant plants of many communities, and certain geophytes possess contractile roots that retract the storage organs of young plants to protect them from fire (Bond and van Wilgen 1996).

Temporal patterns in aboveground net primary production (NPP) are exemplary of the complexity in the prairie. In a wet year, NPP is significantly higher in a burned site than in an unburned one, but in dry years NPP is significantly lower in burned sites than in unburned sites. The primary mechanism by which fire increases NPP is by removal of detritus (see also Chapter 12), which allows more radiant energy to the soil, warming it and the emerging plants and raising NPP—in a wet year. However, the detritus layer also reduces evapotranspiration, so its removal in drought years reduces NPP (Knapp et al. 1998).

Mechanisms for Survival

The mechanisms that allow survival following a fire are often adaptations to a specific fire regime: fire frequency, fire intensity, and season of burning. After a fire, one of the following usually occurs: (a) the plant resumes life from a protected crown comprising perennating buds at the soil surface; or (b) there is increased reproduction from seeds, rhizomes, and stolons, all of which may aid in the survival of grassland plants; or (c) the plant may be locally extirpated.

Hemicryptophytes   The prevailing growth form of grassland species is hemicryptophyte (refer back to Figure 1-2). That means that at least once a year, the aboveground portion of the plant dies back. Prairies are windy places, and the winds tend to dry the aerial portions of the plants. In addition, the plants' low stature allows sunlight penetration, which hastens drying. Thus, when the vegetation is dormant, there is often abundant flammable material at the ground level. Ideal conditions for a fire exist: dry, uncompacted fuels, with plenty of available oxygen.

These fine fuels will be ashed quickly, and high temperatures do not build at ground level. Rather, the flames pass quickly through the plants, with the highest temperatures at the top of the flames, and the heat is dissipated by the prairie wind. Soil surface temperatures in grassland fires may reach only about 300°C, and temperatures 1–5 cm belowground are only about 10–15°C above ambient (Gibson et al. 1990).

There is evidence that 60°C is the lethal temperature for the shoots of most land plants. However, leaf meristems (perennating buds) of many grasses are 40 mm or so below the soil surface, and therefore are not exposed to such high temperatures. Pineland threeawn (*Aristida stricta*) and Curtis dropseed (*Sporobolus curtissi*) have such buried meristems, which are additionally protected by closely packed, persistent leaf sheaths that do not burn readily. The living portion of the plant is virtually unharmed by the fire and will resume growth in the favorable time of the year in a light, moisture, and nutrient regime actually enhanced by fire. The decay-resistant aboveground plant parts contain minerals that are recycled by fire, but most nitrogen is volatilized and lost (Daubenmire 1968b, Knapp and Seastedt 1986, Gilliam et al. 1987, Knapp et al. 1998).

Increased Reproduction from Seeds   Enhanced reproduction may occur as a direct response to fire: Certain grassland species have resistant seeds that require scarification before germination will occur. **Scarification** as used here refers to any agent that ruptures the integrity of the seed coat, leaving it permeable to water and oxygen. Legumes, such as species of *Astragalus* and *Trifolium*, produce seeds that require scarification before seed dormancy will be broken. Bermuda grass (*Cynodon dactylon*) exhibits enhanced reproduction from seed following a fire because of increased seed set. Though many grasses do not possess fire-resistant seeds, some grass species exhibit tolerances that are higher than the temperatures attained by surface fires at the soil surface in grasslands (Daubenmire 1968b).

Biological Invasions and the Grass-and-Fire Cycle

In a global sense, grass invasion promotes a cycle that changes a nonflammable, native-dominated woodland or shrubland into a persistent, nonnative-dominated, flammable grassland (D'Antonio and Vitousek 1992). Historically, this process was deliberately set into motion by ranchers in California who needed grazing sites. In what is called type conversion, vast areas of coastal sage and chaparral were burned and burned again, the burning sometimes augmented with heavy tractors to break up woody debris.

The process works: for all resprouting woody species, some fraction will be killed in each fire, and the pyrogenic grass life form generates the conditions needed for further fire (Freudenberger et al. 1987). A comprehensive article by D'Antonio and Vitousek (1992) gives details as to why the type conversion process works:

1. Grasslands typically support flammable standing dead material.
2. Grasses, due to a large surface:volume ratio, dry quickly.
3. Grasses recover rapidly from fire.
4. The grass canopy itself allows high surface temperatures and large vapor deficits.

Invasion of alien grasses fuels a grass-and-fire cycle that may result in type conversion (D'Antonio and Vitousek 1992). The grasses alter the fire regime itself, with the result that native species are at a continuing disadvantage. In Hawaii Volcanoes National Park, grass invasion has reduced native shrub cover, a reduction that can persist for 20 years even without fire (Hughes and Vitousek 1993).

Mediterranean Climatic Regions

The mediterranean climatic regions are found between 30° and 45° north and south latitudes, on the west coast of continents, and in the Mediterranean Sea region. The mediterranean climate is defined by these characteristics: (a) moderate precipitation concentrated in the cool (not freezing) winter months, (b) summer drought, and (c) marine-moderated atmosphere influence (Keeley and Keeley 1988).

Within that regime, certain areas are occupied by sclerophyllous (hard-leaved) plants and by thin, stony soils with little organic matter; these areas have less than 750 mm of precipitation annually. In the western United States, such sites are occupied by scrub and woodland. The dense scrub is called chaparral (originally a Basque word for scrub oak, *chabarro,* spelled by the Spanish *chaparro,* and used by them to designate the mediterranean type scrub of California; Figure 16-6). In corresponding regions elsewhere are the *matorral* of Chile; the heath, *malee, quongan,* and other sclerophyllous shrublands of Australia; the *fynbos* of the Cape of Good Hope; the *phrygana* of Greece; and the *garrigue* or *garigue* of dry calcareous soils of France, known as *macchia* or *macquis* on siliceous soils (Keeley 1994). In each area, the physiognomy of the vegetation is very similar, as are the plant adaptations to fire. We will discuss several of these adaptations.

In-Soil Seed Storage and Fire-Stimulated Germination

Some species produce seeds that remain dormant in the soil for many years if there is no fire. Such seeds are said to be refractory. Germination depends upon the stimulus of some aspect of fire: intense heat shock, chemicals leached from charred wood, or smoke. The seeds lie—viable but dormant—until some scarifying agent provides the vehicle for resumption of activity. Chamise (*Adenostoma fasciculatum*), a chaparral shrub, responds to fire with phenomenal germination of stored seeds; after a burn, as many as 3000 seedlings occur per square meter.

Seeds of *Ceanothus* are viable after 20 years in laboratory storage, and even longer life may occur in nature (Keeley 1994). Zavitkovski and Newton (1968) wrote that the seeds of snowbrush (*Ceanothus velutinus*) might remain viable in forest litter for up to 575 years between fires.

Herbaceous plants with refractory seeds will often appear in great numbers following a fire (Table 16-3), then decline in the postburn years until another fire.

Figure 16-6   (a) Typical chamise chaparral community. (b) The shrubs are intricately branched, possess volatile oils, and are more flammable than communities not adapted for fire.

(a)

(b)

This pattern has been reported from California, Israel, France, South Africa, and Australia (Trabaud 1987). Seeds of many grasses do not require scarification, so grasses may not predominate in the first postburn year but become very abundant, even dominant, by the third year postfire (Table 16-3; Sweeney 1956, Biswell 1974).

Certain annual species have polymorphic seeds, meaning that some are refractory but a portion germinate without fire-related cues (Keeley et al. 1985). This strategy allows the annuals to take advantage of gaps in the canopy, using generalized gap-responding cues: the refractory seeds germinate following a fire that opens the canopy, the nonrefractory seeds germinate normally without fire (Bond

Table 16-3 Density of certain herbs that appeared 1–4 years after a burn of chaparral in the Highland Springs area of California. Density is given as number of plants per 28 m². From R. J. Sweeney. 1956. *University of California Publications in Botany* 28:143–206. By permission of the University of California Press, Berkeley, CA.

Category	Species	Year 1	Year 2	Year 3	Year 4
Species that peak the first year	*Mimulus bolanderi*	23	3	7	3
	Oenothera micrantha	19	3	4	2
	Antirrhinum vexillocalyculatum	15	4	6	0
	Emmenanthe penduliflora	280	0	3	3
	Silene antirrhina	25	5	12	10
	Malacothrix foccifera	47	8	14	23
	Mentzelia dispersa	31	11	19	14
	Mimulus rattanii	16	0	0	0
Species that peak the second or third year	*Barbarea americana*	9	21	34	17
	Cryptantha torreyana	19	62	844	127
	Gilia capitata	14	7000	0	3
	Lotus humistratus	132	738	33	17
	Bromus mollis	3	43	217	498
Species that peak the fourth year or later	*Bromus rubens*	19	23	147	1499
	Festuca megalura	20	24	417	737
	Chlorogalum pomeridianum	4	5	3	3
Species that remain constant	*Penstemon heterophyllus*	1	1	1	1

and van Wilgen 1996). Keeley (1994) noted that many herbaceous perennials, such as species in the Amaryllidaceae and Liliaceae, have nonrefractory seeds.

What are the fire-related cues that cause germination of refractory seeds? Heat shock is known to loosen cells in the hilum and to weaken the cuticle of seeds in species in diverse families represented in mediterranean shrub communities, including the Rhamnaceae, Fabaceae, and Convolvulaceae, for example (Sampson 1944, Sweeney 1956, Biswell 1974, Trabaud 1987, Borchert and Odion 1995, Keeley and Fotheringham 1997).

The role of charred wood leachate (also called charate) in the germination of refractory seeds has been well studied. Refractory seeds that are known to germinate when incubated with charred wood but not when stimulated by heat shock have been reported from more than 40 chaparral species. The presence of subshrubs (woody at the base) in burned-over areas is due to seed germination stimulated by charate or heat treatment or both (Keeley et al. 1985, Keeley 1994).

Germination can also be induced by smoke in seeds of plants in western Australian heath, Great Basin (Utah) scrub, and in South African fynbos and savanna. Smoke induces germination in 22 species in phylogenetically distant families. In certain species 100% germination can be triggered by sowing seeds on smoke-treated sand or filter paper, by applying smoke-treated water, or even by exposing seeds to vapors from smoke-treated filter paper or sand (Keeley and Fotheringham 1997).

What compounds in smoke trigger germination? Nitrogen dioxide, a component of smoke, induces 100% germination in whispering bells (*Emmenanthe penduliflora*), a chaparral annual fire-follower. Imbibition (the taking up of fluid by a colloidal system), which is altered in heat-stimulated seeds, is not affected by nitrogen oxides. However, nitrogen dioxide in combination with soil water may become HNO_3, a strong acid, which is capable of increasing solute permeability and therefore enhancing germination. The actual mechanism is unknown (Keeley and Fotheringham 1997).

Fire-Induced Flowering

Fire-induced flowering is observed mostly in monocotyledonous plants of the families Amarylidaceae, Iridaceae, Liliaceae, Orchidaceae, Poaceae, and Xanthorrhoeaceae. It has also been reported in dicotyledons, in the families Droseraceae, Loranthaceae, and Protaeaceae. The phenomenon occurs in North America, Australia, South Africa, and southern France (Trabaud 1987). It is important in two ways: (a) those plants that do produce increased numbers of seed would be at a competitive advantage over plants not so adapted; and (b) the seeds fall into an open seedbed, fertilized by mineral ash.

In some mediterranean regions, geophytes resprout from rhizomes, corms, or bulbs, and do not recruit seedlings immediately. However, flowering may be restricted to—or at least enhanced in—the first postfire year. The seeds that are produced are nonrefractory, so they germinate quickly after rains. The second postfire year is vital for seedling recruitment (Keeley 1994). Examples of such geophytes include death camas (*Zygadenus fremontii*), soap plant (*Chlorogalum pomeridianum*), purple-head brodiaea (*Dichelostemma pulchella*) and several species of mariposa lily (*Calochortus* spp.) in the California chaparral.

Bud Protection and Subsequent Resprouting

Buds near the base of the shrub may survive a fire and be released from dormancy by the death of shoots above them. Sprouting is induced by hormones or the onset of winter rains (Figure 16-7). The sprouting may arise along the stem (epicormic) or from a swelling known as a basal burl or lignotuber. These swellings form in the axils of the cotyledonary leaves; they are not a response to injury (Bond and van Wilgen 1996). Protection for the stem bud is provided by the bark, whereas lignotuber buds are protected by the soil.

Lignotubers occur in such families as Ericaceae, Fabaceae, Myrtaceae, Proteaceae, and Rosaceae (Bond and van Wilgen 1996). In the genus *Arctostaphylos*,

Figure 16-7 Stump sprouting. Latent axillary buds are released from inhibition when fire kills the apical meristem.

some species possess basal burls, others do not. Numerous species of sclerophyllous mediterranean plants sprout following fire (Wright and Bailey 1982, Ne'eman et al. 1992). California bay (*Umbellularia californica*) and many oaks (*Quercus* spp.) sprout in response to fire.

It seems apparent that a species which is both a sprouter and a seeder would survive virtually any fire regime, but that is not so. Chamise (*Adenostoma fasciculatum*) can be virtually eliminated from a region if the fire interval is very short, because the resprouting capacity becomes exhausted and all seedlings are killed (Zedler 1995).

Fire Recruiters and Fire Persisters

In mediterranean climate regions, the differential response of plants to disturbance is of particular significance to survival in various fire frequency regimes. Some mediterranean shrubs recruit seedlings only during the first season after a fire. Examples include *Adenostoma fasciculatum* and certain species of *Arctostaphylos* and *Ceanothus*. Keeley has termed these shrubs *fire recruiter species* (1991, 1992). If they also sprout after a fire, they are known as *facultative seeders*. If they do not sprout but only recruit seedlings, they are termed *postfire obligate seeders*. Both groups are *disturbance dependent*. Obligate resprouters, those that resprout but do not recruit seedlings following a fire, are known as *fire persisters*. In terms of seedling recruitment, they are classified as having *disturbance-free* recruitment, because for them seedling recruitment occurs in the absence of disturbance (Keeley 1994).

Obligate resprouters will require a long fire-free period, often three to five years, to mature enough to flower and set seeds; thus the disturbance-free seed-ling recruitment. Such recruitment does not occur in gaps but beneath the canopy, and it is strongly correlated with the biomass and depth of the litter layer (Keeley 1992).

Western and Boreal Coniferous Forest

Coniferous forest dominates the landscape of the montane regions west of the 100th meridian (Johnson 1992, Agee 1993). Deciduous trees are found mostly in linear "forests" that line the waterways. The conifers of the west include the tallest trees (coast redwood, *Sequoia sempervirens*), the most massive trees (giant sequoia, *Sequoiadendron giganteum*), the oldest trees (bristlecone pine, *Pinus aristata*), and the most productive stands of merchantable timber (Pacific Northwest region). Historically, many of these forests experienced frequent surface fires, or very infrequent, intense, stand-replacing fires. Due to fire suppression in the twentieth century, some of the former have experienced a buildup of understory fuels and are more prone to stand-replacing crown fires than in the past (Agee 1996). The aesthetic and ecological value of these forests is incalculable (see also Capter 20), and fire is a significant factor in most of them.

The spatial patterning on a landscape can be strongly influenced by crown fires, which create a patch mosaic of even-age stands. Conversely, the spatial patterns of fuels and terrain affect the extent and rate of spread of a crown fire. The dominant influence is climate, followed by topographic and physiographic features and fuel characteristics. There are always natural fire breaks as well as variations in vegetation, wind speed, and topography that result in a spectrum of fire from low to high intensity, creating and maintaining a mosaic of stands. Rarely does the entire forest burn with equal intensity. The net result is that crown-fire-dominated landscapes are described as nonequilibrium systems (Turner and Romme 1994).

Canopy-Stored Seeds

Serotiny, which literally means "late to open," is characteristic of many species in very diverse families. An excellent example of convergent evolution, serotiny is found in many mediterranean ecosystems, including those of South Africa and Australia (Table 16-4).

There are coniferous plants of other regions that retain viable seeds in cones for many years or even decades. The cones remain on the tree and closed until some mechanism severs the vascular connection with the parent plant, or breaks the resinous bonds that seal the cone scales. Although the plants that exhibit serotiny are phylogenetically diverse, some general principles apply to all of them:

1. Seedling recruitment is disturbance dependent.
2. Seeds are nonrefractory so there is little or no persistent soil seed bank.

Table 16-4 Taxonomic and geographic distribution of serotiny. Approximate fire frequencies refer to crown fires. With the possible exception of *Staberoha* and *Thamnochortus* (Restionaceae), no serotinous monocots have been reported. Modified from W. J. Bond and B. S. van Wilgen. 1996. *Fire and Plants*. Chapman and Hall. Used with permission.

Taxon	Example	Region	Vegetation	Fire frequency (y)
Gymnosperms				
Pinaceae	*Pinus attenuata*	North America	Chaparral	50+
	P. banksiana	North America	Boreal forest	50+
	P. contorta	North America	Conifer forest	50+
	P. rigida	North America	Shrublands, forest	15+
	P. halepensis	Mediterranean	Matorral, woodland	50+
	Picea mariana	North America	Boreal forest	50+
Cupressaceae	*Cupressus*	North America	Chaparral	50+
	Widdringtonia	Africa	Fynbos	15+
	Actinostrobus	Australia	Heathland	15+
	Callitris	Australia	Heathland, woodland	15+
Angiosperms				
Asteraceae	*Helipterum*	Africa	Fynbos	15+
	Phaenocoma	Africa	Fynbos	15+
	Oedera	Africa	Fynbos	15+
Bruniaceae	*Brunia*	Africa	Fynbos	15+
	Berzelia	Africa	Fynbos	15+
	Nebelia	Africa	Fynbos	15+
Casuarinaceae	*Casuarina*	Australia	Woodland	15+
Ericaceae	*Erica sessiliflora*	Africa	Fynbos	15+
Myrtaceae	*Eucalyptus*	Australia	Woodland	15–200+
	Callistemon	Australia	Shrubland, woodland	15+
	Leptospermum and others	Australia	Shrubland, woodland	15+
Proteaceae	*Aulax*	Africa	Fynbos	15+
	Banksia	Australia	Shrubland, woodland	15+
	Dryandra	Australia	Shrubland, woodland	15+
	Hakea	Australia	Shrubland, woodland	15+
	Leucadendron	Africa	Fynbos	15+
	Protea	Africa	Fynbos	15+
	and others	Australia	Shrubland, woodland	15+

3. Seeds are shielded from excessive heat by woody fruits or cones.

4. Late summer or early fall fires are best for seedling survival (Keeley 1994, Stephenson 1997).

Some populations of lodgepole pine require temperatures of 45–50°C (or even higher) to break the resinous bond between the cone scales. Summer soil temperatures may heat the air enough to open low-hanging cones. Knobcone pine, however, requires 200°C for opening the cones. Giant sequoia (*Sequoiadendron giganteum*) cones remain closed due to cone scale turgidity, and open in response to the severance of vascular connections (Vogl 1973, Lotan 1974, Harvey et al. 1980).

Jack Pine

One successful evolutionary response to fire is the development of an aerial or canopy seed bank (Table 16-4). Jack pine (*Pinus banksiana*), widely distributed in North America and a major component of fire-prone boreal forests, is serotinous. Successful seedling recruitment occurs primarily in the postburn environment because the seedlings are shade intolerant, and the trees bear closed cones that require temperatures greater than 50°C to melt the resin. Fire intensity is a selective factor in the degree of serotiny in jack pine populations: low-intensity fires favor nonserotinous trees, which are not fire dependent for seedling recruitment (Gauthier et al. 1996).

Lodgepole Pine

Lodgepole pine (*Pinus contorta*) forms monospecific stands in montane areas from Alberta, Canada, south to Colorado and west to California. Dense stands of more than 100,000 young lodgepole pines per hectare occur in interior British Columbia (Fahey and Knight 1986, Blackwell et al. 1992). "Lodgepole pine has long been known as a fire adapted species because of its ability to produce abundant seed from serotinous cones" (Clements 1910).

Following the Yellowstone fire, branches of lodgepole pine lay on the ground with cones still attached, spilling seeds from opened cone scales. The serotinous cones typical of many populations of lodgepole pine remain on the tree and closed, even for decades, until opened by a crown fire. Seeds from serotinous cones exhibit stimulation of germination by heating, but seeds from nonserotinous cones are *not* stimulated by heating (Despain et al. 1996).

Videos taken of the devastating fire of 1988 in Yellowstone National Park revealed that the length of time needed to completely burn a tree crown was only 15–20 seconds. Thus there are usually enough viable seeds, protected by the cone, to restock a stand even following a severe, stand-replacing fire. In the manner of jack pine, systems restocked after a fire will be mostly serotinous, whereas those originating from nonfire disturbance (windthrow, for instance) will bear a high percentage of normally opening cones (Tinker et al. 1994).

Engelmann spruce (*Picea engelmanii*) often grows with lodgepole pine in the southern Canadian Rocky Mountains. This spruce has no canopy seed bank, no soil seed bank and no juvenile bank. Engelmann spruce reestablishes following fire by dispersal of winged seeds from surviving adult trees: an adult bank in areas adjacent to the burned area. The seeds germinate easily, with no dormancy, but neither Engelmann spruce nor lodgepole pine reproduces effectively without a stand-replacing fire (Johnson and Fryer 1996).

Giant Sequoia Forests

The forests of the Sierra Nevada, especially those that are giant sequoia forest, were characterized primarily by frequent surface fire prior to the time of fire suppression. The fire ecology of giant sequoia (*Sequoiadendron giganteum*) has been studied extensively (Hartesveldt 1964, Biswell 1977, Harvey et al. 1980, SNEP 1996, Stephenson 1996). Fire is essential to the continued vigor of the sequoia–mixed conifer forest, as it plays these primary roles:

1. Removal of competitors
2. Reduction of fuel, decreasing wildfire hazard
3. Recycling of nutrients and preparation of a suitable seedbed on mineral soil in a favorable light regime
4. Maintenance of a subclimax community with a mosaic of vegetation age classes and types
5. Destruction of pathogens such as damping-off fungi
6. Enhancement of habitat for wildlife
7. Opening of serotinous cones

The giant sequoia groves are not pure stands of the big trees but associations of conifers that include white fir (*Abies concolor*), incense-cedar (*Calocedrus decurrens*), and sugar pine (*Pinus lambertiana*) that occur at midelevation on the western slopes of the Sierra Nevada of California. The big trees possess characteristics typical of a fire-adapted species, namely, thick fibrous bark and evanescent lower branches, serotinous cones, and rapidly growing saplings.

Many, if not most, of the mature trees bear fire scars, and many have multiple scars. A 1438-year record of fires in the Mariposa Grove of Yosemite National Park, derived from fire scars and tree rings, revealed that fire intervals during that time ranged from 1–30 years, with fire frequency being greatly reduced after 1860 (Swetnam et al. 1990, Swetnam 1993). Up to 85–95% of the bole of the tree can be burned without killing the tree, though the trees do not stump sprout. A giant sequoia near the entrance station of Sequoia and Kings Canyon National Parks shows evidence of heavy burning but no indication of loss of vigor (Figure 16-8).

Cone production, which begins at around eight years in giant sequoia, is very high: a mature tree might have 40,000 cones, 25,000 of them green—photosynthetic—and closed, that is, serotinous (Figure 16-9). The cones mature within two years, and the seeds remain viable for as long as 22 years. Seeds are very small: it would take over 200,000 giant sequoia seeds to weigh a kilogram.

Figure 16-8   Giant sequoia at the entrance to Sequoia and Kings Canyon National Parks. The fire scar covers more than one-third of the base of the tree and extends nearly 30 m up from the trunk, yet the tree is vigorous and sturdy.

Figure 16-9   A closed cone of the giant sequoia. This cone was cut by a chickaree in Giant Forest, Sequoia National Park. The fleshy scales are eaten by the rodent, and seeds may be dispersed in the process.

Effective seedling recruitment of giant sequoia is disturbance dependent, but there is an almost continual shedding of some seeds due to the activities of various animals. The seeds will germinate, given a suitably moist medium, so—at least theoretically—there would always be several age classes of giant sequoia. Seedling mortality is high in the first year, often more than 86% (Harvey et al. 1980), with desiccation the primary cause of death. Young roots cannot penetrate a thick layer of dry litter to reach water and mineral soil. Heat canker, damping-off fungi, and insect damage can also cause seedling death.

Periodic fire provides optimum conditions for seedling survival, and the dictum "the hotter the better" proposed by Harvey et al. (1980) still stands. The removal of competing species by a hot fire is the single most important effect of fire in giant sequoia forest (Stephenson 1997). In response to recent changes in U.S. Forest Service policy, that agency and the National Park Service are now joint managers of giant sequoia ecosystems. The two agencies jointly evaluate the feasibility of re-creating pre-Euroamerican conditions that will protect, restore, and conserve giant sequoia ecosystems (Stephenson 1996).

Wildland Fire and Management

A wildfire can be of great benefit, or it can be catastrophic, affecting millions of hectares in the United States. Fire suppression and the introduction of alien plants is changing fuel supply and character, and global warming may change the fire regime. Fire can be a tool to attain specified goals at every level of the ecosystem, including reduction of hazardous fuels and restoration or maintenance of natural systems (Agee 1996, Bond and van Wilgen 1996).

Prescribed burning is the skillful application of fire to natural fuels under conditions of weather, fuel moisture, soil moisture, etc., which allow confinement of the fire to a predetermined area and at the same time produce the intensity of heat and rate of spread required to accomplish certain planned benefits to silviculture, wildlife management, grazing, and hazard reduction (Biswell 1977).

Beginning in the 1920s, burning was used in this country for range improvement and for upland game management. Ultimately, it was applied to forestry. By the 1940s, prescribed burning in the southeastern United States was regularly applied as a management tool in longleaf pine (*Pinus palustris*) and slash pine (*P. elliottii*) forests (Figures 16-4, 16-5). In 1951, prescribed burning was used in Everglades National Park, Florida, and by the late 1950s was used as a silvicultural tool for loblolly pine (*P. taeda*) and the shortleaf pines. By 1978, the Forest Service at last reexamined its policy on fire exclusion and determined to include the use of fire in management (van Wagtendonk 1995).

Federal forest plan revisions provide opportunities to resolve issues that involve fire and its role in forest health. Federal wildland fire policy includes appropriate fire management, wildfire suppression, and prescribed fire. When a naturally ignited fire fits the prescription needs of the area, it may be designated as a prescribed natural fire (Schullery 1989, Botti 1995). In this section we will discuss two large wild fires, an urban-wildland interface fire, and management of wildland fire.

Yellowstone National Park   The greater Yellowstone area (GYA) fires of the summer of 1988 were the largest known for that region: 10% or 570,000 ha of GYA, and 45% or 400,000 ha of Yellowstone National Park (YNP) burned (Christensen et al. 1989). It was the driest summer on record. On "Black Saturday," 20 August, flames driven by winds up to 96 km hr^{-1} consumed 64,000 ha. It cost over $120 million to fight the fire and the scars of expensive helicopter landing sites, bulldozer lines, and fire lines will take years to erase. Human efforts did not put out the fire; it was the rains of September that finally put out the fire.

That holocaust was without doubt the most "significant event in the history of the national parks" (Schullery 1989). Could it be considered a natural event? Here are clues that it was: 50 years of fire suppression did not create abnormal fuel loads, as many of these forests typically have centuries-long fire intervals; lightning-caused fires in 1976, 1979, and 1981 were different from the 1988 fires only in degree, not in kind; and even-aged lodgepole pine forests that extend for thousands of hectares must have begun following a very large fire (Figure 16-10; Romme and Despain 1989).

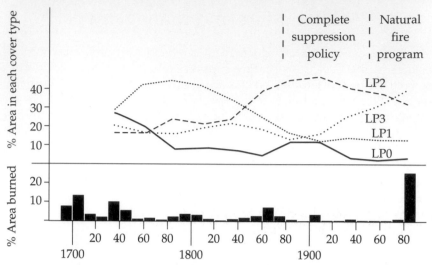

Figure 16-10   Top: Percent of the 129,600-hectare subalpine study area in the greater Yellowstone area covered by each successional stage from 1735 to 1985. The stages: LP0 represents recently burned forests up to the time of canopy closure, about 40 years postfire; LP1 represents stands dominated by dense, young, even-aged lodgepole pines, a stage from 40 to 150 years postfire; LP2 signifies stands of mature even-aged lodgepole pine with a developing understory of Engelmann spruce and subalpine fir, a stage lasting from 150 to 300 years postfire. The LP3 stage, with pine, fir, and spruce of all ages, persists until a stand-replacing fire. The area covered by meadows, water, and other constant features of the landscape are not included in the figure. The reconstructions extend back only to 1735 because extensive fires around 1700 destroyed the evidence necessary to construct earlier landscape mosaics. Bottom: Percent of the study area burned by stand-replacing fires in each decade from 1690 to 1988. From W. H. Romme and D. G. Despain. 1989. "Historical perspective on the Yellowstone fires of 1988." *Bioscience* 39(10):695–699. Used with permission.

The park vegetation and wildlife are recovering. Aesthetically, some areas are still not pleasant to view, but the park is healthy and the plants are vigorously re-clothing the landscape. The responsibilities of keepers of national parks and forests clearly include understanding natural process and putting it to use to achieve and maintain the wilderness character that is desirable. From Christensen et al. (1989) comes this wisdom:

> While accepting that chance exists in the natural world and that management is required even when knowledge is incomplete, managers must also recognize that it is possible to narrow the choice of purposes, to contract the range of ambiguity surrounding objectives, and to shrink the domain of ignorance. Many unknowns can be reduced to uncertainties, and uncertainties to probabilities.

Point Reyes National Seashore   On 3 October 1995, a combination of a hot, windy afternoon and an illegal campfire generated an aggressive fire at Point Reyes National Seashore (PRNS) that eventually burned more than 5000 ha, damaged or destroyed 48 homes, and cost $6.2 million to fight. Although Point Reyes has the foggiest lighthouse on the California coast, in September and October the fog clears, relative humidity and fuel moisture drop, and the temperature rises. Many decades of fire suppression generated fuel buildup in the bishop pine (*Pinus muricata*) forest and in the chaparral below the forest. By 8 October the fire was contained, and by 17 October a plan for recovery was published (U.S. Department of the Interior 1995). The recommendations included postfire mitigation to protect human life and property as well as restoration of the wilderness area of PRNS and the integrity of nonwilderness National Park Service, state park, and private lands within the Seashore.

Wildfire in Mediterranean Climate Regions   The fire-prone nature of the vegetation in mediterranean regions was discussed earlier in this chapter. It is clear that homes set in the middle of such vegetation are at risk from fire. In autumn of 1993, southern California was plagued with massive wildfires that burned more than 80,000 hectares. Firefighters could not contain the flames that erupted in Laguna Beach, Cherry Valley, Pasadena, and Malibu, and more than 1000 homes were destroyed (Keeley et al. 1995, Minnich 1995). To prevent such destructive fires, the homes in such areas must be fire resistant, with nonflammable roof and siding and no overhangs. There must be functional fuel breaks around the home and a gravity-fed water supply if possible (Pyne et al. 1996).

Federal Guidelines   Here are some guidelines for meeting the challenge of wildfire:

1. The first priority in wildland fire management is protection of human life, and property and natural/cultural values are the second.
2. Federal agencies must emphasize education, both of internal and external audiences, about the management of wildland fire.
3. All facets of wildland fire management will involve all partners: federal, state, and local.
4. Valid data and statistics are needed to support responsible fire management decisions.
5. Wildland fire must be reintroduced into the ecosystem.
6. Some form of pretreatment must be used, especially in the wildland/urban interface, if wildland fire cannot be safely reintroduced.
7. Recognizing the "boundaryless behavior" of fire: no one entity can manage and resolve all interface issues; it must be a cooperative effort. (USDA and USDA Forest Service 1997, U.S. Department of the Interior 1997)

"The only way to eliminate wildland fire is to eliminate wildlands" (Christensen et al. 1989).

Summary

Fire has undoubtedly occurred at frequent enough intervals and over a sufficiently long time in most communities to be a major force in the evolution of plant and animal communities. Forests, grasslands, and mediterranean scrub communities show the influence of fire. Only lightning and human activity have the capacity for periodicity and therefore to exert an evolutionary, consistent effect upon the organisms.

Fire may be ground, surface, or crown. It is strongly influenced by meteorological conditions, topography, fuel supply, and fuel moisture. Fire affects biological, chemical, and physical aspects of soil. Soil can help to insulate the macrobiota from fire effects.

Fire functions in several ways to promote southern pine forests. It clears the mineral soil for seed germination and seedling growth, reduces competition from hardwoods and herbaceous plants, eliminates brown needle blight fungus, and opens the forest canopy to allow growth of shade-intolerant young trees without damaging mature trees. In turn, the plants modify fire frequency, fire intensity, and season of burning.

Fire requires fuel, ignition, and favorable conditions to spread, all of which are abundant in grasslands. Fire promotes nutrient cycling and allows sunlight to warm the soil, enhancing productivity. Many grassland species are hemicryptophytes and survive fire without harm. Grassland species evolved with fire and grazing, and benefit from frequent fire. Fire enhances flowering and reproduction.

Mediterranean climate plants exhibit two responses to fire: (1) seeders—fire recruiters—die and come back as seedlings, and (2) sprouters—fire resisters—remain alive and come back as sprouts from burls or root crowns or stems. This division is not a strict dichotomy, for some sprouters are also seeders. Longer fire intervals favor seeders. The resistant seeds are produced and then stored in the soil, remaining dormant until scarification (fire is the most efficient method) or exposure to charred wood or smoke induces germination in a nearly optimal environment. Some fire followers are seen only in the seasons that closely follow fire.

Many coniferous forest trees have serotinous cones that open only following fire: jack pine, lodgepole pine, and giant sequoia only recruit seedlings effectively after fire. Engelmann spruce cones are not serotinous, but the species relies on stand-replacing fire for effective seedling recruitment. Fire removes competing species and opens the forest canopy, as well as reducing litter to mineral ash and reducing populations of pathogens.

Mechanisms that ensure population survival after fire include aerial and soil seed banks, juvenile bank, and adult bank. Landscape pattern and composition in many ecosystems are strongly influenced by spatial heterogeneity in the severity of fire disturbance.

Wildland fire is a vital, natural part of ecosystem functioning. Management in fire-prone areas will require education, cooperation, and active use of prescription fire.

CHAPTER 17

SOIL

I n his classic book, *Foundations of Plant Geography,* Cain (1944) stated that within the constraints placed on plant distribution by climate, soil exerts the most profound influence on the occurrence of plants in a given area. Thus, one cannot truly understand the ecology of terrestrial plants without a firm knowledge of the part of their environment with which they interact most intimately— the soil.

Definition of Soil

We can define soil in various ways, one of which is often used by soil scientists (pedologists) in textbooks of soil science: Soil is a natural product formed from weathered rock by the action of climate and organisms (Thompson and Troeh 1973, Jenny 1980, Rowell 1994, Buol et al. 1997). Such a definition emphasizes the importance of both abiotic and biotic factors in the process of soil formation. This definition, however, overemphasizes only one of several components that make up soil—the mineral component. Furthermore, it fails to recognize that some soils (e.g., those of peat bogs) originate from organic materials, not from rock.

Indeed, as we shall investigate in some detail in this chapter, four components make up soil—mineral particles, organic matter, water, and air. The mineral substrate of soil provides anchorage for plant roots, storage (pore) space for water and air, exchange sites for plant nutrients, and sources of these nutrients via weatherable minerals. Soil organic matter is not only plant and animal detritus at various stages of decomposition, but also cells, tissues, and exudates of soil organisms, such as bacteria, fungi, and earthworms. Organic matter (a) serves as a source and exchange site for nutrient recycling; (b) influences soil structure, pore space, and water-holding capacity; and (c) serves as an energy source for soil microbes and other heterotrophs. Soil water is the solvent for several essential plant nutrients, maintaining an equilibrium between cations and anions that are held on exchange sites, exuded by plant roots, and dissolved in the soil solution itself. Soil air contains O_2 for the aerobic metabolism of plant roots and soil organisms, CO_2 that is released by that metabolic activity and which facilitates weathering, and N_2 for N-fixing soil organisms.

Accordingly, a better definition of soil for our study of the ecology of terrestrial plants might be one that considers soil as a medium for plant growth. Soil is a mixture of mineral and organic materials that is capable of supporting plant life. Implicit in this definition is the intimate link between soil and plants. In fact, one might consider that modern soils and current plant species coevolved, beginning with the Devonian period of the Paleozoic era about 360 million years B.P. (before present) (Richter and Markewitz 1997).

Because of their bias toward agricultural soil, pedologists and plant ecologists alike have considered soil to be rarely more than 1 to 2 m deep. This practical limitation recently has been challenged and gradually is being replaced by the awareness that soil can be as much as 8 m deep and even deeper, including the entire weatherable crust (Richter and Markewitz 1997). It is even likely that many modern soils, especially those of the Great Basin of North America, are superimposed upon several other types of ancient soils. Indeed, this might be said for most locations not affected significantly by glacial processes.

Although materials in the upper fraction of the earth's crust are more intensely exposed to weathering than are the underlying strata, weathering takes place at depths down to and including the bedrock (unweathered rock). A more general term for this weatherable stratum would be *parent material* to include, in addition to consolidated rock, unconsolidated sediments (including organic deposits) as substrate from which soils can be formed.

Sediments are particularly prominent as parent material for soil along the Gulf and Atlantic coasts of North America, and soils of northern South America and Central America (Buol et al. 1997). Examples in the United States would include soils of the Central Valley of California (Jennings 1977), the deep sands of the Adirondacks of New York (Stone and Kszystyniak 1977), and the heavy clays of the lower Coastal Plain of the southeastern United States (Gilliam and Richter 1991).

Unconsolidated sediments are generally of a marine origin at lower elevations close to the coast and of mixed origins (e.g., alluvial-colluvial-deltaic) farther inland (Buol et al. 1997). Regardless of the specific nature of parent materials, they all weather into soil. The depths to and rates at which this occurs vary among types of soil, largely as a function of change through time (Figure 17-1).

Soil Development and the Soil Cycle

Although soil development appears to be a linear and directional process over human time scales, it is an episodic, even cyclic, phenomenon over the more appropriate scale of geologic time. Episodes of soil development are driven primarily

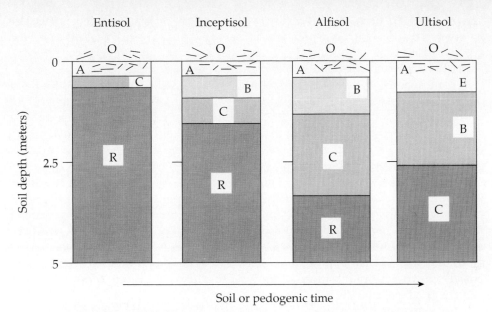

Figure 17-1 Some common soil orders and soil horizons that result from the ecosystem process called soil genesis. The illustrated soils from left to right constitute a general age sequence of soil development (i.e., a chronosequence). Soil horizons such as the O, A, E, B, and C horizons and the underlying partly weathered rock (R) are shown. The soil chronosequence illustrates the gradual formation of the B and C horizons and the deepening of the soil profile through time. From Richter, D. D. and D. Markewitz. "How deep is soil?" *Bioscience* 45:600–609. Copyright 1997 American Institute of Biological Sciences.

by climatic and tectonic perturbations. Cycles of soil development include geologic processes such as (a) sedimentation (deposition) of particles ranging in size from clays to larger fragments, (b) lithification (stone formation), (c) metamorphism and melting from the heat of pressure, (d) crystallization, (e) volcanism, and (f) erosion and transport processes. Soil development may be divided somewhat conveniently into two phases. The first is rock breakdown (weathering) by chemical and physical processes; the second is soil formation.

Rock Weathering

Whatever the substrate, whether a granitic dome, a basaltic lava flow, or a limestone outcrop, it will undergo certain common weathering phenomena. Although we have divided these into two categories (physical and chemical weathering), it should be emphasized that such a distinction is often arbitrary because both types of weathering interact to bring about the breakdown of primary substrates.

Physical Factors One form of physical breakdown, or disintegration, may be attributed to expansion and contraction of rocks as they warm and cool. When outer layers of rock are heated differentially to the inner rock, stress is created. Such stress is relieved by cracking and spreading of the substrate, an effect that may be exacerbated by tectonic activity and gravitational tension. These cracks may fill with water that can exert further physical and even chemical influences.

For example, if this water freezes it will expand. Expansion caused by freezing is quite impressive and may impart pressure of more than 10,000 kg cm^{-2} on unfrozen portions (Figure 17-2). Stresses generated by freezing of water are also significant after initial fragmentation of rocks. The regolith (unconsolidated mantle of weathered rock plus overlying soil material), including the soil at the top, continues to be altered physically by this force.

Exfoliation, the peeling off of layers of rock, results not only from the differential expansion we have discussed but also from expansion that occurs when surface materials are removed by erosion. Igneous intrusions crystallizing far beneath the surface are, in a sense, "spring loaded." Removal of the pressure of overlying material by erosion allows exfoliation, so layers of rock are released much as the modified leaves of an onion bulb peel away one layer at a time. The backside of Half Dome in Yosemite National Park in California and Granite Mountain, Georgia, are good examples of exfoliation.

Physical weathering should be distinguished from erosion and deposition by wind and water. Weathering is the physical and chemical breakdown and recombination of earth materials formed in different temperature, pressure, and/or geochemical environments into forms stable at the surface of the earth. Erosion and deposition describe the physical movement of any earth materials.

Figure 17-2   Cracks in granite caused by the force exerted by the freezing of water, which seeped into minute hairline fractures.

Glaciers transport tremendous amounts of material, and they polish and further reduce rocks and rock fragments with their abrasive materials. Glacial, alluvial (stream-carried), aeolian (wind-carried), and lacustrine (lake-deposited) materials form the basis of parent materials for many soils. Alluvial transport by the Mississippi River is estimated to be in excess of 660,000 kg yr^{-1} of soil into the Gulf of Mexico. The particle size of alluvium may range up to house-sized boulders, depending largely on the volume and velocity of water (Figure 17-3). By contrast, aeolian deposits are usually fine sand to clay-sized particles that can abrade, polish, and further erode rocks in their passing.

Other physical factors that are important in rock weathering include seasonal wet-dry cycles, pressure exerted by plant roots, and planes of weakness inherent in the rock structure.

Chemical Factors   Chemical processes of weathering (decomposition) occur simultaneously with physical disintegration. For example, plant roots exert an important physical force, although their influences are simultaneously of a chemical nature. These processes include several chemical reactions. Carbonization begins when CO_2 comes into equilibrium with H_2O to produce H_2CO_3 (carbonic acid). Carbonic acid partly dissociates into H^+ (hydrogen ion) and HCO_3^- (bicarbonate anion). Hydrogen ions are extremely reactive in chemical decomposition of rock and other parent materials. One source of CO_2 is the atmosphere, making rainfall naturally slightly acidic with a pH of about 5.5, which adds to the chemical weathering of rock surfaces.

The actual amount of CO_2 supplied to the soil from the atmosphere is, however, rather insignificant compared to the amount of CO_2 in the soil generated from aerobic metabolism (respiration) of soil organisms, including plant roots and heterotrophs such as bacteria, fungi, and earthworms. In fact, concentrations of CO_2 in soil air space of forest, pasture, and cultivated soils can be up to 70 times that of

Figure 17-3   An alluvial fan originating at the mouth of a desert canyon. The sparseness of the vegetation allows a clear view of the various sizes of material transported; some of the "particles" are huge boulders. Courtesy of Dr. John S. Shelton.

atmospheric concentrations (Dahlgren et al. 1997, Parfitt et al. 1997). This enhances further chemical weathering of parent materials underlying soil throughout soil formation (Johnson et al. 1977).

Oxidation is generally a disruptive process, causing the breakup of chemical compounds including mineral structures in rocks. This is especially the case for rocks that are high in ferrous (reduced) iron-containing minerals. These minerals disintegrate upon the oxidation of the iron component of the mineral structure.

The most significant chemical weathering is exerted by water in one of three processes: (1) solution, in which the strong dipole moment (polarity) of the water molecule acts as a solvent upon the mineral substrate; (2) hydrolysis, in which minerals are cleaved by combining with H^+ and OH^- ions; and (3) hydration, in which molecules of water become rigidly attached to the material being decomposed.

The products of most of these processes, such as oxidation and hydration, usually have a larger volume than the initial minerals, contributing to the physical forces involved in mineral and rock breakdown. These processes are all influenced by climatic factors, such as temperature and precipitation.

Soil Formation

Soil formation (S) was characterized by Jenny (1941) as dependent on a set of independent variables as follows: climate (*cl*), organisms (*o*), topography or relief (*r*), rock type or parent material (*p*), and time (*t*). Jenny's classic equation is as follows:

$$S = f^{(cl,\, o,\, r,\, p,\, t\,\ldots)}$$

He defined organisms as the biotic potential of a site, not just the existing organisms, which are indeed interdependent or dependent factors. The biotic potential includes consideration of soil organisms, litter, and vegetation type. The soil macrobiota and microbiota interact with the growth medium, altering its capacity to support life. The amount of litter deposited on the soil is determined by the productivity of the community, and the rate of decay or removal of that litter is determined by climate, by the nature of the litter, and by fire frequency. Also, the incorporation of litter-derived materials into the soil, the maintenance of soil porosity and permeability, and ultimately the soil fertility all rely at least in part on the activities of soil organisms.

Vegetation and climate exert a strong influence on the kind of soil that is formed. Spodosols (formerly podzolic soils) develop beneath coniferous forests in cool, moist climates, and have a prominent, ashy gray, bleached upper horizon due to mobilization of organic compounds. Mollisols (formerly Chernozem or "black earth" soils) develop under grasslands in temperate regions and generally have dark surface horizons, rich in organic matter due to rapid turnover of fine roots. As previously noted, a warm, moist environment hastens both disintegration and decomposition. The concepts relating to plant community succession developed in Chapter 11 also apply to soil succession.

Thus, parent materials interacting with soil biota and climate often produce soils with characteristic features. Another important factor in soil formation is time—time for large mineral grains to be disintegrated, decomposed, and changed to secondary minerals; time for finer and finer mineral grains to be redistributed within the soil profile; time for organics and humus to become a significant fraction of the total soil; and also time for finer grains and soluble anions and cations to be leached away. As soils become progressively more weathered and leached over time, the inherent fertility of the soil tends to decrease.

The composition of parent material largely determines the chemical makeup of its derived soils. For instance, acidic soils derive from (a) aluminum-rich bauxites; (b) silica-rich substrates such as quartz-rich sands, slates, and diatomaceous earth; (c) mine tailings of lead and zinc deposits; (d) sulfide-rich, hydrothermally altered volcanic rocks and ash; and (e) sulfur-rich estuarine deposits. Soils with poor fertility may develop from ultrabasic (ultramafic) rock types rich in silicates of iron and magnesium; serpentinite develops from igneous rocks such as peridotite and dunite, which contain silicates of iron and magnesium. Serpentine-derived soils are high in magnesium and low in calcium (Table 17-1). They may also be high in nickel and chromium and deficient in nitrogen and phosphorus. Thus, the chemistry of the parent material results in very infertile soils.

Soil may form from transported material or *in situ* parent material (residual soil). Transported materials tend to have greater surface area, thus developing more quickly than residual bedrock of the same chemical composition. Topography can influence the stability of a soil body in several ways, including its effect on rainfall patterns and the steepness of slopes. Resistant parent material may result in steep slopes, whereas parent rocks or materials that erode or weather more easily will likely produce wide valleys, rounded hills, and often deep residual soils. Steep slopes may continually expose new surfaces to weathering and soil formation, as alluvium and colluvium (material moved by gravity) is moved downslope; resistant soils, if present at all, are young and shallow (Figure 17-3). Gentle slopes restrict transport of materials, often enhancing formation of deep, strongly developed soils.

Table 17-1 Comparison of certain soil nutrients in Serpentine and nonserpentine soils, shown as a ratio.

Nutrient	Nonserpentine:serpentine ratio
Exchangeable calcium	7.4
Exchangeable magnesium	0.5
Total nitrogen, all forms	14.0
Total phosphorus, all forms	65.0
Potassium	3.8
Nickel	0.1
Chromium	0.01

Soil Profiles

Soils, especially the deeper soils of temperate latitudes, develop characteristic layers or horizons, generally differentiated from each other and discernible if a pit or trench is dug into the soil. A section encompassing all the horizons of the soil and extending into its parent material is called a soil profile (Figure 17-1). Rock weathering, as we discussed earlier, is both a chemical and physical process. Soil formation, by contrast, is a largely biochemical process with chemical recombination, along with gradual mixing of organic matter with inorganic soil particles.

As the upper layers of soil are exposed to the roots of plants, to the activities of animals and soil microbes, and to accelerated chemical weathering, mineral grains are altered. Among the mineral grains, the primary minerals are those that were formed initially deep in the earth and are unstable under near-surface conditions. As they weather, many of these minerals break down. By contrast, secondary minerals are those that form at or near the surface from chemical reactions involving weathered products from primary minerals (Table 17-2).

Secondary minerals, abundant organic materials, and associated changes due to interactions between soil and soil biota result in differentiation of soil horizons. Two of the processes are significant enough that they provide names for the principal (master) horizons of mineral soils in which they occur. The A horizon includes the surface layer in which organic matter accumulates and from which salts, clays, and soluble organics are leached. In some soils there is a **zone of eluviation (E)** due to the movement of materials downward from the A horizon, mostly by leaching (Figure 17-1). The B horizon includes the subsurface part of the soil, or subsoil, which may be a **zone of illuviation,** where the materials lost from A and E are accumulated. The illuvial materials may include iron and aluminum compounds, carbonates, clays, salts, and humus. Humus is the dark brown or blackish stable fraction left over after most of the organic residues have been decomposed. Alternatively, the B horizon may be a zone of alteration, where little material accumulates from the A or E but secondary mineral formation or significant changes in structure take place.

The O (organic) horizon and its subdivisions may be largely or totally lacking in certain desert and grassland soils but may be the only horizons existing in organic soils. There is also now a W horizon for soils with water layers in the profile, as in floating bogs or in soils with permafrost. The R layer underlies many, but not all, upland or residual soils, where it may displace the C horizon (Figure 17-1). A young soil may have A and C horizons, but no B horizon. This lack of a B horizon results when there has been too little time for significant illuviation or *in situ* subsoil alteration. These young soils are called Entisols (see the section "Soil Taxonomy") and are the most extensive class of soils. Figure 17-1 depicts temporal change in these horizons during soil development.

The appearance of a soil profile largely determines the fit of a soil into a soil classification scheme. Soils can be identified in the field based on the differentiation of the horizons, the texture, the color of the principal materials, and even which horizons (or portions) are lacking.

Table 17-2   The more important primary and secondary minerals found in soils. The primary minerals are also found abundantly in igneous and metamorphic rocks. Secondary minerals are commonly derived from sedimentary rocks. Modified from N. C. Brady and R. R. Weil. *Nature and Properties of Soils,* 11th ed. Copyright © 1996. Adapted by permission of Prentice-Hall, Inc.

Name	Formula
Primary minerals	
Quartz	SiO_2
Microcline/Orthoclase	$KAlSi_3O_8$
Na plagioclase	$NaAlSi_3O_8$
Anorthite	$CaAl_2Si_2O_8$
Muscovite	$KAl_3Si_3O_{10}(OH)_2$
Biotite	$KAl(Mg, Fe)_3Si_3O_{10}(OH)_2$
Hornblende[a]	$Ca_2Al_2Mg_2Fe_3, Si_6O_{22}(OH)_2$
Augite[a]	$Ca_2(Al, Fe)_4 (Mg, Fe)_4 Si_6O_{24}$
Olivine	$(Mg, Fe)_2SiO_4$
Secondary minerals	
Calcite	$CaCO_3$
Dolomite	$CaMg(CO_3)_2$
Gypsum	$CaSO_4 \cdot 2H_2O$
Apatite	$Ca_5(PO_4)_3 \cdot (Cl, F)$
Goethite	$Fe_2O_3 \cdot H_2O$
Hematite	Fe_2O_3
Gibbsite	$Al_2O_3 \cdot 3H_2O$
Clay minerals	Al silicates

[a]These are approximate formulae only because these minerals are so variable in their composition.

Physical Properties of Soil

Soil Texture

Soil texture is defined as the relative proportion by weight of sand, silt, and clay (the "soil separates" or primary soil particles). That is, only mineral components in the fine-earth fraction (<2 mm) are considered in determining soil texture, even though soil is also composed of organic matter, water, and air. The United States Department of Agriculture (USDA) classification of soil particles places the upper limit for silt at 0.05 mm in diameter and the lower limit at 0.002 mm. Particles greater than 0.05 mm in diameter are considered sand, and particles less than 0.002 mm in diameter are considered clay (Brady 1974, Brady and Weil 1996). Almost any soil contains pebbles and cobbles that are larger than the coarsest sand

(2 mm in diameter), but unless the bulk of the soil is composed of these coarse fragments, they play but a minor role in soil function. If more than 70% of the soil volume consists of coarse fragments, then less than 30% will effectively store water and mineral nutrients.

In general, the structural or skeletal support of the soil is provided by the sand fraction, or at least by the largest soil particles, and the finer particles help to store nutrients and to bind particles together into aggregates. Size classes are assigned arbitrarily, recognizing that minerals (and the building blocks of minerals) form a continuum of sizes, ranging from single tetrahedra (a *tetrahedron* is a three-dimensional figure with four triangular sides, formed chemically with an oxygen ion at each of the four points) to large and very complex mineral crystals. Materials that are without crystalline structure in the fine clay fraction are called amorphous materials. They may consist of opaline silica, aluminum silicates (e.g., allophane) and various iron compounds.

If mineral grains are not spherical, then their diameter is taken to be an average of their maximum and minimum dimensions. Sand grains tend to range from blocky and irregular to subround in shape, and silt particles are similar but smaller. Most clay mineral particles are very unlike sand grains; their platelike shape is reminiscent of mica (Figure 17-4), although particles of any shape less than 0.002 mm in diameter are called clay-sized particles.

In the field, a good estimate of texture can be made by dampening and handling a small amount of the soil. The particles that feel grainy and gritty are sand; those that are silky, like moist talcum powder, are silt. If the sample is sticky and if a self-supporting, flexible ribbon can be extruded between the thumb and finger, there is a high percentage of clay. If it is not sticky or only slightly sticky, and no ribbon can be extruded, there is a low percentage of clay. Different soil textures can be recognized by various combinations of the foregoing.

A more precise determination of soil texture can be made in the laboratory. Particles larger than 2 mm are sieved out of a given sample, which is then treated to oxidize and remove the organic matter. This is usually done with hydrogen peroxide, which also has the effect of rupturing the binding of clays by organic matter. The sample is then dried and weighed. The remaining materials are carefully

Figure 17-4   An idealized group of clay particles representing the card house effect in soil structure, in which positive charges on edge positions are attracted by negative charges on the broad surfaces. The negative charges provide exchange sites for cations and are thus a nutrient storage facility for cations (e. g., K, Ca, Mg). From *Soils and Soil Fertility* by Thompson and Troeh (New York: McGraw-Hill, 1973). Reprinted by permission of Frederick R. Troeh.

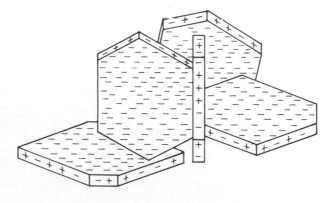

flushed and shaken through a series or nest of sieves that allow particles 1 mm, 0.5 mm, 0.25 mm, 0.1 mm, and 0.05 mm to pass through. The total weight of sand and its distribution can be determined by this process. The fraction passing through the sieves contains silt and clay together, and these can be separated by using a hydrometer or pipette method of analysis (Thompson and Troeh 1973). Such methods, which are based on the differential rate of particle settling due to different particle diameters, allow the separation of the silt and clay fractions.

Once the percentage by weight of separate components has been determined, the use of a soil texture triangle (Figure 17-5) facilitates textural class determination.

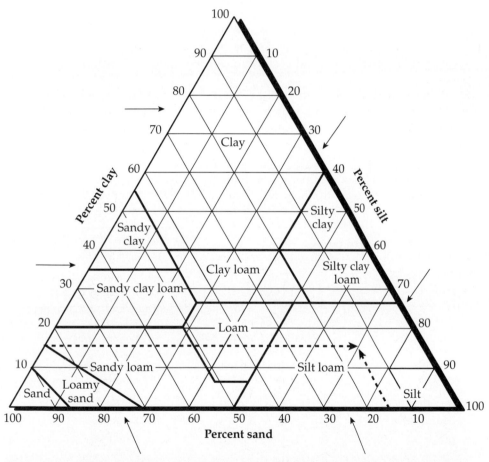

Figure 17-5   Relationship between the particle size distribution of a soil and its textural class name. Depicted is a soil that is 15% sand, 15% clay, and 70% silt: a silt loam. The points corresponding to the percentages of sand and clay present in the soil are located on the sand and clay lines, respectively. Lines are then projected inward from % sand parallel to the silt side of the triangle and inward from % clay parallel to the sand side. Note that the third particle size (in this case, silt) is automatically defined by the two projected lines. The name of the compartment in which the lines intersect is the texture class of the soil in question. Modified from *Nature and Properties of Soils*, 11th ed., by N. C. Brady and R. R. Weil. Copyright © 1996. Adapted by permission of Prentice-Hall, Inc.

Assume, for instance, that a given soil contains 15% sand, 15% clay, and 70% silt. This soil would be called *silt loam*. Notice on the soil triangle that the textural class designation is an area, not just a point on the triangle. Four terms are commonly applied to soil texture: clay, silt, sand, and loam. *Loam* designates a mixture of sand, silt, and clay that exhibits the properties of each fraction about equally. Thus, our sample exhibits the silt fraction more strongly than does loam, but it still has the general properties of a loamy soil, hence silt loam. Table 17-3 lists the general terms used by the USDA to describe soil texture.

To understand the function of these various soil fractions, we must consider two other physical properties of soil, porosity and permeability. **Porosity** refers to the total space that is not occupied by soil particles (Figure 17-6). **Permeability** deals with the ease of passage through soil of gases, liquids, and plant roots. In general, fine-textured soils are dominated by micropores and have greater porosity, with a pore space of as much as 60% of the total soil volume, and sandy soils are dominated by macropores and may have less porosity, with as little as 35% pore space.

Water moves into and drains out of a sandy soil with much greater ease than through a finer-textured soil, which means that sandy soils are more permeable than fine-textured soils. This permeability results from the following facts: water

Table 17-3   General terms used by the U.S. Department of Agriculture to describe soil texture in relation to the basic soil textural class names. Modified from *Nature and Properties of Soils,* 11th ed., by N. C. Brady and R. R. Weil. Copyright © 1996. Adapted by permission of by Prentice-Hall, Inc.

Common names	General terms		Basic soil textural class names
	Texture		
Sandy soils	Coarse		Sand Loamy sand
Loamy soils	Moderately coarse		Sandy loam Fine sandy loam
	Medium		Very fine sandy loam Loam Silt loam Silt
	Moderately fine		Clay loam Sandy clay loam Silty clay loam
Clayey soils	Fine		Sandy clay Silty clay Clay

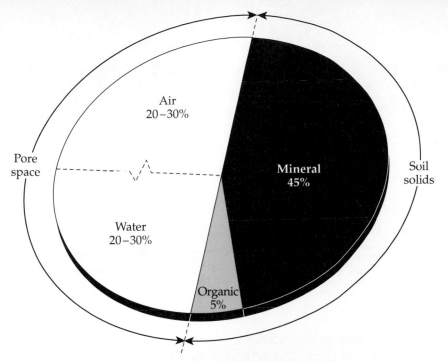

Figure 17-6 Volume composition of a loam surface soil when conditions are good for plant growth. The broken line between water and air indicates that the proportions of these two components fluctuate as the soil becomes wetter or drier. In a fine-textured soil, the pore space would be mainly occupied by micropores; in a coarse-textured soil, the pore space would be mainly occupied by macropores. Modified from *Nature and Properties of Soils*, 11th ed., by N. C. Brady and R. R. Weil. Copyright © 1996. Adapted by permission of Prentice-Hall, Inc.

has strong adhesion and cohesion capacity, and water molecules form a film (**hygroscopic water**) about each clay particle and easily bridge the gap between particles. The cohesion between water molecules is sufficient to maintain the bridge across the micropores. Such soils may be nutritionally rich but can be a problem agriculturally due to low permeability, poor drainage, and poor aeration. Hygroscopic water also forms a tight film about sand grains, but the surface-to-volume ratio is very low, there are fewer grains than in clays, and the large pore size of the macropores does not allow efficient cohesion of water molecules, which therefore drain away under the influence of gravity. Figure 17-6 depicts a loam surface soil with ideal conditions for plant growth. (See Chapter 18 for a full discussion of water potential and soil moisture.)

Clay particles and bits of organic matter within the soil are referred to as **micelles.** Micelles are negatively charged on their surfaces (Figure 17-4). The capacity of micelles to provide the primary soil nutrient storage results from (a) the negative charges on the micelles and (b) their very large surface-to-volume ratio. The hypothetical dicing of a mineral grain (Figure 17-7) allows us to envision the difference in the surface-to-volume ratio of a sand grain with sides 1 mm in length

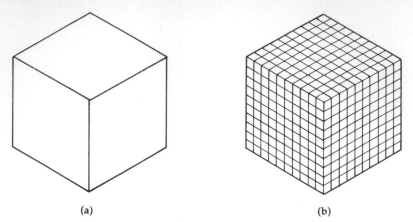

   (a)   (b)

Figure 17-7   (a) A block with sides 1 mm in length has a surface area of 6 mm^2. (b) The same block sliced vertically and horizontally into blocks with sides 0.1 mm in length has a surface area of 60 mm^2. Block b has 10 times the surface area of block a. From *Soils and Soil Fertility* by Thompson and Troeh (New York: McGraw-Hill, 1973). Reprinted by permission of Frederick R. Troeh.

and the same volume of clay particles with sides 0.001 mm in length. The sand grain has a volume of 1 mm^3 and a surface area of 6 mm^2. If we slice this grain into clay particles with sides 0.001 mm in length, the surface area now equals 6000 mm^2, which is 1000 times that of the parent grain.

In summary, the sand fraction provides skeletal (structural) support and aeration and permeability for the soil. The clay fraction provides water-holding and nutrient storage capacities (see also the section "Soil Chemistry"). Silt also provides water-holding capacity and, through weathering, provides plant nutrients and material to form clays, thereby replacing losses from leaching.

Soil Structure

Primary soil particles are arranged into secondary units, called peds (a **ped** is a unit of soil structure such as a block, column, crumb, granule, plate, or prism, formed by natural process). The nature of the arrangement or aggregation of the peds is called soil structure. Soil aeration is greatly enhanced by good soil structure, as is the movement of water and plant roots through the soil.

Aggregate stability, and the shapes and sizes of aggregates, are the most important characteristics of soil structure. *Stability* means the capacity of the peds to retain their structure—that is, to absorb water and not disintegrate. Various agents bind these aggregates. Fungal mycelia, microbial exudates, calcium ions, and certain clays promote stability; sodium ions and expansion and contraction of certain clays promote instability. Chemical additions also affect soil structure, and can reduce soil erosion.

The size and shape of peds are peculiarly significant to the success of plants and other soil-dwelling organisms. Table 17-4 presents illustrations and descrip-

Table 17-4 Classification, description, and illustration of various types of soil structure. Modified from *Nature and Properties of Soils,* 11th ed., by N. C. Brady and R. R. Weil. Copyright © 1996. Adapted by permission of Prentice-Hall, Inc.

Classification	Description	Illustration
Structureless		
Single grain	Each soil particle independent of all others.	
Massive	Entire soil mass clings together, no lines of weakness.	
Structured		
Spheroidal	Granular: Primary soil particles grouped into roughly circular peds, such that there is space between them as they do not fit tightly together. Enhances permeability. Crumb: Crumb structure is a more porous version of granular.	Granular (porous) Crumb
Platelike May be platy-leafy or flaky	Peds with horizontal dimensions greater than vertical. Common at soil surface and often associated with lateral movement of water.	Platy
Blocky	Peds approximately equal in vertical and horizontal dimensions, but fit well together, unlike granular peds. May be angular or subangular.	Angular blocky Subangular blocky
Prismatic	Peds taller than wide, common in B horizons of well-developed soils.	Prismatic
Columnar	Columnar peds may develop from prismatic peds due to old age or high sodium content.	Subangular rounded
Structure destroyed		
Puddled	Soils disturbed when wet, puddled, or run together; structure is destroyed, pores collapse.	

tions of several common soil aggregates. Notice that granular soils are particularly suitable for plant growth because both water and air can circulate easily between the peds. The presence of good structure—granular peds, high degree of aggregate stability—is especially desirable in the topsoil. If the texture of the B horizon is sandy, then structure is of little consequence because drainage and aeration are ensured through macropores. However, certain soils have B horizons with a "claypan" or hardpan, which has a deleterious effect on plant life. An excellent example of such a soil is the Blacklock Spodosol or podzol soil (Typic Duraquod) of the Pygmy Forest in Mendocino County, California. Jenny et al. (1969) noted with some amusement the lack of agreement among naturalists as to the cause and effects:

> [n]aturalists indulged in an apparent *circulus vitiosus*. On field trips the professors of botany would tell their students that the podsol soil is the cause of the unusual assortment of plant species, whereas the visiting professors of pedology (soil science) would attribute—in the light of classical podsol theory—the soil horizon features to the acid-producing vegetation. While it is true that a species individual responds to its soil niche, it is also true that it modifies that niche, which, in turn, reacts upon the individual.

This Blacklock soil is one of the most acid soils known, with a pH of 2.8–3.9 and a characteristic bleached white surface horizon underlaid by an iron-cemented hardpan (Figure 17-8). Leachate from the acid conifer litter cannot drain down through the hardpan but continues to remove nutrients from the already depauperate surface soil, carrying them away in tea-colored rivulets.

Figure 17-8   Soil profile of Blacklock sand, a Spodosol (podzol soil) in the Pygmy Forest, Mendocino County, California. The horizon boundaries are wavy rather than smooth. Note the highly bleached E horizon, beneath which is an iron-cemented hardpan, the Bsm horizon. Rusty, mottled, and weakly cemented sands and sandy loams underlie the B horizon. The entire profile encompasses only about one meter in depth.

O horizon

A horizon

E horizon

Bsm horizon

Bw horizon

Soil Chemistry

One of the most important aspects of soil chemistry is the state of and the relationship between chemical elements in soil systems. The elements may occur as soluble ions in the soil solution, as adsorbed ions on charged particles, and as constituents of mineral and organic particles. **Adsorbed** ions refers to those adhering to the surface of a solid as opposed to **absorbed** ions, those ions that are taken into a cell, tissue, or organ. Soil chemistry deals primarily with reactive materials, which are the chemicals that are of significance to living things.

Reactive materials tend toward equilibria in a changeable environment. Perhaps because life began in the ocean, there is virtually no correlation between the absolute and relative amounts of an element present in the soil and the significance of that element to life. The most abundant, the second most abundant, and the third most abundant elements in the earth's crust are, by weight, oxygen (49.5%), silicon (25.8%), and aluminum (7.5%), respectively. Yet these elements are relatively inaccessible due to the bound nature of their chemical compounds. By contrast, the most important nutrient elements in soil chemistry, listed in Table 17-5, make up less than 16% of the materials in the earth's crust. Soil pH, cation exchange capacity, and the nature of the interface between the solid phase and the liquid phase are all important aspects of soil chemistry.

Soil pH

Typical soils exhibit pH values that range from 4 to 8, although certain soils may have values higher or lower. For instance, saline soils commonly exhibit pH values ranging from 7.3 to 9.5. In southern Death Valley, California, saline soils support a saltbush scrub of desert holly (*Atriplex hymenelytra*), Parry saltbush (*A. parryi*), and honeysweet tidestromia (*Tidestromia oblongifolia*). The distinctly alkaline soils of the broad desert valleys of the Great Basin region of western North America support a shadscale scrub vegetation dominated by shadscale saltbush (*Atriplex confertifolia*) and bud sagebrush (*Artemisia spinescens*). A soil pH of less than 7.0 is considered an acid soil (Brady and Weil 1996).

Certain bog soils only a few hundred years old may be acidic. Some older soils, especially those covered by coniferous forest in humid regions, also have very low pH values. The pygmy cypress (*Cupressus pygmaea*) grows in the Blacklock Spodosol of the Mendocino Pygmy Forest, which we noted earlier has an extremely acid soil of pH 2.8–3.9. The highly weathered Blacklock Spodosols are half a million years old. They consist primarily of very resistant quartz grains and contain very small amounts of nutrients for plant growth. Thus they are similar to highly weathered tropical soils in that most of the nutrients are contained within the living biomass or soil organic matter. In general, leaching tends to remove basic cations and thus lower the pH of the soil. Strong Spodosols with a pH less than 4.0 may develop in as little as 10,000 years in certain glacial materials.

The influence of plants on soil pH is very complex. Nutrients (both cations and anions) are brought up from subsurface soil by plants and then deposited with

Table 17-5   Elements important in soil chemistry and their chemical symbols and principal ions. From *Soils and Soil Fertility* by Thompson and Troeh (New York: McGraw-Hill, 1973). Reprinted by permission of Frederick R. Troeh.

Element	Principal ions	Element	Principal ions
Aluminum	Al^{3+}	Manganese	Mn^{2+}, MnO_4^{2-}
Boron	$B_4O_7^{3-}$	Molybdenum	MoO_4^{2-}
Calcium	Ca^{2+}	Nitrogen	NH_4^+, NO_2^-, NO_3^-
Carbon	CO_3^{2-}, HCO_3^-	Oxygen	With other elements
Chlorine	Cl^-	Phosphorus	$H_2PO_4^-$, HPO_4^{2-}
Cobalt	Co^{2+}	Potassium	K^+
Copper	Cu^{2+}	Sodium	Na^+
Hydrogen	H^+, OH^-	Sulfur	SO_4^{2-}
Iron	Fe^{2+}, Fe^{3+}	Zinc	Zn^{2+}
Magnesium	Mg^{2+}		

litter on the soil surface. Grasses tend to use more bases, and grass cover may act to keep the soil pH from dropping. Coniferous litter, on the other hand, provides acidic leachate, which lowers soil pH. Fewer cations are held on exchange sites, and more cations are released by acid weathering. Figure 17-9 shows the changing availability of plant nutrients in response to changing pH values.

Agricultural soils may develop a low pH due to fertilizer applications that add ammonium-nitrogen and elemental sulfur. For example, ammonium enhances microbial activity, increasing populations of soil organisms that release hydrogen ions (H^+), thus acidifying the soil and displacing stored nutrient cations. Lime (CaO or $CaCO_3$) is often applied to acid soils to make them more basic. Very alkaline soils, on the other hand, may be treated with elemental sulfur to lower pH. In most soils, the pH cannot be changed very much because of effective soil pH buffering.

Although the direct effects of soil pH on plant growth are very limited, the indirect effects are numerous and significant. Toxicity of certain metals, such as aluminum and manganese (which are more soluble at lower pH), would be a direct effect of pH. The most important indirect effect of soil pH is its influence on nutrient availability (Figure 17-9) and on microbial activity. Rates of weathering, availability of nutrients such as nitrogen, phosphorus, and sulfur, and the leachability of nutrients such as potassium are all influenced by soil pH. The availability of nutrients with low solubilities, such as phosphorus, is dramatically influenced by pH changes. Calcium phosphates are less soluble as the pH climbs, and iron and aluminum phosphates are less soluble as the pH drops. A near-neutral pH, between 6.5 and 7.5, is best for phosphorus availability—indeed for the availability of most nutrients needed for plant growth.

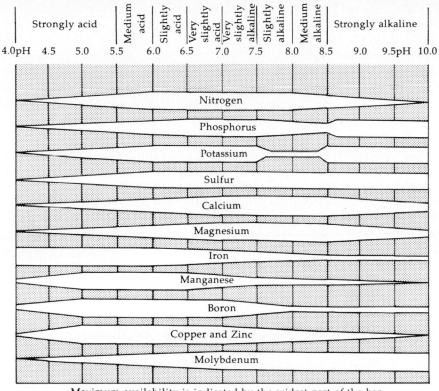

Maximum availability is indicated by the widest part of the bar

Figure 17-9   The relationship between pH and plant nutrient availability. From *Soils and Soil Fertility* by Thompson and Troeh (New York: McGraw-Hill, 1973). Reprinted by permission of Frederick R. Troeh.

Soil Nutrients

As we have seen, one of the important effects soil pH can have on soil chemistry is that of influencing availability of mineral nutrients (Figure 17-9). The distinction between the more general characteristic of soil chemistry and the more specific characteristic of soil fertility is that whereas **soil chemistry** refers to the overall chemical makeup of soil (and the factors which influence that), **soil fertility** refers to the availability of essential nutrients to plants. Several factors other than pH interact to influence the degree to which nutrients become available for uptake by plants. Three of these (humus formation, cation exchange capacity, and soil solution) will be touched on briefly in following sections. Also important are soil particle size (soil texture) and other factors that vary with the soil profile, as discussed previously (Barber 1995, Marschner 1995). Several other aspects of soil nutrients were discussed in the context of ecosystem nutrient cycling in Chapter 13.

Nutrient availability of soil may vary greatly (a) among soil types (even over relatively small spatial scales) and (b) with depth (horizon) within a particular soil type. Table 17-6 shows patterns of nutrient availability with soil horizon for two soils at the University of Michigan Biological Station in Cheboygan County, Michigan. These soils are similar in many respects to the one depicted in Figure 17-8 in that all three soils are Spodosols.

Both Michigan soils support similar stands of largetooth aspen (*Populus grandidentata*) and both are generally sandy textured, having been derived from glacial outwash deposits (sands). Furthermore, the two plots are <10 km apart. These soils, however, show sharply contrasting nutrient characteristics. The Rubicon sand of plot 2 is strongly acidic and low in available (extractable) nutrients; notably, it has a calcareous gravel stratum (note Ca and Mg concentrations for this stratum), which may have been the result of deposits from glacial activity. Note also the high concentration of PO_4 in the B21ir horizon (now referred to as the Bs horizon), the result of adsorption of PO_4 by soil materials, especially iron, which accumulate in this "spodic" horizon; this pattern is also evident in the soil of plot 4. Because parent materials for Rubicon soils are virtually all sandy glacial outwash deposits, water-holding capacity is low; indeed, this site is typical of dry-mesic sites in the area (Roberts and Gilliam 1995b).

Table 17-6   Soil chemical characteristics of two Spodosols from northern lower Michigan. WAI is water availability index, calculated as the amount of water (in %) held by the soil between −0.033 and −1.5 MPa moisture potential (see Chapter 18). OM is organic matter, calculated as twice the percent content of organic carbon. Data were taken from Roberts and Richardson (1985). Nutrients are expressed as extractable levels using Mehlich dilute double-acid extraction, as described in Gilliam and Richter (1991). Data used with permission. Copyright © 1995 National Research Council of Canada.

Horizon	Depth (cm)	pH	WAI (%)	OM (%)	PO_4 ($\mu g/cm^3$)	Ca ($\mu g/cm^3$)	K ($\mu g/cm^3$)	Mg ($\mu g/cm^3$)
Plot 2: Rubicon sand (sandy, mixed, frigid, Entic Haplorthod)								
A1	0–2	4.14	4.0	1.78	2.1	104.8	25.0	15.2
A2	2–7	4.03	4.3	0.96	1.7	36.7	15.5	7.6
B21ir	7–72	4.83	2.4	0.90	14.4	71.6	13.1	11.1
B22ir	72–118	5.13	0.8	0.20	7.0	36.3	6.4	6.5
Gravel	118–138	7.50	1.4		3.0	2340.0	12.4	300.0
C	138+	5.78	0.9	0.16	8.0	306.4	6.4	59.8
Plot 4: Montcalm loamy sand (sandy, mixed, frigid, Alfic Haplorthod)								
A1	0–1	4.54	5.3	5.70	4.0	674.6	40.0	54.7
A2	1–22	4.59	4.0	0.84	2.4	115.2	12.2	17.1
B2ir	22–70	5.11	3.5	1.62	16.5	278.0	21.8	17.7
A'+B'	70–175	5.68	2.6	0.30	8.4	242.9	36.9	42.4

The Montcalm loamy sand overlies glacial till (a mixture of sand, silt, and clay laid down directly by glacial ice), which may be considered an assemblage of buried soil horizons (indicated as A'+B'), as mentioned at the beginning of this chapter. Because of the more loamy texture, water availability is generally much higher for this soil compared to the Rubicon sand (Table 17-6). Higher water availability allows Montcalm soil to support more hardwood vegetation (Roberts and Richardson 1985), resulting in higher soil organic matter, less acidity, and higher nutrient availability than Rubicon soil (Table 17-6). Plot 4 is typical of mesic sites in the region (Roberts and Gilliam 1995b). Such comparisons further emphasize the importance of interactions of numerous abiotic and biotic factors in determining availability of soil nutrients to plants.

Humus Formation

The action of microorganisms on raw organic materials eventually converts them into **humus,** very finely divided particles nearly black in color. By virtue of tiny size, humus can coat soil particles to the point that the soil appears black when it contains as little as 5% organic matter. Root exudates, earthworm excreta, and the like may provide a significant contribution to this nonliving, finely divided organic matter. However, probably 90% of humus is attributable to microorganisms, microbial tissue, and microbial activities on litter.

Humus is itself an intermediate product of decomposition. It may decompose at an annual turnover rate of less than 3% in temperate regions. It comprises not only the resistant fraction of the organic matter but also the microbes that have accomplished the decomposition of the more decomposable fraction. In a climax community or even a stable seral one, the amount of new humus formed each year may approximate that which is removed by decomposition. The significance of slowly decomposing humus is that it provides a long-term supply of continually available nutrients for plant growth.

Although the composition of humus varies, 10–15% of its weight is made up of easily identifiable organic compounds: polysaccharides such as cellulose and its decomposition products; polyphenols such as tannins; proteins and their decomposition products; and a host of hydrocarbons, organic acids, alcohols, esters, and aldehydes. The rest of the humus is not so easily identified but is characterized as humic materials, many of which contain reactive groups—carboxyl, amine, phenolic hydroxyl groups—combined with chains and rings, forming rather large and complex molecules.

Cation Exchange Capacity

Cation exchange capacity (CEC) is a measure of the number of negatively charged sites on soil particles that attract **exchangeable cations,** that is, positively charged ions available for plant uptake. Factors that strongly influence the cation exchange capacity of a soil are its clay content, the kinds of clay minerals or amorphous colloids it contains, its humus content, and its pH. The negative charge of clay particles is a result of substitution of ions with different valences and ionization of hydrogen ions from OH groups.

Soil Solution

Ideally, one must understand soil water *in situ,* in its function as the liquid medium that renders the soil a habitable environment. In reality, soil water is extracted by centrifugation or tension lysimeters and is analyzed in the laboratory (Dahlgren et al. 1997). Readily available ions are present in the soil solution, and they maintain an equilibrium with adsorbed ions (Figure 17-10). These, in turn, maintain an equilibrium with mineral ions at the edge of the micelle. Plant removal of nutrients from the soil solution lowers the concentration of those nutrients in the rhizosphere; ions then move toward the rhizosphere down a chemical gradient. This shift, due to solubility constants of the ions involved, enhances the release of adsorbed ions, and therefore the renewal of the nutrient in the soil solution.

For example, plant uptake of phosphorus establishes a concentration gradient in the soil solution such that the movement of P is toward the root. Under optimum conditions, daily uptake of P in grassland may be as much as 50 times the quantity of P in the solution pool, the pool itself being replenished from the labile pool (that is, from adsorbed ions). Mycorrhizae, discussed in Chapter 7, greatly enhance uptake of P, due to the tremendously enlarged contact zone of root and soil mediated

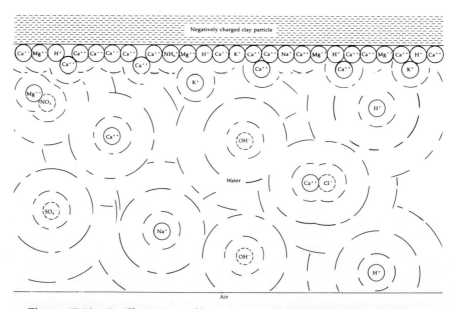

Figure 17-10   An illustration of how ions are distributed in the soil solution near a negatively charged clay particle. From *Soils and Soil Fertility* by Thompson and Troeh (New York: McGraw-Hill, 1973). Reprinted by permission of Frederick R. Troeh.

through hyphal fungi. With continued uptake, the concentration of P in the cytoplasm of root cells may exceed (by up to two orders of magnitude) concentrations in ambient soil solution. When conditions of temperature and water are exactly right, the labile pool has the potential of replacing the solution pool 250 times daily.

Tropical Rain Forest Soils

Once thought to have soils homogeneous enough to allow the term "tropical soil," the tropics contain an extremely high diversity of soil types, thus precluding such a simplistic label for all of them (Richter and Babbar 1991). Let us consider the interaction of vegetation, soil texture, rainfall regime, temperature regime, and soil microorganisms, in reference to cation exchange in a particular tropical ecosystem, the **humid tropical rain forest (TRF).** Many TRF soils may be highly weathered with little or no obvious horizon development. A large proportion of the Oxisols and Ultisols of South America and Africa are acidic and nutrient poor. The litter layer may be extremely thin because decomposition and release of nutrients from organic matter occurs at a very rapid pace in the warm, moist rain forest. Even wood is quickly recycled by subterranean termites.

The luxuriant growth of vegetation (as much as $52.5 \text{ t ha}^{-1} \text{ yr}^{-1}$ gross productivity) belies the poor nutrient content of most TRF soils. The storage for cations resides mainly in the organic component of the soil. The soil is not a significant nutrient pool, and most of the nutrient reserve required by the forest is contained in the phytomass (Walter 1979, Brady and Weil 1996).

What mechanism prevents mobile nutrients from being leached away by the inevitable rains? Work by Went and Stark (1968) showed that feeding roots of TRF trees are directly connected to litter. These rootlets exploit the hyphae of mycorrhizal fungi to obtain nutrients. Mutualistic fungi interact with the roots and thereby increase their host's ability to extract nutrients, which the fungi obtain by absorption, from the forest litter. Due to this highly adaptive relationship there is little chance that free, mobile ions released from litter biomass will be lost through leaching.

Considering that most tropical areas have year-round growing seasons and, in the humid areas, ample rainfall, it is somewhat surprising that those areas have not been more effectively utilized for growing food for humans. Not all soils of tropical areas are depauperate; some are arable and deep. Oxisols in the Philippines and in Hawaii are used successfully for both sugarcane and pineapple crops. But as human populations increase, the demand on the arable land increases, and soils are at risk, as in this scenario:

As long as the rain forest is not disturbed, the lack of a soil nutrient storage compartment is unimportant. The nearly complete recycling of materials allows for stability for hundreds or even thousands of years. Yet deforestation through burning, clearing of the land for cultivation, or logging removes the only significant storage facility, the phytomass, and quickly allows the mobilized ions to move out of the system. In certain areas, exposed soil can become cemented and hard as a pavement. The luxuriant growth of the virgin forest may never (in human terms) be renewed (Brady and Weil 1996, Buol et al. 1997).

Soil Taxonomy

The naming of soils allows one to impart a great deal of information simply by invoking the name of a certain soil. A name is applied to an individual kind of soil that has a definable existence on the landscape. A unit of soil, large enough such that its profile and horizons can be studied and its properties defined, is called a **pedon.** This is the smallest practical volume that can be properly referred to as a soil. A pedon may range from 1 m^2 to 10 m^2 in area, depending on the depth and continuity—or discontinuity—of the soil's horizons within short linear distances (1 to 7 m). Pedons of soils with continuous horizons have the smallest area (1 m^2). Groups of similar adjacent pedons, termed **polypedons,** form the basis for soil survey and soil classification.

To be of value, a classification system must be scrupulously followed and names must be assigned on the basis of certain specified properties. Characteristics that are used to define a soil are termed **differentiating characteristics** or **differentiae.**

Soil scientists strive for a classification scheme that reflects a natural relatedness. The scheme of naming is called soil taxonomy. Any taxonomic system is an arbitrary creation of its author, reflecting certain biases and devised to fill particular needs. In particular, biological taxonomies may have as their basic unit *species* that are discrete—reproductively isolated. But soil taxonomy is similar to vegetation classification schemes in that gradual changes from one type to another may be partitioned into discrete types in spite of the gradient.

History

The term *soil* is very old, and humans have been concerned with soil as a medium for plant growth for millennia. In the 1870s, an innovative and creative Russian school led by V. V. Dukochaev developed and introduced a revolutionary concept of soil and a system of naming soils.

> Soils were . . . independent natural bodies, each with a unique morphology resulting from a unique combination of climate, living matter, earthy parent materials, relief, and age of land form. The morphology . . . reflected the combined effects of the particular set of genetic factors responsible for its development. (U.S. Department of Agriculture 1996).

The Russian soil taxonomy was adopted first in Europe, then in the United States early in the twentieth century. As more soils were described and other information became available, the weaknesses of the Russian system became apparent. In 1938, a system that was essentially Russian in organization, approach, and perspective was developed in the United States and published under the title *Classification of Soils on the Basis of Their Characteristics* (Baldwin et al. 1938). Then, starting in the 1950s, a taxonomic scheme was developed, modified, improved, and presented as a series of approximations, each one more refined and more workable than the last. Six of these approximations were tested by the Soil Survey Staff of the USDA, various U.S. land grant universities, and interested European individuals and agencies.

The seventh approximation was published and issued on a limited basis in 1960 at the Seventh International Congress of Soil Science at Madison, Wisconsin, for wider testing and evaluation (Soil Survey Staff 1960). It was generally referred to as the Seventh Approximation or "the brown book." In 1965, a modified form was officially adopted for soil classification in this country by the National Cooperative Soil Survey. With additional modification, it was published for unrestricted use as a comprehensive system in 1975 by the Soil Survey Staff of the USDA as Agricultural Handbook 436, *Soil Taxonomy—A Basic System of Soil Classification for Making and Interpreting Soil Surveys*. It is this internationally accepted system that we will discuss here.

To be useful, a soil taxonomy should have the following attributes:

1. The definition of each taxon should have the same meaning for each user.
2. The taxonomy should be a multicategoric system, with many taxa in the lower categories.
3. The taxa should refer to real soils, known to occupy geographic areas.
4. Differentiae should be soil properties that can be discerned in the field, inferred from other properties that are observable in the field, or taken from the combined data of soil science and other disciplines.
5. The taxonomy should be modifiable with a minimum of perturbation to the system.
6. The differentiae should maintain pristine soil and manipulated soil that is its equivalent in the same taxon whenever possible.
7. The taxonomy should provide taxa for all the soils of a landscape or at least be capable of such provision. (U.S. Department of Agriculture 1975)

Differentiating Properties of Soils

The properties of a polypedon should serve as differentiae. Soil color and soil horizons have long been used for differentiae. Properties of soil that interact with other properties should be carefully considered when differentiae are selected to define soil taxa.

There are seven defined diagnostic surface horizons or epipedons. An **epipedon** is a horizon, or horizons, forming in the upper part of the soil profile. We will name these diagnostic epipedons but not describe them, due to limitations of space. They are anthropic, histic, melanic, mollic, ochric, plaggen, and umbric (Brady and Weil 1996).

A subsurface horizon (often B, but perhaps part of A) is also of diagnostic value, and the names, descriptions, and genesis or mode of formation of these horizons are also important in soil classification. The diagnostic subsurface horizons include agric, argillic, cambic, duripan, natric, oxic, placic, spodic, and others (Brady and Weil 1996).

Other properties used in defining a soil include an abrupt textural change, lithic contact, microrelief, mineralogical composition, organic soil materials, particle-size classes, soil moisture regimes, soil temperature regimes, and so on.

The Structure of Soil Taxonomy

The USDA soil taxonomy system recognizes a hierarchy of categories. Each category includes a set of taxa. The series is the most basic unit of classification. One of the strengths of the system lies in its capacity for internal modification, as new information is gathered, without disruption of the rest of the system. The hierarchy is as follows:

Categories	*Example of taxa within categories*
Order	Entisol
Suborder	Orthent
Great group	Cryorthent
Subgroup	Typic Cryorthent
Family	Typic Cryorthent loamy-skeletal, carbonatic, frigid
Series	Swift Creek

Most of the formative elements in the names of soil orders are derived from Latin or Greek words appropriate to that soil order, and the soil order name is formed with the suffix *-sol*, from the Latin *solum* for soil. The formative element *ent* is not from Greek or Latin; it is a meaningless syllable. Entisols are often young soils with little or no horizon development. The derivations of the other formative elements are listed in Table 17-7. A new order, Gelisols (soils with permafrost), was added in 1998 (Southard 1998).

The names of suborders are a combination of the formative element of the parent order as the suffix with a prefix that suggests the diagnostic properties of the soil. For example, *orth* (from Greek, *orthos*, "true") combined with *ent* (from Entisol, "recent soil") names a suborder of the Entisol order that is common—*Orthent*. Similarly, a great group name consists of the name of the suborder coupled with a prefix containing one or two formative elements to refer to definitive properties. For instance, *Cryorthent* (from Gr. *kryos*, "coldness") indicates an Entisol with a mean annual temperature $>0°$ but $<8°C$—a very cold soil.

The great group name is combined with one or more modifiers to generate the subgroup name. *Typic* Cryorthent would denote the subgroup thought to typify the great group. Families are named with polynomials consisting of the subgroup name modified by adjectives in a specific order that are names of classes describing the soil's particle size, mineralogy, reaction, and temperature classes, among other factors. Series names are place names, usually taken from the region where the soil was first described. In the field, profiles that do not match the descriptions of officially recognized series may be designated as **variants** of the existing, closely related series. With time and additional field information, a variant may be defined as an official soil series with its own unique name.

It is important for plant ecologists to know that due to limitations of time and money, the definition of soil bodies has only a coarse resolution, and that the common practice of naming variants can be misleading. For example, a soil carries a name and description derived from analysis of that soil in a relatively flat area—a

Table 17-7   Formative elements in the names of soil orders. Modified from *National Soil Survey Handbook 1997*, http://www.statlab.iastate.edu/soils/nssh/. Courtesy of the U.S. Department of Agriculture, Natural Resource Conservation Service.

Name of order	Formative element in name of order	Derivation of formative element	Pronunciation of formative element	Typical vegetation on soils of order
Alfisol	Alf	Al and Fe, symbols for aluminum and iron	Ped*alf*er	Deciduous forest
Andisol	And	J. *ando*, "dark"[a]	*And*esite	Many types
Aridosol	Id	L. *aridus*, "dry"	Ar*id*	Desert scrub
Entisol	Ent	Meaningless syllable	Rec*ent*	Many types
Gelisol[b]	El	L. *gelid*, "cold"	*Gel*atine	Tundra
Histosol	Ist	Gr. *histos*, "tissue"	H*ist*ology	Marsh
Inceptisol	Ept	L. *inceptum*, "beginning"	Inc*ept*ion	Cool grassland
Mollisol	Oll	L. *mollis*, "soft"	M*oll*ify	Grasslands
Oxisol	Ox	F. *oxide*, "oxide"	*Ox*ide	Tropical rain forest
Spodosol	Od	Gr. *spodos*, "wood ash"	*Od*d	Coniferous forest
Ultisol	Ult	L. *ultimus*, "last"	*Ult*imate	Forest
Vertisol	Ert	L. *verto*, "turn"	Inv*ert*	Many types

[a] J. meaning Japanese
[b] Gelisols, a new order (soils with permafrost), added in 1998.

valley, for instance—on which the development has exceeded rates of erosion. The soils on the hillsides surrounding the valley may carry the same name with "slope variant" appended, or they may be designated as a different series. A plant ecologist may note that the plants on the slope and the plants in the valley indicate that the resources for plants are different between the two sites. Water-holding capacity, nutrient storage capacity, and nutrient content may differ significantly in the two adjacent areas. One may find that the slope communities are forever becoming—are seral in nature—and that the flats bear both mature and seral communities.

Summary

Soil can be defined as any mixture of inorganic and organic materials capable of supporting plant growth. The components of soil are mineral grains, organic matter, water, and air. They provide, respectively, plant anchorage and nutrients, intrasystem cycling, solvent medium, and oxygen and nitrogen. The upper layers of soil weather and change in response to abiotic and biotic factors. The first phase of soil development is weathering of parent materials (usually consolidated rock, but sometimes unconsolidated sediments), a process that is facilitated by both physical and chemical processes; the second phase is soil formation.

Soil formation is a function of several environmental factors, including climate, organisms, topography, parent material, and time. The degree of profile development in a soil—the characteristics of the diagnostic horizons—together with the predominant environmental influences on that soil provide the basis for soil classification.

Soil texture refers to the relative proportions by weight of sand, silt, and clay particles of the soil. Skeletal support is provided primarily by the largest particles, such as sand, and the macropores associated with large particles provide permeability; clay provides water and nutrient storage; silt aids in water storage and weathers to produce additional nutrients and clay. The arrangement of peds and stability of aggregates contribute significantly to soil structure.

Micelles (clay and organic colloids) provide the primary storage for readily extractable (available) nutrients in the soil. Equilibrium reactions influence ions dissolved in the soil solution, adsorbed on exchange sites, or bound in weatherable minerals. Most soils exhibit a pH range from about 4 to 8, with most soils becoming more acidic as weathering proceeds. Climate, vegetation, parent material, and relief determine soil pH, which in turn affects vegetation and rates of weathering.

Soil microorganisms convert raw organic matter (plant and animal remains) to humus, a finely divided, nearly black fraction that decomposes at a stable equilibrium rate in climax communities. Cation exchange capacity is influenced by the amount and type of clay minerals and humus. Some tropical soils have little or no obvious horizon development and are nutrient-poor soils with little exchange capacity. The phytomass associated with these soils provides the bulk of their nutrient storage.

Soils are classified by their differentiating characteristics. Soil taxonomy is a classification scheme designed to reflect a natural relatedness among soil groups. To be effective, a taxonomy must have definitions for each taxon that have the same meaning for each user. The system currently used in the United States was developed by the Soil Survey Staff of the USDA Natural Resource Conservation Service. It relies on the properties of the polypedon for differentiating characteristics and has a very complex but flexible multicategoric system. Names for the soil taxa are formed with syllables selected from terms that reflect the nature of the soil. The USDA-NRCS mapping system was not developed in a manner that always allows for direct application to ecological concepts.

CHAPTER 18

PLANT WATER DYNAMICS

T he acquisition of carbon dioxide and nutrients in the face of limiting water resources establishes a dilemma that is a major selective force in most terrestrial plants. In this chapter we examine the nature of water movement and the response of plants to water stress. We will consider the environmental factors that influence water availability in the soil, water loss to the atmosphere, and the characteristics that control plant response to changes in soil and atmospheric water. Terrestrial plants are exposed to water regimes that change seasonally and diurnally. If they are to survive, they must adjust physiologically and/or anatomically, have a broad range of tolerance, or avoid times of limited water. We will also discuss methods of measuring soil and plant water status.

The Water Potential Concept

Originally, water status of plants and their environment was reported in relative values such as percent tissue or soil moisture. Percentage values convey information concerning the amounts of water present but do not tell either the direction of movement or the response of a plant to that amount of water. Now we use the water potential concept, in which the status of water is expressed in thermodynamic terms. All matter tends to move from areas of high free energy (high capacity to do work) to areas of lower free energy. Water potential (ψ) is a measure of the free energy of water in comparison to the free energy of pure water. The ψ of pure water has been arbitrarily assigned a value of zero megapascals (MPa). Megapascal is a pressure unit directly relatable to energy per unit mass (0.1 MPa = 1.0 bar = 0.987 atmosphere = 10^6 ergs g^{-1}). The free energy of water in the biosphere is generally lower than that of pure water; therefore, water potential values are usually negative numbers.

Several factors determine the chemical energy of water. Water potential depends on three predominant forces:

$$\psi_{tot} = \psi_m + \psi_s + \psi_p \qquad\qquad \text{(Equation 18-1)}$$

Matric potential (ψ_m) is a measure of the reduction in free energy caused by the attraction of water molecules to surfaces. The matrix may be macromolecules, soil particles, cell walls, or other surfaces that do not go into solution. Such attractions always reduce the free energy of water, so matric potential is always negative. Osmotic potential (ψ_s) is the reduction in free energy (always negative) attributable to electrical attraction, reorientation, and increased entropy caused by solutes. In living systems, a selectively permeable membrane must be present because no osmotic potential difference could exist without a restriction on the movement of solute molecules (Campbell 1977). Pressure potential (ψ_p) is the positive or negative hydrostatic pressure present in the system. Within living plant cells, ψ_p is usually positive because of turgor pressure, but it may be negative in the xylem when transpiration is occurring. Total water potential is a negative value except in fully turgid cells where ψ_p may balance the negative ψ_s and ψ_m potentials and $\psi_{tot} = 0$ (Kramer 1969).

Humidity and Vapor Pressure

Understanding the processes of evaporation and transpiration (evapotranspiration) is basic to determining the interchange of energy and water between the plant and the environment. Large amounts of energy are involved in any change in the state of water. For example, it takes most of the solar energy falling on 1 cm^2 on a clear summer day to evaporate a cubic centimeter of water. Evaporation of water from plants (transpiration) causes significant cooling of the leaf surface. Plants grown in very high humidity can develop a detrimental heat load because of a reduction in water. Court (1974) estimated that two-thirds of the precipitation falling on the conterminous United States evaporates—this is a massive energy sink. Conversely, when water condenses or freezes, large amounts of energy are released. This energy is in the form of heat, called latent heat. When temperatures reach dew point (the temperature at which atmospheric water condenses), enough latent heat is released to reduce the rate of cooling significantly. Plants growing near bodies of water can often escape frost damage because of this phenomenon (Rosenberg 1974). The freezing of intercellular water also slows cooling.

The rate of evapotranspiration is determined in part by the water vapor content of the atmosphere. At a constant temperature, when pure water evaporates into a closed space an equilibrium will occur at saturation vapor pressure. Saturation vapor pressure is the maximum possible partial pressure of water vapor in the air at a given temperature. Air is seldom saturated, so we need a means of expressing levels below saturation. One simple measurement of water vapor content of the atmosphere can be obtained with psychrometers, which consist of two ventilated thermometers. One is dry and measures air temperature, the other is wet and measures the reduction in temperature caused by evaporative cooling. Figure 18-1 shows the relationship between vapor pressure, relative humidity, and wet and dry

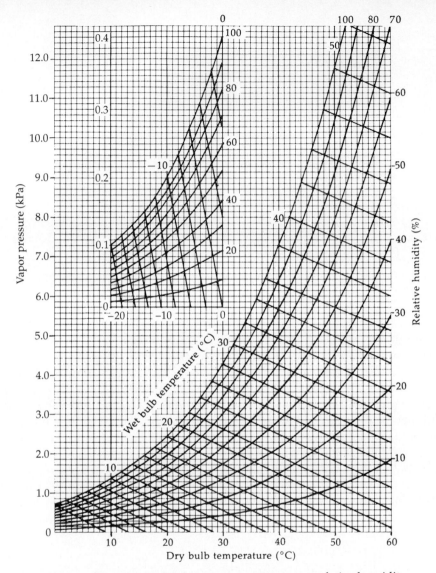

Figure 18-1 The relationship between vapor pressure, relative humidity, and wet and dry bulb temperature at sea level. See inset for values below 0°C. Modified from G. S. Campbell. 1977. *An Introduction to Environmental Biophysics.* By permission of Springer-Verlag, New York.

bulb temperatures. A given value of relative humidity can occur over a range of air temperatures and vapor pressures, making it a meaningless term unless air temperature is specified. Saturation deficit, a more meaningful measure of the evaporative power of the air, is the difference between the actual vapor pressure and the saturation vapor pressure at the same temperature. For example, in Figure 18-1, the saturation vapor pressure (100% relative humidity) at 20°C is about 2.4 kPa (kilo-

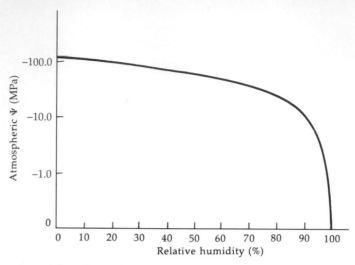

Figure 18-2   The relationship between atmospheric water potential (ψ) and relative humidity at 25°C.

pascals); at 50% relative humidity, vapor pressure is about 1.2 kPa, leaving a saturation deficit of 1.2 kPa. The saturation vapor pressure at 30°C is about 4.3 kPa, and at 50% relative humidity, vapor pressure is about 2.1 kPa, resulting in a 2.2 kPa saturation deficit. The potential evaporation is much greater at 30°C than at 20°C, even though the relative humidity is, in both cases, 50%. It is important to understand that factors determining evaporation are not restricted to the physical state of the atmosphere but also reflect the water potential of the evaporative surface. The importance of the evaporative power of the air is apparent in Figure 18-2, where at 25°C the relative humidity must be above 97% to approach the ψ_{leaf} of even the most xerophytic plants. Consequently, there is usually a large drop in water potential between leaves and the atmosphere.

Plant Hydrodynamics

The general pathway of water through the plant is summarized in Figure 18-3. Water enters plants both through young root tips with root hairs and through cracks in the suberized tissues of the root cortex in older roots. Chung and Kramer (1975) demonstrated that removal of unsuberized roots of loblolly pine caused an insignificant decrease in water absorption. Water enters the young root either by moving from cell to cell or vacuole to vacuole through the symplast or through the extracytoplasmic, apoplastic route in young, unsuberized roots. Eventually, all materials entering the plant are forced to move through cell membranes because of an

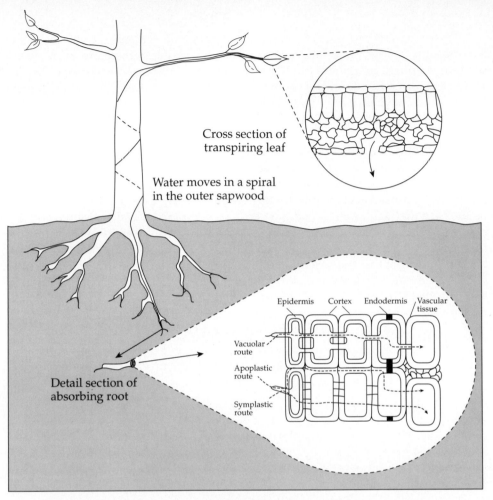

Cross section of
transpiring leaf

Water moves in a spiral
in the outer sapwood

Detail section of
absorbing root

Epidermis   Cortex   Endodermis   Vascular
tissue

Vacuolar
route

Apoplastic
route

Symplastic
route

Figure 18-3   The course of water flow through the soil-plant-atmosphere
continuum.

apoplastic barrier created by the casparian strip of the endodermal cells. Passage
through the endodermis offers considerable resistance, but once the water (and
minerals) enters the vascular cylinder and xylem, resistance is significantly lower.
Vascular strands divide into very fine segments upon entering the leaves and are
distributed throughout the mesophyll so that almost every cell is within one or two
cells of vascular tissue (Esau 1965), but no vascular tissue is exposed directly to the
intercellular spaces adjacent to stomates. From the vascular tissue, water continues
to travel in the liquid phase through the cell walls and between mesophyll cells to
substomatal cavities, where it vaporizes and transpires out through the stomatal
pore into the atmosphere.

Water Movement

The prevailing theory of the mechanism for the movement of water from the soil through the plant and into the atmosphere is based on the concept of a soil-plant-atmosphere continuum (SPAC), often referred to as the cohesion hypothesis. However, Ulrich Zimmerman and others, using pressure probe measurements in living plants, have proposed that SPAC alone cannot explain the pattern of water relations observed in all plants (Zimmerman et al. 1993, Canny 1995).

Transpiration: The Driving Force   According to the SPAC hypothesis, water movement is driven by transpiration and follows a continuous path from the soil, through the root, stem, and leaf, and into the atmosphere. The factors that affect this pathway are shown in Figure 18-4. For water to move through the plant, it must be the case that $\psi_{soil} > \psi_{stem} > \psi_{leaf} > \psi_{air}$, Transpirational water loss reduces ψ_p and the resulting tension (negative pressure) gradient pulls water upward through the xylem. Movement of water by this means requires a continuous column of water kept intact by the cohesive and adhesive properties of water. Even with a suitable ψ gradient, resistance to flow is inherent in the system (Slatyer 1967, Meidner and Sheriff 1976). Slatyer (1967) indicated that a common set of water potential measurements along the gradient might be ψ_{soil} −0.1 MPa, ψ_{stem} −1.0 MPa, ψ_{leaf} −1.5 MPa, $\psi_{atmosphere}$ −100 MPa. The most effective regulation of plant water sta-

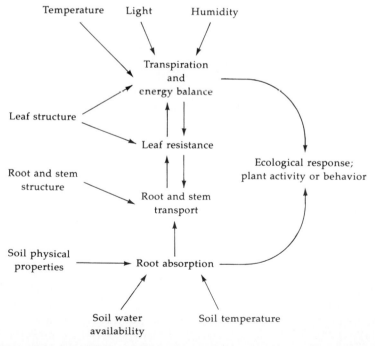

Figure 18-4   Interactions of the most important physical and biotic factors that determine the ecological response of plants to water.

tus is associated with the stomates because of their location in the steepest portion of the gradient. Their ability to open and close represents the major controlling factor of water flux from the plant to the air.

However, several lines of evidence suggest that transpirational pull may not be the only force that moves water in plants. First, movement by negative pressure requires a continuous column of water. Water under negative pressure is metastable and will cavitate (air bubbles will block the xylem) when pressures are sufficiently negative. Experiments using glass tubes show that the cohesive and adhesive forces of water molecules will prevent cavitation in xylem-like conditions only at pressures >-1.0 MPa (Smith 1994). However, plant ecologists have reported negative pressures down to -16.0 MPa in the xylem of transpiring plants and Holbrook et al. (1995) imposed centrifugal force on excised branches and found that xylem sap was able to sustain pressures of <-1.5 MPa without cavitating. It appears, therefore, that xylem elements are not adequately modeled by glass tubes and that negative pressures sufficient to move water exist in xylem (Pockman et al. 1995).

The second line of evidence that questions transpiration as the only driving force for water movement comes from recently developed pressure probes that attempt to measure pressure in functioning xylem cells. The pressure gradient that would be necessary to move water from the roots to the top of tall trees is not found. However, the pressure probe may not reflect the actual pressure of functioning xylem (Sperry et al. 1996). Milburn (1996) believes that the pressures measured with the pressure probe are from immature xylem elements that have varying degrees of turgor pressure.

Other evidence for forces other than transpiration includes the fact that insects with a maximum suction of 0.3 MPa can extract xylem sap from trees (Raven 1983). Also, the presence of gas bubbles in the vessels of transpiring trees suggests that water may not be in a metastable state (Grace 1993).

The Multiforce Hypothesis The multiforce hypothesis is an alternative hypothesis to explain water movement in plants. Part of its contrast to the SPAC hypothesis is that it proposes that plants can obtain transpirational water not only from the soil but also from their own tissues. Each tissue in the plant has a certain capacitance or ability to lose water as transpiration proceeds. Capacitance is reported as the change in tissue water content (m^3) per unit change in water potential. Tissues with a high capacitance deliver large quantities of water with a small change in ψ. The water provided by the tissue is a combination of apoplastic and symplastic water. Plants with a large capacitance in the tissues surrounding the xylem will transpire for longer periods before showing a decrease in water potential than those with a small capacitance (Nilsen et al. 1990). Desert succulents are an example of plants with a very large capacitance. Tissue water would thus keep water potentials from reaching critical levels in the xylem and would be renewed from the soil during times of low transpiration.

A second aspect of the multiforce hypothesis is that if mature xylem sap contains substantial amounts of osmotically active substances, and if the xylem is separated into compartments by solute barrier membranes, the pressure gradient in xylem shown by the pressure probe could be explained (Biles and Abeles 1991; Canny 1993, 1995; Zimmerman et al. 1994). Thus, the movement of water from the roots and surrounding tissues could be at least partially osmotically driven rather than strictly pressure driven as is expected with SPAC. Osmotically active xylem has not been shown to exist and the published data may be explained by other means (Milburn 1996, Richter 1997).

Third, recent pressure probe data have shown that small air bubbles are present on vessel element inner walls. These bubbles induce a circulation of water via the tension that is generated at the liquid-gas interface. This movement along the surface of bubbles is called Marangoni convection. The circulation that is generated could move water through the cell lumen (Nilsen and Orcutt 1996). The reverse osmosis water entering the xylem from the sieve tubes could also contribute to water movement (Milburn 1996). As research proceeds on these and other potential mechanisms for water movement in plants, our concepts of plant hydrodynamics will be refined.

Energy Balance

The energy balance of a leaf depends on absorbed net shortwave and long-wave radiation (R_n, [cal cm^{-2} s^{-1}]), sensible heat transfer* (H), and cooling due to water vaporization (LE):

$$R_n + H + LE = \Delta T_{leaf} \qquad \text{(Equation 18-2)}$$

where L is latent heat of vaporization (590 cal g^{-1}), E is transpiration rate (g cm^{-2} s^{-1}), and ΔT_{leaf} is the change in leaf temperature. The transpiration rate is

$$E = \frac{c^{ias} - c^{air}}{\Sigma r} \qquad \text{(Equation 18-3)}$$

which is the water vapor concentration gradient between the leaf intercellular air spaces (c^{ias}) and the air (c^{air}) divided by the sum of the resistances to flow along the transfer path (Σr). This is related to a form of Fick's law that we used to describe CO_2 flux in Chapter 15.

We can summarize leaf energy balance by considering the fate of 1.4 cal cm^{-2} min^{-1} of energy absorbed by a hypothetical leaf (Figure 18-5). The diagram shows the relative importance to energy dissipation of net radiation, sensible heat transfer, and transpiration. In the rare event that leaf and air temperatures are equal, the absorbed energy is effectively dissipated by transpiration and radiation emission (situation [a] of Figure 18-5). When transpiration is stopped due to stomatal closure (b), leaf temperature increases and energy is dissipated by sensible heat transfer and radiation emission. Situation (c) represents the usual case, where absorbed energy is dissipated by the combined factors. Temperature and water loss are, therefore, intrinsically related. Their relative values, together with the photosynthetic capacity of a plant (which they directly influence), are of central importance in physiological ecology.

---

*Includes heat transfer by both conduction and convection.

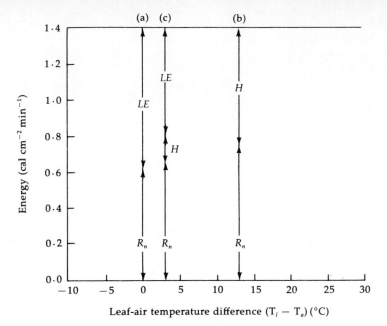

Figure 18-5   Energy exchange diagram for an idealized leaf of downwind width 10 cm, with air temperature 25°C and wind velocity 200 cm sec^{-1}. The total energy absorbed by the leaf is assumed to be 1.4 cal cm^{-2} min^{-1}. In (a) leaf and air temperatures are equal and stomates are open. In (b) stomates are closed and heat is dissipated by sensible heat transfer (H) and radiation emission (R_n). In (c) stomates are open and heat is dissipated by transpiration (LE), H, and R_n. Modified from Slatyer. 1967. *Plant-Water Relationships*. By permission of Academic Press.

When water is limited, there is often a trade-off between CO_2 uptake for photosynthesis and water loss by transpiration. Water use efficiency (WUE) is a measure of this trade-off. WUE is the ratio between net photosynthesis (g CO_2) and the amount of water (kg) released in transpiration. The nature and control of factors influencing WUE are the subject of the remainder of this chapter.

Leaf Resistance

We noted earlier that the steepest water potential gradient in the soil-plant-atmosphere continuum occurs as water leaves the plant in the vapor phase. The most important resistances influencing water loss are, therefore, associated with the leaf (Figure 18-6). **Diffusive resistance*** to water vapor transfer in a leaf (r^{leaf}) may be divided into an external component, referred to as the **boundary layer re-**

---

*Resistance units are s cm^{-1} and express the same resistance to diffusive flow as equivalent path lengths of air of unit cross-sectional area divided by the diffusive coefficient of water in air (2.57×10^{15} m^2 s^{-1} at 20°C). Conductance (g) (cm s^{-1}), the inverse of resistance ($1/r$), is frequently used to evaluate barriers to flow.

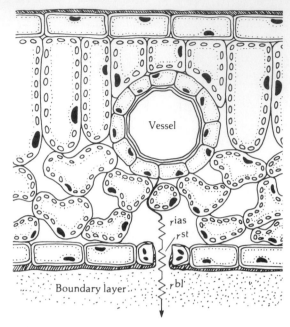

Figure 18-6   Cross section of a leaf showing the resistances associated with water loss. r^{ias} is the resistance to vapor movement in the substomatal cavity, r^{st} is the resistance offered by the stomatal pore and guard cells, and r^{bl} is the resistance offered by the boundary layer on the leaf surface.

sistance (r^{bl}), and an internal component consisting of (a) resistance to movement through intercellular air spaces and the substomatal cavity in the vapor phase (r^{ias}), and (b) resistance to movement through stomatal pore itself (r^{st}). An alternate path of leaf water loss is referred to as **cuticular transpiration.** Here the water travels through the cuticle in the vapor phase, having vaporized at the epidermal cell wall surface. The importance of cuticular transpiration varies with the species, the level of secondary thickening of epidermal cell walls, and characteristics of the waxy cuticle, but resistance is so high that water loss via this path is commonly inconsequential in comparison to stomatal transpiration.

Resistance to vapor movement in intercellular spaces and in the substomatal cavity (r^{ias}) depends on the distance the vapor must travel. The value is more or less constant for a species, depending primarily on stomatal location and density and leaf thickness.

Variation in **stomatal resistance** (r^{st}) is caused by movement of guard cells, which are sensitive to environmental factors. Stomates open and close in response to differences in guard cell turgidity stimulated by variations in light, leaf water potential, CO_2 concentration, and humidity (Figure 18-7). Turgidity changes are related to osmotic potential variations in the guard cells caused by the movement of K^+ into and out of the cells. Movement of K^+ into and out of the cells is driven by a proton pump. When potassium ions enter a guard cell, they cause increased

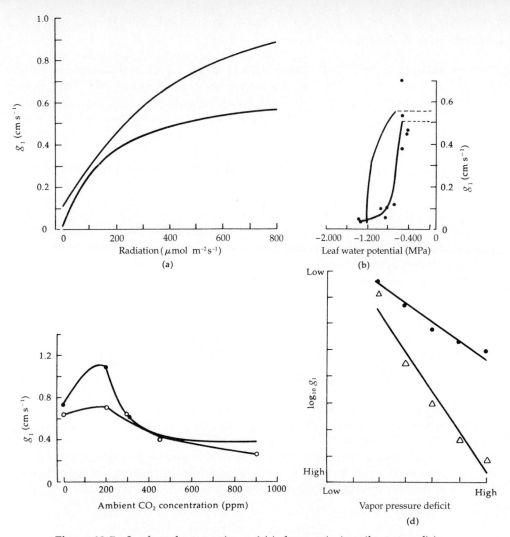

Figure 18-7 Leaf conductance ($g_1 = 1/r$) changes (primarily stomatal) in response to variations in (a) visible radiation, (b) leaf water potential, (c) ambient CO_2 concentration, and (d) vapor pressure deficit between leaf and air for pairs of cultivated plants. Modified from Burrows and Milthorpe. 1976. "Stomatal conductance in the control of gas exchange." In *Water Deficits and Plant Growth*, vol. IV, ed. T. T. Kozlowski. By permission of Academic Press.

guard cell turgidity and opening of stomates. Water stress will stimulate production of abscissic acid (ABA), which inhibits the proton pump that drives K^+ cotransport. K^+ will diffuse from the guard cell, causing stomatal closure. The level of water stress necessary to cause ABA production varies with the drought tolerance of the species. For any species, there is a critical ψ_{leaf} at which the stomates will close, decreasing conductance to zero. Resistance in fully open stomates may range as low as 0.4 s cm^{-1}.

Stomates close when atmospheric vapor pressure goes down (Lange et al. 1971; Schulze et al. 1972, 1973; Camacho-B et al. 1974; Hall and Kaufmann 1975; Smith and Nobel 1977a). The turgor pressure response of guard cells that causes stomatal closure is independent of the water status of surrounding tissues except, perhaps, adjacent subsidiary cells. The mechanism of stomatal response, however, has not been established. A possible explanation was proposed by Lange et al. (1971), based on the concept of **peristomatal transpiration,** wherein guard cells lose water more readily than other epidermal cells and would therefore respond directly to changes in vapor pressure. In *Tradescantia virginiana*, stomatal closure in response to increasing VPD (vapor pressure deficit) is not due to peristomatal transpiration because guard cells and subsidiary cells are covered with cutin both outside and inside the stomatal cavity. Yet guard cells and subsidiary cells lose turgor while other epidermal cells remain turgid. The controlling mechanism appears to be regulation of water transport into subsidiary cells (Nonami et al. 1991). Gradual closing of stomates as evaporative demands of the atmosphere increase may serve to maximize WUE in plants by increasing the time necessary to reach the critical ψ_{leaf}.

Boundary layer resistance (r^{bl}) is considered part of leaf resistance because it is influenced by morphological characteristics of the leaves. Resistance to water vapor diffusion is caused by a layer of air adjacent to the surface of the leaf that does not readily mix with passing air, as with laminar flow over any flat surface. Vapor must pass through this potentially saturated layer by molecular diffusion before moving into the free atmosphere. The thickness of the boundary layer depends on leaf size and wind speed. Greater wind speeds cause turbulence closer to the leaf surface, thus reducing the depth of the boundary layer. Smaller leaves have thinner boundary layers because of increased convective exchange between the surface and the free air. Surface texture, leaf shape, and leaf orientation affect air turbulence near the surface and, therefore, the thickness of the boundary layer.

Optimal Leaf Form

Taylor (1975) and Campbell (1977) considered the interactions of physical factors, such as temperature and light, with biological regulation of water loss (leaf resistance, r^{leaf}) on photosynthesis and water use efficiency of plants. Figure 18-8 shows idealized photosynthetic and WUE responses of leaves of various sizes and with different diffusive resistances when exposed to full sunlight. At high air temperatures (30°C in Figure 18-8a), larger leaves have lower net photosynthetic rates, especially when diffusive resistance is high (e.g., $r^{leaf} = 32$ s cm^{-1}). This is a result of overheating; at low resistances (e.g., $r^{leaf} = 1$ s cm^{-1}), evaporative cooling maintains temperatures that support positive photosynthesis even in leaves 16 cm wide. However, this evaporative cooling requires large amounts of water and the water use efficiency is therefore very low (<4.0 mg g^{-1}). There is a clear advantage for leaves less than 1 cm wide and with a diffusive resistance above 6 s cm^{-1} in a warm, dry environment. Larger leaves have higher photosynthetic rates than smaller

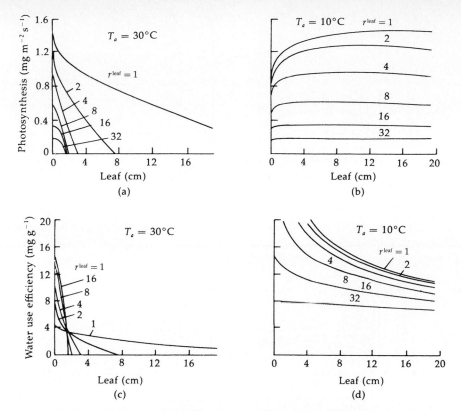

Figure 18-8 Theoretical photosynthetic and water use efficiency responses for C_3 leaves of different sizes and for different levels of resistance (r^{leaf} s cm^{-1}) when exposed to ambient temperatures (T_a) of 30°C and 10°C. WUE = mg CO_2 fixed per g H_2O transpired. From G. S. Campbell. 1977. *An Introduction to Environmental Biophysics.* By permission of Springer-Verlag, New York.

leaves in cool air (10°C in Figure 18-8b), because they are at temperatures above ambient and are therefore closer to the optimum for photosynthesis. Figure 18-8d shows that water use efficiency drops as leaf size increases, but at comparable diffusive resistances, efficiency remains higher than at air temperatures of 30°C (Figure 18-8c). This theoretical model shows that the interactions of temperature, water, and photosynthesis should lead to plants with small leaves in warm, arid environments. This is in fact the case, and may help explain the great success of leguminous trees and small-leaved plants in arid regions. Pinnately compound leaves with small leaflets are common in desert leguminous trees. They are apparently a significant advantage in desert habitats because of reduced heat load and greater WUE. In cooler environments, large leaves are warmer and, with low diffusive resistance, have high photosynthetic rates. There is a gradient of increasing average leaf size from warm, arid to cool, moist habitats.

Theoretical photosynthetic and WUE responses to different light intensities

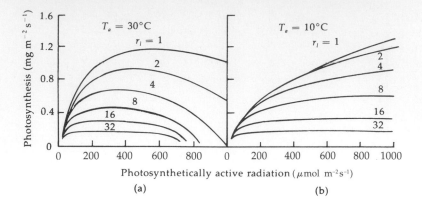

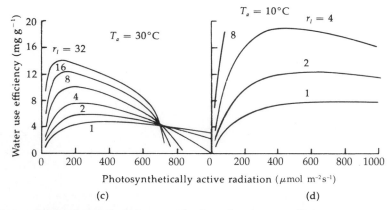

Figure 18-9   Theoretical photosynthetic and water use efficiency responses at different light intensities and for different levels of resistance (r^{leaf}) when exposed to ambient temperatures (T_a) of 30°C and 10°C. All leaves are intermediate sized (3 cm). Leaf temperature is allowed to vary in these simulations. From G. S. Campbell. 1977. *An Introduction to Environmental Biophysics.* By permission of Springer-Verlag, New York.

and diffusive resistances are presented in Figure 18-9 for leaves of intermediate size (3 cm). When ambient temperatures are cool (10°C in Figure 18-9b), maximum photosynthesis is attained at full sunlight (2000 μmol m^{-2} s^{-1}) because leaf temperatures remain moderate. But when temperatures are higher (30°C in Figure 18-9a), leaf temperature increases beyond optimal and intermediate light intensities are the most effective. Full sunlight, in conjunction with a 30°C air temperature, results in very low WUE. In fact, unless diffusive resistance is less than 4 s cm^{-1}, leaf temperature surpasses the tolerance of our hypothetical plant (Figure 18-9c). High WUE is maintained across a range of light intensity when ambient temperatures are low (10°C in Figure 18-9d), again because leaf temperatures remain moderate. Many leaves (especially larger ones) of plants in arid or semiarid habitats are oriented vertically. Vertical orientation allows interception of maximum solar radiation in early morning and late afternoon, when temperatures are lower, and reduces interception at midday, when temperatures are high. Vertical orientation would thus maximize both photosynthesis and WUE in our hypothetical plant.

Although these relationships are for an idealized situation, they correspond in general with observations of real plants. Understanding adaptations of specific plants with regard to these variables will allow us to interpret better the responses to habitat observed in nature.

Diffusive Resistance and Transpiration Measurements

The most popular method of measuring leaf resistance is by the use of a diffusive resistance porometer (Figure 18-10). A porometer measures water loss into a small chamber enclosing part or all of a leaf. Resistance values may be standardized by attaching the porometer to an apparatus with a known diffusive distance (δ). We can then calculate resistance to water vapor transfer as

$$r^{\text{leaf}} = \frac{\delta}{D} \qquad \text{(Equation 18-4)}$$

where D is the diffusion coefficient of water vapor at a given temperature ($cm^2\,s^{-1}$). The units of resistance are $s\,cm^{-1}$. r^{leaf} can also be calculated when transpiration rates (J_{wv}) and ambient humidity are known:

$$r^{\text{leaf}} = \frac{\Delta c_{\text{wv}}}{J_{\text{wv}}} \qquad \text{(Equation 18-5)}$$

Here, Δc is the concentration gradient between mesophyll cell surface (assumed to be saturated) and the air. Humidity is measured by hygrometers with lithium chloride or aluminum oxide sensors. Thermocouples are incorporated because of the interaction between vapor pressure and temperature (Figure 18-1). A fan to circu-

Figure 18-10 A diffusive resistance porometer, the Li-Cor LI-1600 Steady State Porometer. Courtesy of Li-Cor, Inc.

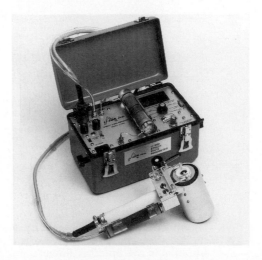

late air reduces the effect of the boundary layer, which may be inordinately large in the small, closed porometer chamber (Nobel 1991). Older diffusive resistance porometers measured the time necessary to increase water content of a dry atmosphere to a predetermined level. Steady-state porometers circulate conditioned air through the chamber and maintain humidity at ambient or other preset levels, thus giving estimates of resistance under ambient conditions rather than maximum water loss into fully dried air. Steady-state measurements are considered to be more accurate (Appleby and Davies 1983). Kanemasu et al. (1969), Beardsell et al. (1972), and Parkinson and Legg (1972) have proposed a variety of porometer designs.

Less expensive estimates of leaf resistance can be made using cobalt chloride paper (Milthorpe 1955, Meidner and Mansfield 1968) or calculation from stomatal dimensions using infiltration techniques or direct measurements. (For a review, see Burrows and Milthorpe 1976.)

Transpiration rates from detached leaves and potted plants are easily made by gravimetric analysis. Repeated weighing of the parts over time gives an estimate of water lost. For detached leaves and whole branches, a potometer can also be used. Figure 18-11 illustrates one form of potometer, which measures water flux through an excised leaf, branch, or shoot system of a plant. More accurate measurements are possible when transpiration measurements are incorporated into a gas exchange system (see Chapter 15) by using a diffusive resistance porometer (Figure 18-10).

Transpiration for a specified leaf area can be calculated and then scaled up to compute whole tree transpiration if we know tree leaf area. For methods of ob-

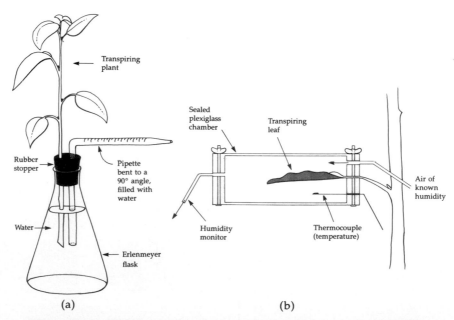

Figure 18-11   (a) Potometer for the measurement of transpiration of plant shoots or leaves. (b) A gas exchange chamber for the measurement of transpiration in continuously flowing air.

taining reliable estimates of leaf area see Yang and Tyree (1994). Transpiration (*E*) is calculated as

$$E = g \, \Delta wv \hspace{4cm} \text{(Equation 18-6)}$$

where *g* is leaf conductance ($1/r$) and Δw is the difference between the vapor pressure of the leaf and the air (Pearcy et al. 1988).

Transpiration can also be estimated by measuring the volume and velocity of water movement in stems and calculating the amount of water that flows through the plant per unit time. Sap flow rate can be determined by measuring the time it takes for heated water in the xylem to move between temperature sensors placed in the sapwood of a tree. Transpiration can then be estimated for individual trees and whole stands by extrapolation using leaf area estimates as described above. See Thorburn et al. (1993), Hinckley et al. 1994, Salama et al. (1994), Jones et al. (1988), and Dawson (1996) as examples.

Stem and Root Transport

Resistance to mass transport of water in vessels and tracheids is low in contrast to the resistances encountered in other phases of the continuum (i.e., root endodermis and mesophyll). Heine (1971) reviewed the development of concepts relating to conductivity in woody plants and stressed the importance of anatomy in determining water flux. Conductivity depends not only on the ψ gradient but also on lumen (central open portion of the cell) area and length of xylem elements. Carlquist (1975) has shown a relationship between element dimensions and the amount of available water. This close relationship between vascular element size and the environment is apparently related to the negative pressures that develop in xylem.

As transpiration proceeds, resistance prevents the instantaneous replacement of transpired water, thereby establishing a negative hydrostatic pressure (tension) in the xylem. This negative pressure is a natural component of vascular water. It is intensified when transpiration demand is high and replacement is slow because of low ψ_{soil} or when high resistance to water flow increases the lag between inflow and outflow. Adhesion of water molecules to cell walls and cohesion between water molecules maintain the integrity of the water column. The more negative the pressure (lower ψ), the greater the stress placed on the cells and the water column. Smaller vascular elements are, therefore, associated with tissues exposed to the greatest stress (Carlquist 1975). Rundel and Stecker (1977) report that this relationship is clearly evident in tracheid diameters of a 90-m-tall nontranspiring Sierra redwood (*Sequoiadendron giganteum*) in Kings Canyon National Park, California (Figure 18-12a). This is not surprising, because the weight of water (gravitational potential) in the conducting tissues of tall trees would lead one to expect that upper xylem tissues would be subjected to the highest levels of stress even in the absence of transpiration (Figure 18-12b).

Water uptake and transport in nonvascular tissues of roots are sensitive to both the water potential gradient and to temperature effects on endodermal cell membrane permeability. Water uptake in response to temperature can be modified

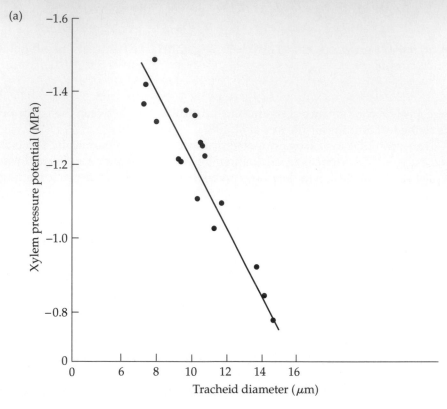

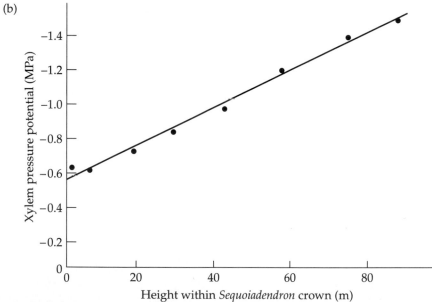

Figure 18-12 (a) The relationship between predawn xylem pressure potential and tracheid diameter in *Sequoiadendron giganteum*. (b) The relationship between height and predawn xylem pressure in *Sequoiadendron giganteum*. From P. W. Rundel and R. E. Stecker. 1977. "Morphological adaptations of tracheid structure to water stress gradients in the crown of *Sequoiadendron giganteum*." *Oecologia* 27:135–139.

by acclimation and varies with species and habitat (Anderson and McNaughton 1973). For example, soil temperature was identified by Fernandez and Caldwell (1975) as a potential stimulus for root activity in Great Basin Desert shrubs. Initiation of rapid growth and water absorption in *Atriplex confertifolia* occurred as the soil profile warmed to progressively deeper levels. Cessation of root activity was associated with soil water depletion, so that activity in any profile zone was limited to one or two weeks of adequate water supply following the attainment of sufficiently warm temperatures. This mechanism prolongs the time that soil water is available by effectively partitioning the water for use during different periods.

Soil Moisture

Soil water potential ultimately determines the availability of water for plant growth. It is important, then, to understand the relationship between the physical properties of soils discussed in Chapter 17 and the availability of water for plants. We will discuss soil properties that influence retention, infiltration, and movement of water, but first we need to define several terms basic to understanding soil water relations.

Moisture Status in Soils: Basic Definitions   Water entering the soil will fill most pore spaces and drain downward through pores in response to gravity. We can add the term ψ_g to our initial water potential formula to represent gravitational potential:

$$\psi_{\text{soil}} = \psi_m + \psi_p + \psi_o + \psi_g \qquad \text{(Equation 18-7)}$$

Gravitational potential is significant only in saturated soils. Water moving through soil in response to gravitational forces is gravitational water. Once gravitational water has moved from the soil or is dispersed in the soil, the soil is left at field capacity (FC). **Field capacity** is the amount of water that can be retained by a soil after gravitational water has been removed. Field capacity for a particular soil is a constant, depending on soil texture, structure, and organic content.

The water available for plant use from soil at field capacity depends on the species, the physiological state of the individual, and environmental conditions. The lower limit of water availability is designated as the **permanent wilting point (PWP).** PWP is reached when ψ_{soil} is equal to or below the minimum osmotic potential of the plant, so that water cannot be removed from the soil. The plant wilts and remains wilted even if placed in a cool, dark, humid chamber. Addition of soil water may allow recovery. PWP for mesophytes is often near -1.5 MPa. The amount of water between FC and PWP is called **available water** or **capillary water** because it is retained against the pull of gravity by capillary action and matric forces. Plants are able to utilize capillary water to varying extents but cannot remove all water from the soil. Below the PWP, water is held so firmly to the soil matrix that it is removable only under conditions surpassing biological tolerance. The most important components of water potential in nonsaturated soils are matric potential and, in some cases such as saline soils, osmotic potential.

The relationship between FC, soil water content, and ψ_{soil} depends to a large

Figure 18-13   Typical water characteristic curves for sand, sandy loam, and clay soils. Modified from Slatyer 1967. *Plant-Water Relationships.* By permission of Academic Press.

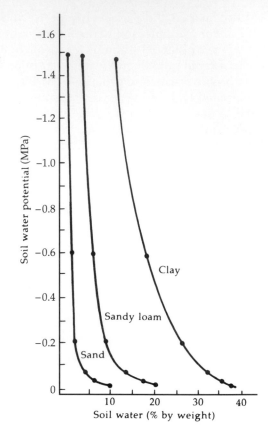

degree on soil texture (Figure 18-13). Clay soils hold large volumes of water at relatively lower water potentials than do coarse-textured soils. This is because of the greater total pore space and greater particle-water contact space in fine-textured soils.

Movement of Soil Water   Availability of water within a soil depends in part on how quickly and how far water will move upward from the water table toward an absorbing root or into the soil during a rainstorm. **Infiltration** and movement of soil water depend on factors that determine soil water potential. Water will infiltrate into a dry soil faster than a wet soil, if they have comparable texture and structure, because the gradient in water potential is greater in the dry soil. The flux density of water through a saturated soil (J_w) is

$$J_w = -K \frac{\Delta \psi_{soil}}{\Delta z}$$

(Equation 18-8)

where K is the **hydraulic conductivity** (the capacity of a soil to transport water, in $g\,cm^{-2}\,s^{-1}$); $\Delta \psi_{soil}$ is the difference in water potential of the soil between two points along which water moves; and z is the distance between the two points in the soil

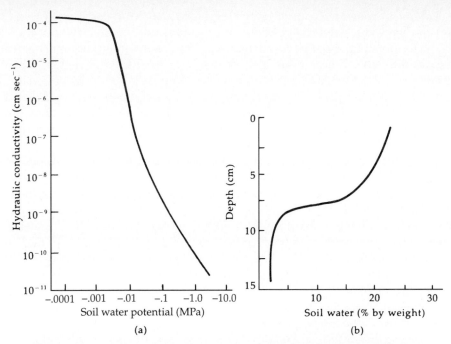

Figure 18-14   (a) The relationship between hydraulic conductivity and soil water potential in a loam soil. (b) The water content profile of an initially dry loam soil at one day following irrigation. After Slatyer. 1967. *Plant-Water Relationships.* By permission of Academic Press.

used to determine $\Delta\psi_{soil}$. As ψ_{soil} drops, there is a significant drop in hydraulic conductivity (Figure 18-14a), especially in sandy soils where the continuity of the water is broken. In finer-textured soils, hydraulic conductivity also decreases, but to a lesser extent because the continuity of water surfaces are maintained at lower ψ_{soil}. After gravitational water has been dispersed, the upper zone of the soil profile is at field capacity and there is a narrow zone of soil with a very steep drop in water content; the soil below that is essentially dry (Figure 18-14b). Similarly, water rises from an underground water table only in limited quantities. The rate of movement depends on the water potential gradient.

Hydraulic lift occurs when the upper soil has a lower water potential than the roots it contains. During the day, water is removed from the soil, drying the upper soil. When the stomates close at night, plant water is replenished from the deep soil. Upper roots attain a higher water potential than the surrounding soil and water moves from the roots to the dry soil. Hydraulic lift occurs in *Artemisia tridentata* (Richards and Caldwell 1987), *Acer saccharum* (Dawson 1993), and *Eucalyptus viminalis* (Phillips and Riha 1994). A growing body of evidence suggests that several, perhaps many, trees and shrubs may lift water from deep soils and pass it to dry, shallower soils. Emerman and Dawson (1996) estimated that a mature sugar maple tree moves >100 L of water upward in the soil each day.

Lateral movement of water through the soil to zones of absorption by roots depends on the water potential gradient and the hydraulic conductivity of the soil. Root water potential must be lower than ψ_{soil} for absorption to occur. This water potential difference can be quite low when ψ_{soil} is -0.5 MPa or higher, whereas with ψ_{soil} at -1.5 MPa, a significantly greater drop in potential is necessary for absorption. This is because hydraulic conductivity decreases as ψ_{soil} drops. The distance through which water will move toward the dry zone surrounding a root depends on the plant's tolerance for stress and the soil properties that influence hydraulic conductivity.

Water infiltration into the soil is a critical factor in areas of steep slopes or intense rainfall. Sandy soils generally have rapid infiltration, but finer soils slow the entrance of water in proportion to clay content. In some cases, organic compounds from plants form "skins" over soil particles so that, rather than penetrating, the water beads up on the soil surface. Such soils are called **hydrophobic** or **water repellent soils.** They have been observed under a wide variety of species and environmental conditions (Bond 1964, Krammes and DeBano 1965, Adams et al. 1970, DeBano and Letey 1969). For example, water repellency increases surface runoff and postfire erosion on chaparral soils in southern California. When chaparral burns, the water repellent zone moves down into the soil (Figure 18-15), leaving a wettable zone at the surface. Postfire rains penetrate and saturate the wettable zone, causing the surface soil to slip downslope. Not all species in a habitat have the same influence on wettability, and few communities have been characterized for wettability patterns. Soil wettability is another environmental factor that affects water availability and therefore regulates the niche relations of plant species.

Development and Significance of Plant Water Stress   Slatyer (1967) devised a hypothetical scheme to show the relationship between ψ_{soil}, ψ_{root}, and ψ_{leaf} as soil water is reduced from field capacity to below the permanent wilting point (Figure 18-16). Significant diurnal patterns of ψ_{plant} are present even when ψ_{soil} is near field capacity. The diurnal range must get larger as the soil becomes drier to maintain sufficient uptake from soils in which both water potential and hydraulic conductivity are decreasing. Following stomatal closure in response to darkness, $\psi_{soil} \cong \psi_{leaf}$. The time necessary for equilibration increases as ψ_{soil} decreases, until $\psi_{soil} \cong \psi_{leaf}$ for only a brief period. Temporary wilting and midday stomatal closure may reduce transpiration at the point along the drying curve where $\psi_{leaf} =$ leaf osmotic potential, ψ_o (the PWP), because a ψ gradient sufficient to extract water from the soil is no longer possible. Osmotic potential is not fixed within an individual plant; a wide range of values is found. We will discuss some examples of osmotic adjustment in the next section.

Plant responses to water stress have been reviewed in detail by Hsiao (1976). We will briefly mention the physiological responses to water stress and later consider some specific adaptations to avoid or deal with stress. Water is involved in and essential to all physiological processes. The photosynthetic decline as water stress increases is primarily due to stomatal closure or other direct effects of water

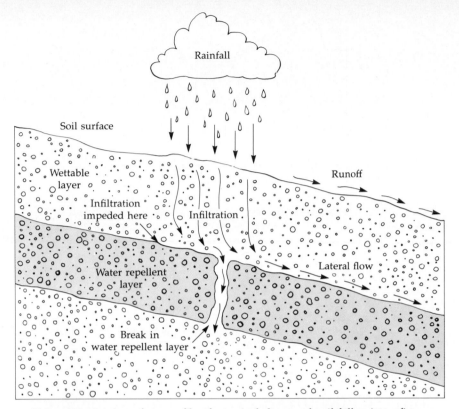

Figure 18-15   A surface profile of a typical chaparral soil following a fire where the surface is wettable but a subsurface water repellent layer causes lateral flow of the surface soil. After J. S. Krammes and J. Osborn. 1969. "Water repellent soils and wetting agents as factors influencing erosion." In *Proceedings of Symposium on Water-Repellent Soils*. L. F. DeBano and J. Letey, eds. University of California, Riverside.

stress on photosynthetic reactions. When stress is great enough, photosynthesis, respiration, protein synthesis, and most other processes involving chemical reactions may be severely reduced because of protein (enzyme) denaturation. Most physiological processes decrease gradually with increasing stress to the point where all turgor is lost, then cease beyond that point. Under moderate stress, biochemical processes may proceed, but growth may be reduced because of lack of sufficient turgor pressure to cause cell enlargement. Lower overall cell size may increase the drought tolerance of the new tissues (Cutler et al. 1977). Newly formed cells may enlarge only when water becomes available, in which case growth much exceeds that expected by the rate of formation of new cells (Oechel et al. 1972). Water stress also increases the viscosity of phloem sap, as well as reducing mass transfer in the xylem, thus limiting transport of nutrients, photosynthate, and hormones.

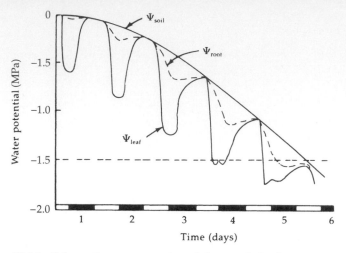

Figure 18-16   Schematic representation of changes in leaf water potential (ψ_{leaf}), root surface potential (ψ_{root}), and soil mass water potential (ψ_{soil}) as transpiration proceeds from a plant rooted in initially wet ($\psi_{root} \approx 0$) soil. The same evaporative conditions are considered to prevail each day. The horizontal dashed line indicates the value of ψ_{leaf} at which wilting occurs. From Slatyer. 1967. *Plant-Water Relationships.* By permission of Academic Press.

Solute Concentration and Plant Water Status

Osmotic concentration of cell sap determines the level of stress necessary to induce wilting. **Osmotic concentration** depends on cell solute content and the amount of intracellular water. Tyree and Hammel (1972) recognized and interpreted osmotic shifts with respect to ecologically meaningful adaptations of plants (Cheung et al. 1975, Roberts and Knoerr 1977, Osonubi and Davies 1978, Bennert and Mooney 1979, Monson and Smith 1982).

Cheung et al. (1975) contrasted *Ginkgo* and *Salix* (Figure 18-17) as an example of plants showing very different osmotic qualities. *Ginkgo* has a lower (more negative) osmotic potential than does *Salix* (point *A* for *Ginkgo*, Figure 18-17) and would consequently remain turgid at much higher stress than would *Salix*. Plasmolysis (turgor pressure = 0 MPa) occurs at point *B*, which is about −2.2 MPa for *Ginkgo* and −1.2 MPa for *Salix*. The steep slope for *Ginkgo* (the more xerophytic of the two species) shows that *Ginkgo* can remove soil water to a much lower ψ_{soil} than *Salix* while maintaining a higher level of tissue hydration. This capacity is partly due to the more rigid (nonelastic) cell walls of *Ginkgo*. Removal of small amounts of water from cells with rigid walls reduces turgor pressure rapidly. In *Salix*, the more flexible cell walls shrink, thus maintaining turgor pressure but creating only a minimal gradient for soil water extraction. *Salix*, therefore, must grow in more mesic environments.

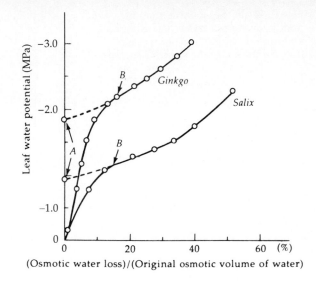

Figure 18-17 A plot of water potential versus osmotic water loss of single *Ginkgo biloba* and *Salix lasiandra* leaves. *A* is the estimate of original osmotic potential; *B* is the osmotic potential at incipient plasmolysis. From Cheung, Tyree, and Dainty (1975). Reproduced by permission of the National Research Council of Canada from the *Canadian Journal of Botany* 53: 1342–1346.

The capacity to reduce turgor pressure to negative values would give xerophytes further ability to extract soil water to very low water potentials. Negative turgor pressures are theoretically possible but have not been shown beyond reasonable doubt to exist (Kyriakopoulos and Richter 1976, Tyree 1976). Such a phenomenon would help explain how some desert plants are able to reduce soil water potentials to < -10 MPa.

Osmotic adjustment is an effective acclimation response to water deficit (Cutler et al. 1977, Osonubi and Davies 1978, Cutler and Rains 1978, Jones and Turner 1978, Bennert and Mooney 1979, Roberts et al. 1980, Monson and Smith 1982). Both stomatal response and wilting point depend on solute concentration, which increases in plants subjected to gradual drying or repeated periods of drought. The chaparral shrub *Ceanothus greggii* shows both seasonal and diurnal adjustments in osmotic potentials that allow it to maintain positive turgor during the extreme drought of late summer (Figure 18-18). In August and September, midday water potentials fall well below the point of zero turgor of early May. Without osmotic adjustment, these plants would experience severe dehydration of tissues during midsummer days (Bowman and Roberts 1985). The increase in solute concentration may be due to increased solute content or to smaller solvent volumes in smaller cells formed under conditions of stress. Osmotic adjustment in *Ceanothus greggii* (Figure 18-18) is due to changes in the volume of osmotic water present in the tissues, whereas other chaparral plants such as *Quercus dumosa* and *Arctostaphylos glandulosa* accumulate solutes as the means of acclimating to the dry season. Regardless of the mechanism operating in a particular species, many plants reflect previous conditions of water availability.

Even drought-sensitive cottonwood (*Populus deltoides*) adjusts osmotic concentration by 0.23 to 0.48 MPa. Almost half of the adjustment is caused by increases

Figure 18-18   Seasonal courses of midday water potential (dashed line) and midday turgor loss point (solid line) for the chaparral shrub *Ceanothus greggii*. Redrawn from Bowman and Roberts. 1985. *Ecology* 66:738–742.

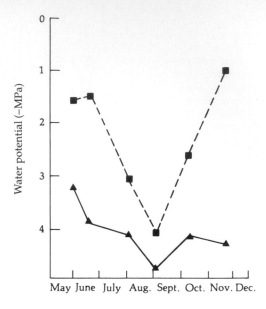

in organic solutes such as sucrose, malic acid, glucose, fructose, etc. (Gebre et al. 1994). Similar levels of osmotic adjustment (0.39 MPa) are found in eastern white cedar (*Thuja occidentalis*; Edwards and Dixon 1995).

Plant and Soil Water Status Measurements

Recent technological advancements have dramatically improved our capabilities of measuring plant and soil water status. Most new techniques are developed around energy differences expressed by rates of evaporation as measured by heat flux from a wet thermocouple, or the negative xylem pressure created by transpiration. Nonelectronic methods based on the direction of water movement and total water content may also be used when more sophisticated equipment is not available.

Pressure Chamber

Scholander et al. (1964) developed the pressure chamber now widely used in ecophysiological research, and Waring and Cleary (1967) showed the great utility of **xylem pressure potentials** in solving ecological problems. This method is based on the fact that negative pressures (tensions) exist in the xylem. When stems are severed, the pressure necessary to force water back to the cut surface is equivalent to the negative pressure in the xylem prior to cutting. A typical pressure chamber apparatus is illustrated in Figure 18-19. Using the pressure chamber, one can mea-

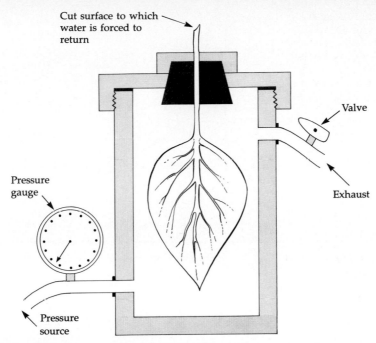

Cut surface to which
water is forced to
return

Valve

Pressure
gauge

Exhaust

Pressure
source

Figure 18-19 Setup of a pressure chamber used to measure plant water potential.

sure many plants in a short time. The apparatus has also been made portable so that data may be taken on plants in their native surroundings. Ritchie and Hinckley (1975) discussed in detail the applications, calibration, and data interpretation of pressure chambers. This review should be consulted before using the pressure chamber technique.

Turgor potential, osmotic potential, and certain tissue characteristics can be determined by examining moisture release curves of plant tissue (also called pressure-volume curves; Figure 18-20). Pressure-volume relationships can be determined with a pressure chamber by placing a fully hydrated, excised shoot into the pressure chamber, incrementally increasing the chamber pressure, and recording the volume of sap expressed at each pressure. Moisture release curves can also be constructed by the repeat pressurization method. In this method, the water potential of a weighed tissue sample is determined using a pressure chamber. The tissue is then removed from the chamber and allowed to desiccate for a short period. It is then reweighed and the water potential of the sample is again determined with the pressure chamber. The process is continued until much of the water is removed from the sample.

Parker and Colombo (1995) developed pressure-volume curves for conifer shoots to compare the sap expression, repeat pressurization, and a composite method of gathering pressure-volume data. The composite method involves exposing tissues to pressure only once instead of the repeated or extended exposure

Figure 18-20 A sample pressure-volume curve. The inverse of the balance pressure is plotted on the ordinate, and the volume of water expressed from the leaf or branch is plotted on the abscissa. *A* is the inverse of the original osmotic potential (0.94 MPa^{-1}). *B* is the inverse of the water potential at incipient plasmolysis (0.66 MPa^{-1}), and *C* is the volume of osmotic water originally in the tissues (0.75 ml). *A* and *C* are determined by extending the linear portion of the curve to the ordinate (*A*) and to the abscissa (*C*). *B* is the point where the curve becomes linear.

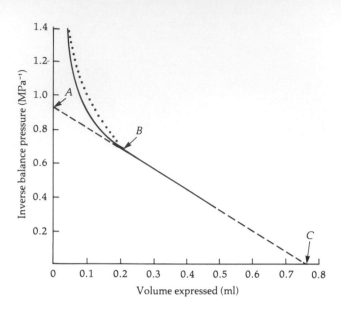

of the other methods. Parker and Colombo found that the curves from sap expression and repeat pressurization were similar to each other but significantly different from the curves generated by the composite method. This suggests that extended or repeated exposure to high pressure may cause unknown changes in membrane permeability (Mastrangelo et al. 1978) or cause cavitation (Cochard et al. 1992) or there may be some response to the changes in tissue temperature associated with exposure to high or rapidly changing pressure (Puritch and Turner 1973). Pallardy et al. (1991) found that the sap expression method gave more accurate results than repeat pressurization in trees.

The moisture release data acquired by any of the above methods are plotted with the inverse of the balance pressure on the ordinate and the volume of water lost on the abscissa. The general form of the resulting moisture release curve is illustrated in Figure 18-20. When the linear portion of the curve is extended, it crosses the ordinate at a value (*A* in Figure 18-20) equal to the inverse of the original osmotic potential. By extending the linear portion of the curve to the abscissa, the original osmotic volume of water can be estimated. The inverse of the tissue water potential at the point of zero turgor is read from the ordinate at the point where the curve becomes linear. Osmotic potential indicates the amount of solutes present per unit volume of cellular water and sets the limit on the magnitude of turgor pressure that can develop. Pressure-volume curves can also be used to determine cell wall elasticity by observing the steepness of the nonlinear portion of the curve. Where cells are very rigid, the slope is very steep (above point *B* in Figure 18-20). Cells that are more elastic will show a more gradual decrease in water potential for each unit of water expressed from the tissue (the dotted line in Figure 18-20). Tyree and Hammel (1972) and Cheung et al. (1975) have interpreted pressure-volume curves in ecological terms. Stress-tolerant plants would be expected to have a higher osmotic potential and less cell wall elasticity than more mesophytic species.

Pressure chamber values represent the total water status of the plant and can be used to determine ψ_{plant} responses to environmental factors, both diurnally and seasonally. These data, in combination with pressure-volume curves, provide an effective means of measuring plant water status and stress tolerance.

Psychrometry and Hygrometry

Psychrometers were developed to measure relative humidity of air. With the development of the water potential concept, it was found that water potential could be determined from measurements of relative humidity. The relationship is

$$\psi = \frac{RT}{V} \ln wv \qquad \text{(Equation 18-9)}$$

where R is the ideal gas constant, T is the absolute temperature, V is the volume of a mole of liquid water, and wv is the relative humidity. **Thermocouple psychrometers** measure temperature depression caused by evaporation from a wet surface. (See Figure 18-1 to review the relationship between wet bulb temperature and humidity.) They can detect small changes in relative humidity in a closed chamber (Spanner 1951, Richards and Ogata 1958). Sample material is placed in a small, sealed chamber containing a thermocouple junction. When the air around the thermocouple is in equilibrium with the sample, the thermocouple is electronically cooled to below the dew point. The film of water collected on the thermocouple will evaporate and cool the thermocouple at a rate inversely proportional to the water potential of the air in the sample chamber. Figure 18-21 is a sketch of a thermocouple psychrometer apparatus that can be used for soil or leaf disc measurements. Various chambers have been developed for specific purposes. Psychrometers with screen or ceramic covers may be inserted into the soil for ψ_{soil} determinations. Thermocouple psychrometers can be inserted into a clamp and attached to leaves or stems as a nondestructive measure of water potential (Campbell and Campbell 1974, Brown and McDonough 1977, Zanstra and Hagenzieker 1977).

It is possible to determine vapor pressure in sample chambers like those used in psychrometry by measuring the dew point. **Dew point hygrometers** contain a thermocouple or mirror with a film of water, just as in a psychrometer, but the hygrometer maintains the film by staying at the dew point temperature. Dew point hygrometry is a continuous measurement of humidity. It is frequently used to measure water vapor in circulating air streams such as those in a gas exchange apparatus (see Chapter 15).

Dye Method

A simple and inexpensive method of measuring plant water potential is based on the change in density of a solution, sometimes referred to as the **Shardokov dye method.** Two series of replicate sugar solutions are prepared to span a range of os-

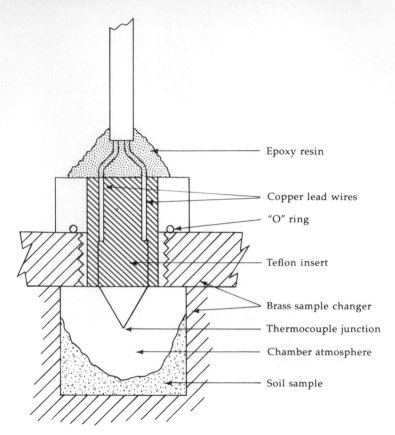

Figure 18-21   The Spanner (Peltier) thermocouple psychrometer and a psychrometer chamber, shown with a soil sample. The entire assembly is immersed in a water bath to maintain constant temperature. From Brown, 1970. "Measurement of water potential with thermocouple psychrometers: Construction and applications." U.S. Dept. of Agriculture Forest Service Research Paper INT-80.

motic potentials. Leaves or leaf discs are placed in one series of solutions. The second series is colored by methylene blue or another appropriate dye. After the tissue has been exposed to the solution for an hour or so, a drop of the colored solution is introduced into the center of the replicate solution that contains the tissue. If the drop moves up, the ψ_{plant} was lower than the osmotic potential of the solution; if it moves down, the ψ_{plant} was higher. The ψ_{plant} is equal to the osmotic potential of the solution when the drop diffuses in all directions. Tables of osmotic potential versus molality of solutions are available. For details and precautions, see Knipling (1967), Knipling and Kramer (1967), and Barrs (1968).

Relative Water Content

Relative water content (RWC) provides a simple means by which information concerning the water status of plants can be obtained. RWC is an expression of the current plant water status relative to what it would be if the plant were fully hydrated.

$$RWC = \frac{\text{fresh weight} - \text{dry weight}}{\text{fully turgid weight} - \text{dry weight}} \times 100 \qquad \text{(Equation 18-10)}$$

Fully turgid weight is obtained by exposing the tissue to water until it no longer increases in weight. RWC is an accurate measure of water content but gives no information concerning plant water status with respect to soil or atmosphere, nor any indication of the plant's response to water stress.

Reflectance of leaves varies with RWC in the range detectable by thematic mappers of some satellites. Thus the RWC of vegetation on a global scale can be tracked by remote means when the leaf water content index (LWCI) is known for common species. Details of the relationship between the reflectance of fully hydrated leaves and LWCI are available in Hunt et al. (1987).

Other Measures of Plant and Soil Water Status

Barrs (1968) has reviewed the extensive literature on water status methodology in plants, and Rawlins (1976) has done the same for soils. Several more useful references and more specific information can be found in Brown (1970) and Brown and van Haveren (1972).

Summary

Water potential is a measure of the free energy of water. Knowing the water potential of a system gives information concerning the potential for movement of water to a neighboring system of known water potential. Water potential is determined by four properties of the system: (a) The matric potential varies with the amount of surface area affecting water molecules. (b) The osmotic potential depends on the solutes in the system. (c) The pressure potential is a measure of energy in the system due to changes in pressure. (d) Gravitational potential, an important component of water potential in saturated soils, is a measure of gravitational pull on water in the system.

Atmospheric water vapor concentrations are important environmental factors for the plant. Evapotranspiration—the loss of water by evaporation and transpiration—is highest in air with a high saturation deficit. Transpiration in plants is controlled by stomates, which are located at the point in the soil-plant-atmosphere system where the most dramatic drop in water potential occurs, the leaf surface.

Evaporative cooling associated with transpiration is an important mechanism of temperature regulation in plants. The resistance to water loss of leaves depends on (a) boundary layer resistance to water vapor diffusion through still air at the leaf surface, (b) stomatal resistance as the stomatal pore is opened and closed in response to internal and external conditions, and (c) resistance in the internal air space of the leaf as water diffuses from the saturated mesophyll cell walls toward the stomate. The size, shape, and orientation of leaves affect their effectiveness in a particular environmental setting. Transpiration and diffusive resistance measurements can be made in gas exchange systems, using steady-state diffusive resistance porometers, or (less expensively) by chemical means.

Transport of water through roots and stems depends on atmospheric water potential, soil water potential, and characteristics of the vascular system. Water potential is highest in the soil and decreases continuously to the atmosphere for an actively transpiring plant. Smaller vessels have greater resistance. They are common in plants of arid habitats and at the tops of tall trees, where new vascular tissues are formed under water stress.

Water held in soil is available for plant growth to varying degrees, depending on soil texture and structure, characteristics of the plant, and soil salinity. Movement of water in the soil depends on the water potential differences along the path of movement and on the hydraulic conductivity of the soil. Plants growing in soils at field capacity show small diurnal fluctuations in tissue water potential, and transpiration exceeds absorption during the midday. As the soil dries, the magnitude of the diurnal fluctuations becomes greater until plants reach wilting during midday. When soil water potential is above plant osmotic potential, tissue water potential will equilibrate with soil water potential during the night. Depletion of soil water to the point of permanent wilting has occurred when soil water potential is equal to tissue osmotic potential.

The osmotic potential of a tissue determines the degree of water stress the plant can tolerate without serious decreases in tissue water. Plants exposed to dry environments tend to have higher solute concentrations in the tissues, allowing them to extract water from the soil under situations of low soil water potential. Some plants acclimate to changes in water availability by actively modifying tissue osmotic potential.

Methods of measuring soil and plant water include the pressure chamber, psychrometry, dew point hygrometry, the dye method, and relative water content. Much remains to be done if we are to understand the wide range of water-related adaptations of terrestrial plants.

CHAPTER 19

WATER: ENVIRONMENT AND ADAPTATIONS

P lant ecologists have expended a great deal of effort studying plant-water relations because no other single environmental factor can be directly related to so many plant responses. We noted the significance of water in determining productivity in Chapter 12, in influencing biogeochemical cycles in Chapter 13, and as an influence on leaf temperature in Chapter 14. If we were to look carefully, we would find that water is an important consideration in almost every chapter in this text. In this chapter, we will get an integrated perspective of the diverse effects of water on plants. This chapter includes the general aspects of water availability to plants, factors determining the distribution of water over the landscape, and vegetational patterns that are correlated with the distribution of water.

Water in the Environment

The differences in water availability between a cypress swamp in Georgia and the Sonoran Desert of Arizona are obvious. Less apparent, yet significant, variations in water availability distinguish most plant communities. Spatial and temporal variations in the hydrologic cycle determine the amount and effectiveness of precipitation in plant communities. The cyclic nature of water is shown in Figure 19-1. Condensation of atmospheric water vapor in the form of precipitation, fog, or dew places liquid water into the biosphere where it is temporarily used or stored in soil or in bodies of water; regardless of its immediate fate, processes of evaporation and transpiration eventually cycle the water molecules back to the vapor state.

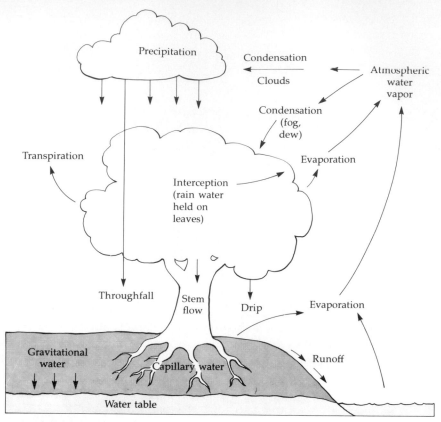

Figure 19-1   Schematic diagram of the hydrologic cycle, showing potential pathways of water flux in an ecosystem.

Fog and Dew

Dew, the condensation of water on objects in the environment, is a source of water for plants, the importance of which has been much debated. In a lucid review of the literature on dew, Stone (1963) reported that there was no evidence suggesting that the presence or absence of dew acted as a limiting factor, but that it was a contributing factor to competitive success. More recent studies support Stone's conclusions and document conditions where the presence of dew may promote survival.

Dew forms when the temperature of a surface falls below the **dew point,** the temperature at which water will condense from air. The formation of dew is promoted under conditions conducive to rapid cooling. Condensation of dew reaches the greatest levels under clear skies (clouds reduce reradiation), low wind speeds, high vapor pressures (which reduce the amount of cooling necessary), and beneath vegetation with open branching and leaf area spread over considerable heights. Lloyd (1961) found that no dew formed under a closed forest canopy, presumably because reradiation was reduced. Dew formation is also enhanced by cool surfaces, poleward facing slopes, in flats and valleys with cold air drainage (Chapter 14), and in areas subjected to cold ocean air when conditions are not such that fog will form. Even under optimal conditions, the theoretical maximum dewfall during a single night is less than 1 mm (Slatyer 1967, Stone 1963).

Soil water may be redistributed by vaporization and condensation, similar to dew formation, within the soil profile (referred to as **distillation** by Stone 1963). Periods of high ambient temperatures heat the upper layers of soil. The heat then moves downward through the soil. When ambient conditions cool, a cool zone of air follows the heated zone into the soil. Air in the interstitial spaces of the soil becomes warmed and water in the heated zone vaporizes and moves upward in the soil. When the warm, moist air encounters the cool soil particles above, the warmed air can be cooled to the dew point. Water from lower in the soil has moved upward in the vapor state and condensed back to the liquid state in the upper zones of the soil. Syvertsen et al. (1975) used this phenomenon to explain daily fluctuations in soil water potential in the rooting zone of desert shrubs, 20–40 cm depth. Vaporization below and condensation in the rooting zone redistributes the soil water to areas where it may be absorbed. It is difficult to distinguish movement of soil water by distillation from water moved by hydraulic lift (see Chapter 18).

Dew may also collect on plant shoots. The amount of dew harvested depends on plant morphology. Shure and Lewis (1973) accidentally noted this difference in ability to harvest dew while introducing radioactive tracers into stems of old-field weeds. Wells placed on stems to hold tracers were full of dew water on common ragweed (*Ambrosia artemisiafolia*) but empty on wild radish (*Raphanus raphanastrum*) and lamb's quarter (*Chenopodium album*). Ragweed collected 2.5–3.2 ml plant^{-1} day^{-1} for an estimated 1100 ml m^{-2} mo^{-1}. This consistent addition of soil water during the growing season may be the factor that gives ragweed a competitive edge during the first year of succession in the northeastern United States.

Stone also found that dew present on leaves had a positive influence on *Pinus ponderosa* seedlings subjected to drought. Stressed seedlings lived an average of three months if sprayed with water, whereas those not sprayed lived only two months. Unsprayed seedlings used twice the soil water of sprayed seedlings. Similar advantages were found for corn stressed to near wilting (Duvdevani 1964). Therefore, dew may be important in reducing water requirements by reducing the leaf-air ψ gradient even when it never enters the soil.

Fog, especially along coastal areas (Figure 19-2), has long been recognized as an important ecological factor. **Fog** forms when warm, moist air (a) passes over cold water or land (as in the coastal communities along the Pacific Ocean); (b) cools as it ascends a mountain (as in tropical cloud forests); or (c) is cooled by rapid reradiation from soil (as in inland valleys). Fog formation is greatly enhanced by the presence of condensation nuclei. Condensation nuclei are frequently salt particles released by the bursting of bubbles from surf, white caps, etc. (Boyce 1951), or dust, pollen, and other particulate matter in the atmosphere.

The importance of **fog precipitation** is most pronounced near the edge of a forest because the water droplets are rapidly removed by obstacles in the environ-

Figure 19-2 Early summer coastal fog in southern California.

ment. This is why fog gauges are rain gauges with a cylindrical vertical screen extending above the opening, forming an obstacle causing water droplets to enter the rain gauge.

The coast redwood forests of northern California are frequently said to exist because of fog. Oberlander (1956) measured summer fog precipitation on the San Francisco peninsula and found that 5–150 cm of water was added to the soil during a single month without rainfall. Azevedo and Morgan (1974) measured 123–388 cm (depending on the species) of fog precipitation in coastal California forests during only 28 summer days. Davis (1966) suggested a relationship between ocean fog and the distribution of spruce-fir forests on the coast of Maine. Vogelmann et al. (1968) measured a 67% increase in precipitation due to fog at 1100 m elevation, but little or no increase below 850 m in the Green Mountains of Vermont; this elevation separated conifer forests (above) from deciduous forests (below). Fog is a potentially significant source of moisture, increasing precipitation 1.5–3 times in many habitats.

Dew and fog may have effects other than providing additional moisture. For example, water removes foliar metabolites from leaves and carries them to the soil, thus speeding up nutrient recycling, especially of K and Ca (Henderson et al. 1977, Tukey and Mecklenberg 1964, Azevedo and Morgan 1974). Tukey (1966) measured carbohydrates, amino acids, and organic acids in plant leachates and del Moral and Muller (1969) measured allelopathic substances in plant leachates.

The direct absorption of water from the surface of leaves into the plant does occur but is greatly restricted by a number of factors. The water potential at the surface of mesophyll cells is relatively high so that only a small energy gradient is established toward the leaf interior even when pure water is on the leaf surface. Stomates are usually closed during periods of dew deposition, severely restricting water vapor entry, especially in xeric-adapted plants (Vaadia and Waisel 1963). Surface tension of water droplets prevents liquid water from moving through the small stomatal openings. Slatyer (1967) concluded that absorption of surface water on leaves could never fulfill more than a small portion of the water requirements of plants; at best it may speed nocturnal recovery of turgor pressure in wilted plants and may delay the onset of stress after sunrise. Fog and dew are the most important source of water for animals, lichens, and some perennial plants in the coastal fog zone of the Namib Desert where fog and dew precipitation provide twice as much water as rainfall (Southgate et al. 1996, Lange et al. 1991).

Precipitation

Water droplets formed around condensation nuclei are called fog when in contact with the earth; they are called clouds when a layer of air with no condensation appears between them and the earth. As these water droplets become large enough to respond to gravity, precipitation will fall as rain. At temperatures below freezing and when conditions are such that the water arrives at the earth's surface as ice, water will fall as snow. Under certain circumstances, sleet or hail may result from summer storms, but these rarer forms of precipitation are of little overall consequence as sources of water for plants (although they may cause considerable physical damage on a local basis).

Condensation of atmospheric moisture in amounts sufficient to result in precipitation is stimulated by any of three sets of atmospheric conditions. Temperature generally decreases with altitude so that any condition where warm, moist air rises may cause precipitation. This type of precipitation is classified as either orographic precipitation (*oreos* = "mountain") or convectional precipitation. **Orographic precipitation** occurs when an air mass is forced upward as it moves across a mountain mass. **Convectional precipitation** occurs during the summer as warm humid air, heated by proximity to the ground, rises abruptly from the earth's surface because of its low density. As this air reaches high altitudes, intense but usually brief summer showers or thunderstorms occur. Cool-season precipitation in temperate areas is usually associated with low-pressure centers moving easterly and along cold polar air masses. **Cyclonic precipitation** occurs as air masses move counterclockwise (in the Northern Hemisphere) around the low-pressure center, become warmed on the advancing front and ascend over cool local air masses, where they cool and cause sustained precipitation over large areas. Because of the seasonally predictable location of cool polar air, high-pressure areas, and pathways of cyclonic storms, seasonal patterns and amounts of precipitation are quite predictable in those areas where the dominant cause of precipitation is cyclonic.

Rainfall The spatial and temporal distribution of rainfall is very complex and is closely related to the productivity, distribution, and life forms of the major terrestrial biomes. The factors determining the distribution and availability of rainfall, and the response of plants to these variations, have occupied plant ecologists for decades.

Vegetation may reflect patterns of orographic precipitation. In California, for example, the deciduous coastal sage scrub community near the coast gets less precipitation than higher elevation chaparral. In very high mountains, such as the Sierra Nevada, air masses continue to move upward after the majority of water has precipitated so that subalpine and alpine communities get much less precipitation than middle-altitude associations. Moisture-laden air masses frequently pass around high peaks through depressions and breaks in mountain ranges, leaving these peaks essentially devoid of orographic precipitation.

Once an air mass crosses the summit of a mountain mass, it falls, expands, and warms, greatly increasing its water-holding capacity. The effect is to increase potential evapotranspiration and to decrease precipitation on the lee side. The leeward area of low precipitation is referred to as a rain shadow. Many of the earth's deserts are caused by rain shadow. The pattern of precipitation in the Cascade Mountains of Washington is influenced primarily by topographic factors (Figure 19-3).

Topography also influences the hydrological cycle by affecting potential evapotranspiration. Chapter 14 describes the variations in solar radiation incident upon slopes of various aspects. Wentworth (1981) incorporated solar radiation with elevation and slope parameters to derive an index to correlate with vegetation mosaics. Significantly less total solar radiation is incident on poleward slopes, leading one to suspect that more water would be present on these slopes because of lowered evapotranspiration. However, several studies (e.g., Griffin 1973, Ng and Miller 1980) show that plants on poleward slopes could be under greater stress during the dry season than those on warmer slopes. Presumably these poleward slope plants have a reduced capacity to restrict water loss compared with those on other slopes, and therefore deplete the available water supply more quickly. Greater LAI (leaf area index) on poleward slopes may also cause more rapid water loss.

Humid air masses may have accumulated significant amounts of water vapor from several sources. The most important source is the ocean, where salt particles form condensation nuclei, resulting in cloud formation. Ocean proximity is thus important in determining amounts of precipitation. Proximity to the ocean does not always ensure a mesic environment. In certain geographic situations, such as in Baja California, Peru, and southwestern Africa, coastal deserts are formed when cold air masses (cooled by passing over cold ocean currents) contact a warm landmass; the air masses warm and cause high rates of evaporation. Conversely, warm ocean currents create warm air masses and are the source of very high levels of coastal precipitation when these heavily water-laden air masses contact cooler landmasses, as in the Pacific Northwest.

Snow   Snowfall has a significant ecological impact in many areas of the world—as the prime source of water to replenish underground supplies, as a source of soil moisture, as an insulator in areas of extreme cold, and as a physical force important in controlling the distribution of vegetation. Snow varies considerably in water content per unit volume, but typically, 8–12 cm of snow is equivalent to 1 cm of

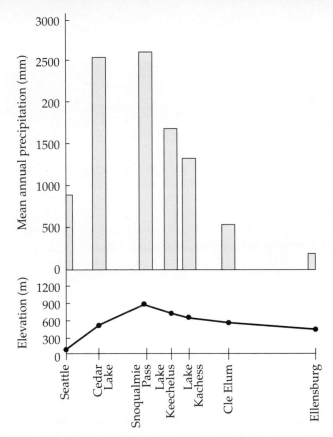

Figure 19-3   Precipitation along a cross section of the Cascade Range in the vicinity of Snoqualmie Pass, Washington (47°25′N. lat.); the distance from Seattle to Ellensburg is approximately 152 km. From Franklin and Dyrness. 1969. *Natural Vegetation of Oregon and Washington,* U.S. Dept. of Agriculture Forest Service Research Paper PNW-80.

rainfall. Increases in soil water from snow depend on the rate of melt. Sudden increases in temperature may cause rapid snowmelt, so that much will be lost as runoff, similar to a very intense rainstorm. Slow melting will increase infiltration and water availability to plants in the immediate vicinity. Soil temperatures under snow seldom fall much below 0°C because the insulating capacity of the snow traps heat radiating from the earth, thus protecting plants from extreme cold. Protection from wind also is important where temperatures are low and winds are desiccating. The physiognomic modification called krummholz (Figure 19-4) is due to physical and desiccation damage caused by wind and blowing ice crystals. In Wyoming, needles of Engelmann spruce (*Picea engelmannii*) and subalpine fir (*Abies lasiocarpa*) directly exposed to winter winds experience greater water stress than leeward needles or those protected by snow. Hadley and Smith (1983) found a high correlation between leaf water status and leaf mortality in these timberline conifers.

Figure 19-4   Krummholz vegetation at timberline in the Rocky Mountains. Flagged appearance is caused by the abrasive and desiccating effects of wind and blowing ice crystals. Photo courtesy of Harold Bradford.

Patterns of vegetation in subalpine forests are also dependent upon snow. Meadows within subalpine forests are often the result of deep snow accumulation and the resulting shorter growing season. Snow avalanches frequently clear large areas of subalpine forest, thereby modifying the vegetational mosaic of steep mountain slopes.

Snow has been more thoroughly studied in alpine tundra habitats than in other communities because of its great importance in determining the distribution and abundance of plants in that habitat. Summer rainfall is typically very low in alpine communities, so snow cover provides the major source of soil moisture for these plants. In fact, most scientists agree that the related factors of snow cover and soil moisture are the primary limiting factors for vascular and nonvascular plants in alpine communities (e.g., Billings and Bliss 1959, Johnson and Billings 1962, Buttrick 1977, Hrapko and LaRoi 1978, Komarkova and Webber 1978, Flock 1978, Stanton et al. 1994, Galen and Stanton 1995). As shown in Figure 19-5, a constant supply of water below snowbeds results in meadow communities in drained areas and sedge-sphagnum communities in concave areas. The higher, steeper slopes above snow accumulations and melt are usually boulder fields with little vegetation because of the combined influence of maximum winter exposure and summer desiccation. The importance of snow as an ecological factor results from both its influence on water supply and its role in physical modification of the habitat. In alpine habitats in Colorado, plant productivity is highest in areas protected by moderate amounts of snow during the winter, where snowmelt occurs early in the

Figure 19-5   (a) Uneven patterns of snow accumulation in a Rocky Mountain alpine community. (b) Relationship between topography, moisture, and snow duration gradients in northern British Columbia alpine communities. (a) Photo courtesy of Harold Bradford. (b) From Buttrick (1977). Reproduced by permission of the National Research Council of Canada from the *Canadian Journal of Botany,* 55:1399–1409.

season, and where soil moisture is high. Snowbanks that persist well into summer retard development of plants even though soil moisture may be high. Flowering time in alpine communities is closely related to persistence of the snow cover (Greenland et al. 1984). The distribution of alpine populations of *Kobresia bellardii* *(myosuroides)* depends on the longer growing season provided by light snow (and early snowmelt) in the Rocky Mountains. Some snow is necessary for protection from wind (Bell and Bliss 1979).

In forest vegetation, snow may suppress seedling survival and tree growth by virtue of its mass. A major ecotone in the Sierra Nevada, at 200 m elevation, correlates strongly with freezing elevation during winter storms (Barbour et al. 1991). Above the ecotone, precipitation tends to fall as snow, whereas below the ecotone it tends to fall as rain. In addition to mechanical stresses, snowpacks often provide habitat for pathogenic fungi and reduce light below the compensation point. Similar studies correlating snowpack with forest distribution have been conducted in British Columbia (Peterson 1969), the Rocky Mountains (Peet 1998), and Japan (Homma 1997).

Precipitation Effectiveness

Knowledge of the total amount of precipitation falling on a community does not always give a clear picture of the availability of water to plants. The season, atmospheric condition, precipitation type, intensity, annual variation, soil condition, and vegetation physiognomy influence the availability of precipitation and its distribution within the habitat. Many attempts have been made to determine **site water balance** (or water budgets) based on measurements of rainfall, runoff, percolation, and evapotranspiration (Rosenberg 1974). However, the accurate measurement of all the parameters necessary to estimate these properties is technologically difficult. Luxmoore (1983) was able to obtain detailed simulations of water fluxes in an oak-hickory stand in Tennessee using a computer model. The database included measurements of soil water, meteorological phenomena, and plant water status.

The conditions discussed earlier that influence atmospheric vapor pressure are important determinants of water loss through evaporation. In fact, measurements of temperature, saturation deficit, and wind have been combined to give a meaningful picture of biologically significant variations in climate. Significant evaporation occurs from free water surfaces, such as droplets intercepted by the vegetation canopy and in slow-draining depressions. Evaporation from exposed soil is limited to the upper 10–20 cm; there is, however, a certain amount of water vapor transfer from lower levels, especially in desert environments. A similar transfer to the vapor state may occur from the surface of snow or ice. This process, called **sublimation,** takes place as ice is transformed to the vapor state without going through a liquid phase. Sublimation, like evaporation, reduces the effectiveness of precipitation.

Seasonal distribution of precipitation determines, in part, how much of the water is available to the vegetation. An annual precipitation rate of 70–80 cm can support either certain deciduous forest communities, or the strongly drought-

adapted broad sclerophyll vegetation of mediterranean climates. The distinguishing factor is that almost all the precipitation in mediterranean climates falls during
the cool season, so that chaparral plants must endure annual extended drought
during the hottest months. In the semiarid mediterranean climate, summer drought
occurs regardless of the amount of winter precipitation or the vegetative cover
(Miller et al. 1983). If the precipitation were more evenly distributed or concentrated in the warm season, more mesic vegetation could be expected. Where a large
portion of the precipitation falls as snow, its effectiveness depends on the melt rate.
When temperatures warm suddenly, a rapid melt will render most of the snow in-
effective—much of the moisture will be lost as runoff, particularly if the soil is
frozen or already saturated.

Intensity of rainfall is an important determinant of runoff and is related to seasonality. Typically, summer showers in temperate areas are of very short duration
and may be so intense that most water will run off the surface before the slower soil
infiltration process can take place.

Predictability of precipitation is also reflected in the flora and vegetation.
Many deserts average enough annual precipitation to support more mesic plants,
but rainfall is very unpredictable and periods between effective precipitation may
extend up to several years. Plants are, therefore, adapted to respond quickly to precipitation and to tolerate periods of drought. Where droughts are predictable, seasonality patterns in plant response are also predictable and little impact on the
vegetation is apparent. However, in mesic environments periodic unpredictable
droughts may have a severe impact on the vegetation.

Many attempts (e.g., Penman 1950, van Bavel 1966, Major 1977) to measure site
water balance have been made. Grier and Running (1977) devised a site water balance index that correlates closely with leaf area in coniferous forest communities
of western Oregon (Figure 19-6). The index was calculated by adding soil water
storage to measured growing season precipitation and then subtracting open pan
evaporation. This approach may be very instructive in those areas where low nu-

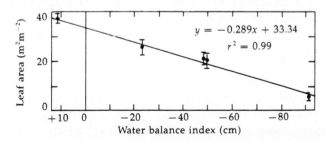

Figure 19-6   Relationship between water balance index and leaf area in
five forest zones of western Oregon. Zones plotted are, from left to right:
Picea sitchensis, Tsuga heterophylla, interior (Willamette) valley, east slope
mixed conifer, and *Juniperus occidentalis.* Bars show range of observed leaf
areas in each vegetation zone. From C. C. Grier and S. W Running. 1977.
"Leaf area of mature northwestern coniferous forests: Relation to site water
balance." *Ecology* 58:893–899. Copyright 1977 by the Ecological Society of
America. Reprinted by permission.

trient levels, severe temperature, physical damage, and high levels of runoff are not important factors. Major (1977) has compiled water balance information for 12 sites on a transect across the Sierra Nevada (Figure 19-7). Water surpluses are greatest at middle elevations where summer water deficits are less than at lower elevation sites in the Central Valley or in the rain shadow of the Sierra Nevada where higher temperatures increase potential evapotranspiration. The ratio of annual precipitation to evaporation *(P/E)* is highly correlated with vegetation. Forests typically have a $P/E > 1.0$, grasslands are 0.7–1.0, scrublands < 0.7, and deserts < 0.1.

Recent advances using stable isotopes of carbon have given plant ecologists insight into site water balance from the plant's perspective (Rundel et al. 1988). In Chapter 15, we described how the ^{13}C isotope of carbon diffuses more slowly from atmosphere to chloroplast and has a lower affinity for RuBP carboxylase than does the ^{12}C isotope of carbon. The difference in diffusion rate causes greater discrimination against ^{13}C in plants with higher water use efficiency (molar ratio of CO_2 absorbed to H_2O lost by transpiration). When plants close their stomates to restrict water loss, they have a more positive $\delta^{13}C$ value because they differentially discriminate against ^{13}C (O'Leary 1988). Therefore, tissue $\delta^{13}C$ values provide an index of water availability integrated over the time that the tissue was produced.

Stewart et al. (1995) used $\delta^{13}C$ values of plants in communities along a rainfall gradient in southern Queensland, Australia, to show that $\delta^{13}C$ values correlated with annual rainfall. The wide range of $\delta^{13}C$ values found in C_3 plants is a reflection of leaf water use efficiency imposed by site water balance.

Ehleringer and Cooper (1988) examined $\delta^{13}C$ ratios of shrubs in a Mojave Desert community in Arizona (Table 19-1). Plants of the same species have increasing values (that is, less negative values) for $\delta^{13}C$ along a gradient of increasing aridity from wash to bajada (the gentle lower slopes of desert mountains; see Chapter 20). The $\delta^{13}C$ values may also reflect patterns of adaptation; for example, long-lived Mojave Desert shrubs had more positive $\delta^{13}C$ values than shorter-lived shrubs living in the same habitat. Carbon isotope ratios provide an integrated view of site water balance.

Stable isotopes are assuming greater importance in ecological studies. For an overview see Rundel et al. (1988).

Floods and Riverine Ecosystems

Riverine ecosystems are located in the zone directly influenced by a river or stream. These **riparian habitats** occur in bands adjacent to active river channels on deposits from past floods or meanderings of the watercourse. The extent of the riverine ecosystem depends on flood frequency and intensity.

During the twentieth century, there has been an enormous decline in riparian communities, especially in the southwestern United States. One cause of this decline is the regulated flow of water from upstream dams. Flood prevention significantly increases salinity and reduces movement and deposition of open habitat for early successional riparian species such as cottonwood (*Populus* spp.) and willow (*Salix* spp.; Scott et al. 1997). Early successional riparian species require open habitats, grow rapidly, have low nutrient needs, and are tolerant of accumulating

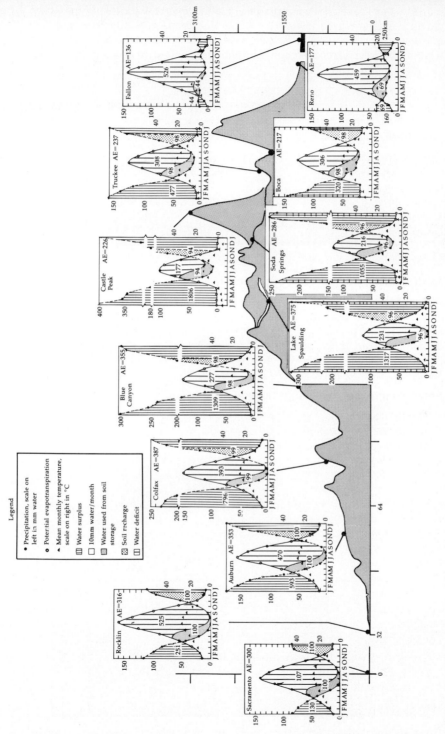

Figure 19-7   A transect over the Sierra Nevada from Sacramento, California (38.5°N), over Donner Summit (39.3°N), to Reno and Fallon, Nevada (39.5°N), showing water balances. Vertical exaggeration 14.5×. From Major. 1977. In *Terrestrial Vegetation of California,* ed. Barbour and Major. Copyright © 1977 John Wiley and Sons, Inc. Reprinted by permission.

Table 19-1   Leaf carbon isotope ratio for individual species in wash, transition, and slope (bajada) microhabitats near Oatman, Arizona. A blank indicates that the species was not present in that microhabitat. Leaf types are dd = drought deciduous, wd = winter deciduous, e = evergreen, and es = evergreen stem. From J. R. Ehleringer and T. A. Cooper. 1988. "Correlations between carbon isotope ratio and microhabitat in desert plants." *Oecologia* 76:562–566.

	Leaf type	Wash	Transition	Slope
Long-lived (50+ years)				
Cercidium floridum	dd	−24.07		
Chilopsis linearis	wd	−25.37		
Chrysothamnus paniculatus	wd	−26.69		
Ephedra viridis	es		−23.30	−23.82
Krameria parvifolia	dd		−24.60	−23.87
Larrea divaricata	e	−24.12	−23.60	−22.67
Lycium andersonii	e	−25.32	−25.09	
Mean		−25.11	−24.15	−23.45
Medium-lived (10–40 years), opportunistic				
Acacia greggii	wd	−27.40	−25.82	
Ambrosia dumosa	dd	−27.37	−26.04	−25.37
Encelia farinosa	dd		−26.13	−25.46
Encelia frutescens	dd	−27.51		
Hymenoclea salsola	dd	−26.45	−23.54	
Mean		−27.18	−25.38	−25.42
Short-lived (1–10 years), opportunistic				
Ambrosia eriocentra	dd	−29.29		
Bebbia juncea	dd	−28.33	−26.66	−25.80
Cassia covesii	dd	−26.03	−26.71	−26.83
Eriogonum fasciculatum	dd		−26.40	−26.53
Eriogonum inflatum	dd	−28.17	−25.82	−25.70
Phoradendron californicum	es	−27.14	−26.96	
Porophyllum gracile	dd	−27.47	−26.75	−26.55
Psilostrophe cooperi	dd		−27.65	−27.02
Sphaeralcea ambigua	dd		−27.38	−27.57
Viguiera laciniata	dd		−26.08	−26.40
Mean		−27.74	−26.71	−26.55

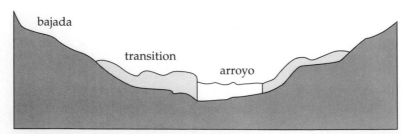

Cross-sectional depiction of the site, illustrating wash, transition, and slope microhabitats.

sediment but are intolerant of salinity. Increased salinity favors establishment of salt cedar *(Tamarix ramosissima),* a rapidly spreading introduced shrub that outcompetes cottonwoods and willows (Busch and Smith 1995). Periodic floods are necessary to establish and maintain the native riparian forests of southwestern North America.

Interception, Throughfall, and Stem Flow

The distribution of rainfall under vegetation canopies depends on how canopy structure influences the amount of water entering the soil at the base of the stem (**stemflow**) and the amount of rainfall coming through the canopy (**throughfall**). Leonard (1961), Lawson (1967), and Henderson et al. (1977) agreed that 10–20% of precipitation is intercepted by forest canopies (somewhat less during the dormant season). Henderson et al. (1977) compared amounts of throughfall in various forest types and found no significant difference between types. The canopies of the desert shrubs creosote bush *(Larrea tridentata),* tarbush *(Flourensia cernua),* and mesquite *(Prosopis glandulosa)* intercept about 44% of rainfall and have 55–65% throughfall (Martinez-Meza and Whitford 1996). Unburned tallgrass prairie intercepts 16% more rainfall than burned prairie. In burned prairie, throughfall averages 78% (Gilliam et al. 1987). Most intercepted water is lost to evaporation because little will be absorbed by the leaves.

Stemflow causes an uneven distribution of soil water by concentrating precipitation near the base of the plant. Geometric shapes that encourage stemflow may be a competitive advantage in water-limited environments. In creosote bush, stem angle and stem length accounted for much of the variability in stemflow, which averaged 10% of total precipitation. Stemflow accounts for 30–40% of precipitation in an Australian woodland and mallee *(Eucalytus* scrubland) (Slatyer 1965, Pressland 1973, Nulsen et al. 1986.) Stemflow may also be a significant nutrient source for plants (see Chapter 13).

Special Adaptations

Osmotic Balance and Toxicity in Saline Habitats

Halophytes are specifically adapted to tolerate widely varying external and internal concentrations of salt. Most saline habitats (Figure 19-8) contain sodium chloride as the primary salt, but sodium, magnesium, and calcium sulfates, magnesium and potassium chlorides, and sodium carbonate are also significant in many situations. **Glycophytes** (plants that are not salt tolerant) generally exclude salt from entering the roots. Excluded salt concentrating at the root surface may soon reverse the chemical energy gradient and cause plant tissues to dehydrate (Greenway and Munns 1980). *Raphanus sativus* (wild radish, a glycophyte) adjusts osmotically to high salinity but suffers long-term salinity damage to processes as-

Figure 19-8 Halophytes growing in a desert depression. Note the salt accumulation at the soil surface. Photograph courtesy of Alan Romspert.

sociated with root conductivity (Ownbey and Mahall 1983). Thus intolerance in some glycophytes may be related to toxicity or nutrient imbalance. Maintaining the chemical energy gradient necessary for water uptake requires that osmotic concentrations in halophytes must be higher than in glycophytes. Although protoplasmic salt tolerance is higher in halophytes than in glycophytes, there is a limit to this tolerance. Mechanisms that restrict cytoplasmic salt concentrations are ultimately necessary to halophyte survival.

Regulation of salt concentrations in the tissues of halophytes may be classified into three basic categories (Albert 1975). First, some halophyte species restrict the rate of salt uptake by increasing rates of ion accumulation only for brief periods or in certain habitats. Ungar (1977) has shown that saltbush (*Atriplex triangularis*) accumulates salt in direct proportion to the salt content of the soil. Salt uptake is restricted or excluded by root membranes in some species (e.g., mangroves; Scholander 1968). When salts are excluded, cellular osmotic potential is adjusted by increasing the amounts of organic metabolites as a means of maintaining the free energy gradient necessary for uptake. Various nitrogen compounds (amino acids and amines), carbohydrates, and organic acids accumulate in cells of halophytes. Such osmotic adjustments occur in ice plant (*Mesembryanthemum crystallinum*), a salt-tolerant succulent that switches from C_3 photosynthesis to CAM and accumulates malate when growing in a saline medium (Luttge et al. as cited in Flowers et al. 1977). Brownell and Crossland (1972, 1974) have shown that some C_4 and CAM plants require or are stimulated by sodium ions in the soil. Such evidence has led to speculation that C_4 and CAM biochemical pathways are adaptations which accumulate organic acids that counterbalance the osmotic influences of saline environments (Laetsch 1974, Flowers et al. 1977).

Salt dilution and salt extrusion are the two other ways that halophytes regulate tissue salt concentrations. **Succulence,** caused by the accumulation of water, serves to dilute salts in cells, thereby reducing solute concentrations (Jennings 1968). Rapid growth of new tissues also has a diluting effect because salts are distributed in greater volumes of tissue (Greenway and Thomas 1965). Salt extrusion, removing excess salts from the plant body, is achieved by various mechanisms. Some species (e.g., *Atriplex* spp.) have salt glands that expel salt onto the surface of leaves where it is washed from the plant by rain, dew, or fog condensation (Figure 19-9). Vesicular bladder hairs are another mechanism whereby salt is deposited on the leaf surface: the hair fills with saltwater until it ruptures, releasing the salt onto the leaf. Salt may also be extruded by the shedding of organs with accumulated salt. Salt evidently induces senility, and as the organ ages, nutrients and stored organics are mobilized to younger parts of the plant, followed by abscission of the organ containing the accumulation of salt.

Any one species usually employs a combination of these mechanisms to reduce the salt load. Further details are in reviews by Waisel (1972), Ranwell (1972), Reimold and Queen (1974), and Flowers et al. (1977).

Anatomical Adaptations

The abundant information concerning anatomical responses to water deficits (e.g., Oppenheimer 1960, Esau 1965, Hanson 1917, Daubenmire 1974) is now being critically reevaluated in light of new technology and plant response information.

Figure 19-9   A scanning electron micrograph of the leaf surface of saltbush (*Atriplex*) showing salt glands. From T. Troughton and L. A. Donaldson. 1972. *Probing Plant Structure.* Reed Methuen Publishers, Auckland, New Zealand.

Xeromorphic leaves are generally small, with reduced cell size; thicker blades; smaller, sunken, and denser stomates located on the lower leaf surface; more pubescence; a thicker cuticle; more heavily lignified epidermal cells; better developed palisade mesophyll; and less intercellular space. It is true that these modifications are found in arid-adapted plants, but we can no longer assume that all plants growing in hot dry environments have similar anatomical adaptations. For example, creosote bush (*Larrea tridentata*), the most common shrub in the warm North American deserts, has no pubescence, its stomates are evenly distributed on both surfaces, and the stomates are not in crypts (Barbour et al. 1974).

Accumulations of surface water on leaves can reduce photosynthesis significantly because CO_2 diffuses through water about 10,000 times slower than through air. Photosynthetic reductions range from 2–95% depending on leaf wettability (Smith and McClean 1989, Brewer and Smith 1994). Trichome (leaf hair) morphology and density are correlated with leaf wettability and habitat in Rocky Mountain montane and subalpine plants. Plants of open meadows where dew deposition is high are less wettable and have shorter retention times for water droplets than adjacent forest understory species (Brewer and Smith 1997). Thus, it appears that one water-related role of trichomes is to increase photosynthesis by reducing leaf wettability in habitats where leaf surface water is a limiting factor (Brewer et al. 1991).

Pubescence, leaf size, and the ratio of mesophyll cell (internal leaf) surface to outer leaf surface (A^{mes}/A) interact to determine water use efficiency in desert plants (Cunningham and Strain 1969; Figure 19-10). Leaf pubescence not only increases the depth of the boundary layer but is even more important as a regulator of light absorptance and, therefore, leaf temperature (Ehleringer et al. 1976, Smith and Nobel 1977b). Small leaves are more efficient energy dissipaters and lose less water per unit leaf area than large leaves. The ratio of mesophyll cell surface to outer leaf surface may be less important in regulating water loss than formerly assumed; its prime importance is in the resistance to CO_2 uptake in the liquid phase (see Chapter 15). Smith and Nobel (1977b) studied the desert shrub *Encelia farinosa* (brittlebush), which develops leaves that vary seasonally in size, pubescence, and A^{mes}/A. Leaves produced in hot, dry conditions are small, highly pubescent, and

Figure 19-10 The influences of pubescence, leaf length, and A^{mes}/A on transpiration and photosynthesis. From W. K. Smith and P. S. Nobel. 1977. "Influences of seasonal changes in leaf morphology on water use efficiency for three desert broadleaf shrubs." *Ecology* 58:1033–1043. Copyright © 1977 by the Ecological Society of America. Reprinted by permission.

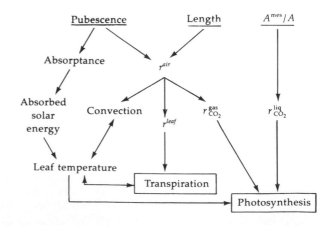

have a high A^{mes}/A ratio. Leaves that develop in cooler, more mesic conditions have decreased A^{mes}/A and pubescence in conjunction with increased size and absorptance. Such morphological changes increase potential photosynthesis over a variety of environmental conditions. Water use efficiency is thus maximized in summer months and decreases during periods of cooler temperatures and more rainfall. Pubescence functions to reduce absorptance, thus lowering leaf temperature in conjunction with reduced leaf size. Minimum r^{leaf} occurs during the wetter periods, maximizing photosynthesis when transpirational loss is not critical. The perennial desert herb *Mirabilis tenuiloba* (white four-o'clock) coexists with brittle-bush but lacks the capacity to modify r^{leaf} seasonally. Active photosynthesis is therefore restricted to a shorter period and the photosynthetic rate is lower during favorable periods (Figure 19-11).

Figure 19-11 (a) Seasonal transpiration, (b) net photosynthesis, and (c) water use efficiency for *Encelia farinosa* (triangles) and *Mirabilis tenuiloba* (circles), based on field measurements. From W. K. Smith and P. S. Nobel. 1977. "Influences of seasonal changes in leaf morphology on water use efficiency for three desert broadleaf shrubs." *Ecology* 58:1033–1043. Copyright © 1977 by the Ecological Society of America. Reprinted by permission.

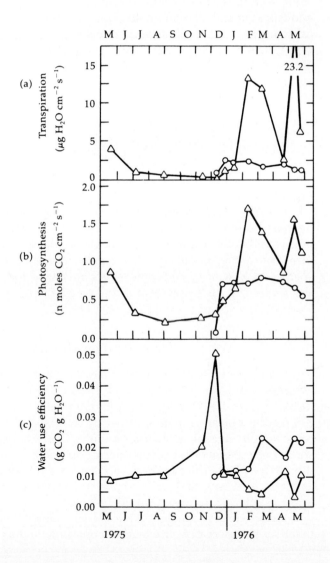

Succulence was discussed in relation to halophytes and with respect to CAM in Chapter 15, but we have not yet addressed the water relations of desert succulents. Perhaps the most important contribution of stored water in succulents lies in the ability of plants to continue opening stomates after soil water potential falls below plant water potential. In the barrel cactus *Ferocactus acanthodes* in the Sonoran Desert, studied by Park Nobel (1977), this occurred at a high value (for a desert plant) of -0.45 to -0.47 MPa. Even after soil water potential fell below this level, stomates continued to open for another 50–60 days, remaining closed only after soil water potential reached -9.0 MPa (more negative values are associated with less available water). Following a seven-month period of no major rainfall, plant water potential fell only to -0.62 MPa (close to the permanent wilting point, as osmotic potential was nearly the same). Such high levels of tissue hydration allowed stomatal response to 19 mm of rainfall within 24 hours, with tissue rehydration and full stomatal opening in 48 hours. Cacti have a high water use efficiency because they are CAM plants and store a volume of water large enough to continue CO_2 uptake for long periods after the soil water potential drops below the permanent wilting point.

Growth Form Responses and Habitat Selection

Evergreen Plants

Plants that maintain leaves during periods of water stress must have a high degree of tolerance for desiccation. The obvious advantage of being evergreen is that when water again becomes available, there is no lag while new tissues are formed; existing tissues quickly rehydrate and become active. Many evergreen species do, however, shed some leaves during periods of water stress; presumably, this serves to reduce surface area and water loss (Oppenheimer 1960, Mooney and Dunn 1970, Evenari et al. 1971, Orshan 1972).

Drought-Deciduous Species

Some species avoid water stress by becoming dormant during the dry season (Mooney and Dunn 1970b, Mooney 1977a). The coastal scrub community of California (Figure 19-12) is an example of an entire community that avoids the hot, dry mediterranean-climate summer by shedding leaves. These plants show maximum activity in cool, wet winter months. Ocotillo (*Fouquieria splendens*) is a desert shrub that ephemerally becomes deciduous, perhaps four or five times a year. Even though the plant grows in the desert, the leaves are not drought tolerant; they drop as stress increases, and new leaves expand within a week after a rain (Cannon 1905a, Scott 1932, see also Chapter 20).

Paloverde (*Cercidium floridum*) is an example of a drought-deciduous plant that lives in desert areas in a leafless condition but has the advantage of green stem

Figure 19-12 Drought-deciduous coastal scrub community of southern California. Photograph courtesy of Ted L. Hanes.

tissue. This gives paloverde and similar desert plants the advantages of ever-greenness and the advantage of increased photosynthetic area during cool, moist periods (Adams and Strain 1968, 1969; Adams et al. 1967).

Phreatophytes

Phreatophytes (literally, "well plants") are usually found in riparian (stream-side) habitats and, even in arid zones, may develop into very large trees (Figure 19-13). They are restricted to habitats with permanent underground water supplies. Cottonwood (*Populus fremontii*), willow (*Salix* spp.), sycamore (*Platanus racemosa*), fan palm (*Washingtonia filifera*) and salt cedar (*Tamarix* spp.) are examples of phreatophyte trees that avoid many of the rigors of arid environments by having roots in constant contact with the fringe of capillary water above a water table. These and many associated shrubs are examples of obligate phreatophytes. Other species are apparently able to take advantage of groundwater when present, but they can tolerate periods of low water availability. These facultative phreatophytes, such as mesquite (*Prosopis glandulosa*) and saltbush (*Atriplex polycarpa*), frequently occur in desert depressions where water and salts accumulate. Because of the high water requirement imposed by arid ecosystems, many facultative and obligate phreatophytic species must be tolerant of rather high levels of salinity. Horton (1977), Horton and Campbell (1974), McDonald and Hughes (1964), Campbell and Dick-Peddie (1964), and Vogl and McHargue (1966) have considered various aspects of phreatophyte ecology.

Mature phreatophyte trees such as bigtoothed maple (*Acer grandidentatum*)

Figure 19-13   Riparian community with large phreatophyte trees in the Chihuahuan Desert of the southwestern United States. Photograph courtesy of Patrick H. Boles.

and boxelder (*Acer negundo*) that tap subsurface flows or bedrock water layers use little surface or stream water even though their roots are in contact with these sources. Young trees use surface and stream water when their root systems are shallow, then switch to deep water as their root systems grow deeper (Dawson and Ehleringer 1991, Kolb et al. 1997). Phreatophytes may require the extra water available in streamside habitats to reach a sufficient rooting depth to tap reliable underground water.

Non-riparian phreatophyte communities dominated by greasewood (*Sarcobatus vermiculatus*), shadscale (*Atriplex confertifolia*), rabbitbrush (*Chrysothamnus nauseosus*), and big sagebrush (*Artemisia tridentata*) occur in the closed basins of the intermountain west. These shrubs are the major natural discharge pathways for groundwater 2–11 m below the surface. Estimates of groundwater removal by transpiration range from 2 to 30 cm yr^{-1} (equivalent to 20,000–300,000 L ha^{-1} yr^{-1}) for these communities (Nichols 1994).

Not all plants associated with watercourses are phreatophytic, but in a given location they are able to survive only in areas with more soil moisture, or they may depend on other attributes of streamside habitats. Riparian plants may also be those able to survive low oxygen conditions in flooded or highly saturated soils, which eliminate upland species from floodplains. In mountain areas a species may be restricted to riparian habitats at lower elevations and extend onto slopes only at higher elevations where more water is available. In desert areas, some species (e.g., *Chilopsis linearis*) are winter deciduous and relatively intolerant of midsummer conditions, so they survive only near watercourses where the additional moisture allows survival between spring and fall growth periods (Odening et al. 1974). Desert leguminous trees such as paloverde (*Cercidium* spp.) require scarification of seeds to initiate germination. Scarification is caused by abrasion of sand and gravel during floods.

Ephemerals

Ephemerals are annual plants that germinate in response to periodic phenomena and are able to complete their life cycle during short periods of mesic conditions. Ephemerals are the most common life form in severe desert situations (Whittaker and Niering 1975), where they are able to survive because they grow only during periods of moderate temperatures and soil water availability. Germination and mortality of desert ephemerals are largely independent of photoperiod because they are controlled by soil water and temperature (Went and Westergaard 1949, Went 1948, Tevis 1958a,b). Beatley (1974, 1969, 1967, 1966) studied the relationship between the ephemerals of the Mojave Desert and the environmental factors that trigger germination and death. C_3 and C_4 plants occupy different temporal habitats, summer ephemerals being typically C_4, and winter ephemerals being typically C_3 in the southwestern deserts (Mulroy and Rundel 1977, Syvertsen et al. 1976).

Summary

The success of plants depends on a multitude of factors. In terrestrial environments, limitations related to the water are always one such factor. Environments differ in the relative amounts of rain, snow, dew, and fog precipitation they receive and in the seasonal distribution of available water. These factors interact with habitat and phenological development of plants to determine the effectiveness of precipitation. The distribution of 1000 mm of precipitation in an environment with a precipitation/evaporation ratio of 2 is presented in Figure 19-14. Large amounts of precipitation are stored in the soil or lost as runoff and drainage. Minor amounts are intercepted by the vegetation canopy, and most of that evaporates. Most of the soil water is vaporized through the vegetation as transpirational water loss. Evapotranspirational water flux depends on the amount of water present and on the evaporative power of the air.

Dew is water that condenses from the air onto objects in the environment with a temperature at or below the dew point. Factors promoting dew formation include clear skies, still air, and high humidity. Some plants collect considerable amounts of water in the soil around the plant by having morphological features that funnel the water to the ground. Dew on the leaves does little to increase leaf water content but does much to reduce the evaporative demands of the atmosphere. A process known as distillation redistributes soil water by vaporization

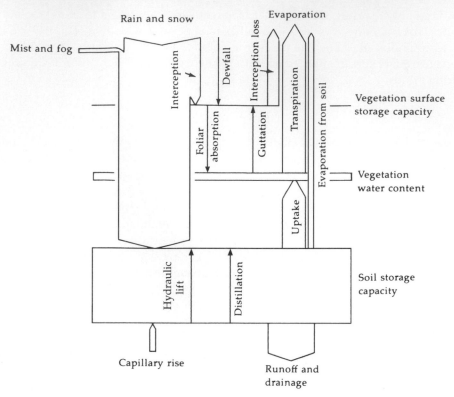

Figure 19-14   Diagram summarizing the hydrological cycle in a vegeta-
tion-soil system; the components are approximately to scale for an annual
rainfall of 1000 mm and evaporation of 500 mm. (Guttation is the exudation of liquid water from leaves.) Modified from Rutter. 1975. "The hydrological cycle in vegetation" in *Vegetation and the Atmosphere,* vol. 1, ed. J. L.
Monteith. By permission of Academic Press, London.

deep in the soil, upward movement of the vapor, then condensation in the upper
areas of the soil during cool periods. This process may be a significant source of
water for desert plants.

Fog forms when warm, moist air cools by moving over cool land areas along
the coast, ascending a mountain mass in the tropics, or rapid reradiation of heat
from soil in inland valleys. Fog precipitation is an important source of water in
coastal forests of the Pacific Northwest, in maritime deserts, and in tropical cloud
forests. Fog drip carries various organic compounds released as foliar metabolites,
contributing to nutrient recycling and the transfer of allelochemics.

Precipitation forms when condensation drops in clouds become large enough
to fall earthward. Precipitation arrives at the earth's surface as rain, hail, or snow
depending on turbulence in the clouds and ambient temperature. There are three
types of precipitation: orographic (associated with air rising over mountains), convectional (associated with convectional air currents rising from the heated ground),

and cyclonic (associated with large air masses circulating around a low-pressure area). Orographic rainfall is highest on the windward side of mountains. The leeward side of mountains is an area of very low rainfall because of the rain shadow caused by the descent of cool air that draws moisture to it as it warms. Convectional storms are usually associated with the summer season. Cyclonic storms are associated with weather fronts moving generally west to east in the Northern Hemisphere. Snow is an important source of soil water and an important insulating factor for plants in very cold environments. Krummholz of timberline are formed by the exposure of plants to the desiccating, abrasive action of wind. Low-growing plants are protected from exposure by the blanket of snow. The distribution and phenology of alpine plants are closely related to the distribution and duration of snow cover.

Many plant adaptations are specifically related to the spatial and temporal distribution of water in the environment. Halophytes are specifically adapted to habitats with high salt content. Glycophytes are nonhalophytic plants. Glycophytes usually exclude salt at the root surface, but when exposed to high salt habitats, they may dehydrate because of osmotic imbalance or die due to nutrient imbalance or toxicity. Halophytes adjust to salt by modifying the internal osmotic concentration to compensate for external salt, diluting tissue salt in high-water-content succulent tissues, or by extruding salt from the vegetative surfaces. Leaves of drought-adapted plants tend to be small, with reduced cell size, thick blades, small and sunken stomates; to be highly pubescent, with a thick cuticle; and to have a high ratio of internal to external surface. Succulence also provides stored water in some xeric plants.

The growth form of many plants is determined by water-related adaptations. Evergreen leaves are generally tolerant of desiccation and can survive dry periods, being ready to photosynthesize immediately upon rehydration. Drought-deciduous species shed their leaves during dry periods. Many of the desert drought-deciduous species have photosynthetic bark to subsidize energy balance during dry periods. Phreatophytes grow where their roots are in constant contact with the fringe of capillary water above a water table. These plants can have deciduous leaves in very low rainfall habitats because of a constant underground water supply. Ephemerals are annual plants adapted to drought by sensitive germination requirements that restrict germination to periods of ample water. Summer desert ephemerals usually have the C_4 photosynthetic pathway, which increases water use efficiency.

CHAPTER 20

MAJOR VEGETATION TYPES OF NORTH AMERICA

The North American continent has a moderately rich flora and a very rich diversity of habitats. The result is a complex mosaic of vegetation types and plant communities. One particular vegetation map for the conterminous United States shows over one hundred community types (Küchler 1964), but surveys of individual states or regions may show 50–100 community types because the level of detail is more refined than that used for the U.S. map. Since we choose to take a very broad survey view, we will discuss this diversity in the context of five vegetation types: *tundra, conifer forest, deciduous forest, grassland,* and *desert scrub,* taking account of regional differences in each case. Our objective is to present enough information on community structure, community dynamics, and ecology of dominants to underscore the uniqueness of each of the five types.

These five vegetation types cover 90% of the 23,500,000 km^2 North American land mass (including Central America). Every reader is likely to find one of these types nearby. Other types, not discussed here, may also be found nearby and may even dominate the local region. Such types include chaparral, freshwater marsh, coastal scrub, woodland, and tropical forest.

Vegetation Types: The Big Picture

The 1990s have seen an explosion of publications about North American vegetation, in particular featuring new approaches to classify it and to relate vegetation patterns to major environmental gradients. The reference section at the end of this chapter lists a dozen major works since 1990, some by federal agencies, some by nongovernmental units, and some by individual researchers who bring international perspectives that enrich our past parochial approaches. At the present time, the U.S. government is fostering a national classification of vegetation. That effort is being coordinated by a multiagency group called the Federal Geographic Data Committee. Within that committee is a vegetation subcommittee. Its efforts are being supported by such nongovernmental units as The Nature Conservancy and the Ecological Society of America. The latter organization has its own Panel on the Vegetation of North America, a group of 20 vegetation ecologists who collectively have continental expertise.

Progress so far has been to describe large, regional units of vegetation (class, subclass, group, formation) on a physiognomic basis, reserving floristic composition details for more local units called alliances (= cover type, series) and associations. **Classes** are based on basic growth form (tree, shrub, herb) of the dominant species and on the amount of ground cover. There are only nine classes in the United States: forest, woodland, sparse woodland, shrubland, sparse shrubland, dwarf shrubland, sparse dwarf shrubland, herbaceous, and sparse vascular/nonvascular. The 33 U.S. **subclasses** are based on leaf traits (evergreen, deciduous, or grasslike) and vegetation height (e.g.: tall grassland, short grassland). Each of the 254 U.S. **formations** adds climatic information (temperate, polar, tropical).

An **alliance** is a physiognomically uniform group of associations sharing one or more diagnostic species. It is equivalent to the "cover type" of the Society of American Foresters and the Society for Range Management or to the "series" of other classification systems. Alliances are named after the diagnostic species and the class. For example, the "*Juniperus osteosperma* sparse woodland" is the technical alliance name for juniper woodland common in semiarid parts of the Great Basin. This woodland, however, occupies a relatively wide range of habitats because the diagnostic species, *Juniperus osteosperma,* has relatively wide environmental tolerances. Within that range lie many more local communities, each with its own unique subset of associated species. These more local communities are technically called **associations.** Association names include the alliance's diagnostic species plus a characteristic species of the local association. For example, some juniper woodlands have an understory that includes needle-and-thread bunchgrass; the technical association name is "*Juniperus osteosperma/Stipa comata* sparse woodland." We do not yet know how many alliances and associations there are in North America nor in any of its component nations. Based on complete surveys of other regions, such as Europe, we can predict that there are more than 3000 alliances and three times as many associations.

Our survey of North American vegetation will not emphasize these classification terms. Instead, we will use less technical, more intuitive and familiar terms. Our purpose in this chapter is to summarize ecological relationships within major vegetation regions rather than to classify vegetation into narrow, technical types.

Finally, we want to make this transcontinental survey as general as possible. That is, at the end of your trip, we hope you can place North America's vegetation in a broader, global context. North America is not that unique. Most of the major vegetation types in North America have counterparts on other continents: tundra, boreal forest, deciduous forest, montane conifer forest, grassland, steppe,

Figure 20-1 (a) Vegetation of North America north of Mexico. Vegetation types that occupy <1% of the North American landmass are not shown. Most ecotones, mosaics, and local subtypes are not indicated. 1a, High arctic tundra and polar desert; 1b, low arctic tundra. 2, Boreal forest. 3a, Mixed mesophytic deciduous forest; 3b, western mixed mesophytic forest; 3c, oak and hickory forest; 3d, oak, hickory, and pine forest; 3e, oak and chestnut forest (now chestnut is absent because of disease); 3f, beech and maple forest; 3g, maple and basswood forest; 3h, northern hardwoods ecotone. 4, High elevation Appalachian Mountains conifer forest. 5, Southeastern Coastal Plain. 6a, Tallgrass prairie including the prairie peninsula and oak savanna ecotone with the eastern deciduous forest; 6b, mixed-grass prairie including aspen parkland ecotone with the boreal forest; 6c, short-grass prairie; 6d, central California grassland; 6e, intermountain grassland or shrub steppe, including Palouse and Columbia Basin areas. 7a, Chihuahuan warm desert scrub; 7b, Sonoran warm desert scrub; 7c, Mojave warm desert scrub; 7d, intermountain (Great Basin, cold desert) scrub. 8, Mediterranean woodlands and scrublands. 9, Pacific coast conifer forest. 10a, Western montane conifer forest of the Rocky Mountains and northern Cascade Range; 10b, western montane conifer forest of the southern Cascade Range, Sierra Nevada, Transverse Range, and Peninsular Range. 11, Mosaic of desert grassland, pinyon-juniper woodland, petran chaparral, and Madrean woodlands and scrublands. 12, Mosaic of desert grassland, warm desert scrub, and Madrean woodlands and scrublands. Adapted from map by W. D. Billings, published in M. G. Barbour and W. D. Billings (1998), with permission from Cambridge University Press.

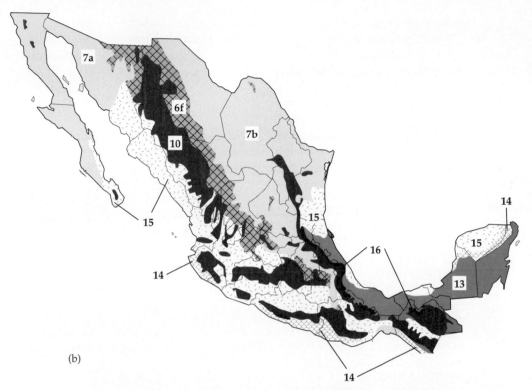

(b)

Figure 20-1 *continued.*   (b) Vegetation of Mexico. Vegetation type 6f, grassland. 7a, Chihuahuan warm desert scrub; 7b, Sonoran warm desert scrub.
10 = Montane conifer-oak forest related to the Rocky Mountains. 13, Lowland evergreen tropical rain forest. 14, Semideciduous tropical rain forest.
15, Thorn scrub. 16, Cloud forest. From Rzedowski (1991).

temperate woodland, and desert scrub. North America lacks only extremely arid
deserts and several tropical grassland, savanna, and forest types.

On a global scale, vegetation is shaped more by present and past abiotic environments than by the genetic constitution of the flora. An example we briefly discussed in Chapter 16 is the analogs of chaparral that occur in mediterranean
climates in many disjunct regions of the world. Unrelated species can arrive at similar forms and functions in similar environments. Good references on global vegetation types and patterns of distribution are by Archibold (1995), Bailey (1996),
Brown et al. (1998), Walter (1984), and Woodward (1987). As shown in Figure 20-1,
most of North America can be mapped in one of the five categories we do discuss.

Tundra

Tundra vegetation occurs in regions beyond timberline that have cold, moist-to-dry climates with short growing seasons. The vegetation is low, often only 10 cm tall, and is dominated by perennial forbs, grasses, sedges, dwarf shrubs, mosses, and lichens (Figure 20-2). In the arctic, on the average, only the top 0.5 m of soil thaws in summer. The permanently frozen subsoil impedes drainage, root growth, and decomposer activity. In the Lapp and Russian languages, *tundra* means a marshy, unforested area, and these are its most prominent characteristics.

There are two major types of tundra in North America. **Arctic tundra** occurs at high latitudes in Canada, Alaska, and Greenland, but at low elevations. **Alpine tundra** occurs at higher elevations in mountain ranges farther south. The vegeta-

(a)

Figure 20-2   Aspect views of two types of arctic tundra. (a) Dwarf heath–cottongrass tussock tundra, a form of low arctic vegetation. Cottongrass, shown flowering here, is *Eriophorum vaginatum*. Heath plants are in the families Ericacea, Empetraceae, and Diapensiaceae. (b) Cushion plant–herb–cryptogam polar semidesert vegetation in the high arctic. Photos courtesy of James B. McGraw.

(b)

tion is similar in both types of tundra. Many species occur in both, but environmental differences (such as maximum summer temperatures, extremes of day length, and intensity of solar radiation) lead to important ecological differences. North of the tundra, or above it in mountains, is a barren region of permanent ice and snow called the **nival** zone. South of the arctic tundra, or below the alpine tundra, is the **boreal conifer forest** (also called the **taiga** or subalpine forest), which will be discussed later in this chapter. Arctic tundra is far more extensive than alpine tundra, and it covers 20% of the North American continent.

Physical Features

The tundra's climate is rigorous and cold, and relatively few species of plants have evolved tolerances to withstand it. The short growing season is 50–90 days on the average, and the mean temperature of the warmest month is less than 10°C. Maximum air temperature, especially in the alpine tundra, can be much higher than 10°C, reaching above 25°C. In addition, leaves near the ground surface can be 10–20°C warmer than air temperature. Summer days in the alpine tundra exhibit greater fluctuations in temperature than in arctic tundra. The thin mountain air and lack of cloud cover allow rapid reradiation of heat at night, permitting frost to occur throughout the growing season. As much as 70% of summer solar radiation is expended on evaporating water, leaving little to warm soil or air.

Day length may be shortened to zero hours in the arctic tundra, and during winter net radiation is negative. Winter conditions are very similar for both tundra types. Minimum temperatures below −50°C, a modest amount of snow cover, and strong winds of 15–30 m s^{-1} (30–60 mph) are common. Annual precipitation varies greatly from region to region, but alpine tundra averages 100–200 cm, which is considerably more than the 10–50 cm arctic areas receive. The latter annual precipitation may seem low, but evapotranspiration is also quite low, so the P/E ratio is between 1 and 2. The arctic tundra may be latitudinally divided into two climatic and vegetational zones: low arctic tundra and high arctic polar semidesert. They differ dramatically in climate, flora, and vegetation (Table 20-1).

Two soil orders dominate the tundra: **histosols** (bog soils, mucks, and organic soils with more than 20% organic matter) and **entisols.** The entisols have little soil profile development and their traits reflect the character of the parent material more than the macroenvironment. A typical entisol profile is shown in Figure 20-3. The top 15–60 cm of soil thaws in summer and freezes in winter. It is coarse-textured and brown from undecomposed organic matter and contains all the root mass of the above. Beneath this topsoil (**talik**) is permanently frozen parent material (**permafrost**) 400–600 m thick. If the vegetation is killed or removed, if the talik is eroded or compacted, or if heat is supplied to the ground, as from a buried pipeline or a house resting on the surface, then the upper part of the permafrost will melt. The result will be unstable soil, further erosion or subsidence, and long-term vegetational change. Even one pass over tundra in summer with a normal wheeled vehicle will leave long-term scars. Bechtel Corporation developed a truck that ran on enormous, broad, air-filled tires to permit nondestructive travel on tundra during construction of the famous oil pipeline from Valdez to the Arc-

Table 20-1 Some differences between high arctic tundra, low arctic tundra, temperate alpine tundra, and tropical alpine tundra. From Bliss (1998), Caldwell et al. (1980), and Smith and Young (1987).

	Arctic tundra		Alpine tundra	
Trait	**High**	**Low**	**Temperate**	**Tropical**
Growing season (mo)	1.5–2.5	3–4	1.5–2.5	12
Maximum solar radiation, growing season (μEi m^{-2} s^{-1})[a]	1000	1500	>2000	>2000
Mean temperature, warmest month (°C)	3–6	8–12	?	6–11
Frost occurrence during growing season	None	None	Common	Nightly
Day length, range during year (hr)	0–24	4–22	6–20	10–14
Typical plant cover (%)	<50	>80	<50	<35
Common growth forms	Cushion plant, rosette plant, grasslike herb	Dwarf shrub, grasslike herb, tussock grass	Cushion plant, rosette plant, grasslike herb	Giant rosette, tussock grass, sclerophyllous shrub, mat, grasslike herb
Lichens and mosses	Abundant	Common	Occasional	Occasional
Root:shoot ratio	1.5:1	3:1	?	<Temperate alpine
Net primary production (g m^{-2} yr^{-1})	5–175	125–400	?	Probably > temperate alpine
Occurrence of fire	Occasional	Rare	Rare	Common

[a]See Table 14-1 for an explanation of the unit einstein.

tic Slope in Alaska. Whenever possible, the pipeline was built aboveground, and care was taken to prevent heat from "leaking" into the ground along the supports. Once damaged, tundra is slow to heal. Roadcuts made in alpine tundra in the Rocky Mountains 50 years ago still show only half the plant cover of nearby, undisturbed tundra.

Freeze/thaw cycles over the course of thousands of years produce a pattern to the landscape that is most visible from above (Figure 20-4). The ground surface is broken into polygons, each 10 cm to 100 m across. The polygon edges may be marked by accumulations of rock, ridges of soil, or dense vegetation. The centers may be depressed and filled with standing water in summer, or they may be raised into hillocks several decimeters high. This **patterned ground** develops most extensively on gentle slopes.

Microrelief affects tundra vegetation markedly because (a) it will determine how late into summer snow will lie on the ground, first covering vegetation, then

Figure 20-3 Profile of a typical tundra soil with permafrost. AO = litter.

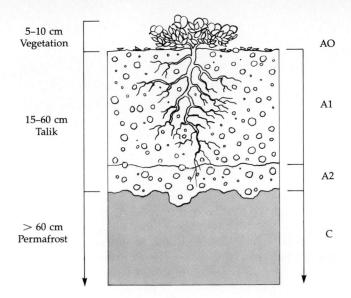

5–10 cm
Vegetation

15–60 cm
Talik

> 60 cm
Permafrost

AO

A1

A2

C

Figure 20-4 Aerial photograph of frost polygons in the tundra. These polygons are up to 100 m across.

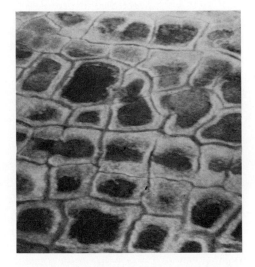

supplying it with water; (b) it will determine the degree of protection that the plants will receive from ice-blasting winter winds; and (c) it will have a bearing on the depth of the talik, and hence on the amount of root space and the adequacy of drainage. Ridge tops support very little plant cover because winter winds remove protecting snow cover. Without the snow, abrasive ice crystals carried by the wind can prune plants back to the ground or, at least, winter winds can desiccate twig tissue. In summer, ridge soils dry rapidly. In contrast, as shown in Figure 20-5, gentle, north-facing slopes may accumulate snow drifts that remain well into summer, providing melt water to **wet meadows** just below them. Wet meadows exhibit 100% ground cover and dominance by grasses and sedges. Rocky talus slopes

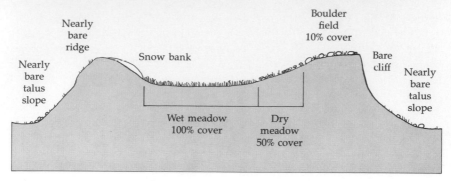

Figure 20-5   Major alpine tundra microenvironments, each with its own characteristic plant community.

and **fell fields,** relatively level boulder-strewn areas, support relatively little cover, scattered in patches where the microenvironment is favorable. Dwarf shrubs, for example, grow tallest in the lee of rocks or other projections. The productivity and biomass of fell fields can be about one-third that of wet meadows, and they resemble the high arctic polar semidesert (see Table 20-1).

Flora and Vegetation

The tundra flora is not rich. The arctic tundra of North America supports only about 700 species, which is only 3% of all angiosperms in the world. Tropical regions of smaller area support tens of thousands of species. The most common families in the tundra are Caryophyllaceae, Asteraceae, Brassicaceae, Cyperaceae, Betulaceae, Diapensiaceae, Empetraceae, Ericaceae, Poaceae, Polygonaceae, Rosaceae, and Salicaceae. Only 1–5% of the species are annual; most are relatively long-lived (20–100 years) herbaceous perennials. Hemicryptophytes—herbaceous perennials with perennating buds just at the soil surface—make up more than 60% of the flora. Trees and tall shrubs are absent. The tundra flora is geologically young, in a regional sense, probably younger than 3 million years. Much of it has been derived from taxa in the highlands of central Asia and the Rocky Mountains.

At its most luxuriant state, as in wet meadows, the vegetation covers 100% of the ground, but it is quite short, 10–40 cm tall, with a leaf area index (LAI) of only 1–2. Most plant biomass is belowground, giving a root:shoot ratio of 3:1. Many perennial forbs are "cushion plants"—their aboveground system is condensed into a compact canopy of small, xerophytic foliage, often with large, showy flowers incongruously scattered over the surface. The xerophytic foliage (small size, sometimes pubescent, sometimes succulent, often light colored, often vertically oriented) may be caused by low soil nitrogen levels more than by high light intensity or late summer drought. Although tundra soils are brown from organic matter, the nutrients are largely unavailable to the plants because decomposer activity is inhibited by the cold and wet environment.

In spring, some plants resume growth while still covered by snow and with root temperatures barely above freezing. The hollow-stemmed forb *Mertensia ciliata* is an example of a subalpine snowbank plant able to conduct photosynthesis in the relatively warm, CO_2-rich air inside its stems even before the leaves have expanded. Some lichens may exhibit positive net photosynthesis while covered by 5 cm of snow. Seeds of tundra plants, on the other hand, require relatively high temperatures, close to 20°C, before they can germinate. Flower buds, formed one or more seasons ago, are induced to open by long days. About two-thirds of the flora is insect-pollinated, so the flowers are large and showy. Some flowers are dish-shaped, heliotropic, and reflect solar radiation into their centers, creating a favorable thermal microenvironment for pollinators such as bumblebees. Pollinator density is low, however, and many species compensate by being facultatively or obligately self-fertilizing. For reasons that are still not clear, most tundra species are polyploids. Seed set is high, the seeds are light, and they have a long life.

During the short growing season, net primary productivity (NPP) is low (see Table 20-1). Much of this is channeled to roots and rhizomes and stored as lipid or protein. Part of the reason that NPP is low is that the ratio of net photosynthesis to respiration is low, only about 2:1 in contrast to a ratio of 10:1 for tropical plants. In addition, the maximum photosynthetic rate of vascular plants is relatively low, roughly equivalent to that of mesic, temperate-zone shade plants. Associated mosses and lichens have maximum photosynthetic rates an order of magnitude (a difference of up to 10 times) lower than those of vascular plants. Temperature optima for photosynthesis is 15–20°C, lower than the 25–30°C optimum of temperate zone plants. Furthermore, there is a broad plateau between 0 and 25°C where photosynthesis is only modestly affected by temperature. UV radiation (wavelength 280–320 nm) may depress photosynthesis and cause epidermal damage, and alpine species or ecotypes appear to be more tolerant of UV exposure than low-elevation arctic taxa. Only a few tropical alpine tundra species have the C_4 photosynthetic pathway.

Dormancy in late summer is triggered by lowered temperatures, increased soil moisture stress, shortened day length, or a change in the solar spectrum that results from a more oblique solar angle (the latter is important only in the arctic tundra). Wet meadow species apparently have little control over stomatal aperture. As the soil dries, plants lose water rapidly, finally becoming dormant at a tissue water potential of only −1 MPa. Dry meadow and fell-field plants can tolerate lower water potentials, to below −4 MPA. Winter snow cover helps to maintain a modest tissue water potential.

Timberline

There is a broad ecotone in Canada and Alaska between the arctic tundra and the boreal forest to the south. In places, it may be 300 km wide. Moving north through the ecotone, the forest first thins, no longer retaining 100% cover, and fruticose lichens cover the ground (Figure 20-6). Then trees become restricted to patches, and finally trees in the patches become dwarfed, twisted, and shrublike (Figure 20-7). The German word **krummholz,** meaning "twisted wood," has been applied to both this life form and to the ecotone "forest"; a synonym is **elfinwood.**

Figure 20-6 Lichen woodland near timberline in central Labrador. The trees are
mainly white spruce (*Picea
glauca*); the understory lichen
is *Cladonia*. From Elliott-Fisk
(1986).

Figure 20-7 Krummholz at timberline at Sonora Pass, 3400 m, in the
Sierra Nevada. The hedgelike trees are whitebark pine (*Pinus albicaulis*),
less than 1 m tall.

Surrounding the patches of trees and krummholz is an increasingly continuous
matrix of shrubby tundra. Finally, even the patches of krummholz disappear and
the tundra proper is entered.

A similar ecotone exists in mountain ranges between alpine tundra and subalpine forest. Ground lichens are not common and the ecotone is much narrower,
extending upward through 150–300 m of elevational change. The exact elevation
of timberline depends on latitude and distance from a maritime climate, as shown

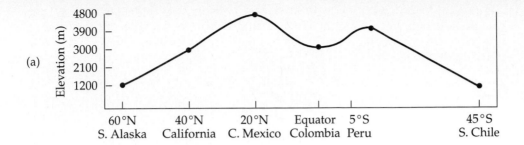

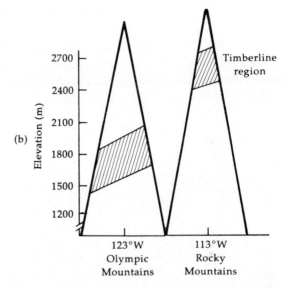

Figure 20-8 (a) The effect of latitude on the elevation of timberline, from the coastal mountains of southern Alaska through the Cascade-Sierran ranges, the Sierra Madre, and the Andes Mountains. (b) The effect of distance inland from oceanic influence at 47°N latitude. The shaded region is the ecotone between closed forest and timberline. (a) Redrawn from R. F. Daubenmire. 1954. "Alpine timberlines in the Americas and their interpretation." *Butler University Botanical Studies* 11:119–136.

in Figure 20-8. In the Rocky Mountains, between 35 and 50°N latitude, timberline falls 100 m for every 1°. Within one mountain range at one point in latitude, timberline will be lower on the temperate, ocean-facing slopes than on the extreme, continental-facing slopes (Figure 20-8b).

The location of timberline is also a function of time, changing in response to changing climate. In some parts of Canada, there is evidence that timberline was 100 km north of its present location 8000 years B.P. (before the present), 300 km north 4000 years B.P., 100 km south 2000 years B.P., and once again 100 km north 1000 years B.P. The ecotone is a dynamic tension zone within which the basic limitation to tree growth seems to be the amount of warmth received during the growing season. Below some critical level of seasonal heat, it becomes impossible to maintain a large amount of woody tissue. Several climatic markers correspond with this heat boundary: (a) North of treeline, the July mean temperature falls below 10–13°C. (b) The yearly heat sum (degree-days above 0°C) falls below 1000. (c) Summers are dominated by the arctic high-pressure system more than 50% of the time. Besides the depression of tree growth, sexual reproduction and successful seed germination become rare. Foliage longevity becomes extreme—up to

30 years for bristlecone pine, with secondary phloem growth throughout that period. In addition, fire is thought to play a large role in the location of timberline in Canada's lichen woodland region, as fire permits the establishment of new tree individuals.

Conservation

Human-caused modifications to the arctic landscape have been modest because of the severity of the climate, instability of the soil, and remoteness of the area from markets. Environmental change has largely resulted from the mining of hydrocarbons (oil) and ores such as zinc and lead. The largest oil fields are located near Prudhoe Bay, Alaska, and the Mackenzie River delta of Canada's Northwest Territories, both facing the Beaufort Sea. There is pressure from oil companies to open the adjacent Arctic National Wildlife Refuge for exploration and drilling.

The Prudhoe oil field covers 500 km^2 and oil is sent south to an all-weather port at Valdez, Alaska, via a pipeline. As mentioned earlier in this chapter, the pipeline is typically elevated above the soil in order to keep the heated oil from melting the permafrost (Figure 20-9). An adjacent service road was built parallel to the pipeline. So far, the road and its dust have posed more of an environmental problem than the pipeline itself. Moose of both sexes and of all ages cross freely underneath the pipeline, but female caribou and calves do not. Long-term studies of vegetation recovery from simulated oil leaks from the pipeline show that wet sedge tundra recovers within 3 years, but upland cottongrass tussock tundra requires as many as 20 years. Revegetation of abandoned disturbed sites, using a combination of native and more southern species, has been successful.

Figure 20-9   The Alaska pipeline, here above the ground and heavily insulated. Care has been taken in planting the supports so that heat is not transferred from the pipe to the ground.

Both Alaska and Canada have committed enormous areas to parks, refuges, and preserves which are predominantly covered with tundra vegetation (Table 20-2). The protected tundra areas in Alaska alone total 40 million hectares, an area equal to the state of California. Canada's contribution is almost as vast. Thus, the impact of humans on North American arctic tundra has been minimal and is likely to remain so into the future so long as global climate does not change. Walter Oechel, who has been studying Alaskan tundra for two decades, has published (1995) what he considered to be evidence that climate change has already affected vegetation at Barrow, Alaska (Figure 20-10). Air temperature during the growing season has significantly increased over a 15-year period, the water table has lowered, and respiration by plants and soil organisms has increased to the point that the ecosystem releases more CO_2 to the atmosphere than it fixes. That is,

Table 20-2   Federal arctic tundra parks, preserves, and reserves in North America. Some of these contain boreal forest as well as tundra. NPS = U.S. National Park Service, FWS = U.S. Fish and Wildlife Service.

Name of unit	Hectares	Authority
Alaska		
Aniakchak Monument and Preserve	217,000	NPS
Arctic Wildlife Refuge	19,351,000	FWS
Bering Land Bridge Preserve	1,000,000	NPS
Cape Krusenstern Monument	178,000	NPS
Denali Park and Preserve	2,400,000	NPS
Gates of the Arctic Park and Preserve	3,300,000	NPS
Glacier Bay Park and Preserve	1,300,000	NPS
Izembek Wildlife Refuge	321,000	FWS
Katmai Park and Preserve	1,700,000	NPS
Kobuk Valley Park	688,000	NPS
Noatak Preserve	2,500,000	NPS
Selawik Wildlife Refuge	2,150,000	FWS
Wrangell–St. Elias Park and Preserve	4,700,000	NPS
Canada		
Aulavik, Banks Island	1,228,000	
Auyuittuq, Baffin Island	2,147,000	
Ellesmere Island	3,950,000	
Ivvavik, Yukon Territory	1,017,000	
Kluane, Yukon Territory	2,000,000	
North Baffin, Baffin Island	2,220,000	
Polar Bear Pass, Bathurst Island	262,000	
Queen Maud Gulf	6,278,200	
Thelon Game Sanctuary	2,396,000	

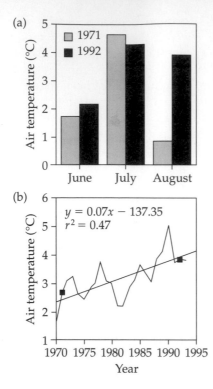

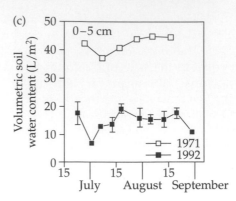

Figure 20-10 Changing climate in the tundra at Barrow, Alaska, between 1971 and 1992. (a) Mean monthly air temperature for the warmest months of the year, 1971 and 1992. Most of the two-decade warming trend is expressed in August. (b) Average temperature for these three summer months for every year between 1971 and 1992, showing a trend. Black boxes are data points shown in part a. (c) Surface soil layers have become much drier in summer as a consequence of climate change. Average summer soil water content has dropped by >50% between 1971 and 1992. From W. C. Oechel et al. (1995).

the past traditional role of arctic tundra as a CO_2 sink has reversed: it has now become a source of CO_2.

Some alpine tundra in mountains to the south (Rocky Mountains, Cascade Range, Sierra Nevada) was grazed by sheep in the nineteenth century. The consequences have included soil compaction, vegetation loss, and erosion. Virtually all alpine tundra today is publicly owned and protected from grazing, but recovery has been slow. The dramatic vistas and clean air attract day hikers and backpackers to the point that local sites can become degraded. Park managers must periodically reroute trails, change the location of camping spots, and limit the number of entry permits.

Conifer Forests

Conifers are not all ecophysiologically alike. They are not equally tolerant of cold, drought, or low nutrients. This is why there are several distinctive types of conifer forest in North America. Those covering the most area will be included in this section.

The most extensive conifer forest type is the **taiga,** which stretches across most of Canada and Alaska and dips south into the Great Lakes states and New England to about 45°N latitude. *Taiga* is a Russian word applied to Eurasian conifer forests described as damp, wild, and scarcely penetrable. A synonym is **boreal forest**. It dominates 25% of the North American landmass, making it the most extensive vegetation type on the continent. The taiga is a monotonous forest, with crowded trees of modest stature and low species diversity in the overstory and understory. This low-elevation forest extends over impressive distances, interrupted often with lakes, ponds, and moss-covered bogs. It can be a quiet forest, not rich in bird activity.

Conifer forests extend south of the taiga at moderate to high elevations in mountain chains. Just below the alpine tundra is a **subalpine forest.** Different species dominate than in the taiga, but its physiognomy is the same. The **montane zones** below this are usually covered by a richer diversity of conifers, and the forests are often more open and the trees larger, more productive, and longer-lived than those of the taiga and subalpine forests. Below the montane zones lie vegetation types dominated by broad-leaved trees, scrub, or grassland. Exceptions to this pattern are the mountain ranges that penetrate north into the taiga; in these areas, the surrounding low-elevation vegetation is still conifer-dominated.

Another magnificent conifer forest dominates a wet, temperate, lowland coastal strip that reaches from northern California to Alaska: the **north coast temperate conifer forest,** sometimes called a **temperate rain forest.** Some of the largest trees in the world, and stands with the greatest biomass, exist in this forest.

Other low-elevation conifer forests, all dominated by pines, are found in smaller areas of coastal California, in the Great Lakes area, in New Jersey, and in the southeastern Coastal Plain of the United States. Some of these grow on droughty, nutritionally poor soils, and many are maintained by fire. These will be discussed briefly at the end of this section.

Physical Features

The taiga is a cold, wet region, with a short growing season only 3–4 months long. During more than 6 months of the year, the mean temperature is below 0°C, snow covers the ground, and net radiation is negative. Maximum summer temperatures reach into the low 20s and winter minimums are in the −50s. The temperature difference between monthly means within a year can be enormous in the interior, continental part of the taiga—a span of 90°C. Annual extremes are dampened near the Atlantic and Pacific coasts (Figure 20-11). Annual precipitation varies between 30 and 85 cm, and snow cover is not deep. Despite the modest amount of precipitation, the *P/E* ratio is still >2. Most precipitation falls in summer. The taiga region experiences 30–120 days with a mean temperature less than 10°C.

The subalpine equivalent to the taiga—in mountains to the south—is 5°C warmer throughout the year than the taiga and it experiences 20–50% more annual precipitation. Extending as they do across many degrees of latitude, montane

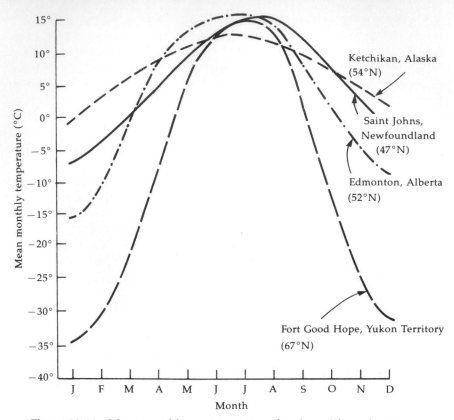

Figure 20-11   Mean monthly temperatures at four boreal forest locations. Two sites are coastal (Ketchikan, Alaska, and Saint Johns, Newfoundland) and exhibit a moderated maritime climate; two others are interior (Edmonton, Alberta, and Fort Good Hope, Yukon Territory) with a continental climate that has extremely cold winters and warmer summers.

conifer forests are much more varied than those of the taiga. Species change with elevation and with latitude in response to complex environmental gradients.

Two important environmental factors are temperature and precipitation. Figure 20-12 shows how average annual temperature and average annual precipitation change with elevation in the Sierra Nevada for west-facing (ocean-facing) and east-facing (desert-facing) slopes. Precipitation increases about 5 cm for every 100 m increase in elevation. The highest elevation shown, 2100 m, corresponds to the red fir forest in the upper montane zone. If higher elevations were shown, you would see that precipitation drops in the subalpine and alpine zones. The east slope is much drier and has a different **lapse rate** (rate of change) because a rain shadow has been created. Winds generally come from the west, drop their moisture on the west slope as air rises, and descend on the east slope as dry winds. The west slope temperature lapse rate shows a drop of 4°C for every 1000 m rise in elevation. Lapse rates differ with different seasons and mountain ranges. Considering

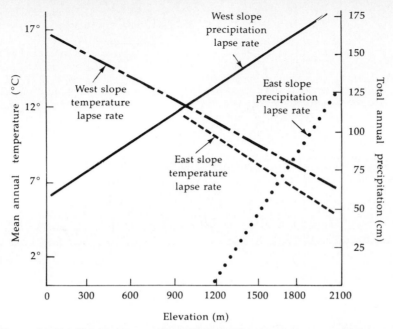

Figure 20-12 Temperature and precipitation lapse rates on west and east slopes of the Sierra Nevada at 38°N latitude. Temperature and moisture lapse rates are steeper on the more arid east side. From Major and Taylor. 1988. In *Terrestrial Vegetation of California,* 2nd ed., eds. Barbour and Major. Copyright © 1988 by the California Native Plant Society.

many factors and locations, montane temperature lapse rates average about 3° per 1000 m. A lapse rate of this nature means that every 500 m climb in elevation is equivalent to moving 800 km north.

Mountains also have latitudinal lapse rates. With increasing distance from the pole, toward lower latitudes, temperatures rise and the growing season lengthens. Consequently there is an upward displacement of a given environment to the south. This displacement is revealed by the upward displacement of vegetation belts. The typical rate of vegetational displacement, in the middle latitudes of the northern temperate zone, is +100 m for every degree south. Not every mountain range has the same displacement: the Appalachian rate is +91, the Rocky Mountain rate is +77, and the Sierra Nevada rate is +172.

The Pacific north coast forest lies in a wet, equable strip. Annual precipitation is generally between 200 and 300 cm, very little falling as snow. Mean January minimums are above freezing, at about 2°C, and mean July maximums are only about 20°C. Diurnal thermoperiods average only 10°C. A pronounced summer dry period exists from Washington to California, but it is moderated by frequent fog that lowers transpiration stress and also drips off the foliage onto the ground, adding perhaps 20 cm of "rain" annually in the Californian part of the range.

Taiga soils mostly belong to the **spodosol** soil order, which has been described

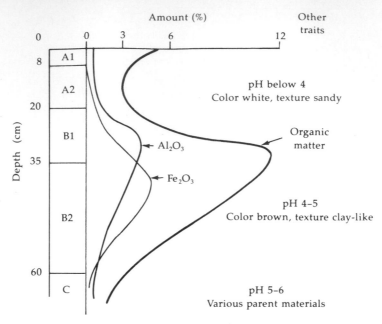

Figure 20-13   Some chemical characteristics of a spodosol soil profile. Note that depths of maximum accumulation of organic matter, aluminum, and iron are slightly different. From the translation of *Plants, the Soil, and Man* by M. G. Stafelt. Copyright © 1972 by John Wiley and Sons, Inc. Reprinted by permission.

in Chapter 17. These are acidic soils (pH 3–5) whose upper layers have been leached of clay, organic matter, and many nutrients (Figure 20-13). Conifers are characterized by shallow root systems (root:shoot ratio = 0.2), which lie above zones of nutrient and clay accumulation in the B2 horizon. However, nutrient uptake is improved by mycorrhizae. All taiga conifers possess mycorrhizae (see Chapter 7), and there is experimental evidence showing that tree growth and survival would be minimal without them. Soils are acidic both because of leaching and because conifer foliage is acidic. The hydrogen ions replace nutrient cations on soil colloids (giving a low percent base saturation), drive some cations into insoluble states, and inhibit bacterial decomposers. Litter tends to accumulate, and as much as 60% of all ecosystem carbon is locked up in humus. Nutrient cycling is further slowed by a layer of feather mosses (typically *Hylocomium splendens* and *Pleurozium schreberi*) up to 30 cm thick, which acts as a nutrient sponge, preventing nutrients from reaching the root zone. Mineral cycling is slow. The half-time for litter decay is about 3–5 years. Permafrost may underlie taiga soils in the northernmost latitudes.

Soils beneath montane and coast forests are generally either podzolic or young and weakly developed (**inceptisol** soil order). If podzolic, they are not as acidic as taiga soils and lack the ashy A2 horizon of taiga soils.

Taiga Vegetation

Taiga vegetation consists of closed forest to the south and more open lichen woodland to the north. Closed forests are dominated by white spruce (*Picea glauca*) and/or black spruce (*P. mariana*). The most common associates are balsam fir (*Abies balsamea*), larch (*Larix laricina*), jack pine (*Pinus banksiana*), lodgepole pine (*P. contorta*), and three broadleaf deciduous trees (*Populus tremuloides, P. balsamifera, Betula papyrifera*). The conifers are closely spaced, with dense canopies (LAI = 9–11), but they are of relatively small stature (up to 20 m tall, 30 cm diameter at breast height [dbh]) and short longevity (100–300 years).

Many species of evergreen and deciduous shrubs and perennial herbs are present, but cover is modest. A ground layer, dominated by feather mosses, lichens, perennial herbs, and ferns, is often continuous (Figure 20-14). The forest, then, is essentially a two-layered community.

Near the edge of bogs, the forest canopy thins and eventually only scattered trees of black spruce and larch (also called tamarack) remain, underlain by a dense shrub canopy of *Vaccinium, Ledum,* and other ericads. Wetter parts of the bog are covered by *Sphagnum* moss. This moss can advance into more upland sites, creating a wetter, more sterile microenvironment, which leads to retrogressive succession. Droughty, sandy soils often support stands of jack pine (*Pinus banksiana*).

The boreal forest is often disturbed by windthrow from winter storms or by fire from summer dry lightning strikes. Black spruce has semiserotinous cones and regenerates well after fire. Succession often proceeds through meadow, then broadleaf tree (*Populus tremuloides, P. balsamifera, Betula papyrifera*) seral stages be-

Figure 20-14   The interior of a taiga forest. Note the nearly continuous ground cover of ferns and other cryptogams.

fore reaching an even-aged spruce-fir climax in a span of about 300 years. The seral broadleaf trees also form an extensive parkland, up to 150 km wide, in an ecotone where the boreal forest meets the central grasslands in Manitoba, Saskatchewan, and Alberta.

Net annual primary productivity is relatively low (480–1280 g m^{-2}), but this is not surprising, given the short, cool growing season and the low nutrient status of the soil. The eastern deciduous forest shows twice that biomass gain, and the tropical rain forests of Central America show more than three times that gain.

The physiological ecology of taiga species has become better known over the past decade. Black spruce, as a representative of overstory conifers, is conservative in nutrient use. Its needles are retained for up to 15 years, and their photosynthetic activity remains high throughout that life span. Optimum temperature for photosynthesis is 15°C, but there is a broad plateau from 9 to 23°C, so temperature during the growing season is not an important limiting factor. Light saturation for photosynthesis is at one-third full sun, and the compensation point is very low, at 1% full sun (12–35 μE m^{-2} s^{-1}). Saplings are therefore very shade tolerant. Maximum photosynthetic rates are low compared to temperate zone plants and are lower than those of most associated growth forms in the forest (Table 20-3). Black spruce is sensitive to water stress, but moisture is not a limiting factor in most years or habitats.

Farther north, black spruce becomes the typical dominant of open lichen woodlands (see Figure 20-6). The understory lichen is either *Stereocaulon paschale* in the west or *Cladonia* (= *Cladinia*) *stellaris* in the east. The climax nature of this woodland is debated; some ecologists believe it is a fire climax rather than a climatic climax. The lichens appear to inhibit tree regeneration mechanically.

Much of the taiga is still botanically and ecologically unexplored. It is largely

Table 20-3 Maximum net photosynthetic rates (μmol CO$_2$ g^{-1} s^{-1}) of taiga species grouped by life form. Modified from W. C. Oechel and W. T. Lawrence. 1985. In *Physiological Ecology of North American Plant Communities,* eds, B. F. Chabot and H. A. Mooney. Chapman and Hall, Publishers.

Life form	Genera	Rate
Deciduous conifer	*Larix*	0.135
Deciduous shrub	*Alnus, Betula, Rubus, Vaccinium*	0.120
Deciduous hardwood	*Alnus, Betula, Populus*	0.090
Sedge	*Carex, Eriophorum*	0.060
Evergreen shrub	*Ledum, Empetrum, Vaccinium*	0.030
Evergreen conifer	*Picea*	0.025
Lichen	*Cladonia* (= *Cladinia*)	0.010
Moss	*Dicranium, Hylocomium, Pleurozium, Polytrichum, Sphagnum*	0.009

undisturbed by any human activities, in fact, and this may be due to small factors as well as large ones. The large factors are climatic hardship and the absence of roads. The small factors are called black flies and mosquitoes.

Although the majority of the taiga has yet to be modified by human activities, about 20% has been subjected to clear-cut logging. This form of logging can have seriously negative consequences for water quality, the water table, soil erosion, and wildlife. Some of the largest clear-cuts in the world have been created in Canada late in the twentieth century. The largest may be in what had been the Gordon Cousins Forest of central Ontario (Figure 20-15). It is so vast that its area (2700 km^2) could only be accurately measured from an orbiting satellite.

Unlike the situation in the United States, most forested land in Canada is under the jurisdiction of each province, rather than the federal government. Provincial governments have traditionally used these forests for immediate harvest, rather than managing them for long-term sustainability. Timber rights have been sold for vast, cumulative areas. For example, in 1989 Manitoba sold timber rights to 108,000 km^2, an area the size of Ohio. In the twentieth century, Alberta has sold timber rights to an area twice that size, or about the area of Great Britain. British Columbia's annual harvest has been averaging 60 million m^3 of timber in the 1990s, and ecologists estimate this volume is more than 30% above the sustainable yield. To put 60 million m^3 in perspective, all forests in the entire Sierra Nevada have had an annual combined harvest, during the same period of time, of only 1–2 million m^3 (about 650 million board feet). British Columbia law requires logging companies to successfully replant 80% of the harvest area with juvenile conifers, but the remaining 20% of the area legally can be left unplanted. Canada's boreal forest has been called the "Brazil of the north" for this reason.

Alaskan and Canadian forest ecologists are focusing on improving revegetation success, selecting the best genotypes for timber production, and developing more ecologically sensitive harvest techniques so as to accommodate wildlife needs (including those of such endangered species as the marten, *Martes americana*).

Subalpine and Montane Vegetation

The Appalachian Mountains   The Appalachian Mountains extend from Maine south to Georgia. These mountains are unique in that the montane zones are largely dominated by hardwoods; conifers are restricted to the subalpine zone. In the north, for example in the Adirondack Mountains of upstate New York, the upper limit of the hardwood montane zone is about 800 m; in the southern Great Smoky Mountains of North Carolina it is about 1500 m.

The taiga dominants, white spruce and balsam fir, do occur in the New England portion of the Appalachians, but they are not dominants. Red spruce (*Picea rubens*) dominates throughout the Appalachians, and it is joined by Fraser fir (*Abies fraseri*) south of Pennsylvania. Broad-leaved subdominant trees include mountain ash (*Sorbus americana*), yellow birch (*Betula lutea*), and maple (*Acer spicatum*).

The boreal forest of New England has been exhibiting dramatic dieback of red spruce for the past 25 years. Spruce decline has been reported for the mountains

Figure 20-15   Part of a vast clear-cut in the Gordon Cousins boreal forest of Ontario. From Devall (1993).

of Vermont, New Hampshire, and New York (and also farther south in North Carolina). Needles become brown and then abscise prematurely, beginning with the youngest and progressing to the oldest. Reproduction stops, basal area growth slows, and trees ultimately die, creating canopy openings that permit windthrow damage to increase. Mortality is apparent in all age classes. Studies of associated fir and of hardwoods at lower elevations do not consistently reveal similar patterns.

Acid deposition has been implicated as a cause of massive conifer dieback in northern and eastern Europe, but the evidence for North America is equivocal. **Acid deposition,** or acid rain, is caused by gaseous SO_x or NO_x emissions becoming strong acids in solution in precipitation. These gases are produced as by-products of internal combustion (automobiles), coal burning, or the refining of certain metals. Normal rain has a pH of 5.6, due to CO_2 becoming weak carbonic acid in solution, but rain in the Northeast has a pH averaging 4.0. Rain of sufficiently low pH can cause leaf lesions and can lower starch and sugar reserves in leaves, increase water stress, reduce nitrogen fixation and nitrification in the soil, inhibit germination, and depress bacterial decomposition. It is not yet clear, however, that the pH of New England rain is alone responsible for tree dieback. Weather patterns may also be implicated.

The Rocky Mountains   Running obliquely from southeast to northwest, the Rocky Mountains cross 40° of latitude and 30° of longitude. There are regional differences in the tree species that associate at particular elevations, but several species range throughout the Rocky Mountains: Engelmann spruce (*Picea engelmannii*) and subalpine fir (*Abies lasiocarpa*) in the subalpine zone, Douglas fir (*Pseudotsuga menziesii*), lodgepole pine (*Pinus contorta*), and white fir (*Abies concolor*) in the upper montane zone, and ponderosa pine (*Pinus ponderosa*) in the lower montane zone.

The elevation of each zone increases to the south. At any given latitude, the zones are highest on the southwest-facing (dry) slopes and lowest on northeast-facing (wet) slopes. In northern Alberta, the upper limit of the subalpine zone is about 2000 m, and the montane zone below begins at 1400 m. The lower montane farther downslope is essentially the same as surrounding low-elevation taiga. Far to the south, in Arizona, the upper limit of the subalpine zone approaches 3000 m and the lower montane begins at 1800 m (Figure 20-16).

Regional associates of spruce-fir in the subalpine zone of the northern Rockies include whitebark pine (*Pinus albicaulis*) and alpine larch (*Larix lyallii*), the latter retaining a tree form higher into timberline than any other species in North America. Subalpine forest in the central Rockies may include whitebark pine, lodgepole pine (*Pinus contorta*), and mountain hemlock (*Tsuga mertensiana*), in addition to spruce-fir. The southern Rockies have limber pine (*Pinus flexilis*), foxtail pine (*P. balfouriana*), and bristlecone pine (*P. longaeva,* which is famous for being the longest-lived tree in the world), in addition to spruce-fir.

The upper montane zone in the Rockies of Montana and Idaho is enriched with coastal tree species because storm tracks coming east from Washington are able to penetrate that far. In this area, only 200 km separate the Rockies from the Cascade Mountains. Dense stands of grand fir (*Abies grandis*), western red cedar (*Thuja plicata*), and western hemlock (*Tsuga heterophylla*) are typical. Only in the drier, eastern part of this region—for example, in Yellowstone National Park—does this zone support the Douglas fir and lodgepole pine forest more typical of the Rocky Mountains as a whole.

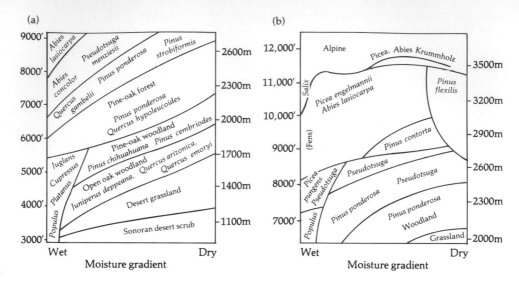

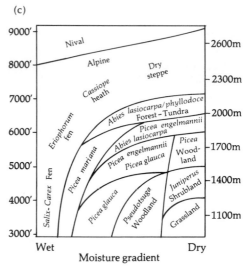

Figure 20-16 Gradient diagrams for major vegetation zones of the Rocky Mountains. (a) Santa Catalina Mountains, Arizona, 32°N latitude; (b) Colorado Front Range, 40°N; (c) Jasper National Park, Alberta, Canada, 53°N. The horizontal axis represents a complex moisture/exposure/aspect gradient and the vertical axis is an elevational gradient (feet on left, meters on right). From Peet (1998).

The Douglas fir–lodgepole pine forest has a natural fire return period of 300–400 years. These fires are hot, stand-replacing crown fires. This is the kind of fire that swept through Yellowstone National Park in 1988, attracting national attention to a "let burn" park management policy. Under that policy, wildfire is understood to be a natural event. Fires started by dry lightning strikes are recognized as having been part of the ecosystem for millions of years. The let-burn policy allows lightning-caused fires that start in areas remote from human habitation to burn, so long as they occur within a particular window of conditions that have to do with wind speed, humidity, and fuel moisture. If these conditions are not met, then the fire would be suppressed. In the 1988 burn, the fire began within the window of acceptable conditions but soon moved outside of them, and by that time the fire could not be contained to remote areas. It entered built-up areas, even privately owned areas outside the park. A national panel of enquiry later exonerated the let-burn policy in general.

The fire return period of forests in the lower montane zone, in contrast, is only 20–40 years and the intensity of those fires is relatively cool, only burning understory vegetation. The open, parklike physiognomy and pure dominance of ponderosa pine are apparently due to this surface fire regime. Fire suppression policies during the 1900s have resulted in denser stands, the invasion of other conifers, more brush cover, and loss of grass cover.

From southern Arizona and southern New Mexico south into the western and eastern Sierra Madres of Mexico, montane vegetation has a pronounced Madrean floristic influence. Mexico is a center of origin for oaks and pines, with more than 100 species of *Quercus* and nearly 40 species of *Pinus* (about half of all *Pinus* species in the world). A single forest stand often has a mix of seven to eight species of each genus. About 15% of the Mexican landmass, largely between 1800 and 3000 m elevation, is covered by pine forest with associated species of *Quercus, Juniperus, Pseudotsuga,* and *Abies*. These forests have been disturbed by extensive woodcutting from the eighteenth century to the present. The introduction of livestock and the practice of annual burning to enhance forage have further degraded the forests to the point where very little old growth remains. Less than 1% of Mexico's area is devoted to the protection of montane vegetation.

The subalpine zone, which lies between 3000 and 4000 m elevation, is not abundant in Mexico. Rocky Mountain *Picea engelmannii* and *Abies lasiocarpa* are absent, replaced with other species, the most widespread being *Picea chihuahuana, P. mexicana, Abies religiosa,* and *A. durangensis*. Harvest of these forests (largely for paper pulp) results in type conversion to alpine bunchgrass vegetation. Once the grasses are in place, conifers are kept from invading by fire and grazing.

The Northern Cascades and the Olympic Mountains   Forest zonation is more complex in this region than in any other mountainous region of North America. Strong gradients of weather and variations in wildfire frequency create this landscape diversity (Figure 20-17). The Olympic Mountains consist of a cluster of peaks between 2200 and 2500 m. The lowland portion of these mountains, below 600 m, lies within the Pacific north coast conifer forest (described on pp. 589–591), which experiences the high annual precipitation and moderate seasonal temperature fluctuations characteristic of temperate maritime climates. About 200 km to the east lie a north-south string of peaks 2200–3300 m in elevation; they form the North Cascade Range. Prevailing winds and storms are from the west, consequently there is a gradient of increasing aridity to the east and the existence of rain shadows on the eastern flanks of both mountain systems.

Once above the low-elevation north coast conifer forest, the west-facing montane zone of the Olympic Mountains is dominated by Pacific silver fir (*Abies amabilis*), associated with western hemlock (*Tsuga heterophylla*) or mountain hemlock (*T. mertensiana*), depending on the elevation. East-facing slopes are dominated by Douglas fir and western hemlock at lower elevations, and by subalpine fir at higher

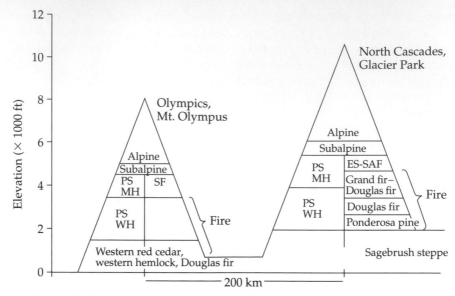

Figure 20-17   Major forest types in the Olympic Mountains and further inland at the same latitude in the North Cascade Range. PS = Pacific silver fir (*Abies amabilis*), WH = western hemlock (*Tsuga heterophylla*), MH = mountain hemlock (*T. mertensiana*), ES = Engelmann spruce (*Picea engelmannii*), SAF = subalpine fir (*Abies lasiocarpa*). Storms come predominantly from the west, so the east-facing (lee) slope of each mountain range is more arid than the west face; also, fire plays a more frequent role on the east face than on the west. Courtesy of Richard Fonda.

elevations. Stand-replacing wildfires occur every two to three centuries within the Douglas fir–western hemlock forest, but they are absent elsewhere. The subalpine zone is dominated by subalpine fir and Alaska cedar (*Chamaecyperis nootkatensis*). Treeline is at 1600 m.

In the Cascade Range, west-facing slopes within the montane zone support the same Pacific silver fir–hemlock forests as in the Olympics, but the east-side montane zone is unique. The lower portion of the montane zone is dominated by Douglas fir with ponderosa pine, whereas the upper portion is dominated by Douglas fir with grand fir (*Abies grandis*). Frequent surface fires characterize these east-side forests. The subalpine zone to the west is dominated by subalpine fir and Englemann spruce, associated with mountain hemlock; to the east whitebark pine (*Pinus albicaulis*) replaces mountain hemlock. This subalpine spruce-fir forest is very similar to the subalpine forest found farther east in the Rocky Mountains. Treeline is at 1900 m, higher than in the Olympics and reflecting the more continental position of the Cascades.

The South Cascades and the Sierra Nevada   The Cascade Range extends through 10° of latitude, from British Columbia to northern California. The southern third of the range has vegetation that is transitional to the Sierra Nevada, a mountain chain that extends south for an additional 6° in latitude and ends at the edge of the Mojave Desert. There is a considerable floristic shift along this latitudinal distance—so much so that the northern and southern limits show a Sorensen-type community coefficient of similarity of about 30 (complete identity = 100 and complete dissimilarity = 0).

The subalpine and montane zones of the Sierra Nevada are unique. The subalpine is an open woodland mixture of pines and hemlock, rather than the dense spruce-fir forest of the Cascades and Rockies. This mixed-conifer subalpine woodland contains whitebark pine, mountain hemlock, and lodgepole pine in the north; limber pine (*Pinus flexilis*), foxtail pine (*P. balfouriana*), and lodgepole pine in the south. It lies between 3300 and 2700 m elevation (Figure 20-18). The upper montane zone, between 2700 and 2100 m, is dominated by red fir (*Abies magnifica*) associated with Jeffrey pine (*Pinus jeffreyi*), western white pine (*P. monticola*), and mountain juniper (*Juniperus occidentalis*) on drier sites or by lodgepole pine on wetter sites. The lower montane zone, between 2100 and 1500 m, is dominated by a mixed conifer forest. Dominance is shared by five tree species: ponderosa pine, Douglas fir, white fir, sugar pine (*Pinus lambertiana*), and incense cedar (*Calocedrus decurrens*).

A certain frequency of surface fires seems to maintain the importance of each species in the mixed conifer forest, and a fire suppression policy over the past 80 years has resulted in a major shift from the previous balance. Table 20-4 shows that the sapling community is quite different from the overstory community: White fir and incense cedar are 2–4 times as important in the sapling layer as in the overstory; sugar pine is holding its own; and Douglas fir and ponderosa pine are

Figure 20-18 Gradient diagram of major Sierra Nevada vegetation, about 37°30′N latitude. The horizontal axis represents a complex moisture/exposure/aspect gradient and the vertical axis is an elevational gradient. Redrawn from Vankat (1982).

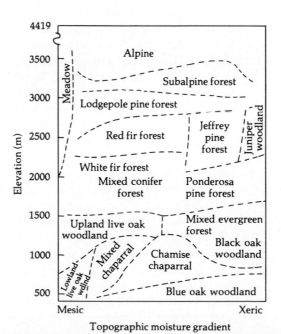

Table 20-4 Importance values (IV) of all conifer species in an old-growth mixed conifer forest, Placer County Big Trees, Sierra Nevada, California, 1600 m elevation. IV = the sum of relative basal area + relative density + relative frequency. Saplings are younger than 40 years of age, trees 3–40 cm diameter at breast height (dbh) are 40–150 years old, and trees >40 cm dbh are 150–400 years old.

Taxon	IV for three size classes		
	<3 cm dbh (saplings)	3–40 cm dbh	>40 cm dbh (overstory adults)
White fir (*Abies concolor*)	132	123	36
Incense cedar (*Calocedrus decurrens*)	62	35	27
Sugar pine (*Pinus lambertiana*)	59	38	69
Ponderosa pine (*P. ponderosa*)	10	47	89
Douglas fir (*Pseudotsuga menziesii*)	37	57	79

2–9 times less important in the sapling layer than in the overstory. The mature forest 100 years from now may be a dense white fir forest, quite unlike the forest of the present and past. One of the reasons for the shift has to do with the seedling requirements of each species. Seedlings of ponderosa pine, for example, are known to do best when the litter layer is thin or absent, and in relatively open sun. Seedlings of white fir can tolerate denser shade and a deeper litter layer.

The mixed conifer forest also dominates mountains south of the Sierra Nevada, including the Transverse and Peninsular Ranges. At these latitudes, Douglas fir is absent and Jeffrey pine replaces ponderosa pine. The Peninsular Range extends into remote regions of Baja California, where fire suppression management has never been instituted. Natural wildfires still occur there, and the structure of the forest is remarkably open (Figure 20-19), reminiscent of anecdotal descriptions of the Sierra Nevada written by travelers in the 1800s. A 60,000 ha national park in the Sierra San Pedro Martir region of the range represents the last landscape-scale remnant of Californian forests as they may have existed prior to Euroamerican contact. Efforts are being made by conservationists from both Mexico and the United States to preserve this unique ecosystem as part of UNESCO's International Biosphere Preserve network.

Scattered within the Sierra Nevada mixed conifer forest are 75 groves of big trees (*Sequoiadendron giganteum*), the most massive organism on earth (Figure 20-20). The largest groves cover more than 800 ha and contain 20,000 trees; the smallest and most northerly grove extends over only 0.4 ha and contains only 6 mature trees. The smaller stands appear to be losing mature trees to death by old age faster than they are gaining saplings. The reasons for the population decline are not known, but the decline has been going on for 500 or more years, so recent human activities are not necessarily the cause (see Chapter 4).

Figure 20-19   Open mixed conifer forest of the Sierra San Pedro Martir, Peninsular Range, Baja California. Elevation is about 2400 m. The most abundant species is Jeffrey pine (*Pinus jeffreyi*).

The lower montane zone is dominated by ponderosa pine throughout the Cascade and Sierra Nevada mountains. It is a relatively open forest and closely resembles the Rocky Mountain lower montane. A well-developed understory of shrubs (*Symphoricarpos, Holodiscus, Chamaebatia, Rosa, Arctostaphylos, Physocarpus, Ceanothus* spp.) or of grasses (*Agropyron, Festuca, Stipa* spp.) can exist. Subdominant oak trees, such as black oak (*Quercus kelloggii*), Oregon white oak (*Q. garryana*), and canyon live oak (*Q. chrysolepis*) are common.

As is true for the Rocky Mountains, many hectares are not covered by forest. In the Sierra Nevada, for example, more than half the land is covered by meadow or brushfield. **Brushfields (montane chaparral)** are often found on hot, rocky, dry exposures where tree growth is limited; they may also invade burned or logged areas on better sites that once supported conifer forest (Figure 20-21). In the latter case, montane chaparral is successional and will ultimately be replaced by a closed forest. The rate of succession is very slow, and management techniques are being developed to suppress brush by selective herbicides that leave the conifers unaffected.

Conservation problems for western conifer forests in general have been highlighted by a recent, large-scale federal study of the Sierra Nevada. The objectives were to (a) assess the current ecological, social, and economic conditions of the Sierra Nevada ecosystem; (b) predict future trends in those areas, given a continuation of current policies and management; and (c) suggest policy or manage-

Figure 20-20   Big tree (*Sequoiadendron giganteum*) in the mixed conifer zone of the Sierra Nevada, about 1400 m.

ment options that might achieve ecological sustainability and social well-being. Major findings included the following:

The human population has risen from a pre-Euroamerican contact average of 1.3 persons per square kilometer to 7.7 in 1990, to a predicted 23.1 in 2040. Urban expansion has been greatest in certain vegetation types, diminishing and fragmenting their remaining area; it has also resulted in the spread of invasive exotic plants and in hydrological changes.

Most of the natural resources generated in the Sierran ecosystem are exported. Economically, the most important resource is water, accounting for 60% of the area's annual $2.2 billion share of the gross national product. In other words, mountains are important sources of water for distant lowlands.

A century of fire suppression management, instituted to protect human habitations, has tripled tree density, shifted dominance from pine to fir, increased fuel biomass, and increased the potential for hot, stand-replacing crown fires. Ironically, fire suppression has made the forest more unsafe for humans. The higher tree density also results in bark beetle epidemics that cause massive tree mortality during the episodic droughts characteristic of this region.

Timber harvest over the past 150 years has removed 75% of the old-growth forest area. Second-growth forests have a simpler physiognomy and age structure, lack debris and snags, and have lower animal species diversity.

Biodiversity has declined. Several plant and animal species have become extinct and dozens are listed as threatened or endangered. The number of nesting

Figure 20-21 Montane chaparral on a disturbed area, east side of the Sierra Nevada, about 1900 m. The dominant shrub is tobacco brush (*Ceanothus velutinus*) and it is suppressing the growth of white fir (*Abies concolor*) coming up beneath it.

birds has declined, including neotropical migrants. Anadromous fish have become nearly extinct due to dams, and native fish are rare. Amphibians at all elevations have severely declined. (Frogs are declining dramatically in lowlands such as the Gulf Coast, as well as in the mountains, and in 140 countries around the world.)

Grazing by domesticated livestock has degraded foothill, riparian, and montane areas. Significant restoration requires decades, and the likelihood of complete restoration is highly unlikely.

Some Sierran airsheds contain such high concentrations of ozone and particulates that their air quality is among the poorest in the nation. Ozone concentration may be greater, especially at night, than in the distant cities where the ozone originates. Federal standards for maximum allowable ozone in the air are too high to protect sensitive conifer species, some of which dominate montane landscapes.

Temperate Low-Elevation Conifer Forests

Pacific North Coast Conifer Forest   This coastal forest extends from the Gulf of Alaska, at 62°N latitude, to the Mendocino coast of California, at 40°N latitude. It is unusual for two reasons. First, hardwoods typically dominate temperate, mesic climates such as this, but here softwood volume is 1000 times hardwood volume.

Second, the size and longevity of the dominants are unprecedented. The largest trees of 10 genera are found here: *Abies, Chamaecyparis, Larix, Calocedrus, Picea, Pinus, Pseudotsuga, Sequoia, Thuja,* and *Tsuga*. Overstory trees are commonly 50–75 m tall and more than 2 m dbh (Table 20-5). Aboveground standing crop averages 2000 t ha^{-1}. Longevity is often well beyond 500 years, and sometimes it is beyond 1000 years.

This is the most productive forest in North America, averaging 1.74 kg m^{-2} yr^{-1}, which is nearly twice that of any other conifer or hardwood forest on the continent. The high productivity is due to mild winter temperatures and high precipitation, especially in winter.

Except for the Californian part of the range, the most common overstory trees are Sitka spruce (*Picea sitchensis*), western hemlock, and western red cedar. Common associates include Douglas fir, grand fir, silver fir, and two species of cedar (*Chamaecyparis*). Douglas fir appears to be a successional species, despite its great size and long life; it does not reproduce well in its own shade and is succeeded by western hemlock, western red cedar, or Pacific silver fir. Broadleaf trees, such as red alder (*Alnus rubra*), may be scattered beneath the conifer canopy, and a lush ground layer of shrubs, perennial herbs, ferns, and cryptogams is typical. Like conifers in other vegetation types, Sitka spruce is noted for striking differences in sun and shade needles, adapting it to full sun as an emergent and to deep shade as a sapling. The needles differ in orientation, size, specific leaf weight, stomatal density, net photosynthetic rate, stomatal conductance, and mesophyll conductance.

In California, coast redwood (*Sequoia sempervirens*) becomes the dominant species. Up to 100 m tall and as old as 2000+ years, redwoods are nevertheless seral species, dependent on periodic disturbance by fire or flood in order to maintain their vigor, reproduction, and dominance.

Table 20-5   Average maximum age, diameter breast height (dbh), and tree height for eight common overstory species of the Pacific north coast forest. Modified from J. F. Franklin and C. E. Halpern, "Pacific Northwest Forests." In *North American Terrestrial Vegetation*, 2nd ed., eds. M. G. Barbour and W. D. Billings. Copyright © 1998 by Cambridge University Press. By permission.

Species	Age	dbh (cm)	Height (m)
Sequoia sempervirens	1250	150–380	75–100
Chamaecyparis nootkatensis	1000	100–150	30–40
Thuja plicata	1000	150–300	60
Pseudotsuga menziesii	750	150–220	70–80
Chamaecyparis lawsoniana	500	120–180	60
Picea sitchensis	500	180–230	70–75
Tsuga heterophylla	400	90–120	50–65
Abies grandis	300	75–125	40–60

Beginning in the 1980s, the Pacific Northwest became a national focus of conservation efforts to preserve old-growth forests. The economy of the region has historically been founded on timber harvest, so the conversion of old-growth forest into second-growth forest has been significant. Old-growth forest is defined in this area as follows: a forest stand, usually >250 years old, with moderate to high canopy closure, having a multilayered, multispecies canopy dominated by large trees (at least some of which have indications of senescence or decadence), with numerous large snags and accumulations of dead wood on the ground.

Such old-growth forests are uniquely suitable for certain species listed as endangered or threatened; one whose range most closely approximates old-growth forest throughout the Pacific Northwest is the northern spotted owl (*Strix occidentalis caurina*). Although at least 17 other listed species occur throughout some or most of the region, the owl became the key species around which conservation plans were developed. In 1993, President Clinton appointed a Forest Ecosystem Management Assessment Team (FEMAT) to develop a policy for maintaining spotted owl populations on a sustained basis for the next 100 years.

FEMAT developed nine different management options, each with a different mix of areas that would permit intensive logging, others that could only be selectively thinned, and reserves that could not be harvested in any way. The economic consequence would be a reduction of timber harvest by about 60%. Even then, FEMAT concluded, the probability of sustaining owl populations unchanged for 100 years was only 60%. Since that time, similar guidelines have been developed for adjacent forests east of the Cascades in Oregon and Washington (using anadromous and resident fish as key species) and for interior Californian forests (using the California spotted owl, *Strix occidentalis occidentalis,* as a key species). To a large degree, implementation of these guidelines began in the late 1990s.

Great Lakes Pine Forest   A mosaic of softwood and hardwood stands spreads over part of the Great Lakes region of the United States, in parts of Michigan, Wisconsin, and Minnesota. Three pines form nearly pure, even-aged stands: red pine (*Pinus resinosa*), eastern white pine (*P. strobus*), and jack pine (*P. banksiana*). The pines are most often restricted to sandy soils, with beech, maple, and yellow birch dominating the surrounding deciduous forest on more mesic soils. This mosaic lies in a tension zone, or ecotone, between the boreal forest and the deciduous forest, and there has been considerable debate about the climax nature of pine stands here. The usual view is that pine is seral to hardwood, and at best the pine forms an edaphic climax on sites too dry to support the hemlock-hardwoods.

These stands were extensively cut for lumber in the nineteenth century, and their area today does not reflect past dominance. Local timber harvesting began in this area in the 1840s, primarily because of the presence of white pine. White pine's range extends east, through much of what had been colonial America, and north into Canada (Figure 20-22), so its value had become well known by the time of westward expansion into the Midwest. It is still considered to be the most valuable timber species in eastern North America. The abundance of white pine in Michigan, Wisconsin, and Minnesota north of 44°N latitude, its tendency to occur in

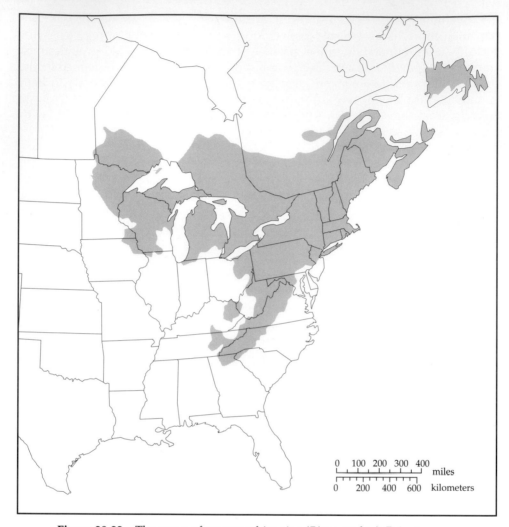

Figure 20-22   The range of eastern white pine (*Pinus strobus*). Prior to extensive logging in the nineteenth century, this species dominated a low-elevation conifer forest around the Great Lakes. From Wendel and Smith (1990).

pure stands, the relatively level and well-drained terrain, and the proximity to Great Lakes shipping routes soon attracted investors and speculators who purchased large tracts of land and usually harvested the forest by clear-cutting.

Clear-cutting advanced through the region at an unprecedented pace. Intensive harvesting began in the 1870s, peaked in 1892, then declined rapidly so that by 1907 local economies had collapsed and the search for timber moved away, to the southeastern United States. Deforestation on this continent has never before or since equaled that experienced by the Lake States: in 20 years, 755 million m^3 (320 billion board feet) of softwoods had been cut, leaving 12.7 million ha (31.3 million acres, or 49,000 square miles) barren. Recovery of this land has been slow because white pine forest grows on coarse, well-drained, glacially derived soils with low nutrient content and low moisture-holding capacity. In addition, the stump- and slash-riddled landscape became subject to spectacular wildfires that set back succession. The growing season is less than 130 days, so not all of the cutover area has been converted to farmland.

New Jersey Pine Barrens   A half million hectares of land in the coastal plain of New Jersey, and limited areas on Long Island and on Cape Cod, have very sandy, acidic soil of low fertility. They support a scrubby forest, about 6 m tall, of pitch pine (*Pinus rigida*) and several oak species. The tree canopy is somewhat open, and there is a well-developed understory of low shrubs, mainly oaks and members of the huckleberry family, and a ground layer of mosses, lichens, and sedges. In part, the poor, droughty soils limit invasion of surrounding oak-hickory deciduous forest, but most ecologists currently believe that a high fire frequency is the major factor favoring the pines. Where fire frequency is 10 years or less, the pines do not reach tree stature, but instead form a shrubby pygmy forest little more than 1 m high. Other parts of the barrens experience a fire frequency of about 20 years.

Despite a large surrounding human population, the pine barrens remain largely a wilderness, a precious resource in the most densely populated state in the United States.

Southeastern Pine Savanna   The Coastal Plain of North Carolina, South Carolina, Georgia, and Florida often supports an open stand of pines underlain with grass. The principal species of pine are longleaf (*Pinus palustris*), shortleaf (*P. echinata*), slash (*P. elliottii*), and loblolly (*P. taeda*). The dependence of this forest type on surface fires or other disturbances is well known. In the absence of disturbance pine is seral to hardwood forest.

This forest became the focus of commercial logging interests immediately after the Lake States forest harvest ended. From about 1905 until 1920, a cumulative area approximately the size of California (40 million ha or 100 million acres) was clear-cut in nine states. About 10% of that area was subsequently converted into farmland, 60% was successfully restocked and managed into relatively healthy second-growth forest, and 30% remained fallow as stunted regrowth, mainly because of poor soil. The South had replaced the Lake States as having the country's worst ecological problem.

Today, remnant stands are being protected, sometimes by unusual partnerships. For example, 188,000 ha of longleaf pine in the Florida panhandle are in the Wade tract under joint management of the Department of Defense (Eglin Air Force Base) and The Nature Conservancy. The vegetation remains critically endangered, however (see Noss et al. 1995). The Wade tract is the largest remaining old-growth stand of longleaf pine. Cavities in older pines provide critical habitat for the endangered red-cockaded woodpecker. As a result of concern about the woodpecker, conservation of longleaf pine ecosystems has become a priority in the southeastern United States.

Deciduous Forests

The deciduous forest offers many contrasts to the boreal forest. The deciduous forest has a lower LAI, permitting more light to reach the forest floor; it has more animal life; the overstory can be very rich in species diversity; and the structure of subdominant canopy layers can be more complex. Because of a longer, warmer growing season, net primary productivity and standing biomass are about twice what they are in the taiga. Litter decay is more rapid, litter half-life being on the order of one year instead of several years. As a result, mineral cycling is more rapid.

Deciduous forests once covered about one-third of all land area in the conterminous United States and about 10% of North America. Some of this area is in the form of narrow fingers of riparian forest that extend west along river courses into nonforested regions, but the great bulk of the forest lies east of 95°W longitude and south of 45°N latitude. Only this eastern portion will be discussed here.

The eastern deciduous forest is not homogeneous throughout its extent. Certain groups of dominant tree species characterize particular regions. Much of what we know about the regional forests comes from the extensive field work done by Dr. Emma Lucy Braun (Chapter 2) prior to World War II. Her gender and diminutive size permitted her to gain the confidence of back-country people, who helped her search for prime, virgin examples of each regional association. Many of the sites she described were logged during World War II, making her 1950 book an important record of what we once had.

Less technical descriptions, recorded by eighteenth-century settlers, also give us a perspective. In Ohio and Pennsylvania, the forest overstory was dominated by 400-year-old oaks, sugar maples, and chestnuts, 1–2 m dbh, with straight trunks rising 25 m before the first side branch. Black walnuts, shagbark hickories, and cottonwoods grew in floodplains near rivers, the latter being large enough for travelers such as Daniel Boone to make into dugout canoes 20 m long and more than 1 m across. The dense overstory canopy, together with grapevines that ran up many trunks, created a deep shade in the understory There was little underbrush, so that it was easy for travelers to wind in and out among the great trunks, even though "one could not shoot an arrow in any direction for more than twenty feet without hitting a tree."

By day, the pristine forest was somber, dark, silent, and gloomy to some. "In the eternal woods it is impossible to keep off a particularly unpleasant, anxious feeling, which is excited irresistibly by the continuing shadow and the confined outlook," wrote one pioneer. The songbirds of our modern forest did not live in that dark forest. At night the forest became unnervingly vocal with calls from wolves, panthers, horned owls, and whippoorwills. "It is clear," wrote John Bakeless in his book on the accounts of early explorers, "that no one lamented the disappearance of the picturesque forest, since there were altogether too many trees for comfort."

Braun described nine regional forest associations (Figure 20-23). In this chapter, we will simplify the picture (and take into account some reinterpretations by other ecologists) by condensing those nine into five: **mixed mesophytic, oak-hickory, southern mixed hardwood, beech-maple,** and **hemlock–northern hardwood.**

Physical Features

Annual precipitation ranges from 80 to 150 cm, generally increasing from the northwest toward the east and south. There is no pronounced seasonal wet or dry period. Except in the southern mixed hardwood forest, snow and hard frost in winter are common. Minimum winter temperatures may drop to −30°C. The growing season is 4–6 months long (longer in the south) and maximum temperatures can reach 38°C. Even moderate temperatures of 30°C are uncomfortable because of the high humidity (as much as 10% of summer rain is intercepted by foliage and evaporates back to the air). The P/E ratio is greater than 1. Despite the wet summer, however, soil moisture deficits develop that are acute enough to affect herb growth and tree seedling survival.

Two soil orders characterize the eastern deciduous forest. The **alfisol** order extends from north of Virginia and Kentucky to well into an ecotone area with the taiga. Alfisols are also known as **gray-brown podzolics;** they are acidic (pH 5–6), but not as acidic as taiga podzols, partly because the pH of deciduous leaves is 5–7 rather than the pH 4 of spruce and pine. There is no bleached horizon, and the fertility (cation exchange capacity and percent base saturation) is relatively high. Diversity and activity of soil biota are high.

The **ultisol** order dominates the southern part of the forest. Ultisols have not been glaciated, hence they are older and more weathered than alfisols. Plowed fields show a B horizon that is bright yellow-orange as a result of oxidized iron. Ultisols are slightly more acidic than alfisols, and considerably lower in fertility.

Physiognomy and Phenology

The modern deciduous forest has been modified by 200 years of uneasy coexistence with farmers, harvesters, and industrialists. Community architecture is no longer so strongly one-layered and impressive as it once was, except in locally protected areas. Today, much of the forest is four-layered (Figure 20-24). The overstory may reach up to 60 m in the best sites, but is more typically 35 m. It is a closed, interlocking canopy, with more than 100% cover. Beneath the overstory is a second tree canopy that is much more open and consists of saplings and genetically smaller trees about 5–10 m tall, such as dogwood (*Cornus*). A third layer is made up of scattered shrubs, often members of the Ericaceae, about 1–2 m tall. A fourth layer consists of seasonally active patches of herbs, mosses, and ferns. Vines such

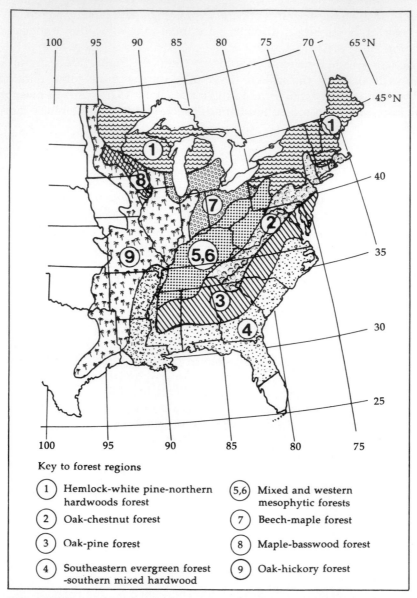

Figure 20-23   Nine regions of the eastern deciduous forest formation as recognized by Braun (1950). These regions are not exactly coincident with those identified and mapped in Figure 20-1a, and the prairie peninsula is not shown. Redrawn from B. Robichaud and M. F. Buell. 1973. *Vegetation of New Jersey.* By permission of Rutgers University Press, New Brunswick, NJ.

as the Virginia creeper (*Parthenocissus quinquefolia*) pass through several layers. Total LAI ranges between 4 and 8, and total biomass averages 30 kg m^{-2}. Annual net primary productivity is about 1200 g m^{-2}.

The aspect of the forest changes dramatically with the seasons because most

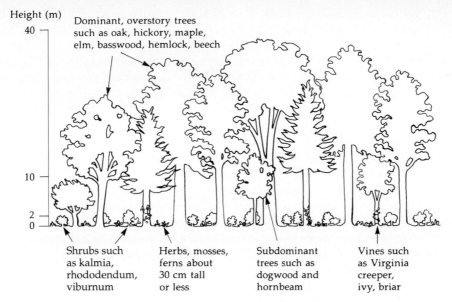

Height (m)

Dominant, overstory trees
such as oak, hickory, maple,
elm, basswood, hemlock, beech

Shrubs such
as kalmia,
rhododendum,
viburnum

Herbs, mosses,
ferns about
30 cm tall
or less

Subdominant
trees such as
dogwood and
hornbeam

Vines such
as Virginia
creeper,
ivy, briar

Figure 20-24   Physiognomy of a typical second-growth eastern deciduous forest. The dominant overstory trees may reach 60 m or more in height, but commonly are 30 m.

species in each stratum are deciduous or die back. If we take New Jersey or Pennsylvania as a reference point, then snowmelt in mid-March coincides with release of spring-flowering herbs from dormancy. At this time, the overstory tree canopy is leafless and more than half of the sunlight reaches the ground. By late March to early April, the ground is green from renewed vegetative activity by herbs that have overwintered as bulbs or rhizomes. These herbaceous geophytes account for about 18% of all species in the forest, a proportion three times Raunkiaer's world average. By mid-April, the herbs are flowering and the overstory trees have resumed meristerm activity and begun to leaf out. The major control of bud break for eastern deciduous trees is a critical heat sum (degree-days) following a winter cold period. In a few exceptional cases (such as beech, *Fagus*), photoperiod is the environmental control. Over a period of 3–4 weeks, the various tree species begin and complete leaf expansion. Species with ring-porous wood and large-diameter vessels (ash, elm, hickory, oak, sumac, walnut) resume cambial activity first, but leaf expansion is delayed. Species with diffuse-porous wood and narrower vessels (alder, birch, cherry, maple, poplar, willow) tend to leaf out first.

As the canopy closes through May, light intensity on the forest floor is reduced to a minimum of 2–5% full sun. Moving light flecks temporarily allow up to 70% full sun to reach ground herbs, and there is some evidence that this light may be responsible for half of herb productivity. By July, many of the spring-flowering herbs have died back and become dormant. This dormancy may be induced by competition for moisture, rather than by competition for light. Other herbs, as described below, grow throughout the summer and flower in August and September. Leaves of shrubs and trees turn color in early October and are shed into November. Length of the growing season, from the time tree leaves are half expanded in early May to the time half their chlorophyll is lost in fall, averages 160 days. Time of radial tree growth averages 105 days. Many phenological events may be triggered by environmental changes in the root zone, rather than in the air.

The guild of understory herbs is floristically rich, with 100 common species. About 95% are perennial, and average life span of a ramet is typically 10–20 years. Large, vegetatively spreading genets may reach ages of 100 years. Sexual reproduction is generally delayed past a juvenile stage of 3+ years. Most species allocate less than 10% of annual photosynthate to sexual reproduction, but a few expend 20–50% (*Claytonia virginica, Erythronium americanum, Hieracium venosum*). Few of these forest floor herbs are wind-pollinated. Seed dispersal is often accomplished by ants, but in other cases there are no special modifications for dispersal, and seeds do not travel far from the parent. Some species rely mainly on vegetative reproduction.

There are at least three phenological herb patterns; **spring ephemeral, summer green,** and **evergreen.** Spring ephemerals complete most of their growth in the short period between snowmelt and tree canopy closure. Leaf emergence of *Erythronium americanum* (Figure 20-25a), for example, begins 5 days after snowmelt, and all leaves have died 50 days later, by late June. Such herbs are heliophytes with a light saturation point of about 25% full sun. It is possible that their flush of growth in spring is ecologically important in that it conserves nutrients that otherwise would be lost in runoff before tree roots become active. Summer green herbs begin growth soon after spring ephemerals, but they maintain leaves and active

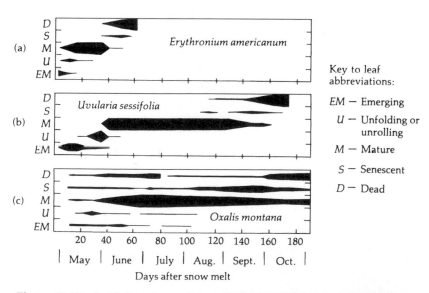

Figure 20-25   Leaf phenology of three understory herbs beneath a northern hardwood forest in New Hampshire. (a) The spring ephemeral *Erythronium americanum;* (b) the summer green *Uvularia sessifolia;* and (c) the evergreen *Oxalis montana.* Thickness of lines indicates relative number of leaves in each state. Modified from Mahall and Bormann (1978).

photosynthesis until fall, when their leaves senesce and drop off (see *Uvularia sessifolia*, Figure 20-25b). Summer greens are shade tolerant, with a light saturation point at 11% of full sun. Their maximum net photosynthetic rate is only about 30 μmol CO_2 g^{-1} s^{-1}, which is half that of spring ephemerals.

Evergreens retain their more sclerophyllous leaves for 1–3 years, but they are dormant during the winter. All new leaves may be produced in spring, or there may be a continuous leaf turnover. Flowering usually occurs in late spring or early summer. *Oxalis montana* (Figure 20-25c) is a good example. Its photosynthetic behavior is intermediate between that of ephemerals and summer greens.

Regional studies reveal only weak associations between particular understory and overstory species. The abundance of any one herb is related more to general physical factors such as depth and texture of litter, soil drainage, and sunflecks, rather than to the effects of a specific overstory tree. As overstory trees change from region to region within the deciduous forest, the understory herb community either shows little change or changes at rates that do not correlate with change in the overstory.

Major Ecotones

Taiga–Deciduous Forest Ecotone: The Hemlock-Hardwoods Forest   In the central part of North America, in Manitoba, Saskatchewan, and Alberta, the taiga passes into a spruce-aspen parkland, then an aspen parkland in a wide ecotone with the grasslands. East of this, in an equally wide ecotone, the taiga passes into the eastern deciduous forest. This ecotone centers on 45°N latitude (southern Maine, northern New York, and northern Michigan), then bends north through northern Wisconsin and northeastern Minnesota. As balsam fir, white spruce, and red spruce become less abundant, a mixture of hardwoods and other conifers assume dominance.

From studies done in the Green Mountains of Vermont, the ecotone's location correlates closely with cloudiness and fog in spring and fall. Other factors are shortness of the frost-free period and soil depth. Above 800 m elevation, fog drip abruptly increases by 60% and the number of frost-free days per year decreases by 40%. Once spruce and fir are established, they modify the site to make it unfavorable for hardwoods: Soil pH drops, litter depth increases, and soil moisture increases.

Ecotone hardwoods include sugar maple (*Acer saccharum*), yellow birch (*Betula allegheniensis*), beech (*Fagus grandifolia*), paper birch (*Betula papyrifera*), and basswood (*Tilia americana*). Conifers include three pines (*Pinus banksiana, P. resinosa, P. strobus*) and hemlock (*Tsuga canadensis*). This **hemlock-hardwoods forest** is the type of forest Thoreau lived in at Walden Pond (Figure 20-26). To the south, it merges into beech-maple or oak-hickory forest. The pines are especially important in the Midwest, and this phase of the ecotone is the Great Lakes pine forest discussed earlier.

Prairie–Deciduous Forest Ecotone: Oak Savanna and Woodland   In an eastward arc extending from Minnesota to Texas, the deciduous forest once gave way to

Figure 20-26   The hemlock-hardwoods forest, an ecotone between the boreal forest and the deciduous forest.

grassland. The ecotone projected east into Illinois, Indiana, Ohio, and Kentucky, permitting a prairie peninsula to cover almost a quarter of a million square kilometers east of the Mississippi River (Figure 20-27). The ecotone was very complex. Sometimes it was abrupt, with a transition from a closed oak forest to an open grassland in a matter of less than 10 m. In other places, the forest gradually thinned to a woodland of more open trees with a grassy understory, then to a savanna of scattered trees (less than 30% canopy cover), and finally to a grassland in the distance of 50 km. The boundary was sinuous, with interdigitations of grassland and forest. It was also a mosaic, with numerous islands of prairie within the forest and outliers of forest in the grassland. Major tree species of the ecotone are bur oak (*Quercus macrocarpa*), post oak (*Q. stellata*), and blackjack oak (*Q. marilandica*).

Many physical and biotic factors change across the ecotone, and it has not been possible for ecologists to determine which is most important. The major factors that correlate with change from forest to grassland are as follows:

1. *P/E* ratio drops below 1 (except in the prairie peninsula).
2. Annual precipitation drops below 60 cm. (Aridity increases to the west, as does prairie.)
3. Yearly variability in precipitation increases and so does the frequency of drought. Low rainfall cannot be the total reason for forest to give way to grasslands, however, because grassland borders farther west support trees under even more arid conditions.

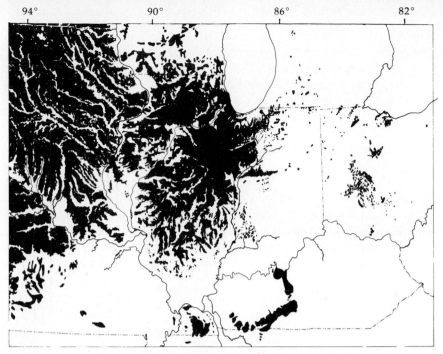

Figure 20-27   The pre-Euroamerican contact location (black) of the prairie peninsula as mapped by Transeau (1935). By permission of the Ecological Society of America.

4. Fire frequency increases, favoring grasses (see next paragraph).

5. Finer, deeper prairie soils favor fibrous roots of grasses.

6. Tree mycorrhizal fungi are absent in prairie soils, and tree saplings are at a competitive disadvantage without mycorrhizae.

7. Heavy grazing by bison until the late nineteenth century may have eliminated tree saplings.

8. The tree line may have been pushed east by the Xerothermic period, about 6000 years ago, and not yet regained its climatic limits.

In the pristine grassland, wildfires were common virtually every autumn. The perennial grasses could recover from rhizomes and root crowns below the surface, but the woody species were killed. Where forest cover was dense enough, ground vegetation was too thin to carry a fire. In this way, fire could maintain an abrupt forest-grassland ecotone. In the absence of fire, some Wisconsin studies show that woody species invade prairie at the average annual rate of 0.3 m. Burning also improves the vigor of the grasses, stimulating productivity and flowering two- to threefold due to removal of litter. In the absence of litter, soil temperatures in the root zone are 5°C higher, which stimulates plant growth and decomposer activity.

Some Regional Forest Associations

The **mixed mesophytic forest association** of the Cumberland Mountains and Allegheny Mountains contains the richest collection of overstory dominants of any forest in North America. As many as 20–25 species of overstory and understory trees can be found in one hectare, and dominance fluctuates so greatly from stand to stand that the community cannot be named by even a handful of species. The most common, characteristic trees include sugar maple (*Acer saccharum*), buckeye (*Aesculus octandra*), beech (*Fagus grandifolia*), tulip tree (*Liriodendron tulipifera*), white oak (*Quercus alba*), northern red oak (*Q. borealis* var. *maxima*), and basswood (*Tilia heterophylla*). Hemlock may sometimes be present, especially in mesic, protected cove forests of the Great Smoky Mountains (Figure 20-28). Understory trees are also diverse and may produce a striking floral display in the spring. They include redbud (*Cercis canadensis*), dogwood (*Cornus florida*), witch hazel (*Hamamelis virginiana*), ironwood (*Carpinus caroliniana*), hornbeam (*Ostrya virginiana*), hackberry (*Celtis occidentalis*), and species of *Rhododendron* and *Magnolia*. Viburnum, sumac, huckleberry, blueberry, and blackberry shrubs are common; midsummer is a berry-picker's delight.

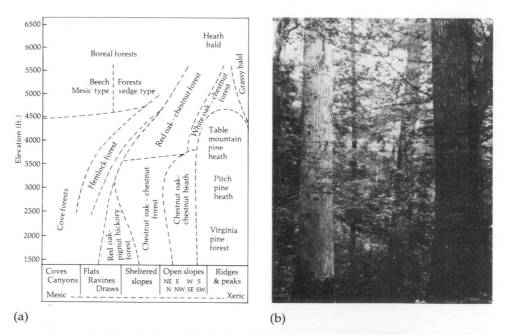

(a)   (b)

Figure 20-28   (a) Gradient diagram of vegetation in the Great Smoky Mountains. The horizontal axis is a complex moisture/exposure/aspect gradient; the vertical axis is elevation (in feet; from Whittaker 1956). (b) The mixed mesophytic forest (hemlock cove forest) occurs in a broad elevational range within the most mesic sites. This particular example is in the Blue Ridge Mountains of North Carolina. Large trunks are >1.5 m dbh.

Nearly surrounding the mixed mesophytic association is the **oak-hickory forest,** which has several species each of *Quercus* and *Carya*. In places, tulip tree may be a codominant. A portion of this forest, just to the east of the mixed mesophytic association, was an area once dominated by chestnut (*Castanea dentata*). Early in the twentieth century, however, a fungal parasite of chestnut was accidentally introduced to the New York area from Europe. By the 1930s, the fungus had spread throughout much of the range of chestnut and had killed all mature trees. Apart from an occasional stump sprout and a few trees, the only remains of chestnut today are "ghosts"—slowly decaying but sometimes still standing skeletons of dead trees. Oaks and hickories now occupy the openings in the canopy vacated by chestnuts.

A **beech-maple forest** extends through much of Ohio and Michigan. Major dominants are sugar maple (*Acer saccharum*) and beech (*Fagus grandifolia*). Common associates include yellow birch (*Betula alleghaniensis*), red maple (*Acer rubrum*), white ash (*Fraxinus americana*), and basswood (*Tilia americana*). The southern limit of this association corresponds to the limit of Wisconsin glaciation. The hemlock-hardwoods forest, already discussed in the ecotone section, lies to the north, through much of Pennsylvania, New York, and New England.

The Southeast is covered by the **southern mixed hardwood forest** (southeastern evergreen forest in map, Figure 20-23). Dominants are many (up to 40 tree species), including evergreens such as southern magnolia (*Magnolia grandiflora*), the semievergreen laurel oak (*Quercus laurifolia*), live oak (*Q. virginiana*), and several pines, and deciduous species such as beech (*Fagus grandifolia*), sweet gum (*Liquidambar styraciflua*), and some hickories. Epiphytes, such as Spanish moss (*Tillandsia usneoides*), are common. A subdominant tree canopy may also be closed, but the shrub and herb strata are patchy. Based on their average importance values, evergreens account for about half of the canopy. From site to site, however, the percent of evergreenness can vary from 0 to 100. In general, evergreen conifers are most abundant on sandy, sterile sites, where retention of leaves may have survival significance.

Historical Changes

All of North American vegetation has undergone enormous latitudinal and areal shifts since the last maximum glacial advance 18,000 years ago. The history of the eastern deciduous forest may be typical (see Figure 20-29). At the glacial peak, climatic gradients from tundra through deciduous forest were steeper than at present, only 800 km separating tundra from oak-hickory-pine. The Appalachian Mountains were covered in tundra. Riverlands and associated bluffs extending north from the Gulf of Mexico were probably refugia for the mixed mesophytic association.

As glaciers retreated 10,000 years B.P. and formed the Great Lakes, grassland and oak savanna encroached from the west. The mixed mesophytic forest dominated its maximal area, through the center of today's eastern deciduous forest. Florida vegetation changed from a sand scrub to an oak savanna. At the end of

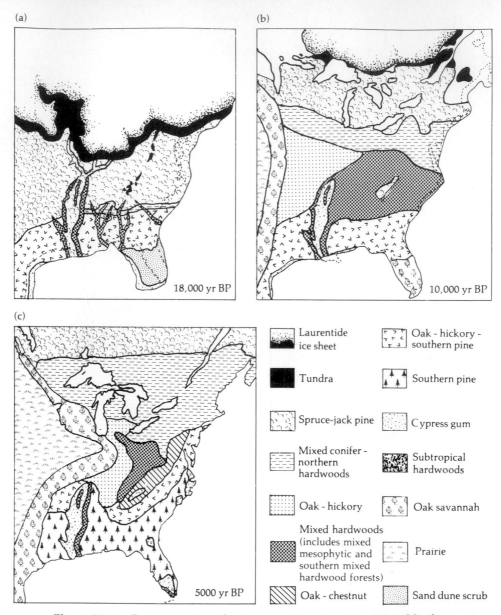

Figure 20-29 Reconstruction of major vegetation types in eastern North America since the last glacial maximum advance: (a) about 18,000 years before the present (B.P.), (b) about 10,000 years B.P., (c) about 5000 years B.P. Redrawn from Delcourt et al. (1983).

the Xerothermic, 5000 years B.P., the southern mixed forest was still more widely established than at present, the prairie peninsula was apparent, the mixed mesophytic forest had contracted, and the northern hardwoods zone was close to its modern limits.

One can now appreciate that modern forests in the northern part of the deciduous forest region are quite young. Sugar maple in the mountains of New Hampshire, for example, has probably been present for fewer than 8000 years. Considering that such a species has a life span averaging 250 years, this is equivalent to 30 generations. Eastern hemlock, with a life span three times as long, has been present there for only 10 generations. It is impressive to recognize, then, that such species have evolved ecotypes. Sugar maples in New Hampshire, separated by less than 1 km in elevation and distance and free of any apparent breeding barriers, exhibit differences in photosynthesis, respiration, and specific leaf weight. High-elevation genotypes have a net photosynthesis rate 40–80% higher than that of low-elevation genotypes.

For the past 10,000 years, Native Americans influenced the vegetation of the eastern deciduous forest in several ways. Prior to Euroamerican contact, Indian population densities averaged 1 person per square kilometer, but ranged as high as 2000 per square kilometer in alluvial bottomlands bordering the Mississippi River, sufficient to have had a marked impact on regional forest structure, species composition, the extent of forest clearings for agriculture and villages, and the frequency of fire. Requirements of land conversion to agriculture and habitation are estimated to have been 1 ha per person, so 10% of eastern North America's landmass would have been converted, not including land abandoned as villages were moved.

As the geographer Michael Williams concluded in his book *Americans and Their Forests* (1989, p. 49), "Indians were a potent ecological factor in the distribution and composition of the [eastern deciduous] forest. The idea of the forest as being in some pristine state of equilibrium with nature, awaiting the arrival of the tranforming hand of the Europeans, has been all too readily accepted as a comforting generalization and as a benchmark against which to measure all subsequent change." Reviews of oak dynamics in the eastern deciduous forest by ecologists such as Marc Abrams have revealed decreasing oak dominance in most areas of the eastern deciduous forest (Table 20-6). He concluded that oaks came to dominance as glaciers retreated and the incidence of fire increased. Indian burning practices prior to Euroamerican contacts further increased oak dominance. Fire scars in old trunks and charcoal deposits in soil indicate that fires in mixed oak stands ca. 1700 may have had a return period of less than 20 years. Oaks in mesic habitats are early successional species and generally give way to maples unless land use managers reinstitute prescribed fires.

The impact of Euroamericans on eastern forests was much greater than that of Native Americans because of greater population densities and technological advances. By 1850, only 40 million ha of eastern deciduous forest had been cleared and converted to farmland, pastures, or settlements, but the pace of conversion intensified after that. The growing needs of industry, transportation (railroads and

Table 20-6 Changing dominance of oak in several eastern deciduous forest types and ecotones. In virtually all these forests, the overstory is dominated by oaks but juvenile tree species in the understory are not oak, suggesting that these oak forests are successional. From Abrams (1992).

Presettlement vegetation	Present overstory species	Present understory species
Savanna western ecotone		
Bur oak	Black oak–white oak	Black cherry–boxelder–elm
White oak–black oak	White oak–shagbark hickory	Shagbark hickory–chokecherry
Blackjack oak–post oak	Blackjack oak–post oak	Blackjack oak–post oak
Oak forest western ecotone		
White oak	White oak–sugar maple	Sugar maple
White oak–black oak	White oak–shagbark hickory	Sugar maple
White oak–black oak	White oak–black oak–red maple	Black cherry–red maple
Mid-Atlantic mixed oak forest		
Oak–chestnut	White oak–black oak	Red maple–sugar maple–birch
White oak	Beech–red maple–white oak	Red maple–sugar maple–cherry
White oak–red oak	White oak–red oak	Beech–blackgum–red maple
Chestnut–red oak	Red oak	Sugar maple–red maple
White oak–white pine	White oak–black oak	Red maple–black cherry
Chestnut oak	Chestnut oak–red oak	Maple–chestnut oak–birch
Northern forest and Lake States ecotones (the latter was logged in the late 1800s)		
White oak–black oak–pine	Black oak–red oak–white pine	Red maple
Conifer–N. hardwoods	Red oak–white oak–red maple	Sugar maple–red maple
White pine–oak	Red oak–white oak–red maple	Sugar maple–red maple
White pine–red pine	Red oak–red maple–paper birch	Red maple
Southern Coastal Plain and piedmont		
Longleaf pine–slash pine– turkey oak	Longleaf pine–slash pine– turkey oak–scrub oak	Turkey oak–scrub oak– scrub hickory
Longleaf pine	Beech–oak–magnolia	Sweetgum–blackgum–beech
White oak–black oak	White oak	Red maple
Oak–chestnut	Red oak–chestnut oak–white oak	Red oak–red maple–hickory

steamships), smelting, charcoal production, and energy generation all required wood as a fuel, and wood was also required as a building material for cities, mines, ships, and furniture. Bark stripped from chestnut, certain oaks (*Quercus prinus*), and eastern hemlock was used for the tanning of leather.

The intensity of timber harvesting continued into the latter half of this century. The cutting of relict old-growth stands became a patriotic thing to do in the 1940s (to support the war effort); the practice of clear-cutting on a rotation of 30–60 years became more common; the scale of disturbance became larger; soil compaction and soil erosion increased as a consequence of modern logging equipment; and re-forestation management produced ecologically unstable monocultures of genetically homogeneous trees.

Grasslands and Steppes

Grasslands are the second most extensive vegetation type in North America, covering 20% of the continent. The central grasslands are locked into the interior of the continent (Figure 20-30; see also Figure 20-37). They occupy a swath 1000 km wide between about 95°W and 105°W, which tapers to a northern limit in Saskatchewan, at 52°N, and to a southern limit in central Texas, at 30°N.

This enormous strip of grassland is not homogeneous. As precipitation decreases from east to west, the height, growth form, and identity of dominant grasses change. The easternmost portion, including the prairie peninsula, is the **tallgrass prairie.** Flowering stalks of some sod-forming grasses here reach 2 m in height. Farther west is the **mixed** (sometimes called the **mid-**) **grass prairie,** where maximum plant height is about 1 m. Still farther west, in the area of lowest rainfall (40 cm), is the **short-grass prairie,** dominated by bunchgrasses less than 0.5 m tall. The spatial relationship of these grasslands to each other and to forests already discussed is summarized for latitude 37°N in Figure 20-31.

Other grasslands include (a) the **intermountain grassland,** extending through Wyoming, Idaho, northern Utah, northern Nevada, and eastern Oregon, and including the Palouse region of eastern Washington (see sagebrush steppe in "Great Basin Vegetation" later in this chapter); (b) the **desert grassland** in Texas, New Mexico, Arizona, and northern Mexico; and (c) the **Central Valley grassland** of California.

All of the North American grasslands have been extensively modified or eliminated, primarily through four activities: cultivation; fire suppression policies; grazing of domestic stock; and the accidental introduction of aggressive, weedy, alien plant species.

Some Definitions

Grasslands are herbaceous communities dominated by **graminoids** (grasses, sedges, rushes), but with **forbs** (non-graminoid herbs) present and sometimes seasonally dominant. Trees are absent except for local sites, such as along water-courses or among rock outcrops.

Grasslands may be dominated by annual grasses, perennial **bunchgrasses,** or perennial **sod-forming** grasses. Annuals are most abundant on dry, overgrazed, or disturbed sites. Some annuals are cool-season types, germinating in late fall and setting seed the following early summer. Cheatgrass (*Bromus tectorum*) in the intermountain area is such an annual. Other annuals are warm-season types. They germinate in spring or summer and complete a much shorter life cycle in a matter

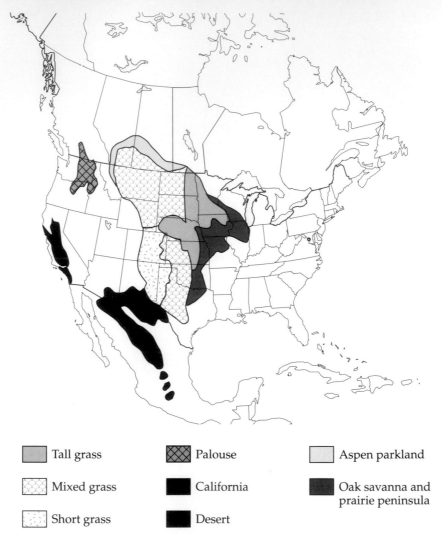

Figure 20-30   Pre-Euroamerican contact distribution of the six major grasslands discussed in the text. A seventh grassland, the intermountain grassland, is later shown as an ecotone in Figure 20-37. Modern limits of these grasslands have been modified by cultivation, fire control, and grazing. For example, much of the desert grassland shown near the Chihuahuan Desert in old versions of this figure is now part of the desert, shown in Figure 20-37. Aspen parkland and oak savanna are ecotones with the boreal forest and eastern deciduous forest, respectively. The prairie peninsula is included here within "oak savanna." For steppes, see Figure 20-37. Redrawn from Sims and Risser (1998), Looman (1983), and Eyre (1963).

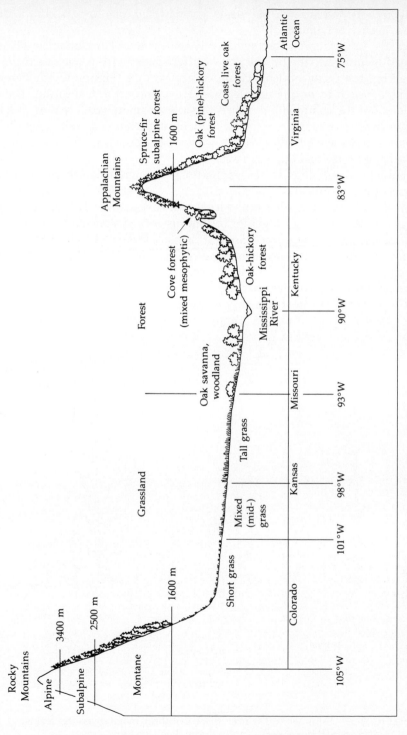

Figure 20-31 Diagrammatic representation of vegetation at 37°N latitude from the Atlantic coast to the Rocky Mountains.

of weeks, rather than months. Six-weeks grama (*Bouteloua barbata*), in the desert grassland, is such an annual.

Annuals do not reproduce vegetatively with runners or rhizomes. Their lateral spread is limited to the production of **tillers** (stems) from buds near the root crown, giving the plant a bushy, clumped appearance (Figure 20-32). Perennial bunchgrasses, such as purple needlegrass (*Stipa pulchra*) in California, also produce tillers, but the continuation of that process for many years results in a large clump a decimeter or more in diameter. Bunchgrasses alone do not generally produce a community with 100% cover, but the spaces between clumps can be seasonally filled by forbs and other grasses.

Perennial sod-forming grasses, such as big bluestem (*Andropogon gerardii*) in the tallgrass prairie, spread laterally by rhizomes. New shoots and roots arise from nodes on the rhizomes in such numbers that a **turf** results. The topsoil is thoroughly penetrated and held together by fibrous root systems, and a continuous sward of shoots covers the surface. Sod-formers are more resistant to grazing than bunchgrasses because they have so many more growing points. Some sod-formers are less aggressive than others, producing only short rhizomes, and some produce rhizomes only under certain environmental conditions.

The term **prairie** is most often applied to a grassland dominated by tall, sod-forming grasses, with 100% cover. **Steppe** is often applied to a grassland dominated by short bunchgrasses interspersed with shrubs and with a ground cover

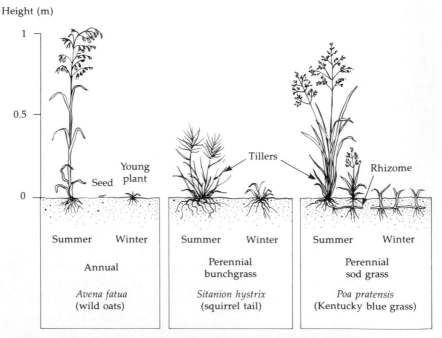

Figure 20-32 Summer and winter aspects of three major grass life forms: annual, perennial bunchgrass, and perennial sod-forming grass.

that may be less than 100%. In this chapter, we will use **grassland** as a general term, applicable to prairie, steppe, and all intermediates. As precipitation increases, grassland gives way to forest. An ecotone vegetation type is an open woodland or **savanna.** Savannas still retain a continuous cover, but trees are regularly present and contribute up to 30% cover. Two major ecotone savannas are shown in Figure 20-30: the northern aspen parkland and the eastern oak savanna. The oak savanna, described earlier in this chapter, is a north-south strip that increases in width from north (50 km wide) to south (hundreds of kilometers wide). Its position corresponds to an aridity gradient on which the *P/E* ratio falls below 1 as one moves west.

The parkland ecotone in Canada is characterized first by increasing aspen (*Populus* spp.) cover, then farther north by increasing aspen and boreal conifer cover. Aspen reproduces vigorously from sucker sprouts that arise from below-ground parts not damaged by fire. The width of the aspen belt ranges from 175 km in the east and north to 75 km in the west. The aspen-conifer belt extends an additional 75–300 km away from the grassland. This ecotone is related to a moisture gradient, but more specifically to the relative amounts of snow or rain that fall in the spring months, April through June. Insufficient precipitation in those months has a negative effect on the reproduction of several dominant grasses. The most characteristic grass in this ecotone is *Festuca campestris*.

Physical Features

Climate of the central grassland is continental, with long, cold winters and long, hot summers. Extreme lows in the winter may dip to $-40°C$, and extreme highs in the summer may reach $+45°C$. An Illinois tallgrass prairie may show a mean monthly temperature for January at $-3°C$, and $24°C$ for July. In the mixed grassland of North Dakota, or in the Palouse region of Washington, mean January temperature can be considerably lower, $-14°C$, while mean July temperature is still high, $21°C$. There are, then, large seasonal temperature changes. Winter snow is common, but it is dry and can be blown into drifts, leaving most of the ground bare and uninsulated.

Annual precipitation for the central grasslands areas is generally 30–80 cm, with a strong summer peak. Evapotranspiration is high, and the *P/E* ratio is less than 1. A pronounced moisture deficit, compared to a minor one in the deciduous forest, characterizes grassland soils in summer. Summer thunderstorms may be violent, accompanied by hail, high winds, and tornadoes. Total rainfall can fluctuate two- to threefold in successive years, and periodic droughts (as in the Dust Bowl years of the 1930s) are to be expected. It has been suggested that the distribution limits of some grassland species correlate with occasional extreme lows in rainfall, rather than with long-term means. Precipitation in the desert, intermountain, and Californian grasslands is lower, about 25–45 cm, and the seasonality also differs from that in central grasslands. The desert area has both summer and winter peaks, while the intermountain and Californian areas have winter peaks.

Gradients of both temperature and precipitation correlate with shifts in photosynthetic pathway of the grasses. North of approximately 44° latitude, cool-season (C_3) grasses predominate; south of this, warm-season (C_4) grasses predominate (Figure 20-33). This ecotone roughly corresponds with the southern borders of Wyoming and Iowa, and along part of that ecotone elevation falls 400 m within a relatively short distance, modifying both temperature and precipitation. Summer temperatures appear to be especially critical to C_4 species: Locations with mean minimum July temperatures below 8°C have few or no C_4 species.

Grassland soils are quite diverse. In the prairie peninsula and in part of the tallgrass area, the prevailing soils are in the ultisol order. A *P/E* ratio of 1 permits weak podzolization to occur. Soil pH is close to neutral, however, and decay of fibrous root systems gives the soil a dark brown color. There is no bleached A horizon. Where precipitation is less than evapotranspiration, **mollisol** soils become prominent. A layer of calcium carbonate accumulates in the subsoil, deposited at shallower and shallower levels as the *P/E* ratio declines. The most mesic mollisol is a **chernozem,** with up to 16% organic matter in the nearly black A horizon, a clay loam texture, high cation exchange capacity, very high percent base saturation, pH slightly basic, and carbonate accumulation only at 100 cm depth. *Chernozem* is a Russian word meaning "black earth." The color is a result of richness in organic matter. Chernozems give way to lighter brown soils as one moves through mixed and short-grass areas. Intermountain grasslands are also on mollisols, but desert grasslands can occur on **aridisols,** which have little profile development and contain less than 1% organic matter in the root zone. Towards the eastern deciduous forest border, prairie can lie on acidic alfisols.

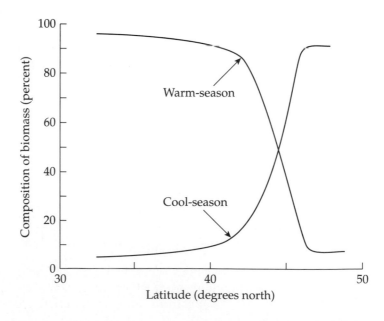

Figure 20-33   Relative percent biomass of C_3 (cool-season) and C_4 (warm-season) grasses as a function of latitude in the mid-prairie region of North America. From Sims and Risser (1998).

Central Grassland Vegetation

The **tallgrass prairie** has a nearly equal mix of bunchgrasses and sod-forming grasses, which together provide a very dense canopy of more than 100% cover and a standing aboveground biomass of 200 g m^{-2}. The LAI is 2–6, a bit less than that in the deciduous forest. Most biomass is belowground, and the root:shoot ratio averages 6–7. Annual aboveground net productivity can be more than 1000 g m^{-2}, the great bulk of which is contributed by warm-season grasses, some of which are C$_4$ species (e.g., little bluestem and blue grama).

The most typical dominants are big bluestem (*Andropogon gerardii*) and Indian grass (*Sorghastrum nutans*), both with flowering shoots up to 2 m tall. Associated with them are shorter grasses, such as little bluestem (*Schizachyrium scoparium*) and *Bouteloua* species, and a variety of perennial forbs. As in all the grasslands, forb species outnumber graminoids by 3 or 4 to 1, but they account for only 10–20% of total community productivity. There is considerable latitudinal variation in the associated species, and even at a particular location major shifts reflect the microenvironment. Prairie cordgrass (*Spartina pectinata*), for example, can be prominent in wet depressions, and Kentucky bluegrass (*Poa pratensis*) can be abundant where the range is heavily grazed.

Most of the tallgrass prairie is gone, replaced by today's corn belt, but the records of early travelers can help us reconstruct it. This is from a journal dated 1837:

> The view from this mound . . . begs all description. An ocean of prairie surrounds the spectator whose vision is not limited to less than 30 or 40 miles. This great sea of verdure is interspersed with delightfully varying undulations, like the vast swells of ocean, and every here and there, sinking in the hollows or cresting the swells, appear spots of trees. . . .

The **mixed-grass prairie** is perhaps the most floristically complex of the central grasslands. The tall sod-formers of the east generally become restricted to locally heavy soils, and the usual dominants are grasses of medium height overtopping shorter bunchgrasses such as little bluestem, hairy grama (*Bouteloua hirsuta*), blue grama (*B. gracilis*), and needlegrass (*Stipa spartea*). To those who have grown up around forests, a grassland such as this may appear homogeneous and monotonous (Figure 20-34), but careful observation reveals a complex mosaic of communities. The mixed-grass prairie is a tension zone or ecotone between the tallgrass and short-grass regions, and it has had a dynamic recent history. Effects of grazing, fire, and drought have been pronounced. It has unique northern and southern provinces. To the north, such C$_3$ grasses of medium height as western wheatgrass (*Pascopyrum smithii*) dominate, associated with needle-and-thread (*Stipa comata*) and the C$_4$ blue grama. Productivity is typically 100–300 g m^{-2} yr^{-1}. To the south, C$_4$ grasses predominate, and productivity is higher. Cover in the mixed grassland is 100%, but standing, live aboveground biomass is only half that of tallgrass prairie, and only half the annual productivity derives from warm-season grasses.

Short-grass communities are dominated by bunchgrasses such as buffalo

Figure 20-34   Mixed grassland on rolling hills in North Dakota.

grass (*Buchlöe dactyloides*), sand bluestem (*Andropogon hallii*), Indian ricegrass (*Oryzopsis hymenoides*), blue grama, and species of *Muhlenbergia*. A few species are sod-formers. Buffalo grass and blue grama, among others, are C_4 species, but about half the annual productivity (depending on latitude) derives from cool-season C_3 species. Aboveground standing crop is only 50 g m^{-2}. When mixed-grass prairie is stressed by heavy grazing or drought, short-grass species tend to increase in abundance. Overgrazing has also contributed to recent invasion by a variety of perennial (*Opuntia, Yucca*) and annual (*Bromus, Festuca, Hordeum, Salsola*) weedy species of low palatability. This is the region (especially southwestern Colorado, southwestern Kansas, and the panhandles of Texas and Oklahoma) that spawned the Dust Bowl.

Desert Grassland

Plateaus above 1000 m elevation at the northern edge of the Sonoran and Chihuahuan Deserts were covered until recently by short bunchgrasses, with desert scrub restricted to ravines, knolls, or other locally poor sites. Important grasses were Indian ricegrass, (*Muhlenbergia*), black grama (*Bouteloua eriopoda*), tobosa (*Hilaria mutica*), and galleta (*H. jamesii*). Its original area amounted to 8% of North American grassland.

A significant fraction of this grassland's acreage has disappeared over the past 100 years. **Desert scrub** has invaded from the locally poor sites and has come to dominate many hectares (Figure 20-35). An 1858 vegetation survey of the Jornada

Figure 20-35 This stand of desert scrub in New Mexico was desert grassland a century ago, according to historical land descriptions.

Experimental Range in south-central New Mexico showed 33,800 ha to have 100% relative cover by grasses. Desert shrubs were absent from those hectares at that time. A resurvey in 1915 showed only 14,400 ha could still be so classified. In 1963, no hectares had 100% relative grass cover (Buffington and Herbel 1965). At the same time, cover by such desert shrubs as creosote bush, mesquite, and tarbush increased twentyfold. A 50⁺-year study of desert grassland on the Santa Rita Experiment Station in southeastern Arizona showed a similar trend, creosote bush alone increasing thirteenfold between 1904 and 1958 (Humphrey and Mehrhoff 1958). Mesquite (*Prosopis juliflora*) has also spread widely.

Some ecologists have hypothesized that the shrub explosion was due to overgrazing of grasses by cattle. However, even after cattle exclosures were erected in 1931 at Jornada and Santa Rita, shrubs continued to spread in both. Possibly shrub invasion is due to erosion of the thin topsoil formerly held in place by the grasses. Other causes suggested for shrub encroachment include fire suppression policies and a slight warming-drying trend in the climate of the southwest region.

Intermountain-Palouse-Willamette Grassland

The northern part of the Great Basin, between the Rocky Mountains and the Cascade Range, is a **shrub steppe** (Figure 20-36). Short bunchgrasses share dominance with cold desert shrubs, especially species of sagebrush (*Artemisia*). Many dominants of this region are also found in the loess-covered Columbia River Basin of eastern Washington (the **Palouse** area), and in another grassland that used to clothe the Willamette Valley of Oregon. As in the Central Valley of California (next section), cultivation has eliminated the Willamette grassland, and we must rely on incomplete records to reconstruct the pristine vegetation.

Dominants of these three regions include the following cool-season bunchgrasses: bluebunch wheatgrass (*Agropyron spicatum*), bluebunch fescue (*Festuca idahoensis*), several other fescues, wild rye (*Elymus* spp.), sacaton (*Sporobolus*

Figure 20-36   *Artemisia* shrub steppe, Wind River area of Wyoming.

wrightii), and California oat grass (*Danthonia californica*). This intermountain grassland has been invaded and changed by annual cheatgrass.

A review of the biogeography of C_4 grasses shows that the intermountain and Californian grasslands are much lower in C_4 species than the central grasslands. Tall and mixed grassland floras are 40% C_4 (grass species only), while the intermountain-Palouse-Willamette-California grassland floras are only 15% C_4.

California Central Valley Grassland

This grassland is located mainly in the Sacramento and San Joaquin valleys, although historically it was also found in the Los Angeles basin and in several other coastal valleys of central and southern California. (An interrupted grassland along the north coast of California is more closely related to the Palouse and Willamette grassland than it is to the Central Valley grassland.) Its area totals only 2% of all North American grasslands.

Two hundred years ago, when first viewed by Europeans, the vegetation may have been dominated by cool-season bunchgrasses such as purple needlegrass (*Stipa pulchra*), nodding needlegrass (*S. cernua*), wild rye (*Elymus* spp.), pine bluegrass (*Poa scabrella*), three awn (*Aristida* spp.), June grass (*Koeleria cristata*), and deer grass (*Muhlenbergia rigens*). These grasses contributed about 50% cover, and the spaces between were filled with great masses of annual and perennial grasses and forbs that flowered in late spring. During the June–October dry period, forbs and grasses alike died back, turning the region golden brown. This grassland once occupied over 5 million hectares, about 13% of the modern state's land area.

Overgrazing, drought, and the introduction of hundreds of weedy annuals from other mediterranean climate regions have changed this grassland into an annual type, with forage of lower nutritional value. Modern dominants are wild oats, bromes, ryegrasses (*Lolium* spp.), foxtails (*Hordeum* spp.), and forbs such as filaree (*Erodium* spp.). Cattle exclosure studies indicate that the native bunchgrasses are very slow to reinvade or to increase in cover, given the presence of these aggressive annuals. However, recent experimentation with spring burns, carefully timed grazing, and the planting of bunchgrass seeds or seedlings has shown that modest acreages can be restored to bunchgrass dominance.

Grassland Habitat Loss

Of all the major vegetation types in North America, grassland has changed the most in the past two centuries. Mixed-grass and short-grass prairies occupy only about 50% of their past range, tallgrass prairie has been reduced by 90%, and the western trio of grasslands (desert, intermountain, and California bunchgrass) have been dramatically changed in species composition by overgrazing and invasion by exotics to the point that natural communities are limited to 1% of their extent prior to Euroamerican contact (Table 20-7). The grasslands have changed because they were easier to manipulate, were more economically valuable when converted to farmland, and because they were dependent on more frequent fire than any other vegetation type. Fire return period in grassland is typically 1–3 years, whereas it is one to two orders of magnitude longer in forest and scrub vegetation. Consequently, the elimination of wildfire by early settlers had an effect within decades instead of centuries.

Currently, 55 grassland species in the United States are formally listed as threatened or endangered, and more than 700 others are candidates for listing. Numerous animal species have been drastically reduced in number, including the bison and prairie dog. The bison population of 200 years ago has been estimated at 12–60 million for North America; today bison number fewer than 100,000. Prairie dog populations have declined 98% in the same period. Bird diversity has significantly dropped. One recent study of 24 relict prairies showed a strong linear relationship between bird species richness and grassland area: 15 species were found in 1000-ha patches and fewer than half that number were found in 10-ha patches. Area accounted for 84% of the variation in bird species richness. We can imagine, then, that fragmentation of the prairie landscape has been the major reason for declines in bird populations.

Restoration activities receive considerable attention at annual meetings of the North American Prairie Conference. The most common techniques for restoration include planting, the use of prescribed fire, and—when necessary—the removal of competitive exotics by herbicide application. A few relict natural prairies of landscape size still exist, and these are important centers for ecological research. One of the best known is the Konza Prairie, 3500 ha of tallgrass prairie in northeastern Kansas. The prairie is owned by The Nature Conservancy and managed in collaboration with Kansas State University. It is part of a national network of long-

Table 20-7   Summary of the estimated current area, historic area, and percent decline of several prairies. N/A indicates data are not available. From Samson and Knopf (1994).

	Historic (ha)	Current (ha)	Decline (%)	Current protected (%)
	Tallgrass			
Manitoba	600,000	300	99.9	N/A
Illinois	8,900,000	930	99.9	<.01
Indiana	2,800,000	404	99.9	<.01
Iowa	12,500,000	12,140	99.9	<.01
Kansas	6,900,000	1,200,000	82.6	N/A
Minnesota	7,300,000	30,350	99.6	<1.0
Missouri	5,700,000	30,350	99.5	<1.0
Nebraska	6,100,000	123,000	98.0	<1.0
North Dakota	1,200,000	1200	99.9	N/A
Oklahoma	5,200,000	N/A	N/A	N/A
South Dakota	3,000,000	449,000	85.0	N/A
Texas	7,200,000	720,000	90.0	N/A
Wisconsin	971,000	4000	99.9	N/A
	Mixed grass			
Alberta	8,700,000	3,400,000	61.0	<.01
Manitoba	600,000	300	99.9	<.01
Saskatchewan	13,400,000	2,500,000	81.3	<.01
Nebraska	7,700,000	1,900,000	77.1	N/A
North Dakota	13,900,000	3,900,000	71.9	N/A
Oklahoma	2,500,000	N/A	N/A	N/A
South Dakota	1,600,000	N/A	N/A	N/A
Texas	14,100,000	9,800,000	30.0	N/A
	Short grass			
Saskatchewan	5,900,000	840,000	85.8	N/A
Oklahoma	1,300,000	N/A	N/A	N/A
South Dakota	179,000	N/A	N/A	N/A
Texas	7,800,000	1,600,000	80.0	N/A
Wyoming	3,000,000	2,400,000	20.0	N/A

term ecological research centers (LTERs) located in a variety of ecosystems and supported by the National Science Foundation.

In some cases, restoration may not be possible. For example, the invasion of cheatgrass (*Bromus tectorum*) into intermountain grassland has modified the fire regime to such an extent that the entire ecosystem has been altered. Cheatgrass is a highly flammable annual which attains such high densities that it can carry fire where previously fire was uncommon. According to D'Antonio and Vitousek (1992), pristine bunchgrass-shrub vegetation of the intermountain region had a fire return period of 60–110 years. After invasion of cheatgrass, the fire return period has shortened to 3–5 years. Consequences—felt over an area of 40 million hectares—include destruction of most plant cover, loss of carrying capacity for livestock, loss of biotic diversity, and increased erosion following intense periods of precipitation. At present, there is no management technique known that would eradicate cheatgrass. Similar ecosystem impacts of grass invasion have been reported for Hawaii, Central America, and Australia.

Desert Scrub

Desert scrub is a vegetation type dominated by shrubs with less than 100% cover and generally restricted to semiarid regions receiving 5–25 cm precipitation a year. In North America, desert scrub occupies about 1.2 million square kilometers, or 5% of the North American landmass. Based on climatic, vegetational, and floristic differences, four desert regions and one ecotone can be recognized (Figure 20-37).

The **Great Basin Desert** (also called the **cold desert**) experiences snow and hard frost in winter. The intermountain grassland, discussed in the last section, extends through the northern part of this desert, but most of the region is thoroughly dominated by big sagebrush (*Artemisia tridentata*, Figure 20-38). The other three deserts are collectively called the **warm** or **hot deserts** because winters are mild. One shrub dominates all three: *Larrea tridentata* (Figure 20-39), variously called creosote bush, greasewood, gobernadora, or hediondilla. Creosote bush is slow-growing but persistent. It is able to spread vegetatively by a splitting of the root crown followed by adventitious rooting at the base of lateral branches beneath the surface. Clonal circles of shrubs develop that may be hundreds to thousands of years old.

When viewed from a low, oblique angle, plant cover seems high, but when the canopies are projected vertically down, only 10–25% of the ground typically lies beneath perennial cover (extremes are 5–50%). Aboveground biomass is low, averaging 700 g m^{-2}, which is nearly the same as tundra standing crop. LAI is less than 1, and during the dry part of the year, plants are either absent, leafless, dormant, or functioning at a low level. Over the course of a year, then, the fraction of solar radiation trapped in photosynthesis is quite low—as low as 0.003%, which is an order or two of magnitude lower than the efficiency of other vegetation types. For all these reasons, net primary productivity is the lowest of all vegetation types examined in this chapter, about 100 g m^{-2} yr^{-1}—50% lower than tundra productivity.

It is a mistake to think that the low community productivity means low rates of photosynthesis for each species. If productivity is expressed in terms of grams fixed per gram of leaf area instead of per unit ground area, then desert shrubs are fully as productive as shrubs in the mesic eastern deciduous forest (e.g., *Larrea* = 0.28 g CO_2 g leaf^{-1} yr^{-1} versus 0.16–0.37 for mesic shrubs). Some desert subshrubs

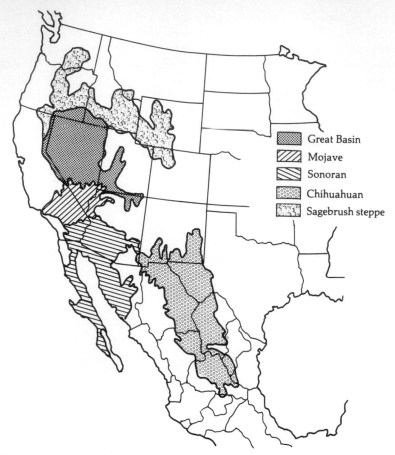

Figure 20-37   Four regions of desert scrub vegetation and one grassland-desert ecotone (sagebrush steppe) in North America. Modified from MacMahon and Wagner (1985) and West (1998).

and annuals have net photosynthesis rates per unit leaf area comparable to or above those of crop plants under irrigated cultivation, with maximum/optimum rates in the range of 60–90 mg CO_2 dm^{-2} hr^{-1}.

Physical Features

Some temperature and precipitation data for each desert are summarized in Table 20-8. Cold winter temperatures in the Great Basin and Chihuahuan Deserts reflect their relatively high elevation on plateaus. The Mojave Desert covers a wide amplitude of elevation, from +1300 to −86 m (the lowest spot in Death Valley). The Sonoran Desert generally is at low elevations. Annual precipitation does not vary much from desert to desert, but its direction, seasonality, and form do. The Great Basin and Mojave have winter peaks from the west and experience moderate to

Figure 20-38   Great Basin Desert, eastern Oregon, dominated by big sagebrush (*Artemisia tridentata*).

Figure 20-39   Creosote bush scrub, showing a clonal ring of creosote bushes. The diameter of the ring is several meters. Courtesy of Frank Vasek.

considerable snowfall. Portions of the Sonoran have winter and summer peaks of equal intensity but no snow, and the Chihuahuan has a pronounced summer peak from the Gulf of Mexico. As in the grassland, fluctuation in rainfall from year to year can be large. The P/E ratio averages 0.3 but may drop below 0.1 at the head of

Table 20-8 Some climatic features of the four desert regions of North America.

	Great Basin	Mojave	Sonoran	Chihuahuan
Area (km²)	409,000	140,000	275,000	453,000
Annual precipitation (mm)	100–300	100–200	50–300	150–300
Precipitation falling in summer (% of total)	30	35	45	65
Snowfall (cm; 10 cm snow = 1 cm rain)	15–30	25–75	trace	trace
Winter mean max/min temperatures (°C)	+8/−8	+15/0	+18/+4	+16/0
Hours of frost (% of total)	5–20	2–5	0–1	2–5
Summer mean max/min temperatures (°C)	34/10	39/20	40/26	34/19
Elevation (m)	>1000	variable	<600	600–1400

the Gulf of California. Summer temperatures are almost equally high in all four deserts.

Some important topographic features common to all deserts are diagrammed in Figure 20-40. The lower slopes of desert mountains are long and gentle. They are given the Spanish name **bajada** (pronounced bah•háh•da). Their average rise is only 1–5%, which is so gradual that one does not realize the elevation gain until one looks back downslope. Bajadas are composed of coarse alluvium eroded from steeper slopes in the montane zone above. They are crisscrossed with channels, called **arroyos** or **washes,** which have been cut by intermittent streams that flow only after heavy rains.

Bajadas support typical desert vegetation, with the highest species diversity of any desert habitat. Evidently, the coarseness of the soil provides a wide range of microhabitats. On a typical Sonoran bajada near Tucson, Arizona, the upper bajada community consisted of 18 perennial species with 35% cover and a species diversity index (H′) of 1.2. Near the base of the bajada, on finer soil, the community consisted of only 6 perennial species with 20% cover and a diversity index of 0.7. Along the same gradient, soil texture changed dramatically (Figure 20-41). Throughout the desert, species diversity correlates better with soil texture than with rainfall.

Arroyos have low plant cover because of frequent disturbance, but the larger ones may have scattered trees. These riparian trees are able to survive because their root systems are in contact with greater supplies of water than are available on the surrounding bajadas. Only in areas of significant summer rainfall, as in the Arizona Sonoran Desert, do these trees occur on the bajadas. Common examples include smoke tree (*Dalea spinosa*), desert ironwood (*Olneya tesota*), mesquite (*Prosopis* spp.), and palo verde (*Cercidium* spp.) All of these are legumes with large seeds that require scarification for germination.

Bajada soils are typical **aridisols,** with little profile development, less than 1%

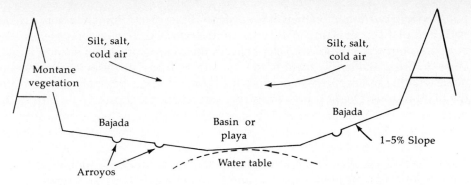

Figure 20-40 Major topographic features in the desert landscape. Typical desert vegetation is restricted to the bajadas.

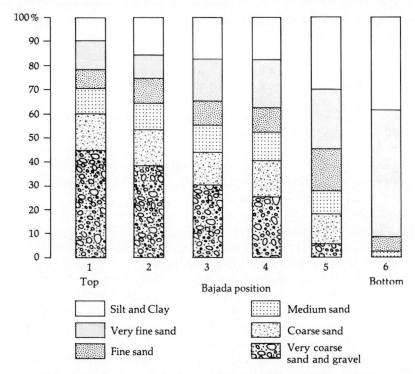

Figure 20-41 Soil particle size distribution for six sites along an Arizona bajada. Redrawn from MacMahon and Wagner (1985).

organic matter, sandy loams to loamy sands, pH 7–8.5, and often with an indurated (hardened) layer of calcium carbonate (called a **caliche** layer) 25–50 cm below the surface. Such a layer impedes root growth, and the shallowness of the caliche affects the species composition of the community above. In some areas, erosional processes produce a **desert pavement**—a surface layer of close-fitting rocks, polished and burnished on their exposed faces. Plant establishment is very difficult here, and plant cover is unusually low. Another surface phenomenon can be a thin but persistent water-repellent layer beneath the canopy of certain shrubs, such as creosote bush. Evidently, material released from living or decaying leaves combines with the soil particles to produce the layer, which may remain for years after the plant is gone. Decay products also accumulate in the topsoil in quantities that differ from species to species. Soils beneath different species may significantly differ in such traits as water-holding capacity, texture, the concentration of organic carbon and nitrogen, and the concentration of various anions and cations such as Na, K, Ca, Mg, Cl, SO_4, Fe, and Mn. Soils in the open, between canopies, naturally differ from soils beneath canopies. In addition, mosses, lichens, and algae may colonize the soil surface beneath canopies, forming a soil crust that affects permeability and nitrogen content (if the lichens or algae are nitrogen-fixers).

Playas (pronounced plý•as) are undrained basins at the base of bajadas. Runoff from bajadas carries fine-textured soil and dissolved salts. This material accumulates in the playa in such a way that there are circular zones of increasing salinity from the edge to the lowest part of the basin. Each zone is characterized by a different community of plants. The innermost zone may be devoid of plants, showing only a crust of salts on the surface. Soil aeration is low because of the fine texture and because a water table may be close to, or at the surface of, the playa bottom. Some research indicates that bajada species are prevented from colonizing playas mainly because of low soil oxygen, rather than because of the high soil salinity (sometimes more than 35,000 ppm salt, pH 9–11).

An additional microenvironmental trait of playas is that they are colder than surrounding slopes. Cold air at night sinks and collects in the basin, and frost can be more common there. One 10-year study in the Mojave Desert of southern Nevada showed that a 60 m drop in elevation from a *Larrea*-dominated bajada slope to an *Atriplex*- and *Lycium*-dominated playa bottom correlated with a drop in mean minimum temperatures from several degrees above freezing to several degrees below freezing. Thus, there are several factors that account for major community differences between playas and bajadas.

Not all community changes can be accounted for by elevational or soil changes. In some level parts of the Great Basin, a mosaic of communities exists with quite narrow ecotones. One community may be strongly dominated by shadscale (*Atriplex confertifolia*), another by winterfat (*Eurotia lanata*), a third by salt sage (*Atriplex nuttallii*), and a fourth by sagebrush. Detailed studies of soil texture, chemistry, and water retention have failed to reveal any changes significant enough to account for the mosaic. On a broader regional scale, these species do tend to separate on gradients of salinity and aridity (Figure 20-42). They may also differ in the amount of snow retained in winter.

Autecology of Major Life Forms

Each vegetation type discussed in this chapter includes species in an array of life forms. Desert vegetation may have a wider array than any other type, depending on how life forms are defined. Forrest Shreve (1942) defined 25 desert life forms. In the following pages, we will briefly discuss a few of these.

Figure 20-42 The general regional distribution of several Great Basin shrubs along gradients of salinity and aridity in Utah. Artr = *Artemisia tridentata*, Cela = *Ceratoides lanata*, Atco = *Atriplex confertifolia*, Atnu = *Atriplex nuttallii*, Save = *Sarcobatus vermiculatus*, Aloc = *Allenrolfea occidentalis*. Redrawn from Caldwell (1985).

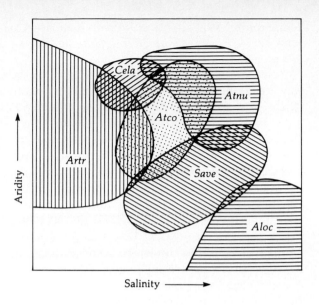

Annuals (Ephemerals) These plants have a life span that is shorter than 8 months. Over 40% of the flora in these four deserts is annual, in contrast to Raunkiaer's world normal of 13%. There are two categories of desert annuals. **Winter annuals** germinate in fall or winter and flower in late spring; **summer annuals** germinate in midsummer and flower in late summer or fall. Rarely does the same species belong to both groups, partly because germination requirements of winter and summer annuals usually differ. If soil temperature is below 18°C, for example, Mojave Desert winter annuals will germinate, but if soil temperature is above 26°C, Mojave summer annuals will germinate. Germination of summer annuals is also enhanced, in some cases, by dry storage for several weeks at 50°C, a requirement that prevents seed germination between the time seeds are shed in fall and a rainy period the following summer.

The diversity of annual species at one season and in one region may be high, and the species may group into distinctive communities that reflect unique microenvironments. Some species grow on bajadas, some at the edge of playas, some in playas; some grow beneath certain canopies and avoid others; some grow only in the open. These patterns may be due to nurse plant requirements (for shade or organic matter), allelopathy, water-repellent layers beneath some shrubs, or salinity tolerances.

Winter annuals germinate after fall or winter rains in excess of 10–15 mm. Below this critical limit, germination is absent. The rain may be necessary to leach inhibitors from the seed coat or to provide moisture beyond some critical inhibitional period of time. The density of annuals correlates with increasing rainfall between 15 and 45 mm. Winter annuals grow slowly through winter, are resistant to frost down to −18°C, then grow rapidly in spring as temperatures rise. Flowering, fruiting, and death occur in late May, for a total life cycle of 5–8 months. Herbivory accounts for some mortality, but most seedlings succumb to drought and high temperature. Most do not survive to maturity. Under the best conditions, density may be 1000 m^{-2}, cover 30%, and biomass 60 g m^{-2}, but typically, density is 100 m^{-2} and biomass 10 g m^{-2}.

Summer annuals germinate in August or September after heavy rains, generally are C$_4$ in metabolism, remain small, and mature by the time of autumn frosts. Their life span is measured in terms of weeks rather than months.

Many winter and summer annuals are able to modify solar radiation striking their leaves by heliotropism. Leaves track the sun's path during the day, such that they remain perpendicular to the sun's rays. Over the course of a day, such leaves receive 38% more radiation than leaves with a fixed horizontal or east-west vertical orientation (see Chapter 15). Some heliotropic leaves fold or cup during periods of drought, and solar radiation is then reduced by an order of magnitude, which in turn reduces leaf temperature and transpiration.

Drought-Deciduous Species   Some drought-deciduous species, such as ocotillo (*Fouquieria splendens*) (Figure 20-43), are able to produce several crops of leaves a year, losing each one as a dry period follows a wet period. Most drought-deciduous species, however, produce only one crop of leaves a year and enter a long dormancy following leaf drop. Their leaves are mesophytic in terms of size, anatomy, and shape, making them energetically inexpensive for the plant to manufacture, compared to evergreen leaves. Their large intercellular spaces and thin blades permit rapid gas exchange (and water loss), so their photosynthetic rate is relatively high, about two to three times the rate for evergreens.

Winter-Deciduous Riparian Species   Most desert riparian species are winter-deciduous, a trait common to riparian species throughout all vegetation types in North America. Onset of stem growth, leaf expansion, and blooming is photoperiodically controlled, in contrast to the moisture controls for other desert life forms.

Figure 20-43   Ocotillo (*Fouquieria splendens*), a drought-deciduous species in full leaf (left) and leafless and dormant (right).

Leaves appear in early summer and exhibit some xeromorphic features. During the hot summer, the water potentials developed within these plants are more negative than those of drought-deciduous species. A considerable amount of photosynthesis is accomplished by young, green twigs in some species.

As already mentioned, many riparian desert species are leguminous trees (Figure 20-44) that produce large seeds with tough seed coats. The scarification they require is easily provided when the seeds are carried for some distance by raging water down an arroyo. Some riparian species are phreatophytes, consequently seedlings produce a root that can grow fast enough to keep ahead of soil drying down from the surface. One of the fastest reported root growth rates is for mesquite: 102 mm day^{-1}, which is an order of magnitude greater than for typical xerophytes or mesophytes.

Succulents   There are many species of **stem** and **leaf succulents** in several families, but the cacti are the best known. Cacti have the CAM photosynthetic pathway, with stomates open at night. Consequently, their water use efficiency is very high, $14-40$ g CO_2 kg^{-1} H_2O, which is an order of magnitude higher than other desert plants. Internal water stress rarely exceeds -0.5 MPa. Cacti possess a shallow root

Figure 20-44   Foothill paloverde (*Cercidium*), a leguminous tree of the Sonoran Desert.

system that is able to absorb water even from light rains, and in dry periods much of this root system dies. In wet periods, water is stored in large parenchyma cells, swelling the stem; in dry periods this water is used and the stem shrinks. Average root depth is only 8 cm, and succulents can show metabolic response to only 18 mm of rain within 24 hours.

Most succulents are not frost resistant, and some ecologists have correlated their distribution limits to isolines of average consecutive hours of frost. Nevertheless, two species of beavertail cactus (*Opuntia* spp.) do extend throughout the colder Great Basin Desert. Several morphological features significantly improve tissue temperature of cacti at marginal northern sites. These include cladode (flattened stem) orientation to face east-west, increase in stem diameter, apical spines or pubescence, apex orientation, and growth beneath nurse plants. *Opuntia polyacantha* in Wyoming can tolerate temperatures down to −17°C, whereas *Opuntia* species in warm deserts to the south can tolerate temperatures only down to −4°C.

Although succulents have a slow growth rate, the largest plants in the desert are cacti: the cardón cactus (*Pachycereus pringlei*) of Baja California and Sonora, Mexico, can reach 18 m and (presumably) hundreds of years in age. The more widely known saguaro cactus of Arizona and Sonora (*Carnegiea gigantea* = *Cereus giganteus*) can reach 10 m and an age of at least 200 years. Some populations of saguaro are clearly senescent, dominated by middle-aged and old plants, possibly because of increasing cattle and rodent populations, both of which graze on young plants.

Evergreens   These are true xerophytes because they grow and transpire throughout even the driest part of the year, at most shedding only a fraction of their leaves. Creosote bush (*Larrea tridentata*) and jojoba (*Simmondsia chinensis*) are good examples. For reasons not yet understood, their cytoplasm is able to resist unusual desiccation. Net photosynthesis is possible even when leaf water potential drops to −5 MPa, and root growth continues to −7 MPa. When moisture is not limiting, their transpiration rates are equivalent to those of mesophytes; when moisture is limiting, their transpiration rates are very low. All desert evergreens have C_3 metabolism.

Their leaves often possess stereotyped morphological features, which ecologists presume are responsible for the low rate of transpiration: for example, thick cuticle, lack of intercellular spaces, sunken stomates, palisade parenchyma beneath both surfaces, small leaves, and vertical leaf orientation. Hairs, salt glands, and waxes on the surface may also serve to reflect light, hence lowering the heat load on the leaf. Many evergreens invest considerable metabolic energy in the production and accumulation of resins and other complex molecules that may have anti-herbivore significance. Some evergreens, such as *Artemisia tridentata* and *Atriplex confertifolia*, exhibit two types of leaves. Small, more xerophytic overwintering leaves are formed in late summer and retained to the following spring, when they are replaced by larger leaves with up to five times higher maximum net photosynthesis rates.

These evergreens give each desert its characteristic appearance, as they dominate the bajadas in terms of cover and biomass. In the warm deserts, CAM succulents and drought-deciduous species may also be important on bajadas, but C_4 plants are significant only in playas. A study of bajada and playa vegetation in the Chihuahuan Desert of New Mexico very dramatically documents this pattern (Table 20-9): 90% of playa biomass was contributed by C_4 species, primarily summer annuals, but only 1% of bajada biomass was contributed by C_4 taxa. Evidently, some aspects of C_4 metabolism are ecologically disadvantageous out on the bajadas, regardless of life form.

Great Basin Vegetation

The cold desert is covered by a one-layered shrub community, about 1 m tall and very low in species diversity. The overwhelming dominant is sagebrush, *Artemisia tridentata* (see Figure 20-38). Although species diversity is low in this desert, there is significant ecotypic variation in several dominant taxa. *Artemisia* is represented by 12 species, and *A. tridentata* alone has 4 ecologically important subspecies (ecotypes). Those ecotypes occupy sites that differ in soil depth, aridity, and temperature (Figure 20-45). Common associates are shadscale (*Artiplex confertifolia*), winterfat (*Ceratoides lanata*), spiny hopsage (*Grayia spinosa*), and Mormon tea (*Ephedra* spp.). Total cover is 15–40%.

To the north, or at higher elevations, is the **sagebrush steppe.** In pristine times, it consisted of scattered sagebrush in a matrix of perennial bunchgrasses and seasonal forbs; total cover could exceed 80%. The grasses included western wheatgrass (*Agropyron smithii*), needlegrass (*Stipa spartea*), Idaho fescue (*Festuca idahoensis*), and bluebunch wheatgrass (*Agropyron spicatum*). Some of this area is now farmland, and most of the rest is changed because of overgrazing, burning, and the invasion of weedy annuals and native short-lived perennials: rabbitbrush (*Chrysothamnus* spp.), cheatgrass, medusahead (*Elymus caput-medusae*), Russian

Table 20-9   Percent C_3, C_4, and CAM of the total species and of the total biomass in three adjacent communities in the Chihuahuan Desert of southern New Mexico. Data were accumulated over the course of several years. From Syvertsen et al. (1976). By permission of *Southwestern Naturalist*.

Community	Attribute	C_3 (%)	C_4 (%)	CAM (%)	Absolute totals	Dominants
Playa bottom	Species	52	48	0	21 species	*Panicum obtusum* (C_4)
	Biomass	10	90	0	1243 kg ha^{-1}	
Playa fringe	Species	54	44	1	77 species	*Hilaria mutica* (C_4),
	Biomass	42	50	8	11,200 kg ha^{-1}	*Prosopis glandulosa* (C_3)
Bajada	Species	66	24	10	144 species	*Larrea tridentata* (C_3), *Yucca*
	Biomass	51	1	48	2088 kg ha^{-1}	*elata* (CAM), *Prosopis glandulosa* (C_3), *Flourensia cernua* (C_3)

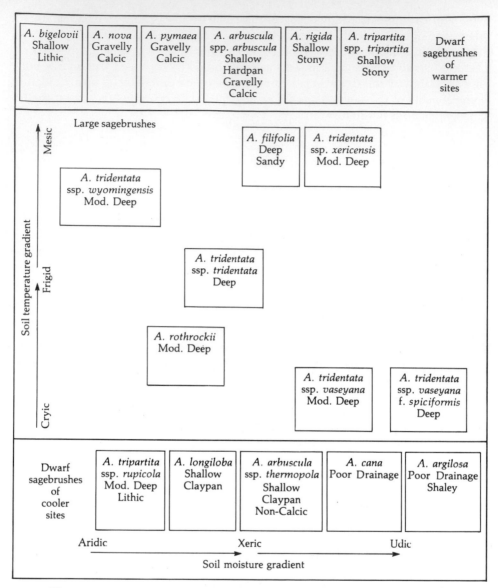

Figure 20-45   Environmental distribution of *Artemisia* species and subspecies in the Great Basin, arranged along gradients of soil moisture, depth, texture, and temperature. From West (1998).

thistle (*Salsola ibirica*), tumble mustard (*Sisymbrium altissimum*), filaree (*Erodium* spp.), and *Halogeton glomeratus*. Sagebrush, which is not palatable, has been purposely reduced in cover by burning (it is not a sprouter). Some of this range has been so degraded that it may not recover even with active management (Fig-

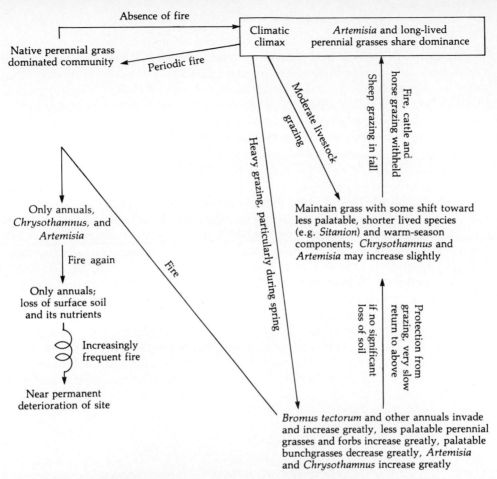

Figure 20-46 Modifications of sagebrush steppe by fire, grazing, and the introduction of weedy annuals in the past 150 years. From West (1998).

ure 20-46). At higher elevations, sagebrush steppe is joined by a 5–15 m tall open canopy of pine and juniper (the **pinyon-juniper woodland**).

In more saline areas is a saltbush-greasewood scrub, dominated by several small shrub species of *Atriplex* and *Sarcobatus vermiculatus*, mainly C$_4$ plants. This vegetation has also been degraded by overgrazing and the invasion of several annual weedy species.

In parts of southern Nevada and eastern Utah, shadscale becomes the dominant, associated with budsage (*Artemisia spinescens*), greasewood (*Sarcobatus baileyi*), and *Ephedra nevadensis*. This has been called the **shadscale zone** of the Great Basin, and its climate is significantly warmer and drier than sagebrush-dominated areas. Ground cover is only 10%.

Warm Desert Vegetation

The **Mojave** is dominated by a two-layered community: an overstory shrub layer, mostly of creosote bush, and an understory subshrub layer, mostly of burro bush (also call bur sage, *Ambrosia dumosa*) (Figure 20-47). It is a monotonous, open community, with only 5–10% cover. At higher elevations (1200–1800 m), creosote bush is displaced by blackbrush (*Coleogyne ramosissima*), Joshua tree (*Yucca brevifolia*), spiny hopsage, matchweed (*Gutierrezia* spp.), and winterfat. The **blackbrush community** is an ecotone between **creosote bush scrub** below and pinyon-juniper woodland above, sharing half of its flora with each.

The **Sonoran Desert** is so complex that it has been divided into as many as seven vegetational regions by some ecologists. We will only consider two extremes here. The **Colorado Desert** of southern California, southwestern Arizona, and the area where the Colorado River empties into the Gulf of California is very arid, and the vegetation is reminiscent of the Mojave. Creosote bush and burro bush dominate. However, there are many differences in annuals and other life forms between this region and the Mojave. Of 545 Mojave plant species, only half are also found in the Sonoran. Many hectares of the Coachella and Imperial Valleys in California are now dominated by citrus, date palms, and sugar beets, irrigated by water diverted from the Colorado River.

The other Sonoran extreme, typical of central Baja California, the **Arizona uplands,** and parts of Sonora, is a complex four-layered community with about 25% ground cover (Figure 20-48). The overstory layer is 3–5 m tall and contains saguaro cactus (*Carnegiea gigantea*), foothill paloverde (*Cercidium microphyllum*), and iron-

Figure 20-47   Typical *Larrea-Ambrosia* community on a Mojave bajada.

Figure 20-48 Arizona uplands community in the Sonoran Desert, with saguaro (*Carnegiea gigantea*) and mesquite (*Prosopis juliflora*) prominent in the overstory.

wood (*Olneya tesota*). A shrub-cactus layer 2–3 m tall has creosote bush, *Acacia* species, ocotillo, and chollas. A third layer is composed of subshrubs 1 m or less in height, such as two species of burro bush (*Ambrosia dumosa* and *A. deltoidea*), brittle bush (*Encelia farinosa*), and a great variety of cacti (barrel, fish hook, cholla, beavertail). There are approximately five times as many species of cacti in the Sonoran Desert as in the Mojave. A fourth layer is dominated by some perennial grasses and many species of summer and winter annuals. Because there are two nearly equal peaks of rainfall in winter and summer in these more mesic parts of the Sonoran, the diversity of species for each annual category is about the same. In contrast, the winter-wet Mojave Desert has about six times as many winter as summer annuals.

The **Chihuahuan Desert** lacks an arboreal element, but it has great species richness in two shrub layers. Beneath an overstory of creosote bush, ocotillo, mesquite, and *Acacia* are tarbush (*Flourensia cernua*), guayule (*Parthenium incanum*), leather plant (*Jatropha* spp.), crucifixion thorn (*Koeberlinia spinosa*), many cacti, and several distinctive members of the agave and lily families with spinescent, basal leaves (Figure 20-49). The latter include century plant (*Agave lechuguilla* and others), sotol (*Dasylerion wheeleri*), Spanish bayonet (*Yucca elata*, *Y. baccata*), and *Nolina*. Ground cover is relatively high, about 20–25%. Summer annuals can be abundant.

This desert is large and complex enough to warrant subdivision into three re-

Figure 20-49   Typical Chihuahuan Desert vegetation, Big Bend National Park, Texas. Prominent plants include ocotillo (*Fouquieria splendens*), creosote bush (*Larrea tridentata*), *Agave* species, and a variety of cacti.

gions. From north to south, these have been called the **Trans-Pecos, Mapimian,** and **Saladin.** Plant cover, species richness, and taller growth forms increase to the south. Most of the preceding details of Chihuahuan vegetation, however, come from the Trans-Pecos region.

Some Concluding Thoughts

We began this chapter by pointing out the vegetational diversity of North America: perhaps 6000 or more associations and representatives of most of the world's major plant formations. As we discussed each vegetation type, however, a theme of habitat loss and biotic loss became prominent. It has been said that 99% of all the species that have ever inhabited this planet have gone extinct, so extinction itself is neither novel nor solely caused by humans. What is unique is the speed with which recent extinctions have happened, and their linkage to preventable human activities. What reasonable prediction can we make regarding our ability to conserve what remains?

The impressions of North American vegetation in this chapter represent a snapshot in time. Photographs do not change, but their subjects do. The landscapes of North America will be different in the future. They will change, even if the half-billion human occupants and all their domesticated livestock were removed and every hectare designated a natural preserve. In the long term—over geologic time—climates will change, continents will move, and floras will evolve. We are learning, however, that climates will also change over the short term. Most climatologists now believe that human activities have changed the earth's atmosphere to the point that a process of climate change has been set in motion and that it will result in measurable biological changes over the entire surface of the globe in a matter of decades (Silver and DeFries 1990, Vemap 1996, Watson et al. 1996). A sobering thought: there is no wilderness left, for even the remotest places on earth are feeling the impact of human activities.

We are not arguing for an earth without humans, only for an earth where humans coexist with as rich a collection of other species as possible. Perhaps our present trend of regaining an environmental sensitivity provides some reason for optimism that we can attain this coexistence and reverse the trend of the last several centuries. Such a reversal can be promoted by the actions and attitudes of those of you who read this book and teach from it. We wish you Godspeed.

Chapter 20 References

This is a very personal list of references. Works cited cover the factual material in the chapter and give a sampling of both current and classical research. This list is neither consistent nor exhaustive, but it does give the reader an entry into the many additional publications not discussed in the text.

References for North America, Overall

Archibold, O.W. 1995. *Ecology of world vegetation.* New York: Chapman and Hall.

Bailey, R.G. 1996. *Ecosystem geography.* New York: Springer-Verlag.

Barbour, M.G. and W.D. Billings (eds.). 1998. *North American terrestrial vegetation,* 2nd ed. New York: Cambridge University Press.

Brown, D.E., C.H. Lowe, and C.P. Pase. 1980. A digitized systematic classification for ecosystems with an illustrated summary of the natural vegetation of North America. Fort Collins, CO: USDA Forest Service, General Technical Report RM-73.

Brown, D.E., F. Reichenbacher, and S.E. Franson. 1998. A classification of North American biotic communities. Salt Lake City: University of Utah Press.

Chabot, B.F. and H.A. Mooney (eds.). 1985. *Physiological ecology of North American plant communities.* New York: Chapman and Hall.

Cowardin, L.M., V. Carter, F.C. Golet, and E.T. LaRoe. 1979. *Classification of wetlands and deepwater habitats of the United States.* Washington, D.C.: U.S. Dept. of the Interior, Fish and Wildlife Service.

Daubenmire, R.F. 1978. *Plant geography, with special reference to North America.* New York: Academic Press.

Driscoll, R.S. et al. 1984. *An ecological land classification framework for the United States.* Washington, D.C.: USDA Forest Service, Miscellaneous Publication 1439.

Ecomap Group. 1993. *National hierarchical framework of ecological units.* Washington, D.C.: USDA Forest Service.

Ecoregions Working Group. 1989. *Ecoclimatic regions of Canada, first approximation.* Ottawa: Canadian Wildlife Service, Ecological Land Classification Series, No. 23.

Eyre, F.H. (ed.). 1980. *Forest cover types of the United States and Canada.* Washington, D.C.: Society of American Foresters.

Flora of North America Editorial Committee. 1993. *Flora of North America north of Mexico.* New York: Oxford University Press.

Grossman, D. (ed.). In press. *International classification of ecological communities: terrestrial vegetation of the United States.* Arlington, VA: The Nature Conservancy.

Knapp, R. 1965. *Die vegetation von Nord- und Mittel-amerika und der Hawaii-Inseln.* Stuttgart, Germany: Gustav Fischer Verlag.

Kuchler, A.W. 1964. *Potential natural vegetation of the conterminous United States.* American Geographical Society, Special Publication 36.

McNab, W.H., and P.E. Avers (eds.). 1994. *Ecological subregions of the United States: section descriptions.* Washington, D.C.: USDA Forest Service.

Noss, R., E.T. LaRoe III, and J.M. Scott. 1995. Endangered ecosystems of the United States: a preliminary assessment of loss and degradation. Washington, D.C.: U.S. Dept. of the Interior, National Biological Service, Biological Report 28.

Peinado, M., J.L. Aguirre, and J. Delgadillo. 1997. Phytosociological, bioclimatic, and biogeographical classification of woody climax communities of western North America. *Journal of Vegetation Science* 8:505–528.

Rivas-Martinez, S. 1997. Syntaxonomical synopsis of the potential natural plant communities of North America, part I. *Itinera Geobotanica* 10:5–148.

Rzedowski, J. 1981. *Vegetación de México.* Mexico, D.F.: Editorial Limusa.

Shiflet, T.N. (ed.). 1994. *Rangeland cover types of the United States.* Denver: Society for Range Management.

Silver, C.S., and R.S. DeFries. 1990. *One earth, one future.* Washington, D.C.: National Academy Press.

Stephenson, N.L. 1990. Climatic control of vegetation distribution: the role of the water balance. *American Naturalist* 135:649–670.

Strong, W.L., E.T. Oswald, and D.J. Downing (eds.). 1990. *The Canadian vegetation classification system: first approximation.* Ottawa: Secretariat, Canada Committee on Ecological Land Classification.

UNESCO. 1973. *International classification and mapping of vegetation: Series 6, Ecology and conservation.* Paris: United Nations Educational, Scientific, and Cultural Organization.

Vankat, J.L. 1979. *The natural vegetation of North America.* New York: Wiley.

———. 1990. A classification of the forest types of North America. *Vegetatio* 88:53–66.

Vegetation Classification Panel. 1998. *Proposed national standards for the floristic levels of vegetation classification in the United States: associations and alliances.* Washington, D.C.: Ecological Society of America.

Vegetation Subcommittee. 1996. *Federal Geographic Data Committee vegetation classification and information standards.* Reston, VA: Federal Geographic Data Committee Secretariat.

Vemap members. 1996. Vegetation/ecosystem modeling and analysis project: comparing biogeography and biogeochemistry models in a continental-scale study of terrestrial ecosystem responses to climate change and CO_2 doubling. *Global Biogeochemical Cycles* 9:407–437.

Walter, H. 1984. *Vegetation of the earth and ecological systems of the geo-biosphere,* 3rd ed. New York: Springer-Verlag.

Watson, R.T., M.C. Zinyowera, and R.H. Moss (eds.). 1996. *Impacts, adaptations, and mitigation of climate change: scientific-technical analyses.* Contribution of Working Group II to the second assessment report of the intergovernmental panel on climate change. New York: Cambridge University Press.

Woodward, F.L. 1987. *Climate and plant distribution.* New York: Cambridge University Press.

Tundra

Billings, W.D. 1974a. Adaptations and origins of alpine plants. *Arctic and Alpine Research* 6:129–142.

———. 1974b. Arctic and alpine vegetation: Plant adaptations to cold summer climates. In *Arctic and alpine environments,* eds. J.D. Ives and R.G. Barry, pp. 403–443. London: Methuen.

Billings, W.D. and P.J. Godfrey. 1967. Photosynthetic utilization of internal carbon dioxide by hollow-stemmed plants. *Science* 158:121–123.

Billings, W.D. and H.A. Mooney. 1968. The ecology of arctic and alpine plants. *Biological Review* 43:481–529.

Black, R.A. and L.C. Bliss. 1980. Reproductive ecology of *Picea mariana* (Mill.) BSP at tree line near Inuvik, Northwest Territories, Canada. *Ecological Monographs* 50:331–354.

Bliss, L.C. 1956. Comparison of plant development in microenvironments of arctic and alpine tundras. *Ecological Monographs* 26:303–337.

———. 1962. Adaptations of arctic and alpine plants to environmental conditions. *Arctic* 15:117–144.

———. 1963. Alpine plant communities of the Presidential range, New Hampshire. *Ecology* 44:687–697.

———. 1966. Plant productivity in alpine microenvironments on Mt. Washington, New Hampshire. *Ecological Monographs* 36:125–155.

———. 1985. Alpine. In *Physiological ecology of North American plant communities,* eds. B.F. Chabot and H.A. Mooney, pp. 41–65. New York: Chapman and Hall.

———. 1987. Arctic tundra and polar desert biome. In *North American terrestrial vegetation,* eds. M.G. Barbour and W.D. Billings, Chap. 1. New York: Cambridge University Press.

Britton, M.E. 1966. *Vegetation of the arctic tundra.* Corvallis: Oregon State University Press.

Caldwell, M.M. 1968. Solar ultraviolet radiation as an ecological factor for alpine plants. *Ecological Monographs* 38: 243–268.

Caldwell, M.M., R. Robberecht, R.S. Nowak, and W.D. Billings. 1982. Differential photosynthetic inhibition by ultra-violet radiation in species from the arctic-alpine life zone. *Arctic and Alpine Research* 14:195–202.

Canaday, B.B. and R.W. Fonda. 1974. The influence of subalpine snowbanks on vegetation pattern, production, and phenology. *Bulletin of the Torrey Botanical Club* 101:340–350.

Chabot, B.F. and W.D. Billings. 1972. Origins and ecology of the Sierran alpine flora and vegetation. *Ecological Monographs* 42:143–161.

Chapin, F.S. III and G.R. Shaver. 1985. Arctic. In *Physiological ecology of North American plant communities,* eds. B.F. Chabot and H.A. Mooney, pp. 16–40. New York: Chapman and Hall.

Chapin, F.S. III, et al. (eds.). 1992. *Arctic ecosystems in a changing climate: An ecological perspective.* New York: Academic Press.

Clausen, J. 1965. Population studies of alpine and subalpine races of conifers and willows in the California high Sierra Nevada. *Evolution* 19:56–68.

Cwynar, L.C. and R.W. Spear. 1991. Reversion of forest to tundra in the central Yukon. *Ecology* 72:202–212.

Daubenmire, R.F. 1954. Alpine timberlines in the Americas and their interpretation. *Butler University Botanical Studies* 11:119–136.

Dawson, T.E. and L.C. Bliss. 1993. Plants as mosaics: leaf, ramet, and gender level variation in the physiology of the dwarf willow, *Salix arctica. Functional Ecology* 7:293–304.

Dennis, J.G. and P.L. Johnson. 1970. Shoot and rhizome-root standing crops of tundra vegetation at Barrow, Alaska. *Arctic and Alpine Research* 2:253–266.

Douglas, G.W. and L.C. Bliss. 1977. Alpine and high subalpine plant communities of the North Cascades Range, Washington and British Columbia. *Ecological Monographs* 47:113–150.

Elliott-Fisk, D.L. 1983. The stability of the northern Canadian tree limit. *Annals of Association of American Geographers* 73:560–576.

———. 1987. The boreal forest. In *North American terrestrial vegetation,* eds. M.G. Barbour and W.D. Billings, Chap. 2. New York: Cambridge University Press.

Ewers, F.W. 1982. Secondary growth in needle leaves of *Pinus longaeva* (bristlecone pine) and other conifers: quantitative data. *American J. Botany* 69:1552–1559.

Forbes, B.C. 1994. The importance of bryophytes in the classification of human-disturbed high arctic vegetation. *Vegetation Science* 5:877–884.

Greller, A.M. 1974. Vegetation of roadcut slopes in the tundra of Rocky Mountain National Park, Colorado. *Biological Conservation* 6(2):84–93.

Griggs, R.F. 1938. Timberlines in the northern Rocky Mountains. *Ecology* 19:548–564.

———. 1946. The timberlines of North America and their interpretation. *Ecology* 27:257–289.

Hoffman, R.S. 1958. The meaning of the word "taiga." *Ecology* 39:540–541.

Hustich, I. 1953. The boreal limits of conifers. *Arctic* 6:149–162.

Johnson, D.A. and M.M. Caldwell. 1976. Water potential components, stomatal function, and liquid phase water transport resistances of four arctic and alpine species in relation to moisture stress. *Physiologia Plantarum* 36:271–278.

Kevan, P.G. 1975. Sun-tracking solar furnaces in high arctic flowers: significance for pollination and insects. *Science* 189:723–726.

Klikoff, L.G. 1965. Microenvironmental influence on vegetational pattern near timberline in the Sierra Nevada. *Ecological Monographs* 35:187–211.

LaMarche, V.C., Jr. and H.A. Mooney. 1972. Recent climatic change and development of the bristlecone pine (*P. longaeva* Bailey) krummholz zone, Mt. Washington, Nevada. *Arctic and Alpine Research* 4:61–72.

Larsen, J.A. 1965. The vegetation of the Ennadai Lake Area, N.W.T. *Ecological Monographs* 35:37–59.

Major, J. and S.A. Bamberg. 1967. Comparison of some North American and Eurasian alpine ecosystems. In *Arctic and alpine environments,* eds. H.E. Wright, Jr. and W.H. Osburn, pp. 89–118. Bloomington: Indiana University Press.

Major, J. and D.W. Taylor. 1977. Alpine. In *Terrestrial vegetation of California,* eds. M.G. Barbour and J. Major, pp. 601–675. New York: Wiley-Interscience.

Marr, J.W. 1948. Ecology of the forest-tundra ecotone on the east coast of Hudson Bay. *Ecological Monographs* 18:117–144.

Moldenke, A.R. 1976. California pollination ecology and vegetation types. *Phytologia* 34:305–361.

Oechel, W.C. and W.T. Lawrence. 1985. Taiga. In *Physiological ecology of North American plant communities,* eds. B.F. Chabot and H.A. Mooney, pp. 66–94. New York: Chapman and Hall.

Oechel, W.C., et al. 1995. Change in arctic CO_2 flux over two decades: effects of climate change at Barrow, Alaska. *Ecological Applications* 5:846–855.

Pewe, T.L. 1966. *Permafrost and its effect on life in the north.* Corvallis: Oregon State University Press.

Polunin, N. 1948. *Botany of the Canadian eastern arctic, III. Vegetation and ecology.* National Museum of Canada Bulletin 104. Ottawa: National Museum of Canada.

Rickard, W.E., Jr. and J. Brown. 1974. Effects of vehicles on arctic tundra. *Environmental Conservation* 1:55–62.

Ritchie, J.C. 1987. *Postglacial vegetation of Canada.* New York: Cambridge University Press.

Shaver, G.R. and W.D. Billings. 1977. Effects of daylength and temperature on root elongation in tundra graminoids. *Oecologia* 28:57–65.

Shaver, G.R. and F.S. Chapin III. 1991. Production:biomass relationships and element cycling in contrasting arctic vegetation types. *Ecological Monographs* 61:1–31.

Stålfelt, M.G. 1960. *Plants, the soil and man.* 1972 translation by M.S. Jarvis and P.G. Jarvis. New York: Wiley.

Teeri, J.A. 1976. Phytotron analysis of a photoperiodic response in a high arctic plant species. *Ecology* 57:374–379.

Tieszen, L.L. (ed.). 1978. *Vegetation and production ecology of an Alaskan arctic tundra.* New York: Springer-Verlag.

Timoney, K.P., et al. 1992. The high subarctic forest-tundra of northwestern Canada: position, width, and vegetation gradients in relation to climate. *Arctic* 45:1–9.

Walker, D.A. 1995. Disturbance and recovery of arctic Alaskan vegetation. In *Landscape function: implications for ecosystem response to disturbance, a case study in arctic tundra,* eds. J.F. Reynolds and J.D. Tenhunen. New York: Springer-Verlag.

Walker, D.A. and M.D. Walker. 1991. History and pattern of disturbance in Alaskan arctic terrestrial ecosystems: a hierarchical approach to analyzing landscape change. *Journal of Applied Ecology* 28:244–276.

Walker, D.A. et al. 1995. Toward a new arctic vegetation map: a review of existing maps. *Vegetation Science* 6:427–436.

Conifer Forests

Agee, J.K. 1993. *Fire ecology of Pacific northwest forests.* Washington, D.C.: Island Press.

Allen, R.B., R.K. Peet, and W.L. Baker. 1991. Gradient analysis of latitudinal variation in southern Rocky Mountain forests. *Biogeography* 18:123–139.

Arno, S.F. and J.R. Habeck. Ecology of alpine larch (*Larix lyallii* Parl.) in the Pacific northwest. *Ecological Monographs* 42:417–450.

Aune, P.S. (ed.). 1992. *Proceedings of the symposium on giant sequoias.* Albany, CA: USDA Forest Service, General Technical Report PSW-151.

Axelrod, D.I. and P.H. Raven. 1985. Origins of the Cordilleran flora. *Biogeography* 12:21–47.

Azvedeo, J. and D.L. Morgan. 1974. Fog precipitation in coastal California forests. *Ecology* 55:1135–1141.

Bahre, C.J. 1991. *Legacy of change: historic human impact on vegetation in the Arizona borderlands.* Tucson: University of Arizona Press.

Baker, W.L. 1992. Structure, disturbance and change in the bristlecone pine forests of Colorado, USA. *Arctic and Alpine Research* 24:17–26.

Baker, W.L. and T.T. Veblen. 1990. Spruce beetles and fire in the nineteenth century subalpine forests of western Colorado, USA. *Arctic and Alpine Research* 22:65–80.

Barbour, M.G. 1987. Californian Upland Forests and Woodlands. In *North American terrestrial vegetation,* ed. M.G. Barbour and W.D. Billings. Chap. 5. New York: Cambridge University Press.

Barbour, M.G., et al. 1991. Snowpack and the distribution of a major vegetation ecotone in the Sierra Nevada of California. *Biogeography* 18:141–149.

Blair, R. (ed.). 1996. The western San Juan Mountains: their geology, ecology, and human history. Niwot: University Press of Colorado.

Bonan, G.B. and H.H. Shugart. 1989. Environmental factors and ecological processes in boreal forests. *Annual Review of Ecology and Systematics* 20:1–28.

Brunstein, F.C. and D.K. Yamaguchi. 1992. The oldest known Rocky Mountain bristlecone pines (*Pinus aristata* Engelm.). *Arctic and Alpine Research* 24:253–256.

Buell, M.F. and W.A. Niering. 1957. Fir-spruce-birch forest in northern Minnesota. *Ecology* 38:602–610.

Chapman, H.H. 1932. Is the longleaf type a climax? *Ecology* 13:328–334.

Cogbill, C.V. 1977. The effect of acid precipitation on tree growth in eastern North America. *Water, Air, and Soil Pollution* 8:89–93.

Cogbill, C.V. and P.S. White. 1991. The altitude-elevation relationship for spruce-fir forest and treeline along the Appalachian Mountain chain. *Vegetatio* 94:153–175.

Cooper, C.F. 1960. Changes in vegetation, structure, and growth of southwestern pine forests since white settlement. *Ecological Monographs* 30:129–164.

Cooper, S.V., K.E. Neiman, and D.W. Robert. 1991. *Forest habitat types of northern Idaho: a second approximation.* USDA Forest Service, General Technical Report INT-236.

Cooper, W.S. 1957. Vegetation of the northwest American province. *Proceedings, 8th Pacific Science Congress* 4:133–138.

Covington, W.W. and M.M. Moore. 1994. Southwestern ponderosa forest structure: changes since Euroamerican settlement. *Journal of Forestry* 92:39–47.

Curtis, J.T. 1959. *The vegetation of Wisconsin.* Madison: University of Wisconsin Press.

Dansereau, P. and F. Segadas-Vianna. 1952. Ecological study of the peat bogs of eastern North America. *Canadian J. of Botany* 30:490–520.

Daubenmire, R.F. 1943a. Soil temperature versus drought as a factor in determining lower altitudinal limits of trees in the Rocky Mountains. *Botanical Gazette* 105:1–13.

———. 1943b. Vegetational zonation in the Rocky Mountains. *Botanical Review* 9:325–393.

Davis, R.B. 1966. Spruce-fir forests of the coast of Maine. *Ecological Monographs* 36:79–94.

Devall, B. 1993. *Clearcut: the tragedy of industrial forestry.* San Francisco: Sierra Club Books and Earth Island Press.

Dick-Peddie, W.A. 1993. New Mexico vegetation: past, present, and future. Albuquerque: University of New Mexico Press.

D'Itri, F.M. 1982. *Acid precipitation: Effects on ecological systems.* Ann Arbor, MI: Science Publishers.

Elliott-Fisk, D.L. 1987. The boreal forest. In *North American terrestrial vegetation,* eds. M.G. Barbour and W.D. Billings, Chap. 2. New York: Cambridge University Press.

Evans, L.S. 1984. Botanical aspects of acidic precipitation. *Botanical Review* 50:449–490.

Fitzsimmons, M.J. 1995. Conserving the boreal forest by shifting the emphasis of management action from vegetation to the atmosphere. *Water, Air, and Soil Pollution* 82:25–34.

Fonda, R.W. and J.A. Bernardi. 1976. Vegetation of Sucia Island in Puget Sound, Washington. *Bulletin of the Torrey Botanical Club* 103:99–109.

Franklin, J.F. 1987. Pacific northwest forests. In *North American terrestrial vegetation,* eds. M.G. Barbour and W.D. Billings, Chap. 4. New York: Cambridge University Press.

Franklin, J.F. and C.T. Dyrness. 1973. *Natural vegetation of Oregon and Washington.* Portland, OR: USDA Forest Service, General Technical Report PNW-8.

Garren, K.H. 1943. Effects of fire on vegetation of the southeastern United States. *Botanical Review* 9:617–654.

Goff, F.G. and P.H. Fedler. 1968. Structural gradient analysis of upland forests in the western Great Lakes area. *Ecological Monographs* 38:65–86.

Graumlich, L.J. 1993. A 1000-year record of temperature and precipitation in the Sierra Nevada. *Journal of Quaternary Research* 39:249–255.

Habeck, J.R. and R.W. Mutch. 1973. Fire-dependent forests in the northern Rocky Mountains. *J. of Quarternary Research* 3:408–424.

Haines, B.L. 1983. Forest ecosystem SO_4-S input-output discrepancies and acid rain: are they related? *Oikos* 41:139–143.

Hartesveldt, R.J., H.T. Harvey, H.S. Shellhammer, and R.E. Stecker. 1975. *The giant sequoia of the Sierra Nevada.* Washington, D.C.: U.S. Dept. of the Interior, National Park Service Publication no. 120.

Hayward, G.D. and J. Verner (eds.). 1994. *Flammulated, boreal, and great gray owls in the United States.* Fort Collins, CO: USDA Forest Service, General Technical Report RM-253.

Heinselman, M.L. 1963. Forest sites, bog processes, and peatland types in the glacial Lake Aggasiz region, Minnesota. *Ecological Monographs* 33:327–374.

Hutchinson, T.C. and M. Havas (eds.). 1980. *Effects of acid precipitation on terrestrial ecosystems.* New York: Plenum.

Jensen, D.B., M.S. Torn, and J. Harte. 1993. *In our own hands: a strategy for conserving California's biological diversity.* Berkeley: University of California Press.

Jones, E.W. 1945. The structure and reproduction of the virgin forests of the north temperate zone. *The New Phytologist* 44:130–148.

Kallend, A.S., A.R.W. Marsh, J.H. Pickles, and M.V. Proctor. 1983. Acidity of rain in Europe. *Atmospheric Environment* 17:127–137.

Knight, D.H. 1994. *Mountains and plains: the ecology of Wyoming landscapes.* New Haven, CT: Yale University Press.

LaRoi, G.H. 1967. Ecological studies in the boreal spruce-fir forests of the North American taiga. *Ecological Monographs* 37:229–253.

Larsen, J.A. 1930. Forest types of the northern Rocky Mountains and their climatic controls. *Ecology* 11:631–672.

———. 1980. *The boreal ecosystem.* New York: Academic Press.

Lassoie, J.P., T.M. Hinckley, and C.C. Grier. 1985. Coniferous forests of the Pacific northwest. In *Physiological ecology of North American plant communities,* eds. B.F. Chabot and H.A. Mooney, pp. 127–161. New York: Chapman and Hall.

Lavoie, C. and S. Payette. 1994. Recent fluctuations of the lichen-spruce forest limit in subarctic Quebec. *Journal of Ecology* 82:725–734.

Lee, J.J. and D.E. Weber. 1982. Effects of sulfuric acid rain on major cation and sulfate concentrations of water percolating through two model hardwood forests. *J. Environmental Quality* 11:57–64.

Lenihan, J.M. 1993. Ecological response surfaces for North American boreal tree species and their use in forest classification. *Vegetation Science* 4:667–680.

Likens, G.E. and T.J. Butler. 1981. Recent acidification of precipitation in North America. *Atmospheric Environment* 15:1103–1109.

Linthurst, R.A. (ed.). 1984. *Direct and indirect effects of acidic deposition on vegetation.* Acid Precipitation Series, Vol. 5. Boston: Butterworth.

Lowe, C.H. and D.E. Brown. 1973. *The natural vegetation of Arizona.* Arizona Resources Information System Cooperative Publication no. 2. Phoenix: Arizona Resources Information System.

Lutz, H.J. 1956. *Ecological effects of forest fires in the interior of Alaska.* Washington, D.C.: U.S. Dept. of Agriculture Technical Bulletin 1133.

Lynham, T.J. and B.J. Stocks. 1991. The natural fire regime of an unprotected section of the boreal forest in Canada. In *Proceedings: Tall Timbers fire ecology conference,* pp. 99–109. Tallahassee, FL: Tall Timbers Research Station.

Martin, P.H. 1996. Will forest preserves protect temperate and boreal biodiversity from climate change? *Forest Ecology Management* 85:335–341.

Maycock, P.F. 1961. The spruce-fir forests of the Keneenaw Peninsula, northern Michigan. *Ecology* 42:357–365.

McCormick, J. 1970. *The pine barrens.* New Jersey State Museum Report no. 2. Trenton: New Jersey State Museum.

McIntosh, R.P. and R.T. Hurley. 1964. The spruce-fir forests of the Catskill Mountains. *Ecology* 45:314–326.

Minnich, R.A., M.G. Barbour, J.H. Burk, and R.F. Fernau. 1995. Sixty years of change in Californian conifer forests of the San Bernardino Mountains. *Conservation Biology* 9:902–914.

Minnich, R.A., M.G. Barbour, J.H. Burk, and J. Sosa-Ramirez. In press. Californian conifer forests under unmanaged fire regimes in the Sierra San Pedro Martir, Baja California, Mexico. *Biogeography.*

Moss, E.H. 1955. The vegetation of Alberta. *Botanical Review* 21:493–567.

Muir, P.S. 1985. Disturbance history and serotiny of *Pinus contorta* forests in western Montana. *Ecology* 66:1658–1668.

———. 1993. Disturbance effects on structure and tree species composition of *Pinus contorta* forests in western Montana. *Canadian Journal of Forest Research* 23:1617–1625.

Oechel, W.C. and W.T. Lawrence. 1985. Taiga. In *Physiological ecology of North American plant communities,* eds. B.F. Chabot and H.A. Mooney, pp. 66–94. New York: Chapman and Hall.

Oosting, H.J. and W.D. Billings. 1951. A comparison of virgin spruce-fir forests in the northern and southern Appalachian system. *Ecology* 32:84–103.

Oosting, H.J. and J.F. Reed. 1944. Ecological composition of pulpwood forests in northwestern Maine. *American Midland Naturalist* 31:181–210.

———. 1952. Virgin spruce-fir of the Medicine Bow Mountains, Wyoming. *Ecological Monographs* 22:69–91.

Osmond, C.B., L.F. Pitelka, and G.M. Hidy (eds.). 1990. *Plant biology of the Basin and Range.* New York: Springer-Verlag.

Pare, D. and Y. Bergeron. 1995. Above-ground biomass accumulation along a 230-year chronosequence in the southern portion of the Canadian boreal forest. *Journal of Ecology* 83:1001–1007.

Parker, A.J. 1994. Latitudinal gradients of coniferous tree species, vegetation, and climate in the Sierran-Cascade axis of northern California. *Vegetatio* 115:145–155.

Peet, R.K. 1981. Forest vegetation of the Colorado Front Range: composition and dynamics. *Vegetatio* 45:3–75.

———. 1987. Conifer forests of the Rocky Mountains. In *North American terrestrial vegetation,* eds. M.G. Barbour and W.D. Billings, Chap. 3. New York: Cambridge University Press.

Perry, J.P., Jr. 1991. *The pines of Mexico and Central America.* Portland: Timber Press.

Potzger, J.E. 1946. Phytosociology of the primeval forest in central and northern Wisconsin and upper Michigan and a brief postglacial history of the Lake Forest formation. *Ecological Monographs* 16:211–250.

Ramamoorthy, T.P., et al. (eds.). 1993. *Biological diversity of Mexico: origins and distribution.* New York: Oxford University Press.

Rennie, P.J. 1977. Forests, muskeg, and organic terrain in Canada. In *Muskeg and the northern environment in Canada,* eds. N.W. Radforth and C.O. Brawner, pp. 167–207. Toronto: University of Toronto Press.

Ritchie, J.C. 1956. The vegetation of northern Manitoba. *Canadian J. of Botany* 34:523–561.

Robichaud, B. and M.F. Buell. 1973. *Vegetation of New Jersey.* New Brunswick, NJ: Rutgers University Press.

Romme, W.H. and D.G. Despain. 1989. Historical perspective on the Yellowstone fires of 1988. *BioScience* 39:695–699.

Ruggiero, L.F., et al. (eds.). 1991. *Wildlife and vegetation of unmanaged Douglas-fir forests.* Portland, OR: USDA Forest Service, General Technical Report PNW-285.

Rundel, P.W. 1971. Community structure and stability in the giant sequoia groves of the Sierra Nevada. *American Midland Naturalist* 85:478–492.

Rundel, P.W., D.J. Parsons, and D.T. Gordon. 1977. Montane and subalpine vegetation of the Sierra Nevada and Cascade Ranges. In *Terrestrial vegetation of California,* eds. M.G. Barbour and J. Major, pp. 559–599. New York: Wiley-Interscience.

Sawyer, J.O. and T. Keeler-Wolf. 1995. *A manual of California vegetation.* Sacramento: California Native Plant Society.

Scott, J.T., T.G. Siccama, A.H. Johnson, and A.R. Briesch. 1984. Decline of red spruce in the Adirondacks, New York. *Bulletin of the Torrey Botanical Club* 111:438–444.

Segura, G. and L.C. Snook. 1992. Stand dynamics and regeneration patterns of a pinyon pine forest in east-central Mexico. *Forest Ecology and Management* 47:175–194.

Shirley, H.L. 1945. Reproduction of upland conifers in the Lake States as affected by root competition and light. *American Midland Naturalist* 33:537–612.

Shugart, H.H., R. Leemans, and G.B. Bonan (eds.). 1992. *A systems analysis of the global boreal forest.* New York: Cambridge University Press.

Siccama, T.G. 1982. Decline of red spruce in the Green Mountains of Vermont. *Bulletin of the Torrey Botanical Club* 109: 162–168.

Sirois, L., G.B. Bonan, and H.H. Shugart. 1994. Development of a simulation model of the forest-tundra transition zone of northeastern Canada. *Journal of Forest Research* 24:687–706.

Smith, W.K. 1985. Western montane forests. In *Physiological ecology of North American plant communities,* eds. B.F. Chabot and H.A. Mooney, pp. 95–126. New York: Chapman and Hall.

Sierra Nevada Ecosystem Project Science Team (eds.). 1996. *Status of the Sierra Nevada.* Final report to Congress of the SNEP. Davis, CA: Wildland Resources Center, Report no. 36, University of California.

Stålfelt, M.G. 1972. *Stålfelt's plant ecology.* (Translated from *Plants, the soil and man,* Stålfelt, 1960.) London: Longman.

Stallard, H. 1929. Secondary succession in the climax forest formation of northern Minnesota. *Ecology* 10:476–548.

Stevens, G.C. and J.F. Fox. 1991. The causes of treeline. *Annual Review of Ecology and Systematics* 22:177–191.

Stewart, R. 1977. *Labrador.* Amsterdam: Time-Life.

Sturtevant, B.R., J.A. Bissonette, and J.N. Long. 1986. Temporal and spatial dynamics of boreal forest structure in western Newfoundland: silvicultural implications for marten habitat management. *Forest Ecology Management* 87:13–25.

Swetnam, T.W. 1993. Fire history and climatic change in giant sequoia groves. *Science* 262:885–890.

Taylor, D.W. 1977. Floristic relationships along the Cascade-Sierran axis. *American Midland Naturalist* 97:333–349.

Ulrich, B. and J. Pankrath (eds.) 1983. *Effects of accumulate of air pollution in forest ecosystems.* Dordrecht, Holland: Reidel Publishing.

USDA Forest Service. 1992. *Old-growth forests* in the southwest and Rocky Mountain regions. Washington, D.C.: USDA Forest Service, General Technical Report RM-213.

———. 1993. *Supplemental environmental impact statement on management of habitat for late-successional and old-growth forest related species within the range of the northern spotted owl.* Portland, OR: USDA Forest Service, General Technical Report PNW.

———. 1995a. *Biodiversity and management of the Madrean archipelago: The sky islands of southwestern United States and northwestern Mexico.* Washington, D.C.: USDA Forest Service, General Technical Report RM-264.

———. 1995b. *Ecology, diversity, and sustainability of the middle Rio Grande Basin.* Washington, D.C.: USDA Forest Service, General Technical Report RM-268.

———. 1996. *Conference on adaptive ecosystem restoration and management: Restoration of Cordilleran conifer landscapes of North America.* Washington, D.C.: USDA Forest Service, General Technical Report RM-278.

Van Cleve, K. and L.A. Viereck. 1981. Forest succession in relation to nutrient cycling in the boreal forest of Alaska. In *Forest succession, concepts and application,* ed. O.C. West, H.H. Shugart, and D.B. Botkin, pp. 185–211. New York: Springer-Verlag.

Vankat, J.L. 1982. A gradient perspective on the vegetation of Sequoia National Park, California. *Madrono* 29:220–214.

Veblen, T.T. and D.C. Lorenz. 1991. *The Colorado Front Range: a century of ecological change.* Salt Lake City: University of Utah Press.

Velazquez, A., V.M. Toledo, and I. Luna. 1998. Temperate vegetation of Mexico. In *North American terrestrial vegetation,* eds. M.G. Barbour and W.D. Billings. New York: Cambridge University Press.

Verner, J. (ed.). 1992. *The California spotted owl.* Albany, CA: USDA Forest Service, General Technical Report PSW-133.

Vogelmann, H.W., G.J. Badger, M. Bliss, and R.M. Klein. 1985. Forest decline on Camels Hump, Vermont. *Bulletin of the Torrey Botanical Club* 112:274–287.

Walsh, S.S., et al. 1994. Influence of snow patterns and snow avalanches on the alpine treeline ecotone. *Journal of Vegetation Science* 5:657–672.

Wang, Z.M., M.J. Lechowicz, and C. Potvin. 1994. Early selection of black spruce seedlings and global change: which genotypes should we favor? *Ecological Applications* 4:604–616.

Wein, R.W. 1990. The importance of wildfire to climate change: hypotheses for the taiga. In *Fire in ecosystem dynamics, Mediterranean and northern perspectives,* eds. J.G. Golammer and M.J. Jenkins, pp. 185–190. The Hague: SPB Academic Publishers.

Welsh, D.A. and L.A. Venier. 1996. Binoculars and satellites: developing a conservation framework for boreal forest wildlife at varying scales. *Forest Ecology and Management* 85:53–65.

Zinke, P.J. 1977. The redwood forest and associated north coast forests. In *Terrestrial vegetation of California,* eds. M.G. Barbour and J. Major, pp. 679–698. New York: Wiley-Interscience.

Deciduous Forests

Abrams, M.D. 1992. Fire and the development of oak forests. *BioScience* 42:346–353.

Arris, L.L. and P.S. Eagleson. 1989. Evidence of a physiological basis for the boreal-deciduous forest ecotone in North America. *Vegetatio* 82:55–58.

Bakeless, J. 1961. *The eyes of discovery.* New York: Dover.

Bierzychudek, P. 1982. Life histories and demography of shade-tolerant temperate forest herbs: a review. *New Phytologist* 90:757–776.

Biondo, B. 1997. In defense of the longleaf pine. *Nature Conservancy* 47(4):10–17.

Bormann, F.H., T.G. Siccama, G.E. Likens, and R.H. Whittaker. 1970. The Hubbard Brook ecosystem study: composition and dynamics of the tree stratum. *Ecological Monographs* 40:373–388.

Boyce, S.G. 1954. The salt spray community. *Ecological Monographs* 24:29–67.

Braun, E.L. 1950. *Deciduous forests of eastern North America.* Philadelphia: Blakiston.

———. 1955. The phytogeography of the eastern United States and its interpretation. *Botanical Review* 21:297–375.

———. 1957. Development of the deciduous forests of eastern North America. *Ecological Monographs* 17:211–219.

Buell, M.F. and W.E. Martin. 1961. Competition between maple-basswood and spruce-fir communities in Itasca Park, Minnesota. *Ecology* 42:428–429.

Cain, S.A. 1943. The tertiary character of the cove hardwood forests of the Great Smoky Mountains. *Bulletin of the Torrey Botanical Club* 70:213–245.

Castello, J.D., D.J. Leopold, and P.J. Smallidge. 1995. Pathogens, patterns, and processes in forest ecosystems. *BioScience* 45:16–24.

Christensen, N.L. 1987. The vegetation of the coastal plain of the southeastern United States. In *North American terrestrial vegetation,* eds. M.G. Barbour and W.D. Billings, Chap. 11. New York: Cambridge University Press.

Clebsch, E.E.C. and R.T. Busing. 1989. Secondary succession, gap dynamics, and community structure in an Appalachian cove forest. *Ecology* 70:728–735.

Core, E.L. 1966. *Vegetation of West Virginia.* Charleston, WV: McClain, Parsons.

Curtis, J.T. 1959. *The vegetation of Wisconsin.* Madison: University of Wisconsin Press.

Daubenmire, R.F. 1936. The "Big Woods" of Minnesota. *Ecological Monographs* 6:233–268.

Delcourt, H.R. and P.A. Delcourt. 1984. Ice age haven for hardwoods. *Natural History* 93(9):22–28.

———. 1987. *Long-term forest dynamics of the temperate zone.* New York: Springer, Ecological Studies No. 63.

———. 1988. Quaternary landscape ecology: relevant scales in space and time. *Landscape Ecology* 2:23–44.

———. 1993. Paleoclimates, paleovegetation, and paleofloras during the Late Quaternary. In *Flora of North America,* Vol. 1, ed. Flora of North America Editorial Committee, pp. 71–94. New York: Oxford University Press.

Delcourt, H.R., P.A. Delcourt, and T. Webb III. 1983. Dynamic plant ecology: the spectrum of vegetational change in space and time. *Quarterly Science Review* 1:153–175.

Dyksterhuis, E.J. 1948. The vegetation of western Cross Timbers. *Ecological Monographs* 18:325–376.

Eagar, C. and M.B. Adams (eds.). 1992. *Ecology and decline of red spruce in the eastern United States.* New York: Springer-Verlag, Ecological Studies No. 96.

Eyre, S.R. 1963. *Vegetation and soils.* Chicago: Aldine.

Forcier, L.K. 1975. Reproductive strategies and the co-occurrence of climax tree species. *Science* 189:808–810.

Frelich, L.E. and C.G. Lorimer. 1991. Natural disturbance regimes in hemlock-hardwood forests of the upper Great Lakes region. *Ecological Monographs* 61:145–164.

Graham, S.A. 1941. Climax forests of the upper peninsula of Michigan. *Ecology* 22:355–362.

Greller, A.M. 1980. Correlation of some climate statistics with distribution of broadleaved forest zones in Florida, USA. *Bulletin of the Torrey Botanical Club* 107:189–219.

———. 1986. Deciduous forest. In *North American terrestrial vegetation,* eds. M.G. Barbour and W.D. Billings, Chap. 10. New York: Cambridge University Press.

———. 1989. Correlation of warmth and temperateness with the distributional limits of zonal forests in eastern North America. *Bulletin of the Torrey Botanical Club* 116:145–163.

Hicks, D.J. and B.F. Chabot. 1985. Deciduous forest. In *Physiological ecology of North American plant communities,* eds. B.F. Chabot and H.A. Mooney, pp. 257–277. New York: Chapman and Hall.

Holt, P.C. (ed.). 1970. *The distributional history of the biota of the southern Appalachians. Part II: Flora.* Blacksburg: Virginia Polytechnic Institute and State University.

Hutchinson, B.A. and D.R. Matt. 1977. The distribution of solar radiation within a deciduous forest. *Ecological Monographs* 47:185–207.

Iverson, L.R., et al. 1989. The forest resources of Illinois: an atlas and analysis of spatial and temporal trends. *Illinois Natural History Survey, Special Publication* 11:1–181.

Keever, C. 1953. Present composition of some stands of the former oak-chestnut forest in the southern Blue Ridge Mountains. *Ecology* 34:44–54.

————. 1973. Distribution of major forest species in southeastern Pennsylvania. *Ecological Monographs* 43:303–327.

Koyama, H. and S. Kawano. 1973. Biosystematic studies on *Maianthemum* (Liliaceae-Polygonatae). VII. Photosynthetic behavior of *M. dialatatum*. *Botanical Magazine of Tokyo* 86:89–101.

Küchler, A.W. 1964. *Potential natural vegetation of the conterminous United States.* New York: American Geographical Society, Special Publication No. 36.

Lechowicz, M.J. 1984. Why do temperate deciduous trees leaf out at different times? Adaptation and ecology of forest communities. *American Naturalist* 124:821–842.

Ledig, F.T. and D.R. Korbobo. 1983. Adaptation of sugar maple populations along altitudinal gradients: photosynthesis, respiration, and specific leaf weight. *American J. Botany* 70:256–265.

Lutz, H.J. 1930. The vegetation of Heart's Content: a virgin forest in northwestern Pennsylvania. *Ecology* 11:1–29.

Maguire, D.A. and R.T.T. Forman. 1983. Herb cover effects on free seedling patterns in a mature hemlock-hardwood forest. *Ecology* 64:1367–1380.

Mahall, B.E. and F.H. Bormann. 1978. A quantitative description of the vegetative phenology of herbs in a northern hardwood forest. *Botanical Gazette* 139:467–481.

Martin, W.H., S.G. Boyce, and A.C. Echternacht (eds.). 1993a. *Biodiversity of the southeastern United States: lowland terrestrial communities.* New York: Wiley.

———— (eds.). 1993b. *Biodiversity of the southeastern United States: upland terrestrial communities.* New York: Wiley.

Miyawaki, A., K. Iwatsuki, and M.M. Grandtner. 1994. *Vegetation in eastern North America.* Tokyo: University of Tokyo Press.

Monk, C.D. 1965. Southern mixed hardwood forest of north-central Florida. *Ecological Monographs* 35:335–354.

————. 1966. An ecological significance of evergreenness. *Ecology* 47:504–505.

Mowbray, T.B. and H.J. Oosting. 1968. Vegetation gradients in relation to environment and phenology in a southern Blue Ridge gorge. *Ecological Monographs* 38:309–344.

Muller, R.N. and F.H. Bormann. 1976. Role of *Erythronium americanum* Ker. in energy flow and nutrient dynamics of a northern hardwood forest ecosystem. *Science* 193:1126–1128.

Oosting, H.J. 1942. An ecological analysis of the plant communities of the Piedmont, North Carolina. *American Midland Naturalist* 28:1–126.

Oosting, H.J. and P.F. Bourdeau. 1955. Virgin hemlock forest segregates in the Joyce Kilmer Memorial Forest of western North Carolina. *Botanical Gazette* 116:340–359.

————. 1959. The maritime live oak forest in North Carolina. *Ecology* 40:148–152.

Paillet, F.L. 1982. The ecological significance of American chestnut (*Castanea dentata* (Marsh.) Borkh.) in the Holocene forests of Connecticut. *Bulletin of the Torrey Botanical Club* 109:457–473.

Potzger, J.E. 1946. Phytosociology of the primeval forest of central-northern Wisconsin and upper Michigan. *Ecological Monographs* 16:211–250.

Potzger, J.E., M.E. Potzger, and J. McCormick. 1956. The forest primeval of Indiana as recorded in the original land surveys and an evaluation of previous interpretations of Indiana vegetation. *Butler University Botanical Studies* 13:95–111.

Quarterman, E. and C. Keever. 1962. Southern mixed hardwood forest: climax in the southeastern coastal plain, USA. *Ecological Monographs* 32:167–185.

Rice, E.L. and W.T. Penfound. 1959. The upland forests of Oklahoma. *Ecology* 40:593–607.

Robichaud, B. and M.F. Buell. 1973. *Vegetation of New Jersey.* New Brunswick: Rutgers University Press.

Rogers, R.S. 1980. Hemlock stands from Wisconsin to Nova Scotia: transitions in understory composition along a floristic gradient. *Ecology* 61:178–193.

————. 1981. Mature mesophytic hardwood forest: community transitions, by layer, from east-central Minnesota to southeastern Michigan. *Ecology* 62:1634–1647.

Rohrig, E. and B. Ulrich (eds.). 1991. *Ecosystems of the world, part 7: temperate deciduous forests.* Amsterdam: Elsevier.

Shane, L.C.K. and E.J. Cushing (eds.). 1991. *Quaternary landscapes.* Minneapolis: University of Minnesota Press.

Siccama, T.G. 1974. Vegetation, soil, and climate on the Green Mountains of Vermont. *Ecological Monographs* 44:325–349.

Siccama, T.G., F.H. Bormann, and G.E. Likens. 1970. The Hubbard Brook ecosystem study: productivity, nutrients, and phytosociology of the herbaceous layer. *Ecological Monographs* 40:389–402.

Skeen, J.N., M.E.B. Carter, and H.L. Ragsdale. 1980. Yellow-poplar: the Piedmont case. *Bulletin of the Torrey Botanical Club* 107:1–6.

Transeau, E.N. 1935. The prairie peninsula. *Ecology* 16:423–437.

USDA Forest Service. 1969. *A forest atlas of the south.* Southern and Southeastern Forest Experiment Stations.

Vogelmann, H.W., T.G. Siccama, D. Leedy, and D.C. Ovitt. 1968. Precipitation from fog moisture in the Green Mountains of Vermont. *Ecology* 49:1205–1207.

Whitney, G.G. 1990. *From coastal wilderness to fruited plain: a history of environmental change in temperate North America from 1500 to the present.* New York: Cambridge University Press.

Whittaker, R.H. 1956. Vegetation of the Great Smoky Mountains. *Ecological Monographs* 26:1–80.

Williams, A.B. 1936. The composition and dynamics of a beech-maple climax forest (Ohio). *Ecological Monographs* 6:319–408.

Williams, M. 1989. *Americans and their forests: an historical geography.* New York: Cambridge University Press.

Woods, F.W. and R.E. Shanks. 1959. Natural replacements of chestnut by other species in the Great Smoky Mountains National Park. *Ecology* 40:349–361.

Wright, H.E., Jr., et al. 1993. *Global climates since the last glacial maximum.* Minneapolis: University of Minnesota Press.

Grasslands

Adams, D.E. and L.L. Wallace. 1985. Nutrient and biomass allocation in five grass species in an Oklahoma tallgrass prairie. *American Midland Naturalist* 113:170–181.

Albertson, F.W. 1937. Ecology of mixed prairie in west central Kansas. *Ecological Monographs* 7:481–547.

Albertson, F.W. and J.E. Weaver. 1947. Reduction of ungrazed mixed prairie to short grass as a result of drought and dust. *Ecological Monographs* 16:449–463.

Anderson, R.C., B.A.D. Hetrick, and G.W.T. Wilson. 1994. Mycorrhizal dependence of *Andropogon gerardii* and *Schizachyrium scoparium* in two prairie soils. *American Midland Naturalist* 132:366–376.

Ayyad, M.A.G. and R.L. Dix. 1964. An analysis of a vegetation-microenvironmental complex on prairie slopes in Saskatchewan. *Ecological Monographs* 34:421–442.

Barry, W.J. 1972. *The central valley prairie.* Sacramento, CA: Dept. of Parks and Recreation.

Beetle, A.A. 1974. Distribution of the native grasses of California. *Hilgardia* 9:309–357.

Bogusch, E.R. 1952. Brush invasion in the Rio Grande Plain of Texas. *Texas J. of Science* 4:85–91.

Borchert, J.R. 1950. The climate of the central North American grassland. *Annals of the Association of American Geographers* 40:1–39.

Branson, F.A. 1985. *Vegetation changes on western rangelands.* Denver: Society for Range Management.

Brown, A.L. 1950. Shrub invasion of southern Arizona desert grassland. *J. of Range Management* 3:172–177.

Buffington, L.C. and C.H. Herbel. 1965. Vegetational changes on a semidesert grassland range from 1858 to 1963. *Ecological Monographs* 35:139–164.

Carpenter, J.R. 1940. The grassland biome. *Ecological Monographs* 10:617–684.

Collins, S.L. and L. Wallace (eds.). 1990. *Fire in North American tallgrass prairies.* Norman: Oklahoma University Press.

Coupland, R.T. 1961. A reconsideration of grassland classification in the northern Great Plains of North America. *J. of Ecology* 49:135–167.

———— (ed.). 1992. *Natural grasslands: introduction and western hemisphere.* Ecosystems of the world, Vol. 81. New York: Elsevier.

D'Antonio, C.M. and P.M. Vitousek. 1992. Biological invasions by exotic grasses, the grass/fire cycle, and global change. *Annual Review of Ecology and Systematics* 23:63–87.

Daubenmire, R.F. 1968. Ecology of fire in grasslands. *Advances in Ecological Research* 5:209–266.

Diamond, D.D. and F.E. Smeins, 1985. Composition, classification, and species response patterns of remnant tallgrass prairie in Texas. *American Midland Naturalist* 113:294–308.

Dyksterhuis, E.J. 1946. The vegetation of the Ft. Worth prairie. *Ecological Monographs* 16:1–29.

Ellison, L. 1960. Influence of grazing on plant succession of rangelands. *Botanical Review* 26:1–78.

Estes, J.R., R.J. Tyrl, and J.N. Brunken (eds.). 1982. *Grasses and grasslands, systematics and ecology.* Norman: University of Oklahoma Press.

Franklin, J.F. and C.T. Dyrness. 1973. *Natural vegetation of Oregon and Washington.* Portland, OR: USDA Forest Service, General Technical Report PNW-8.

Gardner, J.L. 1951. Vegetation of the creosote bush area of the Rio Grande Valley in New Mexico. *Ecological Monographs* 21:379–403.

Gay, C.W. Jr. and D.D. Dwyer. 1965. *New Mexico range plants.* Las Cruces: New Mexico State University, Service Circular 374. Service Cooperative Extension.

Gibbens, R.P., J.M. Tromble, J.T. Hennessy, and M. Cardenas. 1983. Soil movement in mesquite dunelands and former grasslands of southern New Mexico from 1933 to 1980. *J. Range Management* 36:145–148, 370–374.

Gleason, H.A. 1923. The vegetational history of the middle west. *Annals of the Association of American Geographers* 12:39–85.

Hadley, E.B. and R.P. Buccos. 1967. Plant community composition and net primary production within a native eastern North Dakota prairie. *American Midland Naturalist* 77:116–127.

Harris, G.A. 1967. Some competitive relationships between *Agropyron spicatum* and *Bromus tectorum. Ecological Monographs* 37:89–111.

Hastings, J.R. and R.M. Turner. 1965. *The changing mile.* Tucson: University of Arizona Press.

Heady, H.F. 1977. Valley grassland. In *Terrestrial vegetation of California,* eds. M.G. Barbour and J. Major, pp. 491–514. New York: Wiley-Interscience.

Heady, H.F. and R.D. Child. 1994. *Rangeland ecology and management.* Boulder, CO: Westview Press.

Herkert, J.R. 1994. The effects of habitat fragmentation on midwestern grassland bird communities. *Ecological Applications* 4:461–471.

Hitchcock, A.S. 1950. *Manual of the grasses of the United States,* 2nd ed. Washington, D.C.: USDA Forest Service, Miscellaneous Publication 200.

Humphrey, R.R. 1953. The desert grassland: a history of vegetational change and an analysis of causes. *Botanical Review* 24:193–252.

Humphrey, R.R. and L.A. Mehrhoff. 1958. Vegetation changes on a southern Arizona grassland range. *Ecology* 39:720–726.

Johnston, M.C. 1963. Past and present grasslands of southern Texas and N.E. Mexico. *Ecology* 44:456–466.

Jones, C.H. 1944. Vegetation of Ohio prairies. *Bulletin of the Torrey Botanical Club* 71:536–548.

Joyce, L.A. 1993. The life cycle of the range condition concept. *J. of Range Management* 46:132–138.

Knapp, A.K., et al. 1993. Landscape patterns in soil-plant-water relations and primary production in tallgrass prairie. *Ecology* 74:549–560.

Knopf, F.L. 1994. Avian assemblages on altered grasslands. *Studies in Avian Biology* 15:247–257.

Kucera, C.L., R.C. Dahlman, and M.R. Koelling. 1967. Total net productivity and turnover on an energy basis for tallgrass prairie. *Ecology* 48:536–541.

Larson, F. 1940. The role of the bison in maintaining the short grass plains. *Ecology* 21:113–121.

Lauenroth, W.K. and W.A. Laycock (eds.). 1989. *Secondary succession and the evaluation of rangeland condition.* Boulder, CO: Westview Press.

Looman, J. 1980. The vegetation of the Canadian Prairie Provinces, II. The grasslands, part I. *Phytocoenologia* 8:153–190.

———. 1983. Distribution of plant species and vegetation types in relation to climate. *Vegetatio* 54:17–25.

Mack, R.N. 1981. The invasion of *Bromus tectorum* L. into western North America: an ecological chronicle. *Agro-Ecosystems* 7:145–165.

McMillan, D. 1959. The role of ecotypic variation in the distribution of the central grassland of North America. *Ecological Monographs* 29:285–308.

Ojima, D.S., et al. 1994. Long- and short-term effects of fire on nitrogen cycling in tallgrass prairie. *Biogeochemistry* 2:67–84.

Old, S.M. 1969. Microclimates, fire, and plant production in an Illinois prairie. *Ecological Monographs* 39:355–384.

Polley, H.W., H.B. Johnson, and H.S. Mayeux. 1994. Increasing CO_2: comparative response of the C_4 grass *Schizachyrium* and the grassland invader *Prosopis*. *Ecology* 75:976–988.

Redmann, R.E. 1975. Production ecology of grassland plant communities in western North Dakota. *Ecological Monographs* 45:83–106.

Rice, E.L. and R.L. Parenti. 1978. Causes of decreases in productivity in undisturbed tall grass prairie. *American J. of Botany* 65:1091–1097.

Risser, P.G. 1985. Grasslands. In *Physiological ecology of North American plant communities,* eds. B.F. Chabot and H.A. Mooney, pp. 232–256. New York: Chapman and Hall.

———. 1988. Diversity in and among grasslands. In *Biodiversity,* ed. E.O. Wilson. Washington, D.C.: National Academy Press.

Sampson, F. and F. Knopf. 1994. Prairie conservation in North America. *BioScience* 44:418–421.

Schimel, D.S., et al. 1991. Physiological interactions along resource gradients in a tallgrass prairie. *Ecology* 72:672–684.

Schramm, P. 1990. Prairie restoration: a 25-year perspective on establishment and management. In *Proceedings of the twelfth North American Prairie Conference,* pp 169–177. Cedar Falls, IA.

Simms, P.L., W.A. Berg, and J.A. Bradford. 1995. Vegetation of sandhills under grazed and ungrazed conditions. In *Proceeding of the fourteenth North American Prairie Conference,* pp 129–135. Manhattan: Kansas State University.

Simms, P.L. and P. Risser. 1998. Grasslands. In *North American terrestrial vegetation,* 2nd ed., eds. M.G. Barbour and W.D. Billings. New York: Cambridge University Press.

Simms, P.L., J.S. Singh, and W.K. Lauenroth. 1978. The structure and function of ten western North American grasslands. *J. of Ecology* 66:251–285.

Singh, J.S., W.K. Lauenroth, R.K. Heitschmidt, and J.L. Dodd. 1983. Structural and functional attributes of the vegetation of northern mixed prairie of North America. *Botanical Review* 49:117–149.

Sprague, H.B. (ed.). 1959. *Grasslands.* Washington, D.C.: American Association for the Advancement of Science.

Teeri, J.A. and L.G. Stowe. 1976. Climatic patterns and the distribution of C_4 grasses in North America. *Oecologia* 23:1–12.

Tester, J.R. 1989. Effects of fire frequency on oak savanna in east-central Minnesota. *Bulletin of the Torrey Botanical Club* 116:134–144.

Tilman, D. and J.A. Downing. 1994. Biodiversity and stability in grasslands. *Nature* 367:363–365.

Weaver, J.E. 1954. *The North American prairie.* Lincoln, NB: Johnsen.

Weaver, J.E. and F.W. Albertson. 1956. *Grasslands of the Great Plains.* Lincoln, NB: Johnsen.

Weaver, J.E. and T.J. Fitzpatrick. 1932. Ecology and relative importance of the dominants of tall-grass prairie. *Botanical Gazette* 93:113–150.

———. 1934. The prairie. *Ecological Monographs* 4:109–295.

West, N.E. 1993. Biodiversity of rangelands. *Journal of Range Management* 46:2–13.

White, D. 1941. Prairie soil as a medium for tree growth. *Ecology* 22:399–407.

Williams, W.A. 1966. Range improvements as related to net productivity, energy flow, and foliage configuration. *J. of Range Management* 19:29–34.

Young, J.A., R.A. Evans, and J. Major. 1977. Sagebrush steppe. In *Terrestrial vegetation of California*, eds. M.G. Barbour and J. Major, pp. 762–796. New York: Wiley-Interscience.

Young, J.A., R.A. Evans, and P.T. Tueller. 1975. Great Basin plant communities—pristine and grazed. In *Holocene climates in the Great Basin*, ed. R. Elston, pp. 186–215. Reno: Nevada Archeological Survey Occasional Paper.

Zak, D.R., et al. 1994. Plant production and soil microorganisms in late successional ecosystems: a continental-scale study. *Ecology* 75:2333–2347.

Desert Scrub

Adams, S., B.R. Strain, and M.S. Adams. 1970. Water-repellent soil, fire, and annual plant cover in a desert scrub community of southeastern California. *Ecology* 51:696–700.

Anderson, D.J. 1971. Pattern in desert perennials. *J. of Ecology* 59:555–560.

Axelrod, D.I. 1950. Evolution of desert vegetation. In Carnegie Institution of Washington Publication no. 590, pp. 215–306. Washington, D.C.: Carnegie Institution of Washington.

———. 1959. Evolution of the Madro-Tertiary geoflora. *Botanical Review* 24:433–509.

———. 1967. Drought, diastrophism, and quantum evolution. *Evolution* 21:201–209.

———. 1972. Edaphic aridity as a factor in angiosperm evolution. *American Naturalist* 106:311–320.

Bahre, C.J. and M.L. Shelton. 1993. Historic vegetation change, mesquite increases, and climate in southeastern Arizona. *Journal of Biogeography* 20:489–504.

Barbour, M.G. 1969. Age and space distribution of the desert shrub *Larrea divaricata*. *Ecology* 50:679–685.

———. 1973. Desert dogma re-examined: root/shoot productivity and plant spacing. *American Midland Naturalist* 89:41–57.

Beatley, J.C. 1969. Biomass of desert winter annual plant populations in southern Nevada. *Oikos* 20:261–273.

———. 1974. Phenological events and their environmental triggers in Mojave Desert ecosystems. *Ecology* 55:856–863.

———. 1975. Climates and vegetation pattern across the Mojave/Great Basin Desert transition of southern Nevada. *American Midland Naturalist* 93:53–70.

Benson, L. and R.A. Darrow. 1954. *The trees and shrubs of the southwestern deserts*. Tucson: University of Arizona Press.

Benzioni, A. and R.L. Dunstone. 1986. Jojoba: adaptation to environmental stress and the implications for domestication. *Quarterly Review of Biology* 61:177–199.

Betancourt, J.L., T.R. Van Devender, and P.S. Martin (eds.). *Packrat middens, the last 40,000 years of biotic change*. Tucson: University of Arizona Press.

Billings, W.D. 1949. The shadescale vegetation zone of Nevada and eastern California in relation to climate and soils. *American Midland Naturalist* 42:87–109.

———. 1990. *Bromus tectorum*, a biotic cause of ecosystem impoverishment in the Great Basin. In *The earth in transition*, ed. G.M. Woodwell, pp. 301–322. New York: Cambridge University Press.

Björkman, O., R.W. Pearcy, A.T. Harrison, and H.A. Mooney. 1972. Photosynthetic adaptation to high temperatures: a field study in Death Valley, California. *Science* 175:786–789.

Brandt, C.A. and W.H. Rickard. 1994. Alien taxa in the North American shrub-steppe four decades after cessation of livestock. *Biological Conservation* 68:95–105.

Brisson, J. and J.F. Reynolds. 1994. The effect of neighbors on root distribution in a creosotebush (*Larrea tridentata*) population. *Ecology* 75:1693–1702.

Burk, J.H. 1977. Sonoran desert. In *Terrestrial vegetation of California*, eds. M.G. Barbour and J. Major, pp. 869–889. New York: Wiley-Interscience.

Busch, D.E. and S.D. Smith. 1995. Mechanisms associated with decline of woody riparian ecosystems of the southwestern U.S. *Ecological Monographs* 65:347–370.

Caldwell, M. 1985. Cold desert. In *Physiological ecology of North American plant communities*, eds. B.F. Chabot and H.A. Mooney, pp. 198–212. New York: Chapman and Hall.

Cannon, W.A. 1917. Relation of the rate of root growth in seedlings of *Prosopis velutina* to the temperature of the soil. *Carnegie Institution of Washington Yearbook* no. 18.

Capon, B. and W. Van Asdall. 1967. Heat pretreatment as a means of increasing germination of desert annual seeds. *Ecology* 48:305–306.

Chew, R.M. and A.E. Chew. 1965. The primary productivity of a desert shrub (*Larrea tridentata*) community. *Ecological Monographs* 35:353–375.

Chew, R.M. and W.G. Whitford. 1992. A long-term positive effect of kangaroo rats (*Dipodomys spectabilis*) on creosote-bushes (*Larrea tridentata*). *Journal of Arid Environments* 22:375–386.

Clary, W.P. 1975. Ecotypic variation in *Sitanion hystrix*. *Ecology* 56:1407–1415.

Cloudsley-Thompson, J.L. and M.J. Chadwick. 1964. *Life in deserts*. Philadelphia: Dufour.

Cody, M.L. 1989. Growth-form diversity and community structure in desert plants. *Journal of Arid Environments* 17:199–209.

Comstock, J.P. and J.R. Ehleringer. 1992. Plant adaptation in the Great Basin and Colorado Plateau. *Great Basin Naturalist* 52:195–215.

Cooke, R.U. and A. Warren. 1973. *Geomorphology in deserts*. Berkeley: University of California Press.

Danin, A. 1996. *Plants of desert dunes*. New York: Springer-Verlag.

D'Antonio, C.M. and P.M. Vitousek. 1992. Biiological invasions by exotic grasses, the grass/fire cycle, and global changes. *Annual Review of Ecology and Systematics* 23:63–87.

Ehleringer, J. 1985. Annuals and perennials of warm deserts. In *Physiological ecology of North American plant communities*, eds. B.F. Chabot and H.A. Mooney, pp. 162–180. New York: Chapman and Hall.

Ezcurra, E., M. Equihua, and J. Lopez-Portillo. 1987. The desert vegetation of El Pinacate, Sonora, Mexico, *Vegetatio* 71:49–60.

Flowers, S. 1934. Vegetation of the Great Salt Lake region. *Botanical Gazette* 95:353–418.

Freas, K.E. and P.R. Kemp. 1983. Some relationships between environmental reliability and seed dormancy in desert annual plants. *J. Ecology* 71:211–217.

Gardner, J.L. 1951. Vegetation of the creosote bush area of the Rio Grande Valley in New Mexico. *Ecological Monographs* 21:379–403.

Gates, D.H., L.A. Stoddart, and C.W. Cook. 1956. Soil as a factor influencing plant distribution on salt deserts of Utah. *Ecological Monographs* 26:155–175.

Gibson, A.C. and P.S. Nobel. 1986. *The cactus primer*. Cambridge, MA: Harvard University Press.

Guo, Q., et al. 1995. The effects of vertebrate granivores and folivores on plant community structure in the Chihuahuan desert. *Oikos* 73:251–259.

Harper, K.T. and R.L. Pendleton. 1993. Cyanobacteria and cyanolichens: can they enhance availability of essential minerals for higher plants? *Great Basin Naturalist* 53:59–72.

Harper, K.T., et al. 1994. *Natural history of the Colorado Plateau and Great Basin*. Boulder: University Press of Colorado.

Hastings, J.R., R.M. Turner, and D.K. Warren. 1972. *An atlas of some plant distributions in the Sonoran desert*. Tucson: University of Arizona Institute of Atmospheric Physics Technical Report on the Meteorology and Climatology of Arid Regions no. 21.

Hunt, C.B. 1966. *Plant ecology of Death Valley, California*. Washington, D.C.: U.S. Geological Survey Professional Paper no. 509.

Jaeger, E.C. 1957. *The North American deserts*. Stanford, CA: Stanford University Press.

Jensen, M.E., G.H. Simonson, and M. Dosskey. 1990. Correlation between soils and sagebrush-dominated plant communities of northeastern Nevada. *Soil Science Society of America* 54:902–910.

Kelt, D.A. and T.J. Valone. 1995. Effects of grazing on the abundance and diversity of annual plants in Chihuahuan desert scrub habitat. *Oecologia* 103:191–195.

Kemp, P.R. 1983. Phenological patterns of Chihuahuan desert plants in relation to the timing of water availability. *J. Ecology* 71:427–436.

Lowe, C.H. and D.E. Brown. 1973. *The natural vegetation of Arizona*. Phoenix: Arizona Resources Information System Cooperative Publication no. 2.

Ludwig, J.A. 1987. Primary productivity in arid lands: myths and realities. *Journal of Arid Environments* 13:1–7.

Ludwig, J.A., G.L. Cunningham, and P.D. Whitson. 1988. Distribution of annual plants in North American deserts. *Journal of Arid Environments* 15:221–227.

Lunt, O.R., J. Letey, and S.B. Clark. 1973. Oxygen requirements for root growth in three species of desert shrubs. *Ecology* 54:1356–1362.

Mabry, T.J., J.H. Hunziker, and D.R. Difeo, Jr. 1977. *Creosote bush*. Stroudsburg, PA: Dowden, Hutchinson, and Ross.

MacMahon, J.A. 1979. North American deserts: their floral and faunal components. In *Arid-land ecosystems*, Vol. 1, eds. D.W. Goodall and R.A. Perry, pp. 21–82. New York: Cambridge University Press.

MacMahon, J.A. and F.H. Wagner. 1985. The Mojave, Sonoran, and Chihuahuan deserts of North America. In *Hot deserts and arid shrublands*, eds. M. Evenari et al., pp. 105–202. Amsterdam: Elsevier Science.

Marks, J.B. 1950. Vegetation and soil relations in the lower Colorado desert. *Ecology* 31:176–193.

McAuliffe, J.R. 1988. Markovian dynamics of simple and complex desert plant communities. *American Naturalist* 131:459–490.

McClaran, M.P. and T.R. Van Devender (eds.). 1995. *The desert grassland*. Tucson: University of Arizona Press.

McGinnies, W.G., B.J. Goldman, and P. Paylore (eds.). 1968. *Deserts of the world.* Tucson: University of Arizona Press.

Miller, R.F. and P.F. Wigand. 1994. Holocene changes in semiarid pinyon-juniper woodlands. *BioScience* 44:465–474.

Mitchell, J.E., N.E. West, and R.W. Miller. 1966. Soil physical properties in relation to plant community patterns in the shadscale zone of northwestern Utah. *Ecology* 47:627–630.

Monsen, S.B. and S.G. Kitchen (eds.). 1994. *Proceedings of the symposium on ecology and management of annual rangelands.* Ogden, UT: USDA Forest Service, General Technical Report INT-313.

Montana, C. 1990. A floristic-structural gradient related to land forms in the southern Chihuahuan desert. *Journal of Vegetation Science* 1:669–674.

Mozingo, H. 1986. *Shrubs of the Great Basin.* Reno: University of Nevada Press.

Muller, W.H. and C.H. Muller. 1956. Association patterns involving desert plants that contain toxic products. *American J. of Botany* 43:354–361.

Mulroy, T.W. and P.W. Rundel. 1977. Annual plants: adaptations to desert environments. *BioScience* 27:109–114.

Nabhan, G.P. and J.L. Carr (eds.). 1994. Ironwood: an ecological and cultural keystone species of the Sonoran desert. *Occasional Papers in Conservation Biology* no. 1.

Niering, W.A., R.H. Whittaker, and C.H. Lowe. 1963. The saguaro: a population in relation to its environment. *Science* 142:15–23.

Nobel, P.S. 1985. Desert succulents. In *Physiological ecology of North American plant communities,* eds. B.F. Chabot and H.A. Mooney, pp. 181–197. New York: Chapman and Hall.

Odening, W.R., B.R. Strain, and W.C. Oechel. 1974. The effect of decreasing water potential on net CO_2 exchange of intact desert shrubs. *Ecology* 55:1086–1095.

Oppenheimer, H.R. 1960. Adaptation to drought: xerophytism. In *Plant-water relationships in arid and semi-arid conditions, reviews of research,* pp. 105–138. Paris: UNESCO.

Orians, G.H. and O.T. Solbrig. 1977. *Convergent evolution in warm deserts.* Stroudsburg, PA: Dowden, Hutchinson, and Ross.

Pake, C.E. and D.L. Venable. 1996. Seed banks in desert annuals: implications for persistence and coexistence in variable environments. *Ecology* 77:1427–1435.

Pavlik, B.M. 1989. Phytogeography of sand dunes in the Great Basin and Mojave deserts. *Journal of Biogeography* 16:227–238.

Pavlik, P.M. and M.G. Barbour. 1988. Demographic monitoring of endemic sand dune plants, Eureka Valley, California. *Biological Conservation* 46:217–242.

Pearson, L.C. 1966. Primary productivity in a northern desert area. *Oikos* 15:211–228.

Peinado, M., et al. 1995a. Major plant communities of warm North American deserts. *Journal of Vegetation Science* 6:79–94.

Peinado, M., et al. 1995b. Shrubland formations and associations in mediterranean-desert transitional zones of northwestern Baja California. *Vegetatio* 117:165–179.

Polis, G.A. (ed.). 1991. *The ecology of desert communities.* Tucson: University of Arizona Press.

Polunin, N. (ed.). 1986. *Evolution of life-forms.* New York: Wiley.

Rasmuson, D.E., J.E. Anderson, and N. Huntly. 1994. Coordination of branch orientation and photosynthetic physiology in the Joshua tree (*Yucca brevifolia*). *Great Basin Naturalist* 54:204–211.

Rickard, W.H. and J.R. Murdock. 1963. Soil moisture and temperature survey of a desert vegetation mosaic. *Ecology* 44:821–824.

Rogers, G.G. 1982. *Then and now: a photographic history of vegetation change in the central Great Basin desert.* Salt Lake City: University of Utah Press.

Rosentreter, R. and R.G. Kelsey. 1991. Xeric big sagebrush, a new subspecies in the *Artemisia tridentata* complex. *J. of Range Management* 44:330–335.

Rundel, P.W. and A.C. Gibson. 1996. *Ecological communities and processes in a Mojave desert ecosystem: Rock Valley, Nevada.* New York: Cambridge University Press.

Rzedowski, J. 1978. *Vegetación de México.* Mexico City: Editorial Limusa.

Schlesinger, W.H., et al. 1996. On the spatial pattern of soil nutrients in desert ecosystems. *Ecology* 77:364–374.

Schmidt, R.H. 1989. The arid zones of Mexico: climatic extremes and conceptualization of the Sonoran desert. *Journal of Arid Environments* 16:241–256.

Shantz, H.L. and R.L. Piemeisel. 1924. Indicator significance of the natural vegetation of the southwestern desert region. *J. of Agricultural Research* 28:721–802.

Sharp, L.A., K. Sanders, and N. Rimbey. 1990. Forty years of change in a shadscale stand in Idaho. *Rangelands* 12:213–228.

Shreve, F. 1942. The desert vegetation of North America. *Botanical Review* 8:195–246.

Shreve, F. and A.L. Hinkley. 1937. Thirty years of change in desert vegetation. *Ecology* 18:463–478.

Shreve, F. and I.L. Wiggins. 1964. *Vegetation and flora of the Sonoran desert.* Stanford, CA: Stanford University Press.

Smith, S.D., et al. 1995. Soil-plant water relations in a Mojave desert mixed shrub community: a comparison of three geomorphic surfaces. *Journal of Arid Environments* 29:339–351.

Stebbins, R.C. 1974. Off-road vehicles and the fragile desert. *The American Biology Teacher* 36:203–208, 294–304.

Steenbergh, W.F. and C.H. Lowe. 1969. Critical factors during the first years of life of the saguaro (*Cereus giganteus*) at Saguaro National Monument, Arizona. *Ecology* 50:825–834.

Stocker, O. 1960. Physiological and morphological changes in plants due to water deficiency. In *Plant-water relationships in arid and semi-arid conditions, reviews of research*, pp. 63–104. Paris: UNESCO.

Syvertsen, J.P., G.L. Nickell, R.W. Spellenberg, and G.L. Cunningham. 1976. Carbon reduction pathways and standing crop in three Chihuahuan desert plant communities. *Southwestern Naturalist* 21:311–320.

Trimble, S. 1989. *The sagebrush ocean: a natural history of the Great Basin.* Reno: University of Nevada Press.

Tueller, P.T., et al. 1979. *Pinyon-juniper woodlands of the Great Basin: distribution, flora, vegetal cover.* Ogden, UT: USDA Forest Service, Research Paper INT-229.

Turner, R.W. 1963. Growth in four species of Sonoran desert trees. *Ecology* 44:760–765.

———. 1990. Long-term vegetation change at a fully protected Sonoran Desert site. *Ecology* 71:464–477.

Valentine, K.A. and J.J. Norris. 1964. A comparative study of soils of selected creosote bush sites in southern New Mexico. *J. of Range Management* 17:23–32.

Van Devender, T.R., R.S. Thompson, and J.L. Betancourt. 1987. Vegetation history of the deserts of southwestern North America. In *The geology of North America*, vol. K-3, eds. W.F. Ruddiman and H.E. Wright, pp. 323–352. Boulder, CO: Geological Society of America.

Vasek, F.C. and M.G. Barbour. 1977. Mojave desert scrub vegetation. In *Terrestrial vegetation of California*, eds. M.G. Barbour and J. Major, pp. 835–867. New York: Wiley-Interscience.

Vasek, F.C., H.B. Johnson, and G.D. Brum. 1975. Effects of power transmission lines on vegetation of the Mojave desert. *Madroño* 23:114–130.

Vasek, F.C., H.B. Johnson, and D.H. Eslinger. 1975. Effects of pipeline construction on creosote bush scrub vegetation of the Mojave desert. *Madroño* 23:1–13.

Vogl, R.J. and L.T. Mchargue. 1966. Vegetation of California fan palm oases on the San Andreas fault. *Ecology* 47:532–540.

Wallace, A. and E.M. Romney. 1972. *Radioecology and ecophysiology of desert plants at the Nevada Test Site.* Washington, D.C.: U.S. Atomic Energy Commission, TID-25954.

Went, F.W. 1942. The dependence of certain annual plants on shrubs in southern California deserts. *Bulletin of the Torrey Botanical Club* 69:100–114.

———. 1949. Ecology of desert plants. II. The effect of rain and temperature on germination and growth. *Ecology* 30:1–13.

Went, F.W. and M. Westergaard. 1949. Ecology of desert plants. III. Development of plants in the Death Valley National Monument, California. *Ecology* 30:26–38.

Werger, M.J.A., et al. (eds.). 1988. *Plant growth-form strategies and vegetation types in arid environments.* The Hague: SPB Academic Publishing.

West, N.E. 1987. Intermountain deserts, shrubsteppes, and woodlands. In *North American terrestrial vegetation*, eds. M.G. Barbour and W.D. Billings, Chap. 7. New York: Cambridge University Press.

———. Structure and function of microphytic soil crusts in wildland ecosystems of arid to semiarid regions. *Advances in Ecological Research* 20:180–223.

West, N.E. and K.I. Ibrahim. 1968. Soil-vegetation relationships in the shadscale zone of southeastern Utah. *Ecology* 49:445–456.

West, N.E. and M.M. Caldwell. 1983. Snow as a factor in salt desert shrub vegetation patterns in Curlew Valley, Utah. *American Midland Naturalist* 109:376–379.

Whitehead, E.E., et al. (eds.). 1989. *Arid lands: today and tomorrow.* Tucson: University of Arizona Press.

LITERATURE CITED

Aber, J. D., A. Magill, R. Boone, J. M. Melillo, P. A. Steudler, and R. Bowden. 1993b. Plant and soil responses to chronic nitrogen additions at The Harvard Forest, Massachusetts. *Ecological Applications* 3(1): 156–166.

Aber, J. D., A. Magill, S. G. McNulty, R. D. Boone, K. J. Nadelhoffer, M. Downs, and R. Hallett. 1995. Forest biogeochemistry and primary production altered by nitrogen saturation. *Water, Air, and Soil Pollution* 85: 1665–1670.

Aber, J. D., C. Driscoll, C. A. Federer, R. Lathrop, G. Lovett, J. M. Melillo, P. Steudler, and J. Vogelmann. 1993a. A strategy for the regional analysis of the effects of physical and chemical climate change on biogeochemical cycles in notheastern (U.S.) forests. *Ecological Modelling* 67:37–47.

Aber, J.D. and J.M. Melillo. 1991. *Terrestrial ecosystems*. Philadelphia: Saunders College Publishing.

Aber, J. D., J. M. Melillo, and C. A. Federer. 1982. Predicting the effects of rotation length, harvest intensity, and fertilization on fiber yield from Northern Hardwood forests in New England. *Forest Sci.* 28 (1): 31–45.

Aber, J. D., J. M. Melillo, and C. A. McClaugherty. 1990. Predicting long-term patterns of mass loss, nitrogen dynamics, and soil organic matter formation from initial fine litter chemistry in temperate forest ecosystems. *Canadian J. of Botany* 68:2201–2208.

Aber, J. D., K. J. Nadelhoffer, P. Steudler, and J. M. Melillo. 1989. Nitrogen saturation in northern forest ecosystems. *Bio-Science* 39 (6): 378–386.

Abrams, M. D. 1992. Fire and the development of oak forests. *Bioscience* 42:346–353.

Ackerman, E. A. 1941. The Köppen classification of climates in North America. *Geographical Review* 31:105–111.

Adams, M. S. and B. R. Strain. 1968. Photosynthesis in stems and leaves of *Cercidium floridum:* spring and summer diurnal field response and relation to temperature. *Oecologia Pluntarum* 3:285–297.

———. 1969. Seasonal photosynthetic rates in stems of *Cercidium floridum* Benth. *Photosynthetica* 3:55–62.

Adams, S., B. R. Strain, and M. S. Adams. 1970. Water-repellent soil, fire, and annual plant cover in a desert shrub community of southeastern California. *Ecology* 51:696–700.

Adams, S., B. R. Strain, and J. P. Ting. 1967. Photosynthesis in chlorophyllous stem tissue and leaves of *Cercidium floridum:* accumulation and distribution of ^{14}C from $^{14}CO_2$. *Plant Physiology* 42:1797–1799.

Aerts, R. 1995. The advantages of being evergreen. *Trends in Ecology and Evolution* 10:402–407.

Aerts, R., R.G.A. Boot, and P.J.M. van der Aart. 1991. The relation between above- and below-ground biomass allocation patterns and competitive ability. *Oecologia* 87:551–559.

Agee, J. K. 1973. *Prescribed fire effects on physical and hydrologic properties of mixed-conifer forest floor and soil.* Water Resources Center Contribution Report no. 143. Davis, CA: University of California.

———. 1993. *Fire ecology of Pacific Northwest Forests.* Washington D.C.: Island Press.

———. 1996. Achieving conservation biology objectives with fire in the Pacific Northwest. *Weed Technology* 10 (2): 417–421.

Ågren, G.I., R.E. McMurtrie, W.J. Parton, J. Pastor, and H.H. Shugart. 1991. State-of-the-art of models of production—decomposition linkages in conifer and grassland ecosystems. *Ecological Applications* 1:118–138.

Ahlgren, I. F. 1974. The effect of fire on soil organisms. In T. T. Kozlowski and C. E. Ahlgren eds. *Fire and Ecosystems.* New York: Academic Press.

Al-Ani, H. A., B. R. Strain, and H. A. Mooney. 1972. The physiological ecology of diverse populations of the desert shrub *Simmondsia chinesis. J. Ecology* 60:41–57.

Albert, R. 1975. Salt regulation in halophytes. *Oecologia* 21:57–71.

Aldrich, R. J. and R.J. Kremer. 1997. *Principles in Weed Management*, 2nd ed. Ames: Iowa State University Press.

Alexander, M. E. 1982. Calculating and interpreting forest fire intensities. *Canadian J. of Botany* 60:349–357.

Alpert, P., R. Lumaret, and F. DiBiusto. 1993. Population structure inferred from allozyme analysis in the clonal herb *Fragaria chiloensis* (Rosaceae). *American J. of Botany* 80:1002–1006.

Amthor, J.S. 1995. Terrestrial higher-plant response to increasing atmospheric [CO_2] in relation to the global carbon cycle. *Global Change Biology* 1:243–274.

Anderson, J. E. and S. J. McNaughton. 1973. Effects of low soil temperature on transpiration, photosynthesis, leaf relative water content, and growth among elevationally diverse plant populations. *Ecology* 54:1220–1233.

Anderson, K. 1998. From burns to baskets: California Indian women's influence on ecosystems with the use of fire. *Proceedings of Fire in California Ecosystems Conference* U. C. Davis University Extension, San Diego.

Anderson, R. C., J. S. Fralish, and J. Baskin, eds. 1998. *The savanna, barrens, and rock outcrop communities of North America*. Cambridge, England: Cambridge University Press.

Appleby, R. F. and W. J. Davies. 1983. A possible evaporation site in the guard cell wall and the influence of leaf structure on the humidity response by stomata of woody plants. *Oecologia* 56:30–40.

Archibold, O. W. 1995. *Ecology of world vegetation*. New York: Chapman and Hall.

Armstrong, R.A. and R. McGehee. 1980. Competitive exclusion. *Theor. Pop. Bio.* 8:356–375.

Arthur, M.A. and T.J. Fahey. 1990. Mass and nutrient content of decaying boles in an Engelmann spruce–subalpine fir forest, Rocky Mountain National Park, Colorado. *Canadian J. of Forest Research* 20:730–737.

———. 1992. Biomass and nutrients in an Engelmann spruce–subalpine fir forest in north central Colorado: pools, annual production, and internal cycling. *Canadian J. of Forest Research* 20:730–737.

———. 1993. Controls on soil solution chemistry in a subalpine forest in North-Central Colorado. *Soil Sci. Soc. Am. J.* 57:1122–1130.

Ashman, T. 1994. A dynamic perspective on the physiological cost of reproduction in plants. *American Naturalist* 144:300–316.

Ashton, P. S. and P. Hall, 1992. Comparisons of structure among mixed dipterocarp forests of northwestern Borneo. *J. of Ecology* 80:459–481.

Ashton, P.M.S. and G. P. Berlyn. 1994. A comparison of leaf physiology and anatomy of *Quercus* (section Erythrobalanus-Fagaceae) species in different light environments. *American Journal of Botany* 81:589–597.

Augspurger, C.K. 1983. Seed dispersal of the tropical tree *Platypodium elegans,* and the escape of its seedlings from fungal pathogens. *J. of Ecology* 71:759–772.

———. 1984. Seedling survival of tropical tree species: interactions of dispersal distance, light-gaps, and pathogens. *Ecology* 65:1705–1712.

———. 1986. Morphology and dispersal potential of wind-dispersed diaspores of neotropical trees. *American J. of Botany* 73:353–363.

———. 1988. Impact of pathogens on natural plant populations. In A.J. Davy, M.J. Hutchings and A.R. Watkinson, eds., *Plant Population Ecology,* pp. 413–433. Oxford: Blackwell Scientific.

Augspurger, C.K. and K. Katijama. 1992. Experimental studies of seedling recruitment from contrasting seed distributions. *Ecology* 73:1270–1284.

Austin, M. P. and E. M. Adomeit. 1991. Sampling strategies costed by simulation. In *Nature conservation: cost effective biological surveys and data analysis,* C. R. Margules and M. P. Austin (eds.), pp. 167–175. Melbourne, Australia: CSIRO.

Avery, T. E. 1964. To stratify or not to stratify. *J. Forestry* 62:106–108.

Ayala, F. J. 1969. Experimental invalidation of the principle of competitive exclusion. *Nature* 224:176–179.

Ayyad, M. A. G. and R. L. Dix. 1964. An analysis of a vegetation-microenvironmental complex on prairie slopes in Saskatchewan. *Ecological Monographs* 34:421–442.

Azevedo, J. and D. L. Morgan. 1974. Fog precipitation in coastal California forests. *Ecology* 55:1135–1141.

Baker, D.D. and C.R. Schwintzer. 1990. Introduction. In C.R. Schwintzer and J.D. Tjepkema, eds., *The Biology of Frankia and Actinorhizal Plants,* pp. 1–13. San Diego: Academic Press.

Baker, H.G. 1974. The evolution of weeds. *Annual Review of Ecology and Systematics* 5:1–24.

Baker, R. L. and C. E. Thomas. 1983. A point frame for circular plots in southern forest ranges. *J. of Range Management* 14:63–69.

Baldwin, I.T. 1988. The alkaloidal responses of wild tobacco to real and simulated herbivory. *Oecologia* 77:378–381.

———. 1989. The mechanism of damaged-induced alkaloids in wild tobacco. *Journal of Chemical Ecology* 15:1661–1680.

Baldwin, I.T., R.C. Oesch, P.M. Merhige, and K. Hayes. 1993. Damage-induced root nitrogen metabolism in *Nicotiana sylvestris:* testing C/N predictions for alkaloid production. *Journal of Chemical Ecology* 19:3029–3043.

Baldwin, M., C. E. Kellogg, and J. Thorp. 1938. Classification of soils on the basis of their characteristics. In *Soils and Man: 1938 Yearbook of Agriculture*, pp. 979–1001. Washington, D.C.: U.S. Dept. of Agriculture.

Barber, S. A. 1995. *Soil nutrient bioavailability: A mechanistic approach*, 2nd edition. New York: J. Wiley & Sons, Inc.

Barbour, M. G. 1970a. Seedling ecology of *Cakile maritima* along the California coast. *Bulletin of the Torrey Botanical Club* 97:280–289.

———. 1970b. Is any angiosperm an obligate halophyte? *American Midland Naturalist* 84:106–119.

———. 1973a. Chemistry and community composition. In *Air pollution damage to vegetation*, Advances in Chemistry no. 122, ed. M. G. Barbour, pp. 85–100. Washington, D.C.: American Chemical Society.

———. 1973b. Desert dogma re-examined: root/shoot productivity and plant spacing. *American Midland Naturalist* 89:41–57.

———. 1995. Ecological fragmentation in the fifties. In W. Cronon (ed.), *Uncommon ground: rethinking the human place*, pp. 233–255. New York: Norton.

Barbour, M. G., N. H. Berg, T. G. F. Kittel, and M. F. Kunz. 1991. Snow-pack and the distribution of a major vegetation ecotone in the Sierra Nevada of California. *J. of Biogeography* 18:141–149.

Barbour, M. G., G. L. Cunningham, W. C. Oechel, and S. A. Bamberg. 1977. Growth and development, form and function. In *Creosote Bush*, eds. T. J. Mabry, J. H. Hunziker, and D. R. Difeo, Jr. ch. 4 Stroudsburg, PA: Dowden, Hutchinson, and Ross.

Barbour, M. G., D. V. Diaz, and R. W. Breidenbach. 1974. Contributions to the biology of *Larrea* species. *Ecology* 55:1199–1215.

Barbour, M. G., J. A. MacMahon, S. A. Bamberg, and J. A. Ludwig. 1977. The structure and function of *Larrea* communities. *Ecology* 55:1199–1215.

Barbour, M. G. and R. A. Minnich. 1990. The myth of chaparral convergence. *Israel J. of Botany* 39:453–463.

Barnola, J.M., M. Anklin, J. Porcheron, D. Raynaud, J. Schwander, and B. Stauffer. 1995. CO_2 evolution during the last millennium as recorded by Antarctic and Greenland ice. *Tellus* 47B:264–272.

Barradas, V. L. and J. Adem. 1992. Albedo model for a tropical dry deciduous forest in western Mexico. *International J. of Biometeorology* 36:113–117.

Barrs, H. D. 1968. Determination of water deficits in plant tissues. In T. T. Kozlowski, ed., *Water deficits and plant growth*, vol. I, ed. T. T. Kozlowski, pp. 235–268. New York: Academic Press.

Bartholomew, B. 1970. Bare zone between California shrub and grassland communities: the role of animals. *Science* 170:1210–1212.

Baskin, J. M. and C. C. Baskin. 1973. Plant population differences in dormancy and germination characteristics of seeds: heredity or environment? *American Midland Naturalist* 90:493–498.

Bates, L. M. and A. E. Hall. 1981. Stomatal closure with soil water depletion not associated with changes in bulk leaf water status. *Oecologia* 50:62–65.

Bauer, H. L. 1943. The statistical analysis of chaparral and other plant communities by means of transect samples. *Ecology* 24:45–60.

Bazzaz, F. A. 1996. *Plants in changing environments: linking physiological, population, and community ecology*. New York: Cambridge University Press.

———. 1997. *Plant Resource Allocation*. San Diego: Academic Press.

Bazzaz, F.A. and E.D. Fajer. 1992. Plant life in a CO_2-rich world. *Scientific American* 266:68–74.

Beals, E.W. 1984. Bray-Curtis ordination: an effective strategy for the analysis of multivariate ecological data. *Advances in Ecological Research* 14:1–55.

Beard, J. S. 1946. The mora forests in Trinidad, British West Indies. *J. of Ecology* 33:173–192.

Beardsell, M. F., P. G. Jarvis, and B. Davidson. 1972. A null-balance diffusion porometer suitable for use with leaves of many shapes. *J. of Applied Ecology* 9:677–690.

Beare, M. H., R. W. Parmelee, P. F. Hendrix, W. Cheng, D. C. Coleman, and D. A. Crossley Jr. 1992. Microbial and faunal interactions and effects on litter nitrogen and decomposition in agroecosystems. *Ecological Monographs* 62(4): 569–591.

Beatley, J. C. 1966. Winter annual vegetation following a nuclear detonation in the northern Mojave Desert (Nevada Test Site). *Radiation Botany* 6:69–82.

———. 1967. Survival of winter annuals in the northern Mojave Desert. *Ecology* 48:745–750.

———. 1969. Biomass of desert winter annual plant populations in southern Nevada. *Oikos* 20:261–273.

———. 1974. Phenological events and their environmental triggers in Mojave Desert ecosystems. *Ecology* 55:856–863.

Beattie, A. J., D. E. Breedlove, and P. R. Ehrlich, 1973. The ecology of the pollinators and predators of *Frasera speciosa*. *Ecology* 54:81–91.

Becking, R. W. 1957. The Zurich-Montpellier school of phytosociology. *Botanical Review* 23:411–488.

Bell, K. L. and L. C. Bliss. 1979. Autecology of *Kobresia bellardii*: why winter snow accumulation limits local distribution. *Ecological Monographs* 49:377–402.

Belsky, A.J., W.P. Carson, C.L. Jensen, and G.A. Fox. 1993. Overcompensation by plants—herbivore optimization or red herring. *Evolutionary Ecology* 7:109–121.

Bender, M. M., I. Rougani, H. M. Vines, and C. C. Black, Jr. 1973. $^{13}C/^{12}C$ ratio changes in crassulacean acid metabolism plants. *Plant Physiology* 52:427–430.

Bennett, W. H. and H. A. Mooney. 1979. The water relations of some desert plants in Death Valley, California. *Flora* 168:405–427.

Berg, B., G. Ekbohm, M.-B. Johansson, C. McClaugherty, F. Rutigliano, and A. V. de Santo, 1996. Maximum decomposition limits of forest litter types: a synthesis. *Can. J. Bot.* 74:659–672.

Berg, B., C. McClaugherty, A. V. de Santo, M.-B. Johansson, and G. Ekbohm. 1995. Decomposition of litter and soil organic matter—can we distinguish a mechanism for soil organic matter buildup? *Scand. J. For. Res.* 10:108–119.

Bergelson, J. and M.J. Crawley. 1992a. The effects of grazers on the performance of individuals and populations of scarlet gilia, *Ipomopsis aggregata*. *Oecologia* 90:435–444.

———. 1992b. Herbivory and *Ipomopsis aggregata*—the disadvantages of being eaten. *American Naturalist* 139:870–882.

Bergelson, J., T. Juenger, and M.J. Crawley. 1996. Regrowth following herbivory in *Ipomopsis aggregata*—compensation but not overcompensation. *American Naturalist* 148:744–755.

Bergelson, J. and C.B. Purrington. 1996. Surveying patterns in the cost of resistance in plants. *American Naturalist* 148:536–558.

Berner, R. A. 1997. The rise of plants and their effect on weathering and atmospheric CO_2. *Science* 276:544–546.

Berry, J. and O. Björkman. 1980. Photosynthetic response and adaptation to temperature in higher plants. *Annual Review of Plant Physiology* 31:491–543.

Berry, J. A. and G. D. Farquhar. 1978. The CO_2 concentrating function of C_4 photosynthesis: a biochemical model. In D. Hall, J. Coombs, and T. Goodwin, eds., *Proceedings of the 4th International Congress on Photosynthesis*, pp. 119–131. London: Biochemical Society.

Berthet, P. 1960. La mesure de la temperature par determination de la vitesse d' inversion du saccharose. *Vegetatio* 9:197–207.

Bertness, M.D. and R. Callaway. 1994. Positive interactions in communities. *Trends in Ecology and Evolution* 9:191–193.

Bertness, M.D. and S.D. Hacker. 1994. Physical stress and positive associations among marsh plants. *American Naturalist* 144:363–372.

Bertness, M.D. and S.W. Shumway. 1993. Competition and facilitation in marsh plants. *American Naturalist* 142:718–724.

Bertness, M.D. and S.M. Yeh. 1994. Cooperative and competitive interactions in the recruitment of marsh elders. *Ecology* 75:2416–2429.

Biles, C. L. and F. B. Abeles. 1991. Xylem sap proteins. *Plant Physiology* 96:597–601.

Bilger, W., U. Schreiber, and O. L. Lange. 1984. Determination of leaf heat resistance: comparative investigation of chlorophyll fluorescence changes and tissue necrosis methods. *Oecologia* 63:156–162.

Billings, W. D. 1952. The environmental complex in relation to plant growth and distribution. *Quarterly Review of Biology* 27:251–265.

———. 1970. *Plants, man, and the ecosystem.* 2nd ed. Belmont, CA: Wadsworth.

———. 1985. The historical development of physiological plant ecology. In *Physiological ecology of North American plant communities*, ed. B. F. Chabot and H. A. Mooney, pp. 1–15. New York: Chapman and Hall.

Billings, W. D. and L. C. Bliss. 1959. An alpine snowbank environment and its effects on vegetation, plant development, and productivity. *Ecology* 40:388–397.

Billings, W. D., E. E. C. Clebsch, and H. A. Mooney. 1966. Photosynthesis and respiration rates of Rocky Mountain alpine plants under field conditions. *American Midland Naturalist* 75:34–44.

Billings, W. D., P. J. Godfrey, B. F. Chabot, and D. P. Bourque. 1971. Metabolic acclimation to temperature in arctic and alpine ecotypes of *Oxyria digyna*. *Arctic and Alpine Research* 3:277–289.

Biondini, M. E., W. K. Lauenroth, and O. E. Sala. 1991. Correcting estimates of net primary production: are we overestimating plant production in rangelands? *Journal of Range Management* 44:194–198.

Biondo, B. 1997. In defense of the longleaf pine. *Nature Conservancy* 47 (4): 10–17.

Biswell, H. H. 1974. Effects of fire on chaparral. In T. T. Kozlowski and C. E. Ahlgren eds. *Fire and ecosystems*. New York: Academic Press.

———. 1977. *Giant sequoia fire ecology.* University Extension Course X 417.4, Davis, CA: University of California.

Björkman, O. 1968a. Carboxydismutase activity in shape-adapted species of higher plants. *Physiologia Plantarum* 21:1–10.

———. 1968b. Further studies on differentiation of photosynthetic properties of sun and shade ecotypes of *Solidago virgaurea*. *Physiologia Plantarum* 21:84–99.

Björkman, O. and B. Demmig-Adams. 1994. Regulation of photosynthetic light energy capture, conversion, and dissipation in leaves of higher plants. In E.D. Schulze, and M.M. Caldwell, eds., *Ecophysiology of photosynthesis*, pp. 17–47. Berlin: Springer-Verlag.

Björkman, O. and B. Demmig. 1987. Photon yield of O_2-evolution and chlorophyll fluorescence characteristics at 77K among vascular plants of diverse origins. *Planta* 170:489–504.

Blackman, G. E. 1935. A study by statistical methods of the distribution of species in grassland associations. *Annals of Botany* 49:749–778.

Blackwell, B., M. C. Feller, and R. Trowbridge. 1992. Conversion of dense lodgepole pine stands in west-central British Columbia into young lodgepole pine plantations using prescribed fire. 2. Effects of burning treatments on tree seedling establishment. *Canadian J. of Forest Research* 22:572–581.

Bloom, A.J. 1986. Plant economics. *Trends in Ecology and Evolution* 1: 98–100.

Bloom, A.J., F.S. Chapin, III. and H.A. Mooney. 1985. Resource limitation in plants—an economic analogy. *Annual Review of Ecology and Systematics* 16:363–392.

Bocherens, H., E. M. Friis, A. Mariotti, and K. R. Pedersen. 1993. Carbon isotopic abundances in Mesozoic and Cenozoic fossil plants—paleoecological implications. *Lethaia* 26:347–358.

Bock, C. E., J. H. Bock, M. C. Grant, and T. R. Seastedt. 1995. Effects of fire on abundance of *Eragrostis intermedia* in a semi-arid grassland in southeastern Arizona. *J. of Vegetation Science* 6:325–328.

Bolster, K.L., M.E. Martin, and J.D. Aber. 1996. Determination of carbon fraction and nitrogen concentration in tree foliage by near infrared reflectance: a comparison of statistical methods. *Canadian J. of Forest Research* 26:590–600.

Bond, G. 1983. Taxonomy and distribution of non-legume nitrogen-fixing systems. In J.C. Gordon and C.T. Wheeler, eds., *Biological Nitrogen Fixation in Forest Ecosystems: Foundations and Applications*, pp. 55–87. The Hague: Dr. W. Junk.

Bond, R. D. 1964. The influence of the microflora on the physical properties of soils. II. Field studies on water-repellent sands. *Australian J. of Soil Research* 2:123–131.

Bond, W. J. and B. S. van Wilgen. 1996. *Fire and Plants.* New York: Chapman and Hall.

Bonham, C. D. 1989. *Measurement for terrestrial vegetation.* New York: Wiley-Interscience.

Bonnier, G. 1895. Recherches experimentales sur l'adaptation des plantes au climat alpin. *Annales des Sciences Naturelles Botanique.* 7th series 20:217–358.

Booker, F. A., W. E. Dietrich, and L. M. Collins. 1995. The Oakland Hills fire of 20 October: an evaluation of post-fire response. In J. E. Keeley and T. Scott eds. *Brushfires in California: Ecology and Resource Management.* International Association of Wildland Fire.

Borchert, M. I. 1997. Los Padres National Forest, Goleta CA. Personal communication.

Borchert, M. I. and D. C. Odion. 1995. Fire intensity and vegetation recovery in chaparral: a review. In J. E. Keeley and T. Scott eds. *Brushfires in California: Ecology and Resource Management.* International Association of Wildland Fire.

Bormann, F. H. 1953. The statistical efficiency of sample plot size and shape in forest ecology. *Ecology* 34:474–487.

Bormann, F. H. and G. E. Likens. 1967. Nutrient cycling. *Science* 155:242–429.

Bormann, F. H., G. E. Likens, and J. M. Melillo. 1977. Nitrogen budget for an aggrading northern hardwood forest ecosystem. *Science* 196:981–983.

Botti, S. J. 1995. Funding fuels management in the National Park Service: costs and benefits. In D. R. Weise and R. E. Martin, technical coordinators. *The Biswell Symposium: Fire Issues and Solutions in Urban Interface and Wildland Ecosystems.* Walnut Creek, California. Gen. Tech. Report PSW-GTR-158. Albany CA: Pacific Southwest Research Station, Forest Service, USDA.

Botting, D. 1973. *Humboldt and the cosmos.* New York: Harper and Row.

Boucher, D.H. 1985. *The Biology of Mutualism: Ecology and Evolution.* London: Croom Helm.

Boucher, D.H., S. James, and K.H. Keeler. 1982. The ecology of mutualism. *Annual Review of Ecology and Systematics* 13:315–347.

Bowden, R. D., M. S. Castro, J. M. Melillo, P. A. Steudler, and J. D. Aber. 1993b. Fluxes of greenhouse gases between soils and the atmosphere in a temperate forest following a simulated hurricane blowdown. *Biogeochemistry* 21:61–71.

Bowden, R. D., K. J. Nadelhoffer, R. D. Boone, J. M. Melillo, and J. B. Garrison. 1993a. Contributions of aboveground litter, belowground litter, and root respiration to total soil respiration in a temperate mixed hardwood forest. *Canadian J. of Forest Research* 23:1402–1407.

Bowers, J. E. 1988. *A sense of place: the life and work of Forrest Shreve.* Tucson: University of Arizona Press.

Bowman, W. D. and S. W. Roberts. 1985. Seasonal and diurnal water relations adjustments in three evergreen chaparral shrubs. *Ecology* 66:738–742.

Box, E. O. 1996. Plant functional types and climate at the global scale. *J. of Vegetation Science* 7:309–320.

Boyce, S. G. 1951. Source of atmospheric salts. *Science* 113:620–621.

———. 1954. The salt spray community. *Ecological Monographs* 24:29–67.

Boyd, R.S. and M.G. Barbour. 1993. Replacement of *Cakile edentula* by *C. maritima* in the strand habitat of California. *American Midland Naturalist* 130:209–228.

Boyer, J. S. 1976. Water deficits and photosynthesis. In T.T. Kozlowski, ed. *Water deficits and plant growth*, vol. IV, pp. 153–190. New York: Academic Press.

Brady, N. C. 1974. *The nature and properties of soils*, 8th ed. New York: Macmillan.

Brady, N. C. and R. R. Weil. 1996. *The Nature and Properties of Soils*, 11th ed. Prentice Hall: New Jersey.

Braun-Blanquet, J. 1932. *Plant sociology: the study of plant communities.* New York: McGraw-Hill.

Bray, J. R. and J. T. Curtis. 1957. An ordination of the upland forest communities of southern Wisconsin. *Ecological Monographs* 27:325–349.

Brewer, C. A. and W. K. Smith. 1994. Influence of simulated dewfall on photosynthesis and yield in soybean isolines (*Glycine max* [L.] Merr. CV Williams) with different trichome densities. *International J. of Plant Science.* 155:460–466.

———. 1997. Patterns of leaf surface wetness for montane and subalpine plants. *Plant, Cell and Environment* 20:1–11.

Brewer, C. A., W. K. Smith and T. C. Vogelmann. 1991. Functional interaction between leaf trichomes, leaf wettability and the optical properties of water droplets. *Plant Cell and Environment* 14:955–962.

Breymeyer, A. and J.M. Melillo. 1991. The effects of climate change on production and decomposition in coniferous forests and grasslands. *Ecological Applications* 1:111.

Briggs, D. and S. M. Walters. 1969. *Plant variation and evolution.* New York: McGraw-Hill.

Briggs, J. M., and A. K. Knapp. 1991. Estimating aboveground biomass in tallgrass prairie with the harvest method: determining proper sample size using jacknifing and Monte Carlo simulations. *Southwestern Naturalist* 36:1–6.

Bronstein, J.L. 1994. Conditional outcomes in mutualistic interactions. *Trends in Ecology and Evolution* 9:214–217.

Brown, M. J. 1994. A survey of ultraviolet-B radiation in forests. *The J. of Ecology* 82:843–854.

Brown, R. W. 1970. *Measurement of water potential with thermocouple psychrometers: construction and applications.* U.S. Dept. of Agriculture Forest Service Research Paper INT-80. Washington, D.C.: U.S. Dept. of Agriculture.

Brown, R. W. and B. P. van Haveren, eds. 1972. *Psychrometry in water relations research.* Utah Agricultural Experiment Station. Logan, UT: Utah State University.

Brown, R. W. and W. T. McDonough. 1977. Thermocouple psychrometer for *in situ* leaf water potential determinations. *Plant and Soil* 48:5–10.

Brownell, P. F. and C. J. Crossland. 1972. The requirements of sodium as a micronutrient by species having the C_4 dicarboxylic acid photosynthetic pathway. *Plant Physiology* 49:794–797.

———. 1974. Growth responses to sodium by *Bryophyllum tubiflorum* under conditions inducing crassulacean acid metabolism. *Plant Physiology* 54:416–417.

Bryant, J.P., F.D. Provenza, J. Pastor, P.B. Reichardt, T.P. Clausen, and J.T. du Toit. 1989. Interactions between woody plants and browsing mammals mediated by secondary metabolites. *Annual Review of Ecology and Systematics* 22:431–446.

Buchmann, S. L. and G.P. Nabhan. 1996. *The Forgotten Pollinators.* Washington, D.C.: Island Press.

Bunce, J. A. 1977. Nonstomatal inhibition of photosynthesis at low water potentials in intact leaves of species from a variety of habitats. *Plant Physiology* 59:348–350.

Buol, S. W., F. D. Hole, R. J. McCracken, and R. J. Southard. 1997. *Soul genesis and classification,* 4th edition. Ames: Iowa State University Press.

Burd, M. 1994. Bateman's principle and plant reproduction: the role of pollen limitation in fruit and seed set. *Botanical Review* 60:83–139.

Burkholder, P. R. 1952. Cooperation and conflict among primitive organisms. *American Scientist* 40:601–631.

Burleigh, S.H. and J.O. Dawson. 1994. Occurrence of *Myrica*-nodulating *Frankia* in Hawaiian volcanic soils. *Plant and Soil* 164:283–289.

Burrows, F. J. and F. L. Milthorpe, 1976. Stomatal conductance in the control of gas exchange. In T.T. Kozlowski, ed., *Water deficits and plant growth,* vol. IV, pp. 103–152. New York: Academic Press.

Busch, D. E. and S. D. Smith. 1995. Mechanisms associated with decline of woody species in riparian ecosystems of the southwestern U.S. *Ecological Monographs* 65:347–370.

Buttrick, S. C. 1977. The alpine flora of Teresa Island, Atlin Lake, British Columbia, with notes on its distribution. *Canadian J. of Botany* 55:1399–1409.

Byers, J. L. and W. O. Wirtz. 1995. Vegetative characteristics of coastal sage scrub sites used by California gnatcatchers: implications for management in a fire-prone ecosystem. In Greenlee, J. ed. *Proceedings of the Conference on Fire Effects on Threatened and Endangered Species and Habitats,* November 13–16. Coeur D'Alene, Idaho. Fairfield, WA: International Association of Wildland Fire.

Cain, S. A. 1944. *Foundations of plant geography.* New York: Hafner Press.

Cain, S. A. 1947. Characteristics of natural areas and factors in their development. *Ecological Monographs* 17:185–200.

Cajander, A.K. 1909. Über die Waldtypen. *Acta For. Fenn.* 1:1–175.

Caldwell, M. M. and R. Robberecht. 1980. A steep latitudinal gradient of solar ultraviolet-B radiation in the arctic alpine life zone. *Ecology* 61:600–611.

Caldwell, M. M., R. S. White, R. T. Moore, and L. B. Camp. 1977. Carbon balance, productivity, and water use of cold-winter desert shrub communities dominated by C_3 and C_4 species. *Oecologia* 25:275–300.

Callaway R. M., E. H. DeLucia, E. M. Thomas, and W. H. Schlesinger. 1994. Compensatory responses of CO2 exchange and biomass allocation and their effects on the relative growth rate of ponderosa pine in different CO2 and temperature regimes. *Oecologia* 98:159–166.

Callaway, R.M. 1992. Effect of shrubs on recruitment of *Quercus douglasii* and *Quercus lobata* in California. *Ecology* 73:2118–2128.

———. 1995. Positive interactions among plants. *Botanical Review* 61:306–349.

Callaway, R.M., N.M. Nadkarni, and B.E. Mahall. 1991. Facilitation and interference of *Quercus douglasii* on understory productivity in central California. *Ecology* 72:1484–1499.

Camacho-B, S. E., A. E. Hall, and M. R. Kaufmann. 1974. Efficiency and regulation of water transport in some woody and herbaceous species. *Plant Physiology* 54:169–172.

Campbell, C. J. and W. A. Dick-Peddie. 1964. Comparison of phreatophyte communities on the Rio Grande in New Mexico. *Ecology* 45:492–502.

Campbell, G. S. 1977. *An introduction to environmental biophysics.* New York: Springer-Verlag.

Campbell, G. S. and M. D. Campbell. 1974. Evaluation of a thermocouple hygrometer for measuring leaf water potential *in situ. Agronomy J.* 66:24–27.

Canfield, R. H. 1941. Application of the line interception method in sampling range vegetation. *J. of Forestry* 39: 388–394.

Cannon, W. A. 1905*a*. On the transpiration of *Fouquieria splendens. Bulletin of the Torrey Botanical Club* 32:397–414.

Canny, M. J. 1993. The transpiration stream in the leaf apoplast: water and solutes. *Philosophical Transactions of the Royal Society.* B341:87–100. London.

———. 1995. A new theory for the ascent of sap—cohesion supported by tissue pressure. *Annals of Botany* 75:343–357.

Capon, B. and W. Van Asdall. 1966. Heat pretreatment as a means of increasing germination of desert annual seeds. *Ecology* 48:305–306.

Carleton, T.J., R.H. Stitt, and J. Nieppola. 1996. Constrained indicator species analysis (COINSPAN): an extension of TWINSPAN. *J. of Vegetation Science* 7:125–130.

Carlisle, A., A. H. E. Brown, and E. J. White. 1966. The organic matter and nutrient elements in the precipitation beneath a sessile oak (*Quercus petraea*) canopy. *J. of Ecology* 54:87–98.

Carlquist, S. 1975. *Ecological strategies of xylem evolution.* Berkeley, CA: University of California Press.

Carter, R. N. and S. D. Prince, 1981. Epidemic models used to explain biogeographical distribution limits. *Nature* 293:644–645.

———. 1988. Distribution limits from a demographic viewpoint. In A. J. Davy, M. J. Hutchings and A. R. Watkinson, eds., *Plant Population Ecology,* pp. 165–184. Oxford: Blackwell Scientific Publications.

Castro, M. S., P. A. Steudler, and J. M. Melillo. 1995. Factors controlling atmospheric methane consumption by temperate forest soils. *Global Biogeochemical Cycles* 9(1):1–10.

Castro, M. S., P. A. Steudler, J. M. Melillo, J. D. Aber, and S. Millham. 1993. Exchange of N_2O and CH_4 between the atmosphere and soils in spruce-fir forests in the northeastern United States. *Biogeochemistry* 18:119–135.

Caswell, H. 1989. *Matrix Population Models.* Sunderland, MA: Sinauer Associates.

Catana, H. J. 1963. The wandering quarter method of estimating population density. *Ecology* 44:349–360.

Cates, R. G. 1996. The role of mixtures and variation in the production of terpenoids in conifer-insect-pathogen interactions. In J. T. Romeo, J. A. Saunders, and P. Barbosa (eds.), *Phytochemical diversity and redundancy in ecological interactions,* pp. 179–216. New York: Plenum Press.

Cates, R. G., J. Zou, and C. Carlson. 1991. The role of variation in Douglas-fir foliage quality in the silvicultural management of western spruce budworm. In D. Baumgarner and J. Lotan (eds.), *Interior Douglas fir: the species and its management,* pp. 115–127. Pullman: Washington State University.

Causton, D.R. 1988. *Introduction to vegetation analysis.* Boston: Unwin Hyman.

Cavers, P. B. and J. L. Harper. 1967*a*. Germination polymorphism in *Rumex crispus* and *R. obtusifolius. J. of Ecology* 54: 367–382.

———. 1967*b*. The comparative biology of closely related species living in the same area. IX. *Rumex:* the nature of adaptation to a sea-shore habitat. *J. of Ecology* 55:73–82.

Chapin, F.S., III. 1991. Integrated responses of plants to stress. *Bioscience* 41:29–36.

Chapin, F.S., III, A.J. Bloom, C.B. Field and R.H. Waring. 1987. Plant responses to multiple environmental factors. *Bioscience* 37:49–57.

Chapin, F.S., III, and S. McNaughton. 1989. Lack of compensatory growth under phosphorus deficiency in grazing-adapted grasses from the Serengeti Plains. *Oecologia* 79:551–557.

Chapin, F. S., III, and G. R. Shaver. 1985. Arctic. In B.F. Chabot and H.M. Mooney, eds., *Physiological ecology of North American plant communities,* pp. 16–40. New York: Chapman and Hall.

Chapin, F.S., III, L.R. Walker, C.L. Fastie, and L.C. Sharman. 1994. Mechanisms of primary succession following deglaciation at Glacier Bay, Alaska. *Ecological Monographs* 64:149–175.

Chapman, S. B. 1976. Production ecology and nutrient budgets. In *Methods in plant ecology,* ed. S. B. Chapman, pp. 157–228. New York: Halsted Press.

Charnov, E.L. and W.M. Schaffer. 1973. Life history consequences of natural selection: Cole's result revisited. *American Naturalist* 107:791–793.

Chatterton, N. J. 1970. *Physiological ecology of Atriplex polycarpa: growth, salt tolerance, ion accumulation and soil-plant-water relations.* Ph.D. dissertation. Riverside, CA: University of California.

Chazdon, R. L. and R. W. Pearcy. 1986a. Photosynthetic responses to light variation in rainforest species. I. Induction under constant and fluctuating light conditions. *Oecologia* 69:571–523.

———. 1986b. Photosynthetic responses to light variation in rainforest species. I. Carbon gain and photosynthetic efficiency during lightflecks. *Oecologia* 69:524–531.

———. 1991. The importance of sunflecks for forest understory plants. *BioScience* 41:760–766.

Chen, H.Y.H. and K. Klinka. 1997. Light availability and photosynthesis of *Pseudotsuga menziesii* seedlings grown in the open and in the forest understory. *Tree Physiology* 17:23–29.

Chessin, M. and A.E. Zipf. 1990. Alarm systems in higher plants. *Botanical Review* 56:193–235.

Cheung, Y. N. S., M. T. Tyree, and J. Dainty. 1975. Water relations parameters on single leaves obtained in a pressure bomb and some ecological interpretations. *Canadian J. of Botany* 53:1342–1346.

Chew, R. M. 1974. Consumers as regulators of ecosystems: an alternative to energetics. *Ohio J. of Science* 74:359–370.

Chew, R. M. and A. E. Chew. 1965. The primary productivity of a desert shrub (*Larrea tridentata*) community. *Ecological Monographs* 35:353–375.

Christensen, N. L. 1987. The biogeochemical consequences of fire and their effects on the vegetation of the Coastal Plain of the southeastern United States. In L. Trabaud, ed. *The Role of Fire in Ecological Systems.* SPB Academic Publishing.

———. 1996. The report of the Ecological Society of America Committee on the scientific basis for ecosystem management. *Ecological Applications* 6:665–691.

Christensen, N. L., J. K. Agee, P. F. Brussard, et al. 1989. Interpreting the Yellowstone fires of 1988. *Bioscience* 39 (10): 678–685.

Christensen, N. L. and C. H. Muller. 1975. Effects of fire on factors controlling plant growth in *Adenostoma* chaparral. *Ecological Monographs* 45:29–55.

Christensen, N.L. and R.K. Peet. 1984. Convergence during secondary forest succession. *J. of Ecology* 72:25–36.

Chung, H. -H. and P. J. Kramer. 1975. Absorption of water and ³²P through suberized and unsuberized roots of loblolly pine. *Canadian J. of Forest Research* 5:229–235.

Clapham, A. R. 1932. The form of the observational unit in quantitative ecology. *J. of Ecology* 20:192–197.

Clark, D.D. and J.H. Burk. 1980. Resource allocation patterns of two California Sonoran Desert ephemerals. *Oecologia* 46:86–91.

Clark, P. J. and F. C. Evans. 1954. Distance to nearest neighbor as a measure of spatial relationships in populations. *Ecology* 35:445–453.

Clausen, J., D. D. Keck, and W. M. Hiesey. 1940. *Experimental studies on the nature of species. I. The effect of varied environments on western North American plants.* Carnegie Institution of Washington Publ. 520. Washington, D.C.: Carnegie Institution of Washington.

Clayton, W.D. 1981. Evolution and distribution of grasses. *Annals of the Missouri Botanical Garden* 68:5–14.

Clements, E. S. 1960. *Adventures in ecology.* New York: Pageant Press.

Clements, F. E. 1910. The life history of lodgepole pine burn forests. *Bulletin 79*, Washington D.C. United States Department of Agriculture, Forest Service.

———. 1916. *Plant succession: an analysis of the development of vegetation.* Carnegie Institution of Washington Publ. 242. Washington, D.C.: Carnegie Institution of Washington.

———. 1920. *Plant indicators: the relation of plant communities to process and practice.* Carnegie Institution of Washington Publ. 290. Washington, D.C.: Carnegie Institution of Washington.

Clements, F. E. and H. M. Hall. 1921. Experimental taxonomy. *Carnegie Institute, Washington Year Book* 20:395–396.

Cloudsley-Thompson, J.L. 1977. *Man and the biology of arid zones.* London: Edward Arnold.

Cloudsley-Thompson, J.L. and M.J. Chadwick. 1964. *Life in Deserts.* Philadelpia PA: Dufour.

Cochard, H., P. Cruizat, and M. T. Tyree. 1992. Use of positive pressures to establish vulnerability curves. *Plant Physiology* 100:205–209.

Cohen, D. 1966. Optimizing reproduction in a randomly varying environment. *J. of Theoretical Biology* 12:119–129.

Cohen, W.B., M.E. Harmon, D.O. Wallin, and M. Fiorella. 1996. Two decades of carbon flux from forests of the Pacific Northwest. *BioScience* 46:836–844.

Cole, L.C. 1954. The population consequences of life history phenomena. *Quarterly Review of Biology* 29:103–137.

Coleman, D. C., C. P. P. Reid, and C. V. Cole. 1983. Biological strategies of nutrient cycling in soil ecosystems. *Advances in Biological Research* 13:1–55.

Collett, J. L. Jr., B. Oberholzer, L. Mosimann, J. Staehelin, and A. Waldvogel. 1993. Contributions of cloud processes to precipitation chemistry in mixed phase clouds. *Water, Air, and Soil Pollution* 68:43–75.

Collins, S. L. and L. L. Wallace. 1990. *Fire in North American Tallgrass Prairies.* Norman, OK: University of Oklahoma Press.

Conard, H. S. 1935. The plant associations on central Long Island: a study in descriptive sociology. *American Midland Naturalist* 16:433–516.

Condit R., S. P. Hubbell, and R. B. Foster. 1992. Recruitment near conspecific adults and the maintenance of tree and shrub diversity in a neotropical forest. *American Naturalist* 140:261–286.

Connell, J. H. 1971. On the role of natural enemies in preventing competitive exclusion in some marine animals and in rain forest trees. In *Dynamics of population*, ed. P. J. den Boer and G. R. Gradwell, pp. 298–310. Wageningen, Netherlands: Center for Agricultural Publishing and Documentation.

———. 1978. Diversity in tropical rain forests and coral reefs. *Science* 199: 1302–1310.

Cooper, W.S. 1939. A fourth expedition to Glacier Bay, Alaska. *Ecology* 20:130–155.

———. 1957. Sir Arthur Tansley and the science of ecology. *Ecology* 38:658–659.

Cottam, G., F.G. Goff, and R.H. Whittaker. 1978. Wisconsin comparative ordination. In *Ordination of plant communities*, ed. R.H. Whittaker, pp. 185–213. The Hague: Junk.

Court, A. 1974. Water balance estimates for the United States. *Weatherwise* 27:252–256.

Cousens, R. and M. Mortimer. 1995. *Dynamics of Weed Populations*. Cambridge: Cambridge University Press.

Cowles, H.C. 1899. The ecological relations of the vegetation on the sand dunes of Lake Michigan. *Botanical Gazette* 27:95–117, 167–202, 281–308, 361–391.

———. 1901. The physiographic ecology of Chicago and vicinity: a study of the origin, development, and classification of plant societies. *Botanical Gazette* 31:73–108, 145–182.

———. 1909. Present problems in plant ecology: the trend of ecological philosophy. *American Naturalist* 43: 356–368.

———. 1911. the causes of vegetative cycles. *Botanical Gazette* 51:161–183.

Cox, G.W. 1990. *Laboratory Manual of General Ecology*, 6th ed. Dubuque, IA: William C. Brown.

Crafton, W.M. and B.W. Wells. 1934. The old field prisere: an ecological study. *J. of the Elisha Mitchell Science Society* 49:225–246.

Cronk, Q.C.B. and J.L. Fuller. 1995. *Plant Invaders: The Threat to Natural Ecosystems*. London: Chapman & Hall.

Cunningham, G. L. and B. R. Strain. 1969. Irradiance and productivity in a desert shrub. *Photosynthetica* 3:69–71.

Curtis, J. T. and G. Cottam. 1962. *Plant ecology workbook*. Minneapolis, MN: Burgess.

Curtis, J. T. and R. P. McIntosh. 1951. An upland forest continuum in the prairie-forest border region of Wisconsin. *Ecology* 32:476–498.

Cushman, J.H. and A.J. Beattie. Mutualisms—assessing the benefits to hosts and visitors. *Trends in Ecology and Evolution* 6:193–195.

Cutler, J. M. and D. W. Rains. 1978. Effects of water stress and hardening on the internal water relations and osmotic constituents of cotton leaves. *Physiologia Plantarum* 42:261–268.

Cyr, H. and M.L. Pace. 1993. Magnitude and patterns of herbivory in aquatic and terrestrial ecosystems. *Nature* 361:148–150.

D'Antonio, C. M. and P. M. Vitousek. 1992. Biological invasions by exotic grasses, the grass/fire cycle, and global change. *Ann. Rev. Ecol. Syst.* 23:63–87.

Dahlback, A., T. Henriksen, S. H. H. Larson, and K. Stamnes. 1989. Biological UV-doses and effect of an ozone layer depletion. *Photochemistry and Photobiology* 49:621–625.

Dahlgren, R. A., H. J. Percival, and R. L. Parfitt. 1997. Carbon dioxide degassing effects on soil solutions collected by centrifugation. *Soil Science* 162:648–655.

Dansereau, P. 1951. Description and recording of vegetation upon a structural basis. *Ecology* 32:172–229.

———. 1961. Essai de representation cartographique des elements structuraux de la vegetation. In *Methods de la cartographie de la vegetation*, 97th Colloquium, ed. H. Gaussen, pp. 233–255. Paris: Centre National de la Recherche Scientifique.

Darwin, C. R. 1859. *On the origin of species by means of natural selection, or the preservation of favoured races in the struggle for life*. London: John Murray.

Daubenmire, R. F. 1952. Forest vegetation of northern Idaho and adjacent Washington and its bearing on concepts of vegetation classification. *Ecological Monographs* 22:301–330.

———. 1959. A canopy-coverage method of vegetational analysis. *Northwest Science* 33:43–66.

———. 1968a. *Plant communities*. New York: Harper and Row.

———. 1968b. Ecology of fire in grasslands. *Advances in Ecological Research* 5:209–266. London: Academic Press Inc.

———. 1974. *Plants and environment*. 3rd ed. New York: Wiley.

Davis, M.B. 1981. Quaternary history and the stability of forest communities. In D.C. West, H.H. Shugart, eds., *Forest Succession*, pp. 132–153. New York: Springer-Verlag.

Davis, M.B. and C. Zabinski. 1992. Changes in geographical range resulting from greenhouse warming: effects on biodiversity in forests. In R. Peters and T. Lovejoy, eds., *Global Warming and Biological Diversity*, pp. 297–308. New Haven, CT: Yale Univ. Press.

Davis, R. B. 1966. Spruce-fir forests of the coast of Maine. *Ecological Monographs* 36:79–94.

Davis, T. A. W. and P. W. Richards. 1933. The vegetation of Moraballi Creek, British Guiana; an ecological study of a limited area of tropical rain forest. *J. of Ecology* 21:350–385.

Dawson, T. E. 1993. Hydraulic lift and water use by plants: implications for water balance, performance and plant-plant interactions. *Oecologia* 95:565–574.

———. 1996. Determining water use by trees and forests from isotopic, energy balance and transpiration analyses: the roles of tree size and hydraulic lift. *Tree Physiology* 16:263–272.

Dawson, T. E. and J. R. Ehleringer. 1991. Streamside trees that do not use stream water. *Nature* 350:335–337.

De Steven, D. 1991a. Experiments on mechanisms of tree establishment in old-field succession: seedling emergence. *Ecology* 72:1066–1075.

———. 1991b. Experiments on mechanisms of tree establishment in old-field succession: seedling survival and growth. *Ecology* 72:1076–1088.

de Wit, C. D. 1960. *On competition.* Versl. Landbouwkl. Onderzoek. no. 66.8. Wageningen, The Netherlands: Centre for Agricultural Publications and Documentation.

———. 1961. Space relationships within populations of one or more species. In *Mechanisms in biological competition*, ed. F. L. Milthorpe, pp. 314–429. New York: Cambridge University Press.

DeBano, L. F., P. H. Dunn, and C. E. Conrad. 1977. Fire's effect on physical and chemical properties of chaparral soils. In *Proc. Symp. Env. Cons. Fire and Fuel Mangmt. in Medit. Ecosyst.*, pp. 65–74. Washington, D.C.: U.S. Dept. of Agriculture Forest Service.

DeBano, L. F. and J. Letey, eds. 1969. *Proc. Symp. on Water-repellent Soils.* Riverside, CA: University of California.

Deevey, E. S., Jr. 1947. Life tables for natural populations. *The Quarterly Review of Biology* 22:283–314.

del Moral, R. and C. H. Muller. 1969. Fog drip: a mechanism of toxin transport from *Eucalyptus globulus. Bulletin of the Torrey Botanical Club* 96:467–475.

Delacourt, P. A. and H. R. Delacourt. 1987. *Long-term Dynamics of the Temperate Zone: A Case Study of Late-Quaternary Forest History in Eastern North America.* Ecological Study Series 63. New York: Springer-Verlag.

Dennis, J. G. and P. L. Johnson. 1970. Shoot and rhizome-root standing crops of tundra vegetation at Barrow, Alaska. *Arctic and Alpine Research* 2:253–266.

Dept. of the Interior. 1995. Mount Vision Fire Incident Burned Area Emergency Rehabilitation (BAER) Plan. BAER Team North Zone. Washington, D.C.: Department of the Interior.

Despain, D. G., D. L. Clark, and J. J. Reardon. 1996. Simulation of crown fire effects on canopy seed bank in lodgepole pine. *International J. of Wildland Fire* 6 (1): 45–49.

Diamond, J. M. 1975. Assembly of species communities. In M. L. Cody and J. M. Diamond (eds.), pp. 342–444. *Ecology and evolution of communities.* Cambridge, MA: Harvard University Press.

Diaz, S., M. Cabido, and F. Casanoves. 1988. Plant functional traits and environmental filters at a regional scale. *Journal of Vegetation Science* 9:113–122.

Dietvortst, P., E. van der Maarel, and H. van der Putten. 1982. A new approach to the minimal area of a plant community. *Vegetatio* 50:77–91.

Dillenburg, L. R., J. H. Sullivan, and A. H. Teramura. 1995. Leaf expansion and development of photosynthetic capacity and pigments in *Liquidambar styraciflua* (Hamamelidaceae): effects of UV-B radiation. *American J. of Botany* 82:878–885.

Dilworth, J. R. and J. F. Bell. 1978. *Variable probability sampling.* Oregon State Univ. Book Stores, Corvallis.

Dittrich, P. and W. Huber. 1974. Carbon dioxide metabolism in members of the Chlamydospermae. In *Proceedings of the Third International Congress on Photosynthesis*, pp. 1573–1578. Amsterdam: Elsevier.

Dix, R. L. 1961. An application of the point-centered quarter method to the sampling of grassland vegetation. *J. Range Management* 14:63–69.

Downes, R. W. and J. D. Hesketh. 1968. Enhanced photosynthesis at low O_2 concentrations: differential response of temperate and tropical grasses. *Planta* 78:79–84.

Downton, W. J. S. 1975. Checklist of C_4 species. *Photosynthetica* 9:96–105.

Downton, W. J. S. and E. B. Tregunna. 1968. Carbon dioxide compensation—its relation to photosynthetic carboxylation reaction, systematics of the Graminae, and leaf anatomy. *Canadian J. of Botany* 46:207–215.

Drake, B.G. and P. W. Leadley. 1991. Canopy photosynthesis of crops and native plant communities exposed to long-term elevated CO_2. *Plant, Cell and Environment* 14:853–860.

Drude, O. 1890. *Handbuch der Pflanzengeographie.* Stuttgart, Germany: Englemann.

———. 1896. *Deutschlands Pflanzengeographie.* Stuttgart, Germany: Engelhorn.

Dubrovsky, J. G. 1996. *Seed hydration memory in Sonoran desert cacti and its ecological implication. American J. of Botany* 83:624–632.

Dudash, M.R. and C.B. Fenster. 1997. Multiyear study of pollen limitation and cost of reproduction in the iteroparous *Silene virginica. Ecology* 78:484–493.

Dunn, P. H. and L. F. DeBano. 1977. Fire's effect on biological and chemical properties of chaparral soils: In *Proc. Sym. Env. Cons. Fire and Fuel Mangmt. in Medit. Ecosyst.* Washington, DC: USDA Forest Service.

Duvdevani, S. 1964. Dew in Israel and its effect on plants. *Soil Science* 98:14–21.

Easterling, D. R., B. Horton, P. D. Jones, T. C. Peterson, T. R. Karl, D. E. Parker, M. J. Salinger, V. Razuvayev, N. Plummer, P. Jamason and C. K. Folland. 1997. Maximum and minimum temperature trends for the globe. *Science* 277:364–367.

Edwards, D. R. and M. A. Dixon. 1995. Mechanisms of drought response in *Thuja occidentalis* L.: I. Water stress conditioning and osmotic adjustment. *Tree Physiology.* 15:121–127.

Edwards, J.S. and P. Sugg. 1993. Arthropod fallout as a resource in the recolonization of Mount St. Helens. *Ecology* 74:954–958.

Eek, L. and K. Zobel. 1997. Effects of additional illumination and fertilization on seasonal changes in fine-scale grassland community structure. *J. of Vegetation Science* 8:225–234.

Egerton, F. N. 1976. Ecological studies and observations before 1900. In *Evolution of issues, ideas and events in America 1776–1976,* ed. B. J. Taylor, pp. 311–351. Norman, OK: University of Oklahoma Press.

Egler, F.E. 1954. Philosophical and practical considerations of the Braun-Blanquet system of phytosociology. *Castanea* 19:45–60.

Ehleringer, J. and I. Forseth. 1980. Solar tracking by plants. *Science* 210:1094–1098.

Ehleringer, J. R. 1983. Ecophysiology of *Amaranthus palmeri,* a Sonoran Desert summer ephemeral. *Oecologia* 57:107–112.

Ehleringer, J. R. and O. Björkman. 1978. A comparison of photosynthetic characteristics of *Encelia* species possessing glabrous and pubescent leaves. *Plant Physiology* 62:185–190.

Ehleringer, J. R., O. Björkman, and H. A. Mooney. 1976. Leaf pubescence: effects on absorptance and photosynthesis in a desert shrub. *Science* 192:376–377.

Ehleringer, J. R., T. E. Cerling, and B. R. Helliker. 1997. C_4 photosynthesis, atmospheric CO_2, and climate. *Oecologia* 112:285–299.

Ehleringer, J. R. and T. A. Cooper. 1988. Correlations between carbon isotope ratio and microhabitat in desert plants. *Oecologia* 76:562–566.

Ehleringer, J. R. and R. K. Monson. 1993. Evolutionary and ecological aspects of photosynthetic pathway variation. *Annual Review of Ecology and Systematics* 24:411–439.

Eickmeier, W. G. 1978. Photosynthetic pathway distributions along an aridity gradient in Big Bend National Park, and implications for enhanced resource partitioning. *Photosynthetica* 12:290–297.

Ellenberg, H. 1958. Bodenreaktion (einschlieblich Kaltfrage). In *Handbuch der Pflanzenphysiologie,* vol. 4, ed. W. Ruhland, pp. 638–708. Berlin: Springer-Verlag.

Emerman, S. H. and T. E. Dawson. 1996. Hydraulic lift and its influence on the water content of the rhizosphere: an example from sugar maple, *Acer saccharum. Oecologia* 108:273–278.

Engeman, R. M., R. T. Sugihara, L. F. Pank and W. E. Dusenberry. 1994. A comparison of plotless density estimators using Monte Carlo simulation. *Ecology* 75:1769–1779.

Engler, A. 1903. Untersuchungen uber das Wurzelwachstum der Holzarten. *Mitteilungen der schweizer Zentralanstalt fur das forstliche Versuchswesen* 7:247–317.

Erickson, R. O. 1945. The *Clematis fremontii* var. *riehlii* population in the Ozarks. *Annals of the Missouri Botanical Garden* 30:63–68.

Esau, K. 1965. *Plant anatomy.* New York: Wiley.

Evans, G. C. and D. E. Coombe. 1959. Hemispherical and woodland canopy photography and the light climate. *J. of Ecology* 47:103–113.

Evans, J. R. 1989. Photosynthesis and nitrogen relationships in leaves of C_3 plants. *Oecologia* (Berlin) 78:9–19.

Evenari, M., L. Shanan, and N. H. Tadmor. 1971. *The Negev, the challenge of a desert.* Cambridge, MA: Harvard University Press.

Eyre, S. R. 1963. *Vegetation and soils.* Chicago; Aldine.

Faegri, K. and L. van der Pijl. 1971. *The Principles of Pollination Ecology,* 2nd ed. Oxford: Pergamon Press.

Fahey, T. J. and D. H. Knight. 1986. Lodgepole pine ecosystems. *Bioscience* 36 (9): 610–617.

Farman, J. C., B. G. Gardiner and J. D. Shanklin. 1985. Large losses of total ozone in Antarctica reveal seasonal ClO_x/NO_x interaction. *Nature* 315:207–210.

Farquhar, G. D. 1983. On the nature of carbon isotope discrimination in C_4 species. *Australian J. of Plant Physiology* 10:205–226.

Farquhar, G. D., M. C. Ball, S. von Caemmerer, and Z. Roksandic. 1982. Effect of salinity and humidity on the $\delta^{13}C$ value of halophytes— evidence for diffusional isotope fractionalization determined by the ration of intercellular/atmospheric partial pressure of CO_2 under different environmental conditions. *Oecologia* 52:121–124.

Farquhar, G. D., J. R. Ehleringer, and K. T. Hubick. 1989. Carbon isotope discrimination and photosynthesis. *Annual Review of Plant Physiology and Molecular Biology* 40:503–537.

Fasham, M.J.R. 1977. A comparison of nonmetric multidimensional scaling, principal components and reciprocal averaging for the ordination of simulated coenoclines, coenoplanes. *Ecology* 58:551–561.

Fastie, C.L. 1995. Causes and ecosystem consequences of multiple pathways on primary succession at Glacier Bay, Alaska. *Ecology* 76:1899–1916.

Federal Geographic Data Committee. 1996. *FGDC vegetation classification and information standards.* Federal Geographic Data Committee Secretariat, USGS MS 590 National Center, Reston, Virginia.

Feeny, P. 1976. Plant apparency and chemical defense. In J.W. Wallace and R.L. Mansell, eds., *Biochemical Interactions Between Plants and Insects,* pp. 1–40. New York: Plenum Press.

Fenner, M., ed. 1992. *Seeds: The Ecology of Regeneration in Plant Communities.* Wallingford, UK: CAB International.

Fernandez, O. A. and M. M. Caldwell. 1975. Phenology and dynamics of root growth of three cool semi-desert shrubs under field conditions. *J. of Ecology* 63:703–714.

Fiedler, P. L. and S. K. Jain (eds.). 1992. *Conservation biology: the theory and practice of nature conservation, preservation, and management.* New York: Chapman and Hall.

Field, C. and H. A. Mooney. 1986. The photosynthesis-nitrogen relationship in wild plants. In T. Divnish ed., *On the economy of plant form and function,* pp. 25–55. Cambridge: Cambridge University Press.

Field, C. B., J. T. Ball, and J. A. Berry. 1989. Photosynthesis: principles and field techniques. In R. W. Pearcy, J. Ehleringer, H. A. Mooney, and P. W. Rundel, eds. *Plant Physiological Ecology,* pp. 209–253. London: Chapman and Hall.

Findlay, S., M. Carriero, V. Krischik, and C. G. Jones. 1996. Effects of damage to living plants on leaf litter quality. *Ecological Applications* 6 (1): 269–275.

Flock, J. A. W. 1978. Lichen-Bryophyte distributions along a snow-cover-soil-moisture gradient, Niwot Ridge, Colorado. *Arctic and Alpine Research* 10:31–47.

Flowers, T. J., P. F. Troke, and A. R. Yeo. 1977. The mechanism of salt tolerance in halophytes. *Annual Review of Plant Physiology* 28:89–121.

Floyd, D. A. and J. E. Anderson. 1982. A new point interception frame for estimating cover of vegetation. *Vegetatio* 50:185–186.

Fonteyn, P. J. and B. E. Mahall. 1978. Competition among desert perennials. *Nature* 275:544–545.

———. 1981. An experimental analysis of structure in a desert plant community. *J. of Ecology* 69:883–896.

Ford, J. M. and J. E. Monroe. 1971. *Living systems.* San Francisco. CA: Canfield Press.

Forman, R. T. T. 1975. Canopy lichens with blue-green algae: a nitrogen source in a Columbian rain forest. *Ecology* 56:1176–1184.

Forman, R. T. T. and M. Godron. 1981. Patches and structural components for a landscape ecology. *BioScience* 31: 733–740.

Forseth, I. and J. Ehleringer. 1982. Ecophysiology of two solar tracking desert winter annuals. *Oecologia* 54:41–49.

Foster, N. W. and I. K. Morrison. 1976. Distribution and cycling of nutrients in a natural *Pinus banksiana* ecosystem. *Oecologia* 57:110–120.

Fox. B. J. 1989. Small-mammal community pattern in Australian heathland: a taxonomically based rule for species assembly. In *the structure of mammalian communities,* D. W. Morris et al. (eds.), pp. 91–103. Lubbock: Texas Technical University.

Fralish, J. S., R. P. McIntosh, and O. L. Loucks (eds.). *John T. Curtis: fifty years of Wisconsin plant ecology.* Madison: Wisconsin Academy of Sciences, Arts, and Letters.

Frank, D.A and S.J. McNaughton. 1993. Evidence for the promotion of aboveground grassland production by native large herbivores in Yellowstone National Park. *Oecologia* 96:157–161.

Freas, K.E. and P.R. Kemp. 1983. Some relationships between environmental reliability and seed dormancy in desert annuals. *J. of Ecology* 71:217–221.

Frederick, J. E. and H. E. Snell. 1988. Ultraviolet radiation levels during the Antarctic spring. *Science* 241:438–439.

Frelich, L. E., R. R. Calcote, M. B. Davis and J. Pastor. 1993. Patch formation and maintenance in an old-growth hemlock-hardwood forest. *Ecology* 74:513–527.

Freudenberger, D. O., B. E. Fish, and J. E. Keeley. 1987. Distribution and stability of grasslands in the Los Angeles Basin. *Bulletin Southern California Acad. Sci.* 86 (1): 13–26.

Fryer, J. H. and F. T. Ledig. 1972. Microevolution of the photosynthetic temperature optimum in relation to an elevational complex gradient. *Canadian J. of Botany* 50:1231–1235.

Futuyma, D. J. 1986. *Evolutionary biology,* 2nd ed. Sunderland, MA: Sinauer.

Gadgil, M. and W.H. Bossert. 1970. Life historical consequences of natural selection. *American Naturalist* 104:1–24.

Galen, C. 1985. Regulation of seed-set in *Polemonium viscosum:* floral scents, pollination, and resources. *Ecology* 66: 792–797.

Galen, C. and M. L. Stanton. 1995. Responses of snowbed plant-species to changes in growing season length. *Ecology* 76:1546–1557.

Gallardo, A. and W. H. Schlesinger. 1992. Carbon and nitrogen limitations of soil microbial biomass in desert ecosystem. *Biogeochemistry* 18:1–17.

Gamon, J. A. and R. W. Pearcy. 1990. Photoinhibition in *Vitis californica:* interactive effects of sunlight, temperature and water status. *Plant Cell and Environment* 13:267–275.

Garcia-Mendez, G., J. M. Maass, P. A. Matson, and P. M. Vitousek. 1991. Nitrogen transformations and nitrous oxide flux in a tropical deciduous forest in Mexico. *Oecologia* 88:362–366.

Garrison, G. A. 1949. Uses and modifications for the "moosehorn" crown closure estimator. *J. of Forestry* 47:733–735.

Garth, R. E. 1964. The ecology of Spanish moss (*Tillandsia usneoides*). *Ecology* 45:470–481.

Gates. D. M. 1965*a*. Heat transfer in plants. *Scientific American* 213:76–83.

———. 1965b. Radiant energy, its receipt and disposal. *Meteorological Monographs* 6:1–26.

———. 1972. *Man and his environment: climate.* New York: Harper and Row.

Gause, G. F. 1934. *The struggle for existence.* Baltimore, MD: Williams and Wilkins.

Gauthier, S., Y. Bergeron, and J. P. Simon. 1996. Effects of fire regime on the serotiny level of jack pine. *J. of Ecology* 84 (4): 539–458.

Gebre, G. M., M. R. Kuhns, and J. R. Brandle. 1994. Organic solute accumulation and dehydration tolerance in three water-stressed *Populus deltoides* clones. *Tree Physiology* 14:575–587.

Geological Survey Bulletin 965. 1956. Geological investigations in the Paricutin area of Mexico. U.S. Department of the Interior.

Georgiadis, N.J., R.W. Ruess, S.J. McNaughton, and D. Western. 1989. Ecological conditions that determine when grazing stimulates grass production. *Oecologia* 81:316–322.

Geritz, S.A.H. 1995. Evolutionarily stable seed polymorphism and small-scale spatial variation in seedling density. *American Naturalist* 146:685–707.

Gershenzon, J. 1984. Changes in the level of plant secondary metabolites under water and nutrient stress. In B.N. Timmerman, C. Steerlink and E.A. Loewus, eds., *Phytochemical Adaptations to Stress,* pp. 273–321. New York: Plenum Press.

———. 1994. The cost of plant chemical defense against herbivory: A biochemical perspective. In E.A. Bernays, ed., *Insect-Plant Interactions* (vol. 5), pp. 105–173. Boca Raton: CRC Press.

Gibson, et al. 1990. Fire heterogeneity in contrasting fire-prone habitats. *Bulletin Torrey Botanical Club* 117:349–352.

Gilbert, G. S., S. P. Hubbell and R. B. Foster. 1994. Density and distance-to-adult effects of a canker disease of trees in a moist tropical forest. *Oecologia* 98:100–108.

Gilliam, F. S. 1987. The chemistry of wet deposition for a tallgrass prairie ecosystem: inputs and interactions with plant canopies. *Biogeochemistry* 4:203–217.

———. 1988. Interactions of fire with nutrients in the herbaceous layer of a nutrient-poor Coastal Plain forest. *Bulletin of Tor Bot Club* 115 (4): 265–271.

———. 1989. Atmospheric deposition and its potential significance in Southern Appalachian hardwood forest ecosystems. In *Environment in Appalachia. Proceedings, 4th Annual Conference on Appalachia.* J. W. Bagby, ed. Lexington, KY.

———. 1991*a*. The significance of fire in an oligotrophic forest ecosystem. In S. C. Nodvin and T. A. Waldrop eds. *Fire and the environment: ecological and cultural perspectives: Proceedings of an international symposium.* 1990 March 20–24. Knoxville, TN: Gen. Tech. Rep. SE-69. Asheville, NC U.S. Dept. of Agriculture Forest Service S E Forest Expt. Station.

———. 1991*b*. Ecosystem-level significance of acid forest soils. In Wright, R. J. et al. eds. *Plant-soil interactions at low pH.* p. 187–195. The Netherlands: Kluwer Academic Publishers.

Gilliam, F.S. and M.B. Adams. 1995. Plant and soil nutrients in young versus mature central Appalachian hardwood stands. In *Proceedings, 10th Central Hardwood Forest Conference,* ed. K.W. Gottschalk, pp. 109–118. Morgantown, WV. March 1995.

———. 1996. Wetfall deposition and precipitation chemistry for a central Appalachian forest. *J. Air and Waste Management Association* 46:978–984.

Gilliam, F. S., M. B. Adams, and B. M. Yurish. 1996. Ecosystem nutrient responses to chronic nitrogen inputs at Fernow Experimental Forest, West Virginia. *Canadian J. of Forest Research* 26:196–205.

Gilliam, F. S. and N. L. Christensen. 1986. Herb-layer response to burning in pine flatwoods of the lower Coastal Plain of South Carolina. *Bulletin of the Torrey Botanical Club* 113 (1): 42–45.

Gilliam, F. S. and D. D. Richter. 1991. Transport of metal cations through a nutrient-poor forest ecosystem. *Water, Air, and Soil Pollution* 57–58:279–287.

Gilliam, F.S. and M.R. Roberts. 1995. Forest management and plant diversity. *Ecological Applications* 5:911–912.

Gilliam, F. S., T. R. Seastedt, and A. K. Knapp. 1987. Canopy rainfall interception and throughfall in burned and unburned tallgrass prairie. *Southwestern Naturalist* 32 (2): 267–271.

Gilliam, F.S. and N.L. Turrill. 1993. Herbaceous layer cover and biomass in young versus mature stands of a Central Appalachian hardwood forest. *Bulletin of the Torrey Botanical Club* 120:445–450.

———. 1995. Temporal patterns of ozone pollution in West Virginia: implications for high-elevation hardwood forests. *Journal of the Air and Waste Management Association* 45: 621–626.

Gilliam, F.S., N.L. Turrill, and M.B. Adams. 1995. Species composition and patterns of diversity in herbaceous layer and woody overstory of clearcut versus mature central Appalachian hardwood forests. *Ecological Applications* 5:947–955.

Gilliam, F. S., B. M. Yurish, and L. M. Goodwin. 1993. Community composition of an old growth longleaf pine forest: relationship to soil texture. *Bulletin of the Torrey Botanical Club* 120(3): 287–294.

Gillison, A. N. and D. J. Anderson (eds.). 1981. *Vegetation classification in Australia.* Canberra, Australia: CSIRO.

Gillison, A. N. and K. R. W. Brewer. 1985. The use of gradient directed transects or gradsects in natural resource survey. *J. of Environmental Management* 20:103–127.

Gleason, H. A. 1917. The structure and development of the plant association. *Bulletin of the Torrey Botanical Club* 44:463–81.

———. 1926. The individualistic concept of the plant association. *Bulletin of the Torrey Botanical Club* 53:1–20.

Gleeson, S. K. and D. Tilman. 1992. Plant allocation and the multiple limitation hypothesis. *American Naturalist* 139:1322–1343.

Glitzenstein, J. S., W. J. Platt, and D. R. Streng. 1995. Effects of fire regime and habitat on tree dynamics in North Florida longleaf pine savannas. *Ecological Monographs* 65 (4): 441–476.

Goldberg, D. E. and P. A. Werner. 1983. Equivalence of competitors in plant communities: a null hypothesis and a field experimental approach. *American J. Botany* 70:1098–1104.

Goldsworthy, A. 1976. *Photorespiration. Carolina Biology Readers, no. 80.* Burlington, NC: Carolina Biological Suppliers.

Golley, F. B. 1993. *A history of the ecosystem concept in ecology: more than the sum of the parts.* New Haven, CT: Yale University Press.

Good, R. E. 1931. A theory of plant geography. *The New Phytologist* 30:149–203.

———. 1953. *The geography of the flowering plants.* 2nd ed. New York: Longmans, Green, and Co.

Goodall, D. W. 1957. Some considerations in the use of point quadrat methods for the analysis of vegetation. *Australian J. of Biological Science* 5:1–41.

Goodland, R. J. 1975. The tropical origin of ecology: Eugen Warming's jubilee. *Oikos* 26:240–245.

Goodman, L. A. 1969. The analysis of population growth when the birth and death rates depend upon several factors. *Biometrics* 25:659–681.

Gorham, E. 1979. Shoot height, weight, and standing crop in relation to density of monospecific plant stands. *Nature* 279:148–150.

Grace, J. 1993. Consequences of xylem cavitation for plant water deficits. In Smith J. A. C. and H. Griffiths (eds). *Water deficits: plant responses from cell to community,* pp. 109–128. Oxford: Bios Scientific Publishers.

Grace, J.B. 1991. A clarification of the debate between Grime and Tilman. *Functional Ecology* 5:583–587.

Graham, B. F., Jr. and F. H. Bormann. 1966. Natural root grafts. *Botanical Review* 32:255–292.

Gray, A. 1889. *Scientific papers,* selected by C. S. Sargent. New York: Houghton Mifflin.

Greene, D.F. and E.A. Johnson. 1995. Long-distance wind dispersal of tree seeds. *Canadian J. of Botany* 73:1036–1045.

Greenland, D., N. Caine, and O. Pollak. 1984. The summer water budget and its importance in the alpine tundra of Colorado. *Physical Geography* 5:221–239.

Greenway, H. and R. Munns. 1980. Mechanisms of salt tolerance in non-halophytes. *Annual Review of Plant Physiology* 31:149–190.

Greenway, H. and D. A. Thomas. 1965. Plant response to saline substrates. V. Chloride regulation in the individual organs of *Hordeum vulgare* during treatment with sodium chloride. *Australian J. of Biological Science* 18:505–524.

Greer, G.K., R.M. Lloyd, and B.C. McCarthy. 1997. Factors influencing the distribution of pteridophytes in a southeastern Ohio hardwood forest. *J. of the Torrey Botanical Society* 124:11–21.

Gregor, J. W. 1946. Ecotypic differentiation. *The New Phytologist* 45:254–270.

Greig-Smith, P. 1964. *Quantitative plant ecology.* London: Butterworths.

Grier, C. C. and S. W. Running. 1977. Leaf area of mature northwestern coniferous forests: relation to site water balance. *Ecology* 58:893–899.

Griffin, J. R. 1973. Xylem sap tension in three woodland oaks of central California. *Ecology* 54:152–159.

Grime, J.P. 1977. Evidence for the existence of three primary strategies in plants and its relevance to ecological and evolutionary theory. *American Naturalist* 11:1169–1194.

———. 1979. *Plant Strategies and Vegetation Processes.* New York: Wiley.

———. 1982. The concept of strategies: use and abuse. *J. of Ecology* 70:863–865.

———. 1988. A comment on Loehle's critique of the triangular model of primary plant strategies. *Ecology* 69:1618–1620.

———. 1994. The role of plasticity in exploiting environmental heterogeneity. In *Exploitation of Environmental Heterogeneity by Plants,* M.M. Caldwell and R.W. Pearcy, eds., pp. 1–19. San Diego: Academic Press.

———. 1997. Biodiversity and ecosystem function: the debate deepens. *Science* 277:1260–1261.

Grime, J.P., J.G. Hodgson, and R. Hunt. 1990. *The Abridged Comparative Plant Ecology.* London: Unwin Hyman.

Groenewoud, H. 1992. The robustness of Correspondence, Detrended Correspondence, and TWINSPAN Analysis. *J. of Vegetation Science* 3:239–246.

Groffman, P. M., D. R. Zak, S. Christensen, A. Mosier, and J. M. Tiedje. 1993. Early spring nitrogen dynamics in a temperate forest landscape. *Ecology* 74(5): 1579–1585.

Grosenbaugh, L. R. 1952. Plotless timber estimates—new, fast, easy. *J. of Forestry* 50:32–37.

Grubb, P. J., H. E. Green, and R. C. J. Merrifield. 1969. The ecology of chalk heath: its relevance to the calcicole-calcifuge and soil acidification problems. *J. of Ecology* 57:175–212.

Guenther, A. 1997. Seasonal and spatial variations in natural volatile organic compound emissions. *Ecological Applications* 7:34–45.

Hacskaylo, E., ed. 1971. *Proc. of the First North American Conf. on Mycorrhizae, 1969.* U.S. Dept. of Agriculture Forest Service Misc. Publ. 1189. Washington, D.C.: U.S. Dept. of Agriculture Forest Service.

Hadley, J. L. and W. K. Smith. 1983 Influence of wind exposure on needle desiccation and mortality for timberline conifers in Wyoming, USA. *Arctic and Alpine Research* 15:127–135.

Hagen, J. B. 1992. *An entangled bank: the origins of ecosystem ecology.* New Brunswick, NJ: Rutgers University Press.

Haig, D. 1996. The pea and the coconut: seed size and safe sites. *Trends in Evolution and Ecology* 11:1–2.

Haines, B. L., J. B. Waide, and R. L. Todd. 1982. Soil solution nutrient concentrations sampled with tension and zero-tension lysimeters: report of discrepancies. *Soil Society of America J.* 46:658–661.

Hairston, N.G., F.E. Smith, and L.B. Slobodkin. 1960. Community structure, population control, and competition. *American Naturalist* 94:421–425.

Hall, A. E. and M. R. Kaufmann. 1975. The regulation of water transport in the soil-plant-atmosphere continuum. In D. M. Gates and R. B. Schmerl, eds., *Perspectives in biophysical ecology,* pp. 187–202. Berlin: Springer-Verlag.

Hall, A. E., E. D. Schulze, and O. L. Lange. 1976. Current perspectives of steady-state stomatal responses to environment. In O. L. Lange, L. Kappen, and E. D. Schulze, eds., *Water and plant life: problems and modern approaches,* pp. 169–188. Berlin: Springer-Verlag.

Halligan, J. P. 1973. Bare areas associated with shrub stands in grassland: the case of *Artemesia californica. BioScience* 23:429–432.

Halpern, C.B. and T.A. Spies. 1995. Plant species diversity in natural and managed forests of the Pacific Northwest. *Ecological Applications* 5:913–934.

Hanscom, Z., III and P. Ting. 1978. Responses of succulents to plant water stress. *Plant Physiology* 61:327–330.

Hanson, H. C. 1917. Leaf-structure as related to environment. *American J. of Botany* 4:553–560.

———. 1962. *Dictionary of ecology.* New York: Philosophical Library.

Hansson, A-C and O. Andrén. 1986. Belowground plant production in a perennial grass ley (*Festuca pratensis* Huds.) assessed with different methods. *J. of Applied Ecology* 23:657–666.

Harborne, J.B. 1988. *Introduction to Ecological Biochemistry.* 3rd ed. London: Academic Press.

Harcombe, P.A., G.N. Cameron, and E.G. Glumac. 1993. Above-ground net primary productivity in adjacent grassland and woodland on the coastal plain of Texas, USA. *J. of Vegetation Science* 4:521–530.

Harper, J. L. 1964. The individual in the population. *J. of Ecology* 52 (Supplement):149–158.

———. 1967. A Darwinian approach to plant ecology. *J. of Ecology* 55:247–270.

———. 1969. *The biology of mycorrhizae.* 2nd ed. London: Leonard Hill.

———. 1977. *Population biology of plants.* New York: Academic Press.

Harper, J. L. and A. D. Bell. 1979. The population dynamics of growth from in organics with modular construction. In *Population Dynamics,* R. M. Anderson, B. C. Turner, and L. R. Taylor, eds., pp. 29–52. Oxford, England: Blackwell Scientific.

Hartesveldt, R. J. 1964. The fire ecology of the giant sequoias. *Natural History* 73:12–19.

Hartesveldt, R. J. and H. T. Harvey. 1967. The fire ecology of sequoia regeneration. In *Proc. Tall Timbers Fire Ecol. Conf.* no. 7, pp. 65–78. Tallahassee, FL: Tall Timbers Research Station.

Hartsock, T. L. and P. S. Nobel. 1976. Watering converts a CAM plant to daytime CO_2 uptake. *Nature* 262:574–576.

Harvey, H. T., H. S. Shellhammer, and R. E. Stecker. 1980. *Giant Sequoia Ecology: fire and reproduction.* U.S. Dept. Int. NPS Sci. Mono. 12. Washington, D.C.: U.S. Dept. Int. NPS.

Hastings, A. and S. Harrison. 1994. Metapopulation dynamics and genetics. *Annual Review of Ecology and Systematics* V25:167–188.

Hatch, M. D. and C. B. Osmond. 1976. Compartmentation and transport in C_4 photosynthesis. In C.R. Stockings and U. Heber, eds., *Transport in plants. III. Intracellular interactions and transport processes,* pp. 144–184. Berlin-Heidelberg: Springer-Verlag.

Hatch, M. D. and C. R. Slack. 1966. Photosynthesis by sugar-cane leaves: a new carboxylation reaction and the pathway of sugar formation. *Biochemical J.* 101:103–111.

Hayek, L. C. and M. A. Buzas. 1997. *Surveying natural populations.* New York: Columbia University Press.

Hedin, L. O., L. Granat, G. E. Likens, T. A. Buishand, J. N. Galloway, T. J. Butler, and R. H. Rodhe. 1994. Steep declines in atmospheric base cations in regions of Europe and North America. *Nature* 437:351–354.

Heine, R. W. 1971. Hydraulic conductivity in trees. *J. of Experimental Botany* 22:503–511.

Henderson, G. S., W. F. Harris, D. E. Todd, Jr., and T. Grizzard. 1977. Quantity and chemistry of throughfall as influenced by forest type and season. *J. of Ecology* 65:365–374.

Hendrick, R.L. and K.S. Pregitzer. 1992. The demography of fine roots in a northern hardwood forest. *Ecology* 73:1094–1104.

Henry, J. D. and J. M. A. Swan. 1974. Reconstructing forest history from live and dead plant material—an approach to the study of forest succession in southwestern New Hampshire. *Ecology* 55:772–783.

Herms, D.A. and W.J. Mattson. 1992. The dilemma of plants—to grow or defend? *Quarterly Review of Biology* 67: 283–335.

Hickman, J.C. 1975. Environmental unpredictability and plastic energy allocation strategies in the annual *Polygonum cascadense. J. of Ecology* 63:689–701.

Highkin, H. R. 1958. Transmission of phenotypic variability within a pure line. *Nature* 182:1460.

Hill, M.O. 1979b. *TWINSPAN—a FORTRAN program of arranging multivariate data in an ordered two way table by classification of the individuals and the attributes.* Ithaca, NY: Cornell University.

Hinckley, T. M., J. R. Brooks, J. Cermak, R. Ceulemans, J. Kucera, F. C. Meinzer, and D. A. Roberts. 1994. Water flux in a hybrid poplar stand. *Tree Physiology* 14:1005–1018.

Holbrook, N. M., M. J. Burns, and C. B. Field. 1995. Negative xylem pressures in plants: a test of the balancing pressure technique. *Science* 270:1193–1194.

Holland, E. A. and J. K. Detling. 1990. Plant response to herbivory and belowground nitrogen cycling. *Ecology* 71 (3): 1040–1049.

Holm, L. G., D. L. Picknett, J. V. Pancho, and J. P. Herberger. 1977. *The world's worst weeds: distribution and biology.* Honolulu: University Press.

Holmén, K. 1992. The global carbon cycle. In *Global Biogeochemical Cycles*, pp. 239–262. London: Academic Press.

Homma, K. 1997. Effects of snow pressure on growth form and life history of tree species in Japanese beech forest. *J. of Vegetation Science* 8:781–788.

Hopkins, A. D. 1938. *Bioclimatics: a science of life and climatic relations.* U.S. Dept. of Agriculture Misc. Publ. 280. Washington, D.C.: U.S. Dept. of Agriculture.

Horton, J. S. 1977. The development and perpetuation of the permanent Tamarisk type in the phreatophyte zone of the southwest. In U. S. Dept. of Agriculture Forest Service General Technical Report RM-43, Importance, preservation, and management of riparian habitat: symp. proc., pp. 124–127. Washington, D. C.: U. S. Dept. of Agriculture Forest Service.

Horton, J. S. and C. J. Campbell. 1974. Management of phreatophyte and riparian vegetation maximum multiple-use values. USDA Forest Service Research Paper Rm-117.

Horvitz, C.C. and D.W. Schemske. 1988. Demographic cost of reproduction in a neotropical herb: an experimental field study. *Ecology* 69:1741–1745.

Houghton, R.A. 1995. Land-use changes and the carbon cycle. *Global Change Biology* 1:275–287.

Hrapko, J. O. and G. A. La Roi. 1978. The alpine tundra vegetation of Signal Mountain, Jasper National Park. *Canadian J. of Botany* 56:309–332.

Hsiao, T. C. 1973. Plant responses to water stress. *Annual Review of Plant Physiology* 24:519–570.

———. 1976. Stomatal ion transport. In A. Pirson and M. H. Zimmerman, eds., *Encyclopaedia of plant physiology,* vol. 2, part B, pp. 195–217. Berlin: Springer-Verlag.

Hughes, C. E. and B. T. Styles. 1987. The benefits and potential risks of woody legume introductions. *International Tree Crops Journal* 4:209–248.

Hughes, F. and P. M. Vitousek. 1993. Barriers to shrub reestablishment following fire in the seasonal submontane zone of Hawaii. *Oecologia* 93 (4): 557–563.

Hull, R. J. and O. A. Leonard. 1964. Physiological aspects of parasitism in mistletoes (*Arceuthobium* and *Phoradendron*), parts I and II. *Plant Physiology* 39:996–1017.

Humboldt, A. von and A. Bonpland. 1807–1834. *Voyage aux régions equinoxiales du nouveau continent.* 30 vols.

Hume, L. and P. B. Cavers. 1981. A methodological problem in genecology. Seeds versus clones as source material for uniform gardens. *Canadian J. Botany* 59:763–768.

Hunt, E. R., Jr., N. R. Barrett, and P. S. Nobel. 1987. Measurement of relative water content for various species by infrared reflectances. In *International Conference on Measurement of Soil and Plant Water Status*, pp. 9–11. Logan: Utah State University.

Hunt, R. and A.O. Nichols. 1986. Stress and coarse control of root-shoot partitioning in herbaceous plants. *Oikos* 47:149–158.

Hunt, W. H. 1977. A simulation model for decomposition in grasslands. *Ecology* 58:469–484.

Huston, M. A. 1979. A general hypothesis of species diversity. *American Naturalist* 113:81–101.

———. 1994. *Biological diversity: the coexistence of species on changing landscapes.* Cambridge, England: Cambridge University Press.

———. 1997. Hidden treatments in ecological experiments: re-evaluating the ecosystem function of biodiversity. *Oecologia* 110:449–460.

Huston, M.A. and T.M. Smith. 1987. Plant succession: life history and competition. *American Naturalist* 130:168–198.

Hutchison, B. A. and D. R. Matt. 1977. The distribution of solar radiation within a deciduous forest. *Ecological Monographs* 47:185–207.

Ingram, M. 1957. Microorganisms resisting high concentrations of sugars and salts. In *Seventh Symposium of the Society for General Microbiology*, pp. 90–133. Cambridge, England: Cambridge University Press.

Iremonger, S. F. 1990. A structural analysis of three Irish wooded wetlands. *J. of Vegetation Science* 1:359–366.

Jackson, M. T. 1966. Effects of microclimate on spring flowering phenology. *Ecology* 47:407–415.

Janzen, D. H. 1970. Herbivores and the number of tree species in tropical forests. *American Naturalist* 104:501–528.

———. 1975. *Ecology of plants in the tropics.* London: Edward Arnold.

———. 1976. Why bamboos wait so long to flower. *Annual Review of Ecology and Systematics* 7:347–391.

———. 1988. On the broadening of insect-plant research. *Ecology* 69:905.

Jarvis, P. G. and T. A. Mansfield. 1981. *Stomatal physiology.* Cambridge, England: Cambridge University Press.

Jeffree, E. P. 1960. Some long term means from the phenological reports (1891–1948) of the Royal Meteorological Society. *Quarterly J. of the Royal Meteorological Society* 86:95–103.

Jennings, C. W. 1977. Geologic map of California. California geologic data map series, map no. 2. Sacramento, California: Division of Mines and Geology.

Jennings, D. H. 1968. Halophytes, succulence, and sodium in plants—a unified theory. *New Phytologist* 67:899–911.

Jenny H. 1941. *The factors of soil formation.* New York: McGraw-Hill.

———. 1980. *The Soil Resource: Origin and Behavior.* New York: Springer-Verlag.

Jenny, H., R. J. Arkley, and A. M. Schultz. 1969. The pygmy forest-podsol ecosystem and its dune associates of the Mendocino coast. *Madrono* 20:60–74.

Jensen, W. A. and F. B. Salisbury. 1972. *Botany: an ecological approach.* Belmont, CA: Wadsworth.

Johnson, D. A. and M. M. Caldwell. 1975. Gas exchange of four arctic and alpine tundra plant species in relation to atmospheric and soil moisture stress. *Oecologia* 21:93–108.

Johnson, D. W., D. W. Cole, S. P. Gessel, M. J. Singer, and R. V. Minden. 1977. Carbonic acid leaching in a tropical, subalpine, and northern forest soil. *Arctic and Alpine Research* 9:329–343.

Johnson, D. W., R. F. Walker, and J. T. Ball. 1995. Lessons from lysimeters: soil N release from disturbance compromises controlled environment study. *Ecological Applications* 5(2):395–400.

Johnson, E. A. 1992. *Fire and Vegetation Dynamics: Studies from the North American Boreal Forest.* Cambridge: Cambridge University Press.

Johnson, E. A. and G. I. Fryer. 1996. Why Engelmann spruce does not have a persistent seed bank. *Can. J. of For. Res.* 26 (5): 872–878.

Johnson, P. L. and W. D. Billings. 1962. The alpine vegetation of the Beartooth Plateau in relation to cryopedogenic processes and patterns. *Ecological Monographs* 32:105–135.

Jones, C. E. and S. L. Buchmann. 1974. Ultra-violet floral patterns as functional orientation cues in Hymenopteran pollination systems. *Animal Behavior* 22:481–485.

Jones, H. G., P. J. C. Hamer, and K. H. Higgs. 1988. Evaluation of various heat-pulse methods for estimation of sap flow in orchard trees: comparison with micrometeorological estimates of evaporation. *Trees* 2:250–260.

Jones, M. M. and N. C. Turner. 1978. Osmotic adjustment in leaves of sorghum in response to water deficits. *Plant Physiology* 61:122–126.

Jongen, M. M., B. Jones, T. Hebeisen, H. Blum, and G. Hendrey. 1995. The effects of elevated concentrations on the root growth of *Lolium perenne* and *Trifolium repens* grown in a FACE system. *Global Change Biology* 1:361–371.

Jordan, D. F. and J. R. Kline. 1972. Mineral cycling: some basic concepts and their application in a tropical rain forest. *Annual Review of Ecology and Systematics* 3:33–50.

Jordano, P. 1984. Seed weight variation and differential avian dispersal in blackberries *Rubus ulmifolius. Oikos* 43:149–153.

———. 1995. Frugivore-mediated selection on fruit and seed size: birds and St. Lucie's cherry, *Prunus mahaleb. Ecology* 76:2627–2639.

Küchler, A. W. 1964. *The potential natural vegetation of the conterminous United States.* American Geographical Society Special Publ. no. 36. New York: American Geographical Society.

———. 1967. *Vegetation mapping.* New York: Ronald Press.

———. 1977. Potential natural vegetation of California. In *Terrestrial vegetation of California,* ed. M. G. Barbour and J. Major, pp. 909–938 and map. New York: Wiley-Interscience.

Kadmon, R. and A. Shmida. 1990. Patterns and causes of spatial variation in the reproductive success of a desert annual. *Oecologia* 83:139–144.

Kalisz, S. and M. A. McPeak. 1992. Demography of an age-structured annual—resampled projection matrices, elasticity analyses, and seed bank effects. *Ecology* V73:1082–1093.

Kaminsky, R. 1981. The microbial origin of the allelopathic potential of *Adenostoma fasciculatum. Ecological Monographs* 51:365–382.

Kanemasu, E. T., G. W. Thurtell, and C. B. Tanner. 1969. Design, calibration and field use of a stomatal diffusion porometer. *Plant Physiology* 44:881–885.

Kappen, L., G. Shultz, and R. Vanselow. 1995. Direct observations of stomatal movements. In E.-D Schulze, and M. M. Caldwell, eds., *Ecophysiology of photosynthesis,* pp. 231–246. Berlin: Springer-Verlag.

Karban, R. and I.T. Baldwin. 1997. *Induced Responses to Herbivory.* Chicago: University of Chicago Press.

Kattenberg, A., F. Giorgi, H. Grassl, G.A. Meehl, J.F.B. Mithcell, R.J. Stouffer, T. Tokioka, A.J. Weaver, T.M.L. Wigley.

1996. Climate Models—Projections of Future Climate. In J.T. Houghton, L,G, Meira Filho, B.A. Callander, N. Harris, A. Kattenberg, K. Maskell, eds., *Climate Change 1995: The Science of Climate Change*, pp. 285–357. Cambridge: Inter-governmental Panel on Climate Change, Cambridge University Press.

Kauffman, J. B., D. L. Cummings, D. E. Ward, and R. Babbitt. 1995. Fire in the Brazilian Amazon: 1. Biomass, nutrient pools, and losses in slashed primary forests. *Oecologia* 104:397–408.

Kauppi, P.E., K. Mielikainen, and K. Kuusela. 1992. Biomass and carbon budget of European forests. *Science* 256: 70–74.

Kearns, C.A. and D.W. Inouye.1993. *Techniques for Pollination Biologists*. Niwot, CO: University Press of Colorado.

Keeler, K.H. 1985. Cost: benefit models of mutualism. In D.H. Boucher, ed., *The Biology of Mutualism: Ecology and Evolution*, pp. 100–127. London: Croom Helm Publishers.

———. 1992. Local polyploid variation in the native prairie grass *Andropogon gerardii. American J. of Botany* 79:1229–1232.

Keeley, J.E. 1977. Fire-dependent strategies in Arctostaphylos and Ceanothus. In *Proc. Symp. Env. Cons. Fire and Fuel Management in Medit. Ecosystems*, pp. 391–396. Washington, DC: U.S. Dept of Agriculture Forest Service.

———. 1991. Seed germination and life history syndromes in the California chaparral. *Bot. Rev.* 57:81–116.

———. 1992. Demographic structure of California chaparral in the long-term absence of fire. *J. of Vegetation Science* 3 (1): 79–90.

———. 1994. Seed-germination patterns in fire-prone Mediterranean-climate regions. In M. T. K. Arroyo, P. H. Zedler, and M. D. Fox, eds. *Ecology and Biogeography of Mediterranean Ecosystems in Chile, California, and Australia.* New York: Springer-Verlag.

Keeley, J. E. and G. Busch. 1984. Carbon assimilation characteristics of the aquatic CAM plant *Isoetes howellii. Plant Physiology* 76:525–530.

Keeley, J. E., M. Carrington, and S. Trnka. 1995. Overview of management issues raised by the 1993 wildfire of southern California. In J. E. Keeley and T. Scott eds. *Brushfires in California: Ecology and Resource Management.* International Association of Wildland Fire.

Keeley, J. E., D. A. DeMason, R. Gonzalez, and K. R. Markham. 1994. Sediment-based carbon nutrition in tropical alpine *Isoetes.* In P. W. Rundel, A. P. Smith and F. C. Meinzer, eds., *Tropical Alpine Environments: Plant form and function*, pp. 167–194. Cambridge: Cambridge University Press.

Keeley, J. E. and C. D. Fotheringham. 1997. Smoke-induced seed germination in Californian chaparral. *Science* 276:1248–1250.

Keeley, J. E. and S. C. Keeley. 1988. Chaparral. In M. G. Barbour and W. D. Billings, eds. *North American Terrestrial Vegetation.* New York: Cambridge University Press.

Keeley, J. E., B. A. Morton, A. Pedrosa, and P. Trotter. 1985. Role of allelopathy, heat, and charred wood in the germination of chaparral herbs and suffrutescents. *J. of Ecology* 73:445–458.

Keeley, J. E. and P. H. Zedler. Evolution of life histories in *Pinus.* In D. Richardson and R. Cowling, eds., *Ecology and Biogeography of Pines.* Cambridge University Press.

Keeling, C. D., W. G. Mook, and P. P. Tans. 1979. Recent trends in the $^{13}C/^{12}C$ ratio of atmospheric carbon dioxide. *Nature* (London) 277:121–123.

Kent, M. and P. Coker. 1992. *Vegetation description and analysis: a practical approach.* Boca Raton, Florida: CRC Press, Inc.

Kerner, A. 1863. *The plant life of the Danube Basin.* Translated by H. S. Conrad. Republished. Ames, IA: Iowa State College Press. 1951.

———. 1895. *The natural history of plants, their forms, growth, reproduction and distribution.* Translated by F. W. Oliver. London: Blackie.

Kerr, R. A. 1994. Antarctic ozone hole fails to recover. *Science* 266:217.

Kershaw, K. A. 1973. *Quantitative and Dynamic Plant Ecology,* 2nd edition. New York: American Elsevier.

Keys, Jr., J.E., C.A. Carpenter, S.L. Hooks, F.G. Koenig, W.H. McNab, W.E. Russell, and M.L. Smith. 1995. *Ecological units of the eastern United States—first approximation.* U.S. Department of Agriculture, Forest Service, Atlanta, Georgia.

Killingbeck, K. T. 1996. Nutrients in senesced leaves: keys to the search for potential resorption and resorption proficiency. *Ecology* 77 (6): 1716–1727.

King, T. J. 1977. The plant ecology of ant hills in calcareous grasslands. 1. Patterns of species in relation to ant hills in southern England. *J. of Ecology* 65:235–256.

Kinnee, E., C. Geron, and T. Pierce. 1997. United States land use inventory for estimating biogenic ozone precursor emissions. *Ecological Applications* 7:46–58.

Kira, T., H. Ogawa and K. Shinozaki. 1953. Intraspecific competition among higher plants. I. Competition, density-yield interrelationships in regularly dispersed populations. *J. Institute Polytechnic,* Osaka City University D. 4:1–16.

Kirschbaum, M.U.F. 1995. The temperature dependence of soil organic matter decomposition, and the effect of global warming on soil organic C storage. *Soil Biology and Biochemistry* 27:753–760.

Klinka, K., H. Quan, J. Pojar, and D. U. Meidinger. 1966. Classification of natural forest communities of coastal British Columbia. *Vegetatio* 125:149–168.

Kluge, M. 1974. Metabolism of carbohydrates and organic acids. In H. Ellenberg et al., eds., *Progress in botany,* vol. 36, pp. 90–98. Berlin: Springer-Verlag.

Knapp, A. K. 1984. Post-burn differences in solar radiation, leaf temperature, and water stress influencing production in a low-land tallgrass prairie. *American. J. of Botany* 71:220–227.

Knapp, A. K., J. M. Briggs, D. C. Hartnett, and S. L. Collins, eds. 1998. *Grassland Dynamics: Long-Term Ecological Research in Tallgrass Prairie.* Oxford University Press.

Knapp, A.K. and F.S. Gilliam. 1985. Response of Andropogon gerardii (Poaceae) to fire-induced high vs. low irradiance environments in tallgrass prairie: leaf structure and photosynthetic pigments. *American J. of Botany* 72:1668–1671.

Knapp, A. K. and T. R. Seastedt. 1986. Detritus accumulation limits productivity of tallgrass prairie. *Bioscience* 36 (1): 662–668.

Knight, C. L., J. M. Briggs, and M. D. Nellis. 1994. Expansion of gallery forest on Konza Prairie Research Natural Area, Kansas, USA. *Landscape Ecology* 9 (2): 117–125.

Knipling, E. B. 1967. Measurement of leaf water potential by the dye method. *Ecology* 48:1038–1041.

Knipling, E. B. and P. J. Kramer. 1967. Comparison of the dye method with the thermocouple psychrometer for measuring leaf water potentials. *Plant Physiology* 42:1315–1320.

Knops, J. M. H., T. H. Nash III, and W. H. Schlesinger. 1996. The influence of epiphytic lichens on the nutrient cycling of an oak woodland. *Ecological Monographs* 66 (2): 159–179.

Koenig, W.D., R.L. Mumme, W.J. Carmen, and M.T. Stanback. 1994. Acorn production by oaks in central California: variation within and among years. *Ecology* 75:99–109.

Kolb, T. E., S. C. Hart, and R. Amundson. 1997. Boxelder water sources and physiology at perennial and ephemeral stream sites in Arizona. *Tree Physiology* 17:151–160.

Komarek, E. V. Sr. 1964. The natural history of lightning. In *Proceedings of the Tall Timbers Fire Ecology Conference,* no. 2, pp. 181–187. Tallahassee, FL: Tall Timbers Research Station.

———. 1965. Fire ecology—grasslands and man. In *Proceedings of the Tall Timbers Fire Ecology Conference,* no. 4, pp. 169–220. Tallahassee, FL: Tall Timbers Research Station.

———. 1968. Lightning and lightning fires as ecological forces. In *Proceedings of the Tall Timbers Fire Ecology Conference,* no. 8, pp. 169–197. Tallahassee, FL: Tall Timbers Research Station.

Komarkova, V. 1979. *Alpine vegetation of the Indian Peaks area, Front Range, Colorado Rocky Mountains* (2 vols.). Liechtenstein: Cramer, Vaduz.

Komarkova, V. and P. J. Webber. 1978. An alpine vegetation map of Niwot Ridge, Colorado. *Arctic and Alpine Research* 10:1–30.

Kortshak, H. P., C. E. Hartt, and G. O. Burr. 1965. Carbon dioxide fixation in sugar cane leaves. *Plant Physiology* 40: 209–213.

Kozaki, A. and G. Takeba. 1996. Photorespiration protects C_3 plants from photooxidation. *Nature* 384:557–560.

Krajina, V. J. (ed.). 1965. *Ecology of western North America.* Vancouver: University of British Columbia.

Kramer, P. J. 1969. *Plant and soil water relationships: a modern synthesis.* New York: McGraw-Hill.

Krammes, J. S. and L. F. DeBano. 1965. Soil wettability: a neglected factor in watershed management. *Water Resources Research* 1:283–286.

Krebs, C. J. 1972. *Ecology: the experimental analysis of distribution and abundance.* New York: Harper and Row.

———. 1989. *Ecological methodology.* New York: HarperCollins.

Krenzer, E. G., Jr., D. N. Moss, and R. K. Crookston. 1975. Carbon dioxide compensation points of flowering plants. *Plant Physiology* 56:194–206.

Kruckeberg, A. R. 1954. The ecology of serpentine soils. III. Plant species in relation to serpentine soils. *Ecology* 35: 267–274.

Kuhry, P. 1994. The role of fire in the development of *Sphagnum*–dominated peatlands in western boreal Canada, *J. of Ecology* 82 (4): 899–910.

Kuijt, J. 1969. *The biology of parasitic flowering plants.* Berkeley, CA: University of California Press.

Kyriakopoulos, E. and H. Richter. 1976. A comparison of methods for the determination of water status in *Quercus ilex* L. *Zhurnal Pflanzenphysiologie* 82:14–27.

Lachowski, H. 1995. *Guidelines for the use of digital imagery for vegetation mapping.* U.S.D.A. Forest Service, Engineering Staff, Washington, D.C. EM-7140-25.

Laetsch, W. M. 1974. The C_4 syndrome: a structural analysis.. *Annual Review of Plant Physiology* 25:1974.

Lange, O. L. and L. Kappen. 1972. Photosynthesis of lichens from Antarctica. In G. A. Llano, ed., *Antarctic terrestrial biology, Antarctic Research Series, vol. 20,* pp. 83–95. Washington, D.C.: American Geophysical Union.

Lange, O. L., R. Lösch, E. D. Schulze, and L. Kappen. 1971. Responses of stomata to changes in humidity. *Planta* 100:76–86.

Lange, O. L., A. Meyer, I. Ullmann, and H. Zellner. 1991. Microclimate conditions, water-content and photosynthesis of lichens in the coastal fog zone of the Namib Desert–measurements in the fall. *Flora* 185:233–266.

Langlet, O. 1959. A cline or not a cline—a question of scots pine. *Sylvae Genetica* 8:13–22.

Larcher, W. 1995. *Physiological plant ecology: ecophysiology and stress physiology of functional groups,* 3rd edition. New York: Springer-Verlag.

Larcher, W. 1995. Photosynthesis as a tool for indicating temperature stress events. In E.-D Schulze and M. M. Caldwell, eds., *Ecophysiology of photosynthesis,* pp. 261–277. Berlin: Springer-Verlag.

Larrson, S., A. Wiren, L. Lundgren, and T. Ericcson. 1986. Effects of light and nutrient stress on leaf phenolic chemistry in *Salix dasyclados* and susceptibility to *Galerucella lineola* (Coleoptera). *Oikos* 47:205–210.

Laudermilk, J. D. and P. A. Munz. 1938. *Plants in the dung of Nothrotherium from Gypsum Cave, Nevada.* In Carnegie Institute of Washington Publ. 453, pp. 29–37. Washington, D.C.: Carnegie Institute of Washington.

Law, R. 1979. The cost of reproduction effort in clonal plants: a benefit-cost model. *Oikos* 49:199–208.

Lawrence, M. E. 1985. *Senecio* L. (Asteraceae) in Australia: nuclear DNA amounts. *Australian J. Botany* 33:221–232.

Lawson, E. R. 1967. Throughfall and stemflow in a pine-hardwood stand in the Ouachita Mountains of Arkansas. *Water Resources Research* 3:731–735.

Lawton, J.H. and S. McNeill. 1979. Between the devil and the deep blue sea: on the problem of being a herbivore. In K. Anderson, B. Turner and L.R. Taylor, eds., *Population Dynamics,* pp. 223–245. Oxford: Blackwell Scientific.

Leck, M. A., V. T. Parker and R. L. Simpson. 1989. *Ecology of Soil Seed Banks.* San Diego: Academic Press.

Lee, D. W. and J. B. Lowry. 1980. Young leaf anthocyanin and solar ultraviolet. *Biotropica* 12:75–76.

Lee, R. 1969. Chemical temperature determination. *J. of Applied Meteorology* 8:423–430.

Lefkovitch, L. P. 1965. The study of population growth in organisms grouped by stages. *Biometrics* 21:1–18.

Leonard, R. E. 1961. Net precipitation in a northern hardwood forest. *J. of Geophysical Research* 66:2417–2421.

Leslie, P. H. 1945. On the use of matrices in certain population mathematics. *Biometrika* 33:183–212.

Levin, S. A. (ed.) 1996. Economic growth and environmental quality: a forum. *Ecological Applications* 6:12–32.

Levins, R. 1970. Extinctions. In M. Gerstenhaber, ed. *Lectures on Mathematics in the Life Sciences,* pp. 77–107. Providence, RI: American Mathematical Society.

Lewis, E. R. 1972. Delay-line models of population growth. *Ecology* 53:797–807.

Leyval, C. and C. P. P. Reid. 1991. Utilization of microbial siderophores by mycorrhizal and non-mycorrhizal pine roots. *New Phytologist* 119(1): 93–98.

Liebig, J. 1840. *Chemistry in its agriculture and physiology.* London: Taylor and Walton.

Likens, G. E. and F. H. Bormann. 1972. Nutrient cycling in ecosystems. In *Ecosystem structure and function,* ed. J. A. Weins. Corvallis, OR: Oregon State University Press.

Likens, G. E., F. H. Bormann, N. M. Johnson, W. D. Fisher, and R. S. Pierce. 1970. Effects of forest cutting and herbicide treatment on nutrient budgets in the Hubbard-Brook watershed-ecosystem. *Ecological Monographs* 40:23–47.

Likens, G. E., F. H. Bormann, R. S. Pierce, J. S. Eaton, and N. M. Johnson. 1977. *Biogeochemistry of forested ecosystems.* New York: Springer-Verlag.

Likens, G. E., C. T. Driscoll, and D. C. Buso. 1996. Long-term effects of acid rain: response and recovery of a forest ecosystem. *Science* 272:244–246.

Likens, G. E., C. T. Driscoll, D. C. Buso, T. G. Siccama, C. E. Johnson, G. M. Lovett, D. F. Ryan, T. Fahey, and W. A. Reiners. 1994. The biogeochemistry of potassium at Hubbard Brook. *Biogeochemistry* 25:61–125.

Lin, N. H. and V. K. Saxena. 1992a. In-cloud savenging and deposition of sulfates and nitrates: case studies and parameterization. *Atmospheric Environment Part A General Topics* 25 (10): 2301–2320.

———. 1992b. Interannual variability in acidic deposition on the Mt. Mitchell area forest (North Carolina, USA). *Atmospheric Environment Part A General Topics* 25 (2): 517–524.

Lindberg, S. E. and J. G. Owens. 1993. Throughfall studies of deposition to forest edges and gaps in montane ecosystems. *Biogeochemistry* 19:173–194.

Lindberg, S. E., G. M. Lovett, D. D. Richter, and D. W. Johnson. 1986. Atmospheric deposition and canopy interactions of major ions in a forest. *Science* 231:141–145.

Lindsey, A. A. 1955. Testing and line-strip method against full tallies in diverse forest types. *Ecology* 36:485–495.

———. 1956. Sampling methods and community attributes in forest ecology. *Forest Science* 2:287–296.

Lindsey, A. A., J. D. Barton, and S. R. Miles. 1958. Field efficiencies of forest sampling methods. *Ecology* 39: 428–444.

List, R. L. 1951. *Smithsonian meteorological tables.* (6th ed.) Smithsonian Institute Publication No. 4014. Washington, D.C.

Lloyd, D.G. and S.C.H. Barrett. 1996. *Floral Biology: Studies on Floral Evolution in Animal-Pollinated Plants.* New York: Chapman & Hall.

Lloyd, M. G. 1961. The contribution of dew to the summer water budget of northern Idaho. *Bulletin of the American Meteorological Society* 42:572–580.

Loach, K. 1967. Shade tolerance in tree seedlings. I. Leaf photosynthesis and respiration in plants raised under artificial shade. *New Phytologist* 66:607–621.

Loehle, C. 1987. Partitioning of reproductive effort in clonal plants: a benefit-cost model. *Oikos* 49:199–208.

———. 1988a. Problems with the triangular model for representing plant strategies. *Ecology* 69:284–286.

———. 1988b. Tree life history strategies: the role of defenses. *Can. J. of For. Res.* 18:209–222.

Logan, K. T. 1970. Adaptations of the photosynthetic apparatus of sun-and shade-grown yellow birch (*Betula alleghaniensis* Britt.). *Canadian J. of Botany* 48:1681–1688.

Long, S.P. and P.R. Hutchin. 1991. Primary production in grasslands and coniferous forests with climate change: an overview. *Ecological Applications* 1:139–156.

Lonsdale, W. M. 1990. The self-thinning rule: dead or alive? *Ecology* 71:1373–1388.

Lösch, R. and E.-D. Schulze. 1995. Internal coordination of plant responses to drought and evaporational demand. In E.-D. Schulze and M. M. Caldwell, eds., *Ecophysiology of photosynthesis*, pp. 185–204. Berlin: Springer-Verlag.

Lotan, J. E. 1974. Cone serotiny-fire relationships in lodgepole pine. In *Proceedings of the Tall Timbers Fire Ecology Conference*, no. 14, pp. 267–278. Tallahassee, FL: Tall Timbers Research Station.

Loucks, O. L. 1970. Evolution of diversity, efficiency, and community stability. *American Zoologist* 10:17–25.

Lovett, G. M. and P. Tobiessen. 1993. Carbon and nitrogen assimilation in red oaks (*Quercus rubra* L.) subject to defoliation and nitrogen stress. *Tree Physiology* 12:259–269.

Lovett, G. M. and S. E. Lindberg. 1993. Atmospheric deposition and canopy interactions of nitrogen in forests. *Canadian J. of Forest Research* 23:1603–1616.

Lovett-Doust, J. 1989. Plant reproductive strategies and resource allocation. *Trends in Ecology and Evolution* 4:230–234.

Lovett-Doust, J. and L. Lovett-Doust. 1988. *Plant Reproductive Ecology.* New York: Oxford University Press.

Lowry, W. P. 1969. *Weather and life.* New York: Academic Press

Lubchenko, J. et al. 1991. The sustainable biosphere initiative: an ecological research agenda. *Ecology* 72:371–412.

Ludwig, D. 1993. Environmental sustainability: magic, science, and religion in natural resource management. *Ecological Applications* 3:555–558.

Lugwig, J. A. and J. F. Reynolds. 1988. *Statistical ecology.* New York: Wiley-Interscience.

Luken, J.O. 1990. *Directing ecological succession.* London: Chapman and Hall.

Luken, J.O. and R.W. Fonda. 1983. Nitrogen accumulation in a chronosequence of red alder communities along the Hoh River, Olympic National Park, Washington. *Canadian J. of Forest Research* 13:1228–1237.

Luken, J.O. and J.W. Thieret. 1997. *Assessment and management of plant invasions.* New York: Springer-Verlag.

Luo, Y., R. B. Jackson, C. B. Field, and H. A. Mooney. 1996. Elevated CO2 increases belowground respiration in California grasslands. *Oecologia* 108:130–137.

Luxmoore, R. J. 1983. Water budget of an eastern deciduous forest stand. *Soil Science Society of America Journal* 47:785–791.

MacArthur, R. H. 1957. On the relative abundance of bird species. *Proceedings of the National Academy of Science* 45:293–296.

———. 1958. Population ecology of some warblers of northeastern coniferous forests. *Ecology* 39:599–619.

———. 1972. *Geographical Ecology: Patterns in the Distribution of Species.* New York: Harper and Row.

MacArthur, R.H. and E.O. Wilson. 1967. *The Theory of Island Biogeography.* Princeton, NJ: Princeton University Press.

Macior, L. W. 1973. The pollination ecology of *Pedicularis* on Mount Ranier. *American J. of Botany* 60:863–871.

Mack, R. N. and J. L. Harper. 1977. Interference in dune annuals: spatial pattern and neighborhood effects. *J. of Ecology* 65:345–363.

Madronich, S., R. L. McKenzie, M. M. Caldwell and L. O. Bjorn. 1995. Changes in ultraviolet radiation reaching the earth's surface. *Ambio* 24:143–152.

Maguire, D. A. and R. T. T. Forman. 1983. Herb cover effects on tree seedling patterns in a mature hemlock-hardwood forest. *Ecology* 64:1367–1380.

Magurran, A. E. 1988. *Ecological diversity and its measurement.* Princeton, NJ: Princeton University Press.

Mahall, B. E. and F. H. Bormann. 1978. A quantitative description of the vegetative phenology of herbs in a northern hardwood forest. *Botanical Gazette* 139:467–481.

Major, J. 1977. California climate in relation to vegetation. In M. G. Barbour and J. Major, eds., *Terrestrial vegetation of California*, pp. 11–74. New York: Wiley-Interscience.

Manly, B.F.J. 1986. *Multivariate statistical methods: a primer.* New York: Chapman and Hall.

Margulis, L. 1996. Archaeal-eubacterial mergers in the origin of Eukarya: Phylogenetic classification of life. *Proceedings of the National Academy of Sciences* 93:1071–1076.

Marks, G. C. and T. T. Kozlowski, eds. 1973. *Ectomycorrhizae.* New York: Academic Press.

Marschner, H. 1995. *Mineral nutrition of higher plants*, 2nd ed. London: Academic Press.

Marsh, G. P. 1864. *Man and nature.* Reprint 1965. Ed. D. Lowenthal. Cambridge, MA: Harvard University Press.

Marshall, D. R. and S. K. Jain. 1969. Interference in pure and mixed populations of *Avena fatua* and *A. barbata*. *J. of Ecology* 57:251–270.

Martinez-Meza, E. and W. G. Whitford. 1996. Stemflow, throughfall and channelization of stemflow by roots in three Chihuahuan desert shrubs. *J. of Arid Environments* 32:271–287.

Mastrangelo, C. J., J. R. Trudell, and E. N. Cohen. 1978. Antagonism of membrane compression effects by high pressure gas mixtures in a phospholipid bilayer system. *Life Sciences* 22:239–244.

Matson, P. A., and P. M. Vitousek. 1990. Ecosystem approach to a global nitrous oxide budget. *Bioscience* 40:667–672.

Matson, P. A., L. Johnson, C. Billow, J. Miller, and R. Pu. 1994. Seasonal patterns and remote spectral estimation of canopy chemistry across the Oregon transect. *Ecological Applications* 4(2): 280–298.

Matson, P. A., C. Volkmann, K. Coppinger, and W. A. Reiners. 1991. Annual nitrous oxide flux and soil nitrogen characteristics in sagebrush steppe ecosystems. *Biogeochemistry* 14:1–12.

Matthaei, G. L. C. 1905. Experimental researches on vegetable assimilation and respiration. III. On the effect of temperature on carbon-dioxide assimilation. *Philosophical Transactions of the Royal Society of London, Series B* 197:47–105.

May, J. D. and K. T. Killingbeck. 1992. Effects of preventing nutrient resorption on plant fitness and foliar nutrient dynamics. *Ecology* 73(5): 1868–1878.

May, R.M. 1973. *Stability and Complexity in Model Ecosystems.* Princeton: Princeton University Press.

Mayer, E. 1992. A local flora and the biological species concept. *American J. of Botany* 79:222–238.

McCarthy, B.C. 1997. Response of a forest understory community to experimental removal of an invasive nonindigenous plant (*Alliaria petiolata*, Brassicaceae). In *Assessment and management of plant invasions.* ed. Luken, J.O. and J.W. Thieret, 117–130. New York: Springer-Verlag.

McCarthy, B.C. and D.R. Bailey. 1994. Distribution and abundance of coarse woody debris in a managed forest landscape of the central Appalachians. *Canadian J. of Forest Research* 24:1317–1329.

———. 1996. Composition, structure, and disturbance history of Crabtree Woods: and old-growth forest of western Maryland. *Bulletin of the Torrey Botanical Club* 123:350–365.

McCarthy, B.C., C.A. Hammer, G.L. Kauffman, and P.D. Cantino. 1987. Vegetation patterns and structure of an old-growth forest in southeastern Ohio. *Bulletin of the Torrey Botanical Club* 114:33–45.

McCarthy, B.C. and J.A. Quinn. Intra-crown analysis of growth and reproduction. *Oecologia* 91:30–38.

McCauley, D. E., J. E. Stevens, P. A. Peroni, and J. A. Raveill. 1996. The spatial distribution of chloroplast DNA and allozyme polymorphisms within a population of *Silene alba* (Caryophyllaceae). *American J. of Botany* 83:727–731.

McCook, L.J. 1994. Understanding ecological community succession: casual models and theories, a review. *Vegetatio* 110:115–147.

McDonald, C. D. and G. H. Hughes. 1964. *Studies of consumptive use of water by phreatophytes and hydrophytes near Yuma, Arizona.* Geological Survey Professional Paper 486-F. Washington, D. C.: U. S. Geological Survey.

McGraw, J. B. 1985a. Experimental ecology of *Dryas octopetala* ecotypes: relative response to competitors. *New Phytologist* 100:233–241.

———. 1985b. Experimental ecology of *Dryas octopetala* ecotypes. III. Environmental factors and plant growth. *Arctic and Alpine Research* 17:229–239.

McGraw, J. B. and J. Antonovics. 1983. Experimental ecology of *Dryas octopetala* ecotypes. I. Ecotypic differentiation and life-cycle stages of selection. *J. Ecology* 71:879–897.

McGuire, A.D. and L.A. Joyce. 1995. Responses of net primary production to changes in CO_2 and climate. In *Productivity of America's Forests and Climate Change*, ed. L.A. Joyce, pp. 9–45. U.S.D.A. Forest Service General Technical Report RM-271. Fort Collins, CO: U.S. Department of Agriculture, Forest Service.

McGuire, A. D., J. M. Melillo, and L. A. Joyce. 1995. The role of nitrogen in the response of forest net primary production to elevated atmospheric carbon dioxide. *Annual Rev. Ecol. Syst.* 26:437–503.

McInnes, P.F., R.J. Naiman, J. Pastor, and Y. Cohen. 1992. Effects of moose browsing on vegetation and litter of the boreal forest, Isle Royale, Michigan, USA. *Ecology* 73:2059–2075.

McIntosh, R. P. 1974. Plant ecology, 1947–1972. *Annals of the Missouri Botanical Garden* 61:132–165.

———. 1975. H. A. Gleason—individualistic ecologist, 1882–1975: his contribution to ecological theory. *Bulletin of the Torrey Botanical Club* 102:253–273.

———. 1976. Ecology since 1900. In *Evolution of issues, ideas and events in America 1776–1976*, ed. B. J. Taylor, pp. 353–372. Norman, OK: University of Oklahoma Press.

———. 1980a. The background and some current problems of theoretical ecology. *Synthese* 43:195–255.

———. 1980b. The relationship between succession and the recovery process in ecosystems. In *The recovery process in damaged ecosystems*, ed. J. Cairns, Jr., pp. 11–62. Ann Arbor, MI: Ann Arbor Science.

———. 1985. *The background of ecology: concept and theory.* New York: Cambridge University Press.

———. 1987. Pluralism in ecology. *Annual Review of Ecology and Systematics* 18:321–341.

McIntyre, L. 1985. Humboldt's way. *National Geographic* 168(3):318–351.

McLellan, T. M., J. D. Aber, and M. E. Martin. 1991a. Determination of nitrogen, lignin, and cellulose content of decomposing leaf material by near infrared reflectance spectroscopy. *Canadian J. of Forest Research* 21:1684–1688.

McLellan, T. M., M. E. Martin, J. D. Aber, J. M. Melillo, K. J. Nadelhoffer, and B. Dewes. 1991b. Comparison of wet chemistry and near infrared reflectance measurements of carbon-fraction chemistry and nitrogen concentation of forest foliage. *Canadian J. of Forest Research* 21 (11): 1689–1693.

McNaughton, S. J. 1966. Thermal inactivation properties of enzymes from *Typha latifolia* L. ecotypes. *Plant Physiology* 41:1736–1738.

———. 1967. Photosynthetic system II: racial differentiation in *Typha latifolia. Science* 156:1363.

———. 1972. Enzymatic thermal adaptations: the evolution of homeostasis in plants. *The American Naturalist* 106: 165–172.

———. 1983. Compensatory growth as a response to herbivory. *Oikos* 40:329–336.

McNaughton, S.J. and N.J. Georgiadis. 1986. Ecology of African grazing and browsing mammals. *Annual Review of Ecology and Systematics* 17:39–65.

McNeill, S. and T.R.E. Southwood. 1978. The role of nitrogen in the development of insect/plant relationships. In J.B. Harbone, ed., *Biochemical Aspects of Plant and Animal Coevolution*, pp. 77–98. London: Academic Press.

McPherson, J. K. and C. H. Muller. 1969. Allelopathic effects of *Adenostoma fasciculatum*, "chamise," in the California chaparral. *Ecological Monographs* 39:177–198.

Meentemeyer, V., E.O. Box, and R. Thompson. 1982. World patterns and amounts of terrestrial plant litter production. *BioScience* 32:125–128.

Meidner, H. and D. W. Sheriff. 1976. *Water and plants.* New York: Wiley.

Meidner, H. and T. A. Mansfield. 1968. *Physiology of stomata.* New York: McGraw-Hill.

Melillo, J. M., J. D. Aber, and J. F. Muratore. 1982. Nitrogen and lignin control of hardwood leaf litter decomposition dynamics. *Ecology* 63:621–626.

Melillo, J.M., A.D. McGuire, D.W. Kicklighter, B. Moore III, C.J. Vorosmarty, and A.L. Schloss. 1993. Global climate change and terrestrial net primary production. *Nature* 363:234–240.

Merriam, C. H. 1890. Results of a biological survey of the San Francisco mountain region and desert of the Little Colorado, Arizona. *North American Fauna* 3:1–136.

———. 1894. Laws of temperature control of the geographic distribution of animals and plants. *National Geographic* 6:229–238.

———. 1898. Life zones and crop zones of the United States. *U.S. Dept. of Agriculture Biological Survey Division Bulletin* 10:9–79.

Michaels, H.J., B. Benner, A.P. Hartgerink, T.D. Lee, S. Rice, M.F. Willson, and R.I. Bertin. 1988. Seed size variation: magnitude, distribution and ecological correlates. *Evolutionary Ecology* 2:157–166.

Michaelson, G.J., C.L. Ping, and J.M. Kimble. 1996. Carbon storage and distribution in tundra soils of Arctic Alaska, U.S.A. *Arctic and Alpine Research* 28:414–424.

Middleton, E. M. and A. H. Teramura. 1994. Understanding photosynthesis, pigment and growth responses induced by UV-B and UV-A irradiances. *Photochemistry and Photobiology* 60:38–45.

Milburn, J. A. 1996. Sap ascent in vascular plants: challengers to the cohesion theory ignore the significance of immature xylem and the recycling of Münch water. *Annals of Botany* 78:399–407.

Miller, A. and J. C. Thompson. 1975.. *Elements of meteorology.* 2nd ed. Columbus, OH: Merrill.

Miller, G. A. 1994. Functional significance of inflorescence pubescence in tropical alpine species of *Puya. In Tropical Alpine Environments: Plant form and function.* ed. Rundel, P. W., A. P. Smith and F. C. Meinzer. pp. 195–213.

Miller, P. C., D. K. Poole, and P. M. Miller. 1983. The influence of annual precipitation, topography, and vegetative cover on soil moisture and summer drought in southern California. *Oecologia* 56:385–391.

Milne, B. T. 1992. Spatial aggregation and neutral models in fractal landscapes. *American Naturalist* 139:32–57.

Milthorpe, F. L. 1955. The significance of the measurements made by the cobalt chloride paper method. *J. of Experimental Botany* 6:17–19.

Minchin, P.R. 1987. An evaluation of the relative robustness of techniques for ecological ordination. *Vegetatio* 69: 89–107.

Minnich, R. A. 1995. Fuel-driven fire regimes of the California chaparral. In J. E. Keeley and T. Scott eds., *Brushfires in California: Ecology and Resource Management.* International Association of Wildland Fire.

Miyawaki, A., K. Iwatsuki, and M. M. Grandtner. 1994. *Vegetation in eastern North America.* Tokyo: University of Tokyo Press.

Mogensen, H. L. 1996. The hows and whys of cytoplasmic inheritance in seed plants. *American J. of Botany* 83:383–404.

Mohler, C. L. 1983. Effect of sampling pattern on estimation of species distributions along gradients. *Vegetatio* 54: 97–102.

Moldenke, A. R. 1975. Niche specialization and species diversity along a California transect. *Oecologia* 21:219–242.

Monk, C.D. 1966. An ecological significance of evergreenness. *Ecology* 47:504–505.

Monk, C.D., D.W. Imm, R.L. Potter, and G.G. Parker. 1989. A classification of the deciduous forest of eastern North America. *Vegetatio* 80:167–181.

Monserud, R.A., O.V. Denissenko, T.P. Kolchugina, and N.M. Tchebakova. 1995. Change in phytomass and net primary productivity for Siberia from the mid-Holocene to the present. *Global Biogeochemical Cycles* 9:217–226.

Monson, R. K. and S. D. Smith. 1982. Seasonal water potential components of Sonoran Desert plants. *Ecology* 63: 113–123.

Mooney, H. A. 1977*a*. Southern coastal scrub. In M. G. Barbour and J. Major, eds., *Terrestrial Vegetation of California*, pp. 471–489. New York: Wiley Insterscience.

Mooney, H. A. ed. 1977b. *Convergent evolution in Chile and California: Mediterranean climate ecosystems.* Stroudsburg, PA: Dowden, Hutchinson, and Ross.

Mooney, H. A. 1989. Chaparral physiological ecology—Paradigms revisited. In *The California chaparral—paradigms re-examined.* ed. Keeley, S. C. pp. 85–90. Natural History Museum of Los Angeles County, Los Angeles.

Mooney, H. A. and W. D. Billings. 1961. Comparative physiological ecology of arctic and alpine populations of *Oxyria digyna. Ecological Monographs* 31:1–29.

Mooney, H. A. and E. L. Dunn. 1970b. Photosynthetic systems of Mediterranean-climate shrubs and trees of California and Chile. *The American Naturalist* 104:447–453.

Mooney, H. A. and J. R. Ehleringer. 1978. The carbon gain benefits of solar tracking in a desert annual. *Plant Cell and Environment* 1:307–311.

Mooney, H. A., J. Ehleringer, and J. Berry. 1976. High photosynthetic capacity of a winter annual in Death Valley. *Science* 194:322–323.

Mooney, H. A., J. Ehleringer, and O. Björkman. 1977. The energy balance of leaves of the evergreen desert shrub *Atriplex hymenelytra. Oecologia* 29:301–310.

Mooney, H. A. and A. T. Harrison. 1970. The influence of conditioning temperature on subsequent temperature-related photosynthetic capacity in higher plants. In *Prediction and measurement of photosynthetic productivity,* ed. C. T. de Wit, pp. 411–417. Wageningen, The Netherlands: Center for Agricultural Publishing and Documentation.

Mooney, H. A. and M. West. 1964. Photosynthetic acclimation of plants of diverse origin. *American J. of Botany* 51:825–827.

Moore, R. M., ed. 1970. *Australian grasslands.* Canberra, Australia: Australian National University Press.

Mueller-Dombois, D. 1986. Perspectives for an etiology of stand-level dieback. *Annual Review of Ecology and Systematics* 17:221–243.

Mueller-Dombois, D. and H. Ellenberg. 1974. *Aims and Methods of Vegetation Ecology.* New York: Wiley.

Muller, C. H. 1953. The association of desert annuals with shrubs. *American J. of Botany* 40:53–60.

Muller, C. H. 1966. The role of chemical inhibition (allelopathy) in vegetational composition. *Bulletin of the Torrey Botanical Club* 93:332–351.

Muller, W. H. and C. H. Muller. 1956. Association patterns involving desert plants that contain toxic products. *American J. of Botany* 43:354–361.

Mulroy, T. W. and P. W. Rundel. 1977. Annual plants: adaptations to desert environments. *BioScience* 27:109–114.

Murray, B. G. 1975. The cytology of the genus *Briza* L. (Gramineae). *Chromosoma* 49:299–308.

Murty, D., R.E. McMurtrie, and M.G. Ryan. 1996. Declining forest productivity in aging forest stands: a modeling analysis of alternative hypotheses. *Tree Physiology* 16:187–200.

Musil, C. F. 1995. Differential effects of elevated ultraviolet-B radiation on the photochemical and reproductive performance of dicotyledonous and monocotyledonous arid-environment ephemerals. *Plant, Cell and Environment* 18:844–854.

———. 1996. Accumulated effect of elevated ultraviolet-B radiation over multiple generations of the arid-environment annual *Dimorphotheca sinuata* DC. (Asteraceae). *Plant Cell and Environment* 19:1017–1027.

Nadelhoffer, K. J., M. R. Downs, B. Fry, J. D. Aber, A. H. Magill, and J. M. Melillo. 1995. The fate of ^{15}N-labelled nitrate additions to a northern hardwood forest in eastern Maine, USA. *Oecologia* 103:292–301.

Nadkarni, N. 1981. Canopy roots: convergent evolution in rainforest nutrient cycles. *Science* 213:1023–1024.

———. 1983. The effects of epiphytes on nutrient cycles within temperate and tropical rainforest tree canopies. Ph.D. Thesis, University of Washington, Seattle.

———. 1994. Factors affecting the initiation and growth of aboveground adventitious roots in a tropical cloud forest tree: an experimental approach. *Oecologia* 100:94–97.

Nadkarni, N. and R. Primack. 1989. A comparison of mineral uptake by above- and below-ground roots of *Salix syringiana* using gamma spectrometry. *Plant Soil* 113:39–45.

Naeem, S., et al. 1994. Declining biodiversity can alter the performance of ecosystems. *Nature* 368:734–737.

Naeem, S., et al. 1995. Empirical evidence that declining species diversity may alter the performance of terrestrial ecosystems. *Proceedings of the Royal Society of London* B347:249–262.

Nagashima, H. and I. Terashima. 1995. Relationships between height, diameter and weight distributions of *Chenopodium album* plants in stands: effects of dimension and allometry. *Annals of Botany* 75:181–188.

Ne'eman, G., H. Lahav, and I. Izhaki. 1992. Spatial pattern of seedlings one year after fire in a Mediterranean pine forest. *Oecologia* 91:365–370.

Neales, T. F. 1973. The effect of night temperatures on CO_2 assimilation, transpiration, and water use efficiency in *Agave americana* L. *Australian J. of Biological Science* 26:705–714.

Negash, L. 1987. Wavelength-dependence of stomatal closure by ultraviolet radiation in attached leaves of *Eragrostis:* Action spectra under backgrounds of red and blue lights. *Plant Physiology and Biochemistry* 25:753–760.

Neill, C. 1992. Comparison of soil coring and ingrowth methods for measuring belowground production. *Ecology* 73:1918–1921.

Neill, C., B. Fry, J. M. Melillo, P. A. Steudler, J. F. L. Morales, and C. C. Cerri. 1996. Forest- and pasture-derived carbon contributions to carbon stocks and microbial respiration of tropical pasture soils. *Oecologia* (Berlin) 107 (1): 113–119.

Newell, E.A. 1991. Direct and delayed costs of reproduction in *Aesculus californica. J. of Ecology* 79:365–378.

Ng, E. and P. C. Miller. 1980. Soil moisture relations in the southern California chaparral. *Ecology* 61:98–107.

Nichols, W. D. 1994. Groundwater discharge by phreatophyte shrubs in the Great Basin as related to depth to groundwater. *Water Resources Research* 30:3265–3274.

Nicolson, M. 1990. Henry Allan Gleason and the individualistic hypothesis: the structure of a botanist's career. *Botanical Review* 56:91–161.

Nienhuis, J., G. R. Sills, B. Martin, and G. King. 1994. Variance for water-use efficiency among ecotypes and recombinant inbred lines of *Arabidopsis thaliana* (Brassicaeceae). *American J. of Botany* 81:943–947.

Niering, W. A., R. H. Whittaker, and C. H. Lowe. 1963. The saguaro: a population in relation to its environment. *Science* 142:15–23.

Nilsen, E. T. and D. M. Orcutt. 1996. *Physiology of plants under stress: abiotic factors.* New York: Wiley.

Nilsen, E. T., M. R. Sharifi, P. W. Rundel, I. N. Forseth, and J. R. Ehleringer. 1990. Water relations of stem succulent trees in north-central Baja California. *Oecologia* 82:299–303.

Nobel, P.S. 1976. Water relations and photosynthesis of a desert CAM plant, *Agave deserti. Plant Physiology* 58:576–582.

———. 1977. Water relations and photosynthesis of barrel cactus, *Ferocactus acanthodes,* in the Colorado Desert. *Oecologia* 27:117–133.

———. 1980a. Morphology, surface temperatures, and northern limits of columnar cacti in the Sonoran desert. *Ecology* 61:1–7.

———. 1980b. Influences of minimum stem temperatures on ranges of cacti in south-western United States and central Chile. *Oecologia* 47:101–115.

———. 1982a. Orientation of terminal cladodes of platyopuntias. *Botanical Gazette* 143:219–224.

———. 1982b. Interaction between morphology, PAR interception, and nocturnal acid accumulation in cacti. In I. P. Ting and M. Gibbs, eds., *Crassulacean acid metabolism,* pp. 260–277. Rockville, MD: American Society of Plant Physiology.

———. 1982c. Low Temperature tolerance and cold hardening of cacti. *Ecology* 63:1650–1656.

———. 1991. *Physicochemical and Environmental Plant Physiology.* San Diego: Academic Press.

Nonami, H., E. D. Schulze, and H. Ziegler. 1991. Mechanisms of stomatal movement in response to air humidity, irradiance and xylem water potential. *Planta* 183:57–64.

Noy-Meir, I. and E. van der Maarel. 1988. Relations between community theory and community analysis in vegetation science: some historical perspectives. *Vegetatio* 69:5–15.

Nulsen, R. A., K. J. Bligh, I. N. Baxter, E. J. Solin, and D. H. Imrie. 1986. The fate of rainfall in a malle and heath vegetated catchment in southern western Australia. *Australian J. of Ecology.* 11:361–371.

O'Dowd, D.J. and A.M. Gill. 1986. Seed dispersal syndromes in Australian *Acacia.* In D.R. Murray, ed., *Seed Dispersal,* pp. 87–121. San Diego: Academic Press.

O'Lear, H. A, T. R. Seastedt, J. M. Briggs, J. M. Blair, and R. A. Ramundo. 1996. Fire and topographic effects on decomposition rates and nitrogen dynamics of buried wood in tallgrass prairie. *Soil Biology and Biochemistry* 28 (3) : 323–329.

O'Leary, M. H. 1988. Carbon isotopes in photosynthesis. *BioScience* 38:325–336.

O'Leary, M. H. and C. B. Osmond. 1980. Diffusional contributions to carbon isotope fractionation during dark CO_2 fixation in CAM plants. *Plant Physiology* 66:931–934.

Oberlander, G. T. 1956. Summer fog precipitation on the San Francisco peninsula. *Ecology* 37:851–852.

Odening, W. R., B. R. Strain, and W. C. Oechel. 1974. The effect of decreasing water potential on net CO_2 exchange of intact desert shrubs. *Ecology* 55:1086–1095.

Odum, E. P. 1971. *Fundamentals of Ecology.* 3rd ed. Philadelphia PA: W. B. Saunders.

Oechel, W. C., B. R. Strain, and W. R. Odening. 1972. Tissue water potential, photosynthesis, ^{14}C-labeled photosynthetic utilization and growth in the desert shrub *Larrea divaricata. Ecological Monographs* 42:127–141.

Oksanen, L. 1990. Predation, herbivory, and plant strategies along gradients of primary productivity. In J.B. Grace and D. Tilman, eds., *Perspectives on Plant Competition,* pp. 445–474. San Diego: Academic Press.

Olff, H. 1992. Effects of light and nutrient availability on dry matter and N allocation in six successional grassland species. *Oecologia* 89: 412–421.

Ollinger, S. V., J. D. Aber, G. M. Lovett, S. E. Millham, R. G. Lathrop, and J. M. Ellis. 1993. A spatial model of atmospheric deposition for the northeastern U.S. *Ecological Applications* 3 (3): 459–472.

Oppenheimer, H. R. 1960. Adaptation to drought: xerophytism. In *Plant-water relationships in arid and semi-arid conditions,* pp. 105–138. Paris: UNESCO.

Orlóci, L. 1966. Geometric models in ecology. I. The theory and application of some ordination methods. *J. of Ecology* 54:193–215.

Orshan, G. 1972. Morphological and physiological plasticity in relation to drought. In C. M. McKell, J. P. Blaisdell, and

J. R. Goodin, eds., *Wildland shrubs—their biology and utilization*, pp. 245–254. U.S. Dept. of Agriculture Forest Service General Technical Report INT-1. Ogden, UT: U.S. Dept. of Agriculture Forest Service.

Osmond, C. B. 1976. CO_2 assimilation and dissimilation in the light and dark in CAM plants. In R. H. Burris and C. C. Black, eds., *CO_2 metabolism and plant productivity*, pp. 217–233. Baltimore, MD: University Park Press.

Osmond, C. B., O. Björkman, and D. J. Anderson. 1980. Physiological processes in plant ecology: toward and synthesis with *Atriplex. Ecological Studies,* Series #36. Berlin: Springer/Verlag.

Osonubi, O. And W. J. Davies. 1978. Solute accumulation in leaves and roots of woody plants subjected to water stress. *Oecologia* 32:323–332.

Owensby, C.E., P.I. Coyne, and L.M. Auen. 1993. Nitrogen and phosphorus dynamics of a tallgrass prairie ecosystem exposed to elevated carbon dioxide. *Plant, Cell, and Environment* 16:843–850.

Ownbey, R. S. and B. E. Mahall. 1983. Salinity and root conductivity: differential responses of coastal succulent halophyte, *Salicornia virginica* and a weedy glycophyte, *Raphanus sativus. Physiologia Plantarum* 57:189–195.

Paczoski, J. 1921. *Osnowy fitosocjologji.* Cherson, Izd. Stud. Comitet Tech.

Paige, K.N. 1992. Overcompensation in response to mammalian herbivory—from mutualistic to antagonistic interactions. *Ecology* 73:2076–2085.

———. 1994. Herbivory and *Ipomopsis aggregata*—differences in response, differences in experimental protocol—a reply. *American Naturalist* 143:739–749.

Paige, K.N. and T. G. Whitham. 1987. Overcompensation in response to mammalian herbivory: the advantages of being eaten. *American Naturalist* 129:315.

Paine, R. T. 1966. Food web complexity and species diversity. *American Naturalist* 100:65–75.

———. 1969. A note on trophic complexity and community stability. *American Naturalist* 103:91–93.

Pallardy, S. G., J. S. Pereira, and W. C. Parker. 1991. Measuring the state of water in tree systems. In Lassoie, J. P. and T. M. Hinckley, eds., *Techniques and approaches in forest tree ecophysiology*, pp. 27–76. Boca Raton: CRC Press.

Palmer, M.W. 1993. Putting things in even better order: the advantages of canonical correspondence analysis. *Ecology* 74:2215–2230.

Palta, J. 1983. Photosynthesis, transpiration, and leaf diffusive conductance of the cassava leaf in response to water stress. *J. of Botany* 61:373–375.

Parfitt, R. L., H. J. Percival, R. A. Dahlgren, and L. F. Hill. 1997. Soil and soil solution chemistry under pasture and radiata pine in New Zealand. *Plant and Soil* (in press).

Pardo, L. H., C. T. Driscoll, and G. E. Likens. 1995. Patterns of nitrate loss from a chronosequence of clear-cut watersheds. *Water, Air, and Soil Pollution* 95:1659–1664.

Parker, G. G. 1983. Throughfall and stemflow in the forest nutrient cycle. *Advances in Ecological Research* 13:58–134.

Parker, I.M. 1997. Pollinator limitation of *Cytisus scoparius* (Scotch broom), an invasive exotic shrub. *Ecology* 78:1457–1470.

Parker, W. C. and S. J. Colombo. 1995. A critical re-examination of pressure-volume analysis of conifer shoots: comparison of three procedures for generating PV curves on shoots of *Pinus resinosa* Ait. Seedlings. *Journal of Experimental Botany* 46:1701–1709.

Parkinson, K. J. and B. J. Legg. 1972. A continuous flow porometer. *J. of Applied Ecology* 9:669–675.

Parmeter, J. R. Jr. 1977. Effects of fire on pathogens. In *Proc. Symp. Env. Cons. Fire and Fuel Management in Mediterranean Ecosyst.* pp. 58–64. Washington, D.C.: U.S.D.A. Forest Service.

Parrish, J. A. D. and F. A. Bazzaz. 1985. Nutrient content of *Abutilon theophrasti* seeds and the competitive ability of the resulting plants. *Oecologia* 65:247–251.

Paschke, M.W., J.O. Dawson, and B.M. Condon. 1994. *Frankia* in prairie, forest, and cultivated soils of central Illinois, USA. *Pedobiologia* 38:546–551.

Pastor, J., J.D. Aber, C.A. McClaugherty, and J.M. Melillo. 1984. Aboveground production and N and P cycling along a nitrogen mineralization gradient on Blackhawk Island, Wisconsin. *Ecology* 65:256–268.

Pastor, J., B. Dewey, R. J. Naiman, P. F. McInnes, and Y. Cohen. 1993. Moose browsing and soil fertility in the boreal forests of Isle Royale National Park. *Ecology* 74(2): 467–480.

Pastor, J. and W.M. Post. 1986. Influence of climate, soil moisture, and succession on forest carbon and nitrogen cycles. *Biogeochemistry* 2:3–27.

Pearcy, R. W. 1976. Temperature responses of growth and photosynthetic CO_2 exchange rates in coastal and desert races of *Atriplex lentiformis. Oecologia* 26:245–255.

———. 1983. The light environment and growth of C_3 and C_4 tree species in the understory of a Hawaiian forest. *Oecologia* 58:19–25.

Pearcy, R. W. and H. W. Calkin. 1983. The light environment and growth of C_3 and C_4 tree species in the understory of a Hawaiian forest. *Oecologia* 58:19–25.

Pearcy, R. W., J. Ehleringer, H. A. Mooney and P. W. Rundel, eds. 1989. Plant physiological ecology: field methods and instrumentation. New York. Chapman and Hall.

Pearcy, R. W. and A. T. Harrison. 1974. Comparative photosynthetic and respiratory gas exchange characteristics of *Atriplex lentiformis* (Torr.) Wats. in coastal and desert habitats. *Ecology* 55:1104–1111.

Pearcy, R. W., E.-D. Schulze and R. Zimmermann. 1988. Measurements of transpiration and leaf conductance. In

Pearcy, R. W., J. Ehleringer, H.A. Mooney, and P.W. Rundel, eds., *Plant Physiological Ecology: Field Methods and Instrumentation*, pp. 137–160. New York: Chapman and Hall.

Pearson, G. A. 1942. Herbaceous vegetation a factor in natural regeneration of ponderosa pine in the southwest. *Ecological Monographs* 12:315–338.

Peet, R. K. 1998. Forests of the Rocky Mountains. In M. G. Barbour and W. D. Billings, eds., *North American Terrestrial Vegetation*, 2nd ed. New York: Cambridge University Press.

Peet, R.K. and N.L. Christensen. 1980. Succession: a population process. *Vegetatio* 43:131–140.

Peet, R.K., R.G. Knox, J.S. Case, and R.B. Allen. 1988. Putting things in order: the advantages of detrended correspondence analysis. *American Naturalist* 131:924–934.

Peinado, M., J. L. Aguirre, and J. Delgadillo. 1997. Phytosociological, bioclimatic, and biogeographical classification of woody climax communities of western North America. *J. of Vegetation Science* 8:505–528.

Peinado, M., F. Alcaraz, J. L. Aguirre, and J. Delgadillo. 1995. Major plant communities of warm North American deserts. *J. of Vegetation Science* 6:79–94.

Pellmyr, O. and C.J. Huth. 1994. Evolutionary stability of mutualism between yucca and yucca moths. *Nature* 372:257–260.

Pellmyr, O., J. Leebensmack, and C.J. Huth. 1996. Non-mutualistic yucca moths and their evolutionary consequences. *Nature* 380:155–156.

Penman, H. L. 1950. Evaporation over the British Isles. *Quarterly J. of the Royal Meteorological Society* 76:372–383.

Perry, D.A. 1994. *Forest ecosystems.* Baltimore: The Johns Hopkins University Press.

Peterjohn, W. T., M. B. Adams, and F. S. Gilliam. 1996. Symptoms of nitrogen saturation in two central Appalachian hardwood forest ecosystems. *Biogeochemistry* 35:507–522.

Peterjohn, W. T., J. M. Melillo, P. A. Steudler, and K. M. Newkirk. 1994. Responses of trace gas fluxes and N availability to experimentally elevated soil temperatures. *Ecological Applications* 4 (3): 617–625.

Peters, G. A. 1978. Blue-green algae and algal associations. *BioScience* 28:580–585.

Peters, R. and T. Ohkubo. 1990. Architecture and development in *Fagus japonica-F. crenata* forest near Mt. Takahara, Japan. *J. of Vegetation Science* 1:499–506.

Peterson, E. B. 1969. Radiosonde data for characterizing a mountain environment in British Columbia. *Ecology* 50:200–205.

Phillips, D. L. and J. A. MacMahon. 1981. Competition and spacing patterns in desert shrubs. *J. Ecology* 69:97–115.

Phillips, J. 1935. Succession, development, the climax, and the complex organism: an analysis of concepts, part 3. *J. of Ecology* 23:488–508.

Phillips, J. G. and S. J. Riha. 1994. Root growth, water uptake and canopy development in *Eucalyptus viminalis* seedlings. *Australian J. of Plant Physiology* 21:69–78.

Pianka, E. R. 1980. Guild structure in desert lizards. *Oikos* 35:194–201.

Pickett, S.T.A. 1976. Succession: an evolutionary interpretation. *The American Naturalist* 110:107–119.

Pickett, S. T. A. 1980. Non-equilibrium coexistence of plants. *Bulletin of the Torrey Botanical Club* 107:238–248.

Pickett, S.T.A, S.L. Collins, and J.J. Armesto. 1987. Models, mechanisms and pathways of succession. *The Botanical Review* 53:335–371.

Pielou, E. C. 1961. Segregagation and symmetry in two-species populations as studied by nearest neighbor relations. *J. of Ecology* 49:255–269.

———. 1977. *Mathematical Ecology.* New York: Wiley.

———. 1981. The usefulness of ecological models. *Quarterly Review of Biology* 56:17–31.

———. 1984. *The interpretation of ecological data.* New York: J. Wiley and Sons.

Pinero, D., J. Sarukhan, and P. Alberdi. 1982. The costs of reproduction in tropical palm, *Astrocaryum mexicanum.* *J. of Ecology* 70:473–481.

Pirozynski, K.A. 1981. Interactions between fungi and plants through the ages. *Canadian Journal of Botany* 59:1824–1827.

Pirozynski, K.A. and D.W. Malloch. 1975. The origin of land plants: a matter of mycotropism. *Biosystems* 6:153–164.

Pitelka, L.F., S.B. Hansen and J.W. Ashmun. 1985. Population biology of *Clintonia borealis.* I. Ramet and patch dynamics. *J. of Ecology* 73:169–183.

Platenkamp, G. A. and R. G. Shaw. 1993. Environmental and genetic maternal effects on seed characters in Nemophila menziesii. *Evolution* 47:540–555.

Platt, W. J. 1998. Southeastern pine savannas. In R. C. Anderson, J. S. Fralish, and J. Baskin, eds. *The Savanna, Barrens, and Rock Outcrop Communities of North America.* Cambridge, England: Cambridge University Press.

Platt, W. J., G. W. Evans, and S. L. Rathbun. 1988. The population dynamics of a long-lived conifer (*Pinus palustris*). *The American Naturalist* 131:491–525.

Platt, W. J. and I. M. Weiss. 1977. Resource partitioning and competition within a guild of fugitive prairie plants. *American Naturalist* 111:479–513.

———. 1985. An experimental study of competition among fugitive prairie plants. *Ecology* 66:708–720.

Pockman, W. T., J. S. Sperry, and J. W. O'Leary. 1995. Sustained and significant negative water pressure in xylem. *Nature* 378:715–716.

Poore, M. E. D. 1955a. The use of phytosociological methods in ecological investigations. I. The Braun-Blanquet system. *J. of Ecology* 43:226–244.

———. 1955b. The use of phytosociological methods in ecological investigations. II. Practical issues involved in an attempt to apply the Braun-Blanquet system. *J. of Ecology* 43:245–269.

Portnoy, S. and M.F. Willson. 1993. Seed dispersal curves: behavior of the tail of the distribution. *Evolutionary Ecology* 7:25–44.

Powell, J.A. 1992. Inter-relationships of yuccas and yucca moths. *Trends in Ecology and Evolution* 7:10–15.

Power, M. E., et al. 1996. Challenges in the quest for keystones. *BioScience* 46:609–620.

Prentice, I.C. 1977. Non-metric ordination methods in ecology. *J. of Ecology* 65:85–94.

Prescott, C.E., J.P. Corbin, and D. Parkinson. 1989. Biomass, productivity, and nutrient-use efficiency of aboveground vegetation in four Rocky Mountain coniferous forests. *Canadian Journal of Forest Research* 19:309–317.

Pressland, A. J. 1973. Soil moisture redistribution as affected by throughfall and stemflow in an arid zone shrub community. *Australian J. of Botany* 21:235–245.

Price, P.W., G.L. Waring, R. Julkunen-Tiitto, J. Tahvanainen, H.A. Mooney, and T.P. Craig. 1989. Carbon-nutrient balance hypothesis in within-species phytochemical variation of *Salix lasiolepis*. *Journal of Chemical Ecology* 15:1117–1131.

Pulliam, H. R. 1989. Sources, sinks and population regulation. *American Naturalist* 132:652–661.

Puritch, G. S. and N. C. Turner. 1973. Effects of pressure increase and release on temperature within a pressure chamber used to estimate water potential. *J. of Experimental Botany* 24:342–348.

Pyne, S. J., P. L. Andrews, and R. D. Laven. 1996. *Introduction to Wildland Fire,* 2nd ed. New York: Wiley.

Quinn, J. A. 1978. Plant ecotypes: ecological or evolutionary units? *Bulletin of the Torrey Botanical Club* 105:58–64.

———. 1987. Complex patterns of genetic differentiation and phenotypic plasticity versus an outmoded ecotype terminology. In Differentiation patterns in higher plants, pp. 95–113. New York: Academic Press.

Quinn, J. A. and J. C. Colosi. 1977. Separating genotype from environment in germination ecology studies. *American Midland Naturalist* 97:484–489.

Röhrig, E. 1991. Biomass and productivity. In *Ecosystems of the World,* pp. 165–174. Amsterdam: Elsevier Scientific Pub. Co.

Rabotnov, T. A. 1953. L. G. Ramensky. [In Russian.] *Botanisheskii Zhurnal* 38:5.

———. 1978. Structure and method of studying coenotic populations of perennial herbaceous plants. *Soviet J. of Ecology* 9:99–105.

Radosevich, S. R., J. S. Holt and C. Ghersa. 1997. *Weed Ecology: Applications Management.* New York: Wiley.

Raich, J.W., E.B. Rastetter, J.M. Melillo, D.W. Kicklighter, P.A. Steudler, B.J. Peterson, A.L. Grace, B. Moore III, and C.J. Vörösmarty. 1991. Potential net primary productivity in South America: application of a global model. *Ecological Applications* 1:399–429.

Raison, J. K., J. A. Berry, P. A. Armond, and C. S. Pike. 1980. Membrane properties in relation to the adaptation of plants to high and low temperature stress. In N. C. Turner and P. J. Kraemer, eds., *Adaptations of plants to water and high temperature stress,* pp. 261–277. New York: Wiley/Interscience.

Ranwell, D. S. 1972. *Ecology of salt marshes and sand dunes.* London: Chapman and Hall.

Raschke, K. 1976. How stomata resolve the dilemma of opposing priorities. *Philosophical Transactions of the Royal Society of London, Series B* 273:551–560.

Raunkiaer, C. 1934. *The life forms of plants and statistical plant geography.* Oxford: Clarendon Press.

Raven, J. A. 1983. Phytophages of xylem and phloem: a comparison of animal and plant sap-feeders. *Advances in Ecological Research* 13:135–334.

Raven, P. H. and D. I. Axelrod. 1978. *Origin and relationships of the California flora.* Berkeley: University of California Press.

Raven, P. H., R. F. Evert, and S. E. Eichhorn. 1998. *Biology of Plants* 6th ed. New York: Worth Publishers.

Rawlins, S. L. 1976. Measurement of water content and the state of water in soils. In T. T. Kozlowski, ed., *Water deficits and plant growth,* vol. IV, pp. 1–55. New York: Academic Press.

Real, L. (ed.). 1983. *Pollination ecology.* New York: Academic Press.

Reekie, E.G. and F.A. Bazzaz. 1987a. Reproductive effort in plants. 1. Carbon allocation to reproduction. *American Naturalist* 129:876–896.

———. 1987b. Reproductive effort in plants. 2. Does carbon reflect the allocation of other resources? *American Naturalist* 129:897–906.

Reichard, S.H. and C.W. Hamilton. 1997. Predicting invasions of woody plants introduced into North America. *Conservation Biology* 11:193–203.

Reimold, R. J. and W. H. Queen. 1974. *Ecology of halophytes.* New York: Academic Press.

Reiners, W. A. 1992. Twenty years of ecosystem reorganization following experimental deforestation and regrowth suppression. *Ecological Monographs* 62(4):503–523.

Reiners, W. A., A. F. Bouwman, W. F. J. Parsons, and M. Keller. 1994. Tropical rain forest conversion to pasture: changes in vegetation and soil properties. *Ecological Applications* 4(2): 363–377.

Rejmanek, M. 1989. Invasibility of plant communities. In J. A. Drake et al. (eds.), Biological invasions: a global perspective, pp. 369–388. New York: Wiley.

———. 1995. What makes a species invasive? P. Pysek et al. (eds.), *Plant invasions: general aspects and special problems*, pp. 3–13. Amsterdam: Academic Publishing.

———. 1996a. A theory of seed plant invasiveness: the first sketch. *Biological Conservation* 78:171–181.

———. 1996b. Species richness and resistance to invasions. In G. Orians et al. (eds.), Biodiversity and ecosystem processes in tropical forests, pp. 153–172. New York: Springer-Verlag.

Rejmanek, M. and D.M. Richardson. 1996. What attributes make some plant species more invasive? *Ecology* 77: 1655–1661.

Rhoades, D. H. and R. G. Cates. 1976. Toward a general theory of plant antiherbivore chemistry. *Recent Advances in Phytochemistry* 10:168–213.

Richards, J.H. and M.M. Caldwell. 1987. Hydraulic lift: substantial nocturnal water transport between soil layers by *Artemisia tridentata* roots. *Oecologia* 73:486–489.

Richards, L. A. and G. Ogata. 1958. Thermocouple for vapor pressure measurements in biological and soil systems at high humidity. *Science* 128:1089–1090.

Richards, P. W. 1936. Ecological observations on the rain forest of Mount Dulit, Sarawak. *J. of Ecology* 24:1–37 and 340–360.

Richter, D. D. and D. Markewitz. 1997. How deep is soil? *BioScience* 45:600–609.

Richter, D. D. and L. I. Babbar. 1991. Soil diversity in the tropics. *Advances in Ecological Research* 21:315–389.

Richter, H. 1997. Water relations of plants in the field: some comments on the measurement of selected parameters. *J. of Experimental Botany* 48:1–7.

Ricklefs, R. E. 1997. *Economy of Nature* 4th ed. New York: W. H. Freeman.

Risser, P. G. and P. H. Zedler. 1968. An evaluation of the grassland quarter method. *Ecology* 49:1006–1009.

Ritchie, G. A. and T. M. Hinckley. 1975. The pressure chamber as an instrument for ecological research. *Advances in Ecological Research* 9:166–254.

Rivas-Martinez, S. 1997. Syntaxonomical synopsis of the potential natural plant communities of North America, I. *Itinera Geobotanica* 10:5–148.

Robberecht, R. and M. M. Caldwell. 1980. Leaf ultraviolet optical properties along a latitudinal gradient in the arctic alpine life zone. *Ecology* 61:612–619.

Roberts, M.R. and N.L. Christensen. 1988. Vegetation variation among mesic successional forest stands in northern lower Michigan. *Canadian J. of Botany* 66: 1080–1090.

Roberts, M.R. and F.S. Gilliam. 1995a. Patterns and mechanisms of diversity in forested ecosystems: implications for forest management. *Ecological Applications* 5:969–977.

———. 1995b. Disturbance effects on herbaceous layer vegetation and soil nutrients in *Populus* forests of northern lower Michigan. *J. of Vegetation Science* 6:903–912.

Roberts, M.R. and C.J. Richardson. 1985. Forty-one years of population change and community succession in aspen forests on four soil types, northern lower Michigan, U.S.A. *Canadian J. of Botany* 63:1641–1651.

Roberts, S. W. and K. R. Knoerr. 1977. Components of water potential estimated from xylem pressure measurements in five tree species. *Oecologia* 28:191–202.

Roberts, S. W., B. R. Strain, and K. R. Knoerr. 1980. Seasonal patterns of leaf water relations in four co-occurring forest trees species: parameters from pressure volume curves. *Oecologia* 46:330–337.

Rodin, L. E. and N. I. Basilevic. 1967. *Production and mineral cycling in terrestrial vegetation*. Edinburgh, Scotland: Oliver and Boyd.

Rogers, R. S. 1982. Early spring herb communities in mesophytic forests of the Great Lakes region. *Ecology* 63: 1050–1063.

———. 1985. Local coexistence of deciduous forest ground layer species growing in different seasons. *Ecology* 66: 701–707.

Romme, W. H. and D. G. Despain. 1989. Historical perspective on the Yellowstone fires of 1988. *Bioscience* 39 (10): 695–699.

Roodman, D.M. 1994. Global temperature rises slightly. In *Vital Signs 1994*, L.R. Brown, H. Kane, and D.M. Roodman. New York: W.W. Norton and Co.

Rorison, I. H., ed. 1969. *Ecological aspects of the mineral nutrition of plants*. Oxford: Blackwell.

Rosenberg, N. J. 1974. *Micro-climate: the biological environment*. New York: Wiley.

Rosenthal, J.P. and P.M. Kotanen. 1994. Terrestrial plant tolerance to herbivory. *Trends in Ecology and Evolution* 9: 145–148.

Roughgarden, J. 1975. Evolution of marine symbiosis—a simple cost-benefit model. *Ecology* 56:1201–1208.

———. 1979. *Theory of Population Genetics and Evolutionary Ecology: an Introduction*. New York: Macmillan.

Rowe, J. S. 1964. Environmental preconditions with special reference to forestry. *Ecology* 45:399–403.

Rowell, D. L. 1994. *Soil science: methods and applications.* Essex, England: Longman Scientific and Technical.

Ruess, R. W., D. S. Hik, and R. L. Jefferies. 1989. The role of lesser snow geese as nitrogen processors in a sub-arctic salt marsh. *Oecologia* 79:23–29.

Rundel, P. W. and R. E. Stecker. 1977. Morphological adaptations of tracheid structure to water stress gradients in the crown of *Sequioadendron giganteum. Oecologia* 27:135–139.

Rundel, P. W., J. R. Ehleringer, and K. A. Nagy, eds. 1988. *Stable isotopes in ecological research.* New York. Springer-Verlag.

Russell, E. W. B. 1983. Indian-set firest in the forests of the northeastern United States. *Ecology* 64:78–88.

Sage, R.F. 1995. Was low atmospheric CO_2 during the Pleistocene a limiting factor for the origin of agriculture? *Global Change Biology* 1:93–106.

Salama, R. B., G. A. Bartle, and P. Farrington. 1994. Water use of plantation *Eucalyptus camaldulensis* estimated by groundwater hydrograph separation techniques and heat pulse method. *J. of Hydrology* 156:163–180.

Salinger, M. J. 1995. Southwest Pacific temperatures: trends in maximum and minimum temperatures. *Atmospheric Research* 37:87–100.

Sampson, A. W. 1944. Plant succession on burned chaparral lands in northern California. *California Agriculture Experiment Station Bulletin* 685:1–144.

Sampson, R. N., M. Apps, S. Brown, C. V. Cloe, J. Downing, L. S. Heath, D. S. Ojima, T. M. Smith, A. M. Solomon, and J. Wisniewski. 1993. Workshop summary statement: terrestrial biospheric carbon fluxes—quantification of sinks and sources of CO_2. *Water, Air, and Soil Pollution* 70:3–15.

Samson, F. and F. Knopf. 1994. Prairie conservation in North America. *Bioscience* 44(6): 418–421.

Sarukhan, J. and M. Gadgil. 1974. Studies on plant demography: *Ranunculus repens* L., *R. bulbosa* L., and *R. acris* L: III A mathematical model incorporating multiple modes of reproduction. *J. of Ecology* 62:921–936.

Saunier, R. E. and R. F. Wagle. 1965. Root grafting *in Quercus turbinella* Greene. *Ecology* 46:749–750.

Sawyer, J. O. and T. Keeler-Wolf. 1995. *A manual of California vegetation.* Sacramento: California Native Plant Society.

Scheiner, S. M. 1993. Introduction: theories, hypotheses and statistics. Pages 1–13 in (S. M. Scheiner and J. Gurevitch, eds.) Design and Analysis of Ecological Experiments. New York: Chapman and Hall.

Schimel, D.S. 1995. Terrestrial ecosystems and the carbon cycle. *Global Change Biology* 1:77–91.

Schimper, A. F. W. 1898. *Pflanzengeographie auf physiologischer Grundlage.* Jena, Germany: Fisher.

———. 1903. *Plant geography upon a physiological basis.* Translated by W. R. Fisher. Oxford: Clarendon Press.

Schlesinger, W.H. 1997. *Biogeochemistry: An Analysis of Global Change,* 2nd ed. San Diego: Academic Press.

Schmitz, H., H. Bleckmann, and M. Murtz. 1997. Infrared detection in a beetle. *Nature* 386:773–774.

Schneider, S.H. 1989. The changing climate. *Scientific American* 261:70–79.

Scholander, P. F. 1968. How mangroves desalinate seawater. *Physiologia Plantarum* 21:251–261.

Scholander, P. F., H. T. Hammel, E. A. Hemmingsen, and E. D. Bradstreet. 1964. Hydrostatic pressure and osmotic potential in leaves of mangroves and some other plants. *National Academy of Sciences* 52:119–125.

Schreiber, U., W. Bilger, and C. Neubauer. 1995. Chlorophyll fluorescence as a nonintrusive indicator for rapid assessment of *in vivo* photosynthesis. In E.-D. Schulze, and M. M. Caldwell, eds., *Ecophysiology of photosynthesis,* pp. 49 70. Berlin: Springer-Verlag.

Schullery, P. 1989. The fires and fire policy. *Bioscience* 39 (10): 686–694.

Schulze, E.-D. and M. M. Caldwell. 1995. Overview: Perspectives in ecophysiological research of photosynthesis. In E.-D Schulze, and M. M. Caldwell, eds., *Ecophysiology of photosynthesis,* pp. 553–564. Berlin: Springer-Verlag.

Schulze, E.-D., O. L. Lange, U. Buschbom, L. Kappen, and M. Evenari. 1972. Stomatal responses to changes in humidity in plants growing in the desert. *Planta* 108:259–270.

———. 1973. Stomatal responses to changes in temperature at increasing water stress. *Planta* 110:29–42.

Schulze, E.-D. and H. A. Mooney, eds. 1993. *Biodiversity and Ecosystem Function.* Springer-Verlag New York.

Schulze, E.-D., H. Ziegler, and W. Stichler. 1976. Environmental control of crassulacean acid metabolism in *Welwitschia mirabilis* Hook. Fil. in its range of natural distribution in the Namib desert. *Oecologia* 24:323–334.

Schwartz, M.W. 1993. Modelling effects of habitat fragmentation on the ability of trees to respond to climatic warming. *Biodiversity and Conservation* 2:51–61.

———. 1997. Defining indigenous species: an introduction. In J.O. Luken and J.W. Thieret, eds., *Assessment and Management of Plant Invasions,* pp. 7–17. New York: Springer.

Schwartz, M. W. and S. M. Hermann. 1993. The continuing population decline of *Torreya taxifolia* Arn. *Bulletin of Torrey Botanical Club* 120:275–28.

Schwartz, M.W. and J. D. Hoeksema. 1998. Specialization and resource trade: biological markets and a conceptual model for the evolution of mutualisms. *Ecology* (in press).

Scott, F. M. 1932. Some features of the anatomy of *Fouquieria splendens. American J. of Botany* 19:673–678.

Scott, G. D. 1969. *Plant symbiosis.* New York: St. Martin's Press.

Scott, L., G. T. Auble, and J. M. Friedman. 1997. Flood dependency of cottonwood establishment along the Missouri River, Montana, USA. *Ecological Applications* 7:677–690.

Searles, P. S., M. M. Caldwell and K. Winter. 1995. The response of five tropical dicotyledon species to solar ultraviolet-B radiation. *American J. of Botany* 82:445–453.

Seastedt, T. R. 1985. Canopy interception of nitrogen in bulk precipitation by annually burned and unburned tallgrass prairie. *Oecologia* 66:88–92.

———. 1988. Mass, nitrogen, and phosphorus dynamics in foliage and root detritus of tallgrass prairie. *Ecology* 69 (1): 59–65.

Seastedt, T. R., J. M. Briggs, and D. J. Gibson. 1991. Controls of nitrogen limitation in tallgrass prairie. *Oecologia* 87:72–79.

Seastedt, T.R., C.C. Coxwell, D.S. Ojima, and W.J. Parton. 1994. Controls of plant and soil carbon in a semihumid temperate grassland. *Ecological Applications* 4:344–353.

Seastedt, T. R. and A. K. Knapp. 1993. Consequences of nonequilibrium resource availability across multiple time scales: the transient maxima hypothesis. *The American Naturalist* 141:621–633.

Sellers, P.J., F.G. Hall, G. Asrar, D.E. Strebel, and R.E. Murphy. 1992. An overview of the First International Satellite Land Surface Climatology Project (ISLSCP) Field Experiment (FIFE). *J. of Geophysical Research* 97:18345–18371.

Selter, C. M., W. D. Pitts, and M. G. Barbour. 1986. Site microenvironment and seedling survival of Shasta red fir. *American Midland Naturalist* 115:288–300.

Semikhatova, O. A. 1960. The after-effect of temperature on photosynthesis. *Botanisheskii Zhurnal* 45:1488–1501.

Sestak, Z., J. Catsky, and P. G. Jarvis. 1971. *Plant photosynthetic production: manual of methods.* The Hague: Junk.

Shaffer, M. L. 1981. Minimum population sizes for species conservation. *Bioscience* 31:131–134.

Shelford, V. E. 1913. *Animal communities in temperate America.* Chicago: University of Chicago Press.

Shimwell, D. W. 1971. *The description and classification of vegetation.* Seattle, WA: University of Washington Press.

Shipley, B. and R.H. Peters. 1990. A test of the Tilman model of plant strategies: relative growth rate and biomass partitioning. *American Naturalist* 136:139–153.

Shirley, H. L. 1945. Reproduction of upland conifers in the Lake States as affected by root competition and light. *American Midland Naturalist* 33:537–612.

Shmida, A. 1984. Whittaker's plant diversity sampling method. *Israel J. Botany* 33:41–46.

Shotola, S. J., J. R. Guntenspergen, C. P. Dunn, L. A. Leitner and F. Stearns. 1992. Sugar maple invasion in an old-growth oak-hickory forest in southern Illinois. *American Midland Naturalist* 127:125–138.

Shreve, E. B. 1923. Seasonal changes in the water relations of desert plants. *Botanical Gazette* 39:397–408.

Shugart, H. H. 1984. *A theory of forest dynamics.* New York: Springer.

Shumway, S.W. and M.D. Bertness. 1992. Salt stress limitation of seedling recruitment in a salt marsh plant community. *Oecologia* 92:490–497.

Shure, D. J. and A. J. Lewis. 1973. Dew formation and stem flow on common ragweed (*Ambrosia artemisiafolia*). *Ecology* 54:1152–1155.

Siccama, T. G., F. H. Bormann, and G. E. Likens. 1970. The Hubbard Brook ecosystem study: productivity, nutrients, and phytosociology of the herbaceous layer. *Ecological Monographs* 40:389–402.

Silvertown, J., M. Franco and E. Menges. 1996. Interpretation of elasticity matrices as an aid to the management of plant populations for conservation. *Conservation Biology* v10:591–597.

Silvertown, J.W. 1980. The evolutionary ecology of mast seeding in trees. *Biological J. of the Linnean Society* 14:235–250.

Silvertown, J. W. 1982. *Introduction to Plant Population Ecology.* London: Longman.

Silvertown, J. W. and J. L. Doust. 1993. Introduction to Plant Population Biology. Oxford, England: Blackwell Scientific Publications.

Simberloff, D. 1988. The contribution of population and community biology to conservation science. *Annual Review of Ecology and Systematics* 19:473–511.

Simms, E.L. and M. D. Rausher. 1989. The evolution of resistance to herbivory in *Ipomoea purpurea*. II. Natural selection by insects and costs of resistance. *Evolution* 43:573–585.

Singer, M. J. and D. N. Munns. 1996. *Soils: an introduction.* 3rd ed. Upper Saddle River, N.J.: Prentice-Hall, Inc.

Singh, A. and M. Agrawal. 1996. Effects of enhanced UV-B radiation on biomass, net photosynthesis and pigments in four tropical legumes. *International J. of Ecology and Environmental Sciences* 22:23–31.

Sinsabaugh, R. L., R. K. Antibus, A. E. Linkins, C. A. McClaugherty, L. Rayburn, D. Repert, and T. Weiland. 1993. Wood decomposition: nitrogen and phosphorus dynamics in relation to extracelluar enzyme activity. *Ecology* 74(5): 1586–1593.

Slatyer, R. O. 1965. Measurement of precipitation. Interception by an arid plant community (*Acacia aneura* F. Muell.) *Arid Zone Research* 25:181–192.

———. 1967. *Plant-water relationships.* New York: Academic Press.

Slobodkin, L. B. 1974. Comments from a biologist to a mathematician. In *Proc. SIAM-SIMS Conference*, Alta, UT, ed. S. A. Leven, pp. 318–329.

Slobodkin, L.B. and A. Rapoport. 1974. An optimal strategy of evolution. *The Quarterly Review of Biology* 49:181–200.

Smith, A. D. 1940. A discussion of the application of a climatological diagram, the hythergraph, to the distribution of natural vegetation types. *Ecology* 21:184–191.

Smith, A. M. 1994. Xylem transport and the negative pressures sustainable by water. *Annals of Botany* 72:647–651.

Smith, B. N. and S. Epstein. 1971. Two categories of $^{13}C/$ ^{12}C ratios for higher plants. *Plant Physiology* 47:380–384.

Smith, H.B. 1927. Annual vs. biennial growth habit and its inheritance in *Melilotus alba*. *American J. of Botany* 14:129–146.

Smith, J. E. and T. V. Pham. 1996. Genetic diversity of the narrow endemic *Allium aaseae* (Alliaceae). *American J. of Botany* 83:717–726.

Smith, W. H. 1976. Character and significance of forest tree root exudates. *Ecology* 57:324–331.

Smith, W. K. and T. M. McClean. 1989. Adaptive relationship between leaf water repellency, stomatal distribution, and gas exchange. *American J. of Botany* 76:465–469.

Smith, W. K. and P. S. Nobel. 1977a. Temperature and water relations for sun and shade leaves of a deserted broadleaf, *Hyptis emoryi*. *J. of Experimental Botany* 28:169–183.

———. 1977b. Influences of seasonal changes in leaf morphology on water use efficiency for three desert broad leaf shrubs. *Ecology* 58:1033–1043.

Smuts, J. C. 1926. *Holism and evolution*. New York: Macmillan.

SNEP Research Team. 1996. Status of the Sierra Nevada: Sierra Nevada Ecosystem Project report to Congress. Davis, CA: Wildland Resources Center, University of California.

Snook, R.E. and F.P. Day. 1995. Community-level allometric relationships among length, planar area, and biomass of fine roots on a coastal barrier island. *Bulletin of the Torrey Botanical Club* 122:196–202.

Sork, V.L., J. Bramble, and O. Sexton. 1993. Ecology of mast-fruiting in three species of North American deciduous oaks. *Ecology* 74:528–541.

Southard, R. J. 1998. Land, Air, and Water Resources, University of California at Davis. Personal communication.

Southgate, R. I., P. Masters, and M. K. Seely. 1996. Precipitation and biomass changes in the Namib Desert dune ecosystem. *J. of Arid Environments* 33:267–280.

Southwood, T. R. E. 1985. Interactions of plants and animals: patterns and processes. *Oikos* 44:5–11.

Spanner, P. C. 1951. The Peltier effect and its use in the measurement of suction pressure. *J. of Experimental Botany* 2:145–168.

Sperry, J. S., N. Z. Saliendra, W. T. Pockman, H. Cochard, P. Cruiziat, S. D. Davis, F. W. Ewers, and M. T. Tyree. 1996. New evidence for large negative xylem pressures and their measurement by the pressure chamber method. *Plant, Cell and Environment* 19:427–436.

Staehelin, J., A. Waldvogel, J. L. Collett Jr., R. Dixon, R. Heimgartner, W. Heinrich, C. Hsu, L. Li, L. Mosimann, et al. 1993. Contributions of cloud processes to precipitation chemistry in mixed phase clouds. *Water, Air, and Soil Pollution*. 68 (1–2): 1–14.

Stanton, M.L. 1984. Development and genetic sources of seed weight variation in *Raphanus raphanistrum* L. (Brassicaceae). *American J. of Botany* 71:1090–1098.

Stanton, M. L., M. Rejmanek, and C. Galen. 1994. Changes in vegetation and soil fertility along a predictable snowmelt gradient in the Mosquito Range, Colorado, USA. *Arctic and Alpine Research* 26:364–374.

Stearns, S.C. 1977. The evolution of life history traits: a critique of the theory and a review of the data. *Annual Reivew of Ecology and Systematics* 8:145–171.

———. 1992. *The Evolution of Life Histories*. Oxford: Oxford Univ. Press.

Steenbergh, W. F. and C. H. Lowe. 1969. Critical factors during the first years of life of the saguaro (*Cereus giganteus*) at Saguaro National Monument. *Ecology* 50:825–834.

Stephenson, N. L. 1996. Ecology and management of giant sequoia groves. In *Sierra Nevada Ecosystem Project: Final Report to Congress, vol. 2: Assessments and scientific basis for management options*. Davis, CA: University of California, Centers for Water and Wildland Resources.

———. 1997. National Biological Service, Sequoia and Kings Canyon National Parks, Three Rivers, California. personal communication.

Steudler, P. A., J. M. Melillo, R. D. Bowden, and M. S. Castro. 1991. The effects of natural and human disturbances on soil nitrogen dynamics and trace gas fluxes in a Puerto Rican wet forest. *Biotropica* 23(4a): 356–363.

Stewart, G. R., M. H. Turnbull, S. Schmidt, and P. D. Erskine. 1995. ^{13}C natural abundance in plant communities along a rainfall gradient: a biological integrator of water availability. *Australian J. of Plant Physiology* 22:51–55.

Stewart, O. C. 1963. Barriers to understanding the influence of fire by aborigines in vegetation. In *Proceedings of the Tall Timbers Ecol. Conference*, no. 2, pp. 117–126. Tallahassee, FL: Tall Timbers Research Station.

Stohlgren, T. J. 1993. Intra-specific competition crowding of giant sequoias (*Sequoiadendron giganteum*). *Forest Ecology and Management* 59:127–148.

Stohlgren, T. J., G. W. Chong, M. A. Kalkhan and L. D. Schell. 1997. Multiscale sampling of plant diversity: Effects of minimum mapping unit size. *Ecological Applications* 7:1064–1074.

Stohlgren, T. J., M. B. Falkner, and L. D. Schell. 1995. A modified-Whittaker nested vegetation sampling method. *Vegetatio* 117:113–121.

Stone, E. C. 1963. The ecological importance of dew. *Quarterly Review of Biology* 38:328–341.

Stone, E. L. and R. Kszystyniak. 1977. Conservation of potassium in the *Pinus resinosa* ecosystem. *Science* 198:192–194.

Stowe, L. G. and J. A. Teeri. 1978. The geographic distribution of C_4 species of the Dicotyledonae in relation to climate. *American Naturalist* 112:609–623.

Strain, B. R. and V. C. Chase. 1966. Effect of past and prevailing temperatures on the carbon dioxide exchange capacities of some woody desert perennials. *Ecology* 47:1043–1045.

Strong, D.R. J.H. Lawton, and T.R.E. Southwood. 1984. *Insects on Plants: Community Patterns and Mechanisms.* Oxford: Blackwell.

Strong, W. L., E. T. Oswald, and D. J. Downing. 1990. *The Canadian vegetation classification system, first approximation.* Ottawa: Committee on Ecological Land Classification, Sustainable Development Group, Environment Canada.

Stubblefield, S.P., T.N. Taylor, and J.M. Trappe. 1987. Fossil mycorrhizae: a case for symbiosis. *Science* 237:59–60.

Sullivan, J. H., A. H. Teramura and L. R. Dillenburg. 1994. Growth and photosynthetic responses of field-grown sweetgum (*Liquidambar styraciflua:* Hamamelidaceae) seedlings to UV-B radiation. *American J. of Botany* 81:826–832.

Swaine, M. D. 1992. Characteristics of dry forest in West Africa and the influence of fire. *J. of Vegetation Science* 3:365–374.

Sweeney, R. J. 1956. Responses of vegetation to fire. *University of California Publications in Botany* 28:143–206.

Swetnam, T. W. 1993. Fire history and climate change in giant sequoia groves. *Science* 202:885–889.

Swetnam, T. W., R. Touchan, C. H. Baisan, A. C. Caprio, and P. M. Brown. 1990. Giant sequoia fire history in Mariposa Grove, Yosemite National Park. In *Yosemite Centennial Symposium.* El Portal, CA: National Park Service.

Sydes, C. and J. P. Grime. 1981. Effects of tree leaf litter on herbaceous vegetation in deciduous woodland. *J. Ecology* 69:237–262.

Syvertsen, J. P., G. L. Cunningham, and T. V. Feather. 1975. Anomalous diurnal patterns of stem xylem water potentials in *Larrea tridentata. Ecology* 56:1423–1428.

Syvertsen, J. P., G. L. Nickell, R. W. Spellenberg, and G. L. Cunningham. 1976. Carbon reduction pathways and standing crop in three Chihuahuan desert plant communities. *Southwestern Naturalist* 21:311–320.

Tadros, T. M. 1957. Evidence of the presence of an edaphobiotic factor in the problem of serpentine tolerance. *Ecology* 38:14–23.

Taha, F. K., H. G. Fisser, and R. E. Ries. 1983. A modified 100-point frame for vegetation inventory. *J. Range Management* 36:124–125.

Talbot, S. S. and S. L. Talbot. 1994. Numerical classification of the coastal vegetation of Attu Island, Aleutian Islands, Alaska. *J. of Vegetation Science* 5:867–876.

Tansley, A. G. 1914. Presidential address to the first annual general meeting of the British Ecological Society. *J. of Ecology* 2:194–202.

Tansley, A. G. 1947a. Frederic Edward Clements, 1874–1945. *J. of Ecology* 34:194–196.

———. 1947b. The early history of modern plant ecology in Britain. *J. of Ecology* 35:130–137.

Tappeiner, J. C. and A. A. Alm. 1975. Undergrowth vegetation effects on the nutrient content of litterfall and soils in red pine and birch stands in northern Minnesota. *Ecology* 56:1193–1200.

Tateno, M. and F. S. Chapin III. 1997. The logic of carbon and nitrogen interactions in terrestrial ecosystems. *The American Naturalist* 149:723–744.

Taylor, R. J. and R. W. Pearcy. 1976. Seasonal patterns of CO_2 exchange characteristics of understory plants from a deciduous forest. *Canadian J. of Botany* 54:1094–1103.

Taylor, S. E. 1975. Optimal leaf form. In D. M. Gates and R. B. Schmerl, eds., *Perspectives in biophysical ecology,* pp. 73–86. New York: Springer-Verlag.

Teeri, J. A. and L. G. Stowe. 1976. Climatic patterns and the distribution of C_4 grasses in North America. *Oecologia* 23:1–12.

ter Braak, C.J.F. 1986. Canonical correspondence analysis: a new eigenvector technique for multivariate direct gradient analysis. *Ecology* 67:1167–1179.

Teramura, A. H. 1983. Effects of ultraviolet-B radiation on the growth and yield of crop plants. *Physiologia Plantarum* 58:415–427.

Tevis, L., Jr. 1958a. Germination and growth of ephemerals induced by sprinkling a sandy desert. *Ecology* 39:681–687.

———. 1958b. A population of desert ephemerals germinated by less than one inch of rain. *Ecology* 39:688–695.

Thompson, K. and J.P. Grime. 1979. Seasonal variation in the seed banks of herbaceous species in ten contrasting habitats. *J. of Ecology* 67:893–921.

Thompson, L. M. and F. R. Troeh. 1973. *Soils and soil fertility.* New York: McGraw-Hill.

Thorburn, P. J., T. J. Hatton, and G. R. Walker. 1993. Combining measurements of transpiration and stable isotopes of water to determine groundwater discharge from forests. *J. of Hydrology* 150:563–587.

Thoreau, H.D. 1860. The succession of forest trees. In *Excursions* (1893). Boston: Houghton, Mifflin & Co.

Tieszen, L. L. 1978. Photosynthesis in the principal Barrow, Alaska species: A summary of field and laboratory responses. In L. L. Tieszen, ed., *Vegetation and production ecology of an Alaskan arctic tundra*, pp. 241–268. New York: Springer-Verlag.

Tieszen, L. L., B. C. Reed, N. B. Bliss, B. K. Wylie, and D. D. DeJong. 1997. NDVI, C_3 and C_4 production, and distributions in Great Plains grassland land cover classes. *Ecological Applications* 7:59–78.

Tikhomirov, B. A. 1963. *Contributions to the biology of Arctic plants.* Leningrad, USSR: Acad. Nauk.

Tilman, D. 1982. *Resource competition and community structure.* Princeton, NJ: Princeton University Press.

———. 1986. Resources, competition, and the dynamics of plant communities, In M. J. Crawley (ed.), Plant ecology, pp. 51–76. Boston: Blackwell.

———. 1987. On the meaning of competition and the mechanisms of competitive superiority. *Functional Ecology* 1:304–315.

———. 1988. *Plant Strategies and the Structure and Dynamics of Plant Communities.* Princeton, NJ: Princeton Univ. Press.

———. 1994. Community diversity and succession: the roles of competition, dispersal, and habitat modification. In *Biodiversity and ecosystem function,* eds. E.-D. Schulze and H.A. Mooney. Berlin: Springer-Verlag.

———. 1996. Biodiversity: population versus ecosystem stability. *Ecology* 77:350–363.

Tilman, D. and J. A. Downing. 1994. Biodiversity and stability in grasslands. *Nature* 367:363–365.

Ting, I. P. and J. H. Burk. 1983. Aspects of carbon metabolism in *Welwitschia. Plant Science Letters* 32:279–285.

Ting, I. P., and M. Gibbs (eds.). 1982. *Crassulacean acid metabolism.* Rockville, MD: American Society of Plant Physiology.

Tinker, D. B., W. H. Romme, W. W. Hargrove, and M. G. Turner. 1994. Landscape-scale heterogeneity in lodgepole pine serotiny. *Canadian J. of Forest Research* 24 (5): 897–903.

Tobey, R. C. 1981. *Saving the prairies: the life cycle of the founding school of American plant ecology, 1895–1955.* Berkeley: Univ. California Press.

Torrey, J.G. and R.H. Berg. 1988. Some morphological features for generic characterization among the Casuarinaceae. *American Journal of Botany* 75:864–874.

Trabaud, L. 1987. Fire and survival traits of plants. In L. Trabaud, ed. *The Role of Fire in Ecological Systems.* SPB Academic Publishing.

Transeau, E. N. 1926. The accumulation of energy by plants. *Ohio J. of Science* 26:1–10.

Treitz, P. and P. Howarth. 1996. *Remote sensing for forest ecosystem characterization: A review.* Natural Resources Canada, Canadian Forest Service, Sault Ste. Marie, Ontario. NODA/NFP Technical Report TR-12.

Tukey, H. B., Jr. 1966. Leaching of metabolites from aboveground plant parts and its implications. *Bulletin of the Torrey Botanical Club* 93:385–401.

Tukey, H. B., Jr. and R. A. Mecklenberg. 1964. Leaching of metabolites from foliage and subsequent reabsorption and redistribution of the leachate in plants. *American J. of Botany* 51:737–742.

Tuomi, J. T. Hakala, and E. Haukioja. 1983. Alternative concepts of reproductive effort, costs of reproduction and selection in life-history evolution. *American Zoologist* 23:25–34.

Turner, C.L., J.M. Blair, R.J. Schartz, and J.C. Neel. 1997. Soil N and plant responses to fire, topography, and supplemental N in tallgrass prairie. *Ecology* 78:1832–1843.

Turner, D.P., G.J. Koerper, M.E. Harmon, and J.J. Lee. 1995. A carbon budget for forests of the conterminous United States. *Ecological Applications* 5:421–436.

Turner, M. G. and W. H. Romme. 1994. Landscape dynamics in crown fire ecosystems. *Landscape Ecology* 9 (2): 59–77.

Turner, R. M., S. M. Alcorn, G. Olin, and J. A. Booth. 1966. The influence of shade, soil, and water on saguaro seedling establishment. *Botanical Gazette* 127:95–102.

Turrill, W. B. 1946. The ecotype concept, a consideration with appreciation and criticism especially of recent trends. *The New Phytologist* 45:34–43.

Tyree, M. T. 1976. Negative turgor pressure in plant cells: fact or fallacy? *Canadian J of Botany* 54:2738–2746.

Tyree, M. T. and H. T. Hammel. 1972. The measurement of the turgor pressure and the water relations of plants by the pressure-bomb technique. *J. of Experimental Botany* 23:267–282.

United States Congress, Office of Technology Assessment. 1993. Harmful non-indigenous species in the United States. OTA-F-565. Washington, DC: U.S. Government Printing Office.

U.S. Department of Agriculture Forest Service. 1993. *Forest health assessment for the Northeastern Area.* U.S.D.A. Forest Service NA-TP-01-95.

U.S. Department of Agriculture Soil Survey Staff. 1975. *Soil Taxonomy.* Agricultural Handbook 436. Washington D.C.: U.S. Dept. of Agriculture.

U.S. Department of Agriculture Soil Survey Staff. 1996. *Keys to Soil Taxonomy.* 7th ed. Lincoln, NE: U.S. Dept. of Agriculture, Natural Resources Conservation Service.

Ungar, I. A. 1977. The relationship between soil water potential and plant water potential in two inland halophytes under field conditions. *Botanical Gazette* 138:498–501.

Ustin, S. L., R. A. Woodward, M. G. Barbour, and J. L. Hatfield. 1985. Relationships between sunfleck dynamics and red fir seedling distributions. *Ecology* 65:1420–1428.

Vaadia, Y. and Y. Waisel. 1963. Water absorption by the aerial organs of plants. *Physiologia Plantarum* 16:44–51.

Valentini, R., D. Epron, P. De Angelis, G. Matteucci, and E. Dreyer. 1995. *In situ* estimation of net CO_2 assimilation, photosynthetic electron flow and photorespiration in Turkey Oak (*Q. cerris* L.) leaves: diurnal cycles under different levels of water supply. *Plant Cell and Environment* 18:631–640.

Valladares, F. and R. W. Pearcy. 1997. Interactions between water stress, sun-shade acclimation, heat tolerance and photoinhibition in the sclerophyll *Heteromeles arbutifolia*. *Plant Cell and Environment*. 20:25–36.

van Bavel, C. H. M. 1966. Potential evaporation: the combination concept and its experimental verification. *Water Resources Research* 2:445–467.

van der Marrel, L. Orloci, and S. Pignatti (eds.). 1980. *Modern summary of European techniques used in sampling and data analysis*. The Hague: Junk.

Vandermeer, J. 1980. Saguaros and nurse trees: a new hypothesis to account for population fluctuations. *Southwestern Naturalist* 25:357–360.

Vankat, J.L. 1979. *The natural vegetation of North America*. New York: John Wiley & Sons, Inc.

Vankat, J.L. and G.W. Snyder. 1991. Floristics of a chronosequence corresponding to old field-deciduous forest succession in southwestern Ohio. I. Undisturbed vegetation. *Bulletin of the Torrey Botanical Club* 118:365–376.

van der Valk, A. G. 1974. Mineral cycling in coastal foredune plant comunities in Cape Hatteras National Seashore. *Ecology* 55:1349–1358.

van Wagtendonk, J. W. 1995. Dr. Biswell's influence on the development of prescribed burning in California. In D. R. Weise and R. E. Martin, technical coordinators. *The Biswell Symposium: Fire Issues and Solutions in Urban Inteface and Wildland Ecosystems*. Walnut Creek, California: Gen. Tech. Report PSW-GTR-158. Albany, CA: Pacific Southwest Research Station, USDA Forest Service.

Vasek, F. C., H. B. Johnson, and D. H. Eslinger. 1975b. Effects of pipeline construction on creosote bush scrub vegetation of the Mojave Desert. *Madrono* 23:1–13.

Venable, D.L. and L. Lawlor. 1984. Delayed germination and dispersal in desert annuals: escaped in space and time. *Oecologia* 46:272–282.

Viereck, L. E. and W. F. Johnston. 1990. *Picea mariana* (Mill.) BSP, black spruce. In R. M. Burns and B. H. Honkala (eds.) *Silvics of North America*, vol. 1, *Conifers*, 227–237. Agricultural Handbook 654. Washington, D.C.: U.S. Department of Agriculture Forest Service.

Viro, P. J. 1974. Effects of forest fire on soil. In T. T. Kozlowski and C. E. Ahlgren, eds. *Fire and Ecosystems*. New York: Academic Press.

Vitousek, P. M. 1982. Nutrient cycling and nutrient use efficiency. *American Naturalist* 119:553–572.

———. 1997. Biology Department, Stanford University, Stanford California. Personal communication.

Vitousek, P.M., P.R. Ehrlich, A.H. Ehrlich, and P.A. Matson. 1986. Human appropriation of the products of photosynthesis. *BioScience* 36:368–373.

Vitousek, P. M., J. R. Gosz, C. C. Grier, J. M. Melillo, and W. A. Reiners. 1982. A comparative analysis of potential nitrification and nitrate mobility in forest ecosystems. *Ecological Monographs* 52:155–177.

Vitousek, P.M. and R.W. Howarth. 1991. Nitrogen limitation on land and in the sea: how can it occur? *Biogeochemistry* 13:87–115.

Vitousek, P. M. and P. A. Matson. 1985a. Causes of delayed nitrate production in two Indiana forests. *Forest Science* 31:122–131.

Vitousek, P. M. and P. A. Matson. 1985b. Disturbance, nitrogen availability, and nitrogen losses in an intensively managed loblolly pine plantation. *Ecology* 66:1360–1376.

Vitousek, P.M. and W.A. Reiners. 1975. Ecosystem succession and nutrient retention: a hypothesis. *BioScience* 25:376–381.

Vitousek, P.M., D.R. Turner, and K. Kitayama. 1995. Foliar nutrients during long-term soil development in Hawaiian montane rain forest. *Ecology* 76:712–720.

Vogelmann, H. W., T. G. Siccama, D. Leedy, and D. C. Ovitt. 1968. Precipitation from fog moisture in the Green Mountains of Vermont. *Ecology* 49:1205–1207.

Vogl, R. J. 1973. Ecology of knobcone pine in the Santa Ana Mountains, California. *Ecological Monographs* 43:125–143.

———. 1974. Effects of fire on grasslands. In T. T. Kozlowski and C. E. Ahlgren, eds. *Fire and ecosystems*, ch. 5. New York: Academic Press.

Vogl, R. J. and L. T. McHargue. 1966. Vegetation of California fan palm oases on the San Andreas Fault. *Ecology* 47;532–540.

Volterra, V. 1926. Fluctuations in the abundance of a species considered mathematically. *Nature* 118:558–560.

von Leibig, J. 1855. Die Grundsatze der Agriculturchemie. Viewig. Braunschweig.

von Willert, D. J., B. M. Eller, E. Brinckmann, and R. Baash. 1982. CO_2 gas exchange and transpiration of *Welwitschia mirabilis* Hook. Fil. in the central Namib desert. *Oecologia* 55:21–29.

Vose, J. M., K. J. Elliott, D. W. Johnson, R. F. Walker, M. G. Johnson, and D.T. Tingley. 1995. Effects of elevated CO2 and N fertilization on soil respiration from ponderosa pine (*Pinus ponderosa*) in open-top chambers. *Canadian J. of Forest Research* 25:1243–1251.

Wagner, C.A. and T.N. Taylor. 1981. Evidence for endomycorrhizae in Pennsylvanian age plant fossils. *Science* 212:562–563.

Waisel, Y. 1972. *Biology of halophytes*. New York: Academic Press.

Wakimoto, R. H. 1977. Presentation at giant sequoia fire ecology extension course, October 1977, University of California, Davis.

Walker, J., C.H. Thompson, I.F. Fergus, and B.R. Tunstall. 1981. Plant succession and soil development in coastal sand dunes of subtropical eastern Australia. In *Forest Succession: Concepts and Application,* ed. D.C. West, H.H. Shugart, and D.B. Botkin, pp. 107–131. New York: Springer-Verlag.

Walker, L.R. 1995. How unique is primary plant succession at Glacier Bay? In *Proceedings of the Third Glacier Bay Science Symposium, 1993,* ed. D.R. Engstrom. Anchorage, Alaska: National Park Service.

Walker, L.R. and F.S. Chapin, III. 1986. Physiological controls over seedling growth in primary succession on an Alaskan floodplain. *Ecology* 67:1508–1523.

———. 1987. Interactions among processes controlling successional change. *Oikos* 50:131–135.

Walker, L.R. In Press. Patterns and processes in primary succession. In *Ecosystems of disturbed ground,* ed. L.R. Walker. Amsterdam: Elsevier.

Walker, L.R., J.C. Zasada, and F.S. Chapin, III. 1986. The role of life history processes in primary succession on an Alaskan floodplain. *Ecology* 67:1243–1253.

Walker, R. B. 1954. The ecology of serpentine soils. II. Factors affecting plant growth on serpentine soils. *Ecology* 35:259–266.

Walter, H. 1979. The vegetation of the earth and the ecological systems of the geo-biosphere 2nd ed. New York: Springer-Verlag.

Ward, I. 1883. *Dynamic sociology, or applied social science.* Two volumes. Reprinted in 1968 by Greenwood Press, New York.

Wardle, D. A. 1992. A comparative assesment of factors which influence microbial biomass carbon and nitrogen levels in soil. *Biol. Rev.* 67:321–358.

Waring, R. H. and B. D. Cleary. 1967. Plant moisture stress: evaluation by pressure bomb. *Science* 155:1248–1254.

Waring, R.H., A.J.S. McDonald, A.J.S. Larrson, et. al. 1985. Differences in chemical composition of plants grown at constant relative growth rates with stable mineral nutrition. *Oecologia* 66:157–160.

Waring, R.H. and W.H. Schlesinger. 1985. *Forest ecosystems: concepts and management.* Orlando, FL: Academic Press, Inc.

Warming, J. E. B. 1895. *Plantesamfund, prundtraek af den okologiske plantegeografi.* Copenhagen, Denmark: Philipsen.

———. 1909. *Oceology of plants.* London: Oxford University Press.

Warnant, P., L. François, D. Strivay, and J.-C. Gérard. 1994. CARAIB: A global model of terrestrial biological productivity. *Global Biogeochemical Cycles* 8:255–270.

Watkinson, A. R., W. M. Lonsdale, M. H. Andrew. 1989. Modelling the population dynamics of an annual plant *Sorghum intrans* in the wet-dry tropics. *J. of Ecology* 77:162–181.

Weathers, K. C., G. M. Lovett, and G. E. Likens. 1995. Cloud deposition to a spruce forest edge. *Atmospheric Environment* 29 (6): 665–672.

Webb, D.A. 1985. What are the criteria for presuming native status? *Watsonia* 15:231–236.

Webb, S.L. 1986. Potential role of passenger pigeons and other vertebrates in the rapid Holocene migrations of nut trees. *Quaternary Research* 26:367–375.

Weber, R. O., P. Talkner, and G. Stefanicki. 1994. Asymmetric diurnal temperature change in the alpine region. *Geophysical Research Letters* 21:673.

Weiner, J. 1988. In *Plant Reproductive Ecology: Patterns and Strategies.* J. Lovett-Doust and L. Lovett-Doust, eds. New York: Oxford University Press.

———. 1995. Following the growth of individuals in crowded plant populations. *Trends in Ecology & Evolution* 10:389–390.

Weis, E. and J. A. Berry. 1988. Plants and high temperature stress. *Plants and Temperature* 42:329–346.

Weller, D. E. 1987a. A re-evaluation of the -3/2 power rule of self-thinning. *Ecological Monographs* 57:23–43.

———. 1987b. Self-thinning exponent correlated with allometric measures of plant geometry. *Ecology* 68:813–821.

———. 1990. Will the real self-thinning rule please stand up?—a reply to Osawa and Sugita. *Ecology* 71:1204–1207.

Welter, S.C. 1989. In *Plant-Insect Interactions.* Vol. 1. Bernays, E.A., ed., pp. 135–150.

Went, F. W. 1942. The dependence of certain annual plants on shrubs in southern California deserts. *Bulletin of the Torrey Botanical Club* 69:100–114.

———. 1948. Ecology of desert plants. I. Observations on germination in Joshua Tree National Monument. *Ecology* 29:242–253.

Went, F. W. and N. Stark. 1968. Mycorrhiza. *Bioscience* 118:1035–1039.

Went, F. W. and M. Westergaard. 1949. Ecology of desert plants. III. Development of plants in the Death Valley National Monument, California. *Ecology* 30:26–38.

Wentworth, T. R. 1981. Vegetation on limestone and granite in the Mule Mountains, Arizona. *Ecology* 62:469–482.

———. 1983. Distributions of C_4 plants along environmental and composition gradients in southeastern Arizona. *Vegetatio* 52:21–34.

Werner, P.A. 1975. Predictions of fate from rosette size in teasel (*Diplacus fullonum* L.). *Oecologia* 20:197–201.

Werner, P. A. and H. Caswell. 1977. Population growth rates and age versus stage-distribution models for teasel (*Diplacus sylvestris* Huds.) *Ecology* 58:1103–1111.

Wessman, C.A. 1990. Evaluation of canopy biochemistry. In *Remote sensing of biosphere functioning.* pp. 135–156. R.J. Hobbs and H.A. Mooney, eds. New York: Springer-Verlag.

Wessman, C.A., C.A. Bateson, and T.L. Benning. 1997. Detecting fire and grazing patterns in tallgrass prairie using spectral mixture analysis. *Ecological Applications* 7:493–511.

West, N. E. and P. T. Tueller. 1971. Special approaches to studies of competition and succession in shrub communities. In *Wildland shrubs,* U.S. Dept. of Agriculture Forest Service General Technical Report INT-1, ed. C. M. McKell, J. P. Blaisdell, and J. R. Goodin, pp. 165–171. Logan, UT: U.S. Dept. of Agriculture Forest Service.

Westman, W. E., and R. K. Peet. 1982. Robert H. Whittaker (1920–1980): the man and his work. *Vegetatio* 48:97–122.

Westoby, M. 1981. The place of the self-thinning rule in population dynamics. *American Naturalist* 118:581–587.

———. 1984. The self-thinning rule. *Advances in Ecological Research* 14:167–225.

Whelan, R. J. 1995. *The Ecology of Fire.* Cambridge: Cambridge University Press.

White, J. 1979. The plant as a metapopulation. *Annual Review of Ecology and Systematics* 10:109–145.

———. 1980. Demographic factor in populations of plants. In *Demography and Evolution in Plant Populations,* O. T. Solbrig, ed., pp. 21–48. Oxford, England: Blackwell Scientific.

———. 1985. The thinning rule and its application to mixtures of plant populations. In J. White, ed., *Studies on Plant Demography,* pp. 291–309. London: Academic Press.

White, T.C.R.. 1993. *The Inadequate Environment: Nitrogen and the Abundance of Animals.* Berlin: Springer-Verlag.

Whittaker, R. H. 1954. The ecology of serpentine soils. I. Introduction. *Ecology* 35:258–259.

———. 1956. Vegetation of the Great Smoky Mountains. *Ecological Monographs* 26:1–80.

———. 1960. Vegetation of the Siskiyou Mountains of California. *Ecological Monographs* 30:279–338.

———. 1962. Classification of natural communities. *Botanical Review* 28:1–239.

———. 1975. *Communities and ecosystems.* 2nd ed. New York: Macmillan.

Whittaker, R. H. and P. P. Feeny. 1971. Allelochemics: chemical interactions between species. *Science* 171:1757–1770.

Whittaker, R. H. and P. L. Marks. Methods of assessing terrestrial productivity. In *Primary productivity of the biosphere,* ed. H. Lieth and R. H. Whittaker, pp. 55–118. New York: Springer.

Whittaker, R. H. and W. A. Niering. 1975. Vegetation of the Santa Catalina Mountains, Arizona: V. Biomass, production and diversity along the elevation gradient. *Ecology* 56:771–790.

Wilbur, H.M. 1976. Life history evolution in seven milkweed of the genus *Asclepias. J. of Ecology* 64:223–240.

Wildi, O. and L. Orlóci. 1990. *Numerical exploration of community patterns.* The Hague: SPB Academic Publishing.

Williams, K., G.W. Koch and H.A. Mooney. 1985. The carbon balance of flowers of *Diplacus aurantiacus* (Scrophulariaceae). *Oecologia* 66:530–535.

Williams, M., E.B. Rastetter, D.N. Fernandes, M.L. Goulden, G.R. Shaver, and L.C. Johnson. 1997. Predicting gross primary productivity in terrestrial ecosystems. *Ecological Applications* 7:882–894.

Williamson, M. 1972. *The Analysis of Biological Populations.* London: Academic Press.

Willmot, A. and P. D. Moore. 1973. Adaptation to light intensity in *Silene alba* and *S. dioica. Oikos* 24:458–464.

Willson, M.F. 1983. *Plant Reproductive Ecology.* New York: Wiley and Sons.

Wilson, E. O. and W. H. Bossert 1971. *A Primer for Population Biology.* Sunderland, MA: Sinauer and Assoc.

Wilson, J. B., H. Gitah, and A. D. Q. Agnew, 1987. Does niche limitation exist? *Functional Ecology* 1:301–307.

Wilson, J. B. and S. H. Roxburgh. 1994. A demonstration of guild-based assembly rules for a plant community and determination of intrinsic guilds. *Oikos* 60:267–276.

Wilson, J. B., I. Ullman, and P. Bannister. 1996. Do species assemblages ever recur? *J. of Ecology* 84:471–474.

Wilson, J. B. and R. J. Whittaker. 1995. Assembly rules demonstrated in a saltmarsh community. *J. of Ecology* 83:801–807.

Wilson, S.D. and P.A. Keddy. 1986. Species competitive ability and position along a natural stress/disturbance gradient. *Ecology* 67:1236–1242.

Winkler, J. P., R. S. Cherry, and W. H. Schlesinger. 1996. The Q_{10} relationship of microbial respiration in a temperate forest soil. *Soil Biol. Biochem.* 28 (8): 1067–1072.

Winter, K. and U. Lüttge. 1976. Balance between C_3 and CAM pathway of photosynthesis. *Ecological Studies* 19:323–334.

Wishart, D. 1987. *CLUSTAN User Manual (CLUSTAN 3), 4th ed.* Computing Laboratory, University of St. Andrews, Scotland.

Woods, F. W. and K. Brock. 1964. Interspecific transfer of Ca_{45} and P_{32} by root systems. *Ecology* 45:886–889.

Workman, C. 1980. A new chemical method for measurement of mean temperatures with special applicability to cold environments. *Oikos* 35:365–372.

Worster, D. 1985. *Nature's economy: a history of ecological ideas,* 2nd ed. Cambridge, UK: Cambridge University Press.

Wright, E. and R. F. Tarrant. 1957. *Microbial soil properties after logging and slash burning*. U.S.F.S. Pac. N W Forest Range Expt. Sta. Res. Notes 157. Washington, D.C.: U.S. Department of Agriculture Forest Service.

Wright, H. A. and A. W. Bailey. 1982. *Fire Ecology*. New York: John Wiley.

Wyatt, R. 1992. *Ecology and Evolution of Plant Reproduction*. New York: Chapman & Hall.

Yakimchuk, R. and J. Hoddinott. 1994. The influence of ultraviolet-B light and carbon dioxide enrichment on the growth and physiology of seedlings of three conifer species. *Canadian J. of Forest Research*. 24:1–8.

Yang, S. and M. T. Tyree. 1994. Hydraulic architecture of *Acer saccharum and A. rubrum*: comparison of branches to whole trees and the contribution of leaves to hydraulic resistance. *J. of Experimental Botany* 45:179–186.

Yavitt, J. B. and E. L. Smith, Jr. 1983. Spatial patterns of mesquite and associated herbaceous species in an Arizona desert grassland. *American Midland Naturalist* 109:89–93.

Yemm, E. W. and R. G. S. Bidwell. 1969. Carbon dioxide exchanges in leaves. I Discrimination between $^{14}CO_2$ and $^{12}CO_2$ in photosynthesis. *Plant Physiology* 44:1328–1334.

Yin, X., N. E. Foster, and P. A. Arp. 1993. Solution concentrations of nutrient ions below the rooting zone of a sugar maple stand: relations to soil moistsure, temperature, and season. *Canadian J. of Forest Research* 23:617–624.

Yoda, K., T. Kira. H. Ogawa, and K. Kozumi. 1963. Intraspecific competition among higher plants. XI. Self-thinning in overcrowded pure stands under cultivated and natural conditions. *J. of Biology*, Osaka City University 14:107–129.

Young, T.P. 1981. A general model of comparative fecundity for semelparous and iteroparous life histories. *American Naturalist* 118:27–36.

———. 1984. The comparative demography of semelparous *Lobelia telekii* and iteroparous *Lobelia keniensis* on Mount Kenya. *J. of Ecology* 72:637–650.

———. 1985. *Lobelia telekii* herbivory, mortality, and size at reproduction: variation with growth rate. *Ecology* 66:1879–1883.

———. 1990. Evolution of semelparity in Mount Kenya lobelias. *Evolutionary Ecology* 4:157–172.

———. 1994. Population Biology of Mount Kenya lobelias. In P. W. Rundel, A. P. Smith and F. C. Meinzer, eds., *Tropical Alpine Environments: Plant Form and Function*. New York: Cambridge University Press.

Young, T.P. and C.K. Augspurger. 1991. Ecology and Evolution of long-lived semelparous plants. *Trends in Ecology and Evolution* 6:285–289.

Young, T.P. and M.M. Peacock. 1992. Giant senecios and alpine vegetation of Mount Kenya. *J. of Ecology* 80:141–148.

Zangerl, A.R. and F.A. Bazzaz. 1992. Theory and pattern in plant defense allocation. In R.S. Fritts and E.L. Simms eds., *Plant Resistance to Herbivores and Pathogens: Ecology, Evolution and Genetics*, pp. 363–391. Chicago: University of Chicago Press.

Zanstra, P. E. and F. Hagenzieker. 1977. Comments on the psychrometric determination of leaf water potentials *in situ*. *Plant and Soil* 48:347–367.

Zar, J.H. 1996. *Biostatistical analysis, 3rd ed.* Englewood Cliffs, NJ: Prentice Hall.

Zavitkovski, J. and M. Newton. 1968. Ecological importance of snowbrush, *Ceanothus velutinus*, in the Oregon Cascades agricultural ecology. *Ecology* 49:1134–1145.

Zedler, P. H. 1995. Fire frequency in southern California shrublands: biological effects and management options. In J. E. Keeley and T. Scott eds., *Brushfires in California Ecology and Resource Management*. International Association of Wildland Fire.

Zeide, B. 1985. Tolerance and self-tolerance of trees. *Forest Ecology and Management* 13:149–166.

———. 1987. Analysis of the -3/2 power law of self-thinning. *Forest Science* 33:517–537.

Zeiger, E., C. Field, and H.L. Mooney. 1981. Stomatal opening at dawn: possible role of the blue light response. In *Plants and daylight spectrum*, ed. H. Smith. New York: Academic Press.

Zimmerman, M. E. et al. (eds.). 1993. *Environmental philosophy: from animal rights to radical ecology*. Englewood Cliffs, N.J.: Prentice Hall.

Zimmerman, U., A. Haase, D. Langbein, and F. Meinzer. 1993. Mechanisms of long distance water transport in plants: a re-examination of some paradigms in the light of new evidence. *Philosophical Transactions of the Royal Society of London* B341:19–31.

Zimmerman, U., J. J. Zhu, F. Meinzer, G. Goldstein, H. Schneider, G. Zimmermann, R. Benkert, F. Thurmer, P. Melcher, D. Webb, and A. Haase. 1994. High molecular weight organic compounds in the xylem sap of mangroves: implications for long-distance water transport. *Botanica Acta* 107:218–229.

Zobel, D.B. and J.A. Antos. 1997. A decade of recovery of understory vegetation buried by volcanic tephra from Mount St. Helens. *Ecological Monographs* 67:317–344.

INDEX

Page numbers given in bold refer to definitions